Table No.	TABLE CONTENTS	

GREEK ALPHABET

A	α	Alpha	N	ν	Nu
B	β	Beta	Ξ	ξ	Xi
Γ	γ	Gamma	O	o	Omicron
Δ	δ	Delta	Π	π	Pi
E	ϵ	Epsilon	P	ρ	Rho
Z	ζ	Zeta	Σ	σ	Sigma
H	η	Eta	T	τ	Tau
Θ	θ	Theta	Y	υ	Upsilon
I	ι	Iota	Φ	ϕ	Phi
K	κ	Kappa	X	χ	Chi
Λ	λ	Lambda	Ψ	ψ	Psi
M	μ	Mu	Ω	ω	Omega

ELECTRIC

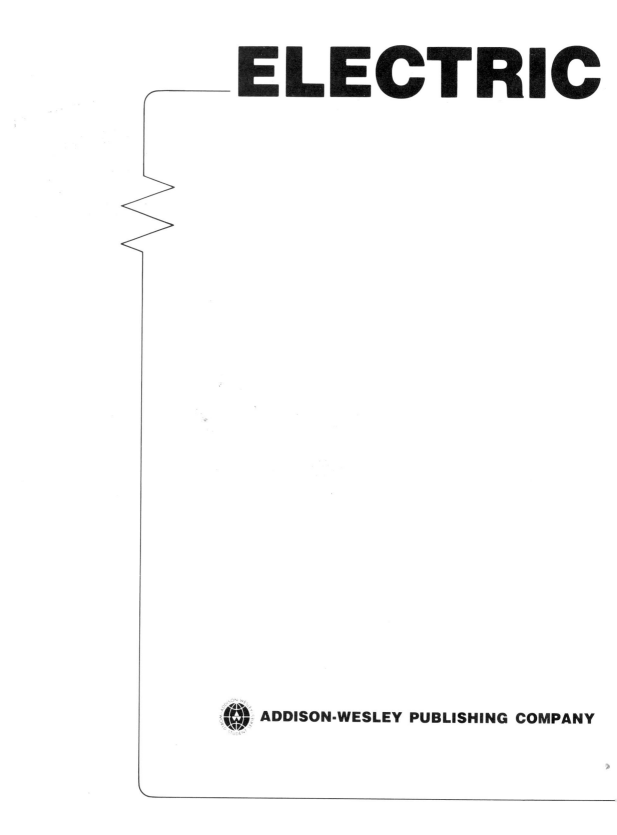

ADDISON-WESLEY PUBLISHING COMPANY

CIRCUITS

JAMES W. NILSSON
IOWA STATE UNIVERSITY

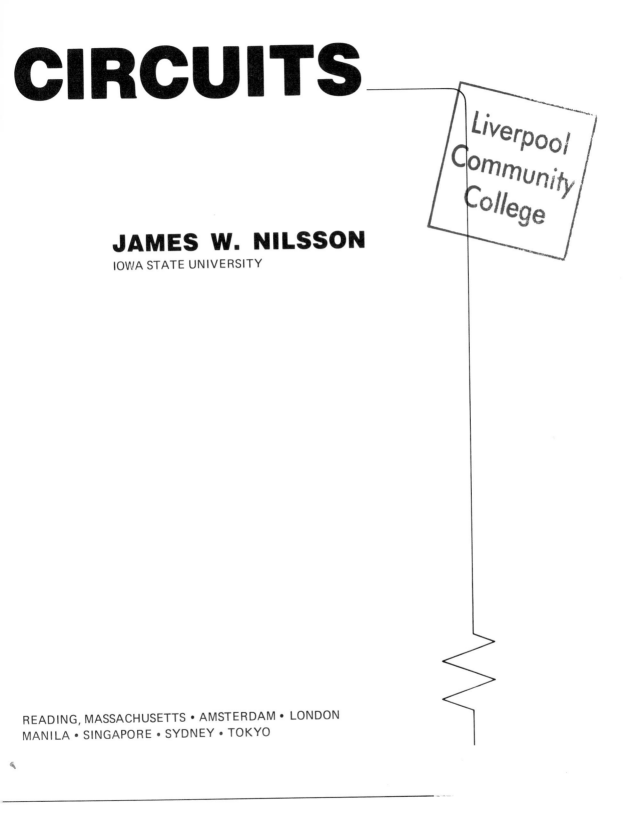

Liverpool
Community
College

READING, MASSACHUSETTS • AMSTERDAM • LONDON
MANILA • SINGAPORE • SYDNEY • TOKYO

This book is in the **Addison-Wesley series in Electrical Engineering**

Sponsoring Editor: *Tom Robbins*
Production Editor: *Doris L. Machado*
Copy Editor: *Martha K. Morong*

Text Designer: *Herb Caswell*
Illustrator: *Parkway Illustrated Press*
Cover Designer: *Ann Scrimgeour Rose*
Art Coordinator: *Susanah H. Michener*

Production Manager: *Karen M. Guardino*
Production Coordinator: *Peter Petraitis*

The text of this book was composed in Century Schoolbook
by Progressive Typographers, Inc.

WORLD STUDENT SERIES

ISBN 0-201-06237-2
ABCDEFGHIJ-DO-89876543

To Anna

PREFACE

Electrical engineers are involved in the creation and operation of a wide variety of systems that are designed to serve the needs and wants of society. These systems range from those that are concerned with the generation, transmission, and control of minute amounts of energy to those that focus on the utilization and control of tremendous amounts of energy. Electrical engineering plays a key role in the development of machines that are used to support human feats of labor and in machines that are used to enhance human feats of computation. Paralleling and supporting the development of "brawn" and "brain" machines is the creation of sophisticated systems of communication, instrumentation, and control. The design and development of all these systems depends in part on the ability to construct mathematical models of electrical components as well as models of interconnected components. These mathematical models are rooted in both electromagnetic field theory and circuit theory. Our purpose, in this text, is to introduce the reader to the theory of circuit analysis.

In writing the first fourteen chapters of the text, I have assumed that the reader has had a course in elementary differential and integral calculus. I have also assumed the reader has had an introductory physics course, at either the high school or university level, that introduces the concepts of energy, power, electric charge, electric current, electric potential, and electromagnetic fields. In writing Chapters 15 through 20, I have assumed the student has had, or is taking, an introductory course in differential equations.

Chapters 1 through 4 introduce the reader, via resistive circuits, to the traditional topics of circuit analysis: Ohm's law, Kirchhoff's laws, node and mesh analysis, Thévenin/Norton equivalents, and the principle of superposition. The d'Arsonval ammeter and voltmeter are introduced early on (in Chapter 3) for two reasons: first, these analog meters continue to find some application outside the laboratory, and second, they give the reader a chance to analyze some simple resistive circuits that are of practical importance. Chapters 5 through 8 introduce the analysis of simple RL, RC, and RLC circuit structures that are disturbed by switching operations.

The operational amplifier circuit is introduced in Chapter 9 for three primary reasons. First, it affords an opportunity, at an elementary level, to discuss what is today a basic building block in circuit design. Second, it gives the reader an opportunity to analyze some circuit structures that are of great practical importance. These circuits serve as a strong motivating topic at the introductory level. Third, the operational amplifier plays an important role in the operation of the digital multimeter, an instrument of utmost importance in today's laboratory. The basic operation of the digital multimeter is included in Chapter 9.

Chapters 10 and 11 introduce analyzing the steady-state behavior of circuits driven by sinusoidal sources. The concept of a phasor and the importance of transferring the analysis from the time domain to the phasor domain is stressed. The meaning of real, reactive, and complex power is also discussed. Chapter 12 is used to introduce the reader to the operation of balanced, three-phase circuits. At the same time, Chapter 12 gives the reader more practice with phasor domain analysis.

The concept of mutual inductance is the topic of Chapter 13. The analysis of both the linear transformer and the ideal magnetic-core transformer are used to illustrate the importance of magnetic coupling.

The first introduction to frequency response is in Chapter 14 where parallel and series resonance are discussed. These simple parallel and series structures are used to introduce the important concepts of resonant frequency, bandwidth, and quality of frequency selectivity.

Chapters 15 and 16 introduce the Laplace transform as a tool for circuit analysis. Functional transforms, operational transforms, partial-fraction expansions, s-domain equivalent circuits, and the Dirac delta function are discussed and illustrated. Chapter 17 expands the usefulness of the Laplace transform approach through the introduction of the transfer function and the convolution integral. The transfer function is also used to broaden the reader's perspective of frequency response via a discussion of Bode amplitude and phase angle plots.

The steady-state response of circuits to periodic driving sources is discussed in Chapter 18. The Fourier series representation of a periodic function is first developed and then the usefulness of the Fourier series in circuit analysis is illustrated. The exponential form of the Fourier series is used as a way of introducing the Fourier transform, which is the topic of Chapter 19. This transition from the Fourier series to the Fourier transform is widely used because it gives good physical insight into the interpretation of the Fourier transform. In discussing the Fourier transform applications to circuit analysis, its relationship to the Laplace transform is also brought out.

The last chapter in the book introduces the reader to the two-port parameters $z, y, a, b, h,$ and g. The relationships among the parameters are given in Table 20.1. The analysis of the terminated two-port in terms of the parameters is summarized in Table 20.2.

Numerical examples are used extensively throughout the text to help the reader understand how the theory is applied to the analysis of a circuit structure. Drill exercises have been introduced into the text to give the readers an opportunity to test their understanding of the material they have just read. The Drill exercises have been printed across the entire page as a way of signaling the reader to stop and solve the exercise before proceeding to the next section.

The text has been designed for use in either a two-semester sequence or a three-quarter sequence. Assuming three recitations, or lectures, per week, the first twelve chapters can be covered during the first semester, leaving the last eight chapters for the second semester. On a quarter schedule, the book can be subdivided into Chapters 1–8, 9–15, and 16–20.

The text can also be used in a single semester introduction to circuit analysis. After covering the first eight chapters, the instructor can choose among Chapters 9 (Operational Amplifiers), 12 (Three-phase circuits), 13 (Mutual Inductance), 14 (Resonance), and 18 (Fourier Series) to develop the desired emphasis.

The introduction of operational amplifier circuits into the text has been done in such a way that this material can be omitted without interfering with the reading of the subsequent chapters. If Chapter 9 is omitted, the instructor simply omits those problems and drill exercises in the chapters following 9 that contain operational amplifiers.

Finally, a few words to the student. Engineering courses have two primary objectives that go hand-in-hand. One is to impart quantitative information about systems and components that reflects the current state-of-the-art. The second is to develop techniques of analysis and synthesis that are applicable to a wide variety of specific problems. Students who develop the habit of thinking in terms of realistic numbers, which quantitatively describe the systems already in existence, and at the same time focus their attention on the principles that underlie the analysis of different systems and components will be in the best position to develop a successful career in a rapidly changing technology. As a practicing engineer, you will not be asked to solve problems that have already been solved. Whether you are trying to improve the performance of an existing system or whether you are creating a new system, you will be working on unsolved problems. As a student, you will devote most of your attention to the discussion of problems already solved. It is through these discussions of how these problems were solved in the past that you will develop the skill to attack and solve the problems of the future.

ACKNOWLEDGMENTS

It is not possible after thirty years of teaching to mention all the people who have contributed to my development as an engineering educator. I do wish to acknowledge those colleagues who have generously offered their advice and counsel during the writing of this text: Professors Charles Cowan, Harry Hale, Edwin Jones, Jr., and Wael Abul-Shohoud who taught from the class notes and Professor Thomas Scott who was kind enough to read and critique several chapters. I also wish to thank the following individuals who reviewed the manuscript and offered many helpful suggestions: Professors N. Bose, James Delansky, J. W. Howze, Gladwyn Lago, S. K. Mitra, and Kalyan Mondal. I am also appreciative of the support I have received from the administration of Iowa State University during the preparation of the manuscript. A special thank you to Dr. J. O. Kopplin, Chairman of the Electrical Engineering Department, and to Dr. David Boylan, Dean of the College of Engineering. Last, but not least, I owe a debt of gratitude to Miss Shellie Siders who typed the original manuscript and did a masterful job of inserting the changes and revisions that went into the final draft.

I should also like to express my sincere appreciation to the people at Addison-Wesley who were given the responsibility of transforming the manuscript into a textbook. Special thanks to Martha Morong, Sue Michener, and Doris Machado, each of whom did an outstanding job. Last but not least, a special thank you to Tom Robbins whose enthusiasm, advice, and encouragement helped make this book possible.

Ames, Iowa   J.W.N.
December 1982

CONTENTS

CHAPTER 4
TECHNIQUES OF CIRCUIT ANALYSIS 71

CHAPTER 5
INDUCTANCE AND CAPACITANCE 132

CHAPTER 10
SINUSOIDAL STEADY-STATE ANALYSIS 276

CHAPTER 11
SINUSOIDAL STEADY-STATE POWER CALCULATIONS 330

CHAPTER **12**
BALANCED THREE-PHASE CIRCUITS 363

CHAPTER **13**
MUTUAL INDUCTANCE 403

CHAPTER **14**
SERIES AND PARALLEL RESONANCE 453

CHAPTER **15**
INTRODUCTION TO THE LAPLACE TRANSFORM   496

CHAPTER **16**
THE LAPLACE TRANSFORM IN CIRCUIT ANALYSIS   538

CHAPTER **17**
THE TRANSFER FUNCTION   580

CHAPTER **18**
FOURIER SERIES   637

CHAPTER **19**
THE FOURIER TRANSFORM   687

CHAPTER **20**
TWO-PORT CIRCUITS 724

APPENDIXES

CIRCUIT VARIABLES

1.1 INTRODUCTION

Electrical engineers are concerned with the design, analysis, and operation of human-made systems involving electrical signals. The number and variety of such systems prohibit any specific enumeration. It is possible, however, to divide the systems into four general groups: (1) communication systems, (2) computer systems, (3) control systems, and (4) power systems. In communication systems, electrical engineers are concerned with the generation, transmission, and distribution of information via electrical signals. In computer systems, electrical engineers focus on the use of electrical signals to carry out computations ranging from business applications to scientific applications. Control systems engineers use electrical signals to control processes varying from the flight control of a space probe to the time and temperature control of a microwave oven. Power systems engineers address the problems involved with the use of electrical signals in the generation, transmission, and distribution of large blocks of power. It is important to recognize that there is considerable interaction between these general types of systems. Thus communication engineers make use of digital computers to control the flow of information. Engineers designing computers must have some understanding of the areas in which the computers are going to be used in order to bring about an intelligent and useful design. Power systems require sophisticated communication components in order to operate in a safe and reliable manner. Thus it is obvious that an engineer with a primary interest in one area must also be knowledgeable in the areas that interact with this interest.

Once we understand that electrical engineering is an extremely diverse field of study involving a wide range of systems and at the same time close interplay between these systems, we can appreciate the problem of deciding how best to introduce a beginning student to this field. Should we start with four introductory courses, one for each area enumerated above? Should we start with one course that surveys the four areas? Should we start with one area and cover it in depth? Is there a subject area supporting all these areas that can be used as a starting point?

The answer to the last question is "yes." The underlying body of knowledge is electromagnetic field theory. A very important special case of field theory is circuit theory. In this book, we have chosen to introduce the beginning student to electrical engineering via circuit theory. There are several reasons for this:

1. Circuit theory provides simple solutions (of sufficient accuracy) to practical problems that would otherwise become hopelessly complicated if approached from the broad point of view of electromagnetic field theory.

2. Many useful electrical systems are less complicated if one understands the electrical phenomena that take place at the terminals of the electrical components and devices that make up the system. Such systems are ade-

quately analyzed, and even synthesized, by the proper application of circuit theory.

3. Circuit theory is an area of study in its own right. A good share of the remarkable development of human-made systems that depend on electrical phenomena can be attributed to the development of circuit theory as a separate discipline of study.

As mentioned above, circuit theory is a special case of field theory. However, even though circuit theory is rooted in electromagnetic field theory, it is possible to introduce circuit concepts before proceeding with an in-depth study of fields. Consequently, a course in electromagnetic field theory is not a prerequisite to understanding the material that follows. We do assume that the student has had an introductory physics course in which electric and magnetic phenomena were introduced.

A final observation about the relationship between circuit theory and field theory is in order before we embark on our study of electric circuits. We are interested in electrical effects caused by electrical charges in motion. These effects will propagate through a system at a finite velocity. The circuit approach is attractive in those circumstances in which the spatial dimensions of the system, in relation to the velocity of propagation, are such that we can assume that the effects occur instantaneously throughout the system. We use the term "lumped-parameter system" to signify that the spatial dimensions of the system do not enter into the circuit model. We can get a quantitative handle on whether or not the physical dimensions of the system are significant by noting that electrical effects propagate by wave phenomena. Thus if the spatial wavelength of the electrical disturbance is large compared to the physical dimensions of the system, we assume that the disturbance occurs instantaneously throughout the system. The wavelength λ is the velocity (which is nearly the velocity of light, $c = 3 \times 10^8$ meters/second) divided by the repetition rate or frequency of the signal; that is, $\lambda = c/f$. The frequency f is measured in cycles/second, or hertz. For example, the frequency of signals used to transmit electrical power is 60 hertz. The spatial wavelength in this case is 5×10^6 meters, or approximately 3100 miles. Radio signals are on the order of 10^9 hertz. At this frequency, the wavelength is 0.3 meter; thus physical dimensions in the order of centimeters are important. Whenever any of the pertinent dimensions of the system approach the wavelength, we must use electromagnetic field theory to analyze the system.

1.2 SYSTEM OF UNITS

All engineers deal in quantitative results. Such results can be communicated in a meaningful way only if all engineers understand the units used. The International System of Units (abbreviated SI) is used by all the major engineering

TABLE 1.1
THE INTERNATIONAL SYSTEM OF UNITS (SI)

Quantity	Basic Unit	Symbol
Length	Meter	m
Mass	Kilogram	kg
Time	Second	s
Electric current	Ampere	A
Thermodynamic temperature	Degree Kelvin	K
Luminous intensity	Candela	cd

societies and hence will be used in this text. It is based on six quantities: (1) length, (2) mass, (3) time, (4) electric current, (5) thermodynamic temperature, and (6) luminous intensity. These quantities, along with the basic unit and symbol for each quantity, are listed in Table 1.1.

In many cases either submultiples or multiples of the basic unit of measurement are more appropriate to use than the unit itself. The International System of Units uses standard prefixes to signify the powers of ten applied to the basic unit. These are listed in Table 1.2. We will find these prefixes very useful in describing the quantitative results we will obtain in the chapters that follow. As a quick example, let us suppose that a time calculation yields a re-

TABLE 1.2
STANDARD PREFIXES TO SIGNIFY POWERS OF TEN

Prefix	Symbol	Power of Ten
atto	a	10^{-18}
femto	f	10^{-15}
pico	p	10^{-12}
nano	n	10^{-9}
micro	μ	10^{-6}
milli	m	10^{-3}
centi	c	10^{-2}
deci	d	10^{-1}
deka	da	10
hecto	h	10^{2}
kilo	k	10^{3}
mega	M	10^{6}
giga	G	10^{9}
tera	T	10^{12}

sult of 10^{-5} second, that is, 0.00001 s. It is much easier to describe this quantity as 10 microseconds, that is, $10^{-5} = 10 \times 10^{-6} = 10$ μs, than as one one hundred-thousandth of a second, or 1/100,000 s.

1.3 CIRCUIT ANALYSIS: AN OVERVIEW

Before we get involved with the details of circuit analysis, let us pause long enough to take a look at circuits from a broad point of view. It is hoped that such an overview will help us keep things in perspective as we study the various segments of circuit theory that make up the whole. Our goal in circuit analysis is to define ideal elements so that when they are connected, the resulting interconnection will quantitatively predict the behavior of the system or device it is intended to represent. For example, we can construct a circuit model that will predict the behavior of the electrical wiring in a residence, or we can construct a circuit model that will predict the behavior of an electronic amplifier. The list is endless. Our point is that ideal circuit elements can be interconnected to represent actual systems. It is this ability to model actual systems with ideal circuit elements that makes circuit theory so useful to engineers.

When we say that the interconnection of ideal circuit elements can be used to predict quantitatively the behavior of a system, we are implying that we can describe the interconnection with mathematical equations. In order for the mathematical equations to be useful, they must be written in terms of measurable quantities. In the case of circuits, the measurable quantities are current and voltage. We will discuss these variables in Section 1.4. For the purposes of our overview, we need know only that they are the variables of primary concern. In general terms, our study of circuit analysis reduces to (1) understanding the behavior of each ideal circuit element in terms of its current and voltage and (2) understanding the constraints imposed on the current and voltage as a result of interconnecting the ideal elements.

1.4 VOLTAGE AND CURRENT

The concept of electrical charge is the basis for describing all electrical phenomena; therefore a brief review of the most important characteristics of this concept is in order. First of all, the charge is bipolar, that is, electrical effects are described in terms of positive and negative charges. Second, electrical charge exists in discrete quantities. Specifically, all quantities of charge are integral multiples of the electronic charge. The electronic charge is 1.6022×10^{-19} coulomb. Third, electrical effects are attributed to both the separation of charge and charges in motion. In circuit theory, we regard the separation of charge as creating an electrical force (voltage) and the motion of charge as creating an electrical fluid (current).

The concepts of voltage and current are useful from an engineering point of view because they can be expressed quantitatively. Whenever positive and negative charges are separated, energy is expended. Voltage is the energy per unit charge that is created by the separation. We express this ratio in differential form; thus

$$v = \frac{dw}{dq},$$ (1.1)

where

$$v = \text{the voltage in volts,}$$
$$w = \text{the energy in joules, and}$$
$$q = \text{the charge in coulombs.}$$

The electrical effects caused by charges in motion depend on the rate of charge flow. Thus this rate of charge flow became a significant variable in scientific work and is known as the electrical current—we have, therefore,

$$i = \frac{dq}{dt},$$ (1.2)

where

$$i = \text{the current in amperes,}$$
$$q = \text{the charge in coulombs, and}$$
$$t = \text{the time in seconds.}$$

Equations (1.1) and (1.2) are definitions for the magnitude of voltage and current, respectively. The bipolar nature of electrical charge requires that we assign polarity references to these variables. We will do so in Section 1.5.

One final comment. In the electrical circuits of interest, the rate of charge flow involves such an enormous number of charge carriers (either positive or negative carriers) that the "graininess" of the charge flow due to the discrete nature of the charge does not enter into the analysis. In other words, it is not necessary to make a detailed study of the progression of individual charge carriers in order to define the current. Thus it is immaterial insofar as the definition of current is concerned whether the charge flow is due to free electrons moving through the crystal lattice structure of a metal or whether it is due to electrons moving within the covalent bonds of a semiconductor material. Note that in order to understand why the circuit model of a copper bus bar differs from the circuit model of a silicon transistor, we must understand the differences associated with these two modes of charge carrier flow. Describing the behavior of an electrical device in terms of the terminal voltage and current masks its internal behavior. Field theory often plays a vital role in the formulation of a circuit model. We can minimize the role of field theory in our present study because we are emphasizing the analysis techniques that take place once the circuit models are derived.

1.5 THE IDEAL BASIC CIRCUIT ELEMENT

At this point, we must define, at least in general terms, what we mean by an ideal basic circuit element. As we introduce each type of circuit element in subsequent chapters, we will discuss its characteristics more extensively. For now, we need only know that an ideal basic circuit element has three attributes: (1) It has only two terminals; (2) it is described mathematically in terms of the circuit variables of current and/or voltage; and (3) it cannot be subdivided into other elements. Figure 1.1 is a representation of an ideal basic circuit element. The box is left blank because we are making no commitment at this time as to the type of circuit element. In Fig. 1.1, the voltage across the terminals of the box is denoted by v and the current in the circuit element is denoted by i. The polarity reference for the voltage is indicated by the plus and minus signs, and the reference direction for the current is shown by the arrow placed alongside the current. The interpretation of these references can be summarized as follows. If the numerical value of v turns out to be a positive number, then there is a drop in voltage in going from terminal 1 to terminal 2. If the numerical value of v is negative, then there is a rise in voltage in going from terminal 1 to terminal 2. If the numerical value of i turns out to be positive, then we can interpret the result as positive charge carriers flowing from terminal 1 to terminal 2 (that is, in the direction of the arrow), or as negative charge carriers flowing from terminal 2 to terminal 1 (that is, opposite the direction of the arrow). If the numerical value of i is negative, then we know that positive charge carriers are flowing from terminal 2 to terminal 1 or that negative charge carriers are flowing from terminal 1 to terminal 2. Note that oppositely charged carriers flowing in opposite directions give the same algebraic sign to the current.

The assignment of the reference polarity for voltage and the reference direction for current is entirely arbitrary. However, once we have assigned the references, we must write all subsequent equations in agreement with the chosen references. The most widely used sign convention applied to these references is called the passive sign convention and will be used throughout this text. The passive sign convention can be stated as follows:

> Whenever the reference direction for the current in an element is in the direction of the voltage drop across the element (as in Fig. 1.1), use a positive sign in the expression that relates the voltage to the current. Otherwise use a negative sign.

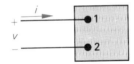

Fig. 1.1 The basic circuit element.

We will apply this sign convention in all the analysis that follows. Our purpose for introducing it even before we have introduced the different types of basic circuit elements is to impress on the student the fact that the selection of polarity references along with the adoption of the passive sign convention is not a function of the basic elements nor the type of interconnections made with the basic elements. We will present the application and interpretation of the passive sign convention in power calculations in Section 1.6.

Before we discuss power (and energy), two more comments about the ideal basic circuit element are in order. We use the adjective "ideal" to imply that a basic circuit element does not exist as a realizable physical component. However, as we implied in Section 1.3, ideal elements can be connected to model actual devices and systems. The word "ideal" does imply that we will be able to formulate a precise mathematical relationship between the terminal voltage and current. We use the adjective "basic" to imply that the circuit element cannot be further reduced or subdivided into other elements. Thus the basic circuit elements form the building blocks for constructing circuit models, but they themselves cannot be modeled with any other type of element.

1.6 POWER AND ENERGY

Power and energy calculations are also important in circuit analysis. One reason for this is that although current and voltage are useful variables in the analysis and design of electrically based systems, the useful output of the system is often nonelectrical and this output is conveniently expressed in terms of power or energy. Another reason is that all practical devices have limitations on the amount of power they can handle. In the design process, therefore, voltage and current calculations by themselves are not sufficient.

Our purpose now is to relate power and energy to the circuit variables of voltage and current and at the same time to discuss the power calculation in relation to the passive sign convention. We begin by first recalling from basic physics that power is the time rate of expending or absorbing energy. (A water pump rated 100 horsepower can deliver more gallons per minute than one rated 10 horsepower.) Mathematically, energy per unit time is expressed in the form of a derivative; thus we have

$$p = \frac{dw}{dt},$$ (1.3)

where

$$p = \text{the power in watts,}$$
$$w = \text{the energy in joules, and}$$
$$t = \text{the time in seconds.}$$

Thus 1 watt is equivalent to 1 joule per second.

The power associated with the flow of charge follows directly from the definition of current and voltage; in other words,

$$p = \frac{dw}{dt} = \left(\frac{dw}{dq}\right)\left(\frac{dq}{dt}\right) = vi, \tag{1.4}$$

where

$p =$ the power in watts,

$v =$ the voltage in volts, and

$i =$ the current in amperes.

From Eq. (1.4) we see that the power associated with a basic circuit element is simply the product of the current in the element and the voltage across the element. Therefore power is a quantity associated with a pair of terminals and we have to be able to tell from our calculation whether power is being delivered to the pair of terminals or extracted from the pair of terminals. This information comes from the correct application and interpretation of the passive sign convention.

If we use the passive sign convention, Eq. (1.4) is correct if the current polarity reference is in the direction of the voltage drop across the terminals. Otherwise, Eq. (1.4) must be written with a minus sign. In other words, if the current is referenced in the direction of a voltage rise across the terminals, the expression for the power is

$$p = -vi. \tag{1.5}$$

The algebraic sign of power is based on the fact that as positive charges move through a drop in voltage, they lose energy, and as they move through a rise in voltage, they gain energy.

The relationship between the polarity references for voltage and current and the expression for power is summarized in Fig. 1.2. Note that in all cases the algebraic sign assigned to the expression for power assumes that we are looking *toward* the terminals from the *outside* of the box.

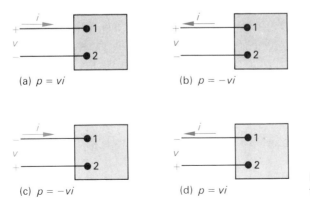

(a) $p = vi$

(b) $p = -vi$

(c) $p = -vi$

(d) $p = vi$

Fig. 1.2   Polarity references and the expression for power.

We can now state the rule for interpreting the algebraic sign of power.

> If the power is positive (that is, if $p > 0$), then power is being delivered to the circuit inside the box. If the power is negative (that is, if $p < 0$), then power is being extracted from the circuit inside the box.

For example, suppose we have selected the polarity references shown in Fig. 1.2(b). Assume further that our calculations for the current and voltage yield the following numerical results:

$$i = 4 \text{ A}$$

and

$$v = -10 \text{ V}.$$

Then the power associated with the terminal pair 1, 2 will be

$$p = -(-10)(4) = 40 \text{ W}.$$

Thus the circuit inside the box is absorbing 40 W.

Our calculations for current can be interpreted as equivalent to positive charge carriers entering terminal 2 and leaving terminal 1 or, alternatively, as negative charge carriers entering terminal 1 and leaving terminal 2. The voltage result tells us that we have a rise of 10 V in going from terminal 1 to terminal 2 or, alternatively, a drop of 10 V in going from terminal 2 to terminal 1.

Let us take our example one step further. Assume that a colleague is solving the same problem but she has chosen the reference polarities shown in Fig. 1.2(c). Her numerical values will be

$$i = -4 \text{ A},$$
$$v = 10 \text{ V},$$
$$p = 40 \text{ W}.$$

Note that when she interprets her results in terms of her reference system, she will reach precisely the same conclusions we have, namely, that the circuit inside the box is absorbing 40 W.

1.7 SUMMARY

The field of electrical engineering is quite diverse. One way of describing such a varied profession is in terms of four major types of systems that depend on electrical phenomena: communication systems, computer systems, control systems, and power systems. The discipline of study underlying the analysis, design, and operation of these systems is electromagnetic field theory. Circuit analysis, the subject of this text, is an extremely important special case of electromagnetic field theory.

The circuits approach to analysis and design focuses on interconnecting ideal basic elements in such a way that the interconnected elements will predict the behavior of actual devices within acceptable limits of accuracy. The definitions of the ideal basic circuit elements, along with descriptions of constraints imposed by the interconnections, depend on being able to define measurable circuit variables. The measurable circuit variables of primary interest are voltage and current.

The concepts of power and energy are also fundamental to the design of electrical systems. For this reason we have introduced the fundamental relationships between power, energy, voltage, and current.

In order for quantitative calculations to have any meaning, they must be based on a system of units that is recognized and acceptable to those working in the field of study. The International System of Units, designated SI, is used in this text because it is more widely accepted than any other as the common language in which scientific and technical results are expressed.

Once the circuit variables are defined, the analysis problem begins to take shape. Basically, the problem reduces to:

1. defining the relationship between the terminal voltage and current for each ideal basic circuit element; and

2. establishing the constraints imposed on the circuit variables by the interconnection.

After we understand the relationships implied in (1) and (2), we can proceed to write and solve the resulting set of equations that describes the circuit. We begin by introducing sources and resistors in Chapter 2.

PROBLEMS

1.1 In electronic circuits it is not unusual to encounter currents in the microampere range. Assume a 10-μA current is due to the flow of electrons.

    a) What is the average number of electrons per second that flow past a fixed reference cross section that is perpendicular to the direction of flow?

    b) Compare the size of this number to the number of centimeters in a straight line between the sun and the earth. You may assume the distance to the sun is 93 million miles.

1.2 How much energy is imparted to an electron as it flows through a 12-V battery from the positive to the negative terminal? Express your answer in attojoules.

1.3 A current of 2000 A exists in a rectangular (0.8-by-10 cm) copper bus bar. The current is due to free electrons moving through the bus bar at an average velocity of v meters/second. If the concentration of free electrons is 10^{29} electrons per cubic meter and if they are uniformly dispersed throughout the bus bar, then what is the average velocity of an electron?

1.4 The voltage and current at the terminals of a circuit element are referenced as shown in Fig. 1.2(a). Assume that the numerical values for v and i are 10 V and -6 A.

a) Calculate the power at the terminals and state whether the power is being absorbed or delivered by the element in the box.
b) Given that the current is due to electron flow, state whether the electrons are entering or leaving terminal 1.
c) Do the electrons gain or lose energy as they pass through the element in the box?

1.5 Two electrical circuits, represented by boxes A and B, are connected as shown in Fig. 1.3. The reference direction for the current, i, in the interconnection and the reference polarity for the voltage, v, across the interconnection are as shown in Fig. 1.3. For each of the following sets of numerical values, calculate the power in the interconnection and state whether the power is flowing from A to B or vice versa.

a) $i = 15$ A, $v = 20$ V   b) $i = -5$ A, $v = 100$ V
c) $i = 4$ A, $v = -50$ V   d) $i = -16$ A, $v = -25$ V

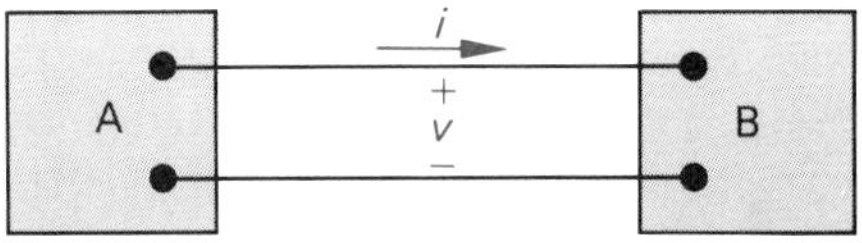

Fig. 1.3 The circuit for Problem 1.5.

CIRCUIT ELEMENTS

2

2.1 INTRODUCTION

We begin our study of circuits with five ideal basic circuit elements:

1. voltage sources,
2. current sources,
3. resistors,
4. inductors,
5. capacitors.

In this chapter we will discuss the characteristics of voltage sources, current sources, and resistors. We begin with these three elements for several reasons. First, ideal sources are used in almost every circuit model of a practical electrical system. Second, the resistive element is described by an algebraic relationship between the terminal voltage and current and this mathematical simplicity makes it an attractive starting point. Third, many practical systems can be modeled by combining only sources and resistors. The second and third reasons combine to give us the fourth reason for starting with sources and resistors, namely, that it enables us to learn the basic techniques of circuit analysis with only algebraic manipulations. When inductors and capacitors are introduced into our circuit models, we will be faced with the solution of integral and differential equations. However, even with inductors and capacitors the basic circuit techniques stay the same. Thus, by the time we start manipulating integral and differential equations, we will be familiar with the methods for writing circuit equations. This approach means that we must temporarily postpone our discussion of inductors and capacitors.

2.2 VOLTAGE AND CURRENT SOURCES

Before we discuss ideal voltage and current sources, it will be helpful to consider first the general nature of electrical sources. By the term "source" we mean a device that is capable of converting nonelectrical energy to electrical energy and vice versa. The conversion process can go either way. A discharging battery converts chemical energy to electrical energy, whereas a battery being charged converts electrical energy to chemical energy. A dynamo is a machine that can convert mechanical energy to electrical energy and vice versa. If operating in the mechanical-to-electrical mode, it is called a generator. If transforming from electrical to mechanical, it is referred to as a motor. The important thing to remember about these sources that are capable of a reversible transformation is that, in a given situation, they can be either delivering or absorbing electrical power. Our problem is to model these practical sources in terms of the ideal basic circuit elements. Now it turns out that practical sources can be described in general terms as devices that tend to maintain either voltage or current. This general behavior of practical devices led to the creation of the ideal voltage source and the ideal current source as basic circuit elements.

Ideal voltage and current sources can be divided into two broad categories: independent sources and dependent sources. When we say that a source is *independent*, we mean that it is independent of any other voltage or current that exists in the circuit to which the source is connected. A *dependent* source, on the other hand, depends on a voltage or current somewhere else in the circuit. These characteristics will become more meaningful after we have discussed some actual circuits. At the moment, we need only be aware that both types of source are used in building circuit models of practical devices. We begin our discussion with independent sources.

The *ideal independent voltage source* is a circuit element that will maintain a prescribed voltage across its terminals regardless of the current in the device. Such a voltage source is capable of generating the prescribed terminal voltage whether the current in the source is zero or finite. Since the terminal voltage is not a function of the current, the ideal independent voltage source is completely defined by the prescribed voltage. The reference polarity for the prescribed voltage must also be given. The graphical, or circuit, symbol for the ideal independent voltage source is shown in Fig. 2.1. The prescribed voltage, with its reference polarity, is given by the symbol v_s.

The *ideal independent current source* is a circuit element that will maintain a prescribed current within its terminals regardless of the voltage across its terminals. Such a current source is capable of generating the prescribed terminal current whether the terminal voltage is zero or finite. Since the current is not a function of the voltage across the terminals, the ideal independent current source is completely defined by the prescribed current and the reference direction. The circuit symbol for the ideal independent current source is shown in Fig. 2.2. The prescribed current is denoted by i_s. The reference direction is given by the arrow inside the circle.

An *ideal dependent,* or *controlled, voltage source* is a source in which the voltage across its terminals is determined by either a voltage or a current existing at some other location in the circuit. Thus we can have either a voltage-controlled voltage source or a current-controlled voltage source. The circuit symbol for a dependent voltage source is shown in Fig. 2.3. The diamond-shaped source is always used to indicate a dependent source. The voltage source v_s is controlled by either a voltage or a current somewhere else in the circuit. If we let v_x, or i_x, symbolize the controlling variable, we have either

$$v_s = \mu v_x$$

or

$$v_s = \rho i_x,$$

where μ and ρ are multiplying constants. Note that μ is dimensionless, whereas ρ carries the dimensions of volts/ampere. It is very important to note that whether we have an independent or a dependent voltage source, the cur-

Fig. 2.1 The circuit symbol for an ideal independent voltage source.

Fig. 2.2 The circuit symbol for an ideal independent current source.

Fig. 2.3 The circuit symbol for a dependent voltage source.

rent in the source cannot be expressed as a function of only its terminal voltage. In other words, knowing the terminal voltage of a voltage source, whether independent or dependent, is not sufficient information to state what current the source may be carrying.

An *ideal dependent,* or *controlled, current source* is a source in which the terminal current is determined by either a voltage or a current existing at some other location in the circuit. Thus we can have either a voltage-controlled current source or a current-controlled current source. The circuit symbol for a dependent current source is shown in Fig. 2.4. The current source i_s is controlled by either a voltage (v_x) or a current (i_x) somewhere else in the circuit; hence we will have either

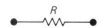

Fig. 2.4 The circuit symbol for a dependent current source.

$$i_s = \alpha v_x$$

or

$$i_s = \beta i_x,$$

where α and β denote multiplying constants. Note that β is dimensionless, whereas α carries the dimension of amperes/volt.

Dependent, or controlled, sources are especially useful in building circuit models of the electronic devices. Note that whether we have an independent or a dependent current source, the voltage across the source cannot be expressed as a function of only the terminal current. Thus knowing the current source, whether independent or dependent, is not sufficient information to state the terminal voltage of the source.

Because both independent and dependent sources are used to model devices that are capable of generating electrical energy, they are also referred to as *active elements.* The other three basic circuit elements (resistors, inductors, and capacitors) are referred to as *passive elements* because when they are used without sources, they can model only devices that cannot generate electrical energy.

2.3 ELECTRICAL RESISTANCE (OHM'S LAW)

Many useful electrical devices are designed to convert electrical energy to thermal energy. Stoves, toasters, irons, and space heaters are examples of household appliances that rely on this conversion process. All these appliances take advantage of the fact that thermal energy arises whenever charge carriers are caused to flow through a metal. The larger the charge flow, the larger the amount of energy converted to heat. This characteristic behavior of metals, such as copper and aluminum, is referred to as the *resistance* of the material to the flow of electrical charge. The circuit element used to model this behavior is the *resistor.* The circuit symbol for the resistor is shown in Fig. 2.5. The letter R is used to denote the resistance value of the resistor.

For purposes of circuit analysis, we must describe the relationship between the terminal voltage and current. There are two ways in which we can

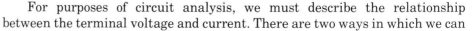

Fig. 2.5 The circuit symbol for a resistor having a resistance of R ohms.

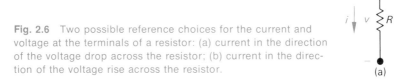

Fig. 2.6 Two possible reference choices for the current and voltage at the terminals of a resistor: (a) current in the direction of the voltage drop across the resistor; (b) current in the direction of the voltage rise across the resistor.

reference the current in the resistor relative to the terminal voltage. The reference for the current can be in the direction of either the voltage drop across the resistor, as shown in Fig. 2.6(a), or the voltage rise across the resistor, as shown in Fig. 2.6(b). If we choose the reference scheme shown in Fig. 2.6(a), then the relationship between the voltage and current is written as

$$v = iR, \tag{2.1}$$

where

$$v = \text{the voltage in volts,}$$
$$i = \text{the current in amperes, and}$$
$$R = \text{the resistance in ohms.}$$

If the reference scheme of Fig. 2.6(b) is used, then we must write

$$v = -iR, \tag{2.2}$$

where v, i, and R are, as before, measured in volts, amperes, and ohms, respectively. The algebraic signs used in Eqs. (2.1) and (2.2) are a direct consequence of the passive sign convention, which we introduced and adopted in Chapter 1. In both Eqs. (2.1) and (2.2), we assume that the resistance parameter itself is positive. There are occasions when a negative resistance appears in the circuit model of a device. The negative resistance implies that the device is a source of electrical energy but the exact physical interpretation is best discussed at the time when the detailed behavior of the device itself is being studied. In our study, we will always assume that R is a positive constant.

Both Eqs. (2.1) and (2.2) are known as *Ohm's law,* after George Simon Ohm, a German physicist who established its validity early in the nineteenth century. Ohm's law is the algebraic relationship to which we referred in Section 2.1. In the International System of Units, resistance is measured in ohms. The Greek letter omega (Ω) is the standard symbol for an ohm. Thus a resistor having a resistance of 8 ohms would appear in a circuit diagram as shown in Fig. 2.7.

Ohm's law is written with the voltage expressed as a function of the current. We will find it convenient in the work that follows to also express the current as a function of the voltage. Thus we have from Eq. (2.1)

$$i = \frac{v}{R}, \tag{2.3}$$

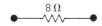

Fig. 2.7 The circuit symbol for an 8-ohm resistor.

or from Eq. (2.2)

$$i = -\frac{v}{R}.$$ (2.4)

The reciprocal of the resistance is referred to as *conductance,* is symbolized by the letter G, and is measured in siemens. The standard abbreviation for siemens is S. Thus

$$G = \frac{1}{R} \text{ S}.$$ (2.5)

An 8-Ω resistor has a conductance value of 0.125 S. We note in passing that in much of the professional literature, the unit used for conductance is the mho (ohm spelled backward) and is symbolized by an inverted omega ($\mho$). Therefore we could also describe an 8-Ω resistor as having a conductance of 0.125 mho, or 0.125 $\mho$.

One must keep in mind several important properties of the ideal resistor. First, the resistance is *constant*. It is not a function of the current in the resistor nor the voltage across the resistor. This ideal resistor is referred to as a *linear time-invariant* resistor. Thus the use of the ideal resistor to model an actual device implies that a linear model is a sufficiently accurate representation of the device. Second, the resistor is *bilateral*. In other words, if the polarity of the voltage reverses, the direction of the current reverses, and vice versa. Not all electrical devices are bilateral, so it is important to keep in mind that the ideal passive circuit elements (resistors, inductors, and capacitors) are bilateral elements. Third, the resistor is a *lumped element,* which means that the circuit element carries no information about spatial dimensions.

We can calculate the power at the terminals of a resistor in several ways. The first approach is to use the defining equation and simply calculate the product of the terminal voltage and current. If we use the reference system of Fig. 2.6(a) we write

$$p = vi.$$ (2.6)

The reference system shown in Fig. 2.6(b) requires that we write

$$p = -vi.$$ (2.7)

We can also express the power at the terminals of a resistor by writing the power in terms of the current and the resistance. Regardless of the reference system, we obtain

$$p = i^2 R.$$ (2.8)

Note that if we start with Eq. (2.6), we have

$$p = vi = (iR)i = i^2 R,$$

whereas if we begin with Eq. (2.7) we obtain

$$p = -vi = -(-iR)i = i^2 R.$$

It is very easy to see from Eq. (2.8) that the power at the terminals of a positive resistor is always positive and therefore a positive resistor always absorbs power from the circuit.

Our third method of expressing the power at the terminals of a resistor is in terms of the voltage and resistance. The expression is independent of the polarity references. We get

$$p = \frac{v^2}{R}. \tag{2.9}$$

Finally, we make the observation that Eqs. (2.8) and (2.9) can also be written in terms of the conductance—thus

$$p = \frac{i^2}{G} \tag{2.10}$$

and

$$p = v^2 G. \tag{2.11}$$

Now that we have introduced the general characteristics of ideal sources and the resistor, let us show how these elements can be used to build the circuit model of a practical system.

2.4 A CIRCUIT MODEL OF A FLASHLIGHT

We have chosen the flashlight as an example of a system that can be modeled by an electric circuit. If the reader is not familiar with the construction of a flashlight, we recommend disassembling one before proceeding with the remainder of this section. A photograph of a widely available flashlight is shown in Fig. 2.8(a). The disassembled flashlight showing the components that make up the system is illustrated in Fig. 2.8(b).

When a flashlight is regarded as an electrical system, the components of primary interest are (1) the batteries; (2) the lamp; (3) the connector; (4) the case; and (5) the switch. We now consider the circuit model for each of these components.

A dry-cell battery will maintain a reasonably constant terminal voltage if the current demand is not excessive. Thus if the dry-cell battery is operating within its intended limits, it can be modeled by an ideal voltage source. The prescribed voltage will be constant and equal to the sum of two dry-cell values. These observations are summarized in Fig. 2.9.

The ultimate output of the lamp is light energy, which is achieved by heating the filament in the lamp to a temperature high enough to cause radiation in the visible range. The filament can be modeled by an ideal resistor. Note in this case that although the resistor will account for the amount of electrical energy converted to thermal energy, it will not predict how much of the thermal energy is converted to light energy. The resistor used to represent the lamp will predict the steady current drain on the batteries, a characteristic of

(a)

reflector

case

lamp

batteries

coiled-wire
connector

(b)

Fig. 2.8   The flashlight viewed as an electrical system: (a) the flashlight; (b) the disassembled flashlight.

(a)

3 V

(b)

Fig. 2.9   The circuit model for two dry cells connected in an additive sense: (a) the actual device; (b) the circuit model.

Fig. 2.10   The circuit model for the flashlight lamp: (a) the bulb; (b) the circuit model.

the system that is also of interest. Figure 2.10 shows the use of a resistor to model the lamp. The lamp resistance is symbolized by R_l.

The connector used in the flashlight serves a dual role. First, it provides an electrical conductive path between the dry cells and the case. Second, it is formed into a springy coil so that it can also be used to apply mechanical pressure to the contacts between the batteries and the lamp. The purpose of this mechanical pressure is to minimize the contact resistance between the two dry cells and between the dry cells and the lamp. This observation brings out an important point: In choosing the wire to form the connectors, we may find that the mechanical properties of the wire will determine our choice of material and the size of the wire. Electrically, the connector can be modeled by an ideal resistor. In Fig. 2.11 we illustrate that the coiled connector is being modeled with a resistor.

The metal case of the flashlight is used to conduct current! That is, the case is one link in the electrical path between the batteries and the lamp. Since it is a metal conductor, its electrical behavior can be modeled by an ideal resistor, as shown in Fig. 2.12. Note that the case also serves two purposes: one electrical and one mechanical. If the flashlight you have disassembled has a plastic case, you will find a metal strip inside the case that connects the coiled connector to the switch. This strip is necessary because the plastic case cannot

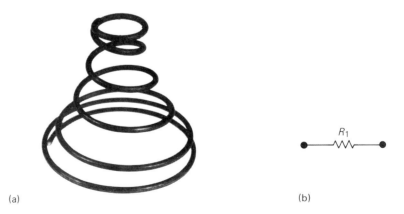

Fig. 2.11   The circuit model for the coiled connector used in the flashlight: (a) the coiled connector; (b) the circuit model.

(a)

(b)

Fig. 2.12 The circuit model for the metal flashlight case: (a) the metal case; (b) the circuit model.

be used as an electrical conductor. The metal strip is also modeled by an ideal resistor.

The final component we must consider is the switch. From an electrical point of view, the switch is a two-state device. It is either ON or OFF. An ideal switch will offer no resistance to the current when it is in the ON state, but will offer infinite resistance to current when it is in the OFF state. These two states represent the limiting values of a resistor—that is, the ON state corresponds to a resistor with a numerical value of zero and the OFF state corresponds to a resistor with a numerical value of infinity. The two extreme values are given the descriptive names "short circuit" ($R = 0$) and "open circuit" ($R = \infty$). Figure 2.13 shows the graphical representation of a short circuit and an open circuit. To represent the fact that a switch can be either a short circuit or an open circuit, depending on the position of its contacts, we use the symbol shown in Fig. 2.14(b).

We are now ready to construct the circuit model of the flashlight. In studying the components of the flashlight, we note that they are connected in tandem, or series. That is, starting with the dry-cell batteries, we see that the positive terminal of one cell is connected to the negative terminal of the second cell (Fig. 2.15). The positive terminal of the second cell is connected to one terminal of the lamp. The other terminal of the lamp makes contact with one side

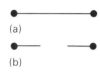

(a)

(b)

Fig. 2.13 The circuit symbol for (a) a short circuit and (b) an open circuit.

(a)

(b)

Fig. 2.14 The circuit symbol for a single-throw, single-pole switch: (a) the flashlight switch; (b) the circuit symbol.

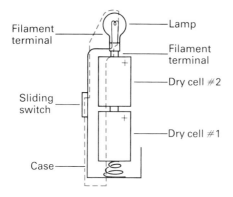

Fig. 2.15 Arrangement of the flashlight components.

of the switch, and the other side of the switch is connected to the metal case. The metal case is then connected to the negative terminal of the first dry cell by means of the metal spring. Note that we have formed a closed path or circuit as we pass through the elements. This closed path is indicated in Fig. 2.15 by the dashed line. The circuit model for the flashlight is shown in Fig. 2.16.

Before we begin the solution of the flashlight circuit, let us make some general observations about our model. First of all, note that we used the ideal resistor to model a lamp, a metal case, and a piece of coiled wire that provides mechanical pressure as well as an electrical connection. The choice of a resistor to model such diverse physical components should alert us to the fact that in selecting a circuit element we must concentrate on the electrical phenomenon that the element is representing. In this device, the resistor is used to model the flow of electrical charge through a metal. We can also see that the resistance of the lamp filament serves a useful function in the system, since it gives rise to the heat that produces the light output of the flashlight. On the other hand, the resistance of the flashlight case as well as that of the coiled connector account for unwanted or parasitic effects. That is, the heat dissipated in the case and connector produces no useful output and at the same time represents a drain on the dry cells. In building circuit models of devices, we must always be alert to these unwanted parasitic effects; otherwise the models we build may not adequately represent our system. Third, note that in building a circuit model for even this simple system we have made approximations. We have as-

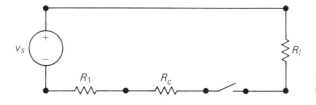

Fig. 2.16 The circuit model for the flashlight.

sumed an ideal switch. In practical switches, contact resistance may be high enough to interfere with the proper operation of the system. Our model will not predict this behavior. We have also assumed that the coiled connector exerts enough pressure to eliminate any contact resistance between the dry cells. Our model does not predict the possible deleterious effect of inadequate pressure. Our use of an ideal voltage source ignores any internal dissipation of energy in the dry cells. This could be accounted for by adding an ideal resistor in series with the source. Our model assumes the internal loss to be negligible.

The skill required to develop the circuit model of a device is every bit as demanding as the skill required to solve the derived circuit. In this text, we are emphasizing the skills required to solve circuits but we must recognize, right at the outset, that other skills are needed in the practice of electrical engineering.

2.5   ANALYSIS OF A FLASHLIGHT CIRCUIT (KIRCHHOFF'S LAWS)

Now that we have some appreciation for the effort involved in developing a circuit model, let us turn our attention to circuit analysis, using the circuit model of the flashlight as our starting point. We have redrawn the circuit (see Fig. 2.17) showing the switch in an ON state. We have also assigned the terminal voltage and current variable for each resistor element. For convenience, we have attached the same subscript to the voltage and current as was previously assigned to the resistor. Note that in assigning the circuit variables, we have also specified their reference polarities.

We say that we have solved a circuit when we know the voltage across and the current in every element. For the circuit at hand, we can identify seven unknowns: i_s, i_1, i_c, i_l, v_1, v_c, and v_l. Recall that v_s is a known voltage, since it represents the sum of the terminal voltages of the two dry cells, that is, a constant voltage of 3 V. How are we going to find these seven unknowns? We know from our study of algebra that to find n unknown quantities we must solve n simultaneous independent equations. From our discussion of Ohm's law in Section 2.3, we know that we can write three of these necessary equations, as follows:

$$v_1 = i_1 R_1, \tag{2.12}$$

$$v_c = i_c R_c, \tag{2.13}$$

$$v_l = i_l R_l. \tag{2.14}$$

Where do we get the other four equations?

The interconnection of circuit elements will impose constraints on the relationships between the terminal voltages and currents. These constraints are referred to as *Kirchhoff's laws*, after Gustav Kirchhoff, who first stated them in a paper published in 1848. The two laws that state the constraints in mathe-

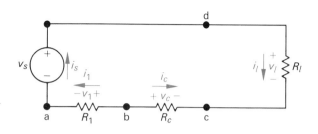

Fig. 2.17 The circuit model of the flashlight with assigned voltage and current variables.

matical form are known as *Kirchhoff's current law* and *Kirchhoff's voltage law*.

Before we can state Kirchhoff's current law, we must first define what we mean by a node. A node is simply a point in a circuit at which two or more circuit elements join. The nodes in the circuit of Fig. 2.17 have been labeled a, b, c, and d. (Observe that node d stretches clear across the top of the diagram. This is done for convenience. *Any uninterrupted line segment in a circuit diagram is always interpreted as a connection having zero resistance.* Note that this is consistent with the circuit representation of a short circuit.) We can now state the following.

> *Kirchhoff's current law:* The algebraic sum of all the currents at any node in a circuit equals zero.

An algebraic sign must be assigned to every current at the node. If we elect to assign a positive sign to a current leaving a node, then we must assign a negative sign to a current entering a node. The converse also applies—that is, if we give a negative sign to a current leaving a node, we must then give a positive sign to a current entering a node.

If we apply Kirchhoff's current law to the four nodes in the circuit shown in Fig. 2.17, using the convention that currents leaving a node are considered positive, we have the following four equations:

$$\text{node a} \qquad i_s - i_1 = 0, \qquad\qquad (2.15)$$

$$\text{node b} \qquad i_1 + i_c = 0, \qquad\qquad (2.16)$$

$$\text{node c} \qquad -i_c - i_l = 0, \qquad\qquad (2.17)$$

$$\text{node d} \qquad i_l - i_s = 0. \qquad\qquad (2.18)$$

Now we pause and make an extremely important observation. The four equations (2.15)–(2.18) are not an independent set, because any one of the four can be derived from the other three. Therefore in any circuit with n nodes, precisely $n - 1$ independent current equations can be derived from Kirchhoff's current law.[†] Let us choose to disregard Eq. (2.18) so that we have six inde-

---

† We will have more to say about this observation in Chapter 4.

pendent equations, namely, Eqs. (2.12)–(2.17). We need one more. We can derive our last equation from Kirchhoff's voltage law.

Before we can state Kirchhoff's voltage law, we must define what we mean by a closed path or loop. We trace a closed path in a circuit by starting at an arbitrarily selected node and then tracing through selected basic circuit elements in such a manner that we return to the original starting node without passing through any intermediate node more than once. In the circuit shown in Fig. (2.17), there is only one closed path or loop. For example, if we choose node d as our starting point and trace in a clockwise direction, we form the closed path by moving through nodes c, b, a, and then back to d. We can now state the following.

> *Kirchhoff's voltage law:* The algebraic sum of all the voltages around any closed path in a circuit equals zero.

The phrase "algebraic sum" implies that we must assign an algebraic sign to each voltage in the loop. As we trace around a closed path, a voltage will appear either as a rise or a drop in our tracing direction. If we assign a positive sign to a voltage rise, then we must assign a negative sign to a voltage drop. Conversely, if we give a negative sign to a voltage rise then we must give a positive sign to a voltage drop.

We now apply Kirchhoff's voltage law to the circuit shown in Fig. 2.17. We elect to trace clockwise around the closed path and at the same time assign a positive algebraic sign to voltage drops. Starting at node d leads to the expression

$$v_l - v_c + v_1 - v_s = 0, \qquad \textbf{(2.19)}$$

which represents the seventh independent equation needed to find the seven unknown circuit variables we mentioned earlier.

The thought of having to solve seven simultaneous equations to find the current delivered by a pair of dry cells to flashlight lamp is not a very pleasant one! Thus we assure you that in the coming chapters you will be introduced to techniques of analysis that will enable you to solve a simple one-loop circuit by writing a single equation! However, before moving on to a discussion of these circuit techniques, we will make several observations that arise from our detailed analysis of the flashlight circuit. These observations are true in general and, therefore, are important to the discussions that will unfold in subsequent chapters. They also support the contention that the flashlight circuit can be solved by defining a single unknown.

First, note that if we know the current in a resistor we also know the voltage across a resistor because current and voltage are directly related through Ohm's law. This means that we can think in terms of associating one unknown variable with each resistor, either the current or the voltage. Let us choose the current as the unknown variable. In other words, if we know the current in a resistor, we can find the voltage across the resistor. We can generalize our observation to say that if we know the current in a passive element, we can find the voltage across the passive element. The relationship between the current and voltage in inductors and capacitors is discussed in Chapter 5. The significance of adopting this point of view is that it greatly reduces the number of simultaneous equations we need to solve. For example, in the flashlight circuit we eliminate the voltages v_c, v_l, and v_1 as unknowns. Thus at the outset we have reduced our analysis problem to thinking in terms of solving four simultaneous equations rather than seven.

The second general observation we can make from our analysis of the flashlight circuit relates to the consequences of connecting only two elements to form a node. It follows directly from Kirchhoff's current law that when only two elements connect to a node, if you know the current in one of the elements, you also know it in the second element. In other words, you need define only one unknown current for the two elements. When just two elements connect together to form a node, the elements are said to be *in series*. (We will have much more to say about the series connection in Chapter 3.) The importance of this second observation is quite apparent when we note that each node in the circuit shown in Fig. 2.17 involves only two elements. Thus only one unknown current needs to be defined. This is substantiated by noting that Eqs. (2.15)–(2.17) lead directly to

$$i_s = i_1 = -i_c = i_e, \tag{2.20}$$

which tells us that if we know any one of the element currents we know them all. For example, if we choose to use i_s as the unknown, we eliminate i_1, i_c, and i_e. Our problem has been reduced to one unknown, namely, i_s.

DRILL EXERCISE
2.1

a) Show that Eq. (2.19) reduces to $i_s R_l + i_s R_c + i_s R_1 - v_s = 0$.

b) Write the explicit expression for i_s in terms of v_s, R_1, R_c, and R_l.

Ans. $i_s = v_s/(R_l + R_c + R_1)$.

DRILL EXERCISE
2.2

For the circuit shown, calculate (a) i_5; (b) v_1; (c) v_2; (d) v_5; (e) the power delivered by the 24-V source.

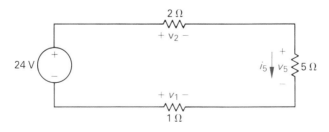

Ans. (a) $i_5 = 3$ A; (b) $v_1 = -3$ V; (c) $v_2 = 6$ V; (d) $v_5 = 15$ V; (e) 72 W.

DRILL EXERCISE Use Ohm's law and Kirchhoff's laws to find the value of R in the circuit shown.
2.3

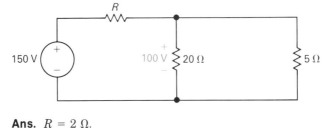

Ans. $R = 2\ \Omega$.

2.6 ANALYSIS OF A CIRCUIT CONTAINING A DEPENDENT SOURCE

We conclude this introduction to elementary circuit analysis with a discussion of a circuit that contains a dependent source (see Fig. 2.18). The circuit is of interest because it represents a structure that is encountered in the analysis and design of transistor amplifiers. The circuit elements inside the shaded box model the transistor. The development of this transistor circuit model need not concern us at this time. Our current interest is to analyze the circuit containing the model; in other words, we will determine the current in each element of the circuit. We assume all the circuit elements are known quantities,

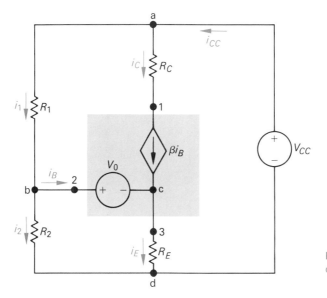

Fig. 2.18 A circuit with a dependent current source.

that is, we assume the values of R_1, R_2, R_C, R_E, V_{CC}, V_0, and β are known. We are also aware that if we know the currents in the circuit, we can find any voltage or power that is of interest. With this in mind, we note that there are six unknown currents, designated in Fig. 2.18 as i_1, i_2, i_B, i_C, i_E, and i_{CC}. In defining these six unknown currents, we have taken advantage of the fact that the resistor R_C is in series with the dependent current source βi_B. Our analysis problem is to derive six independent simultaneous equations involving these six unknowns. Three of these equations can be derived by applying Kirchhoff's current law to any *three* of nodes a, b, c, and d. Let us use a, b, and c. Choosing to label the currents away from a node as positive, we have

$$i_1 + i_C - i_{CC} = 0, \tag{2.21}$$

$$i_B + i_2 - i_1 = 0, \tag{2.22}$$

$$i_E - i_B - i_C = 0. \tag{2.23}$$

We can obtain a fourth independent equation using the constraint imposed by the dependent current source. That is, since R_C is in series with the dependent current source, we have

$$i_C = \beta i_B. \tag{2.24}$$

The remaining two equations can be derived using Kirchhoff's voltage law.

In applying Kirchhoff's voltage law to the circuit shown in Fig. 2.18, it is important to note that the voltage across the dependent current source is unknown and, furthermore, it cannot be expressed as a function of the source current, βi_B. Therefore, in selecting two closed paths, we deliberately avoid any path including the dependent current source. Thus we use the paths bcdb and badb. Choosing voltage drops to be positive, we have

$$V_0 + i_E R_E - i_2 R_2 = 0 \tag{2.25}$$

and

$$-i_1 R_1 + V_{CC} - i_2 R_2 = 0. \tag{2.26}$$

We will not discuss the algebraic manipulations involved in solving these six simultaneous equations, since our goal has been the derivation of the equations. We will report that the solution for i_B is

$$i_B = \frac{\dfrac{V_{CC} R_2}{(R_1 + R_2)} - V_0}{\dfrac{R_1 R_2}{R_1 + R_2} + (1 + \beta) R_E}. \tag{2.27}$$

You will be verifying Eq. (2.27) in Problem 2.8. Note that once i_B is known, we can easily obtain the remaining currents. Problem 2.9 will give you an opportunity to analyze the circuit of Fig. 2.18 when numerical values are assigned to R_1, R_2, R_C, R_E, V_{CC}, V_0, and β.

Let us conclude our discussion of this illustrative circuit with the observation that we defined the number of unknown variables in terms of the number of unknown currents. As with the flashlight circuit, we acknowledged the fact that once the currents are known, any unknown voltages or powers that are of interest can be easily calculated. We would also like to alert you to the fact that we will be using this circuit to illustrate some of the more powerful techniques of circuit analysis that are discussed in subsequent chapters. You, the student, can look forward to the time when you can derive Eq. (2.27) in a single step!

DRILL EXERCISE 2.4

For the circuit shown, find (a) the current i_1 in μA and (b) the voltage v in V.

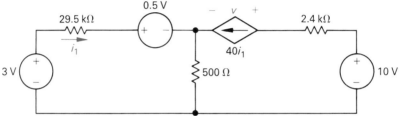

Ans. (a) $i_1 = 50$ μA; (b) $v = 4.175$ V.

DRILL EXERCISE 2.5

The current i_ϕ in the circuit shown is 5 A. Calculate:

a) v_s;

b) the power absorbed by the independent voltage source;

c) the power delivered by the independent current source;

d) the power delivered by the controlled current source;

e) the total power dissipated in the two resistors.

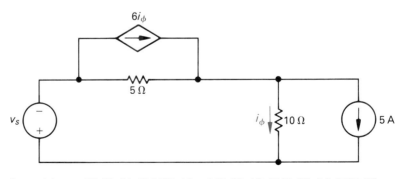

Ans. (a) $v_s = 50$ V; (b) 500 W; (c) -250 W; (d) 3000 W; (e) 2250 W.

2.7 SUMMARY

We have now introduced the characteristics of three basic circuit elements: voltage sources, current sources, and the resistor. We have also shown how an independent voltage source can be combined with some resistor elements to build a circuit model of a flashlight. The development of the equivalent circuit for a flashlight illustrated some of the problems associated with translating the actual system into a circuit model. We next discussed the analysis of the flashlight circuit in terms of the circuit variables current and voltage. This analysis led to the introduction of Kirchhoff's circuit laws, specifically, Kirchhoff's current law and Kirchhoff's voltage law. We further illustrated the application of Kirchhoff's laws with an analysis of a circuit containing a dependent source.

As we move on into circuit analysis, we should keep the following points in mind.

1. The circuit variables are current and voltage.
2. Power can be calculated from current and voltage.
3. Energy can be calculated from power and time.
4. The basic circuit elements define the relationship that exists between the current in the element and the voltage across the element.
5. Kirchhoff's current and voltage laws define the constraints that govern the interconnection of the circuit elements.

Circuit analysis can be described as the judicious application of (4) and (5).

PROBLEMS

2.1   A pair of automotive headlamps are connected to a 12-V battery via the arrangement shown in Fig. 2.19. In the figure, the triangular symbol ▼ is used to indicate that the terminal is connected directly to the metal frame of the car.

    a) Construct a circuit model using resistors and an independent voltage source.
    b) Identify the correspondence between the ideal circuit element and the system component that it represents.

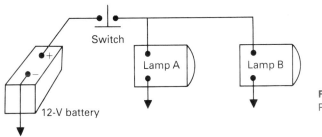

Fig. 2.19   The system for Problem 2.1.

2.2 A simplified circuit model for a residential wiring system is shown in Fig. 2.20.
   a) How many basic circuit elements are there in this model?
   b) How many nodes are there in the circuit?
   c) How many of the nodes connect three or more basic elements?
   d) Identify the circuit elements that form a series pair.
   e) What is the minimum number of unknown currents?
   f) Describe seven closed paths in the circuit.

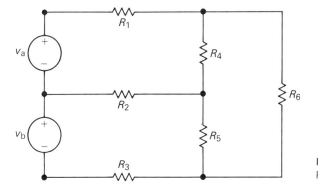

Fig. 2.20 The circuit for Problem 2.2.

2.3 The voltage v_o in the circuit shown in Fig. 2.21 is 30 V. Find (a) i_1; (b) v_g; and (c) the power delivered by the independent voltage source.

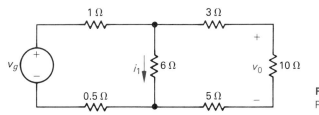

Fig. 2.21 The circuit for Problem 2.3.

2.4 Given the circuit shown in Fig. 2.22, find each of the following.
   a) The value of i_3
   b) The value of i_6
   c) The value of v_s
   d) The power dissipated in each resistor
   e) The power delivered by the 12-A source

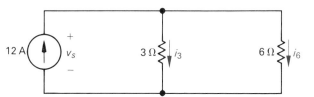
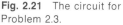

Fig. 2.22 The circuit for Problem 2.4.

2.5 The table in Fig. 2.23(a) gives the relationship between the terminal voltage and current of the practical constant voltage source shown in Fig. 2.23(b).
   a) Plot v_s vs. i_s.
   b) Construct a circuit model of the practical source that is valid for $0 \le i_s \le 10$ A. (Use an ideal voltage source in series with an ideal resistor.)

c) Use your circuit model to predict the current delivered to a 5.8-Ω resistor connected to the terminals of the practical source.

d) Use your circuit model to predict the current delivered to a short circuit connected to the terminals of the practical source.

e) What is the actual short-circuit current?

f) Explain why the answers to (c) and (d) are not the same.

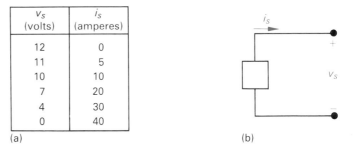

v_S (volts)	i_S (amperes)
12	0
11	5
10	10
7	20
4	30
0	40

(a)

(b)

Fig. 2.23 The circuit for Problem 2.5: (a) v_s and i_s data; (b) practical voltage source.

2.6 The data in the following table pertain to a practical current source.

i_t (terminal current), A	v_t (terminal voltage), V
10	0
8	4,000
6	8,000
4	11,000
2	14,000
0	16,000

a) Plot i_t vs. v_t.

b) Construct a circuit model of this current source that is valid for $0 \leq v_t \leq$ 8000 V. (Use an ideal current source in parallel with an ideal resistor.)

c) Use your circuit model to predict the current delivered to a 500-Ω resistor.

d) Use your circuit model to predict the open-circuit voltage of the current source.

e) What is the actual open-circuit voltage?

f) Explain why the answers to (d) and (e) are not the same.

2.7 Given the circuit shown in Fig. 2.24, calculate each of the following.

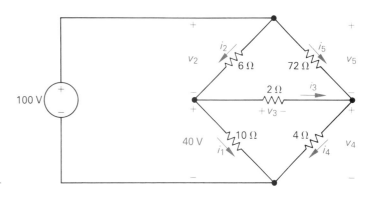

Fig. 2.24 The circuit for Problem 2.7.

a) i_1, i_2, i_3, i_4, and i_5
b) v_2, v_3, v_4, and v_5
c) The power supplied by the source
d) The power dissipated in each resistor
e) Does the power supplied by the source equal the total power dissipated?

2.8 Derive Eq. (2.27). [*Hint:* Use Eqs. (2.23) and (2.24) to express i_E as a function of i_B. Solve Eq. (2.22) for i_2 and substitute the result into both Eqs. (2.25) and (2.26). Solve the "new" Eq. (2.26) for i_1 and substitute this result into the "new" Eq. (2.25). Replace i_E in the "new" Eq. (2.25) and solve for i_B.] Note that since i_{CC} appears only in Eq. (2.21), the solution for i_B involves the manipulation of only five equations.

2.9 For the circuit shown in Fig. 2.18, $R_1 = 60$ kΩ, $R_2 = 40$ kΩ, $R_C = 2$ kΩ, $R_E = 600$ Ω, $V_{CC} = 6$ V, $V_0 = 600$ mV, and $\beta = 19$. Calculate i_B, i_C, i_E, v_{3d}, v_{bd}, i_2, i_1, v_{ab}, i_{CC}, and v_{13}. (*Note:* In the double subscript notation on voltage variables, the first subscript is positive with respect to the second subscript. See Fig. 2.25.)

2.10 Find v_1 and v_g in the circuit shown in Fig. 2.26 when v_o equals 10 V. (*Hint:* Start at the right end of the circuit and work back toward v_g.)

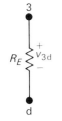

Fig. 2.25 The circuit for Problem 2.9.

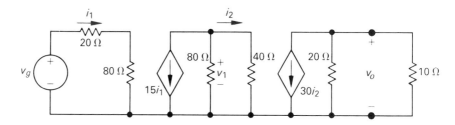

Fig. 2.26 The circuit for Problem 2.10.

SIMPLE RESISTIVE CIRCUITS

3

3.1 INTRODUCTION

At this point in our study, we are in a position to solve some simple circuits by the judicious application of Ohm's law and Kirchhoff's circuit laws. There are two reasons for our choosing to solve these simple circuits before introducing more elegant techniques of circuit analysis. First of all, it will give us a chance to become thoroughly familiar with the laws that underlie the more sophisticated methods, and second, it allows us to introduce some circuits that have important engineering applications. The sources in the circuits discussed in this chapter are limited to voltage and current sources that generate either constant voltages or currents, that is, voltages and currents that are invariant with time. We refer to sources of this type as *direct current,* or *dc,* sources. Historically a direct current was defined as a current produced by a constant voltage. Therefore a constant voltage became known as a direct current, or dc, voltage. One might think that if a constant current is called a direct current, a constant voltage is called a direct voltage. However, the term "direct current voltage," or "dc voltage," is universally used in science and engineering.

3.2 RESISTORS IN SERIES

We have already noted in our discussion of the circuit model of a flashlight that circuit elements may be constrained to carry the same current. *Circuit elements that carry the same current are said to be connected in series.* The resistors in the circuit shown in Fig. 3.1 are connected in series.

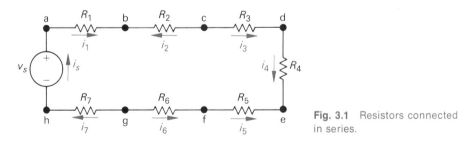

Fig. 3.1 Resistors connected in series.

It is easy to show that these resistors carry the same current by applying Kirchhoff's current law to each node in the circuit. The series interconnection in Fig. 3.1 requires that

$$i_s = i_1 = -i_2 = i_3 = i_4 = -i_5 = -i_6 = i_7, \tag{3.1}$$

which tells us that if we know any one of the seven currents we know them all. Thus Fig. 3.1 can be redrawn as shown in Fig. 3.2, where we have retained the

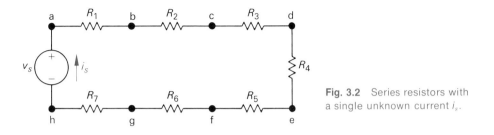

Fig. 3.2 Series resistors with a single unknown current i_s.

identity of the single current i_s. To find i_s, we apply Kirchhoff's voltage law around the single closed loop. Defining the voltage across each resistor as a drop in the direction of i_s, we have

$$- v_s + i_s R_1 + i_s R_2 + i_s R_3 + i_s R_4 + i_s R_5 + i_s R_6 + i_s R_7 = 0, \qquad (3.2)$$

which can be rewritten as

$$v_s = i_s(R_1 + R_2 + R_3 + R_4 + R_5 + R_6 + R_7). \qquad (3.3)$$

The significance of Eq. (3.3) is that insofar as calculating i_s is concerned, the seven resistors can be replaced by a single resistor whose numerical value is the sum of the individual resistors—that is,

$$R_{eq} = R_1 + R_2 + R_3 + R_4 + R_5 + R_6 + R_7 \qquad (3.4)$$

and

$$v_s = i_s R_{eq}. \qquad (3.5)$$

Thus the circuit in Fig. 3.2 can be redrawn as shown in Fig. 3.3.

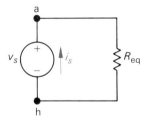

Fig. 3.3 A simplified version of the circuit shown in Fig. 3.2.

It follows directly that if k resistors are connected in series, the equivalent single resistor has a resistance equal to the sum of the k resistances. Mathematically, then,

$$R_{eq} = \sum_{i=1}^{k} R_i = R_1 + R_2 + \cdots + R_k. \qquad (3.6)$$

Another way to think about this concept of an equivalent resistance is to visualize the string of resistors as being inside a "black box." (*Note:* Electrical

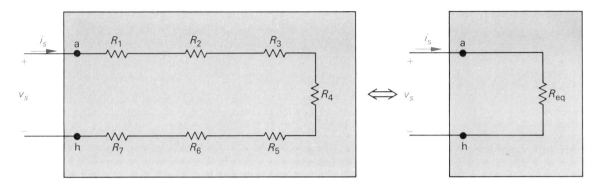

Fig. 3.4   The "black box" equivalent of the circuit shown in Fig. 3.2.

engineers use the term "black box" to imply an opaque container; that is, the contents are hidden from view. The engineer is then challenged to model the contents of the box by studying the relationship between the voltage and current at the terminals of the box.) Insofar as the relationship between the voltage and current at the terminals of the box is concerned, it is impossible to tell whether the box contains k resistors or a single equivalent resistor. This method of studying the circuit in Fig. 3.2 is illustrated in Fig. 3.4.

3.3   RESISTORS IN PARALLEL

Resistors are connected in parallel when the same voltage is applied to their terminals. The circuit in Fig. 3.5 is an illustration of resistors connected in parallel. The parallel resistors can be replaced by a single equivalent resistor, which is related to the individual resistors by the formula

$$\frac{1}{R_{eq}} = \frac{1}{R_1} + \frac{1}{R_2} + \frac{1}{R_3} + \frac{1}{R_4}. \tag{3.7}$$

Equation (3.7) can be derived by a direct application of Ohm's law and Kirchhoff's current law. We proceed as follows. In the circuit shown in Fig. 3.5 let the currents i_1, i_2, i_3, and i_4 be the currents in the resistors R_1 through R_4, respectively. Furthermore, let the positive reference direction for each resistor current be down through the resistor, that is, from node a to node b. From Kirchhoff's current law we have

$$i_s = i_1 + i_2 + i_3 + i_4. \tag{3.8}$$

The parallel connection of the resistors means that the voltage across each resistor must be the same. Hence, from Ohm's law we have

$$i_1 R_1 = i_2 R_2 = i_3 R_3 = i_4 R_4 = v_s. \tag{3.9}$$

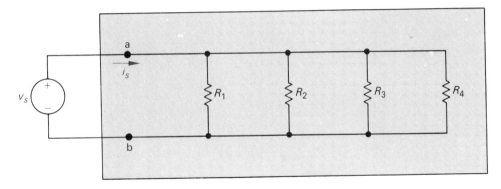

Fig. 3.5 Resistors in parallel.

It follows from Eq. (3.9) that

$$i_1 = \frac{v_s}{R_1},$$

$$i_2 = \frac{v_s}{R_2},$$

$$i_3 = \frac{v_s}{R_3}, \tag{3.10}$$

and

$$i_4 = \frac{v_s}{R_4}.$$

When Eq. (3.10) is substituted into Eq. (3.8) we have

$$i_s = v_s\left(\frac{1}{R_1} + \frac{1}{R_2} + \frac{1}{R_3} + \frac{1}{R_4}\right), \tag{3.11}$$

from which we can write

$$\frac{i_s}{v_s} = \frac{1}{R_{\text{eq}}} = \frac{1}{R_1} + \frac{1}{R_2} + \frac{1}{R_3} + \frac{1}{R_4}. \tag{3.12}$$

Equation (3.12) is what we set out to show. It follows that the four resistors in the circuit shown in Fig. 3.5 can be replaced by a single equivalent resistor. The circuit in Fig. 3.6 illustrates the substitution. For k resistors connected in parallel, Eq. (3.7) becomes

$$\frac{1}{R_{\text{eq}}} = \sum_{i=1}^{k} \frac{1}{R_i} = \frac{1}{R_1} + \frac{1}{R_2} + \cdot \cdot \cdot + \frac{1}{R_k}. \tag{3.13}$$

Sometimes it is more convenient to use conductance when dealing with resistors connected in parallel. In this case Eq. (3.13) becomes

$$G_{\text{eq}} = \sum_{i=1}^{k} G_i = G_1 + G_2 + G_3 + \cdot \cdot \cdot + G_k. \tag{3.14}$$

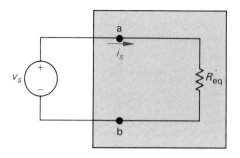

Fig. 3.6 Replacing the four parallel resistors shown in Fig. 3.5 with a single equivalent resistor.

Many times only two resistors are connected in parallel. Figure 3.7 illustrates this special case. The equivalent resistance can be calculated from Eq. (3.13). Thus

$$\frac{1}{R_{eq}} = \frac{1}{R_1} + \frac{1}{R_2} = \frac{R_2 + R_1}{R_1R_2}, \tag{3.15}$$

from which it follows directly that

$$R_{eq} = \frac{R_1R_2}{R_1 + R_2}. \tag{3.16}$$

Thus for *just two* resistors in parallel the equivalent resistance equals the product of the resistances divided by the sum of the resistances. A word of caution: "The product" divided by "the sum" applies *only* for two resistors in parallel.

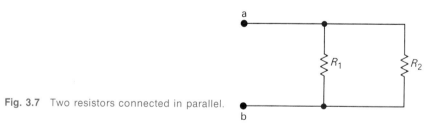

Fig. 3.7 Two resistors connected in parallel.

We conclude our discussion of resistors connected in series or in parallel with a numerical example that will illustrate the usefulness of these results.

Example 3.1 Find i_s, i_1, and i_2 in the circuit shown in Fig. 3.8.

Solution Our purpose is to illustrate how the three specified currents can be found using series–parallel simplifications of the circuit. We begin by noting that the 3-Ω resistor is in series with the 6-Ω resistor and, therefore, this series combination

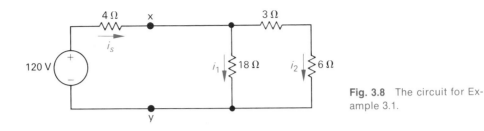

Fig. 3.8 The circuit for Example 3.1.

can be replaced by a 9-Ω resistor. Thus our circuit reduces to that shown in Fig. 3.9(a). Now we observe that the parallel combination of the 9-Ω and 18-Ω resistors can be replaced with a single resistance of $(18 \times 9)/(18 + 9)$, or 6 Ω. This further reduction of the circuit is shown in Fig. 3.9(b). The nodes x and y have been marked on all diagrams to facilitate tracing through the reduction of the circuit.

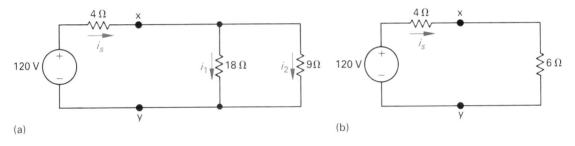

(a)

(b)

Fig. 3.9 A simplification of the circuit shown in Fig. 3.8.

From Fig. 3.9(b) you can verify that i_s will equal $120/10$, or 12 A. At this point in the analysis, we have the result shown in Fig. 3.10, in which we have added the voltage v_1 to help clarify our subsequent discussion. Using Ohm's law we can compute the value of v_1. Thus we have

$$v_1 = (12)(6) = 72 \text{ V}. \tag{3.17}$$

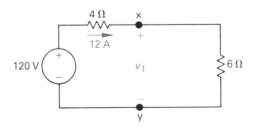

Fig. 3.10 The circuit in Fig. 3.9(b) showing the numerical value of i_s.

But v_1 is the voltage drop from node x to node y. Thus we can return to the circuit in Fig. 3.9(a) and again use Ohm's law to calculate i_1 and i_2. We get

$$i_1 = \frac{v_1}{18} = \frac{72}{18} = 4 \text{ A} \qquad (3.18)$$

and

$$i_2 = \frac{v_1}{9} = \frac{72}{9} = 8 \text{ A}. \qquad (3.19)$$

We have therefore found the three specified currents by using series–parallel reductions in combination with Ohm's law. ■

Before leaving this numerical example, we suggest that you take time to show that the solution satisfies Kirchhoff's current law at every node and Kirchhoff's voltage law around every closed path. (Note that there are three closed paths that can be tested.) It is also informative to show that the power delivered by the voltage source equals the total power dissipated in the resistors. See Problems 3.1 and 3.2.

DRILL EXERCISE
3.1

For the circuit shown, find (a) the voltage v; (b) the power delivered to the circuit by the current source; and (c) the power dissipated in the 10-Ω resistor.

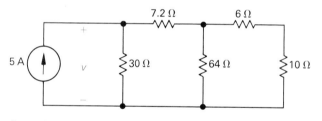

Ans. (a) 60 V; (b) 300 W; (c) 57.6 W.

3.4 THE VOLTAGE DIVIDER CIRCUIT

There are times, especially in electronic circuits, when it is necessary to develop more than one voltage level from a single voltage supply. This can be done by the circuit shown in Fig. 3.11, which is known as a *voltage divider circuit*.

This circuit can be analyzed by a direct application of Ohm's law and Kirchhoff's laws. To facilitate the analysis, we introduce the currents i and i_o, as shown in Fig. 3.11(b). We begin by assuming that the load current i_o is zero. Thus from Kirchhoff's current law we have R_1 and R_2 carrying the same cur-

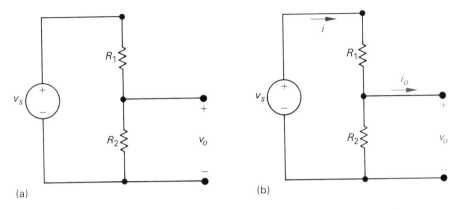

Fig. 3.11   (a) The voltage divider circuit. (b) The divider circuit currents i and i_o.

rent. Applying Kirchhoff's voltage law around the closed loop yields

$$v_s = iR_1 + iR_2 \tag{3.20}$$

or

$$i = \frac{v_s}{R_1 + R_2}. \tag{3.21}$$

Now we can use Ohm's law to calculate v_o:

$$v_o = iR_2 = v_s \frac{R_2}{(R_1 + R_2)}. \tag{3.22}$$

From Eq. (3.22), we see that v_o is a fraction of v_s, where the fraction is the ratio of R_2 to $R_1 + R_2$. Obviously this ratio is always less than 1.0; thus the output voltage v_o is less than the source voltage v_s.

If v_o and v_s are specified, there are an infinite number of combinations of R_1 and R_2 that will yield the proper ratio. For example, suppose that v_s equals 15 V and v_o is to be 5 V. Then $v_o/v_s = \frac{1}{3}$ and from Eq. (3.22), we find that this ratio will be satisfied whenever $R_2 = (\frac{1}{2})R_1$. Other factors that enter into the selection of R_1, and hence R_2, are (1) the power loss in the voltage divider, and (2) the value of the load resistor that will parallel R_2.

If the load on the voltage divider circuit is symbolized by R_l, as shown in Fig. 3.12, the expression for the output voltage becomes

$$v_o = \frac{R_{eq}}{R_1 + R_{eq}} v_s, \tag{3.23}$$

where

$$R_{eq} = \frac{R_2 R_l}{R_2 + R_l}. \tag{3.24}$$

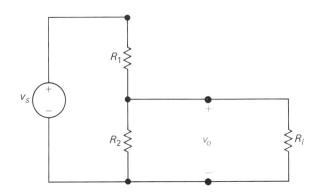

Fig. 3.12 A voltage divider connected to a load of R_l ohms.

When Eq. (3.24) is substituted into Eq. (3.23), we get

$$v_o = \frac{R_2}{R_1[1 + (R_2/R_l)] + R_2} v_s. \qquad \textbf{(3.25)}$$

Note that Eq. (3.25) reduces to Eq. (3.22) as R_l approaches infinity, as it should. The significance of Eq. (3.25) is that it shows that so long as $R_l \gg R_2$, the voltage ratio v_o/v_s is essentially undisturbed by the addition of the load on the divider.

DRILL EXERCISE
3.2

a) Find the no-load value of v_o.

b) Find v_o when R_l is 450 kΩ.

c) How much power is dissipated in the 30-kΩ resistor if the load terminals are accidently short-circuited?

d) What is the maximum power dissipated in the 50-kΩ resistor?

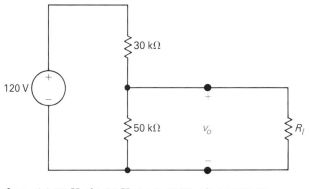

Ans. (a) 75 V; (b) 72 V; (c) 0.48 W; (d) 0.1125 W.

3.5 THE CURRENT DIVIDER CIRCUIT

The *current divider circuit,* shown in Fig. 3.13, consists of two resistors connected in parallel across a current source. The current divider is designed to
divide the current i between R_1 and R_2. The relationship between the current i
and the current in each resistor (that is, i_1 and i_2) can be found by a direct application of Ohm's law and Kirchhoff's current law. The voltage across the parallel resistors is

$$V = i_1 R_1 = i_2 R_2 = \frac{i R_1 R_2}{R_1 + R_2}. \tag{3.26}$$

From Eq. (3.26) we have

$$i_1 = \frac{i R_2}{R_1 + R_2}, \tag{3.27}$$

$$i_2 = \frac{i R_1}{R + R_2}. \tag{3.28}$$

Equations (3.27) and (3.28) tell us that the current divides between two resistors in parallel such that the current in either resistor equals the current
entering the parallel pair multiplied by the resistance of the other branch and
divided by the sum of the resistors. Example 3.2 illustrates the use of the current divider equation.

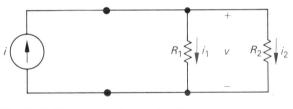

Fig. 3.13   The current divider circuit.

Example 3.2   Find the power dissipated in the 6-Ω resistor shown in Fig. 3.14.

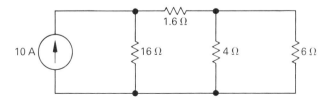

Fig. 3.14   The circuit for Example 3.2.

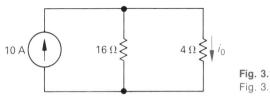

Fig. 3.15 A simplification of the circuit in Fig. 3.13.

Solution   We first must find the current in the resistor by simplifying the circuit using series–parallel reductions. Thus the circuit shown in Fig. 3.14 can be reduced to that shown in Fig. 3.15. The current i_o can be found using the formula for current division. Thus

$$i_o = \frac{(10)(16)}{(16 + 4)} = 8 \text{ A}.$$

Note that i_o is the current in the 1.6-Ω resistor in Fig. 3.14. Now i_o can be further divided between the 6-Ω and 4-Ω resistors. The current in the 6-Ω resistor is

$$i_6 = \frac{(8)(4)}{10} = 3.2 \text{ A},$$

and the power dissipated in the 6-Ω resistor is

$$p = (3.2)^2(6) = 61.44 \text{ W}. \quad \blacksquare$$

3.6   THE d'ARSONVAL METER MOVEMENT

Many instruments for electrical engineering measurements use a *d'Arsonval meter movement*, which is best described with the aid of Fig. 3.16. The movement consists of a movable coil placed in the field of a permanent magnet. The current in the coil creates a torque on the coil, which then rotates until this torque is exactly balanced by a restoring spring. As the coil rotates, it moves a pointer across a calibrated scale. The movement is designed so that the *deflection of the pointer is directly proportional to the current in the movable coil.* From a circuit point of view, the coil is described in terms of a voltage and current rating. For example, one commercially available meter movement is rated at 50 mV and 1 mA. The significance of these ratings is as follows: When the coil is carrying its rated current, the voltage drop across the coil is the rated coil voltage and the pointer is deflected to its full-scale position. The current and voltage ratings of the coil also specify the resistance of the coil. Thus a 50-mV, 1-mA movement has a resistance of 50 Ω.

The important thing to remember about instruments that use the d'Arsonval movement as a read-out mechanism is that the pointer deflection is governed by the current in the coil. The significance of the deflection traces back to what the coil current represents. In the next several sections, we will

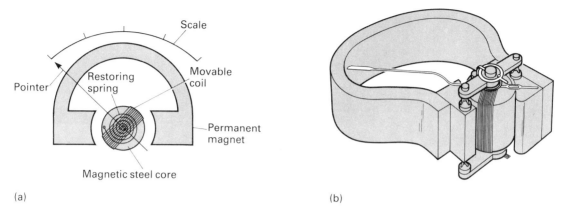

(a)

(b)

Fig. 3.16 The basic parts of a d'Arsonval meter: (a) a schematic diagram; (b) a pictorial diagram.

show how the d'Arsonval movement is used as an ammeter, a voltmeter, an ohmmeter, and a null detector.

3.7 THE AMMETER CIRCUIT

The direct current *ammeter circuit* consists of a d'Arsonval movement in parallel with a resistor, as shown in Fig. 3.17. The purpose of the shunting resistor R_A is to control the amount of current that passes through the meter movement. The shunting resistor R_A and the meter movement form a current divider. Thus for a given d'Arsonval movement, the full-scale reading of the ammeter is determined by R_A. Example 3.3 illustrates the calculations involved to determine R_A.

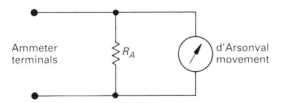

Fig. 3.17 A direct-current ammeter circuit.

Example 3.3 a) A 50-mV, 1-mA d'Arsonval movement is to be used in an ammeter where the full-scale reading is 10 mA. Determine R_A.

b) Repeat (a) if the full-scale reading is 1 A.

c) How much resistance is added to the circuit when the 10-mA ammeter is inserted to measure current?

d) Repeat (c) for the 1-A ammeter.

Solution   a) From the statement of the problem, we know that when the current at the terminals of the ammeter is 10 mA, then 9 mA must be diverted through R_A. At the same time we know that when the movement carries 1 mA, the drop across its terminals is 50 mV. Ohm's law requires that

$$9 \times 10^{-3}R_A = 50 \times 10^{-3},$$

or

$$R_A = 50/9 = 5.555 \ \Omega.$$

b) When the full-scale deflection of the meter is 1 A, R_A must carry 999 mA when the movement carries 1 mA. In this case, then, we have

$$999 \times 10^{-3}R_A = 50 \times 10^{-3},$$

or

$$R_A = \frac{50}{999} \cong 50.05 \ \text{m}\Omega.$$

c) Let R_m represent the equivalent resistance of the meter. For the 10-mA meter

$$R_m = \frac{50 \ \text{mV}}{10 \ \text{mA}} = 5 \ \Omega,$$

or, alternatively,

$$R_m = \frac{(50)(50/9)}{50 + 50/9} = 5 \ \Omega.$$

d) We have

$$R_m = \frac{50 \times 10^{-3}}{1} = 0.050 \ \Omega,$$

or, alternatively,

$$R_m = \frac{(50)(50/999)}{50 + 50/999} = 0.050 \ \Omega. \ \blacksquare$$

DRILL EXERCISE
3.3   a) Find the current in the circuit shown.

b) If the milliammeter in Example 3.3(a) is used to measure the current, what will it read?

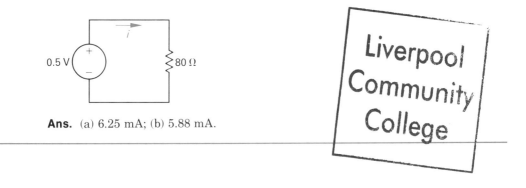

Ans. (a) 6.25 mA; (b) 5.88 mA.

3.8 THE VOLTMETER CIRCUIT

The direct-current *voltmeter circuit* consists of a d'Arsonval movement in series with a resistor, as shown in Fig. 3.18. The purpose of the series resistor R_v is to limit the voltage applied to the meter movement. The series resistor R_v and the d'Arsonval movement form a voltage divider. Thus for a given d'Arsonval movement, the full-scale reading of the voltmeter is determined by R_v. Example 3.4 illustrates the calculations involved to determine R_v.

Fig. 3.18 A direct-current voltmeter circuit.

Example 3.4
a) A 50-mV, 1-mA d'Arsonval movement is to be used in a voltmeter where the full-scale reading is 150 V. Determine R_v.
b) Repeat (a) if the full-scale reading is 5 V.
c) How much resistance does the 150-V meter insert into the circuit?
d) Repeat (c) for the 5-V meter.

Solution
a) Full-scale deflection requires 50 mV across the meter movement, and the movement has a resistance of 50 Ω. Therefore, we can apply Eq. (3.22) with $R_1 = R_v$, $R_2 = 50$, $v_s = 150$, and $v_o = 50$ mV. Thus

$$50 \times 10^{-3} = \frac{150(50)}{R_v + 50}.$$

Solving for R_v, we get

$$R_v = 149{,}950 \ \Omega.$$

b) For a full-scale reading of 5 V, we have

$$50 \times 10^{-3} = \frac{5(50)}{R_v + 50},$$

or

$$R_v = 4950 \ \Omega.$$

c) Let R_m represent the equivalent resistance of the meter. Then

$$R_m = \frac{150}{10^{-3}} = 150,000 \ \Omega,$$

or, alternatively,

$$R_m = 149,950 + 50 = 150,000 \ \Omega.$$

d) We have

$$R_m = \frac{5}{10^{-3}} = 5000 \ \Omega,$$

or, alternatively,

$$R_m = 4950 + 50 = 5000 \ \Omega. \quad \blacksquare$$

It is important to keep in mind that the insertion of either an ammeter or a voltmeter into a circuit disturbs the circuit in which the measurement is being made. An ammeter adds resistance in the branch in which the current is being measured, whereas a voltmeter adds resistance across the terminals, where the voltage is being measured. *How much the meters disturb the circuit in which the measurements are being made depends on the resistance of the meters in comparison to the resistances of the circuit.* If the resistance of the branch without the ammeter is large in comparison to the meter resistance, the insertion of the ammeter will have a negligible effect. If, however, the resistance of the branch is of the same order of magnitude as the ammeter resistance, the insertion of the meter could have a significant effect on the current in the branch. In the latter case, the current measured by the ammeter would not be the same as the current in the branch without the ammeter.

The loading effect of a voltmeter depends on the resistance of the voltmeter in comparison with the resistance that the voltmeter shunts in the circuit. The higher the total resistance of the voltmeter circuit, the smaller the loading effect. Some commercial voltmeters are given a sensitivity rating in ohms/volt so that the user can quickly determine the total resistance that the voltmeter adds to the circuit. For example, the 150-V voltmeter circuit discussed in Example 3.4 would be given a sensitivity rating of 1000 Ω/V, since the total resistance of the voltmeter is 150,000 Ω and the full-scale rating of the meter is 150 V. Direct current voltmeters that use the d'Arsonval meter movement can have sensitivity ratings ranging from 100 Ω/V to 20,000 Ω/V.

Problems 3.10, 3.12, and 3.13 are designed to illustrate why an awareness of possible meter loading effects is important. In concluding this introduction to the effects of meter loading, we hasten to point out that these effects are not peculiar to d'Arsonval meter movements. In any system in which we are making physical measurements, we must extract energy from the system in the process of making the desired measurements. The more energy we extract in relation to the amount of energy available in the system, the more severely we disturb what we are trying to measure. Therefore we must always be conscious of the burden that the measuring device imposes on the system being measured.

DRILL EXERCISE 3.4

a) Find the voltage v across the 75-kΩ resistor.

b) If the 150-V voltmeter of Example 3.4(a) is used to measure the voltage, what will it read?

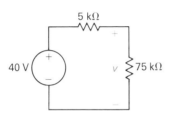

Ans. (a) 37.5 V; (b) 36.36 V.

3.9 THE OHMMETER CIRCUIT

The d'Arsonval *ohmmeter circuit,* shown in Fig. 3.19, consists of a meter movement in series with a battery and a regulating resistor. The operation of the ohmmeter circuit is as follows. The ohmmeter terminals are short-circuited

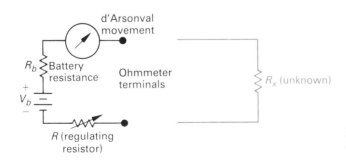

Fig. 3.19 A d'Arsonval ohmmeter circuit.

and the regulating resistor is adjusted to give full-scale deflection of the meter. This is labeled zero resistance on the meter scale. When an unknown resistance is connected to the ohmmeter, the deflection will be less than full scale. A calibrated scale (reading from right to left) can be constructed by connecting a series of known resistors across the ohmmeter and noting the deflection for each resistor. When an unknown resistance is connected to the ohmmeter, its value can be read from the calibrated scale. One of the disadvantages of the d'Arsonval ohmmeter circuit is the inherently nonuniform resistance scale. With a little thought, it should be apparent that the resistance scale will be cramped at the high-resistance end of the scale.

The successful operation of the d'Arsonval ohmmeter is dependent on a stable dc supply. The regulating resistor is used to compensate for changes in the internal resistance of the battery. That is, the regulating resistor enables $R + R_b$ to be held constant. Then so long as V_b is constant, the ohmmeter stays in calibration.

Although this ohmmeter circuit is not a precision instrument, it is an extremely useful one because it is so simple to use. Frequently, this type of ohmmeter is used to check the continuity of a circuit (that is to determine whether $R_x < \infty$).

DRILL EXERCISE
3.5

In the ohmmeter circuit shown, the voltage drop across the 50-μA ammeter and the internal resistance of the 1.5-V battery are negligible. The variable resistor R is adjusted to give full-scale deflection of the microammeter when R_x is zero.

a) What is the numerical value of R?

b) What is the midscale reading of the ohmmeter in ohms?

c) If the variable resistor R can be reduced to zero, how low can the battery voltage drop before the ohmmeter cannot be adjusted to read zero ohms when R_x is zero?

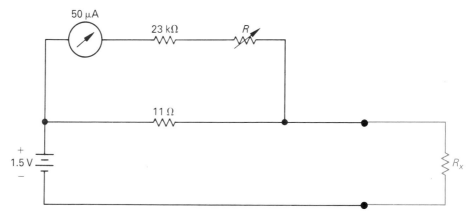

Ans. (a) 7 kΩ; (b) 11 Ω; (c) 1.15 V

3.10 THE POTENTIOMETER CIRCUIT

We have already noted that a dc voltmeter using a d'Arsonval movement possesses a finite resistance between its terminals. For instance, in Example 3.4(d) we noted that when a 50-mV, 1-mA movement is used to construct a voltmeter with a full-scale deflection of 5 V, the meter resistance is 5000 Ω. Suppose we used this meter to measure the terminal voltage of a dc voltage source that has an internal resistance of 3000 Ω. The circuit is shown in Fig. 3.20, in which it is apparent that the meter current is

$$i_m = \frac{5}{3000 + 5000} = \frac{5}{8} \text{ mA},$$

and hence the meter reading will correspond to $\frac{5}{8}$ of full scale, or

$$V_{\text{reading}} = \left(\frac{5}{8}\right)(5) = 3.125 \text{ V}.$$

The meter has loaded the source to the point where the meter reading is off by 37.5%. One way to reduce the effect of meter loading is to use a more sensitive d'Arsonval movement. (For example, if our 5-V voltmeter had been constructed using a 1-mV, 10-μA movement, the meter resistance would have been 500 kΩ and the meter would have indicated the source voltage to be (500/503)5, or 4.97 V.)

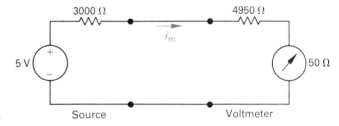

Fig. 3.20 The loading effect of a dc voltmeter.

The *potentiometer circuit* has also been developed to measure the terminal voltage of sources with high internal resistance. It is designed to overcome the loading effect by using a sensitive d'Arsonval movement as a null detector in a voltage comparison circuit. Shown in Fig. 3.21, the circuit consists of (1) a working battery (V); (2) a rheostat (R_r); (3) a slide-wire resistor (R_{ab}); (4) a d'Arsonval movement; (5) a standard cell (E); (6) a switch; and (7) a momentary contact switch. The ability of the potentiometer to measure an unknown voltage accurately is predicated on three assumptions: (1) The standard cell voltage is known; (2) the slide-wire resistance is uniformly distributed along the wire; and (3) the d'Arsonval movement is capable of detecting very small currents, typically in the microampere range.

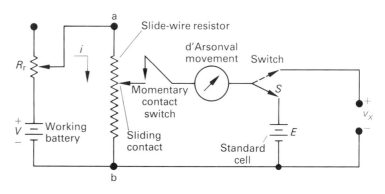

Fig. 3.21   A basic potentiometer circuit.

The operation of the potentiometer can be explained as follows. The voltage drop along the slide wire is calibrated by means of the standard cell. For purposes of discussion, let us assume that the standard cell voltage is 1 V and the slide wire is 200 cm long. Now we set the slide-wire contact at the center of the wire and adjust the rheostat R_r until the d'Arsonval movement indicates no current when it is momentarily connected to the slide-wire contact. At this point we have used the working battery and the rheostat to establish a 2-V drop along the slide wire. Specifically, we have established a voltage scale of 10 mV per centimeter. Now we switch the d'Arsonval detector to the unknown voltage V_x. We then move the slide-wire contact along the slide-wire resistor until the detector indicates zero current during a momentary connection with the slide wire. The position of the slide-wire contact is the value of V_x. Note that the current from the unknown voltage source is zero when the slide-wire contact is positioned to read its voltage. Thus there is no load on the unknown voltage source at the time when its voltage is read from the slide-wire resistance. (Also note that the standard cell current is zero at the time when the voltage calibration is established.) If V_x falls outside the range of the slide-wire voltage scale, we can use a precision voltage divider circuit to bring V_x within range of the potentiometer. Observe that the precision voltage divider circuit will not be loaded by the potentiometer.

One final comment about the potentiometer circuit. The d'Arsonval movement is constructed to read zero in the center of the scale so that it can deflect either up-scale or down-scale during the nulling procedure. This facilitates converging to a null reading. The scale on the detector is not calibrated to read volts (or amps) since only the null reading is of interest. When used in this fashion, the d'Arsonval movement is also referred to as a galvanometer.

DRILL EXERCISE
3.6

In the potentiometer circuit shown, the slide-wire branch consists of 15 discrete resistors, each having a resistance of 10 Ω, in series with a slide-wire resistor of 10 Ω. The rheostat R_r is adjusted so that the 4-V working battery delivers 10 mA after the potentiometer has been calibrated. Both the slide-wire dial and the discrete resistor dial (main dial) are calibrated in volts. The smallest voltage that can be read on the slide-wire dial is 0.5 mV. The main dial is calibrated in steps of 0.1 V, or 100 mV. The standard cell voltage is 1.15 V and the galvanometer movement has a resistance of 50 Ω.

a) After the potentiometer has been calibrated, what is the resistance of the rheostat R_r?

b) At the time when the potentiometer is calibrated, what is the setting of the main dial? The slide-wire dial?

c) The potentiometer is used to measure the voltage of a source that has an open-circuit voltage of 1.4865 V and an internal resistance of 56.6 Ω. What is the galvanometer current when the potentiometer dials are set to 1.4860 V?

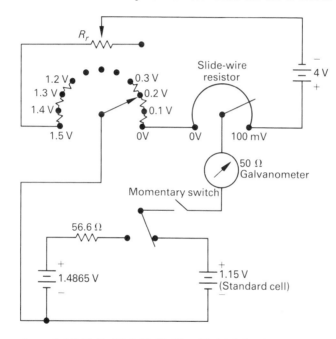

Ans. (a) 240 Ω; (b) 1.10 V, 50 mV; (c) 2.5 μA.

3.11 THE WHEATSTONE BRIDGE

The Wheatstone bridge circuit is a method used for the precision measurement of resistances of medium values, that is, resistances in the range of 1 Ω to 1 MΩ. In commercial models of the Wheatstone bridge, accuracies on the order of ± 0.1% are possible. The bridge circuit consists of four resistance branches, a dc voltage source (usually a battery), and a detector. The detector is generally a d'Arsonval galvanometer in the microamp range. The circuit arrangement of the resistances, battery, and detector is shown in Fig. 3.22, where R_1, R_2, and R_3 are known resistors and R_x is the unknown resistor.

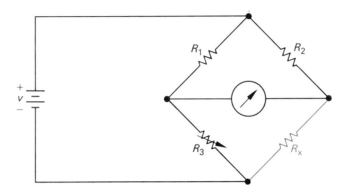

Fig. 3.22   The Wheatstone bridge circuit.

To find the value of R_x, we adjust the variable resistor, R_3, until there is no current in the microammeter branch of the bridge. The unknown resistor is calculated from the simple expression

$$R_x = \frac{R_2}{R_1} \cdot R_3. \tag{3.29}$$

The derivation of Eq. (3.29) follows directly from the application of Kirchhoff's circuit laws to the bridge circuit. The bridge circuit has been redrawn in Fig. 3.23 to show the branch currents that are appropriate to the derivation of Eq. (3.29). In studying the bridge circuit of Fig. 3.23, we note that when i_g is zero, that is, when the bridge is balanced, Kirchhoff's current law requires that

$$i_1 = i_3 \tag{3.30}$$

and

$$i_2 = i_x. \tag{3.31}$$

Now, since i_g is zero, there is no voltage drop across the detector, and therefore

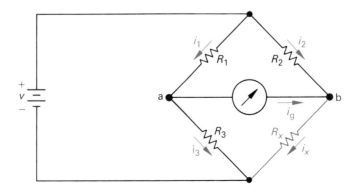

Fig. 3.23   A balanced Wheatstone bridge ($i_g = 0$).

points a and b are at the same potential. Thus when the bridge is balanced, Kirchhoff's voltage law requires that

$$i_3 R_3 = i_x R_x \tag{3.32}$$

and

$$i_1 R_1 = i_2 R_2. \tag{3.33}$$

Combining Eqs. (3.30) and (3.31) with Eq. (3.32) gives us

$$i_1 R_3 = i_2 R_x. \tag{3.34}$$

We obtain Eq. (3.29) by first dividing Eq. (3.34) by Eq. (3.33) and then solving the resulting expression for R_x; thus

$$\frac{R_3}{R_1} = \frac{R_x}{R_2}, \tag{3.35}$$

from which we have

$$R_x = \frac{R_2}{R_1} \cdot R_3. \tag{3.36}$$

Now that we have verified the validity of Eq. (3.29), several comments about the result are in order. First, note that if the ratio R_2/R_1 is unity, the unknown resistor R_x is equal to R_3. In this case, the bridge resistor R_3 must vary over a range that includes the value R_x. For example, if the unknown resistance were 1000 Ω and R_3 could be varied from 0 to 100 Ω, the bridge could never be balanced. Thus in order to cover a wide range of unknown resistors, we must be able to vary the ratio R_2/R_1. In a commercial Wheatstone bridge, R_1 and R_2 consist of decimal values of resistances that can be switched into the bridge circuit. Normally, the decimal values are 1, 10, 100, and 1000 Ω so that the ratio R_2/R_1 can be varied from 0.001 to 1000 in decimal steps. The variable resistor R_3 is usually adjustable in integral values of resistance from 1 to 11,000 Ω.

Although Eq. (3.29) implies that R_x could vary from zero to infinity, there are second-order effects that enter into the operation of the bridge and limit the practical range of R_x from approximately 1 Ω to 1 MΩ. Very low resistances are difficult to measure on the standard Wheatstone bridge because of thermoelectric voltages generated at the junctions of dissimilar metals or because of thermal heating effects—that is, $i^2 R$ effects. Very high resistances are difficult to measure accurately because of leakage currents. In other words, if R_x is very large, the current leakage in the electrical insulation may be compariable to the current in the branches of the bridge circuit.

DRILL EXERCISE
3.7
The bridge circuit in Fig. 3.22 is balanced when $R_1 = 100$ Ω, $R_2 = 1000$ Ω, and $R_3 = 150$ Ω. The bridge is energized from a 5-V dc source.

a) What is the value of R_x?

b) If each bridge resistor is capable of dissipating 250 mW, will the bridge be damaged if left in the balanced state?

Ans. (a) 1500 Ω; (b) no; the total power delivered to the bridge is 110 mW.

3.12 DELTA-TO-WYE (OR PI-TO-TEE) EQUIVALENT CIRCUITS

The bridge configuration in Fig. 3.22 introduces an interconnection of resistances that warrants further discussion. If the galvanometer is replaced by its equivalent resistance, R_m, we can draw the circuit as shown in Fig. 3.24. The interesting thing about this circuit is that the interconnected resistors cannot be reduced to a single equivalent resistance across the terminals of the battery if we are resirricted to the simple series or parallel equivalent circuits that were introduced earlier in this chapter. The interconnected resistors can be reduced to a single equivalent resistor by means of a delta-to-wye (Δ-to-Y) or pi-to-tee (π-to-T) equivalent circuit.[†]

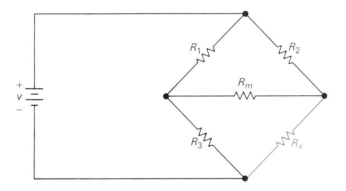

Fig. 3.24 A resistive network generated by a Wheatstone bridge circuit.

The resistors R_1, R_2, and R_m or R_3, R_m, and R_x in the circuit shown in Fig. 3.24 are referred to as a delta (Δ) interconnection because the interconnection looks like the Greek letter Δ. It is also referred to as a pi (π) interconnection because the Δ can be shaped into a π without disturbing the electrical equivalence of the two configurations. The electrical equivalence between the Δ- and the π-interconnections is apparent from Fig. 3.25.

[†] Δ and Y structures occur in a variety of useful circuits (not just resistive networks) and therefore we will find the Δ-to-Y transformation a helpful tool in circuit analysis.

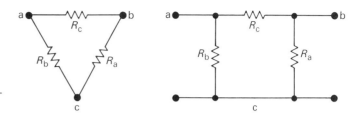

Fig. 3.25 A Δ-configuration viewed as a π-configuration.

The Δ-to-Y equivalent circuit means that the Δ-configuration can be replaced by a Y-configuration such that the terminal behavior of the two configurations will be identical. The Δ-to-Y transformation is illustrated in Fig. 3.26, where the two circuits shown are equivalent at the terminals a, b, and c provided that

$$R_1 = \frac{R_b R_c}{R_a + R_b + R_c},\tag{3.37}$$

$$R_2 = \frac{R_c R_a}{R_a + R_b + R_c},\tag{3.38}$$

$$R_3 = \frac{R_a R_b}{R_a + R_b + R_c}.\tag{3.39}$$

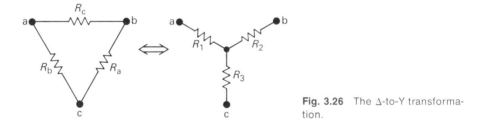

Fig. 3.26 The Δ-to-Y transformation.

The Y-configuration is also referred to as the tee (T) configuration, because the Y-structure can be replaced by the T-structure without disturbing the electrical equivalence of the two structures. The electrical equivalence of the Y- and the T-configurations is apparent from Fig. 3.27.

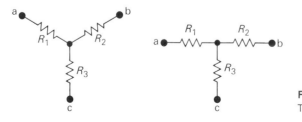

Fig. 3.27 A Y-structure viewed as a T-structure.

Equations (3.37)–(3.39) give the Y-connected resistors as functions of the Δ-connected resistors. It is also possible to reverse the Δ-to-Y (π-to-T) transformation. That is, we can start with the Y-structure and replace it with an equivalent Δ-structure. The expressions for the three Δ-resistors R_a, R_b, and R_c as functions of the three Y-resistors R_1, R_2, and R_3 are

$$R_a = \frac{R_1R_2 + R_2R_3 + R_3R_1}{R_1}, \tag{3.40}$$

$$R_b = \frac{R_1R_2 + R_2R_3 + R_3R_1}{R_2}, \tag{3.41}$$

$$R_c = \frac{R_1R_2 + R_2R_3 + R_3R_1}{R_3}. \tag{3.42}$$

The derivation of Eqs. (3.37)–(3.42) can be accomplished by noting that the two circuits are by definition equivalent with respect to their terminal behavior. Thus if each circuit is placed in a black box we can't tell by external measurements whether the box contains a set of Δ-connected resistors or a set of Y-connected resistors. This can be true only if the resistance between corresponding terminal pairs is the same for each box. For example, we must see the same resistance between terminals a and b whether we use the Δ-connected set or the Y-connected set. It follows directly that

$$R_{ab} = \frac{R_c(R_a + R_b)}{R_a + R_b + R_c} = R_1 + R_2, \tag{3.43}$$

$$R_{bc} = \frac{R_a(R_b + R_c)}{R_a + R_b + R_c} = R_2 + R_3, \tag{3.44}$$

$$R_{ca} = \frac{R_b(R_c + R_a)}{R_a + R_b + R_c} = R_1 + R_3. \tag{3.45}$$

We can obtain Eqs. (3.37)–(3.42) by a straightforward algebraic manipulation of Eqs. (3.43)–(3.45). (See Problem 3.22, which gives you hints on how to start the manipulations.)

Example 3.5 illustrates the use of a Δ-to-Y transformation to simplify the analysis of a circuit.

Example 3.5   Find the current and power supplied by the 40-V source in the circuit shown in Fig. 3.28.

Solution   Since we are interested only in the current and power drain on the 40-V source, our problem is solved once we know the equivalent resistance across the terminals of the source. This equivalent resistance can be found easily after we have replaced either the upper Δ (100 Ω, 125 Ω, 25 Ω) or the lower Δ (40 Ω, 25 Ω, 37.5 Ω) by its equivalent Y. We choose to replace the upper Δ. The three Y-

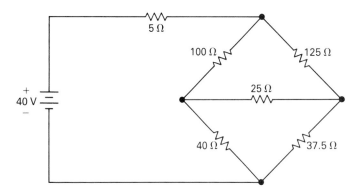

Fig. 3.28   The circuit for Example 3.5.

resistors are defined in Fig. 3.29 and computed from Eqs. (3.37)–(3.39). Thus

$$R_1 = \frac{(100) \times (125)}{250} = 50 \ \Omega,$$

$$R_2 = \frac{125 \times 25}{250} = 12.5 \ \Omega,$$

$$R_3 = \frac{100 \times 25}{250} = 10 \ \Omega.$$

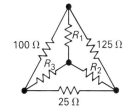

Fig. 3.29   The equivalent Y-resistors.

When the Y-resistors are substituted into the circuit shown in Fig. 3.28, we have the circuit in Fig. 3.30 as a result. From Fig. 3.30, we can easily calculate the resistance across the terminals of the 40-V source by series–parallel

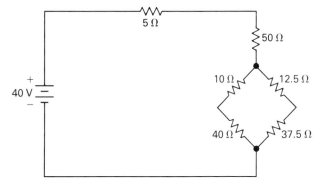

Fig. 3.30   A simplified version of the circuit in Fig. 3.28.

simplifications. Thus we have

$$R_{eq} = 55 + \frac{(50)(50)}{100} = 80 \; \Omega.$$

Our final step is to note that the circuit reduces to an 80-Ω resistor across a 40-V source, as shown in Fig. 3.31, from which it is apparent that the 40-V source will deliver 0.5 A and 20 W to the circuit. ■

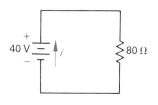

Fig. 3.31 The final step in the simplification of the circuit in Fig. 3.28.

DRILL EXERCISES
3.8

a) Use a delta-to-wye transformation to find the current i in the circuit shown.

b) Find v_1 and v_2. (*Hint:* Use the circuit that exists after the Δ-to-Y transformation has been made.)

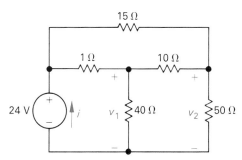

Ans. (a) 1 A; (b) $v_1 = 23.2$ V, $v_2 = 21$ V.

3.13 SUMMARY

In this chapter we have studied some simple resistive circuits that can be analyzed by straightforward application of Ohm's law and Kirchhoff's circuit laws. At the same time we were studying how to use these fundamental laws, we also learned how to simplify combinations of resistors through series–parallel equivalents and delta-to-wye transformations. We analyzed the voltage divider and current divider circuits and introduced their use in simple d'Arsonval meters. We also studied the potentiometer and bridge circuits to show how resistive structures can be used to make measurements on the basis of balancing voltages to create a current null. We are now ready to introduce some elegant methods of circuit analysis that facilitate the study of more complex circuit structures.

PROBLEMS

3.1 a) Show that the solution of the circuit in Fig. 3.8 (see Example 3.1) satisfies Kirchhoff's current law at junctions x and y.

b) Show that the solution of the circuit in Fig. 3.8 satisfies Kirchhoff's voltage law around every closed loop.

3.2 a) Find the power dissipated in each resistor in the circuit shown in Fig. 3.8.

b) Find the power delivered by the 120-V source.

c) Show that the power delivered equals the power dissipated.

3.3 Find the equivalent resistance R_{ab} for each of the circuits in Fig. 3.32.

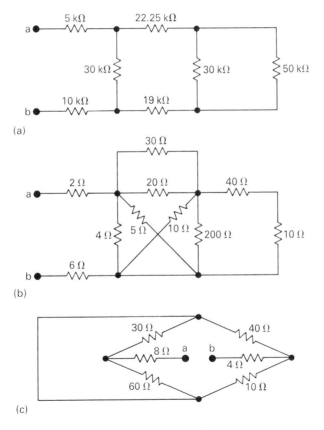

(a)

(b)

(c)

Fig. 3.32 The circuits for Problem 3.3.

3.4 a) Find the voltage drop across the 50-Ω resistor in the circuit shown in Fig. 3.33.

b) Find the voltage drop across the 60-Ω resistor in the circuit shown in Fig. 3.33.

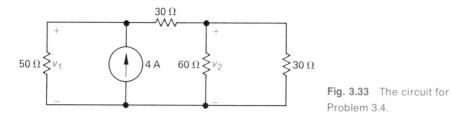

Fig. 3.33 The circuit for Problem 3.4.

3.5 For the circuit shown in Fig. 3.34:

   a) calculate I_0;

   b) calculate the power dissipated in the 14-Ω resistor.

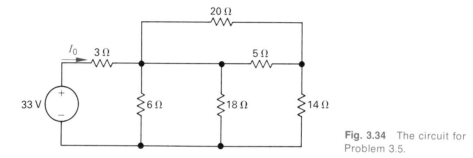

Fig. 3.34 The circuit for Problem 3.5.

3.6 The current in the 2-Ω resistor in the circuit in Fig. 3.35 is 8 A as shown.

   a) Find the voltage v_{ab}.

   b) Find the power dissipated in the 20-Ω resistor.

   c) What percentage of the total power delivered to the terminals a, b is dissipated in the 2-Ω resistor?

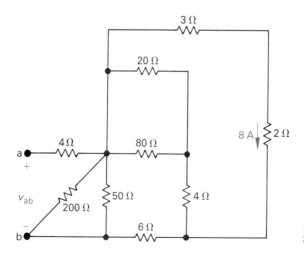

Fig. 3.35 The circuit for Problem 3.6.

3.7 a) Calculate the no-load voltage v_o for the voltage divider circuit shown in Fig. 3.36.

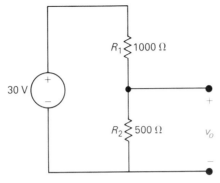

Fig. 3.36 The circuit for Problem 3.7.

b) Calculate the power dissipated in R_1 and R_2.

c) Assume that only 0.25-W resistors are available. The no-load voltage is to be the same as in part (a). Specify the ohmic values of R_1 and R_2.

3.8 The no-load voltage in the voltage divider circuit shown in Fig. 3.37 is 6 V. The smallest load resistor that is ever connected to the divider is 3000 Ω. When the divider is loaded, v_o is not to drop below 5 V.

a) Specify the numerical value of R_1 and R_2.

b) What is the maximum power dissipated in R_1?

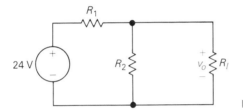

Fig. 3.37 The circuit for Problem 3.8.

3.9 A shunt resistor and a 50-mV, 1-mA d'Arsonval movement are used to build a 25-A ammeter. A resistance of 667 $\mu\Omega$ is placed across the terminals of the ammeter. What is the new full-scale range of the ammeter?

3.10 a) Calculate the current i in the circuit shown in Fig. 3.38(a).

b) What will the 10-mA ammeter of Example 3.3 read if it is inserted in series with the voltage source?

c) Repeat parts (a) and (b) for the circuit in Fig. 3.38(b).

d) Why is it important to know how much resistance the meter inserts into the circuit?

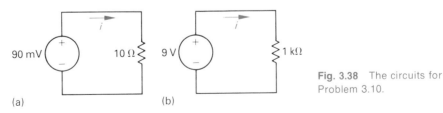

(a)   (b)

Fig. 3.38 The circuits for Problem 3.10.

3.11 a) Show for the ammeter circuit in Fig. 3.39 that the current in the d'Arsonval movement is always one tenth of the current being measured.

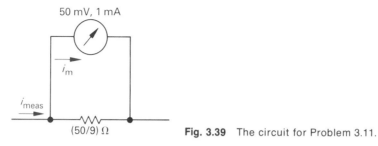

Fig. 3.39   The circuit for Problem 3.11.

b) What would the fraction be if the 50-mV, 1-mA movement were used in a 1-A ammeter?

c) Would you expect a uniform scale on a dc d'Arsonval ammeter?

3.12   A d'Arsonval ammeter having a full-scale reading of 100 mA is used to measure the current in the 3-Ω branch in the circuit shown in Fig. 3.40. The d'Arsonval movement is rated 10 mA and 100 mV.

a) What is the ammeter reading?

b) What is the current in the 3-Ω branch with the ammeter removed?

c) What is the percent error in the measured current based on the actual current?

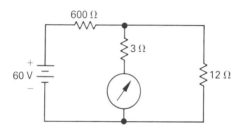

Fig. 3.40   The circuit for Problem 3.12.

3.13   The voltage divider circuit shown in Fig. 3.41 is designed so that the no-load output voltage is six-tenths of the input voltage. A d'Arsonval voltmeter having a sensitivity of 1000 Ω/V and a full-scale rating of 30 V is used to check the operation of the circuit.

a) What will the voltmeter read if it is placed across the 25-V source?

b) What will the voltmeter read if it is placed across the 15-kΩ resistor?

c) What will the voltmeter read if it is placed across the 10-kΩ resistor?

d) Will the voltmeter readings obtained in parts (b) and (c) add to the reading recorded in part (a)? Explain why.

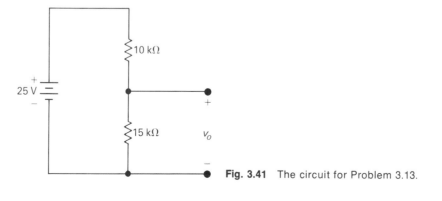

Fig. 3.41   The circuit for Problem 3.13.

3.14 A multirange voltmeter consisting of a 200-mV, 10-mA meter movement and three resistances, R_1, R_2, and R_3, is shown in Fig. 3.42. The desired voltage ranges are 15 V, 150 V, and 300 V. The 15-V, 150-V, and 300-V terminals are as shown in Fig. 3.42.

a) Determine R_1, R_2, and R_3.
b) Assume that a 1500-Ω resistor is connected between the 15-V terminal and the common terminal. The voltmeter is then connected to an unknown voltage using the common terminal and the 300-V terminal. The voltmeter reads 250 V. What is the unknown voltage?

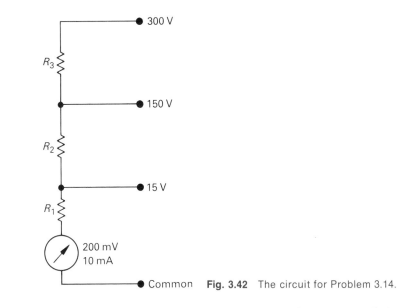

Fig. 3.42 The circuit for Problem 3.14.

3.15 You have been told that the dc voltage of a power supply is about 400 V. When you go to the instrument room to get a dc voltmeter to measure the power supply voltage, you find that there are only two dc voltmeters available. One voltmeter is rated 150 V full scale and has a sensitivity of 100 Ω/V. The second voltmeter is rated 300 V full scale and has a sensitivity of 150 Ω/V.

a) How can you use the two voltmeters to check the power supply voltage?
b) If the power supply voltage is 390 V, what will each voltmeter read?
c) Would you be able to use this technique if the supply were 450 V? Explain.

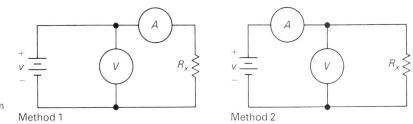

Fig. 3.43 The circuits for Problem 3.16.

Method 1   Method 2

3.16 Figure 3.43 shows two possible ways of connecting a voltmeter and an ammeter in order to determine the unknown resistance R_x. State which method of connecting the voltmeter and the ammeter you would recommend if you knew the following:

The meters are d'Arsonval types with movements rated at 1 mA and 50 mV.
The voltmeter is designed to read 300 V at full scale.
The ammeter is designed to read 5 mA at full scale.
The supply voltage is 150 V.
The resistor is known to be greater than 50 kΩ.
The power rating of the resistor is 0.5 W.

Justify your answer by showing the error arising in each method for (a) $R_x = 50$ kΩ and (b) $R_x = 100$ kΩ.

3.17   The components in the ohmmeter circuit shown in Fig. 3.44 are as follows: A is a 500-μA ammeter; B is a battery with negligible internal resistance and a terminal voltage of 4.5 V; R_1 is a fixed resistor of 2500 Ω; and R is an adjustable resistor.

a) Determine the setting of R so that the microammeter will deflect to its full-scale position when the ohmmeter terminals a, b are short-circuited.
b) Specify the resistance that the ohmmeter should indicate at 100% of full scale; 90% of full scale; 50% of full scale; 10% of full scale; and 0% of full scale.

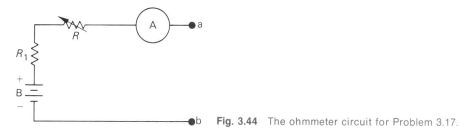

Fig. 3.44   The ohmmeter circuit for Problem 3.17.

3.18   A multirange ohmmeter circuit is shown in Fig. 3.45. Note that the circuit uses a shunting resistor in parallel with the microammeter branch as well as resistors in series with the microammeter. The resistors R_3 through R_{10} have the following numerical values:

$$R_3 = 20 \ \Omega,$$
$$R_4 = 200 \ \Omega,$$
$$R_5 = 2950 \ \Omega,$$
$$R_6 = 29{,}500 \ \Omega,$$
$$R_7 = 990 \ \Omega,$$
$$R_8 = 54 \ \Omega,$$
$$R_9 = 110 \ \Omega,$$
$$R_{10} = 10 \ \Omega.$$

a) Determine the center-scale resistance reading of the ohmmeter for the four positions of the selector switch, that is positions 1, 2, 3, and 4. Assume that the adjustable resistor is set to 500 Ω for all positions of the selector switch.
b) If the center-scale calibration of positions 1, 2, 3, and 4 on the multirange ohmmeter read as 30, 300, 3000, and 30,000, respectively, will the ohmmeter indicate the correct resistance to within ±10%?

3.19   The following data apply to the slide-wire potentiometer shown in Fig. 3.21.

The length of the slide-wire is 200 cm.
The resistance of the slide-wire is 100 Ω.
The working battery voltage is 5 V.

TECHNIQUES OF CIRCUIT ANALYSIS

4

4.1 INTRODUCTION

Up to this point we have been able to analyze relatively simple resistive circuits by the intelligent application of Kirchhoff's laws in combination with Ohm's law. This approach can be used for all circuits, but as the circuits become structurally more complicated and involve more and more elements, we will soon find this direct method quite cumbersome. Our purpose in this chapter is to introduce two powerful techniques of circuit analysis that facilitate the analysis of complex circuit structures: the node-voltage method and the mesh-current method. In addition to these two general methods of analysis, we will also discuss some additional techniques for simplifying circuits. We have already seen how series–parallel reductions and delta-to-wye transformations can be used to simplify a given structure. We will now add source transformations and Thévenin–Norton equivalent circuits to our list of simplification techniques.

Before we begin our discussion of the node-voltage and mesh-current methods of circuit analysis, let us pause and reflect on the groundwork that has been laid to this point. In Chapter 1 we introduced current and voltage as the two variables used to describe the behavior of the basic circuit elements. In Chapter 2 we discussed voltage and current sources along with the circuit parameter of resistance. These three types of basic circuit elements were chosen to start our discussion of circuit analysis because all the basic analytical techniques can be explored using interconnections of just these elements. In both Chapters 2 and 3 we introduced circuit analysis through the direct application of Ohm's law and Kirchhoff's laws. Ohm's law is critical because it describes the relationship between the current and voltage at the terminals of a resistor, and Kirchhoff's laws are important because they describe the constraints imposed on currents and voltages due to the interconnections of the basic elements. We also noted in Section 2.6 in our discussion of both the flashlight circuit and the dependent source circuit that we can simplify the analysis problem by first concentrating on finding the element currents. We are now ready to introduce the node-voltage and mesh-current methods of circuit analysis. Keep in mind that these two methods are of interest because they give us two systematic methods for describing circuits with the minimum number of simultaneous equations.

4.2 TERMINOLOGY

In order to discuss the more elegant methods of circuit analysis, we must define a few basic terms that will enable us to give a clear, concise description of the important features of a given circuit. Thus far all of our circuits have been *planar circuits,* that is, those circuits that can be drawn on a plane such that no branches cross over each other. A circuit that can be drawn with branches crossing over each other is still considered planar if it can be redrawn with no crossover branches. For example, the circuit shown in Fig. 4.1(a) is a planar

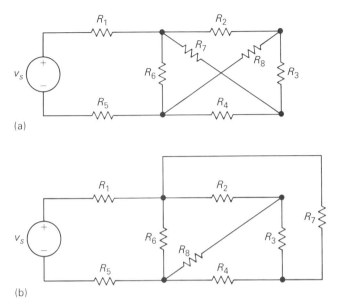

(a)

(b)

Fig. 4.1 (a) A planar circuit. (b) The circuit in part (a) redrawn to verify that it is planar.

circuit since it can be redrawn as shown in Fig. 4.1(b). An example of a nonplanar circuit is shown in Fig. 4.2.

The node-voltage method can be applied to both planar and nonplanar circuits, whereas the mesh-current method is limited to planar circuits. For nonplanar circuits, the mesh-current method is replaced by a technique known as the loop-current method. Although we will not discuss the loop-current method, we wish to call to your attention the fact that once you understand the mesh-current method, the transition to the loop-current method is not difficult.

We have already defined (Section 1.5) what we mean by an ideal basic circuit element. When the basic circuit elements are interconnected to form a circuit, the resulting interconnection is described in terms of nodes, paths, branches, loops, and meshes. We defined both a node and a closed path, or loop, in Section 2.5. Our purpose here is to review those definitions and then expand our vocabulary to include the terms path, branch, and mesh.

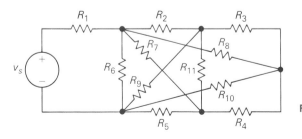

Fig. 4.2 A nonplanar circuit.

A *node* is a point in a circuit where two or more circuit elements join.

A *path* is formed whenever a set of adjoining basic circuit elements is traced, in order, without passing through a connecting node more than once.

A *closed path*, or *loop*, is created by starting at a selected node and then tracing through a set of connected basic circuit elements in such a manner that we return to the original starting node without passing through any intermediate node more than once.

A *branch* is a path that connects two nodes.

A *mesh* is a special type of loop; that is, it does not contain any other loops within it.

These characteristics of a circuit are illustrated with reference to the circuit shown in Fig. 4.3, a careful study of which will reveal there are

1. seven nodes (a, b, c, d, e, f, and g);

2. ten branches (v_1, R_1, R_2, R_3, v_2, R_4, R_5, R_6, R_7, and I); and

3. four meshes (v_1–R_1–R_5–R_3–R_2, v_2–R_2–R_3–R_6–R_4, R_5–R_7–R_6, and R_7–I).

Also note that there are several loops that are not meshes. For example, v_1–R_1–R_5–R_6–R_4–v_2 forms a closed loop, but this is not a mesh because other closed loops can be found within it. A path that is neither closed nor a branch is the trace v_2–R_2–R_3–R_5–R_7.

We will find in the work that follows that it is often convenient to identify only those nodes in the circuit that join *three or more* elements. We will refer to such nodes as *essential nodes*. We also find it convenient to identify only those paths that connect essential nodes *without passing through an essential node*. We will refer to such paths as *essential branches*. In the circuit shown in Fig. 4.3 there are four essential nodes (b, c, e, and g) and seven essential branches (v_1–R_1; R_2–R_3; v_2–R_4; R_5; R_6; R_7; and I). Note that in general the number of essential nodes will be less than or equal to the number of nodes and

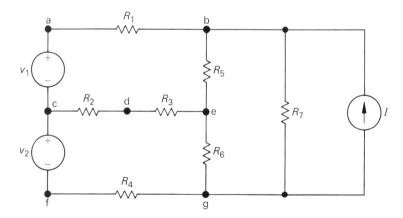

Fig. 4.3 A circuit used to illustrate nodes, branches, meshes, paths, and loops.

the number of essential branches will be less than or equal to the number of branches.

Why are we interested in noting how many nodes, branches, and meshes a given circuit has? It turns out that these attributes of the circuit tell us how many simultaneous equations we must derive in order to solve the circuit. This can be explained as follows. The number of unknown currents in the circuit will equal the number of branches *where the current is not known*. We let b represent this number. For example, in the circuit shown in Fig. 4.3 there are nine branches where the current is unknown. We know from elementary algebra that we must have b independent equations in order to solve a circuit with b unknown currents. If we let n represent the number of nodes in the circuit, we know that we can derive $n - 1$ independent equations by applying Kirchhoff's current law to any set of $n - 1$ nodes. (The application of the current law to the nth node would not generate an independent equation since this equation can be derived from the previous $n - 1$ equations. See Drill Exercise 4.2.) Knowing that we need b equations to describe a given circuit and knowing further that we can obtain $n - 1$ of these equations from Kirchhoff's current law, it follows that the remaining $[b - (n - 1)]$ equations must come from applying Kirchhoff's voltage law to independent loops or meshes.

We see, then, that by counting branches, nodes, and meshes we have established a systematic method for writing the necessary number of equations to solve a given circuit. Specifically, we will apply Kirchhoff's current law to $n - 1$ junctions and Kirchhoff's voltage law to $b - (n - 1)$ independent loops (or meshes). These observations are also valid if we think in terms of essential nodes and essential branches. Thus if we let n_e represent the number of essential nodes and b_e the number of essential branches *where the current is unknown*, then we can apply Kirchhoff's current law at $n_e - 1$ nodes and Kirchhoff's voltage law around $b_e - (n_e - 1)$ loops or meshes.

A circuit may consist of disconnected parts so we must alert you to the fact that the statements pertaining to the number of equations that can be derived from Kirchhoff's current law $(n - 1)$ and voltage law $[b - (n - 1)]$ apply to connected circuits. If a circuit has n nodes and b branches and is made up of s parts, then the current law can be applied $n - s$ times and the voltage law $b - n + s$ times. We also make the observation that any two separate parts can be connected by a *single* conductor. This connection will always cause two nodes to form into one node. Furthermore, no current will exist in the single conductor. This means that any circuit made up of s disconnected parts can always be reduced to a connected circuit.

Let us illustrate this systematic approach to deriving the simultaneous equations that describe a connected circuit in terms of its unknown currents. We will use the circuit shown in Fig. 4.3 and write the equations on the basis of essential nodes and branches. We have already noted that the circuit has four essential nodes and six essential branches where the current is unknown. The circuit in Fig. 4.3 has been redrawn in Fig. 4.4 with the six unknown currents defined by i_1 through i_6.

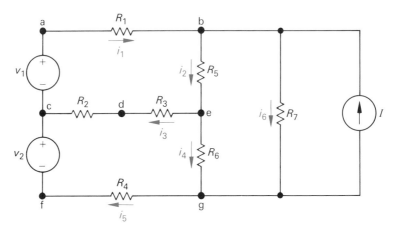

Fig. 4.4 The circuit in Fig. 4.3 with the six unknown branch currents defined.

Three of the six simultaneous equations needed to describe this circuit are derived by applying Kirchhoff's current law to any three of the four essential nodes. We will use the nodes b, c, and e to get

$$-i_1 + i_2 + i_6 - I = 0,$$
$$i_1 - i_3 - i_5 = 0, \qquad \textbf{(4.1)}$$
$$i_3 + i_4 - i_2 = 0.$$

The remaining three equations are derived by applying Kirchhoff's voltage law around three meshes. We have already noted that the circuit has four meshes. We will dismiss the mesh R_7–I because we don't know the voltage across I.† Using the other three meshes in the order in which they are described in item (3) above gives us

$$R_1 i_1 + R_5 i_2 + i_3(R_2 + R_3) - v_1 = 0,$$
$$-i_3(R_2 + R_3) + i_4 R_6 + i_5 R_4 - v_2 = 0, \qquad \textbf{(4.2)}$$
$$-i_2 R_5 + i_6 R_7 - i_4 R_6 = 0.$$

When the six simultaneous equations given by Eqs. (4.1) and (4.2) are rearranged to facilitate their solution, we get the set

$$-i_1 + i_2 + 0i_3 + 0i_4 + 0i_5 + i_6 = I,$$
$$i_1 + 0i_2 - i_3 + 0i_4 - i_5 + 0i_6 = 0,$$
$$0i_1 - i_2 + i_3 + i_4 + 0i_5 + 0i_6 = 0,$$
$$R_1 i_1 + R_5 i_2 + (R_2 + R_3)i_3 + 0i_4 + 0i_5 + 0i_6 = v_1, \qquad \textbf{(4.3)}$$
$$0i_1 + 0i_2 - (R_2 + R_3)i_3 + R_6 i_4 + R_4 i_5 + 0i_6 = v_2,$$
$$0i_1 + R_5 i_2 + 0i_3 - R_6 i_4 + 0i_5 + R_7 i_6 = 0.$$

† We will have more to say about this decision in Section 4.8.

We note in passing that if we sum the current at the nth node (g in this example) we would get the equation

$$i_5 - i_4 - i_6 + I = 0. \tag{4.4}$$

Equation (4.4) is not independent because it can be derived by adding the three equations in Eq. (4.1) and then multiplying the sum by -1. Thus Eq. (4.4) is a linear combination of the equations in Eq. (4.1) and therefore is not independent of them.

Now that we have illustrated a method for deriving the simultaneous equations that describe a circuit in terms of the unknown branch currents, we hasten to point out that we are not content to stop with this systematic formulation of the equations. We wish to carry our procedure one step further. We find that by introducing new variables we will be able to describe a circuit with just $(n - 1)$ equations or just $b - (n - 1)$ equations. Therefore these new variables allow us to obtain a solution by manipulating fewer equations, a desirable goal even if a computer is going to be used to obtain a numerical solution. The new variables are known as *node voltages* and *mesh currents*. The node-voltage method enables us to describe a circuit in terms of $n - 1$ [or $n_e - 1$] equations, whereas the mesh-current method enables us to describe a circuit in terms of $b - (n - 1)$ [or $b_e - (n_e - 1)$] equations. We begin with the node-voltage method.

DRILL EXERCISE
4.1 For the circuit shown, state the numerical value of the number of (a) branches, (b) branches where the current is unknown, (c) essential branches, (d) essential branches where the current is unknown, (e) nodes, (f) essential nodes, and (g) meshes.

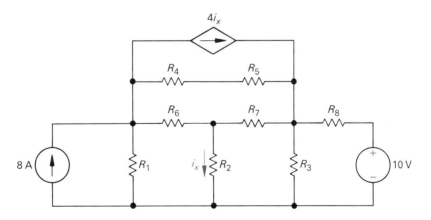

Ans. (a) 11; (b) 9; (c) 9; (d) 7; (e) 6; (f) 4; (g) 6.

DRILL EXERCISE
4.2 A current leaving a node is defined as positive.

a) Sum the currents at each node in the circuit shown.

b) Show that any one of the equations in part (a) can be derived from the remaining two equations.

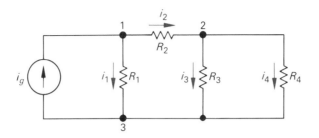

Ans. (a) 1: $i_1 - i_g + i_2 = 0$; 2: $i_3 + i_4 - i_2 = 0$; 3: $i_g - i_1 - i_3 - i_4 = 0$. (b) To derive any one equation from the other two equations, simply add the two equations and then multiply the resulting sum by -1.

DRILL EXERCISE 4.3

a) If only the essential nodes and branches are identified in the circuit of Drill Exercise 4.1, how many simultaneous equations are needed to describe the circuit?

b) How many of these equations can be derived using Kirchhoff's current law?

c) How many need to be derived using Kirchhoff's voltage law?

d) What two meshes should be avoided in applying the voltage law?

Ans. (a) 7; (b) 3; (c) 4; (d) $R_4-R_5-4i_x$ and 8 A$-R_1$.

DRILL EXERCISE 4.4

a) How many separate parts does the circuit shown have?

b) How many nodes?

c) How many independent current equations can be written?

d) How many branches are there?

e) How many branches are there where the current is unknown?

f) How many equations need to be written using the voltage law?

g) Assume that the lower node in each part of the circuit is joined by a single conductor. Repeat the above calculations.

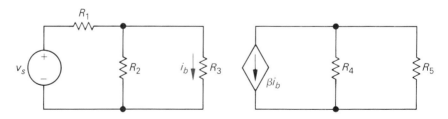

Ans. (a) 2; (b) 5; (c) 3; (d) 7; (e) 6; (f) 3; (g) 1, 4, 3, 7, 6, 3.

4.3 INTRODUCTION TO THE NODE-VOLTAGE METHOD

We will introduce the node-voltage method using the essential nodes of the circuit. To facilitate our discussion, an illustrative circuit is shown in Fig. 4.5. Our first step is to make a neat layout of the circuit so that no branches cross

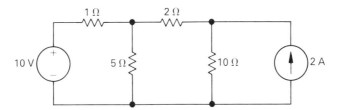

Fig. 4.5   A circuit used in illustrating the node-voltage method of circuit analysis.

over and to mark clearly the essential nodes on the circuit diagram. We note that our circuit has three essential nodes ($n_e = 3$); therefore we need two ($n_e - 1$) node-voltage equations to describe the circuit. The next step in our systematic approach is to select one of the three essential nodes as a reference node. Although theoretically the choice is arbitrary, from a practical point of view, there is often an obvious choice for the reference node. For example, the node with the most branches is usually a good choice. The optimum choice of the reference node (if one exists) will become apparent after you have gained some experience using this method. Since the lower node connects the most branches in the circuit shown in Fig. 4.5, we will use it as the reference node. Once the reference node has been chosen, it is flagged by attaching the symbol ▼, as illustrated in Fig. 4.6.

After the reference node has been selected, the node voltages are defined on the circuit diagram. *A node voltage is defined as the voltage rise from the reference node to a nonreference node.* For the circuit under discussion, we must define two node voltages. These have been denoted as v_1 and v_2 in Fig. 4.6.

We are now ready to write the node-voltage equations. A node-voltage equation is generated by writing the current leaving each branch connected to a nonreference node as a function of the node voltages and then summing these currents to zero in accordance with Kirchhoff's current law. Let us demonstrate this using the circuit shown in Fig. 4.6. The current away from node 1 through the 1-Ω resistor will be the voltage drop across the resistor divided by the resistance (Ohm's law). The voltage drop across the resistor, in the direction of the current away from the node, will be $v_1 - 10$. Therefore the current

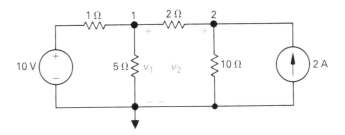

Fig. 4.6   The circuit in Fig. 4.5 showing the reference node and the node voltages.

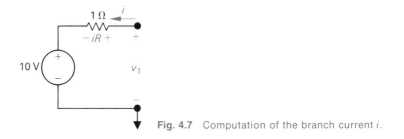

Fig. 4.7 Computation of the branch current i.

in the 1-Ω resistor is $(v_1 - 10)/1$. These observations are readily confirmed after referring to Fig. 4.7, in which the 10-V–1-Ω branch has been drawn with the appropriate voltages and current. Note that summing the voltages around the closed path in accordance with Kirchhoff's voltage law verifies that the voltage drop across the 1-Ω resistor in the direction of i is $(v_1 - 10)$ V.

This same thought process is used to obtain the current in every branch where the current is unknown. Thus the current away from node 1 through the 5-Ω resistor is $v_1/5$ and the current away from node 1 through the 2-Ω resistor is $(v_1 - v_2)/2$. The sum of the three currents leaving node 1 must equal zero; therefore the node-voltage equation derived at node 1 is

$$\frac{v_1 - 10}{1} + \frac{v_1}{5} + \frac{v_1 - v_2}{2} = 0. \tag{4.5}$$

The node-voltage equation written at node 2 is

$$\frac{v_2 - v_1}{2} + \frac{v_2}{10} - 2 = 0. \tag{4.6}$$

In studying Eq. (4.6), note that the first term is the current away from node 2 through the 2-Ω resistor, the second term is the current away from node 2 through the 10-Ω resistor, and the third term is the current away from node 2 through the current source.

Equations (4.5) and (4.6) are the two simultaneous equations that describe the circuit of Fig. 4.6 in terms of the node voltages v_1 and v_2. Solving for v_1 and v_2 yields

$$v_1 = \frac{100}{11} = 9.09 \text{ V}$$

and

$$v_2 = \frac{120}{11} = 10.91 \text{ V}.$$

It is important to note that once the node voltages are known, all the branch currents can be calculated. Once the branch currents are known, the branch voltages and powers can be calculated. Example 4.1 illustrates the use of the node-voltage method to analyze a circuit.

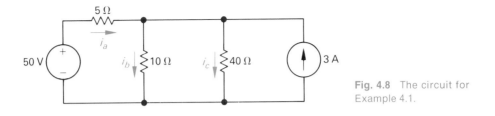

Fig. 4.8 The circuit for Example 4.1.

Example 4.1
a) Use the node-voltage method of circuit analysis to find the branch currents i_a, i_b, and i_c in the circuit shown in Fig. 4.8.

b) Find the power associated with each source, and state whether the source is delivering or absorbing power.

Solution
a) We begin by noting that the circuit has two essential nodes; thus we need to write a single node-voltage expression. We will select the lower node as the reference node and define our unknown node voltage as v_1. These decisions are illustrated in Fig. 4.9. Summing the currents away from node 1 generates the following node-voltage equation:

$$\frac{v_1 - 50}{5} + \frac{v_1}{10} + \frac{v_1}{40} - 3 = 0.$$

Solving for v_1 we obtain

$$v_1 = 40 \text{ V}.$$

It follows directly that

$$i_a = \frac{50 - 40}{5} = 2 \text{ A},$$

$$i_b = \frac{40}{10} = 4 \text{ A},$$

$$i_c = \frac{40}{40} = 1 \text{ A}.$$

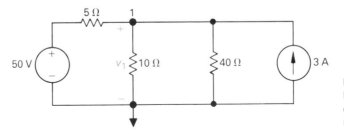

Fig. 4.9 The circuit in Fig. 4.8 showing the reference node and the unknown node voltage v_1.

b) The power associated with the 50-V source is

$$p_{50\,V} = -50i_a = -100 \text{ W} \quad \text{(delivering).}$$

The power associated with the 3-A source is

$$p_{3\,A} = -3v_1 = -3(40) = -120 \text{ W} \quad \text{(delivering).}$$

As a check on our calculations, we note that the total delivered power is 220 W. The total power absorbed by the three resistors is $(4)(5) + 16(10) + 1(40)$, which is 220 W, as it must be. ∎

DRILL EXERCISE
4.5

a) For the circuit shown, use the node-voltage method to find v_1, v_2, and i_1.

b) How much power is delivered to the circuit by the 12-A source?

c) Repeat for the 5-A source.

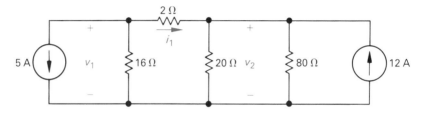

Ans. (a) 48 V, 64 V, -8 A; (b) 768 W; (c) -240 W.

DRILL EXERCISE
4-6

Use the node-voltage method to find v in the following circuit.

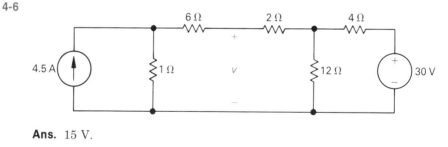

Ans. 15 V.

4.4   THE NODE-VOLTAGE METHOD AND DEPENDENT SOURCES

If the circuit contains dependent sources, the node-voltage equations must be supplemented with the constraint equations imposed by the presence of the dependent sources. Example 4.2 illustrates the application of the node-voltage method to a circuit containing a dependent source.

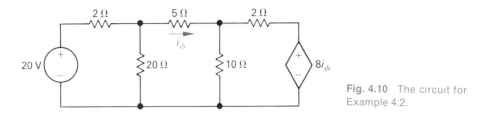

Fig. 4.10   The circuit for Example 4.2.

Example 4.2   Use the node-voltage method to find the power dissipated in the 5-Ω resistor in the circuit shown in Fig. 4.10.

Solution   We begin by noting that our circuit has three essential nodes and, therefore, we will need two node-voltage equations to describe the circuit. Since four branches terminate on the lower node, we will select it as our reference node. The two unknown node voltages are defined on the circuit shown in Fig. 4.11. Summing the currents away from node 1 generates the equation

$$\frac{v_1 - 20}{2} + \frac{v_1}{20} + \frac{v_1 - v_2}{5} = 0.$$

Summing the currents away from node 2 yields

$$\frac{v_2 - v_1}{5} + \frac{v_2}{10} + \frac{v_2 - 8i_\phi}{2} = 0.$$

As written, our two node-voltage equations contain three unknowns, namely, v_1, v_2, and i_ϕ. To eliminate i_ϕ we must express this controlling current in terms of the node voltages. Thus we have

$$i_\phi = \frac{v_1 - v_2}{5}.$$

When this relationship is substituted into the node 2 equation, the two node-voltage equations simplify to

$$0.75v_1 - 0.2v_2 = 10,$$
$$-v_1 + 1.6v_2 = 0.$$

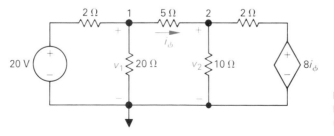

Fig. 4.11   The circuit in Fig. 4.10 with reference node and node voltages.

Solving for v_1 and v_2 gives

$$v_1 = 16 \text{ V} \quad \text{and} \quad v_2 = 10 \text{ V}.$$

It follows directly that

$$i_\phi = \frac{16 - 10}{5} = 1.2 \text{ A}$$

and

$$p_{5\Omega} = (1.44)(5) = 7.2 \text{ W}. \quad \blacksquare$$

In studying the solution of Example 4.2, you might challenge the choice of the reference node. For example, consider the choice of node 2 as the reference node. We still must write two node-voltage equations but if node 2 is used as the reference node we need only solve for one of the unknown node voltages, specifically, the node voltage across the 5-Ω resistor. (See Problem 4.3.)

DRILL EXERCISE 4.7

a) Use the node-voltage method to find the power associated with each source in the following circuit.

b) State whether the source is delivering power to the circuit or extracting power from the circuit.

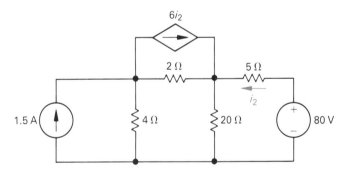

Ans. (a) $p_{1.5\text{ A}} = 15$ W; $p_{6i_2} = 1200$ W; $p_{20\text{ V}} = 320$ W. (b) All sources are delivering power to the circuit.

4.5 THE NODE-VOLTAGE METHOD: SOME SPECIAL CASES

When a voltage source is the only element between two essential nodes, the node-voltage method requires some additional manipulations. The nature of the problem can be seen from the circuit shown in Fig. 4.12. The reference node and the node voltages are as shown in Fig. 4.12. We can see by the figure that we have a problem in writing the expression for the current leaving node 1

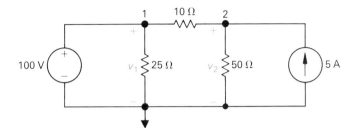

Fig. 4.12   A circuit with a known node voltage.

through the independent voltage source. The problem arises because there is no resistance in series with the 100-V source. At first glance, it may appear that the current in this branch is infinite, as implied by the expression $(v_1 - 100)/0$. However, closer inspection shows that v_1 must be 100 V and therefore we really have the indeterminate form $0/0$. It is the observation that v_1 equals 100 V that allows us to apply the node-voltage method with no further difficulty. That is, once we recognize that v_1 equals 100 V, we see that we have only one unknown node voltage (v_2) and that therefore we can solve this particular circuit by solving a single node-voltage equation. At node 2 we have

$$\frac{v_2 - v_1}{10} + \frac{v_2}{50} - 5 = 0. \tag{4.7}$$

But $v_1 = 100$ V; therefore Eq. (4.7) can be solved for v_2:

$$v_2 = 125 \text{ V}. \tag{4.8}$$

Once v_2 is known, we can calculate the current in every branch. We leave it to the reader to verify that the current into node 1 in the branch containing the independent voltage source is 1.5 A.

It is important to note at this point that when using the node-voltage method any voltage sources that are connected directly between essential nodes reduce the number of unknown node voltages. This follows because whenever a voltage source connects two essential nodes, it constrains the difference between the node voltages at the essential nodes to equal the voltage of the source.

Assume that the circuit shown in Fig. 4.13 is to be analyzed using the node-voltage method. In studying the circuit, we note that there are four essential nodes; thus we anticipate writing three node-voltage equations. However, further study of the circuit reveals that two essential nodes are connected by an independent voltage source and two other essential nodes are tied together via a current-controlled dependent voltage source. Therefore we can deduce that there is really only one unknown node voltage. For example, note that if the voltage across the 50-Ω resistor is known, the voltage across the 100-Ω resistor is also known due to the presence of the dependent voltage source. When

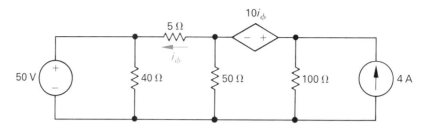

Fig. 4.13 A circuit with a dependent voltage source connected between nodes.

we are choosing which node to use as the reference node, several thoughts come to mind. Either node on each side of the dependent voltage source looks attractive because, if chosen, one of the node voltages would be known to be either $+10i_\phi$ (left node is the reference) or $-10i_\phi$ (right node is the reference). The lower node looks even more attractive because, if chosen, one node voltage is immediately known (50 V) and five branches terminate there. We therefore opt for the lower node as the reference.

The circuit has been redrawn in Fig. 4.14. In addition to flagging the reference node and defining the node voltages, we have introduced the current i, which is needed to support the discussion that follows.

In writing the appropriate node-voltage equation at either node 2 or 3 we cannot express the current in the dependent voltage source branch as a function of the node voltages v_2 and v_3. To solve our dilemma, we introduce the unknown current i and then promptly eliminate it from our equations. Thus at node 2 we have

$$\frac{v_2 - v_1}{5} + \frac{v_2}{50} + i = 0, \tag{4.9}$$

and at node 3,

$$\frac{v_3}{100} - i - 4 = 0. \tag{4.10}$$

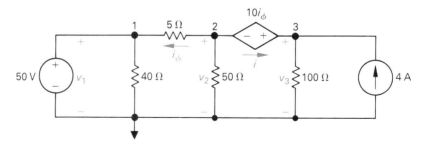

Fig. 4.14 The circuit in Fig. 4.13 defining the selected node voltages.

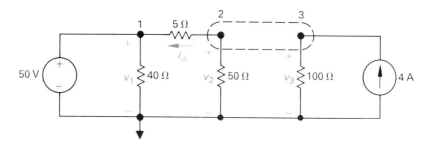

Fig. 4.15   Visualizing nodes 2 and 3 as a "supernode."

We eliminate i by simply adding Eqs. (4.9) and (4.10) to get

$$\frac{v_2 - v_1}{5} + \frac{v_2}{50} + \frac{v_3}{100} - 4 = 0. \tag{4.11}$$

We note in passing that Eq. (4.11) can be written directly, without resorting to the intermediate step represented by Eqs. (4.9) and (4.10). To write Eq. (4.11) directly, we visualize nodes 2 and 3 as comprising a single node and simply sum the currents away from the node in terms of the node voltages v_2 and v_3. This point of view is shown in Fig. 4.15.

Combining nodes 2 and 3 is sometimes referred to as forming a "supernode." Obviously, Kirchhoff's current law must hold for the supernode. Starting with the 5-Ω branch and moving counterclockwise around the supernode, we generate the equation

$$\frac{v_2 - v_1}{5} + \frac{v_2}{50} + \frac{v_3}{100} - 4 = 0, \tag{4.12}$$

which is identical to Eq. (4.11). Therefore the motivation for creating the supernode is apparent. The supernode concept can be used whenever two essential nodes are connected by a voltage source element.

Once Eq. (4.11) (or Eq. 4.12) has been derived, the next step is to reduce the expression to a single unknown node voltage. First we eliminate v_1 from the equation since v_1 is known to be 50 V. Next we can express v_3 as a function of v_2:

$$v_3 = v_2 + 10i_\phi. \tag{4.13}$$

The current controlling the dependent voltage source is now expressed as a function of the node voltages, as follows:

$$i_\phi = \frac{v_2 - 50}{5}. \tag{4.14}$$

Using Eqs. (4.13) and (4.14) along with the fact that v_1 equals 50 V reduces

Eq. (4.11) to

$$v_2 \left[\frac{1}{50} + \frac{1}{5} + \frac{1}{100} + \frac{10}{500} \right] = 10 + 4 + 1$$

$$v_2(0.25) = 15$$

$$v_2 = 60 \text{ V}.$$

It follows directly from Eqs. (4.13) and (4.14) that

$$i_\phi = \frac{60 - 50}{5} = 2 \text{ A}$$

and

$$v_3 = 60 + 20 = 80 \text{ V}.$$

We conclude our introduction to the node-voltage method by using this approach to analyze the circuit first introduced in Section 2.6. For convenience, the circuit has been redrawn in Fig. 4.16. A review of Section 2.6 reveals that when we used the branch-current method of analysis, we were faced with the task of writing and solving six simultaneous equations. We now view this circuit from the vantage point of nodal analysis. The circuit has four essential nodes. Nodes a and d are connected by an independent voltage source, as are nodes b and c. Therefore the problem reduces to finding a single unknown node voltage $[(n_e - 1) - 2]$. Using d as the reference node and combining nodes b and c into a supernode, we have

$$\frac{v_b}{R_2} + \frac{v_b - V_{CC}}{R_1} + \frac{v_c}{R_E} - \beta i_B = 0. \tag{4.15}$$

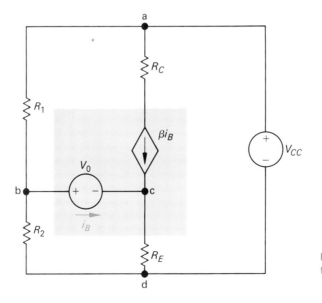

Fig. 4.16 The transistor amplifier circuit of Fig. 2.18.

In writing Eq. (4.15), we have used v_b and v_c to denote the voltage rise from the reference node to nodes b and c, respectively. We have also used the fact that the voltage rise from the reference node to node a is V_{CC}. Now we can eliminate both v_c and i_B from Eq. (4.15) by noting that

$$v_c = (i_B + \beta i_B)R_E \tag{4.16}$$

and

$$v_c = v_b - V_0. \tag{4.17}$$

When Eqs. (4.15) and (4.17) are substituted into Eq. (4.15), we get

$$v_b \left[\frac{1}{R_1} + \frac{1}{R_2} + \frac{1}{(1 + \beta)R_E} \right] = \frac{V_{CC}}{R_1} + \frac{V_0}{(1 + \beta)R_E}. \tag{4.18}$$

Solving Eq. (4.18) for v_b yields

$$v_b = \frac{V_{CC}R_2(1 + \beta)R_E + V_0R_1R_2}{R_1R_2 + (1 + \beta)R_E(R_1 + R_2)}. \tag{4.19}$$

Using the node-voltage method to analyze the circuit in Fig. 4.15 has reduced our problem from manipulating six simultaneous equations to manipulating three simultaneous equations. (See Problem 2.8.) We leave it to the reader to verify that when Eq. (4.19) is combined with Eqs. (4.16) and (4.17) the solution for i_B is identical to Eq. (2.27). (See Problem 4.10.)

**DRILL EXERCISE
4.8**

Use the node-voltage method to find v in the following circuit.

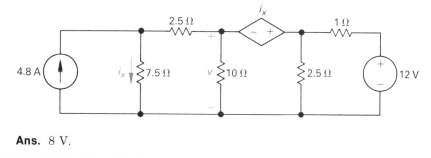

Ans. 8 V.

4.6 INTRODUCTION TO MESH CURRENTS

As stated in Section 4.2, the mesh-current method of circuit analysis enables us to describe a circuit in terms of $b - (n - 1)$ or $[b_e - (n_e - 1)]$ equations. For planar networks, the meshes in the network are identical to the "windows" that are formed when the network is drawn with no branches crossing over each other. The circuit in Fig. 4.1(b) is shown redrawn in Fig. 4.17, where the

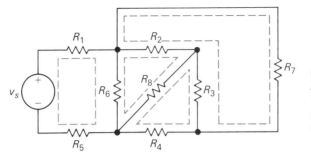

Fig. 4.17 The circuit in Fig. 4.1(b) showing how the "windows" in a planar circuit can be used to identify the meshes in the circuit.

"windows" are identified by the closed dashed paths. Note in Fig. 4.17 that there are seven essential branches where the current is unknown. Since the circuit contains four essential nodes, we need to write four $[7 - (4 - 1)]$ mesh-current equations.

A *mesh current* is defined as the current that exists only in the perimeter of a mesh. It is indicated on a circuit diagram as either a closed solid line or an almost closed solid line that follows the perimeter of the appropriate mesh. The reference direction for the mesh current is indicated by an arrowhead on the solid line. Four mesh currents that are used to describe the circuit in Fig. 4.17 are shown in Fig. 4.18. Note that by definition mesh currents automatically satisfy Kirchhoff's current law. That is, at any node in the circuit, a given mesh current is both into and out of the node.

In studying Fig. 4.18, note that from the definition of a mesh current it is not always possible to identify it in terms of a branch current. For example, in Fig. 4.18 the mesh current i_2 is not equal to any branch current, whereas mesh currents i_1, i_3, and i_4 can be identified with branch currents. Thus it is not always possible to measure a mesh current. For example, in Fig. 4.18 there is no place where an ammeter can be inserted into the circuit to measure the mesh current i_2. The fact that a mesh current can be a fictitious quantity does not mean that it is a useless concept. On the contrary, it is very useful to us in circuit analysis.

The mesh-current method of circuit analysis evolves quite naturally from the branch-current equations. The circuit in Fig. 4.19 can be used to show the

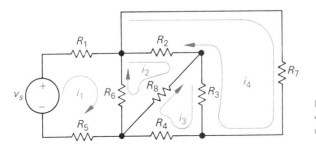

Fig. 4.18 The circuit in Fig. 4.17 with the mesh currents defined.

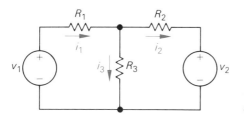

Fig. 4.19 A circuit illustrating the development of the mesh-current method of circuit analysis.

evolution of the mesh-current technique. We begin by using the branch currents (i_1, i_2, and i_3) to formulate the set of independent equations. For this circuit $b_e = 3$ and $n_e = 2$. We can write only one independent current equation; therefore we will need two independent voltage equations. Applying Kirchhoff's current law to the upper node and Kirchhoff's voltage law around the two meshes generates the following set of equations:

$$i_1 = i_2 + i_3, \tag{4.20}$$

$$v_1 = i_1 R_1 + i_3 R_3, \tag{4.21}$$

$$-v_2 = i_2 R_2 - i_3 R_3. \tag{4.22}$$

We can reduce this set of three equations to a set of two equations by solving Eq. (4.20) for i_3 and then substituting this expression for i_3 into Eqs. (4.21) and (4.22). The result is

$$v_1 = i_1(R_1 + R_3) - i_2 R_3, \tag{4.23}$$

$$-v_2 = -i_1 R_3 + i_2(R_2 + R_3). \tag{4.24}$$

We can solve Eqs. (4.23) and (4.24) for i_1 and i_2 to reduce the original problem of solving three simultaneous equations to a problem of solving two simultaneous equations. We have derived Eqs. (4.23) and (4.24) by substituting the $n_e - 1$ current equations into the $b_e - (n_e - 1)$ voltage equations. The value of the mesh-current method lies in the fact that by defining mesh currents we *automatically* eliminate the $n_e - 1$ current equations. Thus the mesh-current method is equivalent to a systematic substitution of the $n_e - 1$ current equations into the $b_e - (n_e - 1)$ voltage equations. The mesh currents for the circuit in Fig. 4.19 that are equivalent to eliminating the branch current i_3 from Eqs. (4.21) and (4.22) are shown in Fig. 4.20.

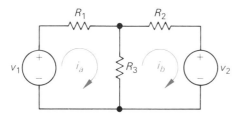

Fig. 4.20 Mesh currents i_a and i_b.

Now we apply Kirchhoff's voltage law around the two meshes, expressing all voltages across resistors in terms of the mesh currents, to get the equations

$$v_1 = i_a R_1 + (i_a - i_b)R_3 \qquad (4.25)$$

and

$$-v_2 = (i_b - i_a)R_3 + i_b R_2. \qquad (4.26)$$

Collecting the coefficients of i_a and i_b in Eqs. (4.25) and (4.26) gives us

$$v_1 = i_a(R_1 + R_3) - i_b R_3 \qquad (4.27)$$

and

$$-v_2 = -i_a R_3 + i_b(R_2 + R_3). \qquad (4.28)$$

When we compare Eqs. (4.27) and (4.28) with Eqs. (4.23) and (4.24), we see that they are identical in form, with the mesh currents i_a and i_b replacing the branch currents i_1 and i_2. By comparing the circuits in Figs. 4.19 and 4.20, we can also see that the branch currents can be expressed in terms of the mesh currents by inspection, hence

$$i_1 = i_a, \qquad (4.29)$$

$$i_2 = i_b, \qquad (4.30)$$

$$i_3 = i_a - i_b. \qquad (4.31)$$

The ability to write Eqs. (4.29)–(4.31) by inspection is crucial to the mesh-current method of circuit analysis. Once we know the mesh currents, we also know the branch currents. And once we know the branch currents, we can compute any voltages or powers of interest.

One final comment about the mesh-current method before we illustrate this approach with a numerical example. Since the meshes have been defined as the "windows" of a planar circuit, we guarantee that the set of mesh-current equations that describe the circuit will be an independent set. Since we state this guarantee without proof, we hasten to assure you that it has been proven.†

Example 4.3   a) Use the mesh-current method to determine the power associated with each voltage source in the circuit shown in Fig. 4.21.

b) Calculate the voltage $[v_0]$ across the 8-Ω resistor.

Solution   a) To calculate the power associated with each source, we need to know the current in each source. In studying the circuit, we see that these source currents will be identical to mesh currents. We also note that our circuit has

---

† See, for example, B. J. Ley, S. G. Lutz, and C. F. Rehberg, *Linear Circuit Analysis*, Ch. 2 (New York: McGraw-Hill, 1959).

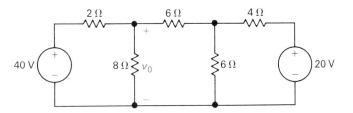

Fig. 4.21 The circuit for Example 4.3.

seven branches where the current is unknown and five nodes. Therefore we need three mesh-current equations to describe the circuit, that is,
$b - (n - 1) = 7 - (5 - 1) = 3$. The three mesh currents that we are using
to describe the circuit in Fig. 4.21 are shown in Fig. 4.22.

If we assume that the voltage drops are positive, the three mesh equations are

$$-40 + 2i_a + 8(i_a - i_b) = 0,$$
$$8(i_b - i_a) + 6i_b + 6(i_b - i_c) = 0, \tag{4.32}$$
$$6(i_c - i_b) + 4i_c + 20 = 0.$$

Equations (4.32) can now be reorganized in anticipation of using Cramer's
method for solving simultaneous equations. We get

$$10i_a - 8i_b + 0i_c = 40,$$
$$-8i_a + 20i_b - 6i_c = 0, \tag{4.33}$$
$$0i_a - 6i_b + 10i_c = -20.$$

The characteristic determinant is

$$\Delta = \begin{vmatrix} 10 & -8 & 0 \\ -8 & 20 & -6 \\ 0 & -6 & 10 \end{vmatrix}$$

$$= 10(200 - 36) + 8(-80)$$
$$= 1640 - 640 = 1000.$$

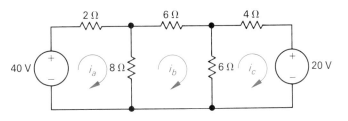

Fig. 4.22 Three mesh currents used to analyze the circuit in Fig. 4.21.

The three mesh currents are

$$i_a = \frac{\begin{vmatrix} 40 & -8 & 0 \\ 0 & 20 & -6 \\ -20 & -6 & 10 \end{vmatrix}}{1000}$$

$$= \frac{40(200 - 36) - 20(48)}{1000}$$

$$= 5.6 \text{ A};$$

$$i_b = \frac{\begin{vmatrix} 10 & 40 & 0 \\ -8 & 0 & -6 \\ 0 & -20 & 10 \end{vmatrix}}{1000}$$

$$= \frac{10(-120) + 8(400)}{1000}$$

$$= 2.0 \text{ A};$$

$$i_c = \frac{\begin{vmatrix} 10 & -8 & 40 \\ -8 & 20 & 0 \\ 0 & -6 & -20 \end{vmatrix}}{1000}$$

$$= \frac{10(-400) + 8(160 + 240)}{1000}$$

$$= -0.80 \text{ A}.$$

Now since the mesh current i_a is identical with the branch current in the 40-V source, the power associated with this source is

$$p_{40\,\mathrm{V}} = -40i_a = -224 \text{ W}.$$

The minus sign tells us that this source is delivering power to the network. The current in the 20-V source is identical to the mesh current i_c; therefore

$$p_{20\,\mathrm{V}} = 20i_c = -16 \text{ W}.$$

The 20-V source is also delivering power to the network.

b) The branch current in the 8-Ω resistor in the direction of the voltage drop v_0 is $i_a - i_b$. Therefore

$$v_0 = 8(i_a - i_b) = 8(3.6) = 28.8 \text{ V.} \quad \blacksquare$$

DRILL EXERCISE

4.9 Use the mesh-current method to find (a) the power delivered to the circuit by the 100-V source, and (b) the power dissipated in the 15-Ω resistor.

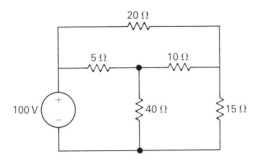

Ans. (a) 600 W; (b) 240 W.

4.7   THE MESH-CURRENT METHOD AND DEPENDENT SOURCES

If the circuit contains dependent sources, the mesh-current equations must be supplemented by the appropriate constraint equations imposed by the presence of the dependent source or sources. Example 4.4 illustrates the application of the mesh-current method when the circuit includes a dependent source.

Example 4.4   Use the mesh-current method of circuit analysis to determine the power dissipated in the 4-Ω resistor in the circuit shown in Fig. 4.23.

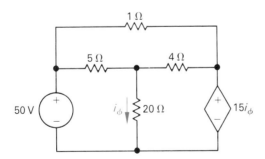

Fig. 4.23   The circuit for Example 4.4.

Solution   Our circuit has six branches where the current is unknown and four nodes; therefore we know that we need three mesh currents to describe the circuit. They are defined on the circuit shown in Fig. 4.24. The three mesh-current

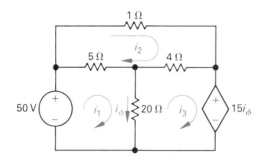

Fig. 4.24 The circuit in Fig. 4.23 showing the three mesh currents.

equations are

$$50 = 5(i_1 - i_2) + 20(i_1 - i_3),$$
$$0 = 5(i_2 - i_1) + 1i_2 + 4(i_2 - i_3), \qquad (4.34)$$
$$0 = 20(i_3 - i_1) + 4(i_3 - i_2) + 15i_\phi.$$

The branch current controlling the dependent voltage source is now expressed in terms of the mesh currents as

$$i_\phi = i_1 - i_3, \qquad (4.35)$$

which is the supplemental equation imposed by the presence of the dependent source. When Eq. (4.35) is substituted into Eqs. (4.34) and the coefficients of i_1, i_2, and i_3 are collected in each equation, we get

$$50 = 25i_1 - 5i_2 - 20i_3,$$
$$0 = -5i_1 + 10i_2 - 4i_3,$$
$$0 = -5i_1 - 4i_2 + 9i_3.$$

The characteristic determinant is

$$\Delta = \begin{vmatrix} 25 & -5 & -20 \\ -5 & 10 & -4 \\ -5 & -4 & 9 \end{vmatrix}.$$

Expanding the characteristic determinant by the first column gives

$$\Delta = 25(90 - 16) + 5(-45 - 80) - 5(20 + 200) = 125.$$

Since we are calculating the power dissipated in the 4-Ω resistor, we compute the mesh currents i_2 and i_3. We have

$$i_2 = \frac{\begin{vmatrix} 25 & 50 & -20 \\ -5 & 0 & -4 \\ -5 & 0 & 9 \end{vmatrix}}{125}$$

$$= \frac{-50(-45 - 20)}{125} = 26 \text{ A}$$

and

$$i_3 = \frac{\begin{vmatrix} 25 & -5 & 50 \\ -5 & 10 & 0 \\ -5 & -4 & 0 \end{vmatrix}}{125}$$

$$= \frac{50(20 + 50)}{125} = 28 \text{ A}.$$

The current in the 4-Ω resistor oriented from left to right is $i_3 - i_2$, or 2 A. Therefore the power dissipated is

$$p_{4\Omega} = (i_3 - i_2)^2(4) = (2)^2(4) = 16 \text{ W}. \quad \blacksquare$$

It is worth noting that if you were not told to use the mesh-current method, you could probably have chosen to use the node-voltage method since the presence of two voltage sources between essential nodes reduces the problem to finding one unknown node voltage. More about making choices later.

DRILL EXERCISE
4.10

a) Use the expression $b - (n - 1)$ to determine the number of mesh-current equations needed to solve the circuit shown.

b) Repeat using $b_e - (n_e - 1)$.

c) Use the mesh-current method to find how much power is being delivered to the dependent voltage source.

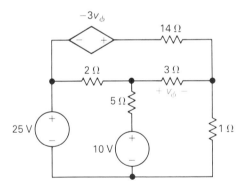

Ans. (a) 3; (b) 3; (c) -36 W.

4.8 THE MESH-CURRENT METHOD: SOME SPECIAL CASES

When a branch includes a current source, the mesh-current method requires some additional manipulations. The nature of the problem can be understood from the circuit shown in Fig. 4.25. The mesh currents i_a, i_b, and i_c, as well as

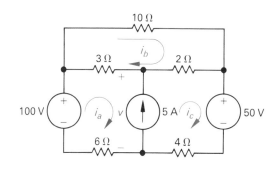

Fig. 4.25 A circuit illustrating mesh analysis when a branch contains an independent current source.

the voltage across the 5-A current source, have been defined in Fig. 4.25 to facilitate the discussion. In analyzing the circuit, we note that there are five essential branches where the current is unknown; furthermore, the circuit has four essential nodes. It follows that we need to write two $[5 - (4 - 1)]$ mesh-current equations in order to solve the circuit. If we use the windows to define the meshes, we see that the three unknown mesh currents reduce to two unknown mesh currents because the current source coupling meshes a and c limits the difference between i_c and i_a to equal 5 A. However, when we attempt to sum the voltages around either mesh a or mesh c, we are forced to introduce the unknown voltage across the 5-A current source into our equations. We can eliminate this unknown voltage by simply introducing it into both mesh equations and then adding the two equations. Thus for mesh a we have

$$100 = 3(i_a - i_b) + v + 6i_a, \tag{4.36}$$

and for mesh c,

$$-50 = 4i_c - v + 2(i_c - i_b). \tag{4.37}$$

Now we add Eqs. (4.36) and (4.37) to obtain

$$50 = 9i_a - 5i_b + 6i_c. \tag{4.38}$$

Summing voltages around mesh b gives us

$$0 = 3(i_b - i_a) + 10i_b + 2(i_b - i_c). \tag{4.39}$$

We can reduce Eqs. (4.38) and (4.39) to two equations and two unknowns using the constraint that

$$i_c - i_a = 5. \tag{4.40}$$

We leave it to the reader to verify that when Eq. (4.40) is combined with Eqs. (4.38) and (4.39) the solutions for the three mesh currents are

$$i_a = 1.75 \text{ A},$$
$$i_b = 1.25 \text{ A},$$

and

$$i_c = 6.75 \text{ A}.$$

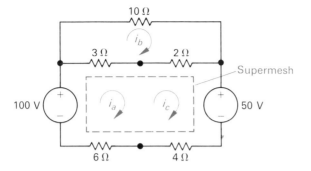

Fig. 4.26 The circuit in Fig. 4.25, illustrating the concept of the supermesh.

Equation (4.38) can be derived without introducing the unknown voltage v by using the concept of a *supermesh*. To create a supermesh, we mentally remove the current source from the circuit by simply avoiding this branch when writing the mesh-current equations. The voltages around the supermesh are expressed in terms of the mesh currents defined by the original windows of the circuit. The supermesh concept is illustrated in Fig. 4.26. When we sum the voltages around the supermesh denoted by the dashed line in Fig. 4.26, we obtain the equation

$$-100 + 3(i_a - i_b) + 2(i_c - i_b) + 50 + 4i_c + 6i_a = 0, \qquad (4.41)$$

which reduces to

$$50 = 9i_a - 5i_b + 6i_c. \qquad (4.42)$$

If we compare Eqs. (4.42) and (4.38), we see that they are identical. Thus the supermesh has eliminated the need for introducing the unknown voltage across the current source into our equations.

We can use the circuit first introduced in Section 2.6 (Fig. 2.18) to illustrate the mesh-current method when a branch contains a dependent current source. The circuit is shown redrawn in Fig. 4.27 with the three mesh currents denoted as i_a, i_b, and i_c. When we study the circuit in Fig. 4.27 with the decision to use the mesh-current method in mind, we note that it has four essential nodes and five essential branches where the current is unknown. Therefore we know that the circuit can be analyzed in terms of two $[5 - (4 - 1)]$ mesh-current equations. Although three mesh currents are defined in Fig. 4.27, we see immediately that the dependent current source forces a constraint between mesh currents i_a and i_c so that we have only two unknown mesh currents. Using the concept of the supermesh, we can redraw the circuit as shown in Fig. 4.28.

Now we sum the voltages around the supermesh in terms of the mesh currents i_a, i_b, and i_c. We get

$$R_1 i_a + V_{CC} + R_E(i_c - i_b) - V_0 = 0. \qquad (4.43)$$

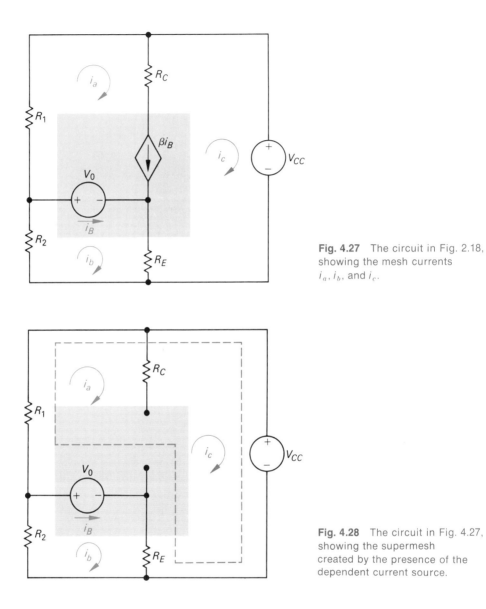

Fig. 4.27 The circuit in Fig. 2.18, showing the mesh currents $i_a, i_b,$ and i_c.

Fig. 4.28 The circuit in Fig. 4.27, showing the supermesh created by the presence of the dependent current source.

The mesh b equation is

$$R_2 i_b + V_0 + R_E(i_b - i_c) = 0. \qquad (4.44)$$

The constraint imposed by the dependent current source is

$$\beta i_B = i_a - i_c. \qquad (4.45)$$

The branch current controlling the dependent current source, expressed as a

function of the mesh currents, is

$$i_B = i_b - i_a. \tag{4.46}$$

From Eqs. (4.45) and (4.46) we have

$$i_c = (1 + \beta)i_a - \beta i_b. \tag{4.47}$$

We can now use Eq. (4.47) to eliminate i_c from Eqs. (4.43) and (4.44). We get

$$[R_1 + (1 + \beta)R_E]i_a - (1 + \beta)R_E i_b = (V_0 - V_{CC}), \tag{4.48}$$

$$-(1 + \beta)R_E i_a + [R_2 + (1 + \beta)R_E]i_b = -V_0. \tag{4.49}$$

We will leave it to the reader to verify that the solution of Eqs. (4.48) and (4.49) for i_a and i_b gives

$$i_a = \frac{V_0 R_2 - V_{CC} R_2 - V_{CC}(1 + \beta)R_E}{R_1 R_2 + (1 + \beta)R_E(R_1 + R_2)}, \tag{4.50}$$

$$i_b = \frac{-V_0 R_1 - (1 + \beta)R_E V_{CC}}{R_1 R_2 + (1 + \beta)R_E(R_1 + R_2)}. \tag{4.51}$$

We also leave it to the reader to verify that when Eqs. (4.50) and (4.51) are used to find i_B, the result is the same as that given by Eq. (2.27).

DRILL EXERCISE   Use the mesh-current method to find the power dissipated in the 2-Ω resistor.
4.11

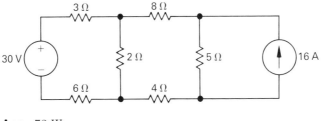

Ans. 72 W.

4.9   THE NODE-VOLTAGE METHOD VERSUS THE MESH-CURRENT METHOD

The greatest advantage of both the node-voltage and mesh-current methods is that they reduce the number of simultaneous equations that must be manipulated. They also require the analyst to be quite systematic in organizing and writing the required simultaneous equations. It is natural to ask, then, "When is the node-voltage method preferred to the mesh-current method, and vice versa?" As one might suspect, there is no clear-cut answer. One possible approach is to compare the number of simultaneous equations required for each method and then to select the one requiring the least number. A second is to analyze the presence and location of voltage and current sources within the circuit structure. Voltage sources may require extra effort in formulating node-voltage equations, whereas current sources may require extra effort in formulating mesh-current equations.

Another point to consider when choosing between the two methods is what information about the circuit being analyzed is of primary interest. In other words, a complete solution of a circuit may not be needed, and therefore the particular piece of information that is of interest may influence what method is used. For example, in the circuit in Fig. 2.18 if only i_{CC} is of interest, the mesh-current method may be selected, whereas if only the voltage across R_2 is of interest, the node-voltage method might be favored.

Perhaps the most important observation to make regarding these two methods of circuit analysis is that in any given situation a little time spent thinking about the problem in relation to the various analytical approaches available will be time well spent.

DRILL EXERCISE   Find the power delivered to the circuit by the 4-A current source.
4.12

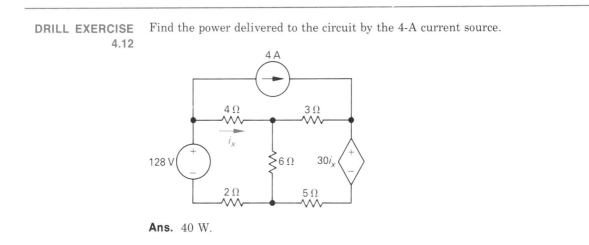

Ans. 40 W.

4.10   SOURCE TRANSFORMATIONS

Even though the node-voltage and mesh-current methods are powerful techniques for solving circuits, we are still interested in methods that can be used to simplify circuits. We begin expanding our list of simplifying techniques with source transformations. A *source transformation,* shown in Fig. 4.29, allows us to replace a voltage source in series with a resistor by a current source in paral-

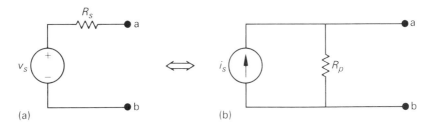

Fig. 4.29   Source transformations.

lel with the same resistor, or vice versa. The double-headed arrow in Fig. 4.29 is used to emphasize that a source transformation is bilateral, that is, we can start with either configuration and derive the other. The two configurations shown in Fig. 4.29 are equivalent with respect to the terminals a, b provided

$$i_s = \frac{v_s}{R_s} \tag{4.52}$$

and

$$R_s = R_p. \tag{4.53}$$

These two equations can be verified by the following arguments. If the two circuits are equivalent with respect to the terminals a, b, they must be equivalent for *all* external values of R connected across a, b. Two extreme values of R that are easy to test are zero and infinity. For zero ohms, or a short circuit, the voltage source will deliver a short-circuit current of v_s/R_s amperes, oriented from terminal a toward b. The short-circuit current delivered by the current source would be i_s, also oriented from terminal a to b. These two short-circuit currents are identical by virtue of Eq. (4.52).

For infinite external resistance, the source arrangement of Fig. 4.29(a) predicts that the voltage from a to b would be v_s with terminal a positive. The voltage across a, b in the circuit in Fig. 4.29(b) is $i_s R_p$, which is equal to v_s by virtue of Eqs. (4.52) and (4.53). Terminal a is also positive, as it must be in order for the two source arrangements to be equivalent.

Observe that if the polarity of v_s is reversed, the orientation of i_s must be reversed in order to maintain equivalence.

The usefulness of making source transformations in order to simplify a circuit analysis problem is illustrated by the following example.

Example 4.5   a) For the circuit shown in Fig. 4.30, find the power associated with the 6-V source.

b) State whether the 6-V source is absorbing or delivering the power calculated in part (a).

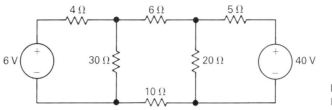

Fig. 4.30 The circuit for Example 4.5.

Solution
a) After studying the circuit in Fig. 4.30 knowing that the power associated with the 6-V source is of interest, several approaches come to mind. First we note that the circuit has four essential nodes and six essential branches where the current is unknown. Thus the current in the branch containing the 6-V source can be found by solving either three [6 − (4 − 1)] mesh-current equations or three (4 − 1) node-voltage equations. If we choose the mesh-current approach, we solve the three mesh-current equations for the mesh current that corresponds to the branch current in the 6-V source. If we elect the node-voltage approach, we solve the three node-voltage equations for the voltage across the 30-Ω resistor, from which the branch current in the 6-V source can be calculated. However, since we are focusing on just one branch current, we can first simplify the circuit using source transformations. We must reduce the circuit in a manner that preserves the identity of the branch containing the 6-V source. For the problem at hand, there is no reason to preserve the identity of the branch containing the 40-V source. Beginning with this branch, we can transform the 40-V source in series with the 5-Ω resistor to an 8-A current source in parallel with a 5-Ω resistor, as shown in Fig. 4.31(a). Next, the parallel combination of the 20-Ω and 5-Ω resistors can be replaced with a 4-Ω resistor. This 4-Ω resistor shunts the 8-A source and therefore can be replaced by a 32-V source in series with a 4-Ω resistor, as shown in Fig. 4.31(b). The 32-V source is in series with 20 Ω of resistance and, hence, can be replaced by a current source of 1.6 A in parallel with 20 Ω, as shown in Fig. 4.31(c). The parallel combination of the 1.6-A current source and the 12-Ω resistor transforms to a voltage source of 19.2 V in series with 12 Ω. The result of this last transformation is shown in Fig. 4.31(d), from which we see that the current in the direction of the voltage drop across the 6-V source is (19.2 − 6)/16, or 0.825 A. Therefore the power associated with the 6-V source is

$$p_{6\,V} = (0.825)(6) = 4.95 \text{ W}.$$

b) The voltage source is absorbing power. ■

DRILL EXERCISE
4.13

a) Use a series of source transformations to find the voltage v in the circuit shown.

b) How much power does the 120-V source deliver to the circuit?

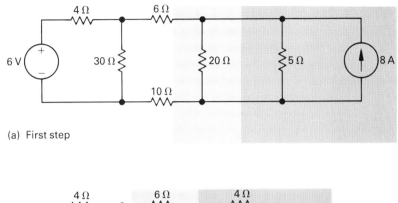

(a) First step

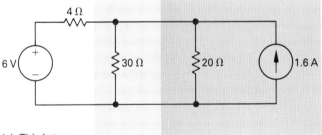

(b) Second step

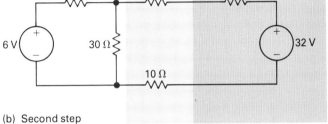

(c) Third step

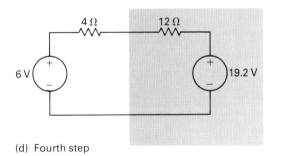

(d) Fourth step

Fig. 4.31 A step-by-step simplification of the circuit in Fig. 4.30.

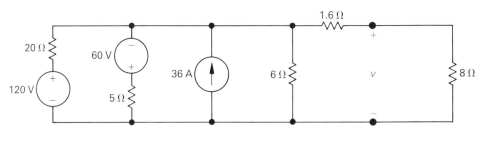

Ans. (a) 48 V; (b) 374.4 W.

4.11 THÉVENIN AND NORTON EQUIVALENTS

There are times in circuit analysis when we wish to concentrate our attention on what happens at a specific pair of terminals in the circuit. For example, when we plug a toaster into an outlet, we are primarily interested in the voltage and current at the terminals of the toaster. We have little or no interest in what effect our connecting the toaster has on voltages or currents elsewhere in the circuit that is supplying the outlet. We can expand this interest in terminal behavior to the case in which we have a whole set of appliances, each requiring a different amount of power, and we are then interested in how the voltage and current delivered at the outlet will change as we change appliances. In other words, we want to focus on the behavior of the circuit supplying the outlet but only at the outlet terminals. Because our interest in circuit behavior is so often focused on a pair of terminals, the Thévenin and Norton equivalent circuits, which we are about to introduce, are extremely valuable aids in analysis. Although at this time we will discuss these equivalent circuits as they pertain to resistive circuits, you should be aware, right at the outset, that Thévenin and Norton equivalent circuits can be used to represent any circuit made up of linear elements.

The significance of the Thévenin equivalent circuit can be described using Fig. 4.32. Figure 4.32(a) represents any circuit made up of sources (both independent and dependent) and resistors. We have identified by the letters a and b the pair of terminals that are of interest. In Fig. 4.32(b), we show the Thévenin equivalent. What we imply by the circuit in Fig. 4.32(b) is that the original interconnection of sources and resistors can be replaced by an independent voltage source V_t in series with a resistor R_t. Furthermore, this series combination of V_t and R_t is equivalent to the original circuit in the sense that if we connect the same load across the terminals a, b of each circuit, we get the same voltage and current at the terminals of the load. This equivalence will hold for *all possible values of load resistance.*

In order to represent the original circuit by its Thévenin equivalent, we must be able to determine the Thévenin voltage V_t and the Thévenin resistance R_t. These two parameters of the Thévenin equivalent can be found as

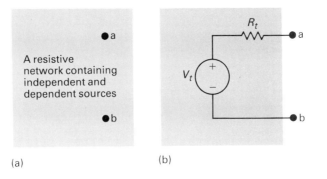

Fig. 4.32 A Thévenin equivalent circuit: (a) a general circuit; (b) the Thévenin equivalent.

follows. First we note that if the load resistance is infinitely large, we have an open-circuit condition. The open-circuit voltage at the terminals a, b in the circuit in Fig. 4.32(b) will be V_t. By hypothesis, this must be the same as the open-circuit voltage at the terminals a, b in the original circuit. Therefore to calculate the Thévenin voltage V_t, we simply calculate the open-circuit voltage in the original circuit.

If the load resistance is reduced to zero, we have a short-circuit condition. If we place a short circuit across the terminals a, b of the Thévenin equivalent circuit, the short-circuit current directed from a to b will be

$$i_{sc} = \frac{V_t}{R_t}. \tag{4.54}$$

By hypothesis, this short-circuit current must be identical to the short-circuit current that exists in a short circuit placed across the terminals a, b of the original network. From Eq. (4.54) we have

$$R_t = \frac{V_t}{i_{sc}}. \tag{4.55}$$

Thus the Thévenin resistance is the ratio of the open-circuit voltage to the short-circuit current. Let us demonstrate with a specific circuit.

To find the Thévenin equivalent circuit of the circuit shown in Fig. 4.33, we first calculate the open-circuit voltage v_{ab}. Note that when the terminals a, b are open, there will be no current in the 4-Ω resistor. Therefore the open-circuit voltage v_{ab} will be identical to the voltage across the 3-A current source. This voltage has been labeled v_0 on the circuit in Fig. 4.33. The voltage v_0 can be found by solving a single node-voltage equation. Choosing the lower node as the reference node, we have

$$\frac{v_0 - 25}{5} + \frac{v_0}{20} - 3 = 0. \tag{4.56}$$

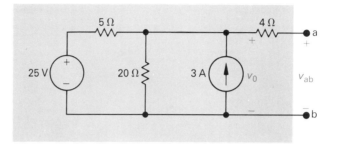

Fig. 4.33 A circuit used in demonstrating a technique for finding a Thévenin equivalent.

Solving for v_0 yields

$$v_0 = 32 \text{ V}. \tag{4.57}$$

It follows that the Thévenin voltage for the circuit in Fig. 4.33 is 32 V.

   The next step in deriving the Thévenin equivalent circuit with respect to the terminals a, b is to place a short circuit across the terminals and calculate the resulting short-circuit current. The circuit with the short in place is shown in Fig. 4.34. Note that the short-circuit current is in the direction of the open-circuit voltage drop across the terminals a, b. (If the short-circuit current is in the direction of the open-circuit voltage rise across the terminals, a minus sign must be inserted in Eq. 4.55.)

   The short-circuit current (i_{sc}) is easily found once v_0 is known. Therefore the problem reduces to finding v_0 with the short in place. Again, if we use the lower node as the reference node, the equation for v_0 becomes

$$\frac{v_0 - 25}{5} + \frac{v_0}{20} - 3 + \frac{v_0}{4} = 0. \tag{4.58}$$

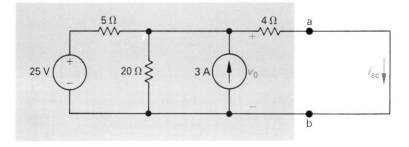

Fig. 4.34 The circuit in Fig. 4.33 with the terminals a, b short-circuited.

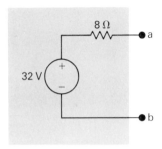

Fig. 4.35 The Thévenin equivalent circuit of the circuit shown in Fig. 4.33.

Solving Eq. (4.58) for v_0 gives us

$$v_0 = 16 \text{ V}. \tag{4.59}$$

It follows that the short-circuit current is

$$i_{sc} = \frac{16}{4} = 4 \text{ A}. \tag{4.60}$$

Now we can find the Thévenin resistance by substituting the numerical values given in Eqs. (4.57) and (4.60) into Eq. (4.55). Thus we have

$$R_t = \frac{V_t}{i_{sc}} = \frac{32}{4} = 8 \ \Omega. \tag{4.61}$$

The Thévenin equivalent circuit for the circuit in Fig. 4.33 is shown in Fig. 4.35.

We will leave it to you to verify that if a 24-Ω resistor is connected across the terminals a, b in the circuit shown in Fig. 4.33, the voltage across the resistor will be 24 V and the current in the resistor will be 1 A. We can see by inspection that the Thévenin circuit in Fig. 4.35 will predict the same voltage and current if a 24-Ω resistor is connected across the terminals a, b.

The Norton equivalent circuit consists of an independent current source in parallel with the Norton equivalent resistance. It can be derived from the Thévenin equivalent circuit simply by making a source transformation. Thus the Norton current equals the short-circuit current at the terminals of interest and the Norton resistance is identical to the Thévenin resistance.

Sometimes we can make effective use of source transformations to derive a Thévenin or Norton equivalent circuit. For example, the Thévenin and Norton equivalent circuits of the circuit shown in Fig. 4.33 can be derived by making the series of source transformations shown in Fig. 4.36. This technique is most useful when the network contains only independent sources. The presence of dependent sources will require retaining the identity of the controlling voltages and/or currents and this constraint will usually prohibit the continued reduction of the circuit via source transformations. In Section 4.12 we will

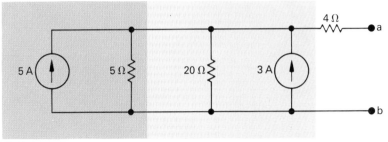

(a) First step

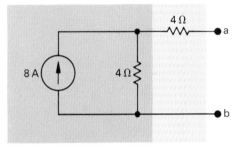

(b) Second step

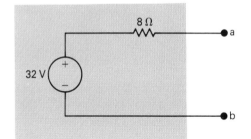

(c) Third step (Thévenin equivalent)

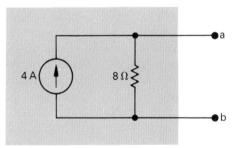

(d) Fourth step (Norton equivalent)

Fig. 4.36 The step-by-step derivation of the Thévenin and Norton equivalent circuits of the circuit shown in Fig. 4.33.

illustrate the problem of finding the Thévenin equivalent when the circuit contains dependent sources.

Find the Thévenin equivalent circuit with respect to the terminals a, b.

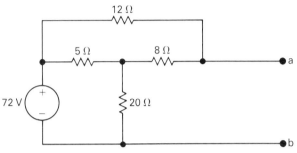

Ans. $V_{ab} = V_t = 64.8$ V, $R_t = 6$ Ω.

Find the Norton equivalent circuit with respect to the terminals a, b.

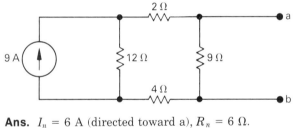

Ans. $I_n = 6$ A (directed toward a), $R_n = 6$ Ω.

4.12 MORE ON DERIVING A THÉVENIN EQUIVALENT

The technique for determining R_t that we discussed and illustrated in the preceding section is not always the easiest method. There are two other methods that are generally simpler to implement. The first is useful only if the network contains independent sources. To calculate R_t for such a network, we first deactivate all independent sources and then calculate the resistance seen looking into the network at the designated terminal pair. *A voltage source is deactivated by replacing it with a short circuit. A current source is deactivated by replacing it with an open circuit.* As an example, consider the circuit shown in Fig. 4.33. Once the independent sources have been deactivated, the circuit in Fig. 4.33 simplifies to that shown in Fig. 4.37. The resistance seen looking into

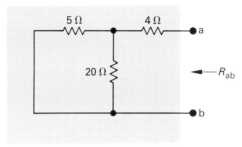

Fig. 4.37   The circuit in Fig. 4.33 after deactivating the independent sources.

the terminals a, b is denoted by R_{ab} in Fig. 4.37. We can see from this circuit that R_{ab} consists of the 4-Ω resistor in series with the parallel combination of the 5- and 20-Ω resistors; thus

$$R_{ab} = R_t = 4 + \frac{5 \times 20}{25} = 8 \; \Omega. \tag{4.62}$$

Note that the derivation of R_t via Eq. (4.62) is much simpler than the derivation of R_t via Eq. (4.61).

    If the network contains dependent sources, an alternative procedure for finding the Thévenin resistance R_t is as follows (see Fig. 4.38). We first deactivate all *independent* sources. We then apply either a test voltage source or a test current source to the Thévenin terminals a, b. The Thévenin resistance will equal the ratio of the voltage across the test source to the current delivered by the test source. In studying the circuit shown in Fig. 4.38, note that it contains an independent 5-V source, a voltage-controlled voltage source, and a current-controlled current source. Also note the controlling signals. The dependent voltage source is controlled by the voltage across the 25-Ω resistor and the dependent current source is controlled by the current in the 2-kΩ resistor. We will first derive the Thévenin equivalent circuit with respect to the terminals a, b using the open-circuit voltage and short-circuit current calcula-

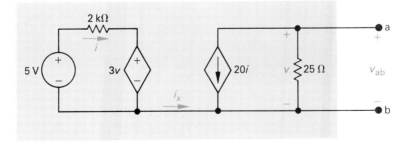

Fig. 4.38   A circuit illustrating the derivation of the Thévenin equivalent circuit when the circuit contains dependent sources.

tions. We will then illustrate the alternative method for finding the Thévenin resistance.

The first step in analyzing the circuit in Fig. 4.38 is to recognize that the current labeled i_x must be zero. (Note that there is no return path for i_x to enter the left-hand portion of the circuit.) The open-circuit, or Thévenin, voltage will be the voltage across the 25-Ω resistor. Since i_x is zero, it follows directly that

$$V_t = v_{ab} = (-20i)(25) = -500i. \tag{4.63}$$

Now we calculate the current i:

$$i = \frac{5 - 3v}{2000} = \frac{5 - 3V_t}{2000}. \tag{4.64}$$

In writing Eq. (4.64), we recognize that the Thévenin voltage is identical to the control voltage. When we substitute Eq. (4.64) into Eq. (4.63), we find that

$$V_t = -5 \text{ V}. \tag{4.65}$$

To calculate the short-circuit current, we place a short circuit across a, b. Observe that when the terminals a, b are shorted together, the control voltage v is reduced to zero. Therefore, with the short in place, the circuit in Fig. 4.38 becomes as shown in Fig. 4.39. With the short circuit shunting the 25-Ω resistor, all the current from the dependent current source will appear in the short; thus

$$i_{sc} = -20i. \tag{4.66}$$

Since the voltage controlling the dependent voltage source has been reduced to zero, the current controlling the dependent current source is

$$i = \frac{5}{2} = 2.5 \text{ mA}. \tag{4.67}$$

When Eq. (4.67) is substituted into Eq. (4.66), we get a short-circuit current of

$$i_{sc} = -20(2.5) = -50 \text{ mA}. \tag{4.68}$$

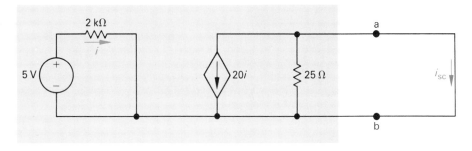

Fig. 4.39 The circuit in Fig. 4.38 with the terminals a, b short-circuited.

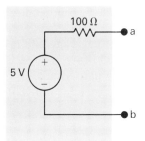

Fig. 4.40 The Thévenin equivalent for the circuit in Fig. 4.38.

It follows directly from Eqs. (4.65) and (4.68) that

$$R_t = \frac{V_t}{i_{sc}} = \frac{-5}{-50} \times 10^3 = 100 \ \Omega. \tag{4.69}$$

The Thévenin equivalent circuit for the circuit shown in Fig. 4.38 is illustrated in Fig. 4.40. Note that the reference polarity marks on the Thévenin voltage source in Fig. 4.38 agree with Eq. (4.65).

Let us now consider the alternative technique for finding the Thévenin resistance R_t. We first deactivate the independent voltage source from the circuit and then excite the circuit from the terminals a, b with either a test voltage source or a test current source. In deciding which type of source to use, we note from the circuit that if we apply a test voltage source we will know the voltage of the dependent voltage source and, hence, the controlling current i. Therefore, in this circuit we opt for the test voltage source. The circuit for computing the Thévenin resistance is shown in Fig. 4.41. The externally applied test voltage source is denoted by v_T and the current that it delivers to the circuit is labeled i_T. To find the Thévenin resistance, we simply solve the circuit shown in Fig. 4.41 for the ratio of the voltage to the current at the test source, that is, $R_t = v_T/i_T$. From Fig. 4.41 we note that

$$i_T = \frac{v_T}{25} + 20i \tag{4.70}$$

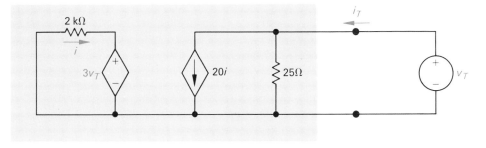

Fig. 4.41 An alternative method for computing the Thévenin resistance.

and

$$i = \frac{-3v_T}{2} \text{ mA.} \tag{4.71}$$

We then substitute Eq. (4.71) into Eq. (4.70) and solve the resulting equation for the ratio v_T/i_T:

$$i_T = \frac{v_T}{25} - \frac{60v_T}{2000},$$

$$\frac{i_T}{v_T} = \frac{1}{25} - \frac{6}{200} = \frac{50}{5000} = \frac{1}{100}. \tag{4.72}$$

It follows from Eqs. (4.72) that

$$R_t = \frac{v_T}{i_T} = 100 \ \Omega. \tag{4.73}$$

In general, the computations involved in this alternative method for finding the Thévenin resistance are easier than those involved in computing the short-circuit current. We should also point out that in a network containing only resistors and dependent sources, the alternative method must be used because the ratio of the Thévenin voltage to the short-circuit current is indeterminate, that is, it is the ratio 0/0. (See Problem 4.21.)

DRILL EXERCISE
4.16

Find the Thévenin equivalent circuit with respect to the terminals a, b.

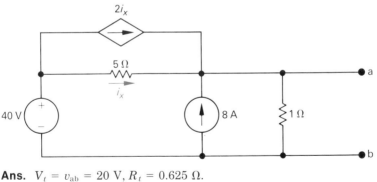

Ans. $V_t = v_{ab} = 20 \text{ V}, R_t = 0.625 \ \Omega.$

There are times when a Thévenin equivalent can be used to reduce one portion of a larger circuit so that the analysis of the larger network is greatly simplified. Let us return to the circuit first introduced in Section 2.6 and subsequently analyzed in Sections 4.5 and 4.8. To facilitate our discussion of using Thévenin's theorem to analyze this circuit, we have redrawn the circuit in Fig. 4.42 and identified the branch currents of interest. Before using a

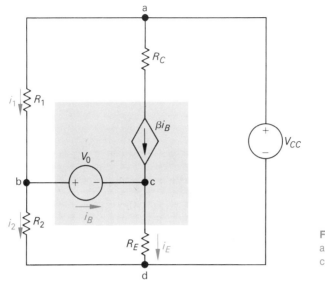

Thévenin equivalent to effect a solution, we make the observation that once we know i_B, we can easily obtain the other branch currents. We argue as follows. The current i_E is simply $(1 + \beta)i_B$. When i_E is known, the voltages v_{cd} and, hence, v_{bd} are known since $v_{bd} = v_{cd} + V_0$. When we know the voltage v_{bd}, we can quickly compute the branch currents i_1 and i_2. Thus $i_2 = v_{bd}/R_2$ and $i_1 = i_2 + i_B$. Realizing that i_B is the key to finding the other branch currents, we redraw the circuit as shown in Fig. 4.43. With a little thought, you should

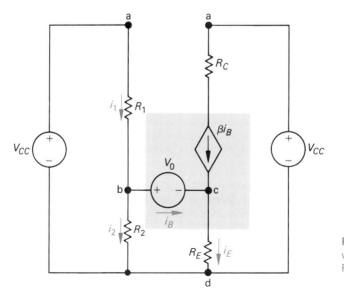

Fig. 4.43 A modified version of the circuit in Fig. 4.42.

be able to observe that this modification will have no effect on the branch currents i_1, i_2, i_B, and i_E.

Now we replace the circuit made up of V_{CC}, R_1, and R_2 with a Thévenin equivalent. The equivalent is made with respect to the terminals b, d. The Thévenin voltage and resistance are

$$V_t = \frac{V_{CC}R_2}{R_1 + R_2} \tag{4.74}$$

and

$$R_t = \frac{R_1 R_2}{R_1 + R_2}. \tag{4.75}$$

With the Thévenin equivalent, the circuit in Fig. 4.43 becomes as shown in Fig. 4.44.

We can now derive an equation for i_B simply by summing the voltages around the left mesh. In writing this mesh equation, we use the fact that $i_E = (1 + \beta)i_B$. Thus

$$V_t = R_t i_B + V_0 + R_E(1 + \beta)i_B, \tag{4.76}$$

from which we can write

$$i_B = \frac{V_t - V_0}{R_t + (1 + \beta)R_E}. \tag{4.77}$$

When we substitute Eqs. (4.74) and (4.75) into Eq. (4.77), we obtain the same

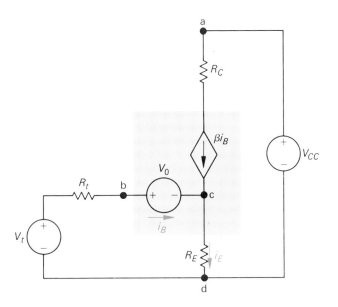

Fig. 4.44   The circuit in Fig. 4.43 modified by a Thévenin equivalent.

expression denoted by Eq. (2.27). Note that once we have incorporated the Thévenin equivalent into the original circuit, we can obtain the solution for i_B by writing a single equation!

4.13 MAXIMUM POWER TRANSFER

Circuit analysis plays an important role in the analysis of systems that are designed to transfer power from a source to a load. The general problem of power transfer can be discussed in terms of two basic types of systems. One emphasizes the efficiency of the power transfer, and the other emphasizes the amount of the power transfer. Power utility systems are a good example of the first type since they are concerned with the generation, transmission, and distribution of large quantities of electrical power. Communication and instrumentation systems are good examples of the second type, since they are designed to transmit information via electrical signals. In the transmission of information, or data, via electrical signals, the power available at the transmitter or detector is limited and it becomes desirable to transmit as much of this power as possible to the receiver, or load. In such applications the amount of power being transferred is small so that the efficiency of transfer is not of primary concern. We now consider the problem of maximum power transfer in systems that can be modeled by a purely resistive circuit.

The problem of maximum power transfer can be described with the aid of the circuit shown in Fig. 4.45. We assume that we are given a resistive network containing independent and dependent sources and a designated pair of terminals (a, b) to which a load (R_l) is to be connected. Our problem is to determine the value of R_l such that maximum power will be delivered to R_l. The first step to finding the critical value of R_l is to recognize that the given resistive network can always be replaced by its Thévenin equivalent. Therefore we redraw the circuit in Fig. 4.45 as shown in Fig. 4.46. Once we have replaced the original network by its Thévenin equivalent, we have greatly simplified the

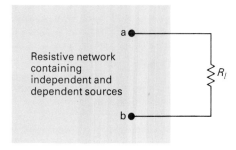

Fig. 4.45 A circuit describing maximum power transfer.

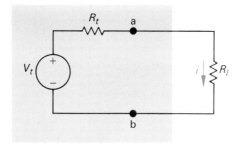

Fig. 4.46 A circuit used to determine the value of R_l for maximum power transfer.

problem of finding R_l. The derivation for R_l requires expressing the power dissipated in R_l as a function of the three circuit parameters V_t, R_t, and R_l. Thus

$$p = i^2 R_l = \left[\frac{V_t}{(R_t + R_l)}\right]^2 R_l. \tag{4.78}$$

Next, we recognize that for a given circuit, V_t and R_t will be fixed; therefore the power dissipated is a function of the single variable R_l. To find the value of R_l that maximizes the power, we use elementary calculus; that is, we solve for the value of R_l where dp/dR_l equals zero. We have

$$\frac{dp}{dR_l} = V_t^2 \left[\frac{(R_t + R_l)^2 - R_l \cdot 2(R_t + R_l)}{(R_t + R_l)^4}\right]. \tag{4.79}$$

Now the derivative will be zero when

$$(R_t + R_l)^2 = 2R_l(R_t + R_l). \tag{4.80}$$

Solving Eq. (4.80) yields

$$R_l = R_t. \tag{4.81}$$

Thus maximum power transfer occurs when the load resistance R_l equals the Thévenin resistance R_t. To find the maximum power delivered to R_l, we simply substitute Eq. (4.81) into Eq. (4.78) to get

$$p_{\max} = \frac{V_t^2 R_l}{(2R_l)^2} = \frac{V_t^2}{4R_l}. \tag{4.82}$$

Example 4.6 a) For the circuit shown in Fig. 4.47, find the value of R_l that results in maximum power being transferred to R_l.

b) Calculate the maximum power that can be delivered to R_l.

c) When R_l is adjusted for maximum power transfer, what percentage of the power delivered by the 360-V source reaches R_l?

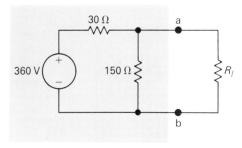

Fig. 4.47 The circuit for Example 4.6.

Solution a) The Thévenin voltage for the circuit to the left of the terminals a, b is

$$V_t = \frac{360}{180} \times 150 = 300 \text{ V}.$$

The Thévenin resistance is

$$R_t = \frac{(150)(30)}{180} = 25 \ \Omega.$$

Replacing the circuit to the left of the terminals a, b with its Thévenin equivalent gives us the circuit shown in Fig. 4.48, from which we see that R_l must equal 25 Ω for maximum power transfer.

b) The maximum power that can be delivered to R_l is

$$p_{\max} = \left(\frac{300}{50}\right)^2 (25) = 900 \text{ W}.$$

c) When R_l equals 25 Ω, the voltage v_{ab} is

$$v_{ab} = \left(\frac{300}{50}\right)(25) = 150 \text{ V}.$$

We can see from Fig. 4.47 that when v_{ab} equals 150 V, the current in the

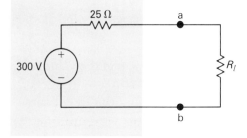

Fig. 4.48 A reduction of the circuit in Fig. 4.47 by means of a Thévenin equivalent.

voltage source in the direction of the voltage rise across the source will be

$$i_s = \frac{360 - 150}{30} = \frac{210}{30} = 7 \text{ A}.$$

Therefore the source is delivering 2520 W to the circuit, that is,

$$p_s = -i_s(360) = -2520 \text{ W}.$$

The percentage of the source power delivered to the load is

$$\frac{900}{2520} \times 100 = 35.71\%. \quad \blacksquare$$

DRILL EXERCISE
4.18

a) Find the value of R that will enable the circuit to deliver maximum power to the terminals a, b.

b) Find the maximum power delivered to R.

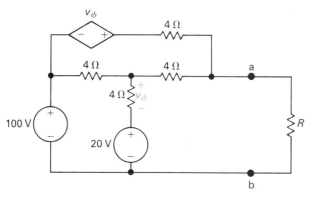

Ans. (a) 3 Ω; (b) 1.2 kW.

DRILL EXERCISE
4.17

Assume that the circuit in Drill Exercise 4.17 is delivering maximum power to the load resistor R.

a) How much power is the 100-V source delivering to the network?

b) Repeat part (a) for the dependent voltage source.

c) What percentage of the total power generated by these two sources is delivered to the load resistor R?

Ans. (a) 3000 W; (b) 800 W; (c) 31.58%.

4.14 SUPERPOSITION

The most distinguishing characteristic of a linear system is the principle of *superposition*, which states that whenever a linear system is excited, or driven, by more than one independent source of energy, we can find the total response

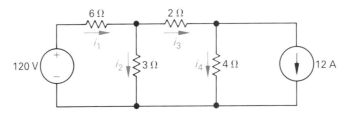

Fig. 4.49 A circuit used to demonstrate the use of superposition in circuit analysis.

by finding the response to each independent source separately and then summing the individual responses. Since we are dealing with circuits made up of interconnected linear-circuit elements, we can apply the principle of superposition directly to the analysis of such circuits when they are driven by more than one independent energy source. At present, we will restrict our illustration to simple resistive networks; however, it is important to bear in mind that the principle is applicable to circuits containing inductance and capacitance as well as resistance. As a matter of fact, it is applicable to any linear system.

We will demonstrate the use of the superposition principle by using it to find the branch currents in the circuit shown in Fig. 4.49. We begin by finding the branch currents due to the 120-V voltage source. We will denote with a prime the component of the branch current that is due to the voltage source. The ideal current source is deactivated by opening the branch in which it appears. Figure 4.50 shows the circuit with the current source removed. We see there the prime notation applied to the branch currents to indicate that the currents in the circuit are due only to the voltage source.

We now note that we can easily find the branch currents in the circuit in Fig. 4.50 once we know the node voltage across the 3-Ω resistor. If we denote this voltage as v_1, we can write

$$\frac{v_1 - 120}{6} + \frac{v_1}{3} + \frac{v_1}{2 + 4} = 0, \tag{4.83}$$

from which it follows that

$$v_1 = 30 \text{ V}. \tag{4.84}$$

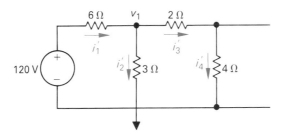

Fig. 4.50 The circuit in Fig. 4.49 with the current source deactivated.

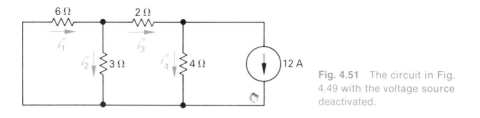

Fig. 4.51 The circuit in Fig. 4.49 with the voltage source deactivated.

Now we can write the expressions for the branch currents i_1'' through i_4'' directly; thus

$$i_1'' = \frac{120 - 30}{6} = 15 \text{ A}, \tag{4.85}$$

$$i_2'' = \frac{30}{3} = 10 \text{ A}, \tag{4.86}$$

$$i_3'' = i_4'' = \frac{30}{6} = 5 \text{ A}. \tag{4.87}$$

The ideal voltage source in Fig. 4.49 is deactivated by replacing it with a short circuit. Thus to find the component of the branch currents due to the current source, we must solve the circuit shown in Fig. 4.51. The double-prime notation for the currents in Fig. 4.51 indicates that these currents are the components of the total current due to the ideal current source.

We will solve for the branch currents in the circuit in Fig. 4.51 by first solving for the node voltages across the 3-Ω and 4-Ω resistors, respectively. The two node voltages are defined as shown in Fig. 4.52, from which it follows that the two node-voltage equations that describe the circuit are

$$\frac{v_3}{3} + \frac{v_3}{6} + \frac{v_3 - v_4}{2} = 0, \tag{4.88}$$

$$\frac{v_4 - v_3}{2} + \frac{v_4}{4} + 12 = 0. \tag{4.89}$$

When we solve Eqs. (4.88) and (4.89) for v_3 and v_4 we get

$$v_3 = -12 \text{ V} \tag{4.90}$$

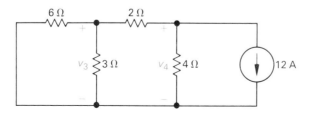

Fig. 4.52 The circuit in Fig. 4.49 showing the node voltages v_3 and v_4.

and

$$v_4 = -24 \text{ V}. \tag{4.91}$$

Now we can write the branch currents i_1'' through i_4'' directly in terms of the node voltages v_3 and v_4 as follows:

$$i_1'' = \frac{-v_3}{6} = \frac{12}{6} = 2 \text{ A}, \tag{4.92}$$

$$i_2'' = \frac{v_3}{3} = \frac{-12}{3} = -4 \text{ A}, \tag{4.93}$$

$$i_3'' = \frac{v_3 - v_4}{2} = \frac{-12 + 24}{2} = 6 \text{ A}, \tag{4.94}$$

$$i_4'' = \frac{v_4}{4} = \frac{-24}{4} = -6 \text{ A}. \tag{4.95}$$

To find the branch currents in the original circuit, that is, the currents i_1, i_2, i_3, and i_4 in the circuit shown in Fig. 4.49, we simply add the currents given by Eqs. (4.92)–(4.95) to the currents given by Eqs. (4.85)–(4.87). Thus

$$i_1 = i_1' + i_1'' = 15 + 2 = 17 \text{ A}, \tag{4.96}$$

$$i_2 = i_2' + i_2'' = 10 - 4 = 6 \text{ A}, \tag{4.97}$$

$$i_3 = i_3' + i_3'' = 5 + 6 = 11 \text{ A}, \tag{4.98}$$

$$i_4 = i_4' + i_4'' = 5 - 6 = -1 \text{ A}. \tag{4.99}$$

We leave it to the reader to verify that the currents given by Eqs. (4.96)–(4.99) are the correct values for the branch currents in the circuit in Fig. 4.49.

DRILL EXERCISE
4.19

a) Use the principle of superposition to find the voltage v in the circuit shown.
b) Find the power dissipated in the 40-Ω resistor.

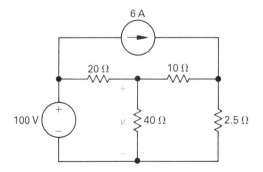

Ans. (a) 40 V; (b) 40 W.

DRILL EXERCISE   Use the principle of superposition to find the voltage v in the following circuit.
4.20

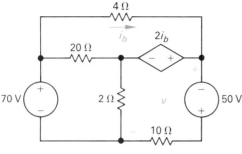

Ans. 30 V.

4.15 SUMMARY

The purpose of this chapter has been to introduce some extremely useful techniques of circuit analysis. The node-voltage and mesh-current techniques are important because they enable us to minimize the number of simultaneous equations needed to describe a circuit. These two techniques are also valuable because they force us to take a systematic approach to writing circuit equations. Source transformations and Thévenin–Norton equivalent circuits, two additional methods of simplifying circuits, are useful when the performance of a circuit focuses on its behavior at a specific pair of terminals. The principle of superposition is an important concept when two or more sources are present in the circuit, because it enables us to isolate the effect of each source on the total response of the circuit.

It is also important to keep in mind that although we have used resistive networks to illustrate the various analytical techniques, the techniques themselves are not limited to resistive circuits. In Chapter 5, we begin discussing the analysis of linear, lumped-parameter circuits that contain inductance and capacitance as well as resistance.

PROBLEMS

4.1   Use the node-voltage method to find the branch currents i_1 through i_5 for the circuit in Fig. 4.53.

4.2   Use the node-voltage method to find the branch currents i_a through i_d in the circuit in Fig. 4.54.

4.3   Use the node-voltage method to find the power associated with the 5-Ω resistor in the circuit in Fig. 4.10 (Example 4.2). Use the node formed by the 5-Ω, 2-Ω, and 10-Ω resistors as the reference node.

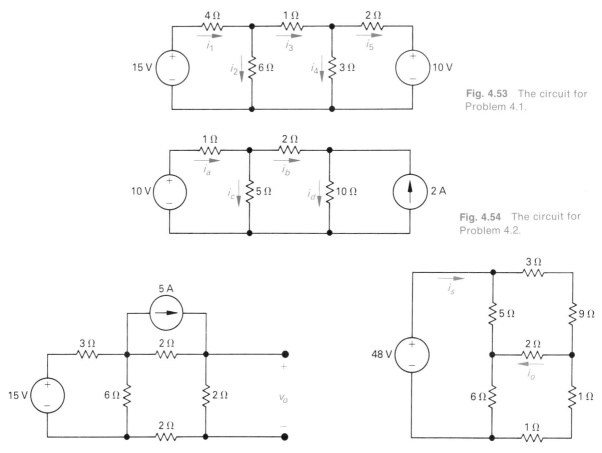

Fig. 4.53 The circuit for Problem 4.1.

Fig. 4.54 The circuit for Problem 4.2.

Fig. 4.55 The circuit for Problem 4.4.

Fig. 4.56 The circuit for Problem 4.5.

4.4 Use the node-voltage method to find the output voltage v_o in the circuit in Fig. 4.55.

4.5 Use the node-voltage method to find the currents i_o and i_s in the circuit in Fig. 4.56.

4.6 a) Use the node-voltage method to find v in the circuit in Fig. 4.57.
    b) Find the power associated with each source in the circuit and state whether the source is delivering power to, or absorbing power from, the circuit.

4.7 Use the node-voltage method to find v in the circuit in Fig. 4.58.

4.8 Use the node-voltage method to find the power associated with the 125-V source in the circuit in Fig. 4.59.

4.9 Use the node-voltage method to find the power associated with each circuit element in the circuit in Fig. 4.60.

4.10 Show that when Eqs. (4.16), (4.17), and (4.19) are solved for i_B, the result is identical to Eq. (2.27).

4.11 Use the mesh-current method to find the power associated with the 45-V source in the circuit in Fig. 4.61. State whether the source is delivering or absorbing power.

4.12 Use the mesh-current method to find the branch current i_o in the circuit in Fig. 4.62.

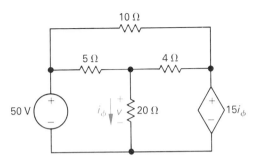

Fig. 4.57 The circuit for Problem 4.6.

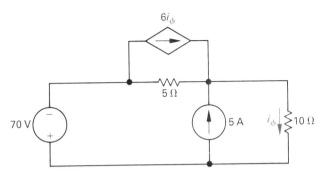

Fig. 4.60 The circuit for Problem 4.9.

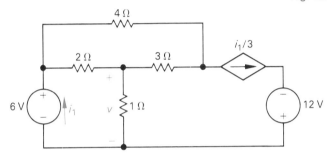

Fig. 4.58 The circuit for Problem 4.7.

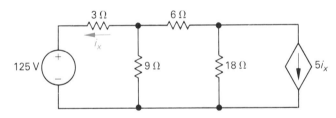

Fig. 4.61 The circuit for Problem 4.11.

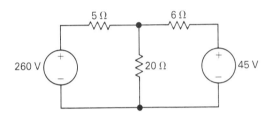

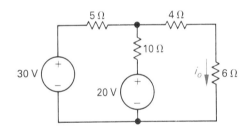

Fig. 4.62 The circuit for Problem 4.12.

Fig. 4.59 The circuit for Problem 4.8.

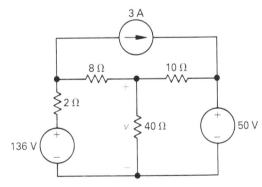

Fig. 4.63 The circuit for Problem 4.13.

4.13 a) Use the mesh-current method to find the voltage v in the circuit in Fig. 4.63.

  b) Find the power associated with each source in the circuit.

  c) State whether the source is extracting power from, or delivering power to, the circuit.

4.14   Use the mesh-current method to find the voltage across the dependent current source in the circuit in Fig. 4.64.

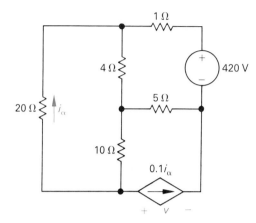

Fig. 4.64   The circuit for Problem 4.14.

4.15 a) Find the branch currents i_0 through i_4 for the circuit shown in Fig. 4.65.
   b) Show that the total power delivered in the circuit equals the total power absorbed.

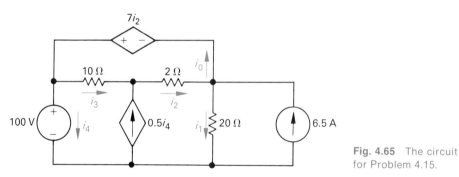

Fig. 4.65   The circuit for Problem 4.15.

4.16   The circuit in Fig. 4.66 is a direct current version of a typical three-wire distribution system. The resistors R_1, R_2, and R_3 represent the resistances of the three conductors that connect the three loads R_a, R_b, and R_c to the 125/250-V voltage supply. The resistors R_a and R_b represent loads connected to the 125-V circuits, and R_c represents a load connected to the 250-V circuit.

   a) Calculate v_a, v_b, and v_c.
   b) Calculate the power delivered to R_a, R_b, and R_c.
   c) Calculate the power delivered by each source.
   d) What percentage of the source power is delivered to the loads?
   e) The R_2 branch represents the neutral conductor in the distribution circuit. What adverse effect occurs if the neutral conductor is opened? (*Hint:* Calculate v_a and v_b and note that appliances or loads designed for use in this circuit would have a nominal voltage rating of 125 V.)

4.17   Find the current in the 38-kΩ resistor in the circuit of Fig. 4.67 by making a succession of appropriate source transformations.

4.18   Determine i_o and v_o in the circuit shown in Fig. 4.68 when R is 0, 2, 4, 6, 10, 18, 24, 42, 90, and 186 Ω.

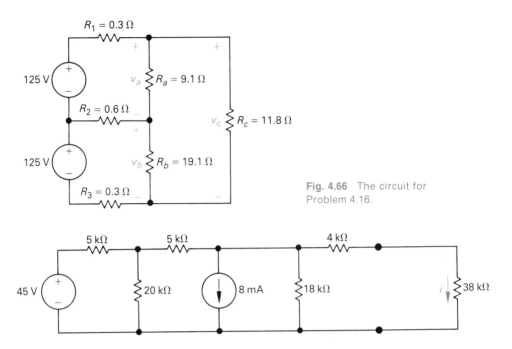

Fig. 4.66 The circuit for Problem 4.16.

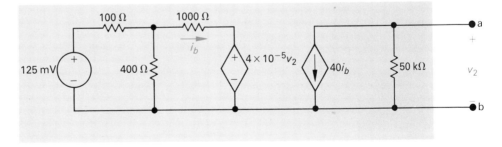

Fig. 4.67 The circuit for Problem 4.17.

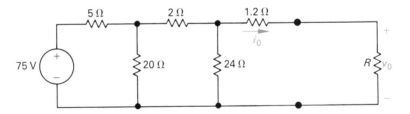

Fig. 4.68 The circuit for Problem 4.18.

Fig. 4.69 The circuit for Problem 4.19.

4.19 Determine the Thévenin equivalent with respect to the terminals a, b for the circuit in Fig. 4.69.

4.20 The Wheatstone bridge in the circuit shown in Fig. 4.70 is balanced when R_3 equals 500 Ω. If the galvanometer has a resistance of 50 Ω, how much current will the galvanometer detect when the bridge is unbalanced by setting R_3 to 501 Ω? (*Hint:* Find the Thévenin equivalent with respect to the galvanometer terminals when $R_3 =$ 501 Ω.) Note that once we have found the Thévenin equivalent with respect to the galvanometer terminals, it is easy to find the amount of unbalanced current in the galvanometer branch for different galvanometer movements.

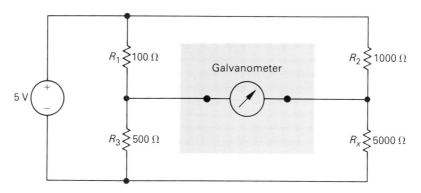

Fig. 4.70 The circuit for Problem 4.20.

4.21 Find the Thévenin equivalent with respect to the terminals a, b for the circuit shown in Fig. 4.71.

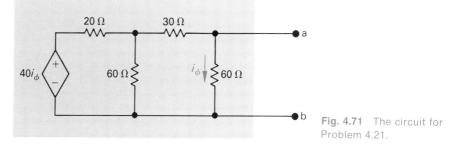

Fig. 4.71 The circuit for Problem 4.21.

4.22 Laboratory measurements on a dc voltage source yield a terminal voltage of 20 V with no load connected to the source and 16 V when loaded with a 400-Ω resistor.

a) What is the Thévenin equivalent with respect to the terminals of the dc voltage source?

b) Show that the Thévenin resistance of the source is given by the expression

$$R_t = \left(\frac{V_t}{V_o} - 1 \right) R_L,$$

where

V_t = the Thévenin voltage,
V_o = the terminal voltage corresponding to the load resistance R_L.

4.23 a) Calculate the power delivered to each resistor in Problem 4.18.

b) Plot the power delivered versus the resistance.

c) At what value of R is the power maximum?

4.24 The resistor R in the circuit shown in Fig. 4.72 is adjusted until maximum power is delivered to the resistor.

   a) What is the value of R?
   b) What is the power delivered to R in milliwatts?
   c) What percentage of the total power delivered by the sources in the circuit is delivered to R?

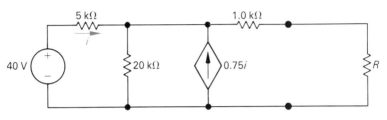

Fig. 4.72 The circuit for Problem 4.24.

4.25 Determine the maximum power that the circuit in Fig. 4.73 can deliver to a resistive load connected to the terminals a, b.

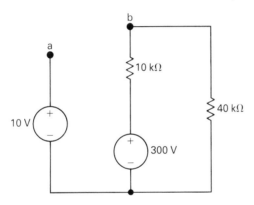

Fig. 4.73 The circuit for Problem 4.25.

4.26 Use the principle of superposition to find the current i_o in the circuit shown in Fig. 4.74.

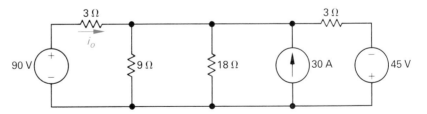

Fig. 4.74 The circuit for Problem 4.26.

5

INDUCTANCE AND CAPACITANCE

5.1 INTRODUCTION

We are now ready to introduce the last two ideal circuit elements mentioned in Chapter 2, namely, inductors and capacitors. First, let us assure you that the circuit analysis techniques introduced in Chapters 3 and 4 are applicable to circuits containing inductors and capacitors. Therefore, once we understand the terminal behavior of inductors and capacitors in terms of current and voltage, we can use Kirchhoff's laws to describe any interconnections with the other basic elements.

Recall from Chapter 1 that one of the advantages of circuit theory is that it provides a relatively simple description of practical components that is unnecessarily complicated if given in terms of electromagnetic field theory. Inductors and capacitors are circuit elements that are easier to describe in terms of circuit variables than field variables. However, before we focus on the circuit descriptions, a brief review of the field concepts underlying these basic elements is in order.

Inductors are circuit elements based on phenomena associated with magnetic fields. The source of the magnetic field is charge in motion, or current. If the current is varying with time, the magnetic field is varying with time. A time-varying magnetic field induces a voltage in any conductor that is linked by the field. The circuit parameter of inductance relates the induced voltage to the current. We will discuss this quantitative relationship in Section 5.2.

Capacitors are circuit elements based on phenomena associated with electric fields. The source of the electric field is separation of charge, or voltage. If the voltage is varying with time, the electric field is varying with time. A time-varying electric field produces a displacement current in the space occupied by the field. The circuit parameter of capacitance relates the displacement current to the voltage. The displacement current is equal to the conduction current at the terminals of the capacitor; therefore we can use capacitance to relate the circuit current to the voltage. We will discuss this quantitative relationship in Section 5.3.

Energy can be stored in both magnetic and electric fields. Knowing this, we should not be too surprised to learn that inductors and capacitors are capable of storing energy. For example, energy can be stored in an inductor and then released to "fire" a spark plug. Energy can be stored in a capacitor and then released to "fire" a flashbulb. At this point it is important to note that we are talking about energy storage, not energy generation. Therefore in ideal inductors and capacitors, we can extract only as much energy as has been stored. Because inductors and capacitors cannot generate energy, they are also classified as passive elements. Energy storage is not unique to electrical systems. Two of the most common examples of mechanical devices used to store energy are springs and flywheels.

We are now ready to describe the behavior of inductors and capacitors in terms of current and voltage.

5.2 THE INDUCTOR

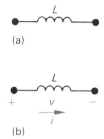

(a)

(b)

Fig. 5.1 The graphical symbol for an inductor with an inductance of L henries.

Whereas resistance is the circuit parameter used to describe a resistor, inductance is the circuit parameter used to describe an inductor. Inductance is symbolized by the letter L, is measured in henries (H), and is represented graphically as a coiled wire, which reminds us that inductance is a consequence of a conductor linking a magnetic field. The inductor is shown in Fig. 5.1(a). If we assign the reference direction of the current in the inductor in the direction of the voltage drop across the terminals of the inductor, as shown in Fig. 5.1(b), then

$$v = L\frac{di}{dt}, \qquad (5.1)$$

where v is measured in volts, L in henries, i in amperes, and t in seconds. If the current is in the direction of the voltage rise across the inductor, Eq. (5.1) is written with a minus sign.

We see from Eq. (5.1) that the voltage across the terminals of an inductor is proportional to the time rate of change of current in the inductor. We can make two important observations at this time. First, if the current is constant, the voltage across the ideal inductor is zero. Thus the inductor looks like a short circuit to a constant, or dc, current. Second, there cannot be an instantaneous change of current in an inductor; that is, the current cannot change by a finite amount in zero time. Equation (5.1) tells us that this would require an infinite voltage, and infinite voltages are not possible. For example, when we open the switch on an inductive circuit in a physically realizable system, the switch will arc over, thus preventing the current from dropping to zero instantaneously. (We note in passing that switching inductive circuits is an important engineering problem because the arcing and voltage surges that can arise must be controlled in order to prevent damage to equipment. The first step to understanding the nature of this problem is to master the introductory material presented in Chapters 5, 6, and 7.) The application of Eq. (5.1) to a simple circuit is illustrated in the following example.

Example 5.1 The independent current source in the circuit in Fig. 5.2 generates zero current for $t < 0$ and a pulse $10te^{-5t}$ for $t > 0$.

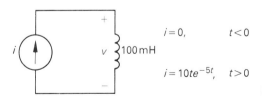

$$i = 0, \qquad t < 0$$

$$i = 10te^{-5t}, \quad t > 0$$

Fig. 5.2 The circuit for Example 5.1.

a) Sketch the current waveform.

b) At what instant of time is the current maximum?

c) Express the voltage across the terminals of the 100-mH inductor as a function of time.

d) Sketch the voltage waveform.

e) Is the voltage maximum when the current is maximum?

f) At what instant of time does the voltage change polarity?

g) Is there ever an instantaneous change in voltage across the inductor? If so, at what time?

Solution   a) The current waveform is shown in Fig. 5.3.

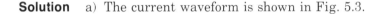

Fig. 5.3 Current waveform for Example 5.1.

b) $\dfrac{di}{dt} = 10(-5te^{-5t} + e^{-5t}) = 10e^{-5t}(1 - 5t)$;

$\dfrac{di}{dt} = 0$ when $t = \frac{1}{5}$ second.   (See Fig. 5.3.)

c) $v = L\dfrac{di}{dt} = (0.1)10e^{-5t}(1 - 5t) = e^{-5t}(1 - 5t)$ V, $t > 0$;

$v = 0$, $t < 0$

d) The voltage waveform is shown in Fig. 5.4.

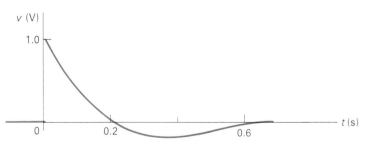

Fig. 5.4 Voltage waveform for Example 5.1.

e) No; the voltage is proportional to di/dt, not i.

f) At 0.2 second, which corresponds to the moment when di/dt is passing through zero and changing sign.

g) Yes, at $t = 0$. Note that there can be an instantaneous change in the voltage across the terminals of an inductor. ∎

Equation (5.1) expresses the voltage across the terminals of the inductor as a function of the current in the inductor. It is also desirable to be able to express the current in the inductor as a function of the voltage. To find i as a function of v, we start by multiplying both sides of Eq. (5.1) by a differential time dt to get

$$v \, dt = L \left(\frac{di}{dt} \right) dt. \tag{5.2}$$

Now we recognize that multiplying the rate at which i varies with t by a differential change in time will generate a differential change in i; therefore, Eq. (5.2) can be written as

$$v \, dt = L \, di. \tag{5.3}$$

To find i as a function of v, we simply integrate both sides of Eq. (5.3). For convenience, we interchange the two sides of the equation and write

$$L \int_{i(t_0)}^{i(t)} dx = \int_{t_0}^{t} v \, d\tau, \tag{5.4}$$

in which we have introduced x and τ as symbols of integration. It follows directly from Eq. (5.4) that

$$i(t) = \frac{1}{L} \int_{t_0}^{t} v \, d\tau + i(t_0), \tag{5.5}$$

where $i(t)$ is the current corresponding to t and $i(t_0)$ is the value of the inductor current at the time when we initiate the integration, namely, t_0. In many practical applications, t_0 is zero and Eq. (5.5) becomes

$$i(t) = \frac{1}{L} \int_{0}^{t} v \, d\tau + i(0). \tag{5.6}$$

Both Eqs. (5.1) and (5.5) give the relationship between the voltage and current at the terminals of an inductor. Equation (5.1) expresses the voltage as a function of current, whereas Eq. (5.5) expresses the current as a function of voltage. Both equations are written with the reference direction for the current in the direction of the voltage drop across the terminals. Note that $i(t_0)$ will carry its own algebraic sign. If the initial current is in the same direction as the reference direction for i, it will be a positive quantity. On the other hand, if the initial current is in the opposite direction from the reference direction for i, it will be a negative quantity. Example 5.2 illustrates the application of Eq. (5.5).

Example 5.2 The voltage pulse applied to the 100-mH inductor shown in Fig. 5.5 is 0 for $t < 0$ and is given by the expression

$$v(t) = 20te^{-10t}$$

for $t > 0$.

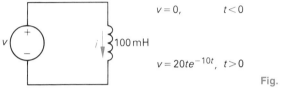

$v = 0,$ $t < 0$

100 mH

$v = 20te^{-10t}, \ t > 0$

Fig. 5.5 The circuit for Example 5.2.

a) Sketch the voltage as a function of time.
b) Find the inductor current as a function of time.
c) Sketch the current as a function of time.

Solution a) The voltage as a function of time is shown in Fig. 5.6.

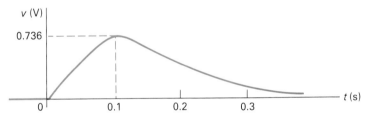

Fig. 5.6 Voltage waveform for Example 5.2.

b) Since $v(t) = 0$ for all $t < 0$, the current in the inductor will be 0 at $t = 0$. Therefore the current for $t > 0$ will be

$$i = \frac{1}{0.1} \int_0^t 20\tau e^{-10\tau} \, d\tau + 0$$

$$= 200 \left[\frac{-e^{-10\tau}}{100} (10\tau + 1) \right] \Big|_0^t$$

$$= 2[1 - 10te^{-10t} - e^{-10t}] \, \text{A}, \quad t > 0.$$

Fig. 5.7 Current waveform for Example 5.2.

c) The current as a function of time is shown in Fig. 5.7. ∎

Note in Example 5.2 that i approaches a constant value of 2 A as t increases. We will have more to say about this result after we have discussed the energy stored in an inductor.

The power and energy relationships for the inductor can be derived directly from the current and voltage relationships. If we assume that the current reference is in the direction of the voltage drop across the terminals of the inductor, the power is

$$p = vi. \tag{5.7}$$

Remember that power is in watts when the voltage is given in volts and the current in amperes. If we express the inductor voltage as a function of the inductor current, Eq. (5.7) becomes

$$p = Li\frac{di}{dt}. \tag{5.8}$$

We can also express the current in terms of the voltage:

$$p = v\left[\frac{1}{L}\int_{t_0}^{t} v\, d\tau + i(t_0)\right]. \tag{5.9}$$

Equation (5.8) is most useful in expressing the energy stored in the inductor. Using the fact that power is the time rate of expending energy, we have

$$p = \frac{dw}{dt} = Li\frac{di}{dt}. \tag{5.10}$$

When we multiply both sides of Eq. (5.10) by a differential time, we get the differential relationship

$$dw = Li\, di. \tag{5.11}$$

Both sides of Eq. (5.11) are integrated with the understanding that the reference for zero energy will be chosen to correspond to zero current in the inductor. Thus

$$\int_0^w dx = L \int_0^i y \, dy,$$

$$w = \frac{1}{2} Li^2.$$

(5.12)

As before, we have introduced different symbols of integration to avoid confusion with the limits placed on the integrals. In Eq. (5.12), the energy is in joules when the inductance is given in henries and the current in amperes. To illustrate the application of Eqs. (5.7) and (5.12), we return to Examples 5.1 and 5.2 by Example 5.3.

Example 5.3
a) For Example 5.1, plot i, v, p, and w versus time. In making the plots, line them up vertically so that for a given time interval we can easily assess the behavior of each variable.

b) In what time interval is energy being stored in the inductor?

c) In what time interval is energy being extracted from the inductor?

d) What is the maximum energy stored in the inductor?

e) Evaluate the integrals

$$\int_0^{0.2} p \, dt \quad \text{and} \quad \int_{0.2}^\infty p \, dt$$

and comment on their significance.

f) Repeat parts (a) through (d) for Example 5.2.

g) In Example 5.2, why is there a sustained current in the inductor as the voltage approaches zero?

Solution
a) The plots of i, v, p, and w follow directly from the expressions for i and v obtained in Example 5.1. In particular, $p = vi$ and $w = (\frac{1}{2})Li^2$. See Fig. 5.8.

b) Energy is being stored in the time interval between 0 to 0.2 s, that is, in the interval when $p > 0$.

c) Energy is being extracted in the time interval between 0.2 s and ∞, that is, in the interval when $p < 0$.

d) $w_{max} = 27.07$ mJ

e) From Example 5.1, we have

$$i = 10te^{-5t} \text{ A} \quad \text{and} \quad v = e^{-5t}(1 - 5t) \text{ V}.$$

Therefore

$$p = vi = 10te^{-10t} - 50t^2 e^{-10t} \text{ W}.$$

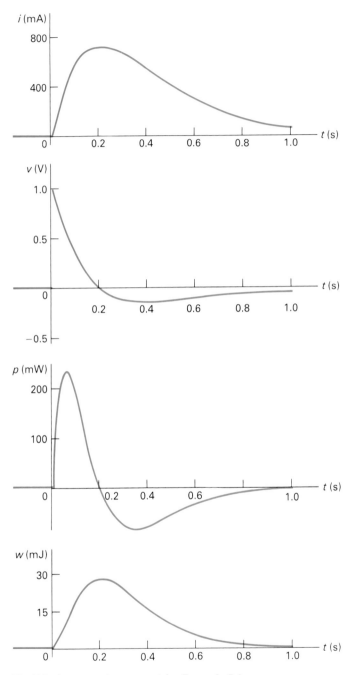

Fig. 5.8 $i, v, p,$ and w versus t for Example 5.1.

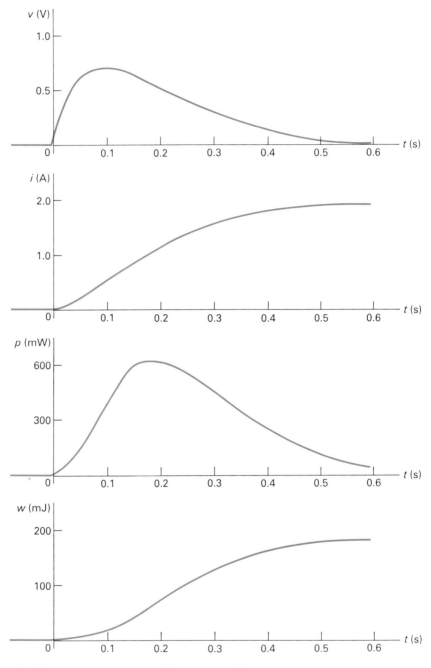

Fig. 5.9 $v, i, p,$ and w versus t for Example 5.2.

Thus

$$\int_0^{0.2} p \, dt = 10 \left[\frac{e^{-10t}}{100} (-10t - 1) \right]_0^{0.2} - 50 \left\{ \frac{t^2 e^{-10t}}{-10} + \frac{2}{10} \left[\frac{e^{-10t}}{100} (-10t - 1) \right] \right\}_0^{0.2}$$

$$= 0.2e^{-2} = 27.07 \text{ mJ};$$

$$\int_{0.2}^{\infty} p \, dt = 10 \left[\frac{e^{-10t}}{100} (-10t - 1) \right]_{0.2}^{\infty} - 50 \left\{ \frac{t^2 e^{-10t}}{-10} + \frac{2}{10} \left[\frac{e^{-10t}}{100} (-10t - 1) \right] \right\}_{0.2}^{\infty}$$

$$= -0.2e^{-2} = -27.07 \text{ mJ}.$$

From the definition of p it follows that the area under the plot of p vs. t represents the energy expended over the interval of integration. Hence, the integration of the power between 0 and 0.2 s represents the energy stored in the inductor during this time interval. The integral of p over the interval from 0.2 s to ∞ is the energy extracted from the inductor. Note that in this time interval all the energy originally stored is removed. That is, after the current pulse has passed, there is no energy stored in the inductor.

f) The plots of v, i, p, and w follow directly from the expressions for v and i as given in Example 5.2. See Fig. 5.9. Note that in this case the power is always positive and hence, energy is always being stored in the inductor during the duration of the voltage pulse.

g) The application of the voltage pulse stores energy in the inductor. Since the inductor is ideal, there is no way for this energy to dissipate after the voltage subsides to zero. Therefore a sustained current circulates in the circuit. A lossless inductor is obviously an ideal circuit element. Practical inductors require the insertion of a resistor in the circuit model. (More about this later.) ■

Examples 5.1, 5.2, and 5.3 have illustrated the application of the basic equations that relate the voltage and current at the terminals of an inductor and the subsequent calculations of power and energy. We are now ready to study the corresponding relationships for the capacitor.

DRILL EXERCISE 5.1 The current source in the circuit shown generates the current pulse

$$i_g(t) = 0, \quad t < 0;$$
$$i_g(t) = 5e^{-200t} - 5e^{-800t}, \quad t \geq 0.$$

Find (a) $v(0)$; (b) the instant of time, greater than zero, when the voltage v passes through zero; (c) the expression for the power delivered to the inductor; (d) the instant when the power delivered to the inductor is maximum; (e) the maximum power; (f) the maximum energy stored in the inductor; and (g) the instant of time when the stored energy is maximum.

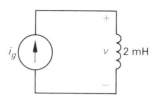

Ans. (a) 6 V; (b) 2.31 ms; (c) $50e^{-1000t} - 10e^{-400t} - 40e^{-1600t}$ W; (d) 616.58 μs; (e) 4.26 W; (f) 5.58 mJ; and (g) 2.31 ms.

5.3 THE CAPACITOR

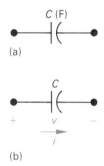

Fig. 5.10 The circuit symbol for a capacitor.

The circuit parameter of capacitance is represented by the letter C, is measured in farads (F), and is symbolized by two short parallel conductive plates, as shown in Fig. 5.10(a). The farad is an extremely large denomination of capacitance; thus submultiples of the farad are found in practical circuits and devices. The most frequently encountered values lie in the picofarad (pF) to microfarad (μF) range. The graphical symbol for the capacitor is designed to remind us that capacitance arises whenever electrical conductors are separated by a dielectric, or insulating, material. The fact that the conductors are separated by a dielectric material implies that electrical charge is not transported through the capacitor. Although the application of a voltage to the terminals of the capacitor cannot cause a movement of charge through the dielectric, it can cause a displacement of charge within the dielectric. As the voltage varies with time, the displacement of charge within the dielectric varies with time, causing what is known as the *displacement current*. From the point of view of the terminals, the displacement current is indistinguishable from a conduction current. The current at the terminals of a capacitor is proportional to the rate at which the voltage across the capacitor varies with time. Mathematically this is written as

$$i = C \frac{dv}{dt}, \qquad (5.13)$$

where i is measured in amperes, C in farads, v in volts, and t in seconds.

In writing Eq. (5.13), we have assumed the passive sign convention shown in Fig. 5.10(b); that is, the current is referenced in the direction of the voltage drop across the capacitor. If the current is referenced in the direction of the voltage rise, we write Eq. (5.13) with a minus sign.

Two important observations follow from Eq. (5.13). First, there cannot be an instantaneous change in the voltage across the terminals of a capacitor. Equation (5.13) indicates that such a change would produce infinite current, a physical impossibility. Second, if the voltage across the terminals is constant, the capacitor current will be zero. This is a direct consequence of the fact that a conduction current cannot be established in the dielectric material of the capacitor. Only a time-varying voltage can produce a displacement current. We can conclude that a capacitor appears as an open circuit to a constant current.

Equation (5.13) gives the capacitor current as a function of the capacitor voltage. It is also useful to express the voltage as a function of the current. We can accomplish this by multiplying both sides of Eq. (5.13) by a differential time dt and then integrating the resulting differentials:

$$i \, dt = C \, dv,$$

or

$$\int_{v(t_0)}^{v(t)} dx = \frac{1}{C} \int_{t_0}^{t} i \, d\tau.$$

Carrying out the integration of the left-hand side of the equation gives us

$$v(t) = \frac{1}{C} \int_{t_0}^{t} i \, d\tau + v(t_0). \tag{5.14}$$

In many practical applications of Eq. (5.14), the initial time is usually zero, that is, $t_0 = 0$; thus Eq. (5.14) becomes

$$v(t) = \frac{1}{C} \int_{0}^{t} i \, d\tau + v(0). \tag{5.15}$$

We can easily derive the power and energy relationships for the capacitor. From the definition of power we have

$$p = vi = Cv \frac{dv}{dt}, \tag{5.16}$$

or

$$p = i \left[\frac{1}{C} \int_{t_0}^{t} i \, d\tau + v(t_0) \right]. \tag{5.17}$$

Combining the definition of energy with Eq. (5.16) we obtain

$$dw = Cv \, dv,$$

from which we have

$$\int_{0}^{w} dx = C \int_{0}^{v} y \, dy,$$

or

$$w = \frac{1}{2} Cv^2 \tag{5.18}$$

In deriving Eq. (5.18), we have chosen the reference for zero energy to correspond to zero voltage.

Examples 5.4 and 5.5 have been designed to illustrate the application of the current, voltage, power, and energy relationships for the capacitor.

Example 5.4   The voltage pulse described by the following equations is impressed across the terminals of a 0.5-μF capacitor:

$$v(t) = 0, \qquad\qquad t \leq 0,$$
$$v(t) = 4t, \qquad\qquad 0 \leq t \leq 1,$$
$$v(t) = 4e^{-(t-1)}, \quad 1 \leq t \leq \infty.$$

a) Derive the expressions for the capacitor current, power, and energy.

b) Sketch the voltage, current, power, and energy as functions of time. In making these sketches, follow the same instructions as set forth in Example 5.3.

c) Specify the interval of time when energy is being stored in the capacitor.

d) Specify the interval of time when energy is being delivered by the capacitor.

e) Evaluate the integrals

$$\int_0^1 p\, dt \quad \text{and} \quad \int_1^\infty p\, dt$$

and comment on their significance.

Solution   a) From Eq. (5.13) we have

$$i = (0.5 \times 10^{-6})(0) = 0, \qquad\qquad\qquad t < 0,$$
$$i = (0.5 \times 10^{-6})(4) = 2\ \mu\text{A}, \qquad\qquad 0 < t < 1,$$
$$i = (0.5 \times 10^{-6})[-4e^{-(t-1)}] = -2e^{-(t-1)}\ \mu\text{A}, \quad 1 < t < \infty.$$

The expression for the power can be derived from Eq. (5.16); thus

$$p = 0, \qquad\qquad\qquad\qquad\qquad\qquad t < 0,$$
$$p = (4t)(2) = 8t\ \mu\text{W}, \qquad\qquad\qquad 0 \leq t < 1,$$
$$p = [4e^{-(t-1)}][-2e^{-(t-1)}] = -8e^{-2(t-1)}\ \mu\text{W}, \quad 1 < t \leq \infty.$$

The energy expression follows directly from Eq. (5.18):

$$w = 0, \qquad\qquad\qquad\qquad\qquad\qquad t < 0,$$
$$w = \left(\frac{1}{2}\right)(0.5)16t^2 = 4t^2\ \mu\text{J}, \qquad\qquad 0 \leq t \leq 1,$$
$$w = \frac{1}{2}(0.5)16e^{-2(t-1)} = 4e^{-2(t-1)}\ \mu\text{J}, \quad 1 \leq t \leq \infty.$$

b) The voltage, current, power, and energy as functions of time are shown in Fig. 5.11.

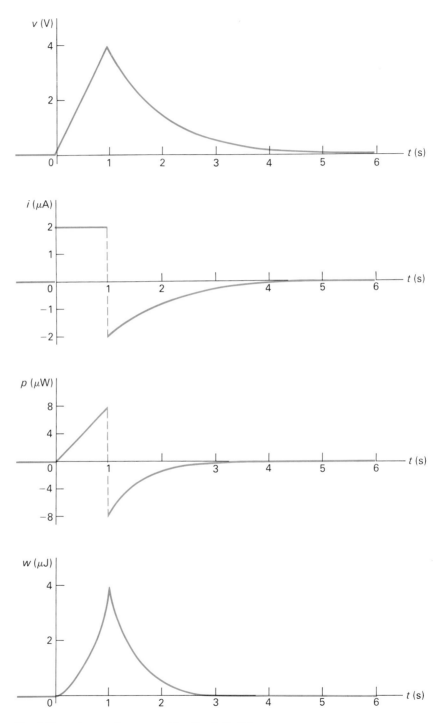

Fig. 5.11 $v, i, p,$ and w versus t for Example 5.4.

c) Energy is being stored in the capacitor whenever the power is positive. Hence, energy is being stored in the interval from 0 to 1 s.

d) Energy is being delivered by the capacitor whenever the power is negative. Thus energy is being delivered by the capacitor for all t greater than 1 s.

e) The integral of $p \, dt$ is the energy associated with the time interval corresponding to the limits on the integral. Thus the first integral represents the energy stored in the capacitor between 0 and 1 s, whereas the second integral represents the energy returned, or delivered, by the capacitor in the interval from 1 s to ∞.

$$\int_0^1 p \, dt = \int_0^1 8t \, dt = 4t^2 \Big|_0^1 = 4 \ \mu J;$$

$$\int_1^\infty p \, dt = \int_1^\infty [-8e^{-2(t-1)}] \, dt$$

$$= (-8) \frac{e^{-2(t-1)}}{-2} \Big|_1^\infty = -4 \ \mu J.$$

Since the voltage applied to the capacitor returns to zero as time increases without limit, the energy returned by this ideal capacitor must equal the energy stored. ◼

Example 5.5 An uncharged 0.2-μF capacitor is driven by a triangular current pulse. The current pulse is described by the following relationships:

$$i(t) = 0, \qquad\qquad t \le 0,$$
$$i(t) = 5000t, \qquad\qquad 0 \le t \le 20 \ \mu s,$$
$$i(t) = 0.2 - 5000t, \quad 20 \le t \le 40 \ \mu s,$$
$$i(t) = 0, \qquad\qquad t \ge 40 \ \mu s.$$

a) Derive the expressions for the capacitor voltage, power, and energy for each of the four time intervals needed to describe the current.

b) Plot i, v, p, and w versus t. Align the plots as specified in the previous examples.

c) Why is there a voltage on the capacitor after the current returns to zero?

Solution a) For $t \le 0$, v, p, and w are all zero. For $0 \le t \le 20 \ \mu s$,

$$v = 5 \times 10^6 \int_0^t 5000\tau \, d\tau + 0 = 12.5 \times 10^9 t^2 \ V,$$

$$p = vi = 62.5 \times 10^{12} t^3 \ W,$$

$$w = \frac{1}{2} Cv^2 = 15.625 \times 10^{12} t^4 \ J.$$

For 20 μs $\leq t \leq$ 40 μs,

$$v = 5 \times 10^6 \int_{20 \mu s}^{t} (0.2 - 5000\tau) \, d\tau + 5.$$

(Note that 5 V is the voltage on the capacitor at the end of the previous interval.) We then have

$$v = (10^6 t - 12.5 \times 10^9 t^2 - 10) \text{ V},$$
$$p = vi = (62.5 \times 10^{12} t^3 - 7.5 \times 10^9 t^2 + 2.5 \times 10^5 t - 2) \text{ W},$$
$$w = \frac{1}{2} C v^2$$
$$= (15.625 \times 10^{12} t^4 - 2.5 \times 10^9 t^3 + 0.125 \times 10^6 t^2 - 2.0 \times 10^{-6} t + 10^{-5}) \text{ J}.$$

For 40 μs $\leq t$,

$$v = 10 \text{ V},$$
$$p = vi = 0,$$
$$w = \frac{1}{2} C v^2 = 10 \ \mu\text{J}.$$

b) The excitation current and the resulting voltage, power, and energy are plotted in Fig. 5.12.

c) Note that the power is always positive during the duration of the current pulse, which means that energy is continuously being stored in the capacitor by the current pulse. When the current returns to zero, the energy stored in the capacitor is trapped since the ideal capacitor offers no means for dissipating energy. Thus a voltage exists on the capacitor after i returns to zero. ■

You are encouraged to work Problem 5.7, which further pursues the solution to this example.

DRILL EXERCISE
5.2

The voltage at the terminals of the 0.5-μF capacitor is 0 for $t < 0$ and $100e^{-20,000t} \sin 40,000t$ V for $t > 0$. Find (a) $i(0)$; (b) the power delivered to the capacitor at $t = \pi/80$ ms; and (c) the energy stored in the capacitor at $t = \pi/80$ ms.

0.5 μF

$+$ v $-$

i

Ans. (a) 2 A; (b) -20.79 W; (c) 519.20 μJ.

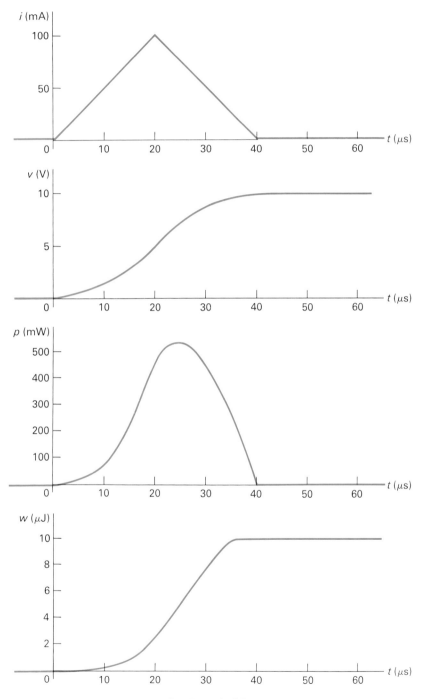

Fig. 5.12 $i, v, p,$ and w versus t for Example 5.5.

The current in the capacitor of Drill Exercise 5.2 is 0 for $t < 0$ and 2 cos 50,000t A for $t \geq 0$. Find (a) $v(t)$; (b) the maximum power delivered to the capacitor at any one instant of time; and (c) the maximum energy stored in the capacitor at any one instant of time.

Ans. (a) 80 sin 50,000t V; (b) 80 W; (c) 1.6 mJ.

5.4   SERIES–PARALLEL COMBINATIONS OF INDUCTANCE AND CAPACITANCE

Just as series–parallel combinations of resistors can be reduced to a single equivalent resistor, series–parallel combinations of inductors or capacitors can be reduced to a single inductor or capacitor. Let us begin our discussion with inductors in series, as shown in Fig. 5.13. In the series connection, the inductors are forced to carry the same current; thus only one current is defined for the series combination. The voltage drops across the individual inductors are

$$v_1 = L_1 \frac{di}{dt},$$

$$v_2 = L_2 \frac{di}{dt},$$

and

$$v_3 = L_3 \frac{di}{dt}.$$

The voltage across the series connection is

$$v = v_1 + v_2 + v_3 = (L_1 + L_2 + L_3) \frac{di}{dt},$$

from which it should be apparent that the equivalent inductance of series-connected inductors is the sum of the individual inductances. For n-inductors in series,

$$L_{eq} = L_1 + L_2 + L_3 + \cdots + L_n. \tag{5.19}$$

If the original inductors carry an initial current of $i(t_0)$, the equivalent inductor carries the same initial current. The equivalent circuit for series inductors carrying an initial current is shown in Fig. 5.14.

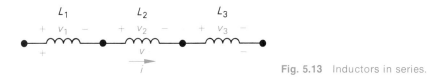

Fig. 5.13 Inductors in series.

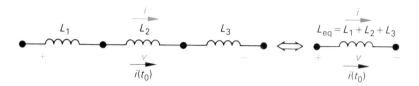

Fig. 5.14 An equivalent circuit for inductors in series carrying an initial current $i(t_0)$.

Inductors in parallel have the same terminal voltage. The equivalent circuit is obtained by writing the current in each inductor as a function of the terminal voltage and the initial current in the inductor. The procedure is illustrated for three inductors in parallel in Fig. 5.15. The currents for the individual inductors are

$$i_1 = \frac{1}{L_1} \int_{t_0}^{t} v \, d\tau + i_1(t_0),$$

$$i_2 = \frac{1}{L_2} \int_{t_0}^{t} v \, d\tau + i_2(t_0), \tag{5.20}$$

$$i_3 = \frac{1}{L_3} \int_{t_0}^{t} v \, d\tau + i_3(t_0).$$

The current at the terminals of the three parallel inductors is the sum of the inductor currents; thus

$$i = i_1 + i_2 + i_3. \tag{5.21}$$

When we substitute Eq. (5.20) into Eq. (5.21), we get

$$i = \left(\frac{1}{L_1} + \frac{1}{L_2} + \frac{1}{L_3} \right) \int_{t_0}^{t} v \, d\tau + i_1(t_0) + i_2(t_0) + i_3(t_0). \tag{5.22}$$

Now we can interpret Eq. (5.22) in terms of a single inductor. This becomes apparent if we rewrite Eq. (5.22) as

$$i = \frac{1}{L_{eq}} \int_{t_0}^{t} v \, d\tau + i(t_0). \tag{5.23}$$

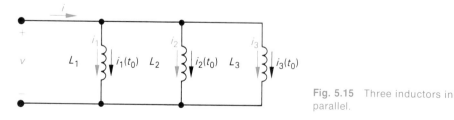

Fig. 5.15 Three inductors in parallel.

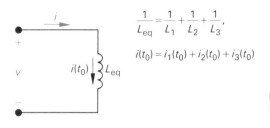

$$\frac{1}{L_{eq}} = \frac{1}{L_1} + \frac{1}{L_2} + \frac{1}{L_3},$$

$$i(t_0) = i_1(t_0) + i_2(t_0) + i_3(t_0)$$

Fig. 5.16 An equivalent circuit for three inductors in parallel.

Comparing Eq. (5.23) with (5.22) yields

$$\frac{1}{L_{eq}} = \frac{1}{L_1} + \frac{1}{L_2} + \frac{1}{L_3} \tag{5.24}$$

and

$$i(t_0) = i_1(t_0) + i_2(t_0) + i_3(t_0). \tag{5.25}$$

Thus the equivalent circuit for the three parallel inductors of Fig. 5.15 is as shown in Fig. 5.16.

The results expressed in Eqs. (5.24) and (5.25) can be extended to n inductors in parallel, namely,

$$\frac{1}{L_{eq}} = \frac{1}{L_1} + \frac{1}{L_2} + \cdots + \frac{1}{L_n} \tag{5.26}$$

and

$$i(t_0) = i_1(t_0) + i_2(t_0) + \cdots + i_n(t_0). \tag{5.27}$$

Capacitors connected in series can be reduced to a single equivalent capacitor. The reciprocal of the equivalent capacitance is equal to the sum of the reciprocals of the individual capacitances. If each capacitor carries its own initial voltage, the initial voltage on the equivalent capacitor will be the algebraic sum of the initial voltages on the individual capacitors. These observations are summarized in Fig. 5.17 and the following equations:

$$\frac{1}{C_{eq}} = \frac{1}{C_1} + \frac{1}{C_2} + \cdots + \frac{1}{C_n}, \tag{5.28}$$

$$v(t_0) = v_1(t_0) + v_2(t_0) + \cdots + v_n(t_0). \tag{5.29}$$

The derivation of the equivalent circuit for series-connected capacitors is left as an exercise for the reader. (See Problem 5.10.)

The equivalent capacitance of capacitors connected in parallel is simply the sum of the capacitances of the individual capacitors, as shown in Fig. 5.18 and the following equation:

$$C_{eq} = C_1 + C_2 + \cdots + C_n. \tag{5.30}$$

Capacitors connected in parallel must carry the same voltage. Therefore if there is an initial voltage across the original paralleled capacitors, this same

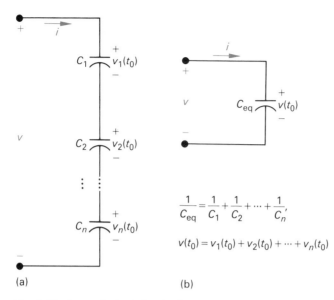

$$\frac{1}{C_{eq}} = \frac{1}{C_1} + \frac{1}{C_2} + \cdots + \frac{1}{C_n},$$

$$v(t_0) = v_1(t_0) + v_2(t_0) + \cdots + v_n(t_0)$$

(a)   (b)

Fig. 5.17 An equivalent circuit of capacitors connected in series: (a) series capacitors; (b) the equivalent circuit.

initial voltage appears across the equivalent capacitance C_{eq}. The derivation of the equivalent circuit for capacitors connected in parallel is left as an exercise for the reader. (See Problem 5.11.)

We will have more to say about these series–parallel equivalent circuits of inductors or capacitors in Chapter 6, where we will be in a position to interpret results based on their use.

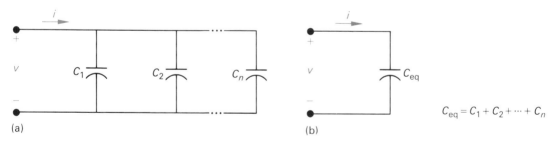

$$C_{eq} = C_1 + C_2 + \cdots + C_n$$

(a)   (b)

Fig. 5.18 Capacitors connected in parallel: (a) capacitors in parallel; (b) the equivalent circuit.

DRILL EXERCISE
5.4

The initial values of i_1 and i_2 in the following circuit are -2 A and $+4$ A, respectively. The voltage at the terminals of the paralleled inductors for $t \geq 0$ is $-40e^{-5t}$ V.

a) If the paralleled inductors are replaced by a single inductor, what is its inductance?

b) What is the initial current and its reference direction in the equivalent inductor?

c) Use the equivalent inductor to find $i(t)$.

d) Find $i_1(t)$ and $i_2(t)$. Check that the solutions for $i_1(t)$, $i_2(t)$, and $i(t)$ satisfy Kirchhoff's current law.

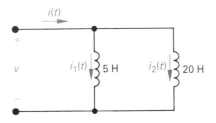

Ans. (a) 4 H; (b) 2 A, down; (c) $2e^{-5t}$ A; (d) $i_1(t) = 1.6e^{-5t} - 3.6$ A, $i_2(t) = 0.4e^{-5t} + 3.6$ A.

DRILL EXERCISE
5.5
The current at the terminals of the two capacitors shown is $240e^{-10t}$ μA for $t \geq 0$. The initial values of v_1 and v_2 are -10 V and -5 V, respectively. Calculate the total energy trapped in the capacitors as t approaches infinity.

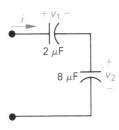

Ans. 20 μJ.

5.5 SUMMARY

We have now introduced two ideal circuit elements—inductors and capacitors—both of which are passive elements that are capable of storing energy. Because the elements are ideal, we note that it is possible to trap energy in the element. Thus an inductor can sustain a current (Example 5.3), and a capacitor can sustain a voltage (Example 5.5). The instantaneous power at the terminals of an inductor, or capacitor, can be either positive or negative, depending on whether energy is being delivered to or extracted from the element. This is in contrast with the fact that the instantaneous power is always positive at the terminals of a resistor because the resistor can only dissipate electrical energy. The salient characteristics of the inductor and capacitor are summarized in Table 5.1. The equations entered in this table are based on the passive sign convention, namely, that the current reference direction is in the direction of the voltage drop across the terminals of the element.

TABLE 5.1

IMPORTANT CHARACTERISTICS OF IDEAL INDUCTORS AND CAPACITORS

Inductors		Capacitors	
1. $v = L\dfrac{di}{dt}$	(V)	1. $v = \dfrac{1}{C}\displaystyle\int_{t_0}^{t} i\,d\tau + v(t_0)$	(V)
2. $i = \dfrac{1}{L}\displaystyle\int_{t_0}^{t} v\,d\tau + i(t_0)$	(A)	2. $i = C\dfrac{dv}{dt}$	(A)
3. $p = vi = Li\dfrac{di}{dt}$	(W)	3. $p = vi = Cv\dfrac{dv}{dt}$	(W)
4. $w = \dfrac{1}{2}Li^2$	(J)	4. $w = \dfrac{1}{2}Cv^2$	(J)
5. Will not permit an instantaneous change in its terminal current.		5. Will permit an instantaneous change in its terminal current.	
6. Will permit an instantaneous change in its terminal voltage.		6. Will not permit an instantaneous change in its terminal voltage.	
7. Appears as a short circuit to constant (dc) currents.		7. Appears as an open circuit to constant (dc) currents.	

The equations appearing in this table are based on the passive sign convention.

Table 5.1 is an especially useful reference for Chapters 6 and 7, in which we discuss what happens to the currents and voltages in a circuit after a switching operation.

PROBLEMS

5.1 Evaluate the integral

$$\int_{0}^{\infty} p\,dt$$

for Example 5.2. Comment on the significance of the result.

5.2 The triangular current pulse shown in Fig. 5.19 is applied to a 20-mH inductor.

    a) Write the expressions that describe $i(t)$ in the four intervals $t < 0$, $0 \le t \le 5$ ms, 5 ms $\le t \le 10$ ms, and $t > 10$ ms.

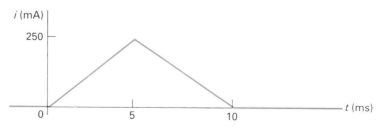

Fig. 5.19 The current pulse for Problem 5.2.

   b) Derive the expressions for the inductor voltage, power, and energy. Use the passive sign convention.

5.3 a) Find the inductor current in the circuit in Fig. 5.20 if $v = 40 \sin 1000t$ V, $L = 16$ mH, and $i(0) = 2.5$ A.
   b) Sketch v, i, p, and w vs. time. In making these sketches, use the format used in Fig. 5.8. Plot over one complete cycle of the voltage waveform.
   c) In the time interval between 0 and 2π ms, describe the subintervals when power is being absorbed by the inductor. Repeat for the subintervals when power is being delivered by the inductor.

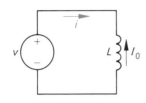

Fig. 5.20   The circuit for Problem 5.3.

5.4 The current in a 160-mH inductor is known to be

$$i = [3 \cos 500t + 1.5 \sin 500t]e^{-250t} \text{ A}, \qquad t \geq 0.$$

What is the maximum magnitude of the inductor voltage? Assume the passive sign convention.

5.5 The voltage across the terminals of a 0.125-μF capacitor is

$$v = 100 \text{ V}, \qquad\qquad\qquad t \leq 0,$$
$$v = A_1 t e^{-2000t} + A_2 e^{-2000t} \text{ V}, \quad t \geq 0.$$

The initial current in the capacitor is -150 mA. Assume the passive sign convention.

   a) What is the initial energy stored in the capacitor?
   b) Evaluate the coefficients A_1 and A_2.
   c) What is the expression for the capacitor current?

5.6 The rectangular-shaped current pulse shown in Fig. 5.21 is applied to a 2.5-μF capacitor. The initial voltage on the capacitor is a 6-V drop in the reference direction of the current. Assume the passive sign convention. Specify the capacitor voltage at 5, 10, 20, and 30 μs.

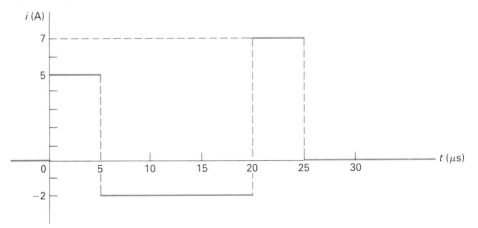

Fig. 5.21   The current pulse for Problem 5.6.

5.7 The expressions for voltage, power, and energy derived in Example 5.5 involved both integration and manipulation of algebraic expressions. As an engineer, you cannot accept such results on faith alone. That is, you should develop the habit of asking yourself, "Do these results make sense in terms of the known behavior of the circuit they purport to describe?" With these thoughts in mind, test the expressions of Example 5.5 by performing the following checks.

   a) Check the expressions to see whether the voltage is continuous in passing from one time interval to the next.
   b) Check the power expression in each interval by selecting a time within the interval and see whether it gives the same result as the corresponding product of v and i. For example, test at 10 μs and 30 μs.
   c) Check the energy expression within each interval by selecting a time within the interval and see whether the energy equation gives the same result as $\frac{1}{2}Cv^2$. Use 10 μs and 30μs as test points.

5.8 Assume that the initial energy stored in the inductors of Fig. 5.22 is zero. Find the equivalent inductance with respect to the terminals a, b.

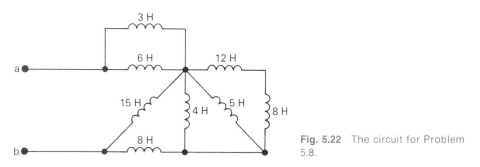

Fig. 5.22 The circuit for Problem 5.8.

5.9 The two parallel inductors in Fig. 5.23 are connected across the terminals of a black box at $t = 0$. The resulting voltage v for $t \geq 0$ is known to be $12e^{-t}$ V.

   a) Replace the original inductors with an equivalent inductor and find $i(t)$ for $t \geq 0$.
   b) Find $i_1(t)$ for $t \geq 0$.
   c) Find $i_2(t)$ for $t \geq 0$.
   d) How much energy is delivered to the black box in the time interval $0 \leq t \leq \infty$?
   e) How much energy was initially stored in the parallel inductors?
   f) How much energy is trapped in the ideal inductors?
   g) Do your solutions for i_1 and i_2 agree with the answer obtained in part (f)?

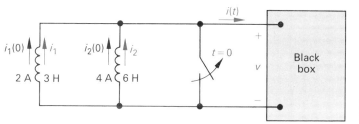

Fig. 5.23 The circuit for Problem 5.9.

5.10 Derive the equivalent circuit for a series connection of ideal capacitors. Assume that each capacitor has its own initial voltage. Denote these initial voltages as $v_1(t_0)$, $v_2(t_0)$, . . . , etc. (*Hint:* Sum the voltages across the string of capacitors, recognizing that the series connection forces the current in each capacitor to be the same.)

5.11 Derive the equivalent circuit for a parallel connection of ideal capacitors. Assume that the initial voltage across the paralleled capacitors is $v(t_0)$. (*Hint:* Sum the currents into the string of capacitors, recognizing that the parallel connection forces the voltage across each capacitor to be the same.)

5.12 The two series-connected capacitors in Fig. 5.24 are connected to the terminals of a black box at $t = 0$. The resulting current $i(t)$ for $t \geq 0$ is known to be $20e^{-t}$ μA.

    a) Replace the original capacitors with an equivalent capacitor and find $v(t)$ for $t \geq 0$.

    b) Find $v_1(t)$ for $t \geq 0$.

    c) Find $v_2(t)$ for $t \geq 0$.

    d) How much energy is delivered to the black box in the time interval $0 \leq t \leq \infty$?

    e) How much energy was initially stored in the series capacitors?

    f) How much energy is trapped in the ideal capacitors?

    g) Do the solutions for v_1 and v_2 agree with the answer obtained in part (f)?

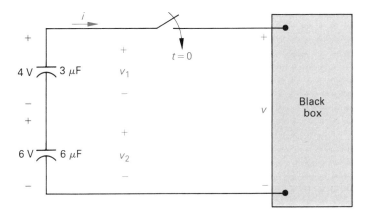

Fig. 5.24 The circuit for Problem 5.12.

6

THE NATURAL RESPONSE OF *RL* AND *RC* CIRCUITS

6.1   INTRODUCTION

In Chapter 5 we discussed the terminal behavior of ideal inductors and capacitors. We noted that one of the important attributes of these circuit elements is their ability to store energy. In this chapter we will determine the currents and voltages that arise when the energy stored in either an inductor or a capacitor is released to a purely resistive network. It is important to recognize at the outset that we are limiting the present analysis to circuits that can be reduced to a single equivalent inductor (or capacitor) and a single equivalent resistor. Thus if more than one inductor, or capacitor, exists in the circuit, they must be interconnected so that, by series–parallel combinations, they can be replaced by a single equivalent element. The same observation must also apply to multiple resistors in the circuit. Such circuits are referred to as *single time constant circuits*, a descriptive phrase that will be explained later. In short, all the circuits analyzed in this chapter are reducible to the forms shown in Fig. 6.1, where the current I_0 represents the initial current in the inductor and signifies that energy has been stored in the inductor and the voltage V_0 represents the initial voltage on the capacitor and signifies that energy has been stored in the capacitor.

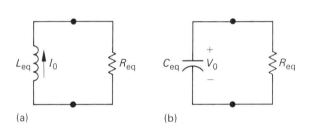

Fig. 6.1   Equivalent forms of the single time constant circuit.

Another important observation to make at this time is that we are assuming that when the stored energy is released there are no sources in the circuit. Thus the sources that were originally used to store energy in either the inductor or the capacitor are removed from the circuit at the time when the stored energy is released. The currents and voltages that are generated by the released energy are referred to as the *natural response* of the circuit. The term natural response emphasizes the fact that the currents and voltages are determined by the nature of the circuit itself and not by external sources of excitation.

In this chapter we will discuss the natural response of single time constant circuits.

6.2 THE NATURAL RESPONSE OF AN *RL* CIRCUIT

The natural response of the *RL* Circuit can be described in terms of the circuit shown in Fig. 6.2. We begin by assuming that the independent current source generates a constant current of *I* amperes and that the switch has been in a closed position for a long time. Thus only constant, or dc, currents can exist in the circuit just prior to opening the switch. This means that the inductor appears as a short circuit [$L \, di/dt = 0$] and therefore all of the current *I* appears in the inductive branch. Note that if the voltage across the inductive branch is zero, there can be no current in either R_0 or *R*. Our problem is to find the voltage and current at the terminals of the resistor after the switch has been opened. If we let $t = 0$ denote the instant when the switch is opened, then we describe our problem as that of finding $v(t)$ and $i(t)$ for $t \geq 0$. For $t \geq 0$ the circuit in Fig. 6.2 reduces to that shown in Fig. 6.3.

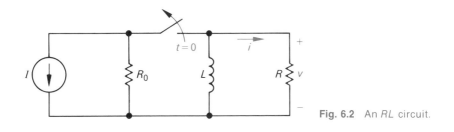

Fig. 6.2 An *RL* circuit.

To find $i(t)$ we use Kirchhoff's voltage law to obtain an expression involving *i*, *R*, and *L*. Summing the voltages around the closed loop, we obtain

$$L \frac{di}{dt} + Ri = 0, \tag{6.1}$$

where we have used the passive sign convention. Equation (6.1) is known as a *first-order ordinary differential equation,* so called because it contains terms involving the ordinary derivative of the unknown, that is, di/dt. The highest-order derivative appearing in the equation is one, hence the term first-order.

Fig. 6.3 The circuit in Fig. 6.2 for $t \geq 0$.

We can go one step further in our description of this equation. The coefficients in the equation (R and L) are constants; that is, they are not functions of either the dependent variable i or the independent variable t. Thus the equation can also be described as an ordinary differential equation with constant coefficients. (The solution of ordinary differential equations is a well-established practice in mathematics. However, we assume that you have not yet studied the solutions of such equations and therefore, whenever possible, we will solve our equations using ordinary calculus.) To solve Eq. (6.1), we divide through by the coefficient of the derivative (L), transpose the term involving i to the right-hand side of the equation, and then multiply both sides by a differential time dt. The result is

$$\left(\frac{di}{dt}\right) dt = -\frac{R}{L} i \, dt. \tag{6.2}$$

Next we recognize the left-hand side of Eq. (6.2) as a differential change in the current i, that is, di. Now we divide through by i, getting

$$\frac{di}{i} = -\frac{R}{L} \, dt. \tag{6.3}$$

We can get an explicit expression for i as a function of t by integrating both sides of Eq. (6.3). Using x and y as symbols of integration, we have

$$\int_{i(t_0)}^{i(t)} \frac{dx}{x} = -\frac{R}{L} \int_{t_0}^{t} dy, \tag{6.4}$$

in which $i(t_0)$ is the current corresponding to the time t_0 and $i(t)$ is the current corresponding to the time t. For the problem at hand, t_0 is zero; therefore carrying out the indicated integration gives

$$\ln \frac{i(t)}{i(0)} = -\frac{R}{L} t. \tag{6.5}$$

It follows directly from the definition of the natural logarithm that

$$i(t) = i(0)e^{-(R/L)t}. \tag{6.6}$$

We noted earlier that the current in the inductor just before opening the switch is I. Now recall from our discussion in Chapter 5 that there cannot be an instantaneous change of current in an inductor. Therefore in the first instant after the switch has been opened, the current in the inductor remains at I. If we use 0^- to denote the time just prior to switching and 0^+ the time immediately following switching, then for the inductor

$$i(0^-) = i(0^+) = I = I_0,$$

where, as in Fig. 6.1, I_0 is used to denote the initial current in the inductor. We also note that the initial current in the inductor is oriented in the same direc-

Fig. 6.4 The current response for the circuit in Fig. 6.3.

tion as the reference direction of i. Therefore Eq. (6.6) becomes

$$i(t) = I_0 e^{-(R/L)t}, \qquad t \geq 0, \tag{6.7}$$

from which we see that the current starts from an initial value of I_0 amperes and decreases exponentially toward zero as t increases. This response is shown graphically in Fig. 6.4.

The rate at which the current approaches zero is determined by the coefficient of t, namely, R/L. The reciprocal of this ratio is defined as the *time constant* of the circuit and is denoted by the Greek letter tau; thus

$$\tau = \text{time constant} = L/R. \tag{6.8}$$

If we use the concept of the time constant, we can write Eq. (6.7) as

$$i(t) = I_0 e^{-t/\tau}, \qquad t \geq 0. \tag{6.9}$$

Since the time constant is an important parameter for this type of circuit, it is worth mentioning several of its characteristics. First, note that it is convenient to think of the time elapsed after the switching has occurred in terms of integral multiples of τ. Thus one time constant after the inductor has begun to release its stored energy to the resistance, the current is reduced to e^{-1} or approximately 0.37 of its initial value. Table 6.1 gives the value of $e^{-t/\tau}$ for inte-

TABLE 6.1
VALUES OF $e^{-t/\tau}$ FOR t EQUAL TO INTEGRAL MULTIPLES OF τ

t	$e^{-t/\tau}$	t	$e^{-t/\tau}$
τ	3.6788×10^{-1}	6τ	2.4788×10^{-3}
2τ	1.3534×10^{-1}	7τ	9.1188×10^{-4}
3τ	4.9787×10^{-2}	8τ	3.3546×10^{-4}
4τ	1.8316×10^{-2}	9τ	1.2341×10^{-4}
5τ	6.7379×10^{-3}	10τ	4.5400×10^{-5}

gral multiples of τ up to 10. It is apparent from the table that when the elapsed time exceeds five time constants, the current is less than one percent of its initial value. We sometimes use this fact to say that, for example, five time constants after the switching has occurred the currents and voltages have, for all practical purposes, reached their final values.

A second characteristic of the time constant is that it gives the time it would take the current to reach its final value if it continued to change at its initial rate. To see this, we evaluate di/dt at 0^+ and assume that the current continues to change at this rate. Thus

$$\frac{di}{dt}(0^+) = -\frac{R}{L}I_0 = -I_0/\tau. \tag{6.10}$$

Now if i starts at I_0 and decreases at a constant rate of I_0/τ amperes per second, the expression for i becomes

$$i = I_0 - \frac{I_0}{\tau}t. \tag{6.11}$$

From Eq. (6.11) we see that i would reach its final value of zero in τ seconds. This characteristic of the time constant is shown in Fig. 6.5. This graphical interpretation of the time constant is sometimes useful in estimating the time constant of a circuit from a cathode-ray oscilloscope trace of its natural response.

The voltage across the resistor in the circuit in Fig. 6.3 can be derived from a direct application of Ohm's law. Hence

$$v = iR = I_0 R e^{-(R/L)t} = I_0 R e^{-t/\tau}, \qquad t \geq 0. \tag{6.12}$$

The power dissipated in the resistor can be derived from any one of the following expressions:

$$p = vi = i^2 R = v^2/R. \tag{6.13}$$

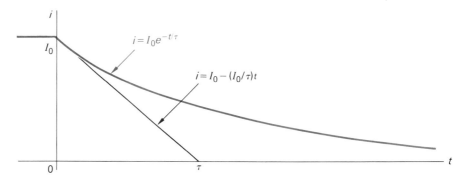

Fig. 6.5 A graphical interpretation of the time constant of the *LR* circuit in Fig. 6.3.

Whichever form is used, the resulting expression can be reduced to

$$p = I_0^2 R e^{-2t/\tau}, \qquad t \geq 0. \tag{6.14}$$

The energy delivered to the resistor in any interval of time after the switch has been opened is

$$w = \int_0^t p \, dx = \int_0^t I_0^2 R e^{-2x/t} \, dx$$

$$= \frac{\tau}{2} I_0^2 R [1 - e^{-2t/\tau}]$$

$$= \frac{1}{2} L I_0^2 (1 - e^{-2t/\tau}), \qquad t \geq 0. \tag{6.15}$$

Note from Eq. (6.15) that as t becomes infinite, the energy dissipated in the resistor approaches the initial energy stored in the inductor.

One final comment about the natural response of the *RL* circuit. We have mentioned that the current starts at an initial value of I_0 amperes and then decays exponentially to zero. Although our ideal circuit components require an infinitely long period of time for the current to decay to zero, we have already noted that after about five time constants the current is a negligible fraction of the initial value. Thus the existence of current in the *RL* circuit shown in Fig. 6.1(a) is a momentary event, and is therefore also referred to as the *transient response* of the circuit. The response that exists a long time after the switching has taken place is called the *steady-state response*. In our circuit, the steady-state response is zero. Note that prior to opening the switch in Fig. 6.2, we assumed that the circuit was operating in a steady-state mode. That is, before opening the switch we assumed that the switch had been in a closed position for a long time so that the current in the inductor had reached its steady-state value of I_0 amperes. In Chapter 7 we will discuss the problems of determining how the current in the inductor reached this steady-state value.

At this point in our discussion, it should be apparent that the key to analyzing the natural response of the *RL* circuit shown in Fig. 6.1(a) is to find the initial current in the inductor and the time constant of the circuit. All subsequent calculations follow from knowing $i(t)$. Example 6.1 illustrates the numerical calculations associated with the natural response of the *RL* circuit.

Example 6.1   Refer to the circuit shown in Fig. 6.6. The initial currents in the inductors L_1 and L_2 have been established by sources not shown. The switch opens at $t = 0$.

a) Find i_1, i_2, and i_3 for $t \geq 0$.

b) Calculate the initial energy stored in the parallel inductors.

c) Determine how much energy is trapped in the inductors as $t \to \infty$.

d) Show that the total energy delivered to the resistive network equals the difference between (b) and (c).

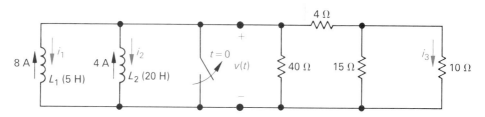

Fig. 6.6   The circuit for Example 6.1.

Solution   a) In studying the circuit, we see that the key to finding the currents i_1, i_2, and i_3 lies in knowing the voltage $v(t)$. We can easily find $v(t)$ if we reduce the circuit in Fig. 6.6 to the equivalent form of Fig. 6.1(a). The paralleled inductors simplify to an equivalent inductance of 4 H, carrying an initial current of 12 A. The resistive network reduces to a single resistance of 8 Ω. Therefore the first step in the solution for the three unknown currents is to replace the circuit in Fig. 6.6 with its equivalent form, as shown in Fig. 6.7. From the circuit in this figure we see immediately that the initial value of $i(t)$ is 12 A and the time constant is $\frac{4}{8}$, or 0.5 s. Therefore we can write

$$i(t) = 12e^{-2t} \text{ A}, \qquad t \geq 0.$$

Now $v(t)$ is simply the product $8i$, so we have

$$v(t) = 96e^{-2t} \text{ V}, \qquad t \geq 0.$$

Once we have $v(t)$ we can implement the calculations for i_1, i_2, and i_3:

$$i_1 = \frac{1}{5} \int_0^t 96e^{-2x} \, dx - 8$$
$$= 1.6 - 9.6e^{-2t} \text{ A}, \qquad t \geq 0;$$

$$i_2 = \frac{1}{20} \int_0^t 96e^{-2x} \, dx - 4$$
$$= -1.6 - 2.4e^{-2t} \text{ A}, \qquad t \geq 0;$$

$$i_3 = \frac{v(t)}{(10)} \cdot \frac{(15)}{25} = 5.76e^{-2t} \text{ A}, \qquad t \geq 0.$$

Fig. 6.7   A simplification of the circuit shown in Fig. 6.6.

b) The initial energy stored in the inductors is

$$w = \frac{1}{2}(5)(64) + \frac{1}{2}(20)(16) = 320 \text{ J}.$$

c) As $t \to \infty$, $i_1 \to 1.6$ A and $i_2 \to -1.6$ A. Therefore a long time after the switch has been opened, the energy stored in the two inductors is

$$w = \frac{1}{2}(5)(1.6)^2 + \frac{1}{2}(20)(-1.6)^2 = 32 \text{ J}.$$

d) The total energy delivered to the resistive network is obtained by integrating the expression for the instantaneous power from zero to infinity. We have, then,

$$w = \int_0^\infty p \, dt = \int_0^\infty 1152e^{-4t} \, dt$$

$$= 1152 \left. \frac{e^{-4t}}{-4} \right|_0^\infty = 288 \text{ J}.$$

We see directly that this is the difference between the initially stored energy (320 J) and the energy trapped in the parallel inductors (32 J). We note in passing that the equivalent inductor for the parallel inductors (which predicts the *terminal* behavior of the parallel combination) has an initial energy of 288 J. That is, the energy stored in the equivalent inductor represents the amount of energy that will be delivered to the resistive network at the terminals of the original inductors. ■

DRILL EXERCISE
6.1

The switch in the circuit shown has been closed for a long time and is opened at $t = 0$.

a) Calculate the initial value of i.

b) Calculate the initial energy stored in the inductor.

c) What is the time constant of the circuit for $t > 0$?

d) What is the numerical expression for $i(t)$ for $t \geq 0$?

e) What percentage of the initial energy stored has been dissipated in the 4-Ω resistor 5 ms after the switch has been opened?

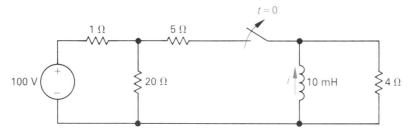

Ans. (a) -16 A; (b) 1.28 J; (c) $\tau = 2.5$ ms; (d) $-16e^{-400t}$ A; (e) 98.17%.

6.3   THE NATURAL RESPONSE OF AN *RC* CIRCUIT

The natural response of the *RC* circuit shown in Fig. 6.1(b) can be described by the circuit in Fig. 6.8. We begin by assuming that the switch has been in position A for a long period of time so that the loop made up of the dc voltage source V_g, the resistor R_1, and the capacitor C has reached a steady-state condition. When this is the case, the constant voltage source cannot sustain a current in the capacitor; therefore the capacitor must be charged to the source voltage of V_g volts. (Recall from Chapter 5 that a capacitor behaves as an open circuit to a constant current.) We note in passing that the problem of determining how the voltage on the capacitor builds up to V_g volts will be discussed in Chapter 7. The important observation to make at this time is that when the switch is moved from position A to position B (at $t = 0$), the voltage on the capacitor is V_g volts. Since there cannot be an instantaneous change in the voltage at the terminals of a capacitor, our problem reduces to the solution of the circuit shown in Fig. 6.9.

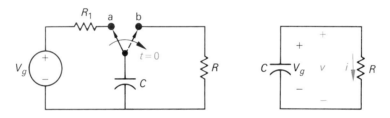

Fig. 6.8   An *RC* circuit.

Fig. 6.9   The circuit in Fig. 6.8 shown after switching.

We can easily find the voltage $v(t)$ by thinking in terms of node voltages. Using the lower junction between R and C as the reference node and summing the currents away from the upper junction between R and C gives

$$C\frac{dv}{dt} + \frac{v}{R} = 0. \tag{6.16}$$

When we compare Eq. (6.16) with Eq. (6.1) we see that the same mathematical techniques can be used to obtain the solution for $v(t)$. We will leave it to the reader to show that

$$v(t) = v(0)e^{-t/RC}, \qquad t \geq 0. \tag{6.17}$$

As we have already noted, the initial voltage on the capacitor equals the voltage source voltage V_g, that is,

$$v(0^-) = v(0^+) = V_g = V_0, \tag{6.18}$$

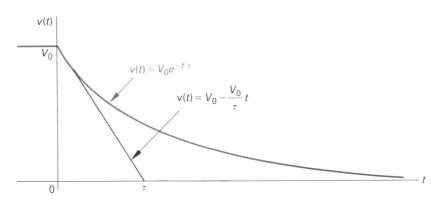

Fig. 6.10 The natural response of the *RC* circuit.

where, as in Fig. 6.1, V_0 denotes the initial voltage on the capacitor. The time constant for the *RC* circuit equals the product of the resistance and capacitance, namely,

$$\tau = RC. \tag{6.19}$$

When we substitute Eqs. (6.18) and (6.19) into Eq. (6.17), we get

$$v(t) = V_0 e^{-t/\tau}, \qquad t \geq 0, \tag{6.20}$$

which tells us that the natural response of the *RC* circuit is an exponential decay of the initial voltage. The rate of decay is governed by the time constant *RC*. The graphical plot of Eq. (6.20) is given in Fig. 6.10, where we also show the graphical interpretation of the time constant.

Once we know $v(t)$, we can easily derive the expressions for i, p, and w. The results are as follows:

$$i(t) = \frac{v(t)}{R} = \frac{V_0}{R} e^{-t/\tau}, \qquad t \geq 0;$$

$$p = vi = \frac{V_0^2}{R} e^{-2t/\tau}, \qquad t \geq 0; \tag{6.22}$$

$$w = \int_0^t p \, dx = \int_0^t \frac{V_0^2}{R} e^{-2x/\tau} \, dx$$

$$= \frac{1}{2} CV_0^2 (1 - e^{-2t/\tau}), \qquad t \geq 0. \tag{6.23}$$

The key to analyzing the natural response of the *RC* circuit shown in Fig. 6.1(b) is to find the initial voltage across the capacitor and the time constant of the circuit. All subsequent calculations follow from knowing $v(t)$. Example 6.2 illustrates the numerical calculations associated with the natural response of the *RC* circuit.

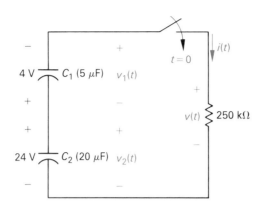

Fig. 6.11   The circuit for Example 6.2.

Example 6.2   The initial voltages on the capacitors C_1 and C_2 in the circuit shown in Fig. 6.11 have been established by sources not shown. The switch closes at $t = 0$.

a) Find $v_1(t)$, $v_2(t)$, $v(t)$, and $i(t)$ for $t \geq 0$.

b) Calculate the initial energy stored in the capacitors C_1 and C_2.

c) Determine how much energy is trapped in the capacitors as $t \to \infty$.

d) Show that the total energy delivered to the 250-kΩ resistor equals the difference between (b) and (c).

Solution   a) After studying the circuit, we can see that once we know $v(t)$, we can obtain the current $i(t)$ from Ohm's law. Once we know $i(t)$, we can calculate $v_1(t)$ and $v_2(t)$, since we know that the voltage across a capacitor is a function of the capacitor current. To find $v(t)$ we replace the series-connected capacitors with an equivalent capacitor. The equivalent capacitor has a capacitance of 4 μF and is charged to a voltage of 20 V. Therefore the circuit in Fig. 6.11 reduces to that shown in Fig. 6.12, from which we see that the initial value of $v(t)$ is 20 V and the time constant of the circuit is $(4)(250) \times 10^{-3}$, or 1 s. Thus the expression for $v(t)$ is

$$v(t) = 20e^{-t} \text{ V}, \qquad t \geq 0.$$

The current $i(t)$ is

$$i(t) = \frac{v(t)}{250,000} = 80e^{-t} \ \mu\text{A}, \qquad t \geq 0.$$

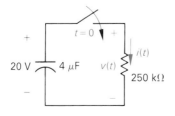

Fig. 6.12   A simplification of the circuit in Fig. 6.11.

Once we know $i(t)$, we can calculate the expressions for $v_1(t)$ and $v_2(t)$. Thus

$$v_1(t) = -\frac{10^6}{5} \int_0^t 80 \times 10^{-6} e^{-x} \, dx - 4$$

$$= (16e^{-t} - 20) \text{ V}, \qquad t \ge 0;$$

$$v_2(t) = -\frac{10^6}{20} \int_0^t 80 \times 10^{-6} e^{-x} \, dx + 24$$

$$= (4e^{-t} + 20) \text{ V}, \qquad t \ge 0.$$

b) The initial energy stored in C_1 is

$$w_1 = \frac{1}{2} (5 \times 10^{-6})(16) = 40 \ \mu\text{J}.$$

The initial energy stored in C_2 is

$$w_2 = \frac{1}{2} (20 \times 10^{-6})(576) = 5760 \ \mu\text{J}.$$

The total initial energy stored in the two capacitors is

$$w_0 = 40 + 5760 = 5800 \ \mu\text{J}.$$

c) As $t \to \infty$ we have

$$v_1 \to -20 \text{ V} \quad \text{and} \quad v_2 \to +20 \text{ V}.$$

Therefore the energy trapped in the two capacitors will be

$$w_\infty = \frac{1}{2} (5 + 20) \times 10^{-6}(400) = 5000 \ \mu\text{J}.$$

d) The total energy delivered to the 250-kΩ resistor is

$$w = \int_0^\infty p \, dt = \int_0^\infty \frac{400e^{-2t}}{250,000} \, dt = 800 \ \mu\text{J}.$$

Comparing the results obtained in parts (b) and (c) we note that

$$800 \ \mu\text{J} = (5800 - 5000) \ \mu\text{J}.$$

We should also point out that the energy stored in the equivalent capacitor in Fig. 6.12 is $\frac{1}{2}(4 \times 10^{-6})(400)$, or $800 \ \mu\text{J}$. Since the equivalent capacitor predicts the *terminal* behavior of the original series-connected capacitor, the energy stored in the equivalent capacitor is the energy that is delivered to the 250-kΩ resistor. ■

DRILL EXERCISE 6.2   The switch in the circuit shown has been closed for a long period of time and opens at $t = 0$. Find each of the following.

a) The initial value of $v(t)$.

b) The time constant for $t > 0$.

c) The numerical expression for $v(t)$ after the switch has been opened.

d) The initial energy stored in the capacitor.

e) The length of time it takes to dissipate 75% of the initially stored energy.

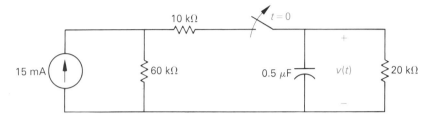

Ans. (a) 200 V; (b) 10 ms; (c) $200e^{-100t}$; (d) 10 mJ; (e) 6.93 ms.

6.4 SUMMARY

Our discussion has centered on determining the natural response of single time constant, source-free circuits, which are either *RL* or *RC* circuits that can be reduced to a single equivalent resistance shunted by either a single equivalent inductor or capacitor. The natural response consists of either the current or voltage that arises as a consequence of releasing energy that has been stored in either the inductor or capacitor to the shunting resistor. We found that in all cases the natural response is an exponentially decaying function. We can determine the initial value of the response by either the initial current in an inductor or the initial voltage across a capacitor. The rate of decay in either type of circuit is governed by the circuit parameters. For the *RL* circuit, the time constant (τ) equals the ratio of inductance to resistance (L/R). For the *RC* circuit, the time constant is given by the product of the resistance and capacitance (RC). In either circuit, the natural responses approach their final value in approximately five time constants.

PROBLEMS

6.1 In the circuit shown in Fig. 6.13, the current and voltage expressions for $t \geq 0$ are

$$i = 8e^{-20t} \text{ A}, \quad t \geq 0;$$
$$v = 240e^{-20t} \text{ V}, \quad t \geq 0.$$

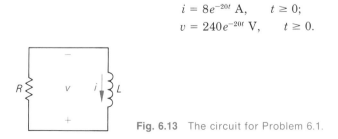

Fig. 6.13 The circuit for Problem 6.1.

Find (a) R, (b) L, (c) τ (ms), (d) the initial energy stored in the inductor, and (e) the amount of energy that has been dissipated in the resistor 100 ms after the current has begun to decay.

6.2 In the circuit shown in Fig. 6.14 the switch makes contact with position b just before breaking contact with position a. This is known as a "make-before-break" switch and is designed so that the switch does not interrupt the current in an inductive circuit. The interval of time between "making" and "breaking" is assumed to be negligible. The switch has been in the a position for a long time. At $t = 0$ the switch is thrown from position a to position b.

   a) Determine the initial current in the inductor.
   b) Determine the time constant of the circuit for $t > 0$.
   c) Find i, v_1, and v_2 for $t \geq 0$.
   d) What percentage of the energy stored in the inductor is dissipated in the 10-Ω resistor 50 ms after the switch is thrown from position a to position b?

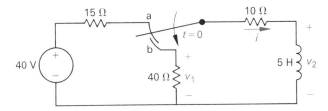

Fig. 6.14 The circuit for Problem 6.2.

6.3 The switch in the circuit in Fig. 6.15 has been open for a long time. At $t = 0$ the switch is closed.

   a) Determine $i(t)$ for $t \geq 0$.
   b) How many milliseconds after the switch has been closed will the current in the switch equal 2 A?

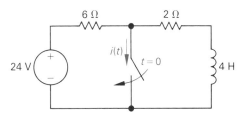

Fig. 6.15 The circuit for Problem 6.3.

6.4 The switch in the circuit shown in Fig. 6.16 has been in position a for a long time. At $t = 0$ the switch is moved to position b, where it remains for 10 ms. The switch is then moved to position c, where it remains indefinitely.

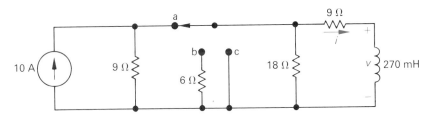

Fig. 6.16 The circuit for Problem 6.4.

a) Find $i(0^+)$.
b) Find $i(8 \text{ ms})$.
c) Find $i(16 \text{ ms})$.
d) Find $v(10^- \text{ ms})$.
e) Find $v(10^+ \text{ ms})$.

6.5 Switches 1 and 2 in the circuit shown in Fig. 6.17 are mechanically linked. When switch 1 is in position a, switch 2 is open. When switch 1 is in position b, switch 2 is closed. At $t = 0$ switch 1 is thrown from position a to position b.

a) For $t \geq 0$, find v, i_1, and i_2.
b) How much energy is trapped in the circuit when switch 1 is left in position b?
c) What is the total energy dissipated in the 6-Ω resistor for $t > 0$?

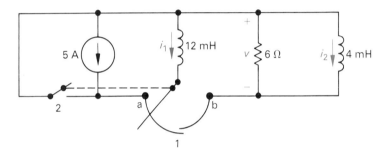

Fig. 6.17 The circuit for Problem 6.5.

6.6 The 600-V, 6-Ω source in the circuit in Fig. 6.18 is inadvertently short-circuited at its terminals a, b. At the time the fault occurs, the circuit is in a steady-state operating condition.

a) What is the initial value of the current in the short-circuit connection between terminals a, b?
b) What is the final value of the current in the short-circuit connection?
c) How many milliseconds after the short circuit has occurred is the current in the short equal to 92 A?

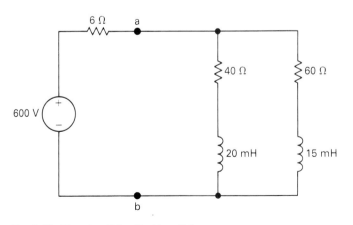

Fig. 6.18 The circuit for Problem 6.6.

6.7 In the circuit in Fig. 6.19, the voltage and current expressions for $t \geq 0$ are

$$v = 8e^{-5t} \text{ V}, \qquad t \geq 0;$$
$$i = 20e^{-5t} \text{ μA}, \qquad t \geq 0.$$

Find (a) R, (b) C, (c) τ (ms), (d) the initial energy stored in the capacitor, and (e) the amount of energy that has been dissipated in the resistor 100 ms after the voltage has begun to decay.

Fig. 6.19 The circuit for Problem 6.7.

6.8 The switch in the circuit in Fig. 6.20 has been in position a for a long time. At $t = 0$ the switch is thrown to position b.

  a) Find $v(t)$ for $t \geq 0$.
  b) What percentage of the initial energy stored in the capacitor is dissipated in the 80-kΩ resistor 4 ms after the switch has been thrown?

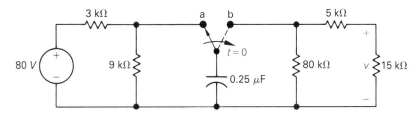

Fig. 6.20 The circuit for Problem 6.8.

6.9 The 5-μF capacitor in the circuit in Fig. 6.21 has been charged to 30 V. At $t = 0$ switch 1 closes, causing the capacitor to discharge into the resistive network. Switch 2 closes 50 ms after the capacitor starts to discharge. Find the magnitude and direction of the current in switch 2 125 ms after switch 1 closes.

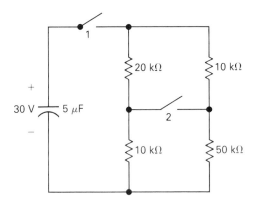

Fig. 6.21 The circuit for Problem 6.9.

6.10 The switch in the circuit in Fig. 6.22 has been in position a for a long time. At $t = 0$ the switch is thrown to position b.

a) Calculate i, v_1, and v_2 for $t \geq 0$.
b) Calculate the energy stored in the capacitor at $t = 0$.
c) Calculate the energy trapped in the circuit and the total energy dissipated in the 25-kΩ resistor if the switch remains in position b indefinitely.

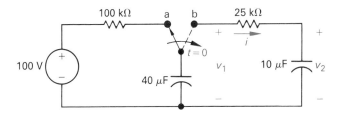

Fig. 6.22 The circuit for Problem 6.10.

7

STEP RESPONSE OF FIRST-ORDER *RL* AND *RC* CIRCUITS

7.1 INTRODUCTION

In Chapter 6 we discussed the natural response of first-order RL and RC circuits. In those circuits, we assumed that energy had been stored initially in the inductors or capacitors and the response was caused by the release of energy to a resistive circuit. We now discuss the problem of finding currents and voltages that are generated in RL and RC circuits when either dc voltage or current sources are suddenly applied to the circuit. The response of a circuit to the sudden application of a constant voltage or current source is referred to as the *step response* of the circuit.

As in Chapter 6, we assume that if more than one inductor or capacitor is present in the circuit, they are interconnected in such a way that the interconnection can be reduced to a single equivalent inductor or capacitor. We begin with the step response of an RL circuit.

7.2 THE STEP RESPONSE OF AN *RL* CIRCUIT

The circuit shown in Fig. 7.1 can be used to describe the step response of an RL circuit. Energy stored in the inductor at the time when the switch is closed is given in terms of a nonzero initial current $i(0)$. Our problem is to find the expression for the current in the circuit as well as the expression for the voltage across the inductance after the switch has been closed. Our procedure is the same as the one we used in Chapter 6: We use circuit analysis to derive the differential equation that describes the circuit in terms of the variable of interest and then use the techniques of elementary calculus to solve the equation.

After the switch in the circuit in Fig. 7.1 has been closed, Kirchhoff's voltage law requires that

$$V_s = Ri + L\frac{di}{dt}, \tag{7.1}$$

which can be solved for the current using the same technique introduced in Chapter 6; that is, we separate the variables i and t and then integrate. The

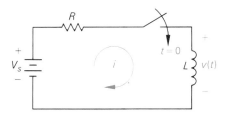

Fig. 7.1   A circuit used to introduce the step response of a first-order *RL* circuit.

first step in this approach is to solve Eq. (7.1) for the derivative di/dt:

$$\frac{di}{dt} = \frac{-Ri + V_s}{L} = \frac{-R}{L}\left(i - \frac{V_s}{R}\right). \tag{7.2}$$

Next, we multiply both sides of Eq. (7.2) by a differential time, dt. This reduces the left-hand side of the equation to a differential change in the current. Thus

$$\left(\frac{di}{dt}\right)dt = \frac{-R}{L}\left(i - \frac{V_s}{R}\right)dt,$$

or

$$di = \frac{-R}{L}\left(i - \frac{V_s}{R}\right)dt. \tag{7.3}$$

The variables in Eq. (7.3) can now be separated. We get

$$\frac{di}{i - (V_s/R)} = \frac{-R}{L}\,dt, \tag{7.4}$$

and then integrate both sides of Eq. (7.4). Using x and y as symbols of integration, we have

$$\int_{I_0}^{i(t)} \frac{dx}{[x - (V_s/R)]} = \frac{-R}{L}\int_0^t\,dy, \tag{7.5}$$

where I_0 is the current at $t = 0$ and $i(t)$ is the current at any t greater than zero. Performing the integration called for in Eq. (7.5) generates the following expression:

$$\ln\frac{[i(t) - (V_s/R)]}{[I_0 - (V_s/R)]} = \frac{-R}{L}\,t, \tag{7.6}$$

from which it follows directly that

$$\frac{[i(t) - (V_s/R)]}{[I_0 - (V_s/R)]} = e^{-(R/L)t},$$

or

$$i(t) = \frac{V_s}{R} + \left(I_0 - \frac{V_s}{R}\right)e^{-(R/L)t}. \tag{7.7}$$

For the case when the initial energy in the inductor is zero, I_0 is zero and Eq. (7.7) reduces to

$$i(t) = \frac{V_s}{R} - \frac{V_s}{R}e^{-(R/L)t}. \tag{7.8}$$

Equation (7.8) tells us that after the switch has been closed, the current increases exponentially from zero to a final value of V_s/R amperes. The rate of increase is determined by the time constant ($\tau = L/R$) of the circuit. One time constant after the switch has been closed, the current will have reached approximately 63% of its final value, that is,

$$i(\tau) = \frac{V_s}{R} - \frac{V_s}{R} e^{-1} \cong 0.6321 \frac{V_s}{R}. \tag{7.9}$$

We also observe that if the current were to continue to increase at its initial rate, it would reach its final value in τ seconds. That is, since

$$\frac{di}{dt} = \frac{-V_s}{R} \left(\frac{-1}{\tau} \right) e^{-t/\tau} = \frac{V_s}{L} e^{-t/\tau}, \tag{7.10}$$

the initial rate at which i is increasing is

$$\frac{di}{dt}(0) = \frac{V_s}{L} \text{ A/s}. \tag{7.11}$$

If the current continued to increase at this rate, the expression for i would be

$$i = \frac{V_s}{L} t, \tag{7.12}$$

from which we see that at $t = \tau$

$$i = \frac{V_s}{L} \cdot \frac{L}{R} = \frac{V_s}{R}. \tag{7.13}$$

These observations relative to Eq. (7.8) are shown graphically in Fig. 7.2.
Since the voltage across an inductor is $L \, di/dt$, it follows from Eq. (7.7)

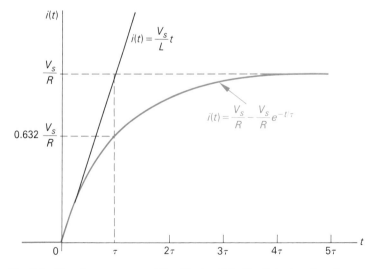

Fig. 7.2 The step response of the *RL* circuit in Fig. 7.1 when $I_0 = 0$.

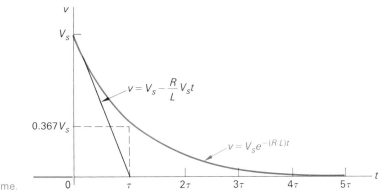

Fig. 7.3 Inductor voltage versus time.

that

$$v = L\left(\frac{-R}{L}\right)\left[I_0 - \frac{V_s}{R}\right]e^{-(R/L)t} = [V_s - I_0R]e^{-(R/L)t}. \tag{7.14}$$

For the case where the initial inductor current is zero, Eq. (7.14) simplifies to

$$v = V_s e^{-(R/L)t}. \tag{7.15}$$

The voltage across the inductor is zero prior to closing the switch. Equation (7.14) tells us that the inductor voltage jumps to $[V_s - I_0R]$ the instant when the switch is closed and then decays exponentially to zero. Does this make sense in terms of the circuit? Since the initial current is I_0 and the inductor prevents an instantaneous change in current, the current will be I_0 the first instant after the switch has been closed. At this instant, the voltage drop across the resistor will be I_0R and the voltage impressed across the inductor will be the source voltage minus this voltage, that is, $V_s - I_0R$. If the initial current is zero, the voltage across the inductor jumps to V_s. We also expect the inductor voltage to approach zero as t increases, since the current in the circuit is approaching the constant value of V_s/R. Equation (7.15) is plotted in Fig. 7.3, where we also show the relationship between the time constant ($\tau = L/R$) and the initial rate at which the inductor voltage is decreasing.

If there is an initial current in the inductor, the solution for the current is given by Eq. (7.7). The algebraic sign of I_0 is positive if the initial current is in the same direction as i; otherwise I_0 carries a negative sign. Example 7.1 illustrates the application of Eq. (7.7) to a specific circuit.

Example 7.1 The switch in the circuit in Fig. 7.4 has been in position a for a long time. At $t = 0$, the switch moves from position a to position b. The switch is a "make-before-break" type so there is no interruption of the inductor current.

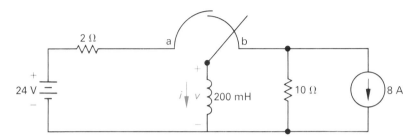

Fig. 7.4 The circuit for Example 7.1.

a) Find the expression for $i(t)$ for $t \geq 0$.

b) What is the initial voltage across the inductor after the switch has been moved to position b?

c) Does this initial voltage make sense in terms of circuit behavior?

d) How many milliseconds after the switch has been put in position b does the inductor voltage equal 24 V?

e) Plot both $i(t)$ and $v(t)$ vs. t.

Solution   a) Since the switch has been in position a for a long time, the 200-mH inductor is a short circuit across the 8-A current source. Therefore the inductor carries an initial current of 8 A. This initial current is oriented opposite to the reference direction for i; thus I_0 is -8 A. When the switch is in position b, the final value of i will be $\frac{24}{2}$, or 12 A. The time constant of the circuit is $\frac{200}{2}$, or 100 ms. Substituting these values into Eq. (7.7) gives us

$$i = 12 + (-8 - 12)e^{-t/0.1}$$
$$= 12 - 20e^{-10t} \text{ A}, \qquad t \geq 0.$$

b) The voltage across the inductor is

$$v = L\frac{di}{dt} = 0.2[200e^{-10t}] = 40e^{-10t} \text{ V}, \qquad t \geq 0.$$

The initial inductor voltage is

$$v(0) = 40 \text{ V}.$$

c) Yes; in the first instant after the switch has been moved to position b, the inductor sustains a current of 8 A in a counterclockwise direction around the newly formed closed path. This current causes a 16-V drop across the 2-Ω resistor. This voltage drop is additive with respect to the drop across the source; therefore a 40-V drop appears across the inductor.

d) The time at which the inductor voltage equals 24 V is found by solving the expression

$$24 = 40e^{-10t}$$

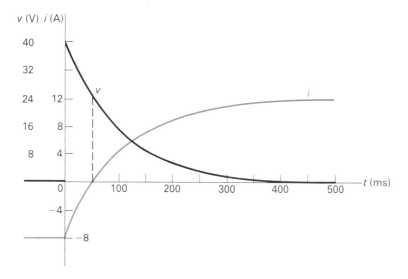

Fig. 7.5 Current and voltage waveform for Example 7.1.

for t. We get

$$t = \frac{1}{10} \ln \frac{40}{24} = 51.08 \times 10^{-3} = 51.08 \text{ ms}.$$

e) Figure 7.5 shows the graphs of $i(t)$ and $v(t)$ vs. t. Note that the instant of time when the current equals zero corresponds to the instant of time when the inductor voltage equals the source voltage of 24 V. ∎

If the series combination of the voltage source and resistor in Fig. 7.1 is converted via a source transformation to a parallel combination of a current source and a resistor, the step response of an *RL* circuit can be described in terms of the circuit shown in Fig. 7.6. We will find it informative to repeat the derivation of the step response of an *RL* circuit using the circuit in Fig. 7.6. To

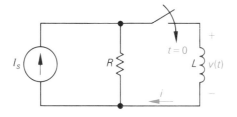

Fig. 7.6 An alternative form of the *RL* circuit.

find $v(t)$, we use the node-voltage method of circuit analysis. Summing the current away from the top node yields

$$I_s = \frac{v}{R} + \frac{1}{L}\int_0^t v \, dy + I_0. \tag{7.16}$$

Equation (7.16) is now differentiated with respect to t. In taking the derivative we note that both I_s and the initial value of i, that is, I_0, are constants. We get

$$0 = \frac{1}{R}\frac{dv}{dt} + \frac{v}{L}. \tag{7.17}$$

Multiplying Eq. (7.17) by R and solving for dv/dt gives

$$\frac{dv}{dt} = \frac{-R}{L}v, \tag{7.18}$$

where we then multiply by dt and separate the variables to get

$$\frac{dv}{v} = \frac{-R}{L}dt. \tag{7.19}$$

Using x and y as symbols of integration, we have

$$\int_{v(0)}^{v(t)} \frac{dx}{x} = \frac{-R}{L}\int_0^t dy, \tag{7.20}$$

where $v(0)$ is the initial voltage across the inductor. After integrating, we obtain

$$v(t) = v(0)e^{-(R/L)t}. \tag{7.21}$$

To find the initial value of the inductor voltage, $v(0)$, we reason as follows. The inductor will hold its initial current to I_0. Therefore the instant when the switch is closed, the current in the resistor will be $I_s - I_0$. The voltage drop across the resistor is identical to the voltage across the inductor; consequently we have

$$v(0) = R[I_s - I_0]. \tag{7.22}$$

Since RI_s must be the same as V_s by virtue of the source transformation, we see that Eq. (7.22) becomes

$$v(t) = [V_s - I_0R]e^{-(R/L)t} \tag{7.23}$$

Since the circuits are equivalent, Eq. (7.23) is identical to Eq. (7.14).

To find the inductor current in the circuit shown in Fig. 7.6, we could integrate the inductor voltage. That is, we can express the current in the inductor as a function of the inductor voltage. Thus

$$i = \frac{1}{L}\int_0^t v(\tau) \, d\tau + I_0. \tag{7.24}$$

Although the integration called for in Eq. (7.24) is not difficult, we can avoid it altogether by noting from the circuit in Fig. 7.6 that

$$i = I_s - \frac{v(t)}{R}.$$ (7.25)

When Eq. (7.23) is substituted in Eq. (7.25) we get

$$i = I_s - \frac{1}{R}[V_s - I_0R]\, e^{-(R/L)t}$$

$$= \frac{V_s}{R} - \left[\frac{V_s}{R} - I_0\right]e^{-(R/L)t}$$

$$= \frac{V_s}{R} + \left[I_0 - \frac{V_s}{R}\right]e^{-(R/L)t}$$ (7.26)

We see that Eq. (7.26) and Eq. (7.7) are the same.

At this point in our discussion of the step response of an *RL* circuit, we make a general observation that will prove helpful later in our work. When we analyzed the circuit in Fig. 7.1, we derived a differential equation for the inductor current, namely, Eq. (7.1). For the purposes of our present discussion, we rewrite Eq. (7.1) in the following form:

$$\frac{di}{dt} + \frac{R}{L}\,i = \frac{V_s}{L}.$$ (7.27)

When we analyzed the circuit in Fig. 7.6, we derived the differential equation (Eq. 7.17) for the inductor voltage, which can be rewritten as

$$\frac{dv}{dt} + \frac{R}{L}\,v = 0.$$ (7.28)

We observe now that Eqs. (7.27) and (7.28) are of the same form. Specifically, each equation equates the sum of the first derivative of the variable and a constant times the variable to a constant value. In Eq. (7.28), the constant on the right-hand side happens to be zero. We also note that in both equations the constant multiplying the dependent variable is the reciprocal of the time constant, that is, $R/L = 1/\tau$. We will return to this observation after discussing the step response of an *RC* circuit.

Assume that the switch in the circuit in Fig. 7.4 has been in position b for a long time and at $t = 0$ it moves to position a. Find (a) $i(0^+)$; (b) $v(0^+)$; (c) τ, $t > 0$; (d) $i(t)$, $t \geq 0$; and (e) $v(t)$, $t \geq 0$.

Ans. (a) 12 A; (b) -200 V; (c) 20 ms; (d) $-8 + 20e^{-50t}$ A, $t \geq 0$; (e) $-200e^{-50t}$ V, $t \geq 0$.

7.3 THE STEP RESPONSE OF AN *RC* CIRCUIT

The step response of a first-order *RC* circuit can be described in terms of either of the circuits in Fig. 7.7. Using the circuit of Fig. 7.7(a), we see that after the switch has been closed, Kirchhoff's voltage law requires that

$$V_s = Ri + \frac{1}{C} \int_0^t i \, d\tau + v_C(0). \tag{7.29}$$

The solution of Eq. (7.29) for i follows the same procedure as that used to solve Eq. (7.16) for v. Thus we begin by differentiating Eq. (7.29) with respect to t. We get

$$R\frac{di}{dt} + \frac{i}{C} = 0,$$

or

$$\frac{di}{dt} + \frac{1}{RC} i = 0. \tag{7.30}$$

To solve Eq. (7.30) for i, we again use separation of variables. Since this procedure has already been detailed in our discussion of the step response of an *RL* circuit, we simply report the result, hence

$$i = I_0 e^{-t/RC}. \tag{7.31}$$

Since there cannot be an instantaneous change in the voltage across the capacitor, the initial current in the circuit will be determined by the difference between V_s and the initial voltage on the capacitor V_0—specifically,

$$i(0) = \frac{V_s - V_0}{R}. \tag{7.32}$$

Substituting Eq. (7.32) into Eq. (7.31) yields

$$i = \frac{[V_s - V_0]}{R} e^{-t/RC}. \tag{7.33}$$

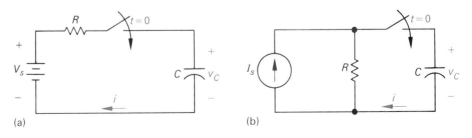

(a)　(b)

Fig. 7.7 Circuits used to derive the step response of an *RC* circuit.

We can find the voltage across the capacitor by noting from the circuit in Fig. 7.7(a) that

$$v_C = V_s - iR. \tag{7.34}$$

Substituting Eq. (7.33) into Eq. (7.34) gives

$$v_C = V_s + [V_0 - V_s]e^{-t/RC}. \tag{7.35}$$

If the initial voltage on the capacitor is zero, Eqs. (7.33) and (7.35) reduce to the following, respectively:

$$i = \frac{V_s}{R} e^{-t/RC}, \tag{7.36}$$

$$v_C = V_s - V_s e^{-t/RC}. \tag{7.37}$$

From Eq. (7.36), we see that when a constant voltage source is suddenly applied through a resistance to an uncharged capacitor, the current in the circuit jumps to an initial value of V_s/R amperes and then decays exponentially to zero. The rate of decay is determined by the time constant of the *RC* circuit. Equation (7.37) gives us the corresponding expression for the capacitor voltage. We see that the voltage builds up exponentially from zero to a final value of V_s volts. Again, the time constant for the buildup is RC seconds. The current and voltage responses are plotted in Fig. 7.8(a) and (b), respectively, in which we also relate the time constant ($\tau = RC$) to the initial rate at which the current and voltage are changing.

It is also informative to derive Eqs. (7.33) and (7.35) by starting with the circuit in Fig. 7.7(b). Thinking of v_C as a node voltage and summing the currents away from the top node generates the following differential equation:

$$C\frac{dv_C}{dt} + \frac{v_C}{R} = I_s. \tag{7.38}$$

Division of Eq. 7.38 by C gives us

$$\frac{dv_C}{dt} + \frac{v_C}{RC} = \frac{I_s}{C}. \tag{7.39}$$

When we compare Eq. (7.39) with Eq. (7.27), we can see that the form of the solution for v_C will be the same as that for the current in the inductive circuit, namely, Eq. (7.7). Therefore, by simply substituting the appropriate variables and coefficients, we can write the solution for v_C directly. The translation requires that I_s replace V_s, C replace L, $1/R$ replace R, and V_0 replace I_0. We get

$$v_C = I_s R + [V_0 - I_s R]e^{-t/RC}. \tag{7.40}$$

We see that Eqs. (7.40) and (7.35) are equivalent because the source transformation requires that $I_s R$ equal V_s.

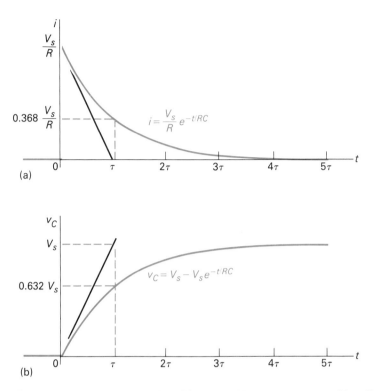

Fig. 7.8   The step response of an *RC* circuit: (a) current response ($V_0 = 0$); (b) voltage response ($V_0 = 0$).

The capacitor current is found directly from Eq. (7.40) by differentiating v_C with respect to time and multiplying by C; thus

$$i = C\frac{dv_C}{dt} = C\left[\left(\frac{-1}{RC}\right)(V_0 - I_s R)e^{-t/RC}\right] = \left(I_s - \frac{V_0}{R}\right)e^{-t/RC}. \qquad (7.41)$$

We see that Eqs. (7.41) and (7.33) are equivalent since $V_s = I_s R$.

Example 7.2 illustrates how to determine the step response of an *RC* circuit.

Example 7.2   The switch in the circuit in Fig. 7.9 has been in position a for a long time. At $t = 0$ the switch is moved to position b.

a) What is the initial value of v_C?

b) What is the final value of v_C?

c) What is the time constant of the circuit when the switch is in position b?

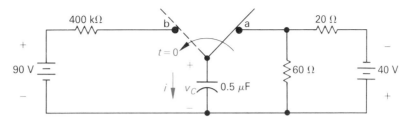

Fig. 7.9   The circuit for Example 7.2.

d) What is the expression for $v_C(t)$ when $t \geq 0$?

e) What is the expression for $i(t)$ when $t \geq 0$?

f) How long after the switch is in position b is the capacitor voltage passing through zero?

g) Plot $v_C(t)$ and $i(t)$ vs. t.

Solution   a) Since the switch has been in position a for a long time, the capacitor will look like an open circuit. Therefore the voltage across the capacitor will be the voltage across the 60-Ω resistor. The voltage across the 60-Ω resistor will be $(40/80) \times 60$, or 30 V. Since the reference for v_C is positive at the upper terminal of the capacitor, we have $V_0 = -30$ V.

b) After the switch has been in position b for a long time, the capacitor will look like an open circuit to the 90-V source. Thus the final value of the capacitor voltage will be $+90$ V.

c) The time constant is

$$\tau = RC = (400 \times 10^3)(0.5) \times 10^{-6} = 0.2 \text{ s.}$$

d) Substituting the appropriate values for V_s, V_0, and RC into Eq. (7.35), we have

$$\begin{aligned} v_C(t) &= 90 + (-30 - 90)e^{-5t} && (t \geq 0) \\ &= 90 - 120e^{-5t} \text{ V} && (t \geq 0). \end{aligned}$$

e) Substituting the appropriate values for V_s, V_0, R, and RC into Eq. (7.33) gives

$$\begin{aligned} i(t) &= \frac{90 - (-30)}{0.4} e^{-5t} \ \mu\text{A} && (t \geq 0) \\ &= 300e^{-5t} \ \mu\text{A} && (t \geq 0). \end{aligned}$$

f) To find how long the switch must be in position b before the capacitor voltage is zero, we solve the equation derived in part (d) for the time when $v_C(t) = 0$. Thus

$$120e^{-5t} = 90$$

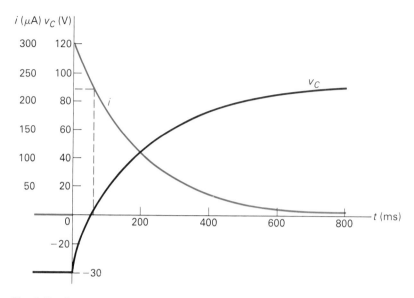

Fig. 7.10   Current and voltage waveforms for Example 7.2.

or

$$e^{5t} = \frac{120}{90}$$

$$t = \frac{1}{5} \ln \left(\frac{4}{3}\right)$$

$$= 57.54 \text{ ms.}$$

Note that when v_C is zero, i is 225 μA and the voltage drop across the 400-kΩ resistor is 90 V.

g) Figure 7.10 shows the graphs of $v_C(t)$ and $i(t)$ vs. t.   ■

We conclude our introduction to the step response of an *RC* circuit by making an observation similar to that made concerning the step response of an *RL* circuit. That is, the capacitor current and voltage both satisfy first-order differential equations with the same form. Specifically, the capacitor current must satisfy the differential equation denoted as Eq. (7.30), while the capacitor voltage must satisfy Eq. (7.39). Note that in each equation the constant multiplying the dependent variable is the reciprocal of the time constant, that is, $1/\tau = 1/RC$. We also call to your attention that these two equations have the same form as the differential equations that describe the current and voltage at the terminals of an inductor, namely, Eqs. (7.27) and (7.28). These observations lead us to our discussion in Section 7.4 of a straightforward technique for deriving the step response of either a first-order *RL* or *RC* circuit.

Assume that the switch in the circuit in Fig. 7.9 has been in position b for a long time and at $t = 0$ it is moved to position a. Find (a) $v_C(0^+)$; (b) $v_C(\infty)$; (c) τ for $t > 0$; (d) $i(0^+)$; (e) v_C, $t \geq 0$; and (f) i, $t \geq 0$.

Ans. (a) 90 V; (b) -30 V; (c) 7.5 μs; (d) -8 A; (e) $(-30 + 120e^{-(400,000/3)t})$ V, $t \geq 0$; (f) $-8e^{-(400,000/3)t}$ A, $t \geq 0$.

7.4 FINDING THE STEP RESPONSE OF A FIRST-ORDER *RL* OR *RC* CIRCUIT

Before developing the general approach to finding the step response of an *RL* or *RC* circuit, let us make sure we understand the type of circuit to which this approach is limited. First, we assume that our circuit contains either a single inductor or a single capacitor. If there are multiple inductors or capacitors, they must be interconnected so that they can be reduced to either a single inductor or a single capacitor. Second, all sources within the network must be constant voltage and/or constant current. Third, we assume that the current and voltage at the terminals of either the inductor or capacitor are changing with time because of a switching operation somewhere in the circuit. Since each circuit contains the equivalent of a single energy storage element, all our circuits can be reduced to one of the four forms illustrated in Fig. 7.11, which follow directly from our knowing that any network connected to the terminals of the inductor or capacitor can be replaced by its Thévenin or Norton equivalent.

To generalize the solution of these four possible circuits, we will let $x(t)$ represent the unknown quantity. Thus $x(t)$ has four possible values. It can represent either the current or voltage at the terminals of an inductor or it can represent either the current or voltage at the terminals of a capacitor. On the basis of the observations we have previously made with respect to Eqs. (7.27), (7.28), (7.30), and (7.39), we know that the differential equation that describes any one of the four circuits in Fig. 7.11 will be of the form

$$\frac{dx}{dt} + \frac{x}{\tau} = K, \tag{7.42}$$

where the value of the constant K can be zero.

Before proceeding with the solution of Eq. (7.42), we must make one more observation. Since the sources in the circuit are constant voltages and/or currents, the final value of our unknown x will be constant. The final value must satisfy Eq. (7.42). When x has reached its final value, the derivative dx/dt must be zero. Hence we can conclude

$$x_f = K\tau, \tag{7.43}$$

where x_f represents the final value of our variable.

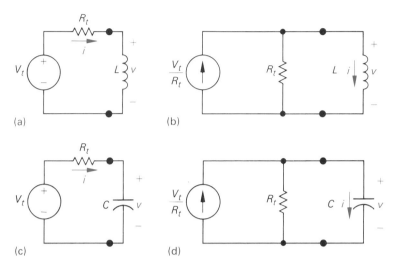

Fig. 7.11   The four possible circuits involved in finding the step response of *RL* and *RC* circuits.

We can solve Eq. (7.42) by separating the variables. We begin by solving for the first derivative:

$$\frac{dx}{dt} = \frac{-x}{\tau} + K = \frac{-(x - K\tau)}{\tau} = \frac{-(x - x_f)}{\tau}. \tag{7.44}$$

In writing Eq. (7.44), we have used the observation contained in Eq. (7.43). Now we multiply both sides of Eq. (7.44) by dt, which allows us to write

$$\frac{dx}{(x - x_f)} = \frac{-1}{\tau} dt. \tag{7.45}$$

Equation (7.45) is now integrated. Using u and v as symbols of integration, we have

$$\int_{x(0)}^{x(t)} \frac{du}{u - x_f} = -\frac{1}{\tau} \int_0^t dv. \tag{7.46}$$

Carrying out the integration called for in Eq. (7.46) gives us

$$x(t) = x_f + [x(0) - x_f]e^{-t/\tau}. \tag{7.47}$$

The importance of Eq. (7.47) is more apparent if we write it out in verbal form:

$$\left\{ \begin{array}{c} \text{The unknown} \\ \text{variable as a} \\ \text{function of time} \end{array} \right\} = \left\{ \begin{array}{c} \text{The final} \\ \text{value of the} \\ \text{variable} \end{array} \right\}$$

$$+ \left[\left\{ \begin{array}{c} \text{The initial} \\ \text{value of the} \\ \text{variable} \end{array} \right\} - \left\{ \begin{array}{c} \text{The final} \\ \text{value of the} \\ \text{variable} \end{array} \right\} \right] \times e^{-t/(\text{time constant})} \tag{7.48}$$

Equation (7.48) reduces the problem of finding the step response of a single time constant RL or RC circuit to a circuits problem involving the computation of three quantities: (1) the final value of the variable; (2) the initial value of the variable; and (3) the time constant.

Examples 7.3 and 7.4 illustrate how Eq. (7.48) can be used to find the step response of an RC or RL circuit.

Example 7.3   The switch in the circuit in Fig. 7.12 has been in the open position for a long time. The initial charge on the capacitor is zero. At $t = 0$, the switch is closed.

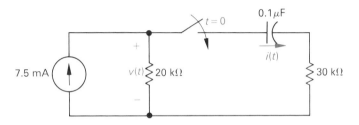

Fig. 7.12   The circuit for Example 7.3.

a) Find the expression for $i(t)$ when $t \geq 0$.

b) Find the expression for $v(t)$ when $t \geq 0$.

Solution   a) Since the initial voltage on the capacitor is zero, at the instant when the switch is closed the current in the 30-kΩ branch will be

$$i(0) = \frac{(7.5)(20)}{(50)} = 3 \text{ mA}.$$

The final value of the capacitor current will be zero because the capacitor will eventually appear as an open circuit to dc current. Thus $i_f = 0$. The time constant of the circuit will equal the product of the Thévenin resistance seen from the capacitor terminals and the capacitor. Therefore $\tau = (20 + 30)10^3(0.1) \times 10^{-6} = 5$ ms. Substituting these values into Eq. (7.48) generates the following expression:

$$i(t) = 0 + (3 - 0)e^{-t/5 \times 10^{-3}} = 3e^{-200t} \text{ mA}, \qquad t \geq 0.$$

b) To find $v(t)$, we note from the circuit that it equals the sum of the voltage across the capacitor and the voltage across the 30-kΩ resistor. To find the capacitor voltage (which is a drop in the direction of the current), we note that its initial value is zero and its final value is (7.5)(20), or 150 V. The time constant is the same as before: 5 ms. Therefore we can use Eq. (7.48) to write

$$v_C(t) = 150 + (0 - 150)e^{-200t} = [150 - 150e^{-200t}] \text{ V}, \qquad t \geq 0.$$

It follows that the expression for the voltage $v(t)$ is

$$v(t) = 150 - 150e^{-200t} + (30)(3)e^{-200t} = [150 - 60e^{-200t}] \text{ V}, \qquad t \geq 0.$$

As one check on our expression for $v(t)$ we note that it predicts that the initial value of the voltage across the 20-kΩ resistor is $150 - 60$, or 90 V. We can see from the circuit that the instant when the switch is closed, the current in the 20-kΩ resistor will be $(7.5)(\frac{30}{50})$, or 4.5 mA. This current will produce a 90-V drop across the 20-kΩ resistor and hence confirms the value predicted by our solution. ■

Example 7.4 The switch in the circuit shown in Fig. 7.13 has been open for a long time. At $t = 0$ the switch is closed.

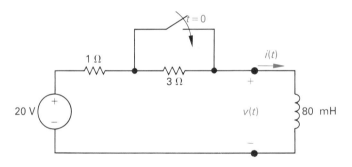

Fig. 7.13 The circuit for Example 7.4.

a) Find the expression for $v(t)$ when $t \geq 0$.

b) Find the expression for $i(t)$ when $i \geq 0$.

Solution a) Since the switch has been open for a long time, the initial current in the inductor will be 5 A oriented from top to bottom. Immediately after the switch has been closed, the current will still be 5 A and therefore the initial voltage across the inductor will become $20 - 5(1)$, or 15 V. The final value of the inductor voltage will be 0 V. With the switch closed, the time constant will be 80/1 ms. We can now use Eq. (7.48) to write down the expression for $v(t)$; thus

$$v(t) = 0 + (15 - 0)e^{-t/80 \times 10^{-3}} = 15e^{-12.5t} \text{ V}, \qquad t \geq 0.$$

b) We have already noted that the initial value of the inductor current is 5 A. After the switch has been closed for a long time, the inductor current will reach 20/1, or 20 A. The circuit time constant is 80 ms; therefore the expression for $i(t)$ is

$$i(t) = 20 + (5 - 20)e^{-12.5t} = [20 - 15e^{-12.5t}]\text{A}, \qquad t \geq 0.$$

We can check that our solutions for $v(t)$ and $i(t)$ are in agreement by noting that

$$v(t) = L\frac{di}{dt} = 80 \times 10^{-3}[15(12.5)e^{-12.5t}] = 15e^{-12.5t} \text{ V}, \qquad t \geq 0. \quad \blacksquare$$

DRILL EXERCISE
7.3

The switch in the circuit shown has been in position a for a long time. At $t = 0$ the switch is moved to position b. Calculate (a) the initial voltage on the capacitor; (b) the final voltage on the capacitor; (c) the time constant (in μs) for $t > 0$; and (d) the length of time (in μs) that it takes the capacitor voltage to reach zero after the switch is moved to position b.

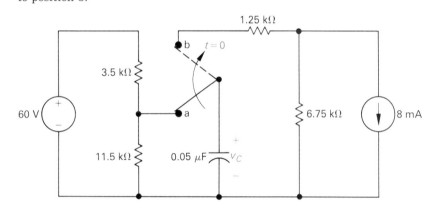

Ans. (a) 46 V; (b) -54 V; (c) 400 μs; (d) 246.47 μs.

DRILL EXERCISE
7.4

After the switch in the circuit shown has been open for a long time, it is closed at $t = 0$. Calculate (a) the initial value of i; (b) the final value of i; (c) the time constant for $t > 0$; and (d) the numerical expression for $i(t)$ when $t \geq 0$.

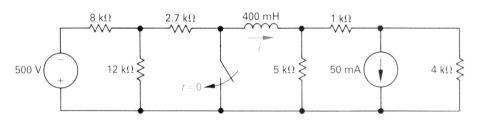

Ans. (a) -20 mA; (b) 40 mA; (c) 160 μs; (d) $i = [40 - 60e^{-6250t}]$ mA, $t \geq 0$.

7.5 SEQUENTIAL SWITCHING

Sequential switching occurs whenever a single switch is switched in sequence to two or more alternative positions or when two or more switches are opened and closed in sequence. The voltages and currents generated by such a switching sequence can be found using the general technique described in Section 7.4. The expressions for $v(t)$ and $i(t)$ are derived for a given position of the switch or switches, and these solutions are then used to determine the initial conditions for the next position of the switch or switches. The following example illustrates the technique.

Example 7.5 The uncharged capacitor in the circuit in Fig. 7.14 is initially switched to terminal a of the three-position switch. At $t = 0$ the switch is moved to position b, where it remains for 15 ms. After the 15-ms delay, the switch is set to position c, where it remains indefinitely.

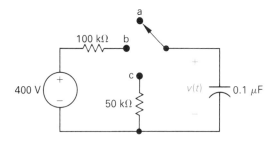

Fig. 7.14 The circuit for Example 7.5.

a) Derive the numerical expression for the voltage across the capacitor.

b) Make a plot of the capacitor voltage vs. time.

c) When will the voltage on the capacitor equal 200 V?

Solution a) At the instant when the switch is moved to position b, the initial voltage on the capacitor is zero. If the switch were to remain in position b, the capacitor would eventually charge to 400 V. The time constant of the circuit when the switch is in position b is 10 ms. With these three observations in mind, we can use Eq. (7.48) to write the expression for the capacitor voltage:

$$v = 400 + (0 - 400)e^{-100t} = [400 - 400e^{-100t}] \text{ V}, \qquad (0 \le t \le 15 \text{ ms}).$$

Since the switch remains in position b for only 15 ms, we note that our expression is valid only for the time interval from 0 to 15 ms. We have emphasized this with the parenthetical notation next to our expression. After the switch has been in position b for 15 ms, the voltage on the capacitor will be

$$v(15 \text{ ms}) = 400 - 400e^{-1.5} = 310.75 \text{ V}.$$

Therefore at the time when the switch is moved to position c, the initial voltage on the capacitor is 310.75 V. With the switch in position c, the final value of the capacitor voltage is zero and the time constant is 5 ms. Again, we use Eq. (7.48) to write down the expression for the capacitor voltage:

$$v = 0 + (310.75 - 0)e^{-200(t-0.015)} = 310.75e^{-200(t-0.015)} \text{ V}, \qquad 15 \text{ ms} \le t.$$

Note that in writing the expression for v, we recognize that it is valid only when $t \ge 15$ ms. This is reflected in the exponent of e by writing $t -- 0.015$ instead of t. Thus when $t = 15$ ms, $v = 310.75$ V.

b) The plot of v vs. t is shown in Fig. 7.15.

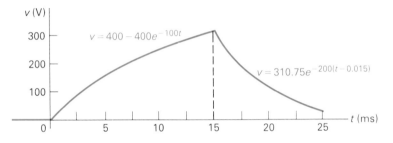

Fig. 7.15 Capacitor voltage in Example 7.5.

c) It is apparent from the plot in Fig. 7.15 that the capacitor voltage will equal 200 V at two different times: once in the interval between 0 and 15 ms and once after 15 ms. The first time $v = 200$ V is found from solving the expression

$$200 = 400 - 400e^{-100t_1},$$

which yields $t_1 = 6.93$ ms. The second time $v = 200$ V is found from solving the expression

$$200 = 310.75e^{-200(t_2-0.015)}$$

In this case, we find that $t_2 = 17.20$ ms.   ■

**DRILL EXERCISE
7.5**

Switch a in the circuit shown has been open for a long time and switch b has been closed for a long time. Switch a is closed at $t = 0$. After remaining closed for one second, switches a and b are opened simultaneously and remain open indefinitely. Determine the expression for the inductor current i that is valid when (a) $0 \le t \le 1$ s and (b) $t \ge 1$ s.

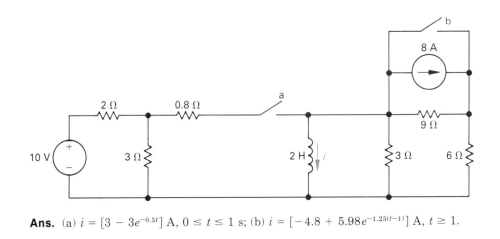

Ans. (a) $i = [3 - 3e^{-0.5t}]$ A, $0 \le t \le 1$ s; (b) $i = [-4.8 + 5.98e^{-1.25(t-1)}]$ A, $t \ge 1$.

7.6 SUMMARY

We have now introduced the step response of an important class of *RL* and *RC* circuits, namely, first-order *RL* and *RC* circuits. First-order circuits are circuits that can be reduced to an equivalent resistive circuit driving a single equivalent inductor or capacitor. The circuits are energized only by constant current and/or voltage sources. Finding the step response of a first-order circuit reduces to finding the current or voltage, or both, at the terminals of the inductor or capacitor. The current and voltage are changing with time as a result of an abrupt switching operation occurring somewhere in the circuit.

We found in all cases that the current or voltage makes a simple exponential transition from its value at the time when the switching occurs to its final value. We also found that the rate at which the exponential transition takes place is determined by the time constant of the circuit. The time constant of the inductive circuit equals the equivalent inductance divided by the Thévenin resistance seen from the terminals of the inductor. The time constant of the capacitive circuit equals the equivalent capacitance times the Thévenin resistance seen from the terminals of the capacitor.

Although this chapter has been restricted to a special type of circuit, the results of our discussion are useful in many practical engineering problems involving switching operations. Some applications are brought out in the following problems.

PROBLEMS

7.1 The current and voltage at the terminals of the inductor in the circuit in Fig. 7.1 are

$$i(t) = [10 - 10e^{-200t}] \text{ A} \qquad (t \ge 0),$$
$$v(t) = 500e^{-200t} \text{ V} \qquad (t \ge 0).$$

a) Specify the numerical values of V_s, R, and L.
b) How many milliseconds after the switch has been closed does the energy stored in the inductor reach one fourth of its final value?

7.2   The switch in the circuit in Fig. 7.16 has been in the OFF position for a long time. At $t = 0$ the switch is moved to the ON position.

a) Find the numerical expression for $i(t)$ when $t \geq 0$.
b) Find the numerical expression for $v(t)$ when $t \geq 0$.
c) Find the numerical expression for $i_s(t)$ when $t \geq 0$.

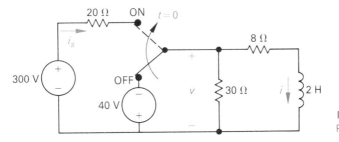

Fig. 7.16   The circuit for Problem 7.2.

7.3   The switch in the circuit in Fig. 7.17 has been closed for a long time. At $t = 0$ the switch is opened. Find the numerical expression for $i(t)$ when $t \geq 0$.

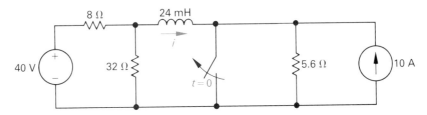

Fig. 7.17   The circuit for Problem 7.3.

7.4   The switch in the circuit in Fig. 7.18 has been closed for a long time. The switch opens at $t = 0$. Find the numerical expressions for $i(t)$ and $v(t)$ when $t \geq 0$.

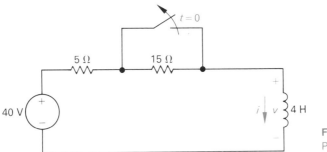

Fig. 7.18   The circuit for Problem 7.4.

7.5   The switch in the circuit in Fig. 7.19 has been closed for a long time. The switch opens at $t = 0$. For $t \geq 0$:

a) Find $v(t)$ as a function of V_{bb}, R_1, R_2, and L.
b) Verify your expression by using it to find $v(t)$ in the circuit in Fig. 7.18.
c) Explain what happens to $v(t)$ as R_2 gets larger and larger.

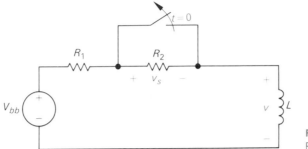

Fig. 7.19   The circuit for Problem 7.5.

d) Find $v_s(t)$ as a function of V_{bb}, R_1, R_2, and L.

e) Explain what happens to $v_s(t)$ as R_2 gets larger and larger.

7.6   The switch in the circuit in Fig. 7.20 has been closed for a long time. A student abruptly opens the switch and reports to her instructor that when the switch opened, an electric arc with noticeable persistence was established across the switch and at the same time the voltmeter placed across the coil was damaged. On the basis of your analysis of the circuit in Problem 7.5, can you explain to the student why this happened?

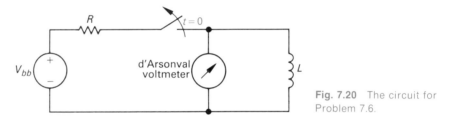

Fig. 7.20   The circuit for Problem 7.6.

7.7   The current and voltage at the terminals of the capacitor in the circuit in Fig. 7.7(b) are

$$i(t) = 5e^{-50t} \text{ mA} \qquad (t \geq 0),$$
$$v(t) = [200 - 200e^{-50t}] \text{ V} \qquad (t \geq 0).$$

a) Specify the numerical values of I_s, R, C, and τ.

b) How many milliseconds after the switch has been closed does the energy stored in the capacitor reach 36% of its final value?

7.8   The switch in the circuit in Fig. 7.21 has been open for a long time. The switch closes at $t = 0$.

a) Find the numerical expression for the voltage across the terminals of the capacitor for $t \geq 0$.

b) Find the numerical expression for the source current $i_s(t)$ for $t \geq 0$.

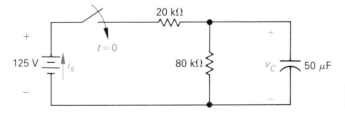

Fig. 7.21   The circuit for Problem 7.8.

7.9   The switch in the circuit in Fig. 7.22 has been in position a for a long time. At $t = 0$ the switch is moved to position b. Find $v_C(t)$ when $t \geq 0$.

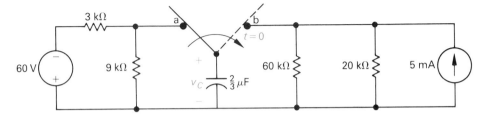

Fig. 7.22   The circuit for Problem 7.9.

7.10   The switch in the circuit in Fig. 7.23 has been closed for a long time. At $t = 0$ the switch is opened.

a) What is the initial value of $i(t)$?
b) What is the final value of $i(t)$?
c) What is the time constant of the circuit for $t \geq 0$?
d) What is the numerical expression for $i(t)$ when $t > 0$?
e) What is the numerical expression for $v(t)$ when $t > 0$?

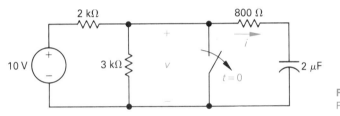

Fig. 7.23   The circuit for Problem 7.10.

7.11   The voltage waveform shown in Fig. 7.24(a) is applied to the circuit in Fig. 7.24(b). The initial voltage on the capacitor is zero.

a) Calculate $v_o(t)$.
b) Make a sketch of $v_o(t)$ vs. t.

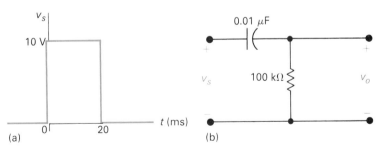

Fig. 7.24   The circuit for Problem 7.11: (a) input voltage; (b) circuit.

7.12   The voltage waveform shown in the circuit in Fig. 7.25(a) is applied to the circuit in Fig. 7.25(b). The initial current in the inductor is zero.

a) Calculate $v_o(t)$.
b) Make a sketch of $v_o(t)$.

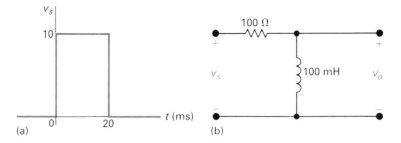

(a)   (b)

Fig. 7.25   The circuit for Problem 7.12: (a) input voltage; (b) circuit.

7.13   The switch in the circuit in Fig. 7.26 closes at $t = 0$ after being open for a long time.

    a) Find the numerical expression for $v_o(t)$ when $t \geq 0$.
    b) Find the numerical expression for $i_1(t)$.

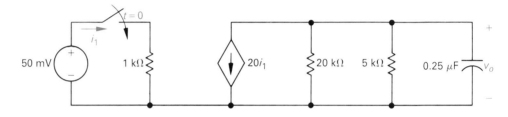

Fig. 7.26.   The circuit for Problem 7.13.

7.14   The gap in the circuit in Fig. 7.27 will arc over whenever the voltage across the gap reaches 12,000 V. The initial current in the inductor is zero. How many milliseconds after the switch has been closed will the gap arc over?

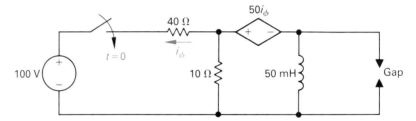

Fig. 7.27   The circuit for Problem 7.14.

7.15   The switch in the circuit in Fig. 7.28 has been closed for a long time. Find the capacitor voltage $v_C(t)$ for $t \geq 0$.

7.16   The circuit shown in Fig. 7.29 is used to close the switch between a and b for a predetermined length of time. The electric relay holds its contact arms down so long as the voltage across the relay coil exceeds 20 V. When the coil voltage equals 20 V, the relay contacts return to their initial position by a mechanical spring action. The switch between a and b is initially closed by momentarily pressing the pushbutton. Assume that the capacitor is fully charged when the pushbutton is first pushed down. The resistance of the relay coil is 10 kΩ, and the inductance of the coil is negligible.

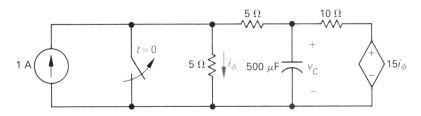

Fig. 7.28 The circuit for Problem 7.15.

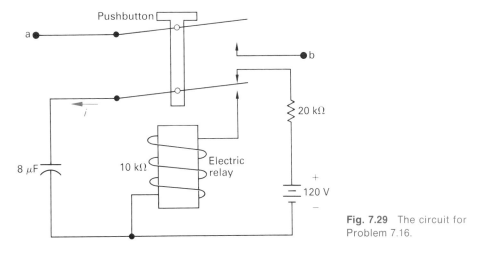

Fig. 7.29 The circuit for Problem 7.16.

a) How long will the switch between a and b remain closed?
b) Write the numerical expression for i from the time when the relay contacts first close to the time when the capacitor is completely charged.

7.17 In the circuit in Fig. 7.30, the lamp starts to conduct whenever the lamp voltage reaches 8 V. During the time when the lamp conducts, it can be modeled as a 5-kΩ resistor. Once the lamp conducts, it will continue to conduct until the lamp voltage drops to 1 V. When the lamp is not conducting, it appears as an open circuit. Assume that the circuit has been in operation for a long time. Let $t = 0$ at the instant when the lamp stops conducting.

a) Derive the expression for the voltage across the lamp for one full cycle of operation.
b) How many times per minute will the lamp turn on?
c) The 500-kΩ resistor is replaced with a variable resistor R. The resistance is adjusted until the lamp "flashes" 12 times per minute. What is the value of R?

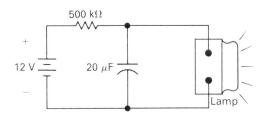

Fig. 7.30 The circuit for Problem 7.17.

8

NATURAL AND STEP RESPONSES OF *RLC* CIRCUITS

8.1 INTRODUCTION

At this point in our study, our discussion of the natural and step responses of circuits containing both inductors and capacitors will be limited to two simple structures: the parallel *RLC* circuit and the series *RLC* circuit. The natural response of a parallel *RLC* circuit consists of finding the voltage across the parallel branches that arises due to the release of energy that has been stored in the inductor or capacitor or both. The problem is defined in terms of the circuit shown in Fig. 8.1, where the initial energy stored in the capacitor is represented by the initial voltage on the capacitor V_0 and the initial energy stored in the inductor is represented by the initial current on the inductor I_0. If the individual branch currents are of interest, they can be found once the terminal voltage is known.

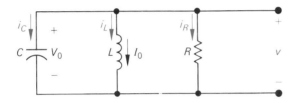

Fig. 8.1 A circuit used to introduce the natural response of the parallel *RLC* circuit.

The step response of a parallel *RLC* circuit is defined by the circuit shown in Fig. 8.2, where we are interested in the voltage that appears across the parallel branches as a result of the sudden application of a dc, or constant, current source. There may or may not be energy stored in the circuit at the time when the current source is applied to the circuit.

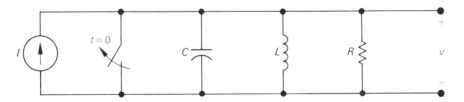

Fig. 8.2 A circuit used to introduce the step response of the parallel *RLC* circuit.

The natural response of a series *RLC* circuit consists of finding the current in the series-connected elements that arises due to the release of initially stored energy in either the inductor or capacitor or both. The problem is repre-

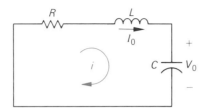

Fig. 8.3 A circuit used to introduce the natural response of the series *RLC* circuit.

sented by the circuit shown in Fig. 8.3. As before, the initially stored energy is represented by the initial inductor current I_0 and the initial capacitor voltage V_0. If any of the individual element voltages are of interest, they can be found once the current is known.

The step response of a series *RLC* circuit can be described in terms of the circuit shown in Fig. 8.4, where we are interested in the current that arises due to sudden application of the dc voltage source. There may or may not be energy stored in the circuit at the time the switch is closed.

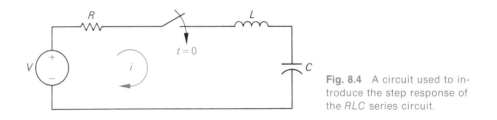

Fig. 8.4 A circuit used to introduce the step response of the *RLC* series circuit.

Although lack of formal training in the solution of ordinary differential equations may make the discussion of the natural and step responses of parallel and series *RLC* circuits a bit hard to follow, the results are important enough to warrant our attention at this time. We begin with the natural response of a parallel *RLC* circuit.

8.2 INTRODUCTION TO THE NATURAL RESPONSE OF A PARALLEL *RLC* CIRCUIT

The first step in finding the natural response of the circuit shown in Fig. 8.1 is to derive the differential equation that the voltage v must satisfy. We can easily do this by summing the currents away from the top node. Each current is expressed as a function of the unknown voltage v. We have

$$\frac{v}{R} + \frac{1}{L}\int_0^t v \, d\tau + I_0 + C\frac{dv}{dt} = 0. \tag{8.1}$$

The integral in Eq. (8.1) can be eliminated by differentiating once with respect to t, hence

$$\frac{1}{R}\frac{dv}{dt} + \frac{v}{L} + C\frac{d^2v}{dt^2} = 0. \tag{8.2}$$

We now divide through Eq. (8.2) by the capacitor C and arrange the derivatives in descending order:

$$\frac{d^2v}{dt^2} + \frac{1}{RC}\frac{dv}{dt} + \frac{v}{LC} = 0. \tag{8.3}$$

In comparing Eq. (8.3) with the differential equations derived in Chapters 6 and 7, we note immediately that they differ by the presence of the term involving the second derivative. Equation (8.3) is described as an "ordinary, *second-order,* differential equation with constant coefficients." Whenever a circuit contains both types of storage elements, the differential equation describing the circuit will be of second order or higher.

We cannot solve Eq (8.3) by separating the variables and integrating as we were able to do with the first-order equations we found in Chapters 6 and 7. The classical approach to solving Eq. (8.3) is to assume that the solution is of exponential form, that is, to assume that the voltage will be of the form

$$v = Ae^{st}, \tag{8.4}$$

where A and s are unknown constants.

Before showing how this assumption leads to the solution of Eq. (8.3), we pause long enough to point out that the assumption contained in Eq. (8.4) is a rational one. The strongest argument we can make in favor of Eq. (8.4) is to note from Eq. (8.3) that the second derivative of the solution plus a constant times the first derivative plus a constant times the solution itself must add to zero *for all values of t.* This can happen only if higher-order derivatives of the solution have the same form as the solution. The exponential function satisfies this criterion. A second argument can be made in favor of Eq. (8.4) by noting that the solutions of all the first-order equations we derived in Chapters 6 and 7 were exponential. It seems reasonable to assume that the solution of the second-order equation will also involve the exponential function.

To show that the assumption contained in Eq. (8.4) leads to the solution of Eq. (8.3), we proceed as follows. If Eq. (8.4) is a solution to Eq. (8.3), it must satisfy Eq. (8.3) for all values of t. When we substitute Eq. (8.3) into Eq. (8.4) we generate the following expression:

$$As^2e^{st} + \frac{As}{RC}e^{st} + \frac{Ae^{st}}{LC} = 0,$$

or

$$Ae^{st}\left(s^2 + \frac{s}{RC} + \frac{1}{LC}\right) = 0, \tag{8.5}$$

which can be satisfied for all values of t only if A is zero or the parenthetical term is zero. We cannot use $A = 0$ as a general solution because this implies that the voltage is zero for all time; this is a physical impossibility if there is energy stored in either the inductor or capacitor. Therefore in order for Eq. (8.4) to be a solution of Eq. (8.3), the parenthetical term in Eq. (8.5) must be zero; thus

$$s^2 + \frac{s}{RC} + \frac{1}{LC} = 0. \tag{8.6}$$

Equation (8.6) is called the *characteristic equation* of the differential equation because the roots of this quadratic equation will determine the mathematical character of $v(t)$.

The two roots of Eq. (8.6) are

$$s_1 = -\frac{1}{2RC} + \sqrt{\left(\frac{1}{2RC}\right)^2 - \frac{1}{LC}} \tag{8.7}$$

and

$$s_2 = -\frac{1}{2RC} - \sqrt{\left(\frac{1}{2RC}\right)^2 - \frac{1}{LC}}. \tag{8.8}$$

If either root is substituted into Eq. (8.4), the assumed solution will satisfy the given differential equation, that is, Eq. (8.3). Note from Eq. (8.5) that this is true regardless of the value of A; therefore

$$v = A_1 e^{s_1 t}$$

or

$$v = A_2 e^{s_2 t}$$

will each satisfy Eq. (8.3). If we denote these two solutions as v_1 and v_2, respectively, we can show that their sum is also a solution. Specifically, if we let

$$v = v_1 + v_2 = A_1 e^{s_1 t} + A_2 e^{s_2 t}, \tag{8.9}$$

then

$$\frac{dv}{dt} = A_1 s_1 e^{s_1 t} + A_2 s_2 e^{s_2 t} \tag{8.10}$$

and

$$\frac{d^2 v}{dt^2} = A_1 s_1^2 e^{s_1 t} + A_2 s_2^2 e^{s_2 t} \tag{8.11}$$

When we substitute Eqs. (8.9), (8.10), and (8.11) into Eq. (8.3), we obtain

$$A_1 e^{s_1 t} \left(s_1^2 + \frac{1}{RC} s_1 + \frac{1}{LC}\right) + A_2 e^{s_2 t} \left(s_2^2 + \frac{1}{RC} s_2 + \frac{1}{LC}\right) = 0. \tag{8.12}$$

But each parenthetical term is zero since by definition s_1 and s_2 are roots of the characteristic equation.

We conclude that the natural response of the parallel *RLC* circuit shown in Fig. 8.1 will be of the form

$$v = A_1 e^{s_1 t} + A_2 e^{s_2 t}. \tag{8.13}$$

The roots of the characteristic equation (s_1 and s_2) will be determined by the circuit parameters R, L, and C. The initial conditions will determine the values of the constants A_1 and A_2.†

The behavior of $v(t)$ will depend on the roots s_1 and s_2. Therefore the first step in finding the natural response is to determine the roots of the characteristic equation. Thus we return to Eqs. (8.7) and (8.8) and rewrite the equations using a notation that is widely used in the literature:

$$s_1 = -\alpha + \sqrt{\alpha^2 - \omega_0^2} \tag{8.14}$$

and

$$s_2 = -\alpha - \sqrt{\alpha^2 - \omega_0^2}, \tag{8.15}$$

where

$$\alpha = \frac{1}{2RC} \tag{8.16}$$

and

$$\omega_0 = \frac{1}{\sqrt{LC}}. \tag{8.17}$$

Since the exponent of e must be dimensionless, both s_1 and s_2 (and hence α and ω_0) must have the dimension of s^{-1}. The reciprocal of time is called *frequency*. To distinguish among the frequencies s_1, s_2, α, and ω_0 we use the following terminology: s_1 and s_2 are referred to as *complex frequencies,* α is called the *neper frequency,* and ω_0 is the *resonant radian frequency.* The full significance of this terminology will unfold as we move through the remaining chapters of the book. All these frequencies have the dimension of radians per second (rad/s).

The nature of the roots s_1 and s_2 will depend on the values of α and ω_0. There are three possible outcomes. First, if $\omega_0^2 < \alpha^2$, both roots will be real and distinct. For reasons to be discussed later, the voltage response is said to be *overdamped* when $\omega_0^2 < \alpha^2$. Second, if $\omega_0^2 > \alpha^2$, both s_1 and s_2 will be complex and furthermore they will be conjugates of each other. In this situation, the voltage response is said to be *underdamped.* The third possible outcome is that ω_0^2 equals α^2. In this case, s_1 and s_2 will be real and equal. Here the voltage response is said to be *critically damped.* We will discuss each case separately in Section 8.3.

Example 8.1 a) Find the roots of the characteristic equation that governs the transient behavior of the voltage shown in Fig. 8.1 if $R = 100\ \Omega$, $L = 40$ mH, and $C = 0.25\ \mu$F.

b) Will the response be over-, under-, or critically damped?

---

† The form of Eq. (8.13) must be modified if the two roots s_1 and s_2 are equal. The modification is discussed in Section 8.3(c).

c) Repeat parts (a) and (b) given that R is increased to 400 Ω.

d) What value of R will cause the response to be critically damped?

Solution   a) For the given values of R, L, and C we have

$$\alpha = \frac{1}{2RC} = \frac{10^6}{(200)(0.25)} = 2 \times 10^4,$$

$$\omega_0^2 = \frac{1}{LC} = \frac{(10^3)(10^6)}{(40)(0.25)} = 10^8.$$

From Eqs. (8.14) and (8.15) we have

$$
\begin{aligned}
s_1 &= -2 \times 10^4 + \sqrt{4 \times 10^8 - 10^8} \\
&= (-2 + \sqrt{3})10^4 \\
&\cong -0.268 \times 10^4, \\
s_2 &= -2 \times 10^4 - 10^4\sqrt{3} \\
&\cong -3.732 \times 10^4.
\end{aligned}
$$

b) The voltage response is overdamped since $\omega_0^2 < \alpha^2$.

c) For $R = 400 \ \Omega$,

$$\alpha = \frac{10^6}{(2)(400)(0.25)} = 5000,$$

$$\alpha^2 = 25 \times 10^6 = 0.25 \times 10^8.$$

Since ω_0^2 remains at 10^8,[†]

$$
\begin{aligned}
s_1 &= -5000 + j10^4\sqrt{0.75} \\
&\cong -5000 + j8660.25, \\
s_2 &\cong -5000 - j8660.25.
\end{aligned}
$$

In this case, the voltage response is underdamped since $\omega_0^2 > \alpha^2$.

d) For critical damping $\alpha^2 = \omega_0^2$. Thus

$$\left(\frac{1}{2RC}\right)^2 = \frac{1}{LC} = 10^8$$

or

$$\frac{1}{2RC} = 10^4$$

and

$$R = \frac{10^6}{(2 \times 10^4)(0.25)} = 200 \ \Omega. \quad \blacksquare$$

---

† In electrical engineering, the imaginary number $\sqrt{-1}$ is represented by the letter j since the letter i is reserved to represent current.

DRILL EXERCISE
8.1 The resistance and inductance in the circuit in Fig. 8.1 are 200 Ω and 10 mH.

a) Find the values of C that will make the voltage response critically damped.

b) If C is adjusted to give a neper frequency of 10 kr/s, find the value of C and the roots of the characteristic equation.

c) If C is adjusted to give a resonant frequency of 50 kr/s, find the value of C and the roots of the characteristic equation.

Ans. (a) $C = 62.5$ nF; (b) $C = 0.25~\mu$F, $s_1 = -10,000 + j17,320.51$ rad/s, $s_2 = -10,000 - j17,320.51$ rad/s; (c) $C = 40$ nF, $s_1 = -25,000$ rad/s, $s_2 = -100,000$ rad/s.

8.3 THE NATURAL RESPONSE OF A PARALLEL *RLC* CIRCUIT

8.3(a) Overdamped Voltage Response   When the roots of the characteristic equation are real and distinct, the voltage response of the circuit in Fig. 8.1 is said to be *overdamped*. The solution for the voltage is of the form

$$v = A_1 e^{s_1 t} + A_2 e^{s_2 t}, \tag{8.18}$$

where s_1 and s_2 are the roots of the characteristic equation. The constants A_1 and A_2 are determined by the initial conditions. We can derive expressions for A_1 and A_2 in terms of the initial voltage on the capacitor and the initial current in the inductor. From the circuit in Fig. 8.1, we see that $v(0)$ must equal V_0 since there cannot be an instantaneous change in the voltage at the terminals of a capacitor. If follows directly from Eq. (8.18) that

$$V_0 = A_1 + A_2. \tag{8.19}$$

We also note from the circuit that since the sum of the currents leaving the top node is zero for all values of t, then at $t = 0$ this generates the expression

$$\frac{V_0}{R} + I_0 + C\frac{dv}{dt}(0) = 0. \tag{8.20}$$

From Eq. (8.20) we have

$$\frac{dv(0)}{dt} = -\frac{V_0}{RC} - \frac{I_0}{C}. \tag{8.21}$$

Evaluating the initial value of dv/dt from Eq. (8.18) gives us

$$\frac{dv}{dt}(0) = s_1 A_1 + s_2 A_2. \tag{8.22}$$

It follows from Eqs. (8.21) and (8.22) that

$$-\frac{V_0}{RC} - \frac{I_0}{C} = s_1 A_1 + s_2 A_2. \tag{8.23}$$

Now we can solve Eqs. (8.19) and (8.23) for A_1 and A_2. Thus

$$A_1 = \frac{V_0 s_2 + (1/C)[(V_0/R) + I_0]}{(s_2 - s_1)}$$ (8.24)

and

$$A_2 = -\frac{V_0 s_1 + (1/C)[(V_0/R) + I_0]}{(s_2 - s_1)}$$ (8.25)

At this point, we know both the roots of the characteristic equation and the coefficients A_1 and A_2. Therefore to obtain the expression for the over-damped response we simply substitute these known values into Eq. (8.18). Example 8.2 illustrates the numerical calculations involved in finding the overdamped response of a parallel *RLC* circuit.

Example 8.2 The initial voltage across the capacitor in the circuit in Fig. 8.1 is 50 V and the initial current in the inductor is 2 A. The circuit element values are 0.25 μF, 40 mH, and 100 Ω, respectively.

a) Find the initial current in each branch of the circuit.

b) Find the initial value of dv/dt.

c) Find the expression for $v(t)$.

d) Sketch $v(t)$ in the interval $0 \le t \le 150$ μs.

Solution a) Since the inductor prevents an instantaneous change in its current, the initial value of the inductor current is 2 A, that is, $i_L(0) = I_0 = 2$ A. The capacitor will hold the initial voltage across the parallel elements to 50 V; thus the initial current in the resistive branch will be 50/100, or 0.5 A. Kirchhoff's current law requires the sum of the currents leaving the top node to equal zero at every instant; thus

$$i_C(0) = -i_L(0) - i_R(0) = -2.5 \text{ A}.$$

b) Since $i_C = C(dv/dt)$ it follows that $(dv/dt)(0) = i_C(0)/C$. Therefore

$$\frac{dv}{dt}(0) = \frac{-2.5}{0.25} \times 10^6 = -10^7.$$

c) The numerical values of the circuit elements are identical to those of Example 8.1(a). Therefore the roots of the characteristic equation are

$$s_1 = (-2 + \sqrt{3})10^4 \cong -2679.49,$$
$$s_2 = (-2 - \sqrt{3})10^4 \cong -37,320.51.$$

Since the roots are real and distinct, we know that response is overdamped and hence has the form given by Eq. (8.18). The coefficients A_1 and A_2 can

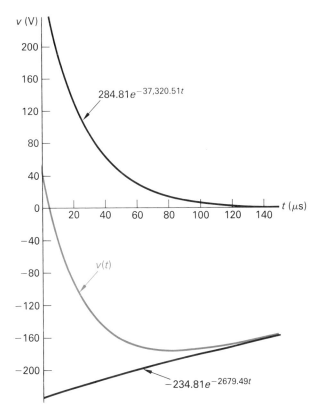

Fig. 8.5 The voltage response for Example 8.2.

be calculated from Eqs. (8.24) and (8.25). We get

$$A_1 = \frac{50(-2 - \sqrt{3})10^4 + 4 \times 10^6(2.5)}{-2\sqrt{3} \cdot 10^4} = 25(1 - 6\sqrt{3}) \cong -234.81$$

and

$$A_2 = -\frac{50(-2 + \sqrt{3})10^4 + 4 \times 10^6(2.5)}{-2\sqrt{3} \cdot 10^4} = 25(1 + 6\sqrt{3}) \cong 284.81.$$

The overdamped voltage response is

$$v(t) = [-234.81e^{-2679.49t} + 284.81e^{-37,320.51t}]\,\text{V}, \qquad t \geq 0.$$

d) A plot of $v(t)$ vs. t over the interval $0 \leq t \leq 150\ \mu$s is shown in Fig. 8.5. ∎

DRILL EXERCISE
8.2
The element values in the circuit in Fig 8.1 are $R = 400\ \Omega$, $L = 50$ mH, and $C = 50$ nF. The initial current in the inductor is -4 A and the initial voltage on the capacitor is 0 V. The output signal is the voltage v. Find (a) $i_R(0)$; (b) $i_C(0)$; (c) $dv(0)/dt$; (d) A_1; (e) A_2; and (f) $v(t)$ when $t \geq 0$.

Ans. (a) 0;   (b) 4 A;   (c) 8×10^7 V/s;   (d) 8000/3;   (e) $-8000/3$;
(f) $(8000/3)[e^{-10,000t} - e^{-4000t}]$ V when t $\geq$ 0.

8.3(b) Underdamped Voltage Response   When the roots of the characteristic equation are complex, $\omega_0^2 > \alpha^2$ and the response is *underdamped*. When we are discussing the underdamped response, it is convenient to express the roots s_1 and s_2 as

$$s_1 = -\alpha + \sqrt{-(\omega_0^2 - \alpha^2)}$$
$$= -\alpha + j\sqrt{\omega_0^2 - \alpha^2}$$
$$= -\alpha + j\omega_d \tag{8.26}$$

and

$$s_2 = -\alpha - j\omega_d, \tag{8.27}$$

where

$$\omega_d = \sqrt{\omega_0^2 - \alpha^2} \tag{8.28}$$

and ω_d is known as the *damped radian frequency*. The reason for this terminology will be explained later.

The underdamped voltage response of a parallel *RLC* circuit can be written as

$$v(t) = B_1 e^{-\alpha t} \cos \omega_d t + B_2 e^{-\alpha t} \sin \omega_d t, \tag{8.29}$$

which follows directly from Eq. (8.18) as we will now show. In making the transition from Eq. (8.18) to Eq. (8.29), we use the Euler identity:

$$e^{\pm j\theta} = \cos \theta \pm j \sin \theta. \tag{8.30}$$

Thus

$$v(t) = A_1 e^{(-\alpha + j\omega_d)t} + A_2 e^{-(\alpha + j\omega_d)t}$$
$$= A_1 e^{-\alpha t} e^{j\omega_d t} + A_2 e^{-\alpha t} e^{-j\omega_d t}$$
$$= e^{-\alpha t}[A_1 \cos \omega_d t + jA_1 \sin \omega_d t + A_2 \cos \omega_d t - jA_2 \sin \omega_d t]$$
$$= e^{-\alpha t}[(A_1 + A_2) \cos \omega_d t + j(A_1 - A_2) \sin \omega_d t].$$

At this point in the transition from Eq. (8.18) to (8.29), we replace the arbitrary constants $A_1 + A_2$ and $j(A_1 - A_2)$ with new arbitrary constants denoted as B_1 and B_2, to get

$$v = e^{-\alpha t}[B_1 \cos \omega_d t + B_2 \sin \omega_d t]$$
$$= B_1 e^{-\alpha t} \cos \omega_d t + B_2 e^{-\alpha t} \sin \omega_d t.$$

In Eq. (8.29), the two arbitrary constants B_1 and B_2 are determined by the initial energy stored in the circuit, whereas α and ω_d are fixed by the circuit parameters R, L, and C.

Before deriving expressions for B_1 and B_2 as functions of the initial conditions, let us pause and reflect on the general nature of the underdamped response. First note that the trigonometric functions tell us that the under-damped response is oscillatory; that is, the voltage alternates between positive and negative values. The rate at which the voltage oscillates is fixed by ω_d. Second, note that the amplitude of the oscillation decreases exponentially. The rate at which the amplitude falls off is determined by α. Because α determines how quickly the oscillations subside, it is also referred to as the *damping factor,* or *damping coefficient.* It should now be apparent why ω_d is called the *damped radian frequency.* If there is no damping, α is zero and the frequency of oscillation will be ω_0. Whenever there is a dissipative element (R) in the circuit, α is nonzero and the frequency of oscillation (ω_d) is less than ω_0. Thus when α is nonzero, the frequency of oscillation is said to be damped.

The oscillatory behavior is possible because of the presence of the two types of energy storage elements in the circuit: the inductor and capacitor. (A mechanical analogy of our electrical circuit is that of a mass suspended on a spring, where oscillation is possible because energy can be stored in both the spring and the moving mass.) We will have more to say about the characteristics of the underdamped response after working a numerical example, but first we need to obtain expressions for B_1 and B_2 in terms of the initial conditions.

To obtain expressions for B_1 and B_2, we proceed as follows. From Eq. (8.29), we see that the initial value of v equals B_1; thus

$$v(0) = B_1 = V_0 \tag{8.31}$$

To find B_2 we evaluate the initial value of dv/dt. It follows from Eq. (8.29) that

$$\frac{dv}{dt} = e^{-\alpha t}[(\omega_d B_2 - \alpha B_1) \cos \omega_d t - (\omega_d B_1 + \alpha B_2) \sin \omega_d t]. \tag{8.32}$$

Evaluating Eq. (8.32) at $t = 0$ yields

$$\frac{dv}{dt}(0) = \omega_d B_2 - \alpha B_1. \tag{8.33}$$

We have already noted (Eq. 8.21) that the circuit forces the initial value of dv/dt to equal $-(1/C)[(V_0/R) + I_0]$; therefore

$$\omega_d B_2 - \alpha B_1 = -\frac{1}{C}\left(\frac{V_0}{R} + I_0\right). \tag{8.34}$$

Solving Eq. (8.34) for B_2 gives

$$B_2 = \frac{\alpha}{\omega_d} V_0 - \frac{1}{\omega_d C}\left(\frac{V_0}{R} + I_0\right) = -\frac{\alpha}{\omega_d}(V_0 + 2I_0 R). \tag{8.35}$$

When Eqs. (8.31) and (8.35) are substituted into Eq. (8.29), we get the expression for the underdamped voltage response in terms of the initial conditions, that is,

$$v(t) = V_0 e^{-\alpha t} \cos \omega_d t - \frac{\alpha}{\omega_d}(V_0 + 2I_0 R)e^{-\alpha t} \sin \omega_d t. \tag{8.36}$$

The following example is designed to illustrate the calculations involved in finding the natural underdamped voltage response of the parallel *RLC* circuit.

Example 8.3   In the circuit in Fig. 8.1, $V_0 = 0$ and $I_0 = -12.25$ mA. The circuit parameters are $R = 20$ kΩ, $L = 8$ H, and $C = 0.125$ μF.

a) Calculate the roots of the characteristic equation.

b) Calculate the voltage response for $t \geq 0$.

c) Plot $v(t)$ vs. t for the time interval $0 \leq t \leq 11$ ms.

Solution   a) Since

$$\alpha = \frac{1}{2RC} = \frac{10^6}{2(20)10^3(0.125)} = 200,$$

$$\omega_0^2 = \frac{1}{LC} = \frac{10^6}{(8)(0.125)} = 10^6, \qquad \omega_0^2 > \alpha^2,$$

the response is underdamped. Now $\omega_d = \sqrt{\omega_0^2 - \alpha^2} = \sqrt{10^6 - 4 \times 10^4} = 100\sqrt{96} = 979.80$ r/s, $s_1 = -\alpha + j\omega_d = -200 + j979.80$, and $s_2 = -\alpha - j\omega_d = -200 - j979.80$.

b) We have

$$v(t) = B_1 e^{-\alpha t} \cos \omega_d t + B_2 e^{-\alpha t} \sin \omega_d t,$$

$$B_1 = V_0 = 0,$$

$$B_2 = \frac{\alpha V_0}{\omega_d} - \frac{1}{\omega_d C}\left(\frac{V_0}{R} + I_0\right) = -\frac{I_0}{\omega_d C}$$

$$= \frac{12.25 \times 10^{-3} \times 10^6}{(100\sqrt{96})(0.125)} \cong 100.$$

Substituting the numerical values of α, ω_d, B_1, and B_2 into the expression for $v(t)$ gives

$$v(t) = 100e^{-200t} \sin 979.80t \text{ V}, \qquad t \geq 0.$$

c) The plot of $v(t)$ vs. t for the first 11 ms after the stored energy is released is shown in Fig. 8.6, which clearly indicates the damped oscillatory nature of the underdamped response. The voltage $v(t)$ approaches its final value, alternating between values that are greater than and less than the final value. Furthermore, these swings about the final value are decreasing exponentially with time. ■

There are several characteristics of this type of response that we now wish to point out. First, as the dissipative losses in the circuit decrease, the persistence of the oscillations increases and the frequency of the oscillation approaches ω_0. To see this, note that as R approaches infinity the dissipation in

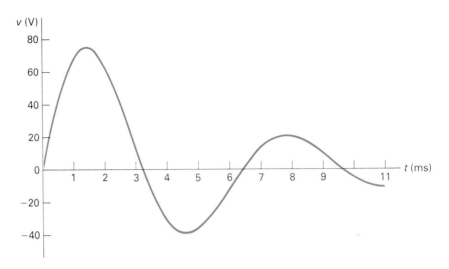

Fig. 8.6 The voltage response for Example 8.3.

the circuit in Fig. 8.1 approaches zero since $p = v^2/R$. As R approaches infinity, α approaches zero and this in turn tells us that ω_d approaches ω_0. When α is zero, the maximum amplitude of the voltage remains constant; thus we have a sustained oscillation at ω_0 rad/s. In Example 8.3, if R is increased to infinity, the solution for $v(t)$ becomes

$$v(t) = 98 \sin 1000t \text{ V}, \qquad t \geq 0.$$

Thus in this case we have a sustained oscillation where the maximum amplitude of the voltage is 98 V and the frequency of oscillation is 1000 rad/s.

The *period* of the oscillation is the length of time it takes the response to go through one complete cycle—say, from one positive maximum value to the next positive maximum value. This time interval, denoted as T_d, can be calculated directly from ω_d—specifically,

$$\omega_d = \frac{2\pi}{T_d} \text{ rad/s} \tag{8.37}$$

or

$$T_d = \frac{2\pi}{\omega_d} \text{ s}. \tag{8.38}$$

The reciprocal of the period of oscillation is known as the *frequency* and is measured in cycles/second, or hertz:

$$f_d = \frac{1}{T_d} \text{ Hz}. \tag{8.39}$$

In Example 8.3, the period of oscillation is

$$T_d = \frac{2\pi}{100\sqrt{96}} = 6.41 \text{ ms}$$

and the frequency of oscillation in hertz is

$$f_d = \frac{10^3}{6.41} = 155.94 \text{ Hz.}$$

We can now describe the difference between an underdamped and an overdamped response. In an underdamped system, the response oscillates or "bounces" about its final value. This oscillation is also referred to as *ringing*. In an overdamped system, the response approaches its final value without ringing or in what is sometimes described as a "sluggish" manner.

DRILL EXERCISE
8.3

A 10-mH inductor, a 1-μF capacitor, and a variable resistor are connected in parallel as shown in Fig. 8.1. The resistor is adjusted so that the roots of the characteristic equation are $-8000 \pm j6000$ sec^{-1}. The initial voltage on the capacitor is 10 V and the initial current in the inductor is 80 mA. Find (a) R; (b) $dv(0)/dt$; (c) B_1 and B_2 in the solution for v; and (d) $i_L(t)$.

Ans. (a) 62.5 Ω;   (b) $-240{,}000$ V/s;   (c) $B_1 = 10$, $B_2 = -80/3$;   (d) $i_L(t) = 10e^{-8000t}[8 \cos 6000t + (82/3) \sin 6000t]$ mA when $t \geq 0$.

8.3(c) Critically Damped Voltage Response   The second-order circuit in Fig. 8.1 is critically damped when $\omega_0^2 = \alpha^2$, or $\omega_0 = \alpha$. When the circuit is critically damped, the response is just on the verge of oscillating. For the critically damped case, the two roots of the characteristic equation are real and equal, that is,

$$s_1 = s_2 = -\alpha = -\frac{1}{2RC}. \tag{8.40}$$

When the roots of the characteristic equation are real and equal, the solution for the voltage no longer has the form given by Eq. (8.18). We can see that Eq. (8.18) breaks down because if $s_1 = s_2 = -\alpha$, it predicts that

$$\begin{aligned} v &= (A_1 + A_2)e^{-\alpha t} \\ &= A_0 e^{-\alpha t}, \end{aligned} \tag{8.41}$$

where A_0 is an arbitrary constant. Equation (8.41) cannot satisfy two independent initial conditions $[V_0, I_0]$ with only one arbitrary constant A_0. Remember that α is fixed by the circuit parameters R and C.

Our dilemma can be traced back to the assumption that the solution will be of the form given by Eq. (8.18). It turns out that in the case where the roots of the characteristic equation are equal, the solution for the differential equation takes the form

$$v(t) = D_1 t e^{-\alpha t} + D_2 e^{-\alpha t} \tag{8.42}$$

Thus in the case of a repeated root, the solution involves a simple exponential term plus the product of a linear term and an exponential term. The justification of Eq. (8.42) is left for an introductory course in differential equations. Our goal is to show how to use this solution in order to satisfy the differential equation in the special case in which $s_1 = s_2 = -\alpha$.

It follows directly from Eq. (8.42) that

$$\frac{dv}{dt} = e^{-\alpha t}[(1 - \alpha t)D_1 - \alpha D_2]. \tag{8.43}$$

When we combine Eqs. (8.42) and (8.43) considering the constraints imposed by the circuit, we get

$$v(0) = D_2 = V_0, \tag{8.44}$$

$$\frac{dv}{dt}(0) = D_1 - \alpha D_2 = -\frac{1}{C}\left(\frac{V_0}{R} + I_0\right). \tag{8.45}$$

Equations (8.44) and (8.45) can be solved for D_1 and D_2:

$$D_1 = \alpha V_0 - \frac{1}{C}\left(\frac{V_0}{R} + I_0\right)$$

$$= -\alpha(V_0 + 2I_0 R) \tag{8.46}$$

and

$$D_2 = V_0. \tag{8.47}$$

Example 8.4 illustrates the calculations involved in finding the critically damped response of the parallel *RLC* circuit.

Example 8.4   a) For the circuit in Example 8.3, find the value of R that will result in a critically damped voltage response.

b) Calculate $v(t)$ for $t \geq 0$.

c) Plot $v(t)$ vs. t for $0 \leq t \leq 7$ ms.

Solution   a) From Example 8.3 we know that $\omega_0^2 = 10^6$. Therefore for critical damping

$$\alpha = 10^3 = \frac{1}{2RC},$$

or

$$R = \frac{10^6}{(2000)(0.125)} = 4000 \ \Omega.$$

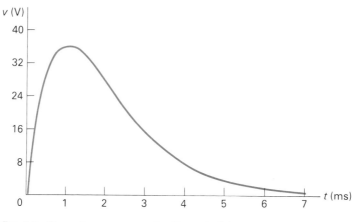

Fig. 8.7 The voltage response for Example 8.4.

b) From Eq. (8.46) we have

$$D_1 = -\alpha(V_0 + 2I_0 R)$$
$$= -1000[0 + 2(-12.25 \cdot 10^{-3})(4000)]$$
$$= 98,000,$$

and from Eq. (8.47),

$$D_2 = V_0 = 0.$$

When we substitute these values for α, D_1, and D_2 into Eq. (8.42) we get

$$v(t) = 98,000t \cdot e^{-1000t}, \qquad t \geq 0.$$

c) A plot of $v(t)$ vs. t in the interval $0 \leq t \leq 7$ ms is shown in Fig. 8.7. ■

DRILL EXERCISE
8.4

The resistor in the circuit shown in Fig. 8.1 is adjusted for critical damping. The inductance and capacitance values are 0.4 H and 10 μF. The initial energy stored in the circuit is 25 mJ and is distributed equally between the inductor and capacitor. Find (a) R; (b) V_0; (c) I_0; (d) D_1 and D_2 in the solution for v; and (e) i_R, $t \geq 0$.

Ans. (a) 100 Ω; (b) 50 V; (c) 250 mA; (d) $-50,000$, 50; (e) $i_R = -500t \cdot e^{-500t} + 0.50e^{-500t}$ A, $t \geq 0$.

We conclude our discussion of the natural response of the parallel *RLC* circuit with a brief summary of the results. The first step in finding the natural response is to calculate the roots of the characteristic equation. Once we have determined the roots, we know immediately whether the response is overdamped, underdamped, or critically damped.

If the roots are real and distinct $(\omega_0^2 < \alpha^2)$, the response is overdamped and the voltage is given by the expression

$$v(t) = A_1 e^{s_1 t} + A_2 e^{s_2 t},$$

where

$$s_1 = -\alpha + \sqrt{\alpha^2 - \omega_0^2},$$
$$s_2 = -\alpha - \sqrt{\alpha^2 - \omega_0^2},$$
$$A_1 = \frac{V_0 s_2 + (1/C)[(V_0/R) + I_0]}{(s_2 - s_1)},$$

and

$$A_2 = -\frac{V_0 s_1 + (1/C)[(V_0/R) + I_0]}{(s_2 - s_1)}.$$

If the roots are complex $(\omega_0^2 > \alpha^2)$, the response is underdamped and the voltage is given by the expression

$$v(t) = B_1 e^{-\alpha t} \cos \omega_d t + B_2 e^{-\alpha t} \sin \omega_d t,$$

where

$$\alpha = \frac{1}{2RC},$$
$$\omega_d = \sqrt{\omega_0^2 - \alpha^2},$$
$$B_1 = V_0,$$

and

$$B_2 = -\frac{\alpha}{\omega_d}(V_0 + 2I_0 R).$$

If the roots of the characteristic equation are real and equal $(\omega_0^2 = \alpha^2)$, the voltage response is given by the expression

$$v(t) = D_1 t e^{-\alpha t} + D_2 e^{-\alpha t},$$

where

$$\alpha = \frac{1}{2RC},$$
$$D_1 = -\alpha(V_0 + 2I_0 R),$$
$$D_2 = V_0.$$

8.4 THE STEP RESPONSE OF A PARALLEL *RLC* CIRCUIT

Finding the step response of a parallel *RLC* circuit involves finding the voltage that appears across the parallel branches, or the current that appears in the individual branches, as a result of the sudden application of a constant, or dc,

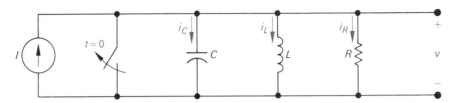

Fig. 8.8 A circuit describing the step response of the parallel *RLC* circuit.

current source. There may or may not be energy stored in the circuit at the time when the current source is applied. Our problem is represented by the circuit in Fig. 8.8.

In order to develop a general approach to finding the step response of a second-order circuit—that is, a circuit described by a second-order differential equation—we will focus on finding the current in the inductive branch (i_L). This current is of particular interest because it does not approach zero as t increases. Thus after the switch has been open for a long time, the inductor current will equal the dc source current I. Since we wish to focus on the technique for finding the step response, we will assume that the initial energy stored in the circuit is zero. This will simplify our calculations and make it easier to concentrate on how to find the step response. (The introduction of initially stored energy will enter into calculations involving the determination of the arbitrary constants, but it will not alter the basic thought process involved in finding the step response.)

To find the inductor current i_L, we must solve a second-order differential equation equated to the forcing function I. We can derive this differential equation as follows. From Kirchhoff's current law, we have

$$i_L + i_R + i_C = I,$$

or

$$i_L + \frac{v}{R} + C\frac{dv}{dt} = I. \qquad \textbf{(8.48)}$$

Now we use the fact that

$$v = L\frac{di_L}{dt} \qquad \textbf{(8.49)}$$

and therefore.

$$\frac{dv}{dt} = L\frac{d^2i_L}{dt^2}. \qquad \textbf{(8.50)}$$

When we substitute Eqs. (8.49) and (8.50) into Eq. (8.48), we get

$$i_L + \frac{L}{R}\frac{di_L}{dt} + LC\frac{d^2i_L}{dt^2} = I. \qquad \textbf{(8.51)}$$

For convenience, we divide through by LC and rearrange the terms as follows:

$$\frac{d^2 i_L}{dt^2} + \frac{1}{RC}\frac{di_L}{dt} + \frac{i_L}{LC} = \frac{I}{LC}. \tag{8.52}$$

When we compare Eq. (8.52) with Eq. (8.3), we see that our problem is altered by the presence of a nonzero term on the right-hand side of the equation. Before showing how we can solve Eq. (8.52) by a direct approach, let us obtain the solution indirectly. Once we know what the solution of Eq. (8.52) is, it is easier to explain the direct approach.

We can solve for i_L indirectly by first finding the voltage v. We can find v by the techniques introduced in Section 8.3 because the differential equation that v must satisfy is identical to Eq. (8.3). To see this, we simply return to Eq. (8.48) and express i_L as a function of v; thus

$$\frac{1}{L}\int_0^t v \, d\tau + \frac{v}{R} + C\frac{dv}{dt} = I. \tag{8.53}$$

Now when Eq. (8.53) is differentiated once with respect to t, the right-hand side reduces to zero since I is a constant. Thus

$$\frac{v}{L} + \frac{1}{R}\frac{dv}{dt} + C\frac{d^2 v}{dt^2} = 0,$$

or

$$\frac{d^2 v}{dt^2} + \frac{1}{RC}\frac{dv}{dt} + \frac{v}{LC} = 0. \tag{8.54}$$

As we discussed in Section 8.3, the solution for v will depend on the roots of the characteristic equation. Thus the three possible solutions are

$$v = A_1 e^{s_1 t} + A_2 e^{s_2 t}, \tag{8.55}$$

$$v = B_1 e^{-\alpha t} \cos \omega_d t + B_2 e^{-\alpha t} \sin \omega_d t, \tag{8.56}$$

and

$$v = D_1 t e^{-\alpha t} + D_2 e^{-\alpha t}. \tag{8.57}$$

To find the three possible solutions for i_L we substitute Eqs. (8.55)–(8.57) into Eq. (8.48). You should be able to verify that when this has been done, the three solutions for i_L will be of the forms

$$i_L = I + A_1' e^{s_1 t} + A_2' e^{s_2 t}, \tag{8.58}$$

$$i_L = I + B_1' e^{-\alpha t} \cos \omega_d t + B_2' e^{-\alpha t} \sin \omega_d t, \tag{8.59}$$

and

$$i_L = I + D_1' t e^{-\alpha t} + D_2' e^{-\alpha t}, \tag{8.60}$$

where A_1', A_2', B_1', B_2', D_1', and D_2' are arbitrary constants.

In each case, the primed constants can be found in terms of the arbitrary constants associated with the voltage solution (see Problem 8.9). However, this is a cumbersome approach. It is much easier to find the primed constants directly in terms of the initial values of the response function. For the circuit under discussion, we would find the primed constants from $i_L(0)$ and $di_L(0)/dt$. We will illustrate this approach in the examples that follow.

It is important to note at this point that the solution for the second-order differential equation with a constant forcing function is equal to the forced response plus a response function that is identical in *form* to the natural response. Thus we can always write the solution for the step response in the form

$$i = I_f + \left\{\begin{matrix} \text{function of the same form as} \\ \text{the natural response} \end{matrix}\right\} \tag{8.61}$$

or

$$v = V_f + \left\{\begin{matrix} \text{function of the same form} \\ \text{as the natural response} \end{matrix}\right\}, \tag{8.62}$$

where I_f and V_f represent the final value of the response function. The final value may be zero, as, for example, the final value of the voltage v in the circuit in Fig. 8.8 was.

The technique of finding the step response of a parallel *RLC* circuit is illustrated in the following examples.

Example 8.5   The initial energy stored in the circuit in Fig. 8.8 is zero. At $t = 0$, a dc current source of 24 mA is applied to the circuit. The circuit elements are a 400-Ω resistor, a 25-mH inductor, and a 25-nF capacitor.

a) What is the initial value of i_L?

b) What is the initial value of di_L/dt?

c) What are the roots of the characteristic equation?

d) What is the numerical expression for $i_L(t)$ when $t \geq 0$?

Solution   a) Since there is no energy stored in the circuit prior to applying the dc current source, the initial current in the inductor is zero. The inductor will prohibit an instantaneous change in inductor current; therefore $i_L(0)$ is zero immediately after the switch has been opened.

b) The initial voltage on the capacitor is zero before the switch has been opened; therefore it will be zero immediately after the switch has been opened. Now since $v = L\, di_L/dt$, it follows that

$$\frac{di_L}{dt}(0) = 0.$$

c) From the circuit elements, we have

$$\omega_0^2 = \frac{1}{LC} = \frac{10^{12}}{(25)(25)} = 16 \times 10^8$$

and

$$\alpha = \frac{1}{2RC} = \frac{10^9}{(2)(400)(25)} = 5 \times 10^4,$$

or

$$\alpha^2 = 25 \times 10^8.$$

Since $\omega_0^2 < \alpha^2$, the roots of the characteristic equation are real and distinct; thus

$$s_1 = -5 \times 10^4 + 3 \times 10^4 = -20,000 \text{ s}^{-1}$$

and

$$s_2 = -5 \times 10^4 - 3 \times 10^4 = -80,000 \text{ s}^{-1}.$$

d) Because the roots of the characteristic equation are real and distinct, the inductor current response will be overdamped. Thus $i_L(t)$ will have the form given by Eq. (8.58), namely,

$$i_L = I + A_1' e^{s_1 t} + A_2' e^{s_2 t}.$$

It follows directly from this solution that

$$i_L(0) = I + A_1' + A_2' = 0$$

and

$$\frac{di_L}{dt}(0) = s_1 A_1' + s_2 A_2' = 0.$$

Solving for A_1' and A_2' gives us

$$A_1' = \frac{-s_2 I}{s_2 - s_1} = -\frac{4}{3}(24 \times 10^{-3}) = -32 \text{ mA}$$

and

$$A_2' = \frac{s_1 I}{s_2 - s_1} = \frac{1}{3}(24 \times 10^{-3}) = 8 \text{ mA}.$$

The numerical solution for $i_L(t)$ is

$$i_L(t) = [24 - 32e^{-20,000t} + 8e^{-80,000t}] \text{ mA}, \qquad t \geq 0. \quad \blacksquare$$

Example 8.6 The resistor in the circuit of Example 8.5 is increased to 625 Ω. Find $i_L(t)$ for $t \geq 0$.

Solution Since L and C remain fixed, ω_0^2 has the same value as in Example 8.4, that is, $\omega_0^2 = 16 \times 10^8$. Increasing R to 625 Ω decreases α to 3.2×10^4. Since $\omega_0^2 > \alpha^2$, the roots of the characteristic equation are complex. Thus we have

$$s_1 = -3.2 \times 10^4 + j2.4 \times 10^4 \text{ s}^{-1}$$

and

$$s_2 = -3.2 \times 10^4 - j2.4 \times 10^4 \text{ s}^{-1}.$$

The current response is now underdamped and given by Eq. (8.59); thus

$$i_L(t) = I + B_1' e^{-\alpha t} \cos \omega_d t + B_2' e^{-\alpha t} \sin \omega_d t.$$

For the problem at hand, α is 32,000 s^{-1}, ω_d is 24,000 rad/s, and I = 24 mA.
As in Example 8.5, B_1' and B_2' are determined from the initial conditions; thus

$$i_L(0) = I + B_1' = 0$$

and

$$\frac{di_L}{dt}(0) = \omega_d B_2' - \alpha B_1' = 0.$$

It follows that

$$B_1' = -I = -24 \times 10^{-3} = -24 \text{ mA}$$

and

$$B_2' = \frac{\alpha}{\omega_d}(-I) = \frac{4}{3}(-24) = -32 \text{ mA}.$$

The numerical solution for $i_L(t)$ is

$$i_L(t) = (24 - 24e^{-32,000t} \cos 24,000t - 32e^{-32,000t} \sin 24,000t) \text{ mA}, \quad t \geq 0. \quad \blacksquare$$

Example 8.7   The resistor in the circuit in Example 8.5 is set at 500 Ω. Find $i_L(t)$ for $t \geq 0$.

Solution   We know that ω_0^2 remains at 16×10^8. With R set at 500 Ω, α becomes 4 $\times$ 10^4 s^{-1}. This corresponds to critical damping and therefore the solution for $i_L(t)$ takes the form of Eq. (8.60):

$$i_L(t) = I + D_1' t e^{-\alpha t} + D_2' e^{-\alpha t}.$$

Again D_1' and D_2' are computed from initial conditions; thus

$$i_L(0) = I + D_2' = 0$$

and

$$\frac{di_L}{dt}(0) = D_1' - \alpha D_2' = 0.$$

It follows that

$$D_1' = -\alpha I = -40,000(24) = -960,000 \text{ mA},$$
$$D_2' = -I = -24 \text{ mA}.$$

The numerical expression for $i_L(t)$ is

$$i_L(t) = (24 - 960,000 t e^{-40,000t} - 24e^{-40,000t}) \text{ mA}, \quad t \geq 0. \quad \blacksquare$$

Example 8.8   a) Plot on a single graph, over a range from 0 to 220 μs, the overdamped, underdamped, and critically damped responses derived in Examples 8.5–8.7, respectively.

b) Use the plots of part (a) to find the time it takes i_L to reach 90% of its final value.

c) On the basis of the results obtained in part (b), which response would you describe as being the most "sluggish"?

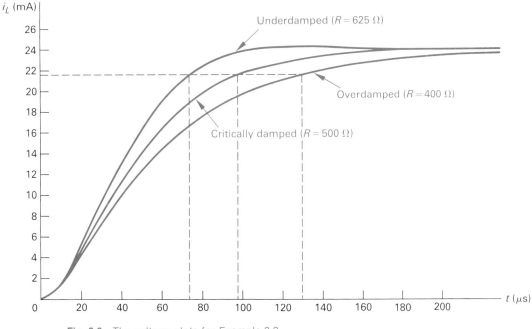

Fig. 8.9   The voltage plots for Example 8.3.

Solution   a) See Fig. 8.9.

b) Since the final value of i_L is 24 mA, we can read the times off the plots corresponding to i_L = 21.6 mA. Thus t_{od} = 130 μs, t_{cd} = 97 μs, and t_{ud} = 74 μs.

c) The overdamped response is the most "sluggish" since it takes the longest time for i_L to reach 90% of its final value.   ■

Example 8.9   Energy is stored in the circuit of Example 8.7 at the instant when the dc current source is applied. The initial current in the inductor is 29 mA and the initial voltage across the capacitor is 50 V. Find (a) $i_L(0)$; (b) $di_L(0)/dt$; (c) $i_L(t)$ for $t \geq 0$.

Solution   a) Since there cannot be an instantaneous change of current in an inductor, the initial value of i_L in the first instant after the dc current source has been applied must be 29 mA.

b) The capacitor will hold the initial voltage across the inductor to 50 V. Therefore

$$L\frac{di_L(0)}{dt} = 50,$$

or

$$\frac{di_L}{dt}(0) = \frac{50}{25} \times 10^3 = 2000 \text{ A/s}.$$

c) From the solution of Example 8.7, we know that the current response is critically damped. Thus

$$i_L(t) = I + D_1' t e^{-\alpha t} + D_2' e^{-\alpha t},$$

where

$$\alpha = \frac{1}{2RC} = 40,000 \text{ s}^{-1} \quad \text{and} \quad I = 24 \text{ mA}.$$

We obtain the constants D_1' and D_2' from the initial conditions. Thus

$$i_L(0) = I + D_2' = 29 \text{ mA},$$

from which we get

$$D_2' = 29 - 24 = 5 \text{ mA}.$$

The solution for D_1' is

$$\frac{di_L}{dt}(0) = D_1' - \alpha D_2' = 2000$$

or

$$\begin{aligned}
D_1' &= 2000 + \alpha D_2' \\
&= 2000 + (40,000)(5 \times 10^{-3}) \\
&= 2200 \text{ A/s} = 2.2 \times 10^6 \text{ mA/s}.
\end{aligned}$$

Thus the numerical expression for $i_L(t)$ is

$$i_L(t) = (24 + 2.2 \times 10^6 t e^{-40,000t} + 5e^{-40,000t}) \text{ mA}, \qquad t \geq 0. \quad \blacksquare$$

DRILL EXERCISE
8.5 In the circuit shown in Fig. 8.8, $R = 250 \ \Omega$, $L = 0.32$ H, $C = 2 \ \mu$F, $I_0 = 0.5$ A, $V_0 = 80$ V, and $I = -1.5$ A. Find (a) $i_R(0^+)$; (b) $i_C(0^+)$; (c) $di_L(0^+)/dt$; (d) s_1, s_2; and (e) $i_L(t)$ for $t \geq 0$.

Ans. (a) 320 mA; (b) -2.32 A; (c) 250 A/s; (d) $(-1000 + j750)$ s^{-1}, $(-1000 - j750)$ s^{-1}; (e) $i_L(t) = [-1.5 + 2e^{-1000t} (\cos 750t + 1.5 \sin 750t)]$ A for $t \geq 0$.

8.5 THE NATURAL AND STEP RESPONSES OF A SERIES *RLC* CIRCUIT

In Section 8.1 we defined the natural and step responses of a series *RLC* circuit with reference to the circuits in Figs. 8.3 and 8.4. The procedures for finding the natural or step responses of a series *RLC* circuit are the same as those used to find the natural or step responses of a parallel *RLC* circuit, since both circuits are described by differential equations that have the same form. For example, the differential equation that describes the current in the circuit in Fig. 8.3 has the same form as the differential equation that describes the voltage in the circuit in Fig. 8.1. We can show this by summing the voltages around the closed path in the circuit in Fig. 8.3; thus

$$Ri + L\frac{di}{dt} + \frac{1}{C}\int_0^t i\,d\tau + V_0 = 0. \tag{8.63}$$

We now differentiate Eq. (8.63) once with respect to t to get

$$R\frac{di}{dt} + L\frac{d^2i}{dt^2} + \frac{i}{C} = 0, \tag{8.64}$$

which we can rearrange as

$$\frac{d^2i}{dt^2} + \frac{R}{L}\frac{di}{dt} + \frac{i}{LC} = 0. \tag{8.65}$$

When we compare Eq. (8.65) with Eq. (8.3), we see that they have the same form. Therefore, to find the solution of Eq. (8.65), we follow the same thought process that led us to the solution of Eq. (8.3).

It follows directly from Eq. (8.65) that the characteristic equation for the series *RLC* circuit is

$$s^2 + \frac{R}{L}s + \frac{1}{LC} = 0. \tag{8.66}$$

The roots of the characteristic equation are

$$s_{1,2} = -\frac{R}{2L} \pm \sqrt{\left(\frac{R}{2L}\right)^2 - \frac{1}{LC}} \tag{8.67}$$

or

$$s_{1,2} = -\alpha \pm \sqrt{\alpha^2 - \omega_0^2}. \tag{8.68}$$

The neper frequency (α) for the series *RLC* circuit is

$$\alpha = \frac{R}{2L}\ \text{s}^{-1}, \tag{8.69}$$

whereas the expression for the resonant radian frequency is the same as that of the parallel *RLC* circuit:

$$\omega_0 = \frac{1}{\sqrt{LC}} \text{ rad/s.} \tag{8.70}$$

The current response will be overdamped, underdamped, or critically damped according to whether $\omega_0^2 < \alpha^2$, $\omega_0^2 > \alpha^2$, or $\omega_0^2 = \alpha^2$, respectively. Thus the three possible solutions for the current are

$$i(t) = A_1 e^{s_1 t} + A_2 e^{s_2 t}, \qquad \text{(overdamped)} \tag{8.71}$$

$$i(t) = B_1 e^{-\alpha t} \cos \omega_d t + B_2 e^{-\alpha t} \sin \omega_d t, \qquad \text{(underdamped)} \tag{8.72}$$

$$i(t) = D_1 t e^{-\alpha t} + D_2 e^{-\alpha t}. \qquad \text{(critically damped)} \tag{8.73}$$

We also note that once we know the natural current response, we can find the natural voltage response across any circuit element.

To verify that the procedure for finding the step response of a series *RLC* circuit is the same as that for finding the step response of a parallel *RLC* circuit, we show that the differential equation that describes the capacitor voltage in the circuit in Fig. 8.10 has the same form as the differential equation that describes the inductor current in the circuit in Fig. 8.8. For convenience, we assume that zero energy is stored in the circuit at the instant when the switch is closed.

Applying Kirchhoff's voltage law to the circuit of Fig. 8.10 gives us

$$V = Ri + L\frac{di}{dt} + v_C. \tag{8.74}$$

The current in the circuit (i) is related to the capacitor voltage (v_C) by the expression

$$i = C\frac{dv_C}{dt}, \tag{8.75}$$

from which it follows that

$$\frac{di}{dt} = C\frac{d^2 v_C}{dt^2}. \tag{8.76}$$

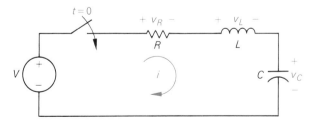

Fig. 8.10   A circuit used in showing the step response of the series *RLC* circuit.

We can substitute Eqs. (8.75) and (8.76) into Eq. (8.74) and write the resulting expression as

$$\frac{d^2 v_C}{dt^2} + \frac{R}{L}\frac{dv_C}{dt} + \frac{v_C}{LC} = \frac{V}{LC}. \tag{8.77}$$

Equation (8.77) has the same form as Eq. (8.52); therefore the procedure for finding v_C parallels that for finding i_L. The overdamped, underdamped, and critically damped solutions for v_C will be

$$v_C = V_f + A_1' e^{s_1 t} + A_2' e^{s_2 t}, \qquad \text{(overdamped)} \tag{8.78}$$

$$v_C = V_f + B_1' e^{-\alpha t} \cos \omega_d t + B_2' e^{-\alpha t} \sin \omega_d t, \qquad \text{(underdamped)} \tag{8.79}$$

$$v_C = V_f + D_1' t e^{-\alpha t} + D_2' e^{-\alpha t}, \qquad \text{(critically damped)} \tag{8.80}$$

where V_f is the final value of v_C. It follows directly from the circuit in Fig. 8.10 that the final value of v_C will be the dc source voltage V.

Examples 8.10 and 8.11 illustrate the mechanics of finding the natural and step responses of a series *RLC* circuit.

Example 8.10   The 0.1-μF capacitor in the circuit in Fig. 8.11 is charged to 100 V. At $t = 0$ the capacitor is discharged through a series combination of a 100-mH inductor and a 560-Ω resistor.

a) Find $i(t)$ for $t \geq 0$.

b) Find $v_C(t)$ for $t \geq 0$.

Solution   a) The first step to finding $i(t)$ is to calculate the roots of the characteristic equation. For the given element values,

$$\omega_0^2 = \frac{1}{LC} = \frac{(10^3)(10^6)}{(100)(0.1)} = 10^8,$$

$$\alpha = \frac{R}{2L} = \frac{560}{2(100)} \times 10^3 = 2800.$$

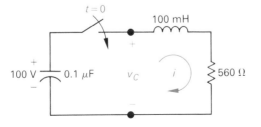

Fig. 8.11   The circuit for Example 8.10.

Next we compare α^2 to ω_0^2 and note that $\omega_0^2 > \alpha^2$, since

$$\alpha^2 = 7.84 \times 10^6 = 0.0784 \times 10^8.$$

At this point, we know that the response is underdamped and the solution for $i(t)$ is of the form

$$i(t) = B_1 e^{-\alpha t} \cos \omega_d t + B_2 e^{-\alpha t} \sin \omega_d t,$$

where $\alpha = 2800$ s^{-1} and $\omega_d = 9600$ rad/s.

The numerical values of B_1 and B_2 are found from the initial conditions. The inductor current is zero before the switch has been closed and, hence, it is zero immediately after the switch has been closed; therefore

$$i(0) = 0 = B_1.$$

To find B_2, we evaluate $di(0)/dt$. From the circuit we note that since $i(0) = 0$ immediately after the switch has been closed, there will be no voltage drop across the resistor and therefore, the initial voltage on the capacitor will appear across the terminals of the inductor. This leads us to the expression.

$$L \frac{di(0)}{dt} = V_0$$

or

$$\frac{di(0)}{dt} = \frac{V_0}{L} = \frac{100}{100} \times 10^3 = 1000 \text{ A/s}.$$

Using the fact that B_1 is zero, we have from the solution

$$\frac{di}{dt} = 400 B_2 e^{-2800t} (24 \cos 9600t - 7 \sin 9600t).$$

Thus

$$\frac{di(0)}{dt} = 9600 B_2$$

and

$$B_2 = \frac{1000}{9600} \cong 0.1042 \text{ A}.$$

The solution for $i(t)$ is

$$i(t) = 0.1042 e^{-2800t} \sin 9600t \text{ A}, \qquad t \geq 0.$$

b) To find $v_C(t)$, we can use either of the following relationships:

$$v_C = -\frac{1}{C} \int_0^t i \, d\tau + 100$$

or

$$v_C = iR + L \frac{di}{dt}.$$

Whichever expression is used (the second is recommended) the result is

$$v_C(t) = [100 \cos 9600t + 29.17 \sin 9600t]e^{-2800t} \text{ V}, \qquad t \geq 0. \quad \blacksquare$$

Example 8.11 There is no energy stored in the 100-mH inductor or the 0.4-μF capacitor at the time when the switch in the circuit in Fig. 8.12 is closed. Find $v_C(t)$ for $t \geq 0$.

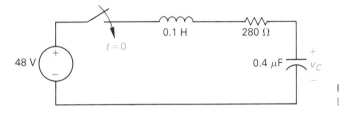

Fig. 8.12 The circuit for Example 8.11.

Solution The roots of the characteristic equation are

$$s_1 = -\frac{280}{0.2} + \sqrt{\left(\frac{280}{0.2}\right)^2 - \frac{10^6}{(0.1)(0.4)}}$$

$$= (-1400 + j4800) \text{ s}^{-1},$$

$$s_2 = (-1400 - j4800) \text{ s}^{-1}.$$

Since the roots are complex, the voltage response is underdamped; thus

$$v_C(t) = 48 + B_1' e^{-1400t} \cos 4800t + B_2' e^{-1400t} \sin 4800t, \qquad t \geq 0.$$

Since there is no initial energy stored in the circuit, both $v_C(0)$ and $dv_C(0)/dt$ are zero. We have, then,

$$v_C(0) = 0 = 48 + B_1'$$

and

$$\frac{dv_C(0)}{dt} = 0 = 4800B_2' - 1400B_1'.$$

Solving for B_1' and B_2' yields

$$B_1' = -48 \text{ V} \quad \text{and} \quad B_2' = -14 \text{ V}.$$

Therefore the solution for $v_C(t)$ is

$$v_C(t) = [48 - 48e^{-1400t} \cos 4800t - 14e^{-1400t} \sin 4800t] \text{ V}, \qquad t \geq 0. \quad \blacksquare$$

DRILL EXERCISE
8.6 The switch in the circuit shown has been in position a for a long time. At $t = 0$ it moves to position b. Find (a) $i(0^+)$; (b) $v_C(0^+)$; (c) $di(0^+)/dt$; (d) s_1, s_2; and (e) $i(t)$ for $t \geq 0$.

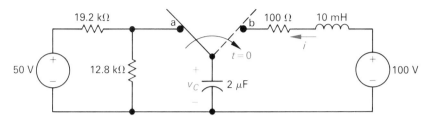

Ans. (a) 0; (b) 20 V; (c) 8000 A/s; (d) $(-5000 + j5000)$ s^{-1}, $(-5000 - j5000)$ s^{-1}; (e) $i(t) = (1.6e^{-5000t} \sin 5000t)$ A for $t \geq 0$.

DRILL EXERCISE 8.7

Find $v_C(t)$ for $t \geq 0$ for the circuit in Drill Exercise 8.6.

Ans. $v_C = [100 - 80e^{-5000t} (\cos 5000t + \sin 5000t)]$ V for $t \geq 0$.

8.6 SUMMARY

We described both the parallel and series *RLC* circuits by second-order differential equations. The first step in the solution of these differential equations is to determine the roots of the characteristic equation. For the parallel structure, the roots are

$$s_{1,2} = -\frac{1}{2RC} \pm \sqrt{\left(\frac{1}{2RC}\right)^2 - \frac{1}{LC}},$$

while for the series circuit the roots are

$$s_{1,2} = -\frac{R}{2L} \pm \sqrt{\left(\frac{R}{2L}\right)^2 - \frac{1}{LC}}.$$

The nature of the characteristic roots tells us whether the solution for any voltage or current in the circuit is overdamped, underdamped, or critically damped. If the roots are real and distinct, the response is overdamped; if the roots are complex, the response is underdamped; and if the roots are real and equal, the response is critically damped.

The natural response of either circuit structure involves finding the time domain expressions of voltages and/or currents that are generated by the release of energy stored in either the inductor, the capacitor, or both. In all cases, finding the natural response involves finding two arbitrary constants whose values depend on the initial value of the variable and the initial value of the first derivative of the variable. The three possible natural response functions take the following forms:

$$f(t) = A_1 e^{s_1 t} + A_2 e^{s_2 t}, \qquad \text{(overdamped)}$$
$$f(t) = B_1 e^{-\alpha t} \cos \omega_d t + B_2 e^{-\alpha t} \sin \omega_d t, \qquad \text{(underdamped)}$$
$$f(t) = D_1 t e^{-\alpha t} + D_2 e^{-\alpha t}, \qquad \text{(critically damped)}$$

where $f(t)$ is either a voltage response or a current response.

The step response of either circuit structure involves finding the time domain expressions of voltages and/or currents that are generated by the sudden application of a dc source. There may or may not be energy stored in the circuit at the time when the dc source is applied. The roots of the characteristic equation are unaffected by the presence of the dc source. The three possible step-function responses take the following forms:

$$f(t) = f_F + A_1' e^{s_1 t} + A_2' e^{s_2 t}, \qquad \text{(overdamped)}$$
$$f(t) = f_F + B_1' e^{-\alpha t} \cos \omega_d t + B_2' e^{-\alpha t} \sin \omega_d t, \qquad \text{(underdamped)}$$
$$f(t) = f_F + D_1' t e^{-\alpha t} + D_2' e^{-\alpha t}. \qquad \text{(critically damped)}$$

In these expressions, $f(t)$ represents either a voltage or current response and f_F is the final value of the desired response.

The terms overdamped, underdamped, and critically damped are used to describe the impact of the dissipative element (R) on the response. The effect of R is reflected in the neper frequency, or damping factor α. If α is large compared to the resonant frequency ω_0, the voltage or current approaches its final value without oscillation and this nonoscillatory response is called overdamped. If α is small compared to ω_0, the response oscillates about its final value and this response is called underdamped. The smaller the value of α is, the longer the oscillation persists. If the dissipative element is removed from the circuit, α equals zero and the voltage or current response becomes a sustained oscillation. The critical value of α occurs when $\alpha = \omega_0$; in this case, the response is on the verge of oscillation, and is called the critically damped response.

PROBLEMS

8.1 The resistance, inductance, and capacitance in a parallel RLC circuit are 2.5 kΩ, 15.625 H, and 0.4 μF, respectively.

  a) Calculate the roots of the characteristic equation that describe the voltage response of the circuit.
  b) Will the voltage response be over-, under-, or critically damped?
  c) What value of inductance will yield a pair of complex conjugate roots equal to $-500 \pm j500$ rad/s?
  d) If $L = 15.625$ H and $C = 0.4$ μF, what value of R will result in a critically damped response?

8.2 The initial voltage on the capacitor in the circuit in Fig. 8.1 is zero. The initial current in the inductor is 18 mA. The voltage response for $t \geq 0$ is

$$v(t) = 15e^{-8000t} - 15e^{-2000t} \text{ V}.$$

  a) Determine the numerical values of R, L, C, α, and ω_0.
  b) Calculate $i_R(t)$, $i_C(t)$, and $i_L(t)$ for $t \geq 0$.

8.3 The circuit elements in the circuit in Fig. 8.1 are $R = 2000$ Ω, $C = 0.25$ μF, and $L = 6.25$ H. The initial inductor current is zero and the initial capacitor voltage is -60 V.

  a) Calculate the initial current in each branch of the circuit.

b) Find $v(t)$ for $t \geq 0$.
c) Find $i_L(t)$ for $t \geq 0$.

8.4 The natural voltage response of the circuit in Fig. 8.1 is

$$v = 10e^{-1000t}(4 \cos 5000t - \sin 5000t) \text{ V}, \qquad t \geq 0,$$

when the capacitor is 0.5 μF. Find (a) R; (b) L; (c) V_0; (d) I_0; and (e) $i_L(t)$.

8.5 In the circuit in Fig. 8.1, $R = 5$ kΩ, $L = 8$ H, $C = 0.125 \mu$F, $V_0 = 30$ V, and $I_0 = 6$ mA.

a) Find $v(t)$ for $t \geq 0$.
b) Find the first three values of t for which dv/dt is zero. Let these values of t be denoted t_1, t_2, and t_3.
c) Show that $t_3 - t_1 = T_d$.
d) Show that $t_2 - t_1 = T_d/2$.
e) Calculate $v(t_1)$, $v(t_2)$, and $v(t_3)$.
f) Sketch $v(t)$ vs. t for $0 \leq t \leq t_2$.

8.6 a) Find $v(t)$ for $t \geq 0$ in the circuit in Problem 8.5 if the 5-kΩ resistor is removed from the circuit.
b) Calculate the frequency of $v(t)$ in hertz.
c) Calculate the maximum amplitude of $v(t)$ in volts.

8.7 In the circuit in Fig. 8.1, a 50-mH inductor is shunted by an 80-pF capacitor; the resistor R is adjusted for critical damping; and $V_0 = 5$ V and $I_0 = -0.4$ mA.

a) Calculate the numerical value of R.
b) Calculate $v(t)$ for $t \geq 0$.
c) Plot $v(t)$ vs. t for $0 \leq t \leq 10 \mu$s.
d) What percentage of the total energy stored in the inductor and capacitor remains 5 μs after the initially stored energy has begun to dissipate?

8.8 The resistor in the circuit in Example 8.3 is changed to 3200 Ω.

a) Find the numerical expression for $v(t)$ when $t \geq 0$.
b) Plot $v(t)$ vs. t for the time interval $0 \leq t \leq 7$ ms. Compare this response with that of Example 8.3 ($R = 20$ kΩ) and Example 8.4 ($R = 4$ kΩ). In particular, compare peak values of $v(t)$ and the times when these peak values occur.

8.9 a) Show by substituting the solution given by Eq. (8.55) into Eq. (8.48) that i_L has the form given by Eq. (8.58).
b) From the results of part (a), show that

$$A_1' = -\left(\frac{A_1}{R} + A_1 C s_1\right)$$

and

$$A_2' = -\left(\frac{A_2}{R} + A_2 C s_2\right).$$

8.10 For the circuit in Example 8.5, find for $t \geq 0$ (a) $v(t)$; (b) $i_R(t)$; and (c) $i_C(t)$.

8.11 Use the solution from Problem 8.10(a) to check the expressions derived in Problem 8.9(b).

8.12 For the circuit in Example 8.6, find for $t \geq 0$ (a) $v(t)$ and (b) $i_C(t)$.

8.13 For the circuit in Example 8.7, find $v(t)$ for $t \geq 0$.

8.14 In the circuit in Fig. 8.8, the initial energy stored in the inductor is zero, and the initial energy stored in the capacitor is 153.6 μJ. The circuit elements are $R = 5$ kΩ, $L = 120$ H, and $C = (10/3) \mu$F.

a) Find $i_L(t)$ for $t \geq 0$ if the dc current source is 4 mA.

b) Find the maximum value of $i_L(t)$.

c) Find the time (ms) at which the maximum occurs.

8.15 The initial energy stored in the 0.1-μF capacitor in the circuit in Fig. 8.13 is 45 μJ. The initial energy stored in the inductor is zero. The roots of the characteristic equation that describes the natural behavior of the current i are -2000 s^{-1} and -8000 s^{-1}.

a) Find the numerical values of R and L.

b) Find the numerical values of $i(0)$ and $di(0)/dt$ immediately after the switch has been closed.

c) Find $i(t)$ for $t \geq 0$.

d) How many microseconds after the switch closes does the current reach its maximum value?

e) What is the maximum value of i in mA?

f) Find $v_L(t)$ for $t \geq 0$.

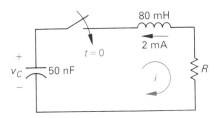

Fig. 8.13   The circuit for Problem 8.15.

8.16 In the circuit in Fig. 8.14, the resistor is adjusted for critical damping. The initial capacitor voltage is zero and the initial inductor current is 2 mA.

a) Find the numerical value of R.

b) Find the numerical values of i and di/dt immediately after the switch is closed.

c) Find $v_C(t)$ for $t \geq 0$.

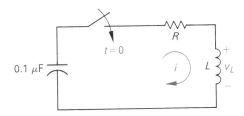

Fig. 8.14   The circuit for Problem 8.16.

8.17 The initial energy stored in the circuit in Fig. 8.15 is zero.

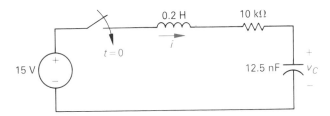

Fig. 8.15   The circuit for Problem 8.17.

a) Find the roots of the characteristic equation.
b) Find the capacitor voltage for $t \geq 0$.
c) Find the inductor current for $t \geq 0$.

8.18 Repeat Problem 8.17 given that the initial voltage on the capacitor is -6 V and the initial value of the inductor current is 3 mA.

8.19 The inductor in the circuit in Problem 8.17 is adjusted until the step response of the inductor current is critically damped.

a) What is the numerical value of L?
b) What is the numerical expression for the inductor current?
c) What is the maximum value of i?
d) How many microseconds after the switch is closed does the maximum current occur?

8.20 Assume that the capacitor voltage in the circuit in Fig. 8.10 is underdamped. Also assume that there is no energy stored in the circuit elements at the time when the switch is closed.

a) Show that $dv_C/dt = (\omega_0^2/\omega_d)Ve^{-\alpha t} \sin \omega_d t$.
b) Show that $dv_C/dt = 0$ when $t = n\pi/\omega_d$, where $n = 0, 1, 2, \ldots$.
c) Let $t_n = n\pi/\omega_d$ and show that $v_C(t_n) = V - V(-1)^n e^{-\alpha n\pi/\omega_d}$.
d) Show that

$$\alpha = \frac{1}{T_d} \ln \frac{[v_C(t_1) - V]}{[v_C(t_3) - V]},$$

where $T_d = t_3 - t_1$.

8.21 The voltage across a 1.6-nF capacitor in the circuit in Fig. 8.10 is described as follows. After the switch has been closed for several seconds, the voltage is constant at 72 V. The first time the voltage exceeds 72 V it reaches a peak of 100.80 V. This occurs $(5\pi/3)$ μs after the switch has been closed. The second time the voltage exceeds 72 V, it reaches a peak of 76.61 V. This second peak occurs 5π μs after the switch has been closed. At the time when the switch is closed, there is no energy stored in either the capacitor or the inductor. Find the numerical values of R and L. (*Hint:* Work Problem 8.20 first.)

8.22 The switch in the circuit in Fig. 8.16 has been in the ON position for a long time. The switch is moved to the OFF position at $t = 0$. Find the numerical expression for $v_C(t)$ when $t \geq 0$.

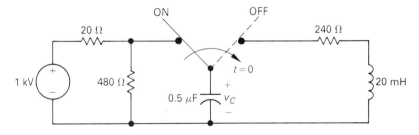

Fig. 8.16   The circuit for Problem 8.22.

8.23 The switch in the circuit in Fig. 8.17 has been closed for a long time. Time is measured from the instant when the switch is opened.

a) Find the numerical expression for $v_L(t)$ when $t \geq 0$.
b) Find the maximum value of $v_L(t)$.

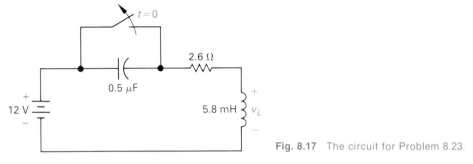

Fig. 8.17 The circuit for Problem 8.23.

8.24 The "make-before-break" switch in the circuit in Fig. 8.18 moves from position a to position b at $t = 0$. Find $v(t)$ for $t \geq 0$.

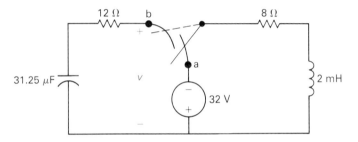

Fig. 8.18 The circuit for Problem 8.24.

9

THE OPERATIONAL AMPLIFIER

9.1 INTRODUCTION

The electronic circuit known as an operational amplifier has become increasingly important in recent years. A detailed analysis of this circuit requires an understanding of such electronic devices as diodes and transistors. You may wonder, then, why we are introducing the circuit before you have achieved an understanding of the electronic components that make up the circuit. There are several reasons. First and foremost is our contention that circuit theory is important because it enables us to design and analyze electrical systems that have important applications and hence commercial value. Second, we can develop an appreciation for how the operational amplifier can be used as a circuit building block by focusing on its terminal behavior. Thus at an introductory level we need not fully understand the operation of the electronic components that govern the terminal behavior. Third, the circuit model of the operational amplifier requires the use of a dependent source. Thus we have a chance to use a dependent source in a practical circuit rather than as an abstract circuit component. Fourth, we can combine the operational amplifier and the passive circuit elements with which we are already familiar to solve differential equations. Since we have just completed our introduction to solving simple circuits that can be described in terms of differential equations, it is appropriate that we now show how electrical components can be used in turn to implement the solutions of such equations. Finally, the operational amplifier plays an important role in the design of the digital multimeter. In order to get some feel for this important laboratory instrument, we must understand the terminal behavior of the operational amplifier.

The fact that we are focusing on the terminal behavior of the operational amplifier implies that we are taking a "black-box" approach to its operation; that is, we are not interested in the internal structure of the amplifier nor the currents and voltages that exist in this internal structure. The important thing to remember is that the internal behavior of the amplifier accounts for the voltage and current constraints imposed at the terminals. (At this introductory level, we ask that you accept these constraints on faith.)

The operational-amplifier circuit first came into existence as a basic building block in the design of analog computers. It was referred to as "operational" because it was used to implement the mathematical operations of integration, differentiation, addition, sign changing, and scaling. In recent years, the range of application has broadened beyond implementing mathematical operations; however, the original name for the circuit persists. Engineers and technicians have a penchant for creating technical jargon, hence the operational amplifier is widely known as an *op amp*.

9.2 OPERATIONAL-AMPLIFIER TERMINALS

Since we are stressing the terminal behavior of the operational amplifier, let us begin by discussing the terminals on a commercially available amplifier. In 1968 Fairchild Semiconductor introduced an operational amplifier that

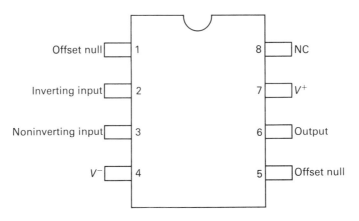

Fig. 9.1 The eight-lead MINIDIP package (top view).

has found widespread acceptance: the μA741. (The μA prefix is used by Fairchild to indicate a microcircuit fabrication of the amplifier.) The amplifier is available in several different packages. For our discussion, we will assume an eight-lead MINIDIP† package. Figure 9.1 shows a top view of the package, and the terminal designations are given alongside the terminals. The terminals of primary interest are:

1. the inverting input,
2. the noninverting input,
3. the output,
4. the positive power supply (V^+), and
5. the negative power supply (V^-).

Before discussing these five terminals, we pause long enough to indicate why the remaining three terminals are of little or no concern. The offset null terminals may be used in an auxiliary circuit to compensate for a degradation in amplifier performance due to aging and due to imperfections that arise in the circuit during its fabrication. However, since the degradation of performance is in most cases negligible, the offset terminals are often unused and play a secondary role in the circuit analysis problem. We will assume that our operational amplifiers require no trimming. Terminal number eight (8) is of no interest simply because it is an unused terminal: NC stands for no connection, which means that the terminal is not connected to the amplifier circuit. (When the μA741 is packaged in a fourteen-lead DIP package, there are seven unused terminals.)

---

† DIP is an abbreviation for Dual In-line Package. This means that the terminals on each side of the package are in line and, at the same time, terminals on opposite sides of the package also line up.

Fig. 9.2   The circuit symbol for an operational amplifier.

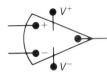

Fig. 9.3 A simplified circuit symbol for the op amp.

Figure 9.2 shows a widely used circuit symbol for the operational amplifier that contains the five terminals of primary interest. Since it is inconvenient to use word labels for the terminals when the operational amplifier is imbedded in a circuit, the terminal designations are simplified in the following way. The noninverting input terminal is labeled with a plus sign (+) and the inverting terminal is labeled with a minus sign (−). The power supply terminals, which are always drawn external to the triangle, are marked V^+ and V^-. The terminal at the apex of the triangular box is always understood to be the output terminal. These simplified designations are summarized in Fig. 9.3.

9.3   TERMINAL VOLTAGES AND CURRENTS

We are now ready to introduce the terminal voltages and currents that are used to describe the behavior of the operational amplifier. Let us begin with the voltage variables. Those used to describe the behavior of an operational amplifier are measured from a common reference node. The voltage variables with their *reference* polarities are shown in Fig. 9.4. All voltages are considered as voltage rises from the common node. This is the same convention that is used in the node-voltage method of analysis. A positive supply voltage (V_{CC}) is connected between V^+ and the common node. A negative supply voltage ($-V_{CC}$) is connected between V^- and the common node. The voltage between the inverting input terminal and the common node is denoted by v_1. The voltage between the noninverting input terminal and the common node is designated as v_2. The voltage between the input terminal and the common node is denoted by v_o.

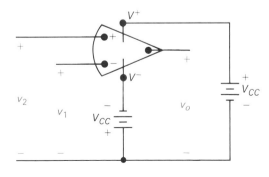

Fig. 9.4   Terminal voltage variables.

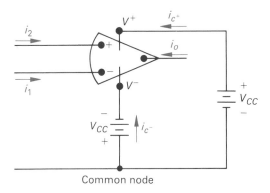

Common node

Fig. 9.5   Terminal current variables.

The current variables with their *reference* directions are shown in Fig. 9.5. In studying that figure, note the following. All the current reference directions are into the terminals of the operational amplifier: i_1 is the current into the inverting input terminal; i_2 is the current into the noninverting input terminal; i_o is the current into the output terminal; i_{c^+} is the current into the positive power supply terminal; i_{c^-} is the current into the negative power supply terminal.

In order to predict the behavior of an operational amplifier when circuit elements are externally connected to its terminals, we must understand the constraints imposed on the terminal voltages and currents by the amplifier itself. Those imposed on the terminal voltages are as follows:

$$v_o = A(v_2 - v_1) \tag{9.1}$$

and

$$-V_{CC} \le v_o \le V_{CC}. \tag{9.2}$$

Equation (9.1) states that the output voltage is proportional to the difference between v_2 and v_1. The proportionality constant A is known as the *open-loop voltage gain*. (The significance of the term "open-loop" will be explained later.) Equation (9.2) states that the output voltage is bounded. In particular, v_o must lie between $\pm V_{CC}$, the power supply voltages. If v_o is at either limiting value, we say that the operational amplifier is saturated. The amplifier is operating in its linear range so long as $v_o < |V_{CC}|$. The significance of Eqs. (9.1) and (9.2) is summarized by the graph shown in Fig. 9.6, where we note that the x-axis variable is $(v_2 - v_1)$, that is, the difference between the two input voltages.

The importance of the voltage constraints lies in knowing the typical numerical values of V_{CC} and A. The dc power supply voltages seldom exceed 20 V and A is rarely less than 10,000, or 10^4. Referring to Fig. 9.6 we see that in the linear range of operation the magnitude of $v_2 - v_1$ must be less than $20/10^4$, or 2 mV. We conclude, then, that when the operational amplifier is operating in its linear region

$$v_1 \approx v_2. \tag{9.3}$$

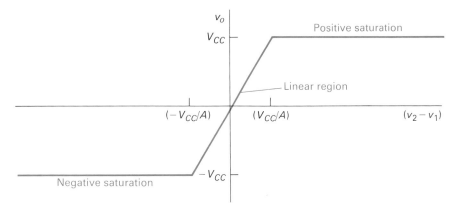

Fig. 9.6 The voltage transfer characteristic of an op amp.

We will soon see that Eq. (9.3) is a valuable aid to predicting the performance of linear op-amp circuits.

From Kirchhoff's current law we know that the sum of the currents entering the operational amplifier is zero; thus

$$i_1 + i_2 + i_o + i_{c^+} + i_{c^-} = 0. \tag{9.4}$$

The constraint imposed on the currents by the operational amplifier is that i_1 and i_2 must be extremely small in comparison to the other terminal currents. In an ideal operational amplifier the input terminal currents are zero. Thus we have

$$i_1 = i_2 \approx 0, \tag{9.5}$$

which tells us that the input resistance to an operational amplifier is very large. Typical values range from hundreds of kilohms to thousands of megohms. We will find Eq. (9.5) very useful in analyzing circuits containing operational amplifiers.

When the constraint given by Eq. (9.5) is substituted into Eq. (9.4) we have

$$i_o = -(i_{c^+} + i_{c^-}). \tag{9.6}$$

The significance of Eq. (9.6) is that even though the curent at the input terminals is negligible, there may still be appreciable current at the output terminal.

Before we start analyzing circuits containing operational amplifiers, let us further simplify our circuit symbol. The first step is to remove the dc power supplies. This can be done by simply indicating the power supply voltage next to the appropriate terminal. The simplified symbol is shown in Fig. 9.7.

We can simplify further when we know that the amplifier is operating within its linear range. In this situation, the dc voltages $\pm V_{CC}$ do not enter

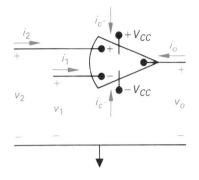

Fig. 9.7 The op-amp symbol with power supplies removed.

into the circuit equations. It is implicitly understood that $v_o < |V_{CC}|$. If we are given linear operation of the op amp, we can remove the power supply terminals from the symbol, as shown in Fig. 9.8. (A word of caution: Because the power supply terminals have been omitted, there is a danger of inferring from the symbol that $i_1 + i_2 + i_o = 0$. We have already noted that this is not the case, that is, $i_1 + i_2 + i_o + i_{c^+} + i_{c^-} = 0$. In other words, the constraint that $i_1 = i_2 \cong 0$ does not imply that $i_o \cong 0$.)

A couple of additional comments are in order. It is not necessary that the positive and negative power supply voltages be equal in magnitude. In the linear operating range, v_o must lie between the two supply voltages. For example, if $V^+ = 15$ V and $V^- = -10$ V, then $-10 \le v_o \le 15$. The open-loop voltage gain (A) is not constant under all operating conditions. At this point in our study, however, we will assume that it is. A discussion of how and why the open-loop gain (A) can change must be delayed until after we understand how the electronic devices and components are used to fabricate the amplifier circuit.

Example 9.1 has been designed to illustrate the analysis of a circuit containing an operational amplifier through the judicious application of Eqs. (9.3) and (9.5). When we use these equations to predict the behavior of a circuit containing an operational amplifier, we are in effect using an idealized model of the amplifier.

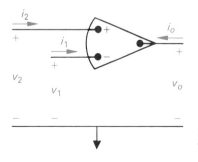

Fig. 9.8 The op-amp symbol with the power supply terminals removed.

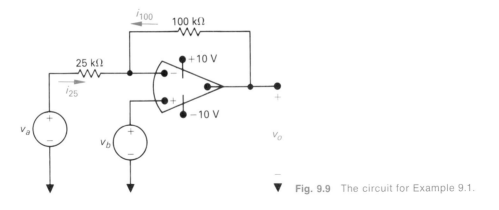

Example 9.1   The operational amplifier in the circuit in Fig. 9.9 is ideal.

a) Calculate v_o if $v_a = 1$ V and $v_b = 0$ V.

b) Repeat part (a) if $v_a = 1$ V and $v_b = 2$ V.

c) If $v_a = 1.5$ V specify the range of v_b such that the amplifier does not saturate.

Solution   a) Since $v_2 = v_b$ by virtue of the external connection and $v_1 = v_2$ because of the constraint imposed by the operational amplifier, we have $v_1 = 0$. Therefore the current in the 25-kΩ resistor from left to right will be $\frac{1}{25}$ mA. The current in the 100-kΩ resistor from right to left will be $v_o/100$ mA. If we apply Kirchhoff's current law to the inverting input junction and remember that the current into the operational amplifier is negligible, we have

$$\frac{1}{25} + \frac{v_o}{100} = 0.$$

It follows directly that v_o is -4 V. We pause to note that v_o lies between ± 10 V, hence the operational amplifier is in its linear range of operation.

b) Using the same thought process as outlined in part (a) we note that

$$v_2 = v_b = v_1 = 2 \text{ V},$$

$$i_{25} = \frac{v_a - v_1}{25} = \frac{1 - 2}{25} = -\frac{1}{25} \text{ mA},$$

$$i_{100} = \frac{v_o - v_1}{100} = \frac{v_o - 2}{100} \text{ mA},$$

$$i_{25} = -i_{100}.$$

Therefore $v_o = 6$ V. Again we note that v_o lies within ± 10 V.

c) As before, $v_1 = v_2 = v_b$ and $i_{25} = -i_{100}$. Since v_a is 1.5 V we have

$$\frac{1.5 - v_b}{25} = -\frac{(v_o - v_b)}{100}.$$

Solving for v_b as a function of v_o gives

$$v_b = \frac{1}{5}(6 + v_o).$$

Now if the amplifier is to be within the linear range of operation, $-10 \leq v_o \leq 10$ V. When we substitute these limits on v_o into our expression for v_b we see that v_b is limited to

$$-0.8 \text{ V} \leq v_b \leq 3.2 \text{ V.} \quad \blacksquare$$

We are now ready to discuss the operation of some important operational amplifier circuits using Eqs. (9.3) and (9.5) to idealize the behavior of the amplifier itself.

DRILL EXERCISE
9.1

Assume that the operational amplifier in the circuit shown is ideal.

a) Calculate v_o for the following values of v_s: 0.4 V, 0.72 V, 2.0 V, -0.6 V, -0.8 V, and -2.0 V.

b) Specify the range of v_s such that the amplifier does not saturate.

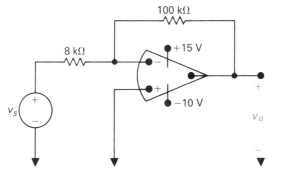

Ans. (a) -5 V, -9 V, -10 V, 7.5 V, 10 V, 15 V; (b) -1.2 V $\leq v_s \leq 0.8$ V.

9.4   THE INVERTING AMPLIFIER CIRCUIT

The inverting amplifier circuit is shown in Fig. 9.10. We are assuming that the operational amplifier is operating in its linear range. In studying the circuit in Fig. 9.10, note that the circuit external to the operational amplifier consists of two resistors (R_f, R_s), a voltage signal source (v_s), and a short circuit connected between the noninverting input terminal and the common node.

We will now analyze this circuit assuming an ideal op amp. Our goal in the analysis is to obtain an expression for the output voltage (v_o) as a function of the source voltage (v_s). First we note that v_2 is zero *because of the external*

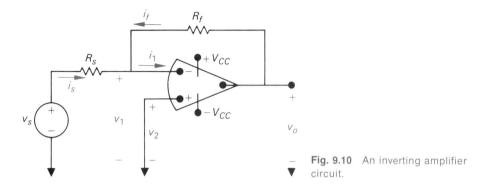

Fig. 9.10 An inverting amplifier circuit.

short circuit. We have already discovered through Eq. (9.3) that the op amp forces v_1 to be very nearly equal to v_2 (within a few millivolts if A is at least 10^4). We conclude, therefore, that v_1 is also near zero. Using the fact that v_1 is, for all practical purposes, zero gives us

$$i_s = \frac{v_s}{R_s} \tag{9.7}$$

and

$$i_f = \frac{v_o}{R_f}. \tag{9.8}$$

Now we invoke the constraint stated in Eq. (9.5), namely, that the terminal current i_1 is negligible. It follows directly that

$$i_f = -i_s. \tag{9.9}$$

When we substitute Eqs. (9.7) and (9.8) into Eq. (9.9) we get the sought-after result, specifically,

$$v_o = \frac{-R_f}{R_s}\, v_s, \tag{9.10}$$

from which we see that the output voltage is an inverted (sign reversal), scaled replica of the input. The scaling factor is the ratio R_f/R_s. The minus sign in Eq. (9.10) is, of course, the reason why the circuit is referred to as an inverting amplifier.

The result given by Eq. (9.10) is valid only if the op amp in the circuit in Fig. (9.10) is ideal, that is, if the open-loop gain (A) is infinite and the input current (i_1) is zero. For a practical op amp, Eq. (9.10) is an approximation. In most cases the approximation is a very good one. (We will have more to say about this later.) Equation (9.10) is an important result because it tells us that if the open-loop gain is large, we can control the gain of the inverting amplifier with the external resistors R_f and R_s. The upper limit on the gain (R_f/R_s) will be determined by the power supply voltage and the value of the signal volt-

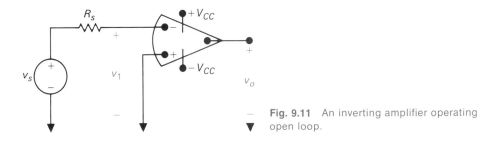

Fig. 9.11 An inverting amplifier operating open loop.

age v_s. If we assume equal power supply voltages, that is, $V^+ = -V^- = V_{CC}$, we have

$$|v_o| < V_{CC},$$

$$\left|\frac{R_f}{R_s} v_s\right| < V_{CC},$$

$$\frac{R_f}{R_s} < \left|\frac{V_{CC}}{v_s}\right|.$$

For example, if $V_{CC} = 15$ V and $v_s = 10$ mV, then the ratio R_f/R_s must be less than 1500.

The inverting amplifier circuit in Fig. 9.10 can be used to explain the term "open-loop gain." Note in Fig. 9.10 that the resistor R_f ties the output terminal to the inverting input terminal. This connection provides what is called a *feedback* path in the circuit. If R_f is removed, the feedback path is opened and the amplifier is said to be operating open loop. The open-loop operation is shown in Fig. 9.11.

Opening the feedback path drastically changes the behavior of the circuit. First we note that the output voltage is now

$$v_o = -Av_1, \tag{9.11}$$

assuming as before that $V^+ = -V^- = V_{CC}$; then $|v_1| < V_{CC}/A$ for linear operation. Since the inverting input current is zero for an ideal op amp, the voltage drop across R_s is zero and the inverting input voltage equals the signal voltage v_s, that is, $v_1 \approx v_s$. We can conclude, therefore, that the amplifier can operate open loop in the linear mode only if $|v_s| < V_{CC}/A$. If $|v_s| > V_{CC}/A$, the op amp simply saturates. In particular, if $v_s < -V_{CC}/A$ the op amp saturates at $+V_{CC}$, and if $v_s > V_{CC}/A$ the op amp saturates at $-V_{CC}$.

DRILL EXERCISE
9.2

The source voltage v_s in the circuit of Drill Exercise 9.1 is -640 mV. The 100-kΩ feedback resistor is replaced by a variable resistor R_x. What range of R_x will allow the inverting amplifier to operate in its linear range?

Ans. $0 \le R_x \le 187.5$ kΩ.

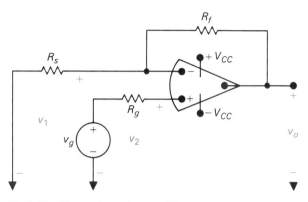

Fig. 9.12   The noninverting amplifier.

9.5   THE NONINVERTING AMPLIFIER CIRCUIT

The noninverting amplifier circuit is shown in Fig. 9.12, where the signal source is represented by v_g in series with the resistor R_g. In deriving the expression for the output voltage as a function of the source voltage, we will assume an ideal operational amplifier operating within its linear range. Thus, as before, we use Eqs. (9.3) and (9.5) as the basis for the derivation. The fact that the op-amp input current is zero allows us to write $v_2 = v_g$ and it follows from Eq. (9.3) that v_1 is also equal to v_g. Now observe that because the input current ($i_1 = i_2 = 0$) is zero, the resistors R_f and R_s form an unloaded voltage divider across v_o. Therefore we can write

$$v_1 = v_g = \frac{v_o R_s}{R_s + R_f}. \tag{9.12}$$

Solving Eq. (9.12) for v_o gives us the sought-after expression

$$v_o = \frac{R_s + R_f}{R_s} v_g. \tag{9.13}$$

Operation in the linear range requires

$$\frac{R_s + R_f}{R_s} < \left| \frac{V_{CC}}{v_g} \right|.$$

Problems 9.6 and 9.7 are designed to illustrate a simple application of the noninverting amplifier circuit.

DRILL EXERCISE
9.3

Assume that the operational amplifier in the circuit shown is ideal.

a) Find the output voltage when the variable resistor is set to 80 kΩ.

b) How large can R_x be before the amplifier saturates?

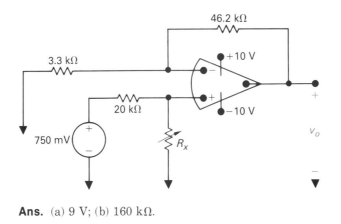

Ans. (a) 9 V; (b) 160 kΩ.

9.6   THE INTEGRATING AMPLIFIER

Using the ideal model of the operational amplifier we can show how the amplifier can be combined with a resistor and capacitor in order to build an integrating circuit. The purpose of the circuit is to develop an output voltage that is proportional to the integral of the input voltage. The integrating amplifier is shown in Fig. 9.13. If we apply the same arguments we used when discussing the inverting amplifier, we can conclude that in the case of an ideal op amp

$$i_f = -i_s \tag{9.14}$$

and

$$v_1 = v_2 = 0. \tag{9.15}$$

It follows directly that

$$C_f \frac{dv_o}{dt} = -\frac{v_s}{R_s},$$

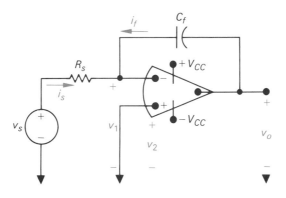

Fig. 9.13   An integrating amplifier.

or

$$\frac{dv_o}{dt} = -\frac{1}{R_sC_f}\,v_s.$$ (9.16)

Multiplying both sides of Eq. (9.16) by a differential time dt and then integrating yields

$$v_o(t) - v_o(t_o) = -\frac{1}{R_sC_f}\int_{t_o}^{t} v_s\,dy.$$ (9.17)

Assuming the voltage on the capacitor is zero at $t = t_o$ gives us

$$v_o(t) = -\frac{1}{R_sC_f}\int_{t_o}^{t} v_s\,dy,$$ (9.18)

where y is used as the variable of integration. Equation (9.18) tells us that for an ideal operational amplifier the output voltage is an inverted (minus sign), scaled $(1/R_sC_f)$ replica of the integral of the input signal voltage.

   In practice, of course, we cannot synthesize a perfect integrator. However, we can design high-quality integrators that, within specified limits, perform the integration function very well. The ability to design high-quality integrators is reflected in the existence of very useful and powerful analog computers. In fact, the need to build better analog computers was the original motivation for developing operational electronic amplifiers.

DRILL EXERCISE
9.4

Assume that the operational amplifier in the circuit shown is ideal. At $t = 0$, the switch is connected to terminal a.

a) If the initial voltage on the capacitor is zero, how many milliseconds does it take to saturate the amplifier?

b) At the instant that the operational amplifier reaches saturation when the switch is in position a, the switch is moved to position b. How many milliseconds after the switch is in position b does the output voltage reach zero?

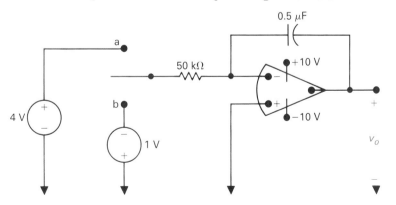

Ans. (a) 62.5 ms; (b) 250 ms.

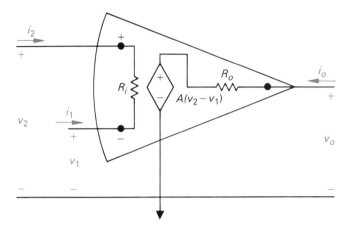

Fig. 9.14 An equivalent circuit for an operational amplifier.

9.7 AN EQUIVALENT CIRCUIT FOR THE OPERATIONAL AMPLIFIER

Let us now consider a more realistic model that predicts the performance of the operational amplifier in its *linear region* of operation. In building a more realistic circuit model we will modify three assumptions used in characterizing an ideal op amp: (1) a finite input resistance (R_i); (2) a finite open-loop gain (A); and (3) a nonzero output resistance (R_o). These modifications are in the circuit diagram shown in Fig. 9.14.

Whenever we use the equivalent circuit in Fig. 9.14 we disregard the assumptions that $v_1 = v_2$ (Eq. 9.3) and $i_1 = i_2 = 0$ (Eq. 9.5). Furthermore, Eq. (9.1) is no longer valid because of the presence of the nonzero output resistance (R_o).

Another way to think about the circuit in Fig. 9.14 is to reverse the thought process. That is, we can say that Fig. 9.14 reduces to the ideal model of the operational amplifier when $R_i \rightarrow \infty$, $A \rightarrow \infty$, and $R_o \rightarrow 0$. For the μA741 op amp, the typical values of R_i, A, and R_o are 2 MΩ, 10^5, and 75 Ω, respectively.

Although the presence of R_i and R_o makes the analysis of circuits containing op amps more cumbersome, the analysis remains very straightforward. To illustrate, we will analyze both the inverting and noninverting amplifiers using the equivalent circuit shown in Fig. 9.14. We begin with the inverting amplifier.

If we use the op-amp circuit in Fig. 9.14, the circuit for the inverting amplifier is that shown in Fig. 9.15. As before, our goal is to express the output voltage (v_o) as a function of the source voltage (v_s). We can obtain the desired expression by writing the two node-voltage equations that describe the circuit and then solving the resulting set of equations for v_o. The two nodes are labeled a and b in Fig. 9.15. Also note that v_2 equals zero by virtue of the external

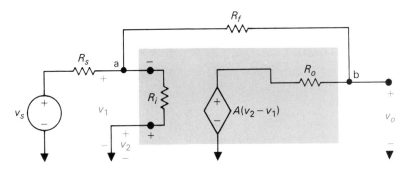

Fig. 9.15 An inverting amplifier circuit.

short-circuit connection at the noninverting input terminal. The two node-voltage equations are

$$\text{Node a:} \qquad \frac{v_1 - v_s}{R_s} + \frac{v_1}{R_i} + \frac{v_1 - v_o}{R_f} = 0, \tag{9.19}$$

$$\text{Node b:} \qquad \frac{v_o - v_1}{R_f} + \frac{v_o - A(-v_1)}{R_o} = 0. \tag{9.20}$$

Equations (9.19) and (9.20) are rearranged so that the solution for v_o by Cramer's method becomes apparent; thus

$$\left[\frac{1}{R_s} + \frac{1}{R_i} + \frac{1}{R_f}\right] v_1 - \left[\frac{1}{R_f}\right] v_o = \left[\frac{1}{R_s}\right] v_s, \tag{9.21}$$

$$\left[\frac{A}{R_o} - \frac{1}{R_f}\right] v_1 + \left[\frac{1}{R_f} + \frac{1}{R_o}\right] v_o = 0. \tag{9.22}$$

Solving for v_o results in the following:

$$v_o = \frac{-A + R_o/R_f}{\dfrac{R_s}{R_f}\left[1 + A + \dfrac{R_o}{R_i}\right] + \left[\dfrac{R_s}{R_i} + 1\right] + \dfrac{R_o}{R_f}} \cdot v_s. \tag{9.23}$$

Note that Eq. (9.23) reduces to Eq. (9.10) as $R_o \to 0$, $R_i \to \infty$, and $A \to \infty$.

We note in passing that if the inverting amplifier in Fig. 9.15 is loaded at its output terminals with a load resistance of R_L ohms, the relationship between v_o and v_s becomes

$$v_o = \frac{-A + R_o/R_f}{\dfrac{R_s}{R_f}\left[1 + A + \dfrac{R_o}{R_i} + \dfrac{R_o}{R_L}\right] + \left[1 + \dfrac{R_o}{R_L}\right]\cdot\left[1 + \dfrac{R_s}{R_i}\right] + \dfrac{R_o}{R_f}} \cdot v_s. \tag{9.24}$$

Problems 9.10, 9.11, and 9.12 are designed to familiarize you with numerical calculations involving Eqs. (9.23) and (9.24).

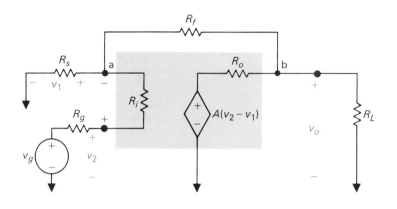

Fig. 9.16   A noninverting amplifier circuit.

When we use the equivalent circuit in Fig. 9.14 to analyze the noninverting amplifier, we obtain the circuit shown in Fig. 9.16, where the signal source is represented by the voltage source v_g in series with the resistance R_g. The load on the amplifier is denoted by the resistor R_L. The analysis of the amplifier consists of deriving an expression for v_o as a function of v_g. This is accomplished by writing the node-voltage equations at nodes a and b. We proceed as follows: At node a we have

$$\frac{v_1}{R_s} + \frac{v_1 - v_g}{R_g + R_i} + \frac{v_1 - v_o}{R_f} = 0, \tag{9.25}$$

and at node b,

$$\frac{v_o - v_1}{R_f} + \frac{v_o}{R_L} + \frac{v_o - A(v_2 - v_1)}{R_o} = 0. \tag{9.26}$$

Now using the fact that the current in R_g is the same current as in R_i, we have

$$\frac{v_2 - v_g}{R_g} = \frac{v_1 - v_g}{R_i + R_g}. \tag{9.27}$$

We use Eq. (9.27) to eliminate v_2 from Eq. (9.26), giving us a pair of equations involving the unknown voltages v_1 and v_o. This algebraic manipulation leads to the following:

$$v_1 \left[\frac{1}{R_s} + \frac{1}{R_g + R_i} + \frac{1}{R_f} \right] - v_o \left[\frac{1}{R_f} \right] = v_g \left[\frac{1}{R_g + R_i} \right], \tag{9.28}$$

$$v_1 \left[\frac{AR_i}{R_o(R_i + R_g)} - \frac{1}{R_f} \right] + v_o \left[\frac{1}{R_f} + \frac{1}{R_o} + \frac{1}{R_L} \right] = v_g \left[\frac{AR_i}{R_o(R_i + R_g)} \right]. \tag{9.29}$$

Solving for v_o yields

$$v_o = \frac{[(R_f + R_s) + (R_s R_o / AR_i)]v_g}{R_s + \dfrac{R_o}{A}(1 + K_r) + \left[\dfrac{R_f R_s + (R_f + R_s)(R_i + R_g)}{AR_i} \right]}, \tag{9.30}$$

where

$$K_r = [(R_s + R_g)/R_i] + [(R_f + R_s)/R_L] + [(R_fR_s + R_fR_g + R_gR_s)/R_iR_L].$$

Observe that Eq. (9.30) reduces to Eq. (9.13) when $R_o \rightarrow 0, A \rightarrow \infty$, and $R_i \rightarrow \infty$. For the unloaded ($R_L = \infty$) noninverting amplifier, Eq. (9.30) simplifies to

$$v_o = \frac{[(R_f + R_s) + R_sR_o/AR_i]v_g}{R_s + \dfrac{R_o}{A}\left[1 + \dfrac{R_s + R_g}{R_i}\right] + \dfrac{1}{AR_i}[R_fR_s + (R_f + R_s)(R_i + R_g)]}. \quad (9.31)$$

Note in deriving Eq. (9.31) from Eq. (9.30) that K_r reduces to $(R_s + R_g)/R_i$. Problem 9.13 is designed to illustrate the effect of R_i, A, and R_o on the performance of a noninverting amplifier.

9.8 THE DIFFERENTIAL MODE

The operational amplifier is also useful as a linear differential amplifier. In the differential mode, the amplifier is meant to produce an amplified replica of the difference between v_1 and v_2. We touched on this possibility in Section 9.4 when we briefly discussed the possibility of the inverting amplifier operating open loop with $v_2 = 0$. We now wish to introduce the differential mode when neither v_1 nor v_2 is set equal to zero. It should be apparent from our previous discussion that the difference between v_1 and v_2 must be extremely small; otherwise the high-gain amplifier will simply saturate. Thus the operational amplifier is useful as a linear differential amplifier in applications where v_1 and v_2 are approximately equal. The bridge circuit in Fig. 9.17 is a good example of such an application. The purpose of the bridge structure is to generate a signal $v_1 - v_2$ when the bridge arm resistor R_x changes from its balance value of R_1R_2/R_4. That is, when ϵ is zero, the bridge is balanced and $v_1 = v_2$ or $v_1 - v_2 = 0$. When R_x changes (due to some phenomenon such as temperature

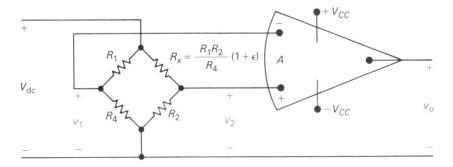

Fig. 9.17 Using an operational amplifier in a differential input mode.

or strain) the voltage $v_1 - v_2$ is nearly proportional to ϵ if ϵ is relatively small. We will leave it to the reader (see Problem 9.14) to show that for an ideal operational amplifier with infinite input resistance

$$v_1 - v_2 = \left[\frac{V_{dc}R_1R_4}{(R_1 + R_4)^2} \right] \cdot \frac{\epsilon}{\left[1 + \dfrac{R_1}{R_1 + R_4}\epsilon \right]}. \tag{9.32}$$

We should also mention that we are assuming the bridge is energized from a regulated power supply so that the bridge voltage V_{dc} is constant for all values of R_x.

For small values of ϵ, Eq. (9.32) simplifies to

$$v_1 - v_2 \approx \frac{V_{dc}R_1R_4}{(R_1 + R_4)^2} \, \epsilon. \tag{9.33}$$

Since R_1, R_4, and V_{dc} are all fixed quantities, the input to the differential amplifier is proportional to ϵ. The output of an ideal operational amplifier is $-A(v_1 - v_2)$; therefore, in the circuit in Fig. 9.17, v_o is

$$v_o = \frac{-V_{dc}R_1R_4A\epsilon}{(R_1 + R_4)^2} \quad \text{for } (|\epsilon| \ll 1). \tag{9.34}$$

It follows from Eq. (9.34) that if the amplifier is to operate in its linear region

$$\frac{V_{dc}R_1R_4A\epsilon}{(R_1 + R_4)^2} \le V_{CC}$$

or

$$|\epsilon| \le \frac{V_{CC}(R_1 + R_4)^2}{V_{dc}R_1R_4A}. \tag{9.35}$$

It should be apparent from the inequality of Eq. (9.35) that a very large open-loop gain (A) means that the amplifier can tolerate only a very small difference in v_1 and v_2 before saturating. Typically $(v_1 - v_2)$ is limited to the μV range.

When ϵ is zero, v_o should be zero since $v_1 = v_2$. However, even with $v_1 = v_2$ there will be some output from the amplifier because in practice it is not possible to get perfect balance between the two channels in the amplifier. The fact that the differential amplifier yields an infinitesimal output when the same signal is applied to the two input terminals (common mode) represents a flaw in the operation of the amplifier. How well an operational amplifier can reject a common-mode signal is referred to as the *common-mode rejection* property of the amplifier. The ability of a differential amplifier to reject a signal that is common to both input terminals is expressed quantitatively by a figure of merit called the *common-mode rejection ratio* (CMRR), which we will discuss in Section 9.9.

In the bridge circuit in Fig. 9.17, the circuit parameters are $V_{dc} = 1.5$ V, $V_{CC} = 20$ V, $R_1 = 10$ kΩ, $R_2 = 2$ kΩ, $R_4 = 1$ kΩ, and $A = 2 \times 10^6$.

a) From $(v_1 - v_2)$ in μV when R_x is larger than its balance value by 1 Ω.

b) Specify the range of R_x such that the operational amplifier does not saturate.

Ans. (a) 6.20 μV; (b) 19,998.387 Ω ≤ R_x ≤ 20,001.613 Ω.

9.9 THE COMMON-MODE REJECTION RATIO

When we are using the operational amplifier in the differential mode, one measure of its effectiveness is its ability to produce zero output if the voltages v_1 and v_2 are identical. That is, since

$$v_o = A(v_2 - v_1) \tag{9.36}$$

we would expect v_o to be zero when v_2 was equal to v_1. In practical operational amplifiers, we find that v_o is not zero when v_1 and v_2 are equal. This inability to produce a zero output when v_1 and v_2 are the same is caused by unavoidable imperfections in the electronic components used in the fabrication of the amplifier. These imperfections make it impossible to have identical amplifying channels. The nature of the problem can be described in terms of the circuit shown in Fig. 9.18.

If both channels through the operational amplifier were identical then, except for sign inversion, the output would be the same regardless of which channel were used. For example, if v_2 were made zero by grounding the noninverting input terminal and v_1 were set equal to v_s, then

$$v_o = A_1 v_s, \tag{9.37}$$

where A_1 represents the gain of the amplifier via the inverting channel.

Now we reverse the process; that is, we ground the inverting input terminal and set v_2 equal to v_s. Then

$$v_o = A_2 v_s, \tag{9.38}$$

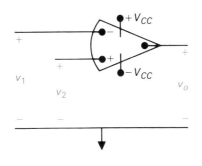

Fig. 9.18 A circuit used in introducing the concept of the common-mode rejection ratio.

where A_2 represents the gain of the amplifier via the noninverting channel. For an ideal operational amplifier we would expect

$$A_1 v_s = -A_2 v_s. \tag{9.39}$$

In terms of the open-loop gain previously used in our analysis, the ideal op amp requires that

$$A = -A_1 = A_2. \tag{9.40}$$

In order to quantitatively discuss the difference between channels we rewrite the expression for the output voltage as a function of the two input voltages:

$$v_o = A_1 v_1 + A_2 v_2. \tag{9.41}$$

Note that this equation reduces to the more familiar form of Eq. (9.36) if we idealize the operational amplifier by assuming that Eq. (9.40) is valid. It is also interesting to note that Eq. (9.41) is a direct application of the principle of superposition. Thus Eq. (9.41) reinforces the fact that we are assuming operation within the linear range of the operational amplifier, that is, $v_o \leq |V_{CC}|$.

The next step in the development of the common-mode rejection ratio is to regard all signal voltages as being made up of two components: a common-mode component and a difference-mode component. We can justify the need for such a point of view by considering the following two sets of signals. Let us assume that the two voltage sets are described in terms of dc voltages. The first set of voltages are $v_1 = -10 \ \mu V$ and $v_2 = +10 \ \mu V$, and the second set are $v_1 = 30 \ \mu V$ and $v_2 = 50 \ \mu V$. Now observe that in both sets the difference between v_2 and v_1 is 20 μV; the sum in the first set is 0, whereas in the second set it is 80 μV. Because the two sets have the same difference we would expect each set to generate the same output from an operational amplifier operating in the differential mode. However, we have already noted that the output will not be the same because of differences in the two channels. We now identify the outputs as different because the two sets have different common-mode components. The *common-mode component* of the two signals is defined as the average value of the two signals; thus

$$v_c = \frac{1}{2}(v_1 + v_2). \tag{9.42}$$

The *differential-mode component* is defined as the difference between v_2 and v_1, namely,

$$v_d = v_2 - v_1. \tag{9.43}$$

When we solve Eqs. (9.42) and (9.43) for v_1 and v_2 we get the desired expressions for the signal voltages as functions of common-mode and differential-mode components; thus

$$v_1 = v_c - \frac{1}{2}v_d, \tag{9.44}$$

and

$$v_2 = v_c + \frac{1}{2} v_d. \tag{9.45}$$

In terms of the two sets of signals mentioned above, we note that the first set has a common-mode component of 0 V and a differential-mode component of 20 μV. The second set has a common-mode component of 40 μV and a differential-mode component of 20 μV. These observations are tabulated in the following equations:

$$\text{Set 1:} \quad v_c = \frac{1}{2}(v_1 + v_2) = \frac{1}{2}(-10 + 10) = 0, \tag{9.46}$$

$$v_d = (v_2 - v_1) = [10 - (-10)] = 20 \ \mu\text{V}; \tag{9.47}$$

$$\text{Set 2:} \quad v_c = \frac{1}{2}(v_1 + v_2) = \frac{1}{2}(30 + 50) = 40 \ \mu\text{V}, \tag{9.48}$$

$$v_d = (50 - 30) = 20 \ \mu\text{V}. \tag{9.49}$$

Using the common-mode and differential-mode components, we can describe the two sets of signals as follows:

$$\text{Set 1:} \quad v_1 = v_c - \frac{1}{2} v_d = 0 - \frac{1}{2}(20) = -10 \ \mu\text{V}, \tag{9.50}$$

$$v_2 = v_c + \frac{1}{2} v_d = 0 + \frac{1}{2}(20) = 10 \ \mu\text{V}; \tag{9.51}$$

$$\text{Set 2:} \quad v_1 = 40 - \frac{1}{2}(20) = 30 \ \mu\text{V}, \tag{9.52}$$

$$v_2 = 40 + \frac{1}{2}(20) = 50 \ \mu\text{V}. \tag{9.53}$$

Now that we have introduced the concept of signals being described in terms of common-mode and differential-mode components, we are in a position to describe quantitatively the operational amplifier's ability to reject the common-mode component. We begin by returning to Eq. (9.41) and writing v_1 and v_2 in terms of their common-mode and differential-mode components. We get

$$v_o = A_1(v_c - \frac{1}{2} v_d) + A_2(v_c + \frac{1}{2} v_d). \tag{9.54}$$

Collecting coefficients of v_c and v_d allows us to rewrite Eq. (9.54) as

$$v_o = (A_1 + A_2)v_c + \frac{1}{2}(A_2 - A_1)v_d, \tag{9.55}$$

from which we see that $A_1 + A_2$ is the gain for the common-mode signal and

$\frac{1}{2}(A_2 - A_1)$ is the gain for the differential-mode signal. We emphasize this observation by letting

$$A_c = A_1 + A_2 \tag{9.56}$$

and

$$A_d = \frac{1}{2}(A_2 - A_1). \tag{9.57}$$

If we substitute Eqs. (9.56) and (9.57) into Eq. (9.55), we get

$$v_o = A_c v_c + A_d v_d. \tag{9.58}$$

Note that for an ideal op amp where $A_2 = -A_1 = A$, we have $A_c = 0$ and $A_d = A$. Thus an operational amplifier that exhibits a very small value for A_c and a very large value for A_d approaches the characteristics of an ideal op amp. The figure of merit that reflects the magnitude of the ratio of A_c/A_d is the common-mode rejection ratio, abbreviated CMRR:

$$\text{CMRR} \triangleq \left| \frac{A_d}{A_c} \right|. \tag{9.59}$$

It is customary to express this ratio in decibels. The use of the decibel as a unit for measuring ratios will be introduced in Chapter 17. For the moment we simply note that when the common-mode rejection ratio is expressed in decibels, we write

$$\text{CMRR} = 20 \log_{10} \left| \frac{A_d}{A_c} \right|. \tag{9.60}$$

Typical values of the CMRR range from 60 dB to 120 dB. From Eq. (9.60) we can see that this translates into ratios that range from 10^3 to 10^6.

One final observation. We can measure the common-mode gain by making $v_1 = v_2$ and measuring the corresponding output. If $v_1 = v_2$ it follows that $v_c = v_1$ and $v_d = 0$. Thus from Eq. (9.58) we have $v_o = A_c v_1$, where A_c is simply the ratio of the measured output to the input. The measurement of A_c is summarized in Fig. 9.19.

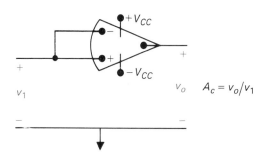

Fig. 9.19   A circuit for the measurement of A_c.

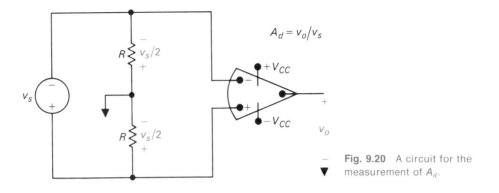

$A_d = v_o/v_s$

Fig. 9.20 A circuit for the measurement of A_d.

The differential-mode gain can be measured by making $v_1 = -v_2$. In this case, v_c will equal zero and v_d will equal twice v_2. The measurement of A_d can be implemented by the circuit shown in Fig. 9.20, where we note that

$$v_2 = \frac{v_s}{2} = -v_1. \tag{9.61}$$

Now since $v_c = 0$ and $v_d = 2v_2$, then

$$v_o = A_d v_d = A_d[2v_s/2] = A_d v_s, \tag{9.62}$$

from which we have

$$A_d = v_o/v_s. \tag{9.63}$$

Once A_c and A_d have been measured, the common-mode rejection ratio can be calculated from Eq. (9.59). The rejection in dB is calculated from Eq. (9.60).

DRILL EXERCISE 9.6

The common-mode and differential-mode gains of the amplifier in Fig. 9.18 are 10^3 and 10^6, respectively. The input voltages to the amplifier are $v_1 = 15.6\ \mu V$ and $v_2 = 4.4\ \mu V$. Assume that the amplifier is operating in its linear region. Compute (a) the common-mode and differential-mode components of v_1 and v_2; (b) the CMRR in decibels; (c) the channel gains A_1 and A_2; and (d) the output voltage v_o. (e) Repeat part (d) if $v_1 = v_2 = 15.6\ \mu V$ and if $v_1 = v_2 = 4.4\ \mu V$.

Ans. (a) $v_c = 10\ \mu V$, $v_d = -11.2\ \mu V$; (b) 60 dB; (c) $A_1 = -999,500$, $A_2 = 1,000,500$; (d) -11.19 V; (e) 15.6 mV, 4.4 mV.

9.10 THE DIGITAL MULTIMETER

The digital multimeter is an instrument that is designed to measure dc voltage and current, ac voltage and current,[†] and resistance. The value of the measured quantity is converted by means of electronic circuits to a digital or

[†] ac, or alternating current, voltage and current signals are introduced in Chapter 10.

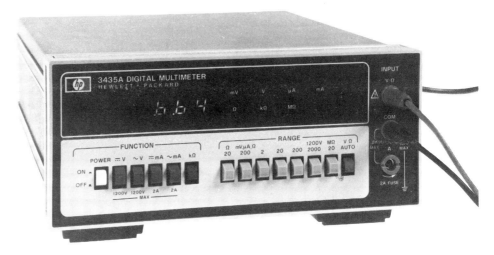

Fig. 9.21 A digital multimeter.

numeric readout. A commercially available digital multimeter is shown in Fig. 9.21. Although we cannot discuss in detail how each electronic circuit carries out its assigned function, we can give you a feel for how the instrument works. In particular, we will show how the operational-amplifier circuit is used in the digital multimeter.

A simplified block diagram of the digital multimeter is shown in Fig. 9.22. The first two blocks indicate that the input variable and its range are guided to the proper internal circuits of the multimeter by manually set switches. (Automatic ranging of ac and dc voltage and resistance is available on some multimeters.) The third block emphasizes that internal circuits of the multimeter convert any selected input variable to a dc voltage V_a. This dc voltage, which is proportional to the input variable, is next converted to digital form. Finally, the numerical value of the measured variable is transferred to a digital display.

An input dc voltage is already in proper form and therefore is guided directly through range-selection operational amplifiers to the analog-to-digital (A-to-D) converter.

An input ac voltage can be converted to a dc voltage by several techniques. One common method is to base the ac voltage calibration of the multimeter on

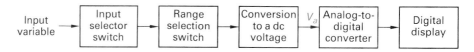

Fig. 9.22 A simplified block diagram of the digital multimeter.

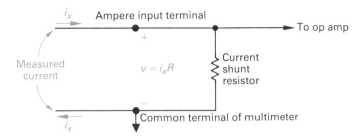

Fig. 9.23   Conversion of current to voltage.

the assumption that the ac voltage is a pure sinusoidal signal. The signal is passed through a full-wave rectifier circuit that inverts the negative portion of the signal. A dc voltage that is proportional to the average value of the rectified signal is generated. This dc voltage is then translated into the rms (root-mean-square) value corresponding to a pure sine wave.† If the input signal is not a sine function, the multimeter does not indicate its true rms value. Therefore another technique is to produce a dc voltage proportional to the rms value of the ac signal. A digital multimeter based on the rms conversion process is referred to as a true rms meter.

To convert a current (either dc or ac) to a voltage, we feed the unknown current through a current shunt resistor, as shown in Fig. 9.23. We then feed the voltage developed across the shunt resistor into an appropriate operational amplifier for further processing. A voltage proportional to an ac current is converted to a dc voltage by using one of the techniques described above.

The circuit shown in Fig. 9.24 is used to convert resistance to a dc voltage. The dc voltage source (V_{dc}) and the reference resistor R_{ref} are part of the internal circuitry of the multimeter. The output voltage of the operational-

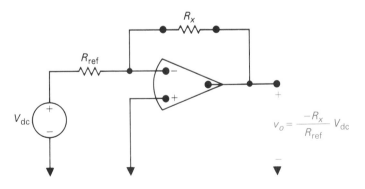

Fig. 9.24   Conversion of resistance to voltage.

† The root-mean-square value of a signal is discussed in Section 11.3.

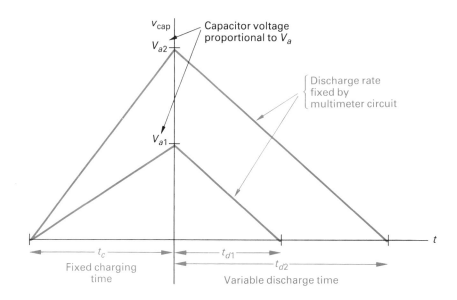

Fig. 9.25 A dual-ramp analog-to-digital system.

amplifier circuit is directly proportional to the unknown resistance R_x, that is,

$$v_o = -\frac{R_x}{R_{\text{ref}}} V_{\text{dc}} = \left(-\frac{V_{\text{dc}}}{R_{\text{ref}}}\right) R_x. \tag{9.64}$$

The most common analog-to-digital conversion process used in the digital multimeter is known as the *dual-ramp system*. The basic idea is described as follows. The dc voltage V_a is used to charge a capacitor for a fixed length of time. At the end of this charging interval, the voltage on the capacitor is proportional to V_a. The capacitor is now discharged at a uniform rate. The time it takes to discharge the capacitor to zero volts is proportional to the initial voltage on the capacitor, which in turn is proportional to V_a. Thus the dual-ramp system converts the dc voltage V_a to an interval of time that is proportional to V_a. This time is measured by an electronic clock; which generates voltage pulses at a uniform rate. The number of pulses generated during the time when the capacitor discharges is counted and this count is therefore proportional to V_a. The dual-ramp system is illustrated in Fig. 9.25.

The circuit used to create an interval of time proportional to V_a uses an integrating amplifier, as shown in Fig. 9.26. The position of the switch S is internally controlled by the multimeter, as is the polarity of the dc reference voltage V_{ref}. The polarity of V_{ref} is always the opposite of that of V_a.

To see how the dual-slope integrating circuit works, assume that V_a is a positive dc voltage. Let t_c be the length of time that the switch S connects the integrator to V_a. After t_c seconds the voltage, v_o, is

$$v_o = -\frac{1}{RC} \int_0^{t_c} V_a \, dt = -\frac{V_a}{RC} t_c. \tag{9.65}$$

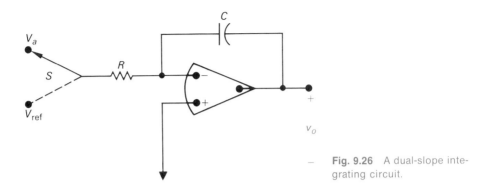

Fig. 9.26   A dual-slope integrating circuit.

The integrator is now switched to the reference voltage V_{ref} and remains in this position until v_o is reduced to zero. Since the polarity of V_{ref} is always the opposite of the polarity of V_a, the capacitor voltage can always be made equal to zero. At the instant when the switch S connects the integrating amplifier to V_{ref}, the voltage on the capacitor is v_o or $-V_a t_c/RC$, as shown in Fig. 9.27.

For the circuit in Fig. 9.27

$$-\frac{V_{ref}}{R} + C\frac{dv_o}{dt} = 0, \tag{9.66}$$

from which it follows that

$$\int_{-V_a t_c/RC}^{v_o} dv = \frac{V_{ref}}{RC}\int_0^t dt,$$

or

$$v_o = -\frac{V_a t_c}{RC} + \frac{V_{ref}}{RC}t. \tag{9.67}$$

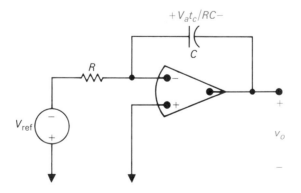

Fig. 9.27   The integrating amplifier connection during the discharge of the capacitor C.

Now we let t_d represent the length of time it takes to reduce v_o to zero. From Eq. (9.67) we have

$$t_d = \frac{V_a t_c}{V_{\text{ref}}}, \tag{9.68}$$

which tells us that the length of time to reduce v_o to zero is directly proportional to V_a since both t_c and V_{ref} are constants set by the designer of the multimeter. The length of time that the electronic clock drives the digital display is denoted by t_d. Since V_a is proportional to the quantity being measured, the digital multimeter has converted the unknown variable to a known numerical value.

As a laboratory instrument, the digital multimeter is now replacing the electromechanical instruments introduced in Chapter 3. However, in some applications the electromechanical instrument is still a viable choice. A more important observation to be made at this point is that electrical measurements represent an area of study in their own right. The correct interpretation of the measurement requires a knowledge of the instrument used. Whether an instrument is designed around electronic circuitry, an electromechanical device, or a combination of both, it disturbs the system into which it is inserted. Thus careful circuit analysis is often required to properly assess the significance of the measurement.

9.11 SUMMARY

Our introduction to circuits using operational amplifiers has, of necessity, been simplistic. There are many practical problems related to the successful application of op amps. A discussion of these problems must be delayed until you have a more extensive knowledge of electronic circuit behavior. Noise generation, bias drift, thermal stability, frequency response, slew rate, and feedback are all topics relevant to a deeper understanding of op amp circuits. The actual behavior of many op-amp configurations is also better taught in a laboratory environment where direct observations and measurements can be made. This introductory material is intended to alert you to some of the exciting things that can be done by simply combining a high-gain amplifier with passive circuit elements such as resistors and capacitors.

PROBLEMS

9.1 The operational amplifier in the circuit in Fig. 9.28 is ideal.
  a) Calculate v_o if $v_a = 4$ V and $v_b = 0$ V.
  b) Calculate v_o if $v_a = 2$ V and $v_b = 0$ V.
  c) Calculate v_o if $v_a = 2$ V and $v_b = 1$ V.
  d) Calculate v_o if $v_a = 1$ V and $v_b = 2$ V.
  e) If $v_b = 1.6$ V, specify the range of v_a such that the amplifier does not saturate.

9.2 The operational amplifier in the circuit in Fig. 9.29 is ideal.

    a) Calculate v_o.

    b) Calculate i_o.

9.3 Assume that the ideal op amp in the circuit in Fig. 9.30 is operating in its linear region.

    a) Find the output voltage as a function of v_a, v_b, and v_c.

    b) Discuss the significance of the result.

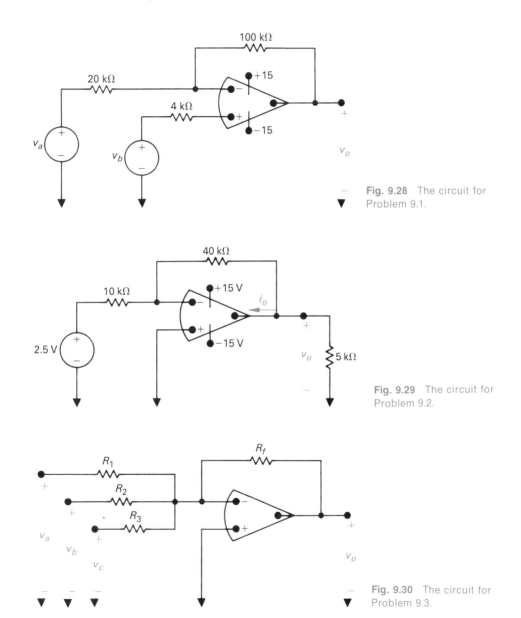

Fig. 9.28 The circuit for Problem 9.1.

Fig. 9.29 The circuit for Problem 9.2.

Fig. 9.30 The circuit for Problem 9.3.

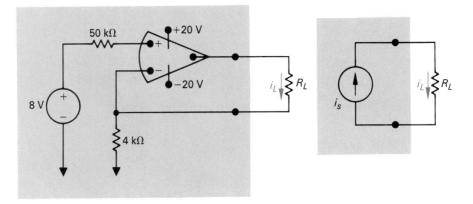

Fig. 9.31   The circuit for Problem 9.4.

9.4   The circuit inside the shaded area in Fig. 9.31 is a constant current source for a limited range of values of R_L.

a) Find the value of i_L for $R_L = 4$ kΩ.
b) Find the maximum value for R_L for which i_L will have the value of part (a).
c) Assume that $R_L = 16$ kΩ. Explain the operation of the circuit. You can assume that $i_1 = i_2 \approx 0$ under all operating conditions.
d) Sketch i_L vs. R_L for $0 \leq R_L \leq 16$ kΩ.

9.5   The op amp in the circuit in Fig. 9.32 is ideal.

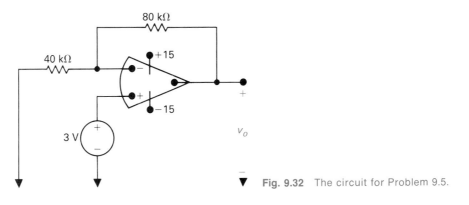

Fig. 9.32   The circuit for Problem 9.5.

a) Calculate v_o.
b) State whether the amplifier is inverting or noninverting.

9.6   Assume that the ideal op amp in the circuit in Fig. 9.33 is operating in its linear region.

a) Show that $v_o = [(R_1 + R_2)/R_1]v_s$.
b) What happens if $R_1 \to \infty$ and $R_2 \to 0$?

9.7   Assume that the ideal op amp in the circuit in Fig. 9.34 is operating in its linear region.

a) Calculate the power delivered to the 600-Ω resistor.
b) Repeat (a) with the operational amplifier removed from the circuit, that is, with

the 600-Ω resistor connected in series with the voltage source and the 29.4-kΩ resistor.

c) Find the ratio of the power found in part (a) to that found in part (b).

d) Does the insertion of the op amp between the source and the load serve a useful purpose? Explain.

9.8 We now wish to illustrate how several op-amp circuits can be interconnected to solve a differential equation.

a) Derive the differential equation for the spring–mass system shown in Fig. 9.35. Assume that the force exerted by the spring is directly proportional to the spring displacement, that the mass is constant, and that the frictional force is directly proportional to the velocity of the moving mass.

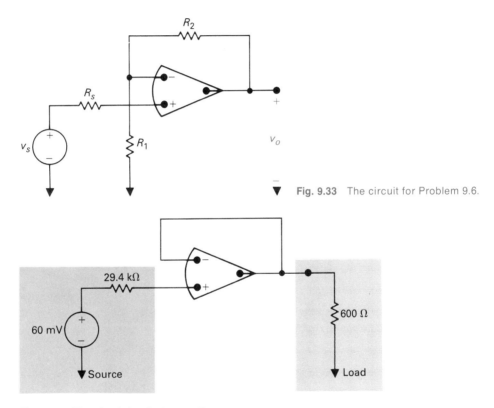

Fig. 9.33 The circuit for Problem 9.6.

Fig. 9.34 The circuit for Problem 9.7.

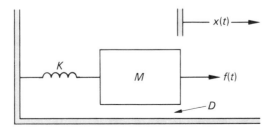

Fig. 9.35 The spring–mass system for Problem 9.8.

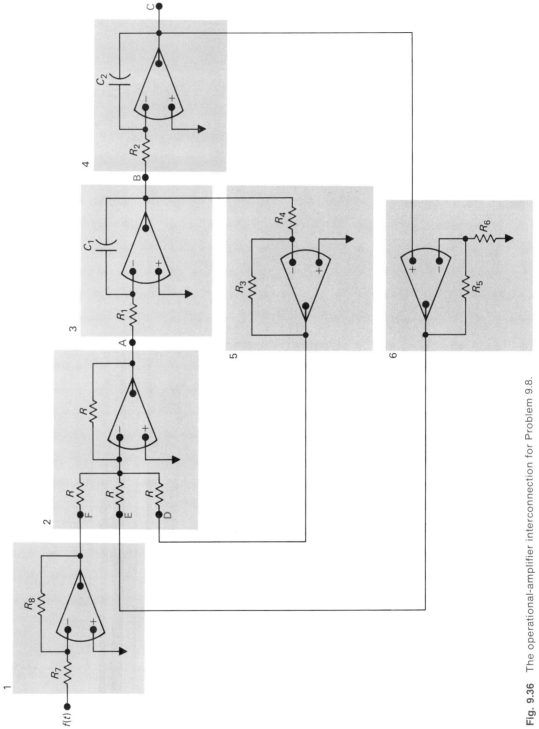

Fig. 9.36 The operational-amplifier interconnection for Problem 9.8.

b) Rewrite the differential equation derived in part (a) so that the highest-order derivative is expressed as a function of all the other terms in the equation. Now assume that a voltage equal to d^2x/dt^2 is available and by successive integrations generate dx/dt and x. We can synthesize the coefficients in the equations by scaling amplifiers and combine the terms required to generate d^2x/dt^2 by a summing amplifier. With these ideas in mind, analyze the interconnection shown in Fig. 9.36. In particular, describe the purpose of each shaded area in the circuit and describe the signal at the points labeled B, C, D, E, and F, assuming the signal at A represents d^2x/dt^2. Also discuss the parameters $R; R_1,$ $C_1; R_2, C_2; R_3, R_4; R_5, R_6$; and R_7, R_8 in terms of the coefficients in the differential equation.

9.9 a) Interchange the capacitor C_f and the resistor R_s in the circuit in Fig. 9.13 and explain why the resulting circuit is called a *differentiating circuit*.

b) What is the gain of the differentiating amplifier?

9.10 The inverting amplifier in the circuit in Fig. 9.37 has an input resistance of 500 kΩ, an output resistance of 5 kΩ, and an open-loop gain of 300,000. Assume that the amplifier is operating in its linear region. Calculate the following:

a) the voltage gain (v_o/v_s) of the amplifier:

b) the value of v_1 in mV when $v_s = 1$ V;

c) the resistance seen by the signal source (v_s).

d) Repeat parts (a), (b), and (c) using the ideal model for the op amp.

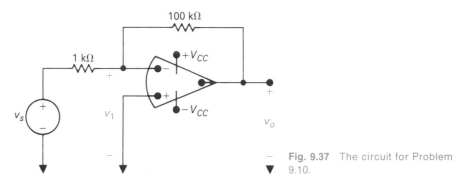

Fig. 9.37 The circuit for Problem 9.10.

9.11 Repeat Problem 9.10 given that the inverting amplifier is loaded with a 1-kΩ resistor.

9.12 Find the Thévenin equivalent circuit with respect to the output terminals a, b for the inverting amplifier in Fig. 9.38.

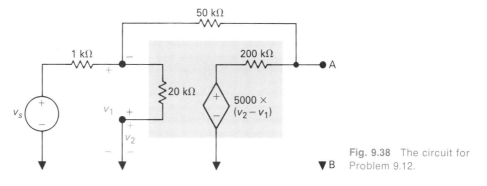

Fig. 9.38 The circuit for Problem 9.12.

a) What is the output resistance of the amplifier?
b) What is the input resistance seen by the source v_s when the load at the terminals a, b is 2000 Ω?

9.13 The operational amplifier in the noninverting amplifier circuit in Fig. 9.39 has an input resistance of 400 kΩ, an output resistance of 5 kΩ, and an open-loop gain of 20,000. Assume that the op amp is operating in its linear region. Calculate the following:

a) the voltage gain (v_o/v_g);
b) the inverting and noninverting input voltages v_1 and v_2 if $v_g = 1$ V;
c) the difference $(v_2 - v_1)$ in μV when $v_g = 1$ V;
d) the current drain in pA on the signal source v_g.
e) Repeat parts (a)–(d) assuming an ideal op amp.

9.14 a) Derive Eq. (9.32). Note that Eq. (9.32) assumes an ideal operational amplifier and a constant bridge voltage of V_{dc} volts.

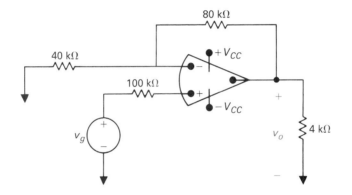

Fig. 9.39 The circuit for Problem 9.13.

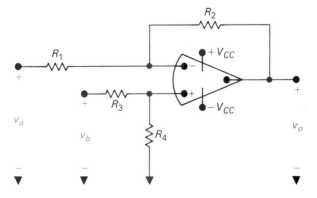

Fig. 9.40 The circuit for Problem 9.15.

b) Verify the approximation given by Eq. (9.33) by using long division to expand

$$\frac{\epsilon}{1 + \dfrac{R_1}{R_1 + R_4}\,\epsilon}$$

into a power series. Simplify the algebra by letting $R_1/(R_1 + R_4) = \alpha$. Note that $\alpha < 1$ by definition. Comment on the significance of higher-order terms for small ϵ.

9.15 The circuit in Fig. 9.40 is an ideal linear differential amplifier in which the gain of the amplifier is a function of the external resistors R_1, R_2, R_3, and R_4.

a) Show that

$$v_o = \frac{R_4(R_1 + R_2)}{R_1(R_3 + R_4)}\,v_b - \frac{R_2}{R_1}\,v_a.$$

b) What is the expression for v_o if $R_3 = R_1$ and $R_4 = R_2$?

9.16 Select values of R_B, R_C, and R_D in the amplifier circuit in Fig. 9.41 such that $v_o = 8(v_b - v_a)$ and the voltage source v_o sees an input resistance of 4500 Ω. Use the ideal model for the operational amplifier.

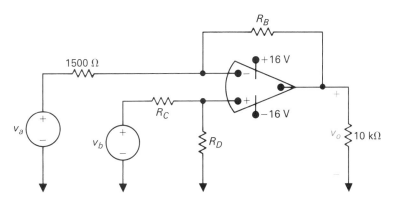

Fig. 9.41   The circuit for Problem 9.16.

10

SINUSOIDAL STEADY-STATE ANALYSIS

10.1   INTRODUCTION

The steady-state behavior of circuits that are energized by sinusoidal sources is an important area of study for several reasons. First, the generation, transmission, distribution, and consumption of electrical energy occurs under essentially sinusoidal steady-state conditions. Second, it is possible to predict the behavior of circuits to nonsinusoidal sources if the sinusoidal behavior is understood. Third, it is often easier to carry out the design of electrical systems on the basis of steady-state sinusoidal behavior. Thus design specifications are spelled out in terms of a desired steady-state sinusoidal response. The circuit, or system, is then designed to meet these sinusoidal characteristics. If these specifications are satisfied, the designer knows that the circuit will respond satisfactorily to nonsinusoidal inputs.

The importance of sinusoidal steady-state behavior cannot be over-emphasized. Most of the topics in the subsequent chapters of this text are based on a thorough understanding of the techniques needed to implement the analysis of circuits driven by sinusoidal sources. We begin our study with a review of the important characteristics of the sinusoidal function.

10.2   THE SINUSOIDAL SOURCE

A *sinusoidal voltage source* (independent or dependent) produces a voltage that varies sinusoidally with time. A *sinusoidal current source* (independent or dependent) produces a current that varies sinusoidally with time. In discussing the sinusoidal function we will review its properties using a voltage source. The observations we make also apply to a current source.

We can express a sinusoidally varying function using either the sine function or the cosine function. There is no clearcut choice for either function. Although either works equally well in sinusoidal steady-state analysis, it is important to recognize that both functional forms cannot be used simultaneously. We will use the cosine function throughout our discussion.

Using the cosine function, we can write a sinusoidally varying voltage as follows:

$$v = V_m \cos (\omega t + \phi). \tag{10.1}$$

To facilitate our discussion of the parameters in Eq. (10.1), we show the voltage vs. time plot in Fig. 10.1. Our first observation is that the sinusoidal function is repetitive. Such a function is called *periodic*. One of the parameters of interest, therefore, is the length of time it takes the sinusoidal function to pass through all of its possible values. This time is referred to as the *period* of the function and is denoted by T. The reciprocal of T gives the number of cycles per second, or frequency, of the sine wave and is denoted by f; thus

$$f = \frac{1}{T}. \tag{10.2}$$

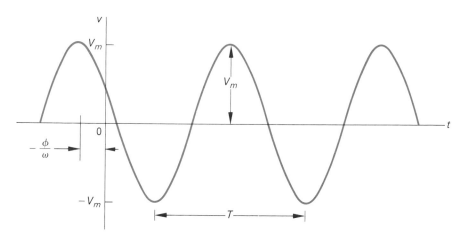

Fig. 10.1 A sinusoidal voltage.

A cycle per second is referred to as a hertz and is abbreviated Hz. (The term "cycles per second" is rarely used in contemporary technical literature.) The numerical value of T or f is contained in the coefficient of t in Eq. (10.1). Omega (ω) is used to represent the angular frequency of the sinusoidal function; thus

$$\omega = 2\pi f = 2\pi/T \quad \text{(radians/second)}. \tag{10.3}$$

Equation (10.3) is derived from the fact that the sine (or cosine) function passes through a complete set of values each time its argument passes through 2π radians (360 degrees). From Eq. (10.3) we see that whenever t is an integral multiple of T the argument ωt increases by an integral multiple of 2π radians.

The maximum amplitude of the sinusoidal voltage is given by the coefficient V_m. Since the cosine function is bounded by ± 1 the amplitude is bounded by $\pm V_m$. This is clearly evident in Fig. 10.1.

The phase angle of the sinusoidal voltage of Eq. (10.1) is the angle ϕ. It determines the value of the sinusoidal function at $t = 0$; therefore it fixes the point on the periodic wave at which we start measuring time. Changing the phase angle ϕ shifts the sinusoidal function along the time axis, but has no effect on either the amplitude [V_m] or the angular frequency [ω]. Note, for example, that if ϕ is reduced to zero the effect on the plot shown in Fig. 10.1 is to shift the sinusoidal function ϕ/ω time units to the right. Also note that if ϕ is positive, the sinusoidal function shifts to the left, whereas if ϕ is negative, the function shifts to the right. (See Problem 10.2.)

A comment with regard to the phase angle is in order. It should be obvious that ωt and ϕ must carry the same units, since they are added together in the argument of the sinusoidal function. Since ωt is expressed in radians, one would expect ϕ to be also. We find, however, that ϕ is normally given in degrees and ωt is converted from radians to degrees before the two quantities are added. We will continue this bias toward degrees by expressing the phase angles in degrees.

It should be apparent at this point in our discussion that the sinusoidal signal is completely specified once its *maximum amplitude, frequency,* and *phase angle* are given. Examples 10.1, 10.2, and 10.3 illustrate these basic properties of the sinusoidal function.

Example 10.1 A sinusoidal current has a maximum amplitude of 20 A. The current passes through one complete cycle in 1.0 ms. The magnitude of the current at zero time is 10 A.

a) What is the frequency of the current in hertz?

b) What is the frequency in radians/second?

c) Write the expression for $i(t)$ using the cosine function. Express ϕ in degrees.

Solution   a) From the statement of the problem we have $T = 1$ ms, hence $f = 1/T = 1000$ Hz.

b) $\omega = 2\pi f = 2000\pi$ rad/s.

c) We have that $i(t) = I_m \cos(\omega t + \phi) = 20 \cos(2000\pi t + \phi)$. But $i(0) = 10$ A; therefore $10 = 20 \cos\phi$, from which we know that $\phi = 60°$. Thus our expression for $i(t)$ becomes

$$i(t) = 20 \cos(2000\pi t + 60°). \quad \blacksquare$$

Example 10.2 A sinusoidal voltage is given by the expression $v = 300 \cos(120\pi t + 30°)$.

a) What is the period of the voltage in ms?

b) What is the frequency in Hz?

c) What is the magnitude of v at $t = 2.778$ ms?

Solution   a) From the expression for v we note that $\omega = 120\pi$ rad/s. Since $\omega = 2\pi/T$ it follows that $T = 2\pi/\omega = \frac{1}{60}$ s, or 16.667 ms.

b) The frequency is $1/T$, or 60 Hz.

c) At $t = 2.778$ ms, ωt is very nearly 1.047 radians, or 60°; therefore $v(2.778 \text{ ms}) = 300 \cos(60° + 30°) = 0$ V. $\quad \blacksquare$

Example 10.3 Since we have adopted the cosine function to represent our sinusoidal signals, it is worth noting that the sine function can be translated to the cosine function by subtracting 90° ($\pi/2$ radians) from the argument of the sine function.

a) Verify that this is so by showing that

$$\sin(\omega t + \theta) = \cos(\omega t + \theta - 90°).$$

b) Use the result of part (a) to express $\sin(\omega t + 30°)$ as a cosine function.

Solution a) Our verification is done by a direct application of the trigonometric identity

$$\cos(\alpha - \beta) = \cos\alpha\cos\beta + \sin\alpha\sin\beta.$$

We let $\alpha = \omega t + \theta$ and $\beta = 90°$. Since $\cos 90° = 0$ and $\sin 90° = 1$, we have

$$\cos(\alpha - \beta) = \sin\alpha = \sin(\omega t + \theta) = \cos(\omega t + \theta - 90°).$$

b) From part (a) we have

$$\sin(\omega t + 30°) = \cos(\omega t + 30° - 90°) = \cos(\omega t - 60°). \quad\blacksquare$$

With these attributes of the sinusoidal function in mind, we are now ready to discuss the problem of finding the steady-state sinusoidal response of a circuit.

DRILL EXERCISE 10.1 A sinusoidal voltage is given by the expression

$$v = 40\cos(2513.27t + 36.87°).$$

Find (a) f in Hz; (b) T in ms; (c) V_m; (d) $v(0)$; (e) ϕ in degrees and radians; (f) the smallest positive value of t at which $v = 0$; and (g) the smallest positive value of t at which $dv/dt = 0$.

Ans. (a) 400 Hz; (b) 2.5 ms; (c) 40 V; (d) 32 V; (e) 36.87°, or 0.6435 radian; (f) 368.96 μs; (g) 993.96 μs.

10.3 THE SINUSOIDAL RESPONSE

Before we focus on the steady-state response to sinusoidal sources, let us pause long enough to look at the problem in broader terms. Such an overview will help us keep the steady-state solution in perspective. The general nature of the problem can be described in terms of the circuit shown in Fig. 10.2, where v_s is a sinusoidal voltage, that is,

$$v_s = V_m\cos(\omega t + \phi). \tag{10.4}$$

For convenience, the initial current in the circuit is assumed to be zero. Time is

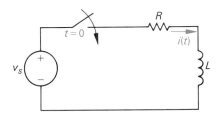

Fig. 10.2 An *RL* circuit excited by a sinusoidal voltage source.

measured from the moment when the switch closes. The problem is to derive the expression for $i(t)$ when $t \geq 0$. Thinking back to Chapter 7 we see that our problem is quite similar to finding the step response of the RL circuit. The only difference is that the voltage source is now a time-varying sinusoidal voltage rather than a constant, or dc, voltage. A direct application of Kirchhoff's voltage law to the circuit in Fig. 10.2 leads us to the ordinary differential equation

$$L \frac{di}{dt} + Ri = V_m \cos(\omega t + \phi), \tag{10.5}$$

the formal solution of which is discussed in an introductory course in differential equations. We ask those of you who have not yet studied differential equations to accept that the solution for i is

$$i = \frac{-V_m}{\sqrt{R^2 + \omega^2 L^2}} \cos(\phi - \theta) e^{-(R/L)t} + \frac{V_m}{\sqrt{R^2 + \omega^2 L^2}} \cos(\omega t + \phi - \theta), \tag{10.6}$$

where the angle θ is defined as the angle whose tangent is $\omega L/R$. Thus for a given circuit driven by a sinusoidal source of known frequency, θ is easily determined.

We can check the validity of Eq. (10.6) by showing that it satisfies Eq. (10.5) for all values of $t \geq 0$. We first show that Eq. (10.6) equals zero for $t = 0$, thus confirming that the initial current is zero, as stipulated by the initial condition. Next, we substitute Eq. (10.6) into the left-hand side of Eq. (10.5) and show that it is equivalent to the right-hand side, thus proving that the solution is valid for all $t > 0$. (See Problem 10.5.)

The first term on the right-hand side of Eq. (10.6) is referred to as the *transient component* of the current because it becomes infinitesimal as time elapses. The second term on the right-hand side of Eq. (10.6) is known as the *steady-state component* of the solution. It will exist as long as the switch remains closed and the source continues to supply the sinusoidal voltage. Our purpose in this chapter is to develop a technique for calculating the steady-state response directly, thus avoiding the problem of solving the differential equation. It should be pointed out that in solving for the steady-state solution directly we forfeit obtaining either the transient component or the total response, which is the sum of the transient and steady-state components.

We now focus on the steady-state portion of Eq. (10.6) because it exhibits the characteristics of the sinusoidal steady-state that are true in general. There are four important observations to make.

1. The steady-state solution is a sinusoidal function.
2. The frequency of the response signal is identical to the frequency of the source signal. This is always true in a linear circuit when the circuit parameters R, L, and C are constant. (If there are frequencies in the response signals that are not present in the source signals, there is a nonlinear element in the circuit.)

3. The maximum amplitude of the steady-state response will, in general, differ from the maximum amplitude of the source. For the circuit under discussion, the maximum amplitude of the response signal is $V_m/\sqrt{R^2 + \omega^2 L^2}$ and the maximum amplitude of the signal source is V_m.

4. The phase angle of the response signal will, in general, differ from the phase angle of the source. For the problem at hand, the phase angle of the current is $(\phi - \theta)$ and that of the voltage source is ϕ.

These observations are important because they help us understand the motivation for the phasor method, which we begin developing in Section 10.4. In particular, observe that once the decision has been made to find only the steady-state response, the problem is reduced to finding the maximum amplitude and phase angle of the response signal. The waveform and frequency of the response are already known.

DRILL EXERCISE
10.2

The voltage applied to the circuit in Fig. 10.2 at $t = 0$ is $100 \cos (400t + 60°)$. The circuit resistance is 40 Ω and the initial current in the 75-mH inductor is zero.

a) Find $i(t)$ for $t \geq 0$.

b) Write the expressions for the transient and steady-state components of $i(t)$.

c) Find the numerical value of i 1.875 ms after the switch has been closed.

d) What are the maximum amplitude, the frequency (in rad/s), and the phase angle of the steady-state current?

e) By how many degrees are the voltage and the steady-state current out of phase?

Ans. (a) $-1.84e^{-533.33t} + 2 \cos (400t + 23.13°)$ A; (b) $-1.84e^{-533.33t}$ A, $2 \cos (400t + 23.13°)$ A; (c) 133.61 mA; (d) 2 A, 400 rad/s, 23.13°; (e) 36.87°.

10.4 THE PHASOR

The phasor is a complex number† that carries the amplitude and phase angle information of a sinusoidal function. The phasor concept is rooted in Euler's identity, which relates the exponential function to the trigonometric function:

$$e^{\pm j\theta} = \cos \theta \pm j \sin \theta. \qquad (10.7)$$

Equation (10.7) is important to the work at hand because it gives us another way of expressing the cosine and sine functions.‡ We can think of the cosine function as the real part of the exponential function and the sine function

---

† The reader who feels a bit uneasy about complex numbers may wish to pause here and peruse Appendix A.

‡ Since the letter i is used for current in electrical engineering literature, the letter j has been adopted to signify the square root of minus one, that is, $j = \sqrt{-1}$.

as the imaginary part of the exponential function, that is,

$$\cos \theta = \mathscr{Re}\{e^{j\theta}\} \tag{10.8}$$

and

$$\sin \theta = \mathscr{Im}\{e^{j\theta}\}, \tag{10.9}$$

where $\mathscr{Re}$ stands for "the real part of" and $\mathscr{Im}$ stands for "the imaginary part of."

Since we have already chosen to use the cosine function in our analysis of the sinusoidal steady state (Section 10.1), we can make direct application of Eq. (10.8). In particular, let us write the sinusoidal voltage function given by Eq. (10.1) in the form suggested by Eq. (10.8). We have

$$\begin{aligned} v &= V_m \cos(\omega t + \phi) \\ &= V_m \mathscr{Re}\{e^{j(\omega t + \phi)}\} \\ &= V_m \mathscr{Re}\{e^{j\omega t}e^{j\phi}\}. \end{aligned} \tag{10.10}$$

We can move the coefficient V_m inside the argument of the "real part of" without altering the result. We can also reverse the order of the two exponential functions inside the argument so that we can write Eq. (10.10) as

$$v = \mathscr{Re}\{V_m e^{j\phi} e^{j\omega t}\}. \tag{10.11}$$

In studying Eq. (10.11) observe that the coefficient of the exponential $e^{j\omega t}$ is a complex number that carries the amplitude and phase angle of the given sinusoidal function. This complex number is by definition the *phasor representation,* or *phasor transform,* of the given sinusoidal function. Thus

$$\mathbf{V} = V_m e^{j\phi} = \mathscr{P}\{V_m \cos(\omega t + \phi)\}, \tag{10.12}$$

where the notation $\mathscr{P}\{V_m \cos(\omega t + \phi)\}$ is read "the phasor transform of $V_m \cos(\omega t + \phi)$." Thus the phasor transform transfers the sinusoidal function from the time domain to the complex-number domain. As in Eq. (10.12), a phasor quantity will be represented throughout the text by a boldface letter.

Equation (10.12) is the polar form of a phasor. A phasor can also be expressed in rectangular form. Thus Eq. (10.12) can be written as

$$\mathbf{V} = V_m \cos \phi + jV_m \sin \phi. \tag{10.13}$$

We will find both polar and rectangular forms useful in circuit applications of the phasor concept.

One additional comment regarding Eq. (10.12) is in order. The frequent occurrence of the exponential function $e^{j\phi}$ has led to a shorthand abbreviation that lends itself to text material. This abbreviation is the angle notation

$$1\underline{/\phi} \equiv 1e^{j\phi}.$$

We will use this notation extensively in the material that follows.

Thus far we have emphasized moving from the sinusoidal function to its phasor transform. However, it should be apparent that we can also reverse the process. That is, given the phasor we can write the expression for the sinusoidal function. Thus if we are given that $\mathbf{V} = 100\underline{/-26°}$, the expression for v is 100 cos $(\omega t - 26°)$, since we have decided to use the cosine function for all sinusoids. Observe that we cannot deduce the value of ω from the phasor. *The phasor carries only amplitude and phase information.* The step of going from the phasor transform to the time-domain expression is referred to as finding the *inverse* phasor transform and is formalized by the equation

$$\mathscr{P}^{-1}\{V_m e^{j\phi}\} = \mathscr{R}e\{V_m e^{j\phi} e^{j\omega t}\}, \tag{10.14}$$

where the notation $\mathscr{P}^{-1}\{V_m e^{j\phi}\}$ is read as "the inverse phasor transform of $V_m e^{j\phi}$." Equation (10.14) tells us that to find the inverse phasor transform we multiply the phasor by $e^{j\omega t}$ and then extract the real part of the product.

The phasor transform is so useful in circuit analysis because it reduces the problem of finding the maximum amplitude and phase angle of the steady-state sinusoidal response to the algebra of complex numbers. We can show this through the following observations.

1. Since the transient component vanishes as time elapses, the steady-state component of the solution must also satisfy the differential equation (see Problem 10.5b).

2. In a linear circuit driven by sinusoidal sources, the steady-state response is also sinusoidal.

3. Using the notation introduced in Eq. (10.8) we can postulate that the steady-state solution will be of the form $\mathscr{R}e\{Ae^{j\beta} e^{j\omega t}\}$, where A is the maximum amplitude of the response and β is the phase angle of the response.

4. When the postulated steady-state solution is substituted into the differential equation, the exponential term $e^{j\omega t}$ cancels out, leaving the solution for A and β in the domain of complex numbers.

Let us illustrate these observations using the circuit in Fig. 10.2.

We know that the steady-state solution for the current i is of the form

$$i_{ss}(t) = \mathscr{R}e\{I_m e^{j\beta} e^{j\omega t}\}, \tag{10.15}$$

where we have used the subscript "ss" to emphasize the fact that we are dealing with the steady-state solution. When we substitute Eq. (10.15) into Eq. (10.5) we generate the expression

$$\mathscr{R}e\{j\omega L I_m e^{j\beta} e^{j\omega t}\} + \mathscr{R}e\{R I_m e^{j\beta} e^{j\omega t}\} = \mathscr{R}e\{V_m e^{j\phi} e^{j\omega t}\}. \tag{10.16}$$

In deriving Eq. (10.16) we have used the fact that both differentiation and multiplication by a constant can be taken inside the "real part of" operation. We have also rewritten the right-hand side of Eq. (10.5) using the notation of Eq. (10.8). The sum of the real parts is the same as the real part of the sum. Therefore the left-hand side of Eq. (10.16) can be reduced to a single term:

$$\mathscr{R}e\{(j\omega L + R)I_m e^{j\beta} e^{j\omega t}\} = \mathscr{R}e\{V_m e^{j\phi} e^{j\omega t}\}, \tag{10.17}$$

which is true only if the arguments are equal; thus

$$(j\omega L + R)I_m e^{j\beta}e^{j\omega t} = V_m e^{j\phi}e^{j\omega t},$$

or

$$I_m e^{j\beta} = \frac{V_m e^{j\phi}}{R + j\omega L} \qquad (10.18)$$

We see from Eq. (10.18) that $e^{j\omega t}$ has been eliminated from the determination of the amplitude (I_m) and phase angle (β) of the response. Thus for this circuit the problem of finding I_m and β involves the algebraic manipulation of the complex quantities $V_m e^{j\phi}$ and $R + j\omega L$. Note that we encountered both polar and rectangular forms.

The phasor transform is also useful in circuit analysis because it applies directly to the sum of sinusoidal functions. Since circuit analysis involves summing currents and voltages, the importance of this observation is obvious. We can formalize this property as follows. If

$$v = v_1 + v_2 + \cdots + v_n, \qquad (10.19)$$

where all the voltages on the right-hand side are sinusoidal voltages of the same frequency, then

$$\mathbf{V} = \mathbf{V}_1 + \mathbf{V}_2 + \cdots + \mathbf{V}_n. \qquad (10.20)$$

Thus the phasor representation of the sum is the sum of the phasors of the individual terms. The development of Eq. (10.20) is discussed in Section 10.6.

Before we apply the phasor transform to circuit analysis, we pause long enough to illustrate its usefulness in solving a problem with which you are already familiar, namely, adding sinusoidal functions via trigonometric identities. Example 10.4 has been designed to show how the phasor transform greatly simplifies this type of problem.

Example 10.4   Given that $y_1 = 20 \cos (\omega t - 30°)$ and $y_2 = 40 \cos (\omega t + 60°)$, express $y = y_1 + y_2$ as a single sinusoidal function.

a) Solve by using trigonometric identities.

b) Solve by using the phasor concept.

Solution   a) First we expand both y_1 and y_2 using the cosine of the sum of two angles to get

$$y_1 = 20 \cos \omega t \cos 30° + 20 \sin \omega t \sin 30°,$$

$$y_2 = 40 \cos \omega t \cos 60° - 40 \sin \omega t \sin 60°.$$

If we then add y_1 and y_2, we obtain

$$y = (20 \cos 30 + 40 \cos 60) \cos \omega t + (20 \sin 30 - 40 \sin 60) \sin \omega t$$

$$= 37.32 \cos \omega t - 24.64 \sin \omega t.$$

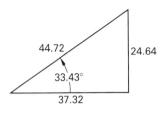

Fig. 10.3   A right triangle used in the solution for y.

To combine these two terms we treat the coefficients of the cosine and sine as sides of a right triangle (Fig. 10.3) and then multiply and divide the right-hand side by the hypotenuse. Our expression for y becomes

$$y = 44.72 \left[\frac{37.42}{44.72} \cos \omega t - \frac{24.64}{44.72} \sin \omega t \right]$$

$$= 44.72[\cos 33.43° \cos \omega t - \sin 33.43° \sin \omega t].$$

We again invoke the identity involving the cosine of the sum of two angles and write

$$y = 44.72 \cos (\omega t + 33.43°).$$

b) The problem can be solved using phasors as follows. Since

$$y = y_1 + y_2$$

it follows from Eq. (10.20) that

$$\mathbf{Y} = \mathbf{Y}_1 + \mathbf{Y}_2$$
$$= 20 \underline{/-30°} + 40 \underline{/60°}$$
$$= (17.32 - j10) + (20 + j34.64)$$
$$= 37.32 + j24.64 = 44.72 \underline{/33.43°}.$$

Once the phasor $\mathbf{Y}$ is known, the corresponding trigonometric function for y can be written by taking the inverse phasor transform. Thus

$$y = \mathcal{P}^{-1}\{44.72e^{j33.43}\} = \mathcal{R}e\,\{44.72e^{j33.43}e^{j\omega t}\} = 44.72 \cos (\omega t + 33.43°).$$

The superiority of the phasor approach to adding sinusoidal functions should be apparent. You should note that the phasor approach requires the ability to move back and forth between the polar and rectangular forms. ∎

DRILL EXERCISE
10.3

Find the phasor transform of each of the following trigonometric functions.

a) $v = 170 \cos (377t - 40°)$ V

b) $i = 10 \sin (1000t + 20°)$ A

c) $i = [5 \cos (\omega t + 36.87°) + 10 \cos (\omega t - 53.13°)]$ A

d) $v = [300 \cos (20{,}000\pi t + 45°) - 100 \sin (20{,}000\pi t + 30°)]$ mV

Ans. (a) $\mathbf{V} = 170\underline{/-40°}$ V; (b) $\mathbf{I} = 10\underline{/-70°}$ A; (c) $\mathbf{I} = 11.18\underline{/-26.57°}$ A; (d) $\mathbf{V} = 339.90\underline{/61.51°}$ mV.

DRILL EXERCISE Find the time-domain expression corresponding to each of the following phasors.

10.4
a) $\mathbf{V} = 86.3\underline{/+26°}$ V
b) $\mathbf{I} = [10\underline{/30°} + 25\underline{/60°}]$ mA
c) $\mathbf{V} = [60 + j30 + 100\underline{/-28°}]$ V

Ans. (a) $v = 86.3 \cos(\omega t + 26°)$ V; (b) $i = 34.03 \cos(\omega t + 51.55°)$ mA; (c) $v = 149.26 \cos(\omega t - 6.52°)$ V.

10.5 THE PASSIVE CIRCUIT ELEMENTS IN THE PHASOR DOMAIN

The systematic application of the phasor transform in circuit analysis requires two steps. First, we must establish the relationship between the phasor current and the phasor voltage at the terminals of the passive circuit elements. Second, we must develop the phasor-domain version of Kirchhoff's laws, which we discussed in Section 10.6. In this section, we will establish the relationship between the phasor current and the phasor voltage at the terminals of the resistor, inductor, and capacitor. We begin with the resistor. We will use the passive sign convention in all our derivations.

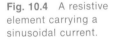

Fig. 10.4 A resistive element carrying a sinusoidal current.

It follows directly from Ohm's law that if the current in a resistor varies sinusoidally with time, that is, if $i = I_m \cos(\omega t + \theta_i)$, then the voltage at the terminals of the resistor (Fig. 10.4) will be

$$v = R[I_m \cos(\omega t + \theta_i)]$$
$$= RI_m[\cos(\omega t + \theta_i)], \tag{10.21}$$

where I_m is the maximum amplitude of the current in amperes and θ_i is the phase angle of the current.

The phasor transform of this voltage is

$$\mathbf{V} = RI_m e^{j\theta_i} = RI_m\underline{/\theta_i}. \tag{10.22}$$

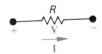

Fig. 10.5 The phasor domain equivalent circuit of a resistor.

But $I_m\underline{/\theta_i}$ is the phasor representation of the sinusoidal current; thus we can write Eq. (10.22) as

$$\mathbf{V} = R\mathbf{I}, \tag{10.23}$$

which tells us that the phasor voltage at the terminals of a resistor is simply the resistance times the phasor current. The circuit diagram for a resistor in the phasor domain is shown in Fig. 10.5.

Equation (10.21) or (10.23) also contains another important piece of information, namely, that at the terminals of a resistor there is no phase shift between the current and voltage. This phase relationship is shown in Fig. 10.6. The signals are said to be in time phase because they both reach corresponding

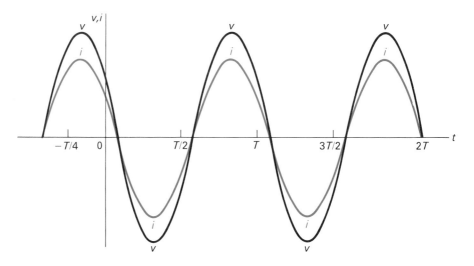

Fig. 10.6 A plot showing that the voltage and current at the terminals of a resistor are in phase ($\theta_i = 60°$).

values on their respective curves at the same time (for example, they are at their positive maxima at the same instant).

The relationship between the phasor current and phasor voltage at the terminals of the inductor is derived by assuming a sinusoidal current and using $L\, di/dt$ to establish the corresponding voltage. Thus given $i = I_m \cos(\omega t + \theta_i)$, we see that the expression for the voltage is

$$v = L\frac{di}{dt} = -\omega L I_m \sin(\omega t + \theta_i). \tag{10.24}$$

Now we can rewrite Eq. (10.24) using the cosine function:

$$v = -\omega L I_m \cos(\omega t + \theta_i - 90°). \tag{10.25}$$

The phasor representation of the voltage given by Eq. (10.25) is

$$\begin{aligned}
\mathbf{V} &= -\omega L I_m e^{j(\theta_i - 90°)} \\
&= -\omega L I_m e^{j\theta_i} e^{-j90} \\
&= j\omega L I_m e^{j\theta_i} \\
&= j\omega L \mathbf{I}
\end{aligned} \tag{10.26}$$

Note that in deriving Eq. (10.26) we used the identity

$$e^{-j90°} = \cos 90° - j\sin 90° = -j.$$

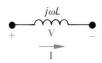

Fig. 10.7 The phasor domain equivalent circuit for an inductor.

From Eq. (10.26) we see that the phasor voltage at the terminals of an inductor equals $j\omega L$ times the phasor current. The phasor-domain equivalent circuit for the inductor is shown in Fig. 10.7.

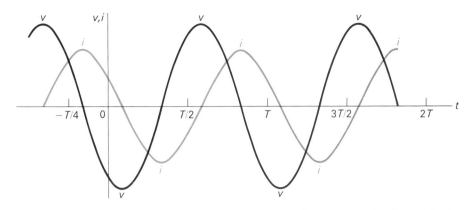

Fig. 10.8   A plot showing the phase relationship between the current and voltage at the terminals of an inductor ($\theta_i = 60°$).

The phase relationship between the current and voltage at the terminals of an inductor is contained in Eq. (10.26). This relationship becomes obvious if we rewrite Eq. (10.26) as

$$\mathbf{V} = (\omega L \underline{/90°})(I_m \underline{/\theta_i})$$
$$= \omega L I_m \underline{/\theta_i + 90°}, \tag{10.27}$$

from which we see that the voltage and current will be out of phase by exactly 90°. In particular, the voltage will lead the current by 90° or, what is equivalent, the current will lag behind the voltage by 90°. This concept of "voltage leading current" or "current lagging voltage" is made clear by the plot shown in Fig. 10.8, where we note that as time increases from zero the voltage comes to a particular value on the sine function ahead of the current. For example, it reaches its negative peak exactly 90° before the current reaches its negative peak. The same observation can be made with respect to the "zero-going-positive" crossing or the positive peak. Thus we say that the voltage *leads* the current by 90° or, what is equivalent, the current *lags* behind the voltage by 90°.

The phase shift can also be expressed in seconds. A phase shift of 90° corresponds to one fourth of a period, hence the voltage leads the current by $T/4$, or $1/4f$ second.

The relationship between the phasor current and phasor voltage at the terminals of a capacitor follows directly from the derivation of Eq. (10.26). That is, if we note that for a capacitor

$$i = C\frac{dv}{dt}$$

and then assume

$$v = V_m \cos{(\omega t + \theta_v)},$$

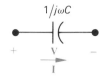

$1/j\omega C$

Fig. 10.9 The phasor domain equivalent circuit for a capacitor.

it follows that

$$\mathbf{I} = j\omega C\mathbf{V}. \qquad (10.28)$$

Now if we solve Eq. (10.28) for the voltage as a function of the current we get

$$\mathbf{V} = \left(\frac{1}{j\omega C}\right)\mathbf{I}. \qquad (10.29)$$

From Eq. (10.29) we see that the equivalent circuit for the capacitor in the phasor domain is as shown in Fig. 10.9.

The voltage across the terminals of a capacitor will *lag* behind the capacitor current by exactly 90°. We can easily see this by rewriting Eq. (10.29) as

$$\mathbf{V} = \left(\frac{1}{\omega C}\right)\underline{/-90°}\ I_m\underline{/\theta_i}$$

$$= \frac{I_m}{\omega C}\ \underline{/\theta_i - 90°}. \qquad (10.30)$$

The alternative way to express the phase relationship contained in Eq. (10.30) is to say that the current *leads* the voltage by 90°. The phase relationship between the current and voltage at the terminals of a capacitor is shown in Fig. 10.10.

We conclude our discussion of the passive circuit elements in the phasor domain with a very important observation. When we compare Eqs. (10.23), (10.26), and (10.29), we note that they are all of the form

$$\mathbf{V} = Z\mathbf{I}, \qquad (10.31)$$

where Z represents the *impedance* of the circuit element. Thus the impedance of a resistor is R, the impedance of an inductor is $j\omega L$, and the impedance of a capacitor is $1/j\omega C$. In all cases, the impedance is measured in ohms. The concept of impedance is very powerful in sinusoidal steady-state analysis and we

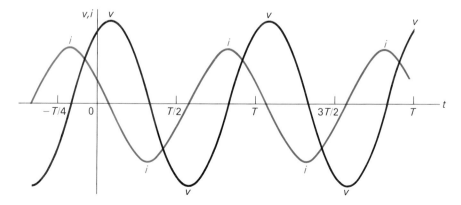

Fig. 10.10 A plot showing the phase relationship between the current and voltage at the terminals of a capacitor ($\theta_i = 60°$).

will have much more to say about its usefulness in subsequent sections. There are a couple of additional terms associated with the impedance of the passive circuit elements that we introduce at this time.

The term *inductive reactance* is used to describe the product ωL. The reactance of an inductor is ωL (Ω), whereas the impedance of an inductor is $j\omega L$ (Ω).

The term *capacitive reactance* is used to describe the quantity $-1/\omega C$. The reactance of a capacitor is $-1/\omega C$ (Ω), whereas the impedance of a capacitor is $1/j\omega C$ or $j(-1/\omega C)$ (Ω).

One final comment about impedance. Note that although impedance is a complex number it is *not* a phasor. It is important to remember that although phasors are complex numbers, not all complex numbers are phasors.

Finally, we remind you that if the reference direction for the current in a passive circuit element is in the direction of the voltage rise across the element, a minus sign must be inserted into the equation that relates the voltage to the current.

DRILL EXERCISE
10.5

The current in the 75-mH inductor is 4 cos (40,000t − 38°) mA. Calculate (a) the inductive reactance; (b) the impedance of the inductor; (c) the phasor voltage **V**; and (d) the steady-state expression for $v(t)$.

Ans. (a) 3000 Ω; (b) j3000 Ω; (c) 12$\underline{/52°}$; (d) 12 cos (40,000t + 52°) V.

DRILL EXERCISE
10.6

The voltage across the terminals of the 0.2-μF capacitor is 40 cos (10^5t − 50°) V. Calculate (a) the capacitive reactance; (b) the impedance of the capacitor; (c) the phasor current **I**; and (d) the steady-state expression for $i(t)$.

Ans. (a) −50 Ω; (b) −j50 Ω; (c) 0.8$\underline{/40°}$; (d) 0.8 cos (10^5t + 40°) A.

10.6 KIRCHHOFF'S LAWS IN THE PHASOR DOMAIN

We pointed out in Section 10.4, with reference to Eqs. (10.19) and (10.20), that the phasor transform is useful in circuit analysis because it applies to the sum of sinusoidal functions. This usefulness was illustrated in Example 10.4. We would now like to formalize this observation by developing Kirchhoff's laws in the phasor domain.

We begin by assuming that v_1 through v_n represent voltages around a closed path in a circuit. Furthermore, we assume that the circuit is operating in a sinusoidal steady state. Thus Kirchhoff's voltage law requires that

$$v_1 + v_2 + \cdots + v_n = 0, \tag{10.32}$$

which in the sinusoidal steady state becomes

$$V_{m_1} \cos (\omega t + \theta_1) + V_{m_2} \cos (\omega t + \theta_2) + \cdots + V_{m_n} \cos (\omega t + \theta_n) = 0. \tag{10.33}$$

Now we use Euler's identity to write Eq. (10.33) as

$$\mathscr{Re}[V_{m_1} e^{j\theta_1} e^{j\omega t}] + \mathscr{Re}\{V_{m_2} e^{j\theta_2} e^{j\omega t}\} + \cdots + \mathscr{Re}\{V_{m_n} e^{j\theta_n} e^{j\omega t}\} = 0. \tag{10.34}$$

From the algebra of complex numbers we know that the sum of the real parts is the same as the real part of the sum; therefore we can rewrite Eq. (10.34) as

$$\mathscr{Re}\{V_{m_1} e^{j\theta_1} e^{j\omega t} + V_{m_2} e^{j\theta_2} e^{j\omega t} + \cdots + V_{m_n} e^{j\theta_n} e^{j\omega t}\} = 0. \tag{10.35}$$

The term $e^{j\omega t}$ can be factored from each term, enabling us to write

$$\mathscr{Re}\{[V_{m_1} e^{j\theta_1} + V_{m_2} e^{j\theta_2} + \cdots + V_{m_n} e^{j\theta_n}] e^{j\omega t}\} = 0,$$

or

$$\mathscr{Re}\{[\mathbf{V}_1 + \mathbf{V}_2 + \cdots + \mathbf{V}_n] e^{j\omega t}\} = 0. \tag{10.36}$$

But $e^{j\omega t}$ cannot equal zero; therefore,

$$\mathbf{V}_1 + \mathbf{V}_2 + \cdots + \mathbf{V}_n = 0, \tag{10.37}$$

which is the statement of Kirchhoff's voltage law as it applies to phasor voltages. That is, if Eq. (10.32) applies to a set of sinusoidal voltages in the time domain, then Eq. (10.37) is the equivalent statement in the phasor domain.

A similar derivation applies to a set of sinusoidal currents. Thus if we are given that

$$i_1 + i_2 + \cdots + i_n = 0, \tag{10.38}$$

then it follows that

$$\mathbf{I}_1 + \mathbf{I}_2 + \cdots + \mathbf{I}_n = 0, \tag{10.39}$$

where $\mathbf{I}_1, \mathbf{I}_2, \ldots, \mathbf{I}_n$ are the phasor representations of the individual currents $i_1, i_2, \ldots, i_n$.

Equations (10.31), (10.37), and (10.39) form the basis for circuit analysis in the phasor domain. Observe that Eq. (10.31) has the same algebraic form as Ohm's law and Eqs. (10.37) and (10.39) state Kirchhoff's laws for phasor quantities. Therefore all the techniques developed for analyzing resistive circuits can be used to find phasor currents and voltages. It should be a comfort to the reader to learn that new analysis techniques are not needed to analyze circuits in the phasor domain. The basic tools of

1. series–parallel simplifications,
2. Δ-to-Y transformations,

3. source transformations,

4. Thévenin–Norton equivalent circuits,

5. superposition,

6. node-voltage analysis, and

7. mesh-current analysis

all can be used in the analysis of circuits in the phasor domain. The problem of learning phasor circuit analysis consists of two fundamental parts. First, you must be able to construct the phasor-domain model of a circuit. Second, you must be able to algebraically manipulate complex numbers and/or quantities to arrive at a solution. These attributes of phasor analysis will be illustrated in the discussion that follows.

Four branches terminate at a common node. The reference direction of each branch current i_1, i_2, i_3, and i_4 is toward the node. If $i_1 = 100 \cos (\omega t + 25°)$ A, $i_2 = 100 \cos (\omega t + 145°)$ A, and $i_3 = 100 \cos (\omega t - 95°)$ A, find i_4.

Ans. $i_4 = 0$.

10.7   SERIES–PARALLEL AND DELTA-TO-WYE SIMPLIFICATIONS

The rules for combining impedances in series, or parallel, and for making delta-to-wye transformations are the same as those for resistors. The only difference is that combining impedances involves the algebraic manipulation of complex numbers.

Impedances in series can be combined into a single impedance by simply adding the individual impedances. The problem is defined in general terms by the circuit shown in Fig. 10.11, where the impedances $Z_1, Z_2, \ldots, Z_n$ are connected in series between the terminals a, b. The impedances are in series because they carry the same phasor current **I**. From Eq. (10.31) we know that the voltage drop across each impedance will be $Z_1\mathbf{I}, Z_2\mathbf{I}, \ldots, Z_n\mathbf{I}$ and from Kirch-

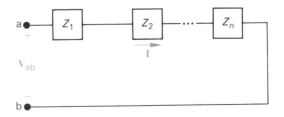

Fig. 10.11   Impedances in series.

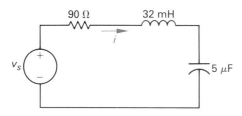

Fig. 10.12   The circuit for Example 10.5.

hoff's voltage law we have

$$\mathbf{V}_{ab} = \mathbf{Z}_1\mathbf{I} + \mathbf{Z}_2\mathbf{I} + \cdots + \mathbf{Z}_n\mathbf{I}$$
$$= (\mathbf{Z}_1 + \mathbf{Z}_2 + \cdots + \mathbf{Z}_n)\mathbf{I}. \tag{10.40}$$

The equivalent impedance between the terminals a, b is

$$\mathbf{Z}_{ab} = \frac{\mathbf{V}_{ab}}{\mathbf{I}} = \mathbf{Z}_1 + \mathbf{Z}_2 + \cdots + \mathbf{Z}_n. \tag{10.41}$$

Example 10.5 illustrates a numerical application of Eq. (10.41).

Example 10.5   A 90-Ω resistor, a 32-mH inductor, and a 5-μF capacitor are connected in series across the terminals of a sinusoidal voltage source, as shown in Fig. 10.12. The steady-state expression for the source voltage v_s is $750 \cos (5000t + 30°)$.

a) Construct the phasor-domain equivalent circuit.

b) Calculate the steady-state current i by the phasor method.

Solution   a) From the expression for v_s we have $\omega = 5000$ rad/s. Therefore the impedance of the 32-mH inductor is

$$\mathbf{Z}_L = j\omega L = j(5000)(32 \times 10^{-3}) = j160 \ \Omega,$$

and the impedance of the capacitor is

$$\mathbf{Z}_C = j\left(\frac{-1}{\omega C}\right) = -j\frac{10^6}{(5000)(5)} = -j40 \ \Omega.$$

The phasor transform of v_s is

$$\mathbf{V}_s = 750\underline{/30°} \text{ V}.$$

The phasor-domain equivalent circuit of the circuit shown in Fig. 10.12 is illustrated in Fig. 10.13.

b) We can compute the phasor current by simply dividing the voltage of the voltage source by the equivalent impedance between the terminals a, b. From Eq. (10.41) we have

$$\mathbf{Z}_{ab} = 90 + j160 - j40$$
$$= 90 + j120 = 150\underline{/53.13°} \ \Omega.$$

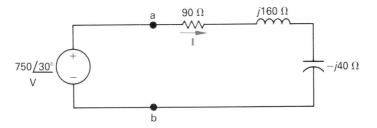

Fig. 10.13 The phasor domain equivalent circuit of the circuit shown in Fig. 10.12.

Thus

$$\mathbf{I} = \frac{750\underline{/30°}}{150\underline{/53.13°}} = 5\underline{/-23.13°} \text{ A.}$$

We can now write the steady-state expression for i directly:

$$i = 5 \cos (5000t - 23.13°) \text{ A.} \quad \blacksquare$$

Impedances connected in parallel can be reduced to a single equivalent impedance by the reciprocal relationship

$$\frac{1}{Z_{ab}} = \frac{1}{Z_1} + \frac{1}{Z_2} + \cdots + \frac{1}{Z_n}. \tag{10.42}$$

The parallel connection of impedances is illustrated in Fig. 10.14. Note that the impedances are in parallel because they have the same voltage across their terminals. The derivation of Eq. (10.42) follows directly from Fig. 10.14. We simply combine Kirchhoff's current law with the phasor-domain version of Ohm's law, that is, Eq. (10.31). From Fig. 10.14 we have

$$\mathbf{I} = \mathbf{I}_1 + \mathbf{I}_2 + \cdots + \mathbf{I}_n$$

or

$$\frac{\mathbf{V}}{Z_{ab}} = \frac{\mathbf{V}}{Z_1} + \frac{\mathbf{V}}{Z_2} + \cdots + \frac{\mathbf{V}}{Z_n}. \tag{10.43}$$

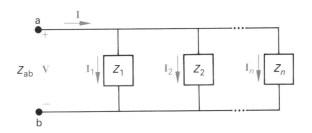

Fig. 10.14 Impedances in parallel.

Once the common voltage term is canceled out of Eq. (10.43) we see that we have Eq. (10.42).

Equation (10.42) can also be expressed in terms of *admittance*, which is defined as the reciprocal of impedance and is denoted by the letter Y. Thus

$$Y = \frac{1}{Z} = G + jB \quad \text{(siemens)}. \tag{10.44}$$

Like conductance, admittance is measured in siemens and commonly expressed in mhos.

When we use Eq. (10.44) in Eq. (10.42) we have

$$Y_{ab} = Y_1 + Y_2 + \cdots + Y_n. \tag{10.45}$$

We also note from Eq. (10.42) that for the special case of just two impedances in parallel,

$$Z_{ab} = \frac{Z_1 Z_2}{Z_1 + Z_2}. \tag{10.46}$$

It is also worth noting the admittance of each of the ideal passive circuit elements. In particular, the admittance of a resistor is $1/R$ mhos. We have already noted in our discussion of resistive circuits that the reciprocal of resistance is called *conductance*. The admittance of an inductor is $1/j\omega L$, or $j(-1/\omega L)$ mhos. The quantity $-1/\omega L$ is referred to as the *inductive susceptance* of the inductor. The admittance of a capacitor is $j\omega C$ mhos. The quantity ωC is referred to as the *capacitive susceptance* of the capacitor.

Example 10.6 illustrates the application of Eqs. (10.44) and (10.45) to a specific circuit.

Example 10.6   The sinusoidal current source in the circuit in Fig. 10.15 produces the current $i_s = 8 \cos 200{,}000t$ A.

a) Construct the phasor domain equivalent circuit.

b) Find the steady-state expressions for v, i_1, i_2, and i_3.

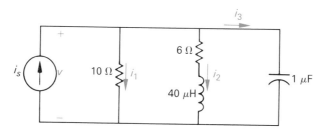

Fig. 10.15   The circuit for Example 10.6.

Solution   a) The phasor transform of the current source is $8\underline{/0°}$; the resistors transfer directly to the phasor domain as 10 Ω and 6 Ω; the 40-μH inductor has an impedance of $j8$ Ω at the given frequency of 200,000 rad/s; and at this frequency the 1-μF capacitor has an impedance of $-j5$ Ω. The phasor domain equivalent circuit is shown in Fig. 10.16, where the phasor transforms of the unknowns are also symbolized.

   b) In studying the circuit in Fig. 10.16 we note that we can easily obtain the voltage across the current source once we know the equivalent impedance of the three parallel branches. We also observe that once **V** is known, we can calculate the three phasor currents $\mathbf{I}_1$, $\mathbf{I}_2$, and $\mathbf{I}_3$ using Eq. (10.31). To find the equivalent impedance of the three branches we first find the equivalent admittance by simply adding the admittances of each branch. The admittance of the first branch is

$$\mathbf{Y}_1 = \frac{1}{10} = 0.1 \ \mho,$$

the admittance of the second branch is

$$\mathbf{Y}_2 = \frac{1}{6 + j8} = \frac{6 - j8}{100} = 0.06 - j0.08 \ \mho,$$

and the admittance of the third branch is

$$\mathbf{Y}_3 = \frac{1}{-j5} = j0.2 \ \mho.$$

The admittance of the three branches is

$$\mathbf{Y} = \mathbf{Y}_1 + \mathbf{Y}_2 + \mathbf{Y}_3 = 0.16 + j0.12 = 0.2\underline{/36.87°} \ \mho.$$

The impedance seen by the current source is

$$\mathbf{Z} = 1/\mathbf{Y} = 5\underline{/-36.87°} \ \Omega.$$

The voltage **V** is

$$\mathbf{V} = \mathbf{ZI} = 40\underline{/-36.87°} \ \text{V}.$$

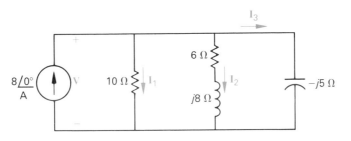

Fig. 10.16   A phasor domain equivalent circuit.

It follows directly that

$$\mathbf{I}_1 = \frac{40\underline{/-36.87°}}{10} = 4\underline{/-36.87°} = 3.2 - j2.4 \text{ A},$$

$$\mathbf{I}_2 = \frac{40\underline{/-36.87°}}{6 + j8} = 4\underline{/-90°} = -j4 \text{ A},$$

and

$$\mathbf{I}_3 = \frac{40\underline{/-36.87°}}{5\underline{/-90°}} = 8\underline{/53.13°} = 4.8 + j6.4 \text{ A}.$$

It is worth noting that we can check our computations at this point by verifying that

$$\mathbf{I}_1 + \mathbf{I}_2 + \mathbf{I}_3 = \mathbf{I}.$$

Specifically,

$$3.2 - j2.4 - j4 + 4.8 + j6.4 = 8 + j0.$$

The corresponding steady-state time domain expressions are

$$v = 40 \cos (200{,}000t - 36.87°) \text{ V},$$
$$i_1 = 4 \cos (200{,}000t - 36.87°) \text{ A},$$
$$i_2 = 4 \cos (200{,}000t - 90°) \text{ A},$$
$$i_3 = 8 \cos (200{,}000t + 53.13°) \text{ A}. \quad \blacksquare$$

DRILL EXERCISE
10.8

A 100-Ω resistor is connected in parallel with a 50-mH inductor. This parallel combination is connected in series with a 10-Ω resistor and a 10-μF capacitor.

a) Calculate the impedance of this interconnection if the frequency is 1 krad/s.

b) Repeat part (a) if the frequency is increased to 4 krad/s.

c) At what finite frequency will the impedance of the interconnection be purely resistive?

d) What is the impedance at the frequency found in part (c)?

Ans. (a) $30 - j60$ Ω; (b) $90 + j15$ Ω; (c) 2 krad/s; (d) 60 Ω.

DRILL EXERCISE
10.9

The interconnection described in Drill Exercise 10.8 is connected across the terminals of a voltage source that is generating the voltage $300 \cos 2000t$ V. What is the maximum amplitude of the current in the 50-mH inductor?

Ans. 3.54 A.

DRILL EXERCISE
10.10

Three branches having impedances of $3 + j4$ Ω, $16 - j12$ Ω and $-j4$ Ω, respectively, are connected in parallel. What are the equivalent (a) admittance, (b) conductance, and (c) susceptance of the parallel connection in m℧? (d) If the paralleled branches are excited from a sinusoidal current source where $i = 8 \cos \omega t$ A, what is the maximum amplitude of the current in the purely capacitive branch?

Ans. (a) $200\underline{/36.87°}$; (b) 160; (c) 120; (d) 10 A.

The Δ-to-Y transformation that we discussed in Section 3.12 with regard to resistive circuits also applies to impedances. The Δ-connected impedances along with the Y-equivalent circuit are defined in Fig. 10.17. The Y-impedances as functions of the Δ-impedances are given by the following:

$$Z_1 = \frac{Z_b Z_c}{Z_a + Z_b + Z_c}, \tag{10.47}$$

$$Z_2 = \frac{Z_c Z_a}{Z_a + Z_b + Z_c}, \tag{10.48}$$

$$Z_3 = \frac{Z_a Z_b}{Z_a + Z_b + Z_c}. \tag{10.49}$$

It is also possible to reverse the Δ-to-Y transformation. That is, we can start with the Y-structure and replace it with an equivalent Δ-structure. The Δ-impedances as functions of the Y-impedances are given by the following:

$$Z_a = \frac{Z_1 Z_2 + Z_2 Z_3 + Z_3 Z_1}{Z_1}, \tag{10.50}$$

$$Z_b = \frac{Z_1 Z_2 + Z_2 Z_3 + Z_3 Z_1}{Z_2}, \tag{10.51}$$

$$Z_c = \frac{Z_1 Z_2 + Z_2 Z_3 + Z_3 Z_1}{Z_3}. \tag{10.52}$$

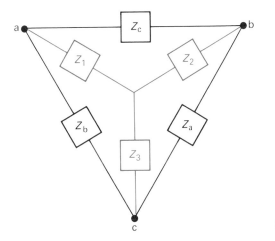

Fig. 10.17 The delta-wye transformation.

The derivation of Eqs. (10.47)–(10.49) or Eqs. (10.50)–(10.52) follows the same thought process as was used to derive the corresponding equations for pure resistive circuits. In fact, if we compare Eqs. (3.37)–(3.39) with Eqs. (10.47)–(10.49) and Eqs. (3.40)–(3.42) with Eqs. (10.50)–(10.52), we can see that the symbol R has been replaced by the symbol Z. You may wish to review Problem 3.22 concerning the derivation of the Δ-to-Y transformation.

The usefulness of the Δ-to-Y transformation in phasor circuit analysis is illustrated in the following example.

Example 10.7   Use a Δ-to-Y impedance transformation to find $\mathbf{I}_0$, $\mathbf{I}_1$, $\mathbf{I}_2$, $\mathbf{I}_3$, $\mathbf{I}_4$, $\mathbf{I}_5$, $\mathbf{V}_1$, and $\mathbf{V}_2$ in the circuit in Fig. 10.18.

Solution   First note that the circuit is not amenable to series–parallel simplifications as it now stands. A Δ-to-Y impedance transformation allows us to solve for all the branch currents without resorting to either the node-voltage method or the mesh-current method. In studying the circuit, we see that if either the upper delta (abc) or the lower delta (bcd) is replaced by its Y-equivalent, we can further simplify the resulting circuit by series–parallel combinations. In deciding which delta to replace, it is worth checking the sum of the impedances around the delta since this quantity forms the denominator for the equivalent Y-impedances. Since the sum around the lower delta is $30 + j40$, we choose to eliminate it from the circuit. The Y-impedance connecting to terminal b is

$$Z_1 = \frac{(20 + j60)(10)}{(30 + j40)} = 12 + j4 \ \Omega,$$

the Y-impedance connecting to terminal c is

$$Z_2 = \frac{10(-j20)}{30 + j40} = -3.2 - j2.4 \ \Omega,$$

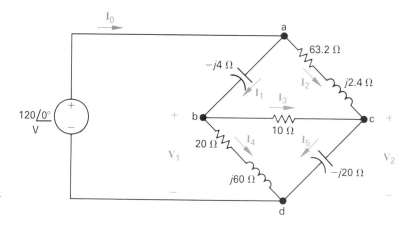

Fig. 10.18   The circuit for Example 10.7.

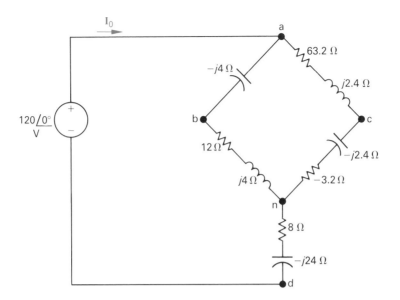

Fig. 10.19   The circuit in Fig. 10.18 with the lower delta replaced by its equivalent wye.

and the Y-impedance connecting to terminal d is

$$Z_3 = \frac{(20 + j60)(-j20)}{30 + j40} = 8 - j24 \ \Omega.$$

When the Y-equivalent impedances are inserted into the circuit, we get the circuit shown in Fig. 10.19, which can now be simplified by series–parallel reductions. The impedance of the abn branch is

$$Z_{abn} = 12 + j4 - j4 = 12 \ \Omega,$$

and the impedance of the acn branch is

$$Z_{acn} = 63.2 + j2.4 - j2.4 - 3.2 = 60 \ \Omega.$$

Now observe that the abn branch is in parallel with the acn branch; therefore these two branches can be replaced with a single branch having an impedance of

$$Z_{an} = \frac{(60)(12)}{72} = 10 \ \Omega.$$

This 10-Ω resistor can be combined with the impedance between n and d so that the circuit in Fig. 10.19 reduces to that shown in Fig. 10.20. It follows directly from the circuit in Fig. 10.20 that

$$\mathbf{I}_0 = \frac{120/0°}{(18 - j24)} = 4\underline{/53.13°} = 2.4 + j3.2 \ \text{A}.$$

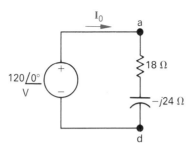

Fig. 10.20 A simplified version of the circuit shown in Fig. 10.19.

Once we know $\mathbf{I}_0$, we can work back through the equivalent circuits to find the branch currents in the original circuit. We begin by noting that $\mathbf{I}_0$ is the current in the branch nd of Fig. 10.19. Therefore

$$\mathbf{V}_{nd} = (8 - j24)\mathbf{I}_0 = 96 - j32 \text{ V.}$$

We can now calculate the voltage $\mathbf{V}_{an}$ since

$$\mathbf{V} = \mathbf{V}_{an} + \mathbf{V}_{nd}$$

and both $\mathbf{V}$ and $\mathbf{V}_{nd}$ are known. Thus

$$\mathbf{V}_{an} = 120 - 96 + j32 = 24 + j32 \text{ V.}$$

We can now compute the branch currents $\mathbf{I}_{abn}$ and $\mathbf{I}_{acn}$.

$$\mathbf{I}_{abn} = \frac{24 + j32}{12} = 2 + j\frac{8}{3} \text{ A,}$$

$$\mathbf{I}_{acn} = \frac{24 + j32}{60} = \frac{4}{10} + j\frac{8}{15} \text{ A.}$$

At this point in our analysis, we note that in terms of the branch currents defined in Fig. 10.18

$$\mathbf{I}_1 = \mathbf{I}_{abn} = 2 + j\frac{8}{3} \text{ A,}$$

$$\mathbf{I}_2 = \mathbf{I}_{acn} = \frac{4}{10} + j\frac{8}{15} \text{ A.}$$

We can check our calculations of $\mathbf{I}_1$ and $\mathbf{I}_2$ by noting that

$$\mathbf{I}_1 + \mathbf{I}_2 = 2.4 + j3.2 = \mathbf{I}_0.$$

To find the branch currents $\mathbf{I}_3$, $\mathbf{I}_4$, and $\mathbf{I}_5$ we must first calculate the voltages $\mathbf{V}_1$ and $\mathbf{V}_2$. Referring to Fig. 10.18 we see that

$$\mathbf{V}_1 = 120\underline{/0°} - (-j4)\mathbf{I}_1 = \frac{328}{3} + j8 \text{ V}$$

and

$$\mathbf{V}_2 = 120\underline{/0°} - (63.2 + j2.4)\mathbf{I}_2 = 96 - j\frac{104}{3} \text{ V.}$$

The branch currents $\mathbf{I}_3$, $\mathbf{I}_4$, and $\mathbf{I}_5$ can now be calculated:

$$\mathbf{I}_3 = \frac{\mathbf{V}_1 - \mathbf{V}_2}{10} = \frac{4}{3} + j\frac{12.8}{3}\ \text{A},$$

$$\mathbf{I}_4 = \frac{\mathbf{V}_1}{20 + j60} = \frac{2}{3} - j1.6\ \text{A},$$

$$\mathbf{I}_5 = \frac{\mathbf{V}_2}{-j20} = \frac{26}{15} + j4.8\ \text{A}.$$

We can check our calculations by noting that

$$\mathbf{I}_4 + \mathbf{I}_5 = \frac{2}{3} + \frac{26}{15} - j1.6 + j4.8 = 2.4 + j3.2 = \mathbf{I}_0,$$

$$\mathbf{I}_3 + \mathbf{I}_4 = \frac{4}{3} + \frac{2}{3} + j\frac{12.8}{3} - j1.6 = 2 + j\frac{8}{3} = \mathbf{I}_1,$$

$$\mathbf{I}_3 + \mathbf{I}_2 = \frac{4}{3} + \frac{4}{10} + j\frac{12.8}{3} + j\frac{8}{15} = \frac{26}{15} + j4.8 = \mathbf{I}_5.\quad\blacksquare$$

DRILL EXERCISE
10.11

Use a Δ-to-Y transformation to find the current $\mathbf{I}$ in the circuit shown.

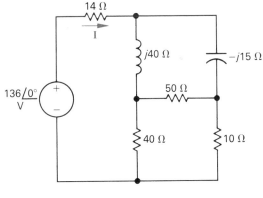

Ans. $\mathbf{I} = 4\underline{/+28.07°}$ A.

10.8 SOURCE TRANSFORMATIONS AND THÉVENIN–NORTON EQUIVALENT CIRCUITS

The source transformations introduced in Section 4.8 and the Thévenin–Norton equivalent circuits discussed in Section 4.9 are analytical techniques that can also be applied to phasor domain circuits. To prove the validity of these techniques, we follow the same thought process that we used in Sections 4.8 and 4.9, except that we substitute impedance (Z) for resistance (R). The source

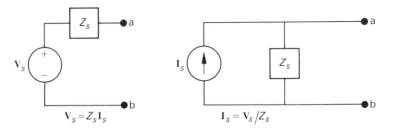

Fig. 10.21   A source transformation in the phasor domain.

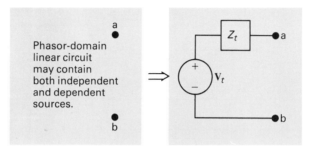

Fig. 10.22   The phasor domain version of the Thévenin equivalent circuit.

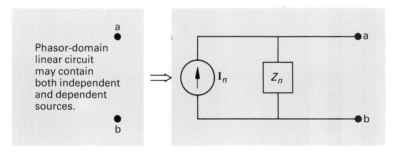

Fig. 10.23   The phasor domain version of the Norton equivalent circuit.

transformation equivalent circuit using the nomenclature of the phasor domain is shown in Fig. 10.21.

The phasor domain version of the Thévenin equivalent circuit is illustrated in Fig. 10.22 and the Norton equivalent circuit is shown in Fig. 10.23. The techniques for finding the Thévenin equivalent voltage and impedance are identical to those used for resistive circuits, except that the phasor domain equivalent circuit involves the manipulation of complex quantities. The same can be said for finding the Norton equivalent current and impedance.

The application of the source transformation equivalent circuit to phasor domain analysis is illustrated in Example 10.8. The details of finding a

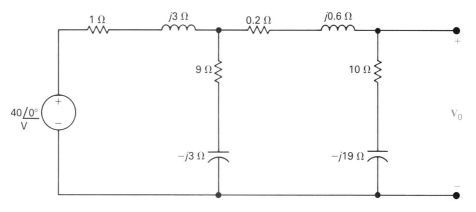

Fig. 10.24   The circuit for Example 10.8.

Thévenin equivalent circuit in the phasor domain are illustrated in Example 10.9.

Example 10.8   Use the concept of source transformations to find the phasor voltage $\mathbf{V}_0$ in the circuit in Fig. 10.24.

Solution   We begin by noting that the series combination of the voltage source [$40\underline{/0°}$] and the impedance of $1 + j3$ Ω can be replaced by the parallel combination of a current source and the $1 + j3$-Ω impedance. The source current is

$$\mathbf{I} = \frac{40}{1 + j3} = \frac{40}{10}(1 - j3) = 4 - j12 \text{ A.}$$

Thus the circuit in Fig. 10.24 can be modified as shown in Fig. 10.25. Note that

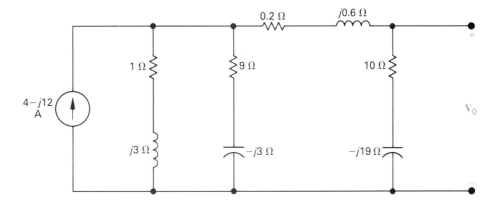

Fig. 10.25   The first step in reducing the circuit in Fig. 10.24.

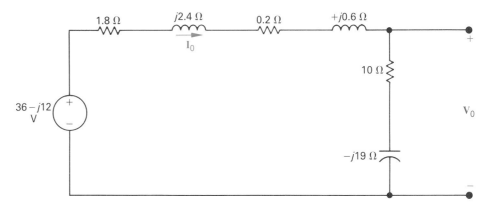

Fig. 10.26 The second step in simplifying the circuit in Fig. 10.24.

the reference direction for $\mathbf{I}$ follows from the given polarity reference of the 40-V source.

Next we combine the two parallel branches into a single impedance,

$$Z = \frac{(1 + j3)(9 - j3)}{10} = 1.8 + j2.4 \ \Omega,$$

which is in parallel with the current source of $4 - j12$ A. Another source transformation will convert this parallel combination to a series combination consisting of a voltage source in series with the impedance of $1.8 + j2.4 \ \Omega$. The voltage of the voltage source will be

$$\mathbf{V} = (4 - j12)(1.8 + j2.4) = 36 - j12 \text{ V.}$$

Using this source transformation, we can redraw the circuit in Fig. 10.25 as shown in Fig. 10.26. Note the polarity of the voltage source. The current $\mathbf{I}_0$ has been added to the circuit to expedite the solution for $\mathbf{V}_0$.

At this point in our solution for $\mathbf{V}_0$ we see that the circuit has been reduced to a simple series circuit. The current $\mathbf{I}_0$ can be calculated by dividing the voltage of the source by the total series impedance. Specifically,

$$\mathbf{I}_0 = \frac{36 - j12}{12 - j16} = \frac{12(3 - j1)}{4(3 - j4)} = \frac{39 + j27}{25} = 1.56 + j1.08 \text{ A.}$$

Now we can obtain the value of $\mathbf{V}_0$ by multiplying $\mathbf{I}_0$ by the impedance $10 - j19$; thus

$$\mathbf{V}_0 = (1.56 + j1.08)(10 - j19) = 36.12 - j18.84 \text{ V.} \quad \blacksquare$$

Example 10.9 Find the Thévenin equivalent circuit with respect to the terminals a, b for the circuit shown in Fig. 10.27.

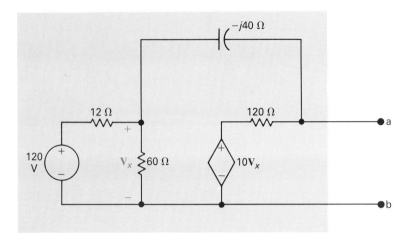

Fig. 10.27 The circuit for Example 10.9.

Solution We begin by first finding the Thévenin equivalent voltage. This voltage is the open-circuit voltage appearing at the terminals a, b. We will choose the reference for the Thévenin voltage as positive at terminal a. First we see that we can make two source transformations relative to the 120-V, 12-Ω, and 60-Ω circuit elements that will simplify this portion of the circuit. At the same time, these transformations must preserve the identity of the controlling voltage $\mathbf{V}_x$ because of the dependent voltage source. We can determine the two source transformations by first replacing the series combination of the 120-V source and 12-Ω resistor with a 10-A current source in parallel with 12 Ω. Next we replace the parallel combination of the 12-Ω and 60-Ω resistors by a single 10-Ω resistor. Finally we replace the 10-A source in parallel with 10 Ω by a 100-V source in series with 10 Ω. The resulting circuit is shown in Fig. 10.28. The current **I** has been added to the circuit in Fig. 10.28 to facilitate further discussion. With a little thought, it should be apparent to you that once we know **I** we can compute the Thévenin voltage. We can find the current **I** by summing the voltages around the closed path in the circuit in Fig. 10.28. It follows that

$$100 = 10\mathbf{I} - j40\mathbf{I} + 120\mathbf{I} + 10\mathbf{V}_x = (130 - j40)\mathbf{I} + 10\mathbf{V}_x.$$

We can relate the controlling voltage $\mathbf{V}_x$ to the current **I** by noting from Fig. 10.28 that

$$\mathbf{V}_x = 100 - 10\mathbf{I}.$$

It follows directly that

$$\mathbf{I} = \frac{-900}{30 - j40} = 18\underline{/-126.87°} \text{ A}.$$

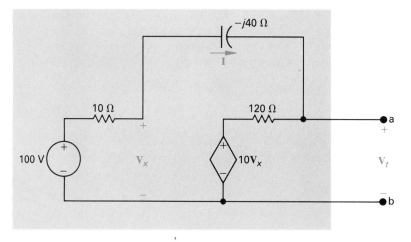

Fig. 10.28 A simplified version of the circuit shown in Fig. 10.27.

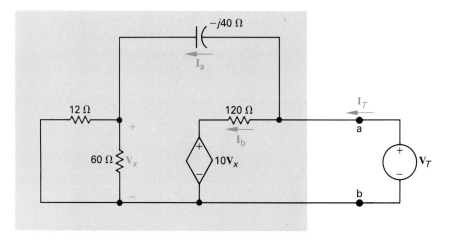

Fig. 10.29 A circuit for calculating the Thévenin equivalent impedance.

Now we calculate $\mathbf{V}_x$:

$$\mathbf{V}_x = 100 - 180\underline{/-126.87°} = 208 + j144 \text{ V.}$$

Finally we note from Fig. 10.28 that

$$\mathbf{V}_t = 10\mathbf{V}_x + 120\mathbf{I}$$
$$= 2080 + j1440 + 120(18)\underline{/-126.87°}$$
$$= 784 - j288 = 835.22\underline{/-20.17°} \text{ V.}$$

To find the Thévenin impedance, we can use any of the techniques previously used to find the Thévenin resistance. We will illustrate the test-source method in this example. Recall that in using the test-source method we deactivate all independent sources from the circuit and then apply either a test voltage source or a test current source to the terminals of interest. The ratio of the voltage to the current at the test source is the Thévenin impedance. This technique as it applies to the circuit in Fig. 10.27 is shown in Fig. 10.29, where we note that we have chosen a test voltage source $\mathbf{V}_T$. Also note that the independent voltage source has been deactivated by an appropriate short circuit and the identity of $\mathbf{V}_x$ has been preserved. The branch currents $\mathbf{I}_a$ and $\mathbf{I}_b$ have been added to the circuit to facilitate the calculation of $\mathbf{I}_T$. By straightforward applications of Kirchhoff's circuit laws, you should be able to verify the following relationships:

$$\mathbf{I}_a = \frac{\mathbf{V}_T}{10 - j40},$$

$$\mathbf{V}_x = 10\mathbf{I}_a,$$

$$\mathbf{I}_b = \frac{\mathbf{V}_T - 10\mathbf{V}_x}{120}$$

$$= \frac{-\mathbf{V}_T(9 + j4)}{120(1 - j4)},$$

$$\mathbf{I}_T = \mathbf{I}_a + \mathbf{I}_b$$

$$= \frac{\mathbf{V}_T}{10 - j40}\left[1 - \frac{(9 + j4)}{12}\right]$$

$$= \frac{\mathbf{V}_T(3 - j4)}{12(10 - j40)},$$

$$Z_t = \frac{\mathbf{V}_T}{\mathbf{I}_T} = 91.2 - j38.4 \ \Omega.$$

The Thévenin equivalent circuit is shown in Fig. 10.30. ∎

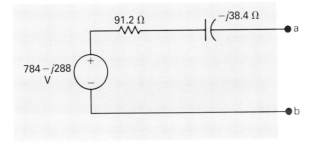

Fig. 10.30 The Thévenin equivalent circuit for the circuit in Fig. 10.27.

Find the Thévenin equivalent with respect to the terminals a, b.

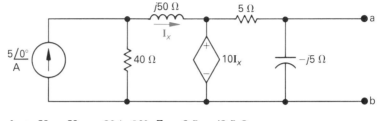

Ans. $\mathbf{V}_t = \mathbf{V}_{ab} = 20\underline{/-90°}$; $Z_t = 2.5 - j2.5\ \Omega$.

10.9 THE NODE-VOLTAGE METHOD

We introduced in Sections 4.2–4.4 the basic concepts behind the node-voltage method of circuit analysis. These same observations apply when the node-voltage method is used to analyze phasor domain circuits. Our purpose here is to simply illustrate, by numerical example, the solution of a phasor domain circuit by the node-voltage technique. Drill Exercise 10.13 and Problems 10.21–10.23 have been designed to give you some experience using the node-voltage method to solve for steady-state sinusoidal responses.

Example 10.10   Use the node-voltage method to find the branch currents $\mathbf{I}_a$, $\mathbf{I}_b$, and $\mathbf{I}_c$ in the circuit in Fig. 10.31.

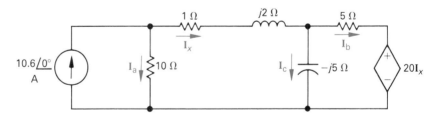

Fig. 10.31   The circuit for Example 10.10.

Solution   The circuit in Fig. 10.31 can be described in terms of two node voltages since the circuit contains three essential nodes. Because four branches terminate at the essential node, which stretches across the bottom of Fig. 10.31, we will use it as the reference node. The remaining two essential nodes are labeled 1 and 2 and the appropriate node voltages are designated as $\mathbf{V}_1$ and $\mathbf{V}_2$. Figure 10.32 reflects the choice of the reference node and the terminal labels. Summing the

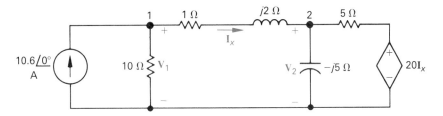

Fig. 10.32 The circuit in Fig. 10.31 with the node voltages defined.

currents away from node 1 yields the equation

$$-10.6 + \frac{\mathbf{V}_1}{10} + \frac{\mathbf{V}_1 - \mathbf{V}_2}{1 + j2} = 0.$$

Multiplying through by $(1 + j2)$ and collecting the coefficients of $\mathbf{V}_1$ and $\mathbf{V}_2$ generates the expression

$$\mathbf{V}_1(1.1 + j0.2) - \mathbf{V}_2 = 10.6 + j21.2.$$

Summing the currents away from node 2 gives us

$$\frac{\mathbf{V}_2 - \mathbf{V}_1}{(1 + j2)} + \frac{\mathbf{V}_2}{-j5} + \frac{\mathbf{V}_2 - 20\mathbf{I}_x}{5} = 0.$$

The controlling current $\mathbf{I}_x$ is

$$\mathbf{I}_x = \frac{\mathbf{V}_1 - \mathbf{V}_2}{1 + j2}.$$

Substituting this expression for $\mathbf{I}_x$ into the node-2 equation, multiplying through by $1 + j2$, and collecting coefficients of $\mathbf{V}_1$ and $\mathbf{V}_2$ produces the equation

$$-5\mathbf{V}_1 + (4.8 + j0.6)\mathbf{V}_2 = 0.$$

The solutions for $\mathbf{V}_1$ and $\mathbf{V}_2$ are

$$\mathbf{V}_1 = 68.40 - j16.80 \text{ V},$$
$$\mathbf{V}_2 = 68 - j26 \text{ V}.$$

The branch currents directly follow:

$$\mathbf{I}_a = \frac{\mathbf{V}_1}{10} = 6.84 - j1.68 \text{ A},$$

$$\mathbf{I}_x = \frac{\mathbf{V}_1 - \mathbf{V}_2}{1 + j2} = 3.76 + j1.68 \text{ A},$$

$$\mathbf{I}_b = \frac{\mathbf{V}_2 - 20\mathbf{I}_x}{5} = -1.44 - j11.92 \text{ A},$$

$$\mathbf{I}_c = \frac{\mathbf{V}_2}{-j5} = 5.2 + j13.6 \text{ A}.$$

As a check on our work we note that

$$\mathbf{I}_{a} + \mathbf{I}_{x} = 6.84 - j1.68 + 3.76 + j1.68 = 10.6 \text{ A}$$

and

$$\mathbf{I}_{x} = \mathbf{I}_{b} + \mathbf{I}_{c} = -1.44 - j11.92 + 5.2 + j13.6 = 3.76 + j1.68 \text{ A.} \quad \blacksquare$$

DRILL EXERCISE   Use the node-voltage method to find the steady-state expression for $v(t)$ in the circuit
10.13   shown. The sinusoidal sources are $i_s = 10 \cos \omega t$ A and $v_s = 100 \sin \omega t$ V, where $\omega = 50$ krad/s.

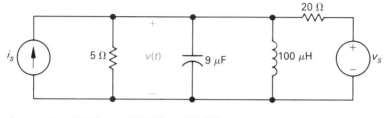

Ans. $v(t) = 31.62 \cos (50,000t - 71.57°)$.

10.10   THE MESH-CURRENT METHOD

The mesh-current method can also be used in the analysis of phasor domain circuits. The procedures used in phasor domain applications of the method are the same as those used in analyzing resistive circuits. We introduced in Sections 4.5–4.7 the basic techniques of the mesh-current method. We will demonstrate the extension of the mesh-current method to phasor domain circuits by Example 10.11.

Example 10.11   Use the mesh-current method of circuit analysis to find the voltages $\mathbf{V}_1$, $\mathbf{V}_2$, and $\mathbf{V}_3$ in the circuit in Fig. 10.33.

Solution   Since the circuit has two windows and a dependent voltage source, we must write two mesh-current equations and a constraint equation. The reference direction for the mesh currents $\mathbf{I}_1$ and $\mathbf{I}_2$ is clockwise, as shown in Fig. 10.34, from which we can see that once we know $\mathbf{I}_1$ and $\mathbf{I}_2$ we can easily find the unknown voltages. Summing the voltages around mesh 1 gives us

$$150 = (1 + j2)\mathbf{I}_1 + (12 - j16)(\mathbf{I}_1 - \mathbf{I}_2)$$

or

$$150 = (13 - j14)\mathbf{I}_1 - (12 - j16)\mathbf{I}_2.$$

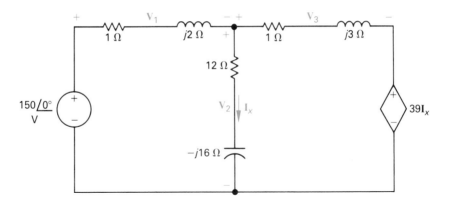

Fig. 10.33 The circuit for Example 10.11.

Summing the voltages around mesh 2 generates the equation

$$0 = (12 - j16)(\mathbf{I}_2 - \mathbf{I}_1) + (1 + j3)\mathbf{I}_2 + 39\mathbf{I}_x.$$

From Fig. 10.34 we see that the controlling current $\mathbf{I}_x$ is the difference between $\mathbf{I}_1$ and $\mathbf{I}_2$. In equation form, our constraint is

$$\mathbf{I}_x = \mathbf{I}_1 - \mathbf{I}_2.$$

When this constraint is substituted into the mesh-2 equation and the resulting expression is simplified, we get

$$0 = (27 + j16)\mathbf{I}_1 - (26 + j13)\mathbf{I}_2.$$

Solving for $\mathbf{I}_1$ and $\mathbf{I}_2$, we get

$$\mathbf{I}_1 = -26 - j52 \text{ A,}$$
$$\mathbf{I}_2 = -24 - j58 \text{ A,}$$
$$\mathbf{I}_x = -2 + j6 \text{ A.}$$

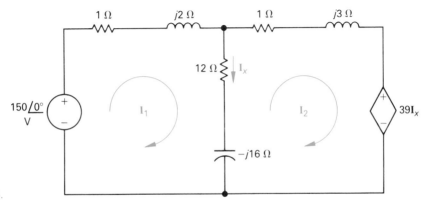

Fig. 10.34 Mesh currents used to solve the circuit in Fig. 10.33.

The three voltages are

$$\mathbf{V}_1 = (1 + j2)\mathbf{I}_1 = 78 - j104 \text{ V},$$
$$\mathbf{V}_2 = (12 - j16)\mathbf{I}_x = 72 + j104 \text{ V},$$
$$\mathbf{V}_3 = (1 + j3)\mathbf{I}_2 = 150 - j130 \text{ V},$$
$$39\mathbf{I}_x = -78 + j234 \text{ V}.$$

We can check our calculations by summing the voltages around closed paths. Thus

$$-150 + \mathbf{V}_1 + \mathbf{V}_2 = -150 + 78 - j104 + 72 + j104 = 0,$$
$$-\mathbf{V}_2 + \mathbf{V}_3 + 39\mathbf{I}_x = -72 - j104 + 150 - j130 - 78 + j234 = 0,$$
$$-150 + \mathbf{V}_1 + \mathbf{V}_3 + 39\mathbf{I}_x = -150 + 78 + j104$$
$$+ 150 - j130 - 78 + j234 = 0. \quad \blacksquare$$

DRILL EXERCISE
10.14

Use the mesh-current method to find the phasor current **I** in the circuit shown.

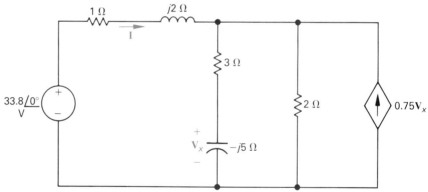

Ans. $\mathbf{I} = 29 + j2 = 29.07\underline{/3.95°}$.

10.11 PHASOR DIAGRAMS

At times when we are using the phasor method to analyze the steady-state sinusoidal operation of a circuit, a graphical diagram of the phasor currents and voltages gives further insight into the behavior of the circuit. A phasor diagram shows the magnitude and phase angle of each phasor quantity in the complex-number plane. Phase angles are measured counterclockwise from the positive real axis, and magnitudes are measured from the origin of the axes. For example, the phasor quantities $10\underline{/30°}$, $12\underline{/150°}$, $5\underline{/-45°}$, and $8\underline{/-170°}$ are drawn as shown in Fig. 10.35.

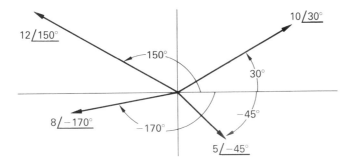

Fig. 10.35 A graphical representation of phasors.

Constructing phasor diagrams of circuit quantities generally involves both currents and voltages. As a result, two different magnitude scales are necessary, one for currents and one for voltages. We also call to your attention that the ability to visualize a phasor quantity on the complex-number plane can be very useful when checking pocket calculator calculations. Since the typical pocket calculator does not offer a printout of the data entered, it is very helpful when entering complex numbers to visualize in which quadrant the number lies. Then when the calculated angle is displayed, you can quickly check whether you have keyed in the appropriate values. For example, suppose the polar form of $-7 - j3$ is to be computed. Without making any calculations you should anticipate a magnitude greater than 7 and an angle in the third quadrant that is more negative than $-135°$ or less positive than $225°$. These observations should be apparent from Fig. 10.36.

To illustrate the usefulness of phasor diagrams in circuit analysis, we construct a phasor diagram for the circuit shown in Fig. 10.37. In this circuit, the parallel combination of R_2 and L_2 represents a load on the output end of a dis-

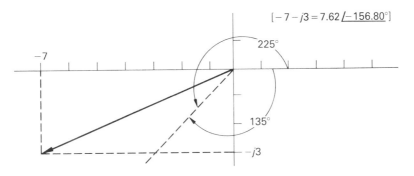

Fig. 10.36 The complex number $-7 - j3$.

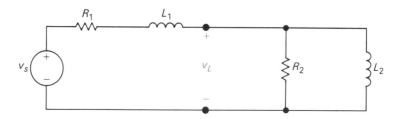

Fig. 10.37 A circuit illustrating the construction of a phasor diagram.

tribution line, which is modeled by the series combination of R_1 and L_1. The sinusoidal voltage source is modeled by v_s.

Our analysis problem can be described as follows. We wish to know what effect the addition of a capacitor across the terminals of the load will have on the amplitude of v_s if we adjust v_s so that the amplitude of v_L remains constant. This is a technique used by utility companies to control the voltage drop on their lines. Let's see how it works.

We begin by assuming zero capacitance. After constructing the phasor diagram for the zero-capacitance case, we can add the capacitor to the circuit and study its effect on the amplitude of v_s assuming the amplitude of v_L is held constant. The phasor domain equivalent circuit of the circuit in Fig. 10.37 is shown in Fig. 10.38. The phasor branch currents $\mathbf{I}$, $\mathbf{I}_a$, and $\mathbf{I}_b$ have been added to Fig. 10.38 to aid in the discussion of the phasor diagram.

The stepwise evolution of the phasor diagram is shown in Fig. 10.39. In relating the phasor diagram to the circuit in Fig. 10.38, we note the following.

1. Since we are holding the amplitude of the load voltage constant, we have chosen $\mathbf{V}_L$ as our reference.

2. We know that $\mathbf{I}_a$ will be in phase with $\mathbf{V}_L$ and that its magnitude will be $|\mathbf{V}_L|/R_2$. (On the phasor diagram, the magnitude scale for the current phasors is independent of the magnitude scale chosen for voltage phasors.)

3. We know that $\mathbf{I}_b$ will lag behind $\mathbf{V}_L$ by 90° and that its magnitude will be $|\mathbf{V}_L|/\omega L_2$.

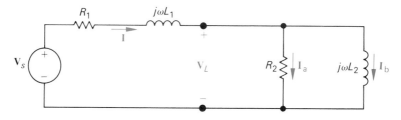

Fig. 10.38 A phasor domain equivalent circuit.

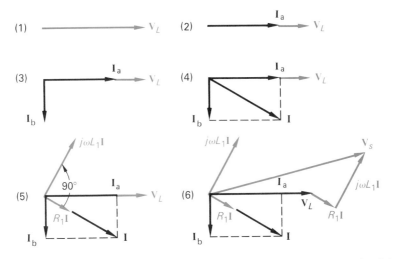

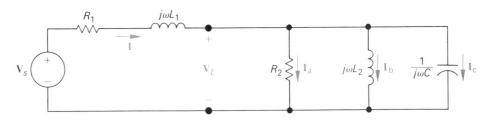

Fig. 10.39 The step-by-step evolution of the phasor diagram for the circuit in Fig. 10.38.

4. The line current $\mathbf{I}$ is equal to the sum of $\mathbf{I}_a$ and $\mathbf{I}_b$.

5. The voltage drop across R_1 will be in phase with the line current and the voltage drop across $j\omega L_1$ will lead the line current by 90°.

6. The source voltage is the sum of the load voltage and the drop along the line, that is, $\mathbf{V}_s = \mathbf{V}_L + (R_1 + j\omega L_1)\mathbf{I}$.

Note that the completed phasor diagram shown in part (6) of Fig. 10.39 shows very clearly the amplitude and phase angle relationships between all the currents and voltages in the circuit in Fig. 10.38.

We now add the capacitor branch as shown in Fig. 10.40. Since we are holding $\mathbf{V}_L$ constant, we can construct the phasor diagram for the circuit in Fig. 10.40 following the same steps as those in Fig. 10.39 except that at step 4 we add the capacitor current $\mathbf{I}_c$ to the diagram. In so doing, we note that $\mathbf{I}_c$ will lead $\mathbf{V}_L$ by 90° and that its magnitude will be $|\mathbf{V}_L|\omega C$. The effect of $\mathbf{I}_c$ on the line current is shown in Fig. 10.41, from which we can see that both the magnitude

Fig. 10.40 The addition of a capacitor to the circuit in Fig. 10.38.

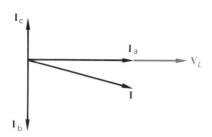

Fig. 10.41 The effect of the capacitor current I_c on the line current **I**.

and phase angle of the line current **I** will change with changes in the magnitude of I_c. As the magnitude and phase angle of **I** change, so do the magnitude and phase angle of the voltage drop along the line. As the drop along the line changes, the magnitude and phase angle of V_s change. These observations are shown in the phasor diagram in Fig. 10.42, where the dotted phasors represent the pertinent currents and voltages before the addition of the capacitor. Thus by comparing the dotted phasors of **I**, $R_1 I$, $j\omega L_1 I$, and V_s with their solid counterparts, we clearly see the effect of adding C to the circuit. In particular, note that by adding the capacitor across the load it is possible to reduce the amplitude of the source voltage and still maintain the amplitude of the load voltage. From a practical point of view, this means that as the load increases (that is, as I_a and I_b increase), capacitors can be added to the system (that is, I_c is increased) so that under heavy load conditions V_L can be maintained without increasing the amplitude of the source voltage.

We will make use of phasor diagrams in our future work whenever such diagrams give us additional insight into the steady-state sinusoidal operation of the circuit under investigation. Problem 10.32 is designed to show how a phasor diagram can help to explain the operation of a phase-shifting circuit.

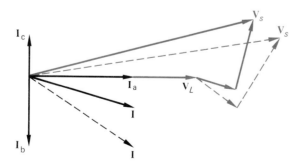

Fig. 10.42 A phasor diagram showing the effect of adding a load-shunting capacitor to the circuit in Fig. 10.38 if V_L is held constant.

DRILL EXERCISE
10.15

The parameters in the circuit in Fig. 10.38 are

$$R_1 = 0.1 \ \Omega, \ \omega L_1 = 0.8 \ \Omega, \ R_2 = 24 \ \Omega, \ \omega L_2 = 32 \ \Omega, \ \text{and} \ \mathbf{V}_L = 240 + j0 \ \text{V}.$$

a) Calculate the phasor voltage $\mathbf{V}_s$.
b) Connect a capacitor in parallel with the inductor, hold $\mathbf{V}_L$ constant, and adjust the capacitor until the magnitude of $\mathbf{I}$ is a minimum. What is the capacitive reactance? What is the value of $\mathbf{V}_s$?
c) Find the value of the capacitive reactance that will keep the magnitude of $\mathbf{I}$ as small as possible and at the same time make $|\mathbf{V}_s| = |\mathbf{V}_L| = 240 \ \text{V}$.

Ans. (a) $247.11 \underline{/1.68°}$ V; (b) $-32 \ \Omega$, $241.13 \underline{/1.90°}$ V; (c) $-26.90 \ \Omega$.

10.12 SUMMARY

The phasor method for finding the steady-state sinusoidal response is extremely important in engineering circuit analysis. Now that we have been introduced to the technique, it is appropriate that we pause and reflect on the important characteristics of the method. Since we are dealing with time-invariant linear circuits, we recognize that the only frequencies that can appear in the current and voltage responses must be present in the driving sources. Our analysis and examples have assumed that all sources are at the same frequency. If sources of different frequencies are present in the circuit, we can invoke the principle of superposition to find the currents and voltages at each frequency. Thus once we learn how to analyze a circuit in which all the sources operate at the same frequency, the extension to circuits containing sources of different frequencies is straightforward. Since in linear circuits the frequency of the response is determined by the frequency of the source, the problem of finding the steady-state sinusoidal response reduces to determining the amplitude and phase angle of the response signal.

The first step in the phasor method is to express all sources in terms of co-sine functions. The second step is to construct the phasor domain equivalent circuit. Sources are represented in the phasor domain by the complex number that carries the amplitude and phase angle of the given sinusoidal function. Resistors transfer to the phasor domain without modification. Each inductor appears in the phasor domain as an impedance of $j\omega L$ ohms, and each capacitor appears as an impedance of $(1/j\omega C)$ ohms. Once we have constructed the phasor domain equivalent circuit, we can use any of the techniques introduced relative to dc circuit analysis to analyze the circuit. There are two reasons why the techniques introduced in dc circuit analysis apply to phasor circuit analysis. First, Kirchhoff's laws hold for phasor voltages and currents. Second, the phasor current and voltage at the terminals of a passive circuit element are related by the equation $\mathbf{V} = Z\mathbf{I}$.

Up to this point in our introduction to the analysis of circuits operating in the sinusoidal steady state, we have focused on the problem of finding the currents and voltages. We are now ready to turn our attention to the power calculations associated with sinusoidal currents and voltages.

PROBLEMS

10.1 Given the sinusoidal voltage

$$v = 40 \cos (100\pi t + 60°) \text{ V.}$$

a) What is the maximum amplitude of the voltage?
b) What is the frequency in rad/s?
c) What is the frequency in Hz?
d) What is the phase angle in degrees?
e) What is the phase angle in radians?
f) What is the period in ms?
g) What is the first time after $t = 0$ that $v = -40$ V?
h) The sinusoidal function is shifted (10/3) milliseconds to the right along the time axis. What is the expression for $v(t)$?
i) What is the minimum number of milliseconds that the function must be shifted to the right if the expression for $v(t)$ is 40 sin $100\pi t$?
j) What is the minimum number of milliseconds that the function must be shifted to the left if the expression for $v(t)$ is 40 sin $100\pi t$?

10.2 In a single graph, sketch

$$v = 100 \cos (\omega t + \phi) \text{ vs. } \omega t$$

for $\phi = 30°, 0°, -30°,$ and $60°$.

a) State whether the voltage function is shifting to the right or left as ϕ becomes more negative.
b) What is the direction of shift if ϕ changes from 0 to $+30°$?

10.3 At $t = -2$ ms a sinusoidal voltage is known to be zero and becoming positive. This occurs next at $+8$ ms. It is also known that the voltage is 80.90 V at $t = 0$.

a) What is the frequency of v in Hz?
b) What is the expression for v?

10.4 A sinusoidal voltage is zero at $t = -2\pi/3$ ms and increasing at a rate of 80,000 V/s. The maximum amplitude is 80 V.

a) What is the frequency of v in rad/s?
b) What is the expression for v?

10.5 a) Verify that Eq. (10.6) is the solution of Eq. (10.5). This can be done by substituting Eq. (10.6) into the left-hand side of Eq. (10.5) and then noting that it equals the right-hand side for all values to $t > 0$. At $t = 0$, Eq. (10.6) should reduce to the initial value of the current.
b) Since the transient component vanishes as time elapses and since our solution must satisfy the differential equation for all values of t, the steady-state component, by itself, must also satisfy the differential equation. Verify this observation by showing that the steady-state component of Eq. (10.6) satisfies Eq. (10.5).

10.6 Use the concept of the phasor to combine the following sinusoidal functions into a single trigonometric expression.

$$y = 100 \cos (800t + 120°) + 60 \cos (800t - 150°)$$
$$y = 50 \cos (300t + 30°) - 20 \sin (300t + 210°)$$

10.7 A 60-Hz sinusoidal voltage with a maximum amplitude of 25 V at $t = 0$ is applied across the terminals of an inductor. The maximum amplitude of the steady-state current in the inductor is 5 A.

    a) What is the frequency of the inductor current?
    b) What is the phase angle of the voltage?
    c) What is the phase angle of the current?
    d) What is the inductive reactance of the inductor?
    e) What is the inductance of the inductor in mH?
    f) What is the impedance of the inductor?

10.8 A 10-kHz sinusoidal voltage has zero phase angle and a maximum amplitude of 200 mV. When this voltage is applied across the terminals of a capacitor, the resulting steady-state current has a maximum amplitude of 1256.64 μA.

    a) What is the frequency of the current in rad/s?
    b) What is the phase angle of the current?
    c) What is the capacitive reactance of the capacitor?
    d) What is the capacitance of the capacitor in μF?
    e) What is the impedance of the capacitor?

10.9 A 30-Ω resistor, a 20-mH inductor, and a 6.25-μF capacitor are connected in series. The series-connected elements are energized by a sinusoidal voltage source whose voltage is 200 cos (4000t + 30°) V.

    a) Draw the phasor-domain equivalent circuit.
    b) Reference the current in the direction of the voltage rise across the source and find the phasor current.
    c) Find the steady-state expression for $i(t)$.

10.10 A 5-Ω resistor and a 2.5-μF capacitor are connected in parallel. This parallel combination is also in parallel with the series combination of a 3-Ω resistor and a 25-μH inductor. These three parallel branches are driven by a sinusoidal current source whose current is 4 cos (160,000t + 60°) A.

    a) Draw the phasor-domain equivalent circuit.
    b) Reference the voltage across the current source as a rise in the direction of the source current and find the phasor voltage.
    c) Find the steady-state expression for $v(t)$.

10.11 Find the impedance Z_{ab} in the circuit in Fig. 10.43. Express Z_{ab} in both polar and rectangular form.

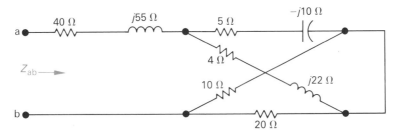

Fig. 10.43 The circuit for Problem 10.11.

10.12 Find the admittance Y_{ab} in the circuit in Fig. 10.44. Express Y_{ab} in both polar and rectangular form.

10.13 a) Show that at a given frequency ω, the circuits in Figs. 10.45 and 10.46 will have the same impedance between the terminals a, b if

$$R_1 = \frac{R_2}{1 + \omega^2 R_2^2 C_2^2} \quad \text{and} \quad C_1 = \frac{1 + \omega^2 R_2^2 C_2^2}{\omega^2 R_2^2 C_2}.$$

b) Find the values of resistance and capacitance that when connected in series will have the same impedance at 8 krad/s as that of a 1-kΩ resistor connected in parallel with a 0.125-μF capacitor.

10.14 a) Show that at a given frequency ω, the circuits in Figs. 10.45 and 10.46 will have the same impedance between the terminals a, b if

$$R_2 = \frac{1 + \omega^2 R_1^2 C_1^2}{\omega^2 R_1 C_1^2} \quad \text{and} \quad C_2 = \frac{C_1}{1 + \omega^2 R_1^2 C_1^2}.$$

(*Hint:* The two circuits will have the same impedance if they have the same admittance.)

b) Find the values of resistance and capacitance that when connected in parallel will give the same impedance at 50 krad/s as that of a 30-kΩ resistor connected in series with a capacitance of 500 pF.

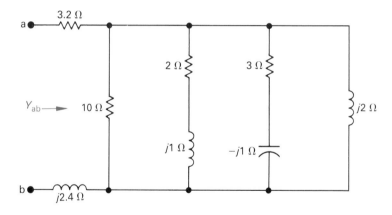

Fig. 10.44 The circuit for Problem 10.12.

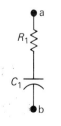

Fig. 10.45 The circuit for Problems 10.13 and 10.14.

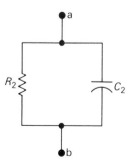

Fig. 10.46 The circuit for Problem 10.14.

10.15 a) Show that at a given frequency ω, the circuits in Figs. 10.47 and 10.48 will have the same impedance between the terminals a, b if

$$R_1 = \frac{\omega^2 L_2^2 R_2}{R_2^2 + \omega^2 L_2^2} \quad \text{and} \quad L_1 = \frac{R_2^2 L_2}{R_2^2 + \omega^2 L_2^2}.$$

b) Find the values of resistance and inductance that when connected in series will have the same impedance at 20 krad/s as that of a 6-kΩ resistor connected in parallel with a 400-mH inductor.

10.16 a) Show that at a given frequency ω, the circuits in Figs. 10.47 and 10.48 will have the same impedance between the terminals a, b if

$$R_2 = \frac{R_1^2 + \omega^2 L_1^2}{R_1} \quad \text{and} \quad L_2 = \frac{R_1^2 + \omega^2 L_1^2}{\omega^2 L_1}.$$

(*Hint:* The Two circuits will have the same impedance if they have the same admittance.)

b) Find the values of resistance and inductance that when connected in parallel will have the same impedance at 250 rad/s as a 20-Ω resistor connected in series with a 160-mH inductor.

10.17 The phasor current $\mathbf{I}_1$ in the circuit in Fig. 10.49 is $3\underline{/0°}$ A.
a) Find $\mathbf{I}_2$, $\mathbf{I}_3$, and $\mathbf{V}_s$.
b) If $\omega = 10^5$ rad/s, write the expressions for $i_2(t)$, $i_3(t)$, and $v_s(t)$.

Fig. 10.47 The circuit for Problems 10.15 and 10.16.

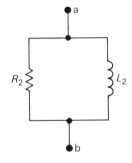

Fig. 10.48 The circuit for Problems 10.15 and 10.16.

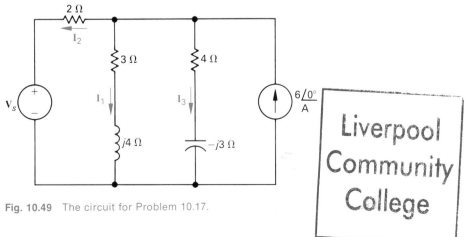

Fig. 10.49 The circuit for Problem 10.17.

10.18 Find $\mathbf{I}_0$ and Z in the circuit in Fig. 10.50 if $\mathbf{V}_s = 40 + j30$ V and $\mathbf{I} = 6\underline{/0°}$ A.

10.19 Use the technique of source transformations to find the steady-state expression for $v(t)$ in the circuit in Fig. 10.51. The sinusoidal voltage sources are given by the expressions

$$v_1 = 180 \cos (400t - 36.87°) \text{ V}$$

and

$$v_2 = 150 \sin 400t \text{ V}.$$

10.20 Find the Thévenin equivalent with respect to the terminals a, b of the circuit shown in Fig. 10.52.

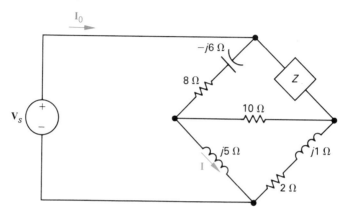

Fig. 10.50 The circuit for Problem 10.18.

Fig. 10.51 The circuit for Problem 10.19.

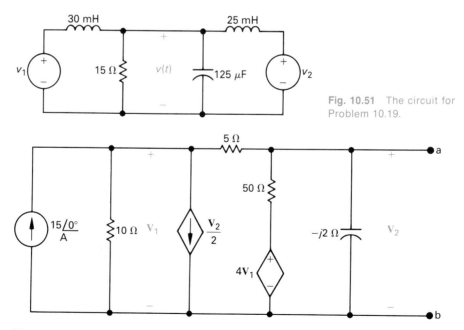

Fig. 10.52 The circuit for Problem 10.20.

10.21 Use the node-voltage method to find the phasor voltage $\mathbf{V}_0$ in the circuit in Fig. 10.53. Express the voltage in both polar and rectangular form.

10.22 Use the node-voltage method to find $\mathbf{V}_2$ and $\mathbf{I}_1$ in the circuit in Fig. 10.54.

10.23 In the circuit in Fig. 10.55, $v_s = 20 \sin 10^{4-} t$ V and $i_s = 5 \cos (10^4 t - 60°)$. Use the node-voltage method to determine the steady-state expressions for v_1 and v_2.

10.24 Use the mesh-current method to find the phasor branch currents $\mathbf{I}_1$, $\mathbf{I}_2$, and $\mathbf{I}_3$ in the circuit in Fig. 10.56.

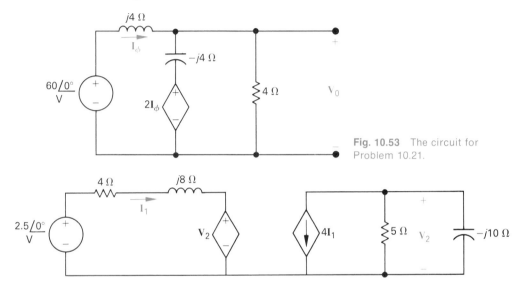

Fig. 10.53 The circuit for Problem 10.21.

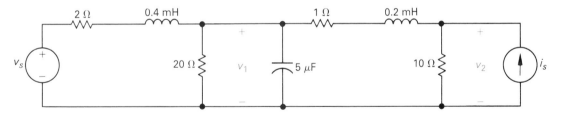

Fig. 10.54 The circuit for Problem 10.22.

Fig. 10.55 The circuit for Problem 10.23.

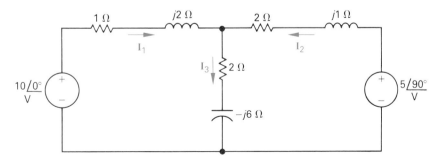

Fig. 10.56 The circuit for Problem 10.24.

10.25 Use the mesh-current method to find the branch currents $\mathbf{I}_1$, $\mathbf{I}_2$, $\mathbf{I}_3$, and $\mathbf{I}_4$ in the circuit in Fig. 10.57.

10.26 In order to introduce you to a circuit configuration that is widely used in residential wiring, we have shown a representative circuit in Fig. 10.58. In this simplified model, the resistor R_3 is used to model a 240-V appliance (such as an electric range), and the resistors R_1 and R_2 to model 120-V appliances (such as a lamp, toaster, and iron). The branches carrying $\mathbf{I}_1$ and $\mathbf{I}_2$ are modeling what electricians refer to as the "hot" conductors in the circuit, and the branch carrying $\mathbf{I}_n$ is modeling the neutral conductor. Our purpose in analyzing the circuit is to show the importance of the neutral conductor in the satisfactory operation of the circuit. You are to choose the method for analyzing the circuit.

a) Show that $\mathbf{I}_n$ is zero if $R_1 = R_2$.
b) Show that $\mathbf{V}_1 = \mathbf{V}_2$ if $R_1 = R_2$.

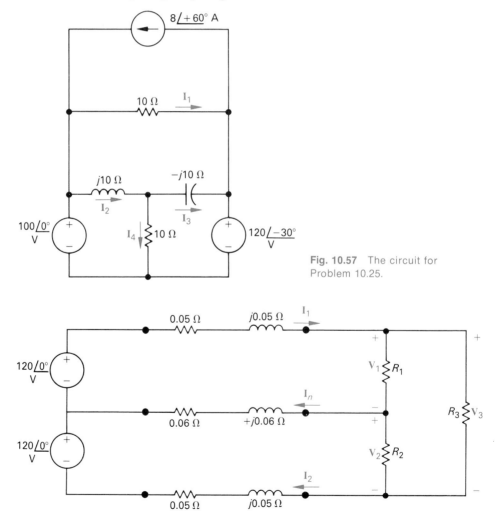

Fig. 10.57 The circuit for Problem 10.25.

Fig. 10.58 The circuit for Problem 10.26.

c) Open the neutral branch and calculate $\mathbf{V}_1$ and $\mathbf{V}_2$ if $R_1 = 72 \; \Omega, R_2 = 720 \; \Omega$, and $R_3 = 8 \; \Omega$.

d) Close the neutral branch and repeat part (c).

e) On the basis of your calculations, explain why the neutral conductor is never fused in such a manner that it could open while the "hot" conductors are energized.

10.27 Find the Norton equivalent circuit with respect to the terminals a, b for the circuit shown in Fig. 10.59.

10.28 Find the Thévenin equivalent circuit with respect to the terminals a, b for the circuit shown in Fig. 10.60.

10.29 Find the Thévenin equivalent circuit with respect to the terminals a, b for the circuit shown in Fig. 10.61.

10.30 Find the Norton equivalent circuit with respect to the terminals a, b for the circuit shown in Fig. 10.62.

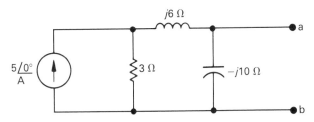

Fig. 10.59 The circuit for Problem 10.27.

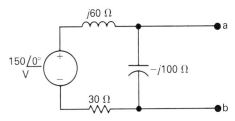

Fig. 10.60 The circuit for Problem 10.28.

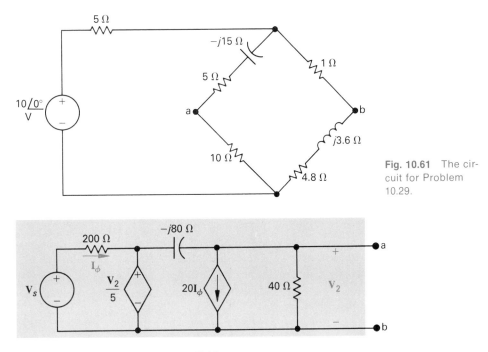

Fig. 10.61 The circuit for Problem 10.29.

Fig. 10.62 The circuit for Problem 10.30.

10.31 a) For the circuit shown in Fig. 10.63, compute $\mathbf{V}_s$ and $\mathbf{V}_l$.
  b) Construct a phasor diagram showing the relationship between $\mathbf{V}_s$, $\mathbf{V}_l$, and the load voltage of $120\underline{/0^\circ}$.
  c) Repeat parts (a) and (b) given that the load resistance changes from 6 Ω to 1.5 Ω and the load reactance changes from 8 Ω to 2 Ω. Assume that the load voltage remains constant at $120\underline{/0^\circ}$. How much must the amplitude of $\mathbf{V}_s$ be increased in order to maintain the load voltage at 120 V?
  d) Repeat part (c) given that at the same time the load resistance and reactance changes, a capacitive reactance of $-2\ \Omega$ is connected across the load terminals.

10.32 Show by using a phasor diagram what happens to the magnitude and phase angle of the voltage v_o in the circuit in Fig. 10.64 as R_x is varied from zero to infinity. The amplitude and phase angle of the source voltage are held constant as R_x varies.

10.33 The operational amplifier in the circuit in Fig. 10.65 is ideal.
  a) Find the steady-state expression for $v_0(t)$.
  b) How large can the amplitude of v_g be before the amplifier saturates?

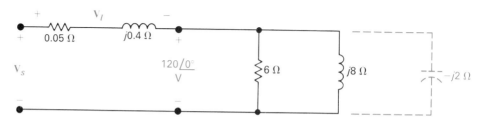

Fig. 10.63 The circuit for Problem 10.31.

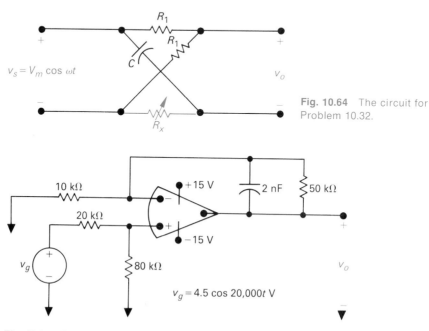

Fig. 10.64 The circuit for Problem 10.32.

$v_g = 4.5 \cos 20{,}000t$ V

Fig. 10.65 The circuit for Problem 10.33.

10.34 The operational amplifier in the circuit in Fig. 10.66 is ideal. Find the steady-state expression for $v_o(t)$ when $i_g = 0.36 \cos 10^5 t$ mA.

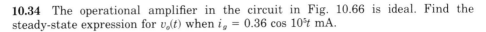

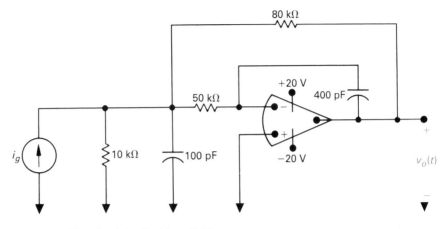

Fig. 10.66 The circuit for Problem 10.34.

11

SINUSOIDAL STEADY-STATE POWER CALCULATIONS

.1 INTRODUCTION

In Chapter 10 we learned how to determine the steady-state voltages and currents in circuits driven by sinusoidal sources. We now concentrate on power calculations associated with steady-state sinusoidal operation. Our primary interest is to determine the average power that is either delivered to, or extracted from, a pair of terminals by a sinusoidal voltage and current. The power associated with a pair of terminals in a circuit operating in the sinusoidal steady-state is an important calculation. The voltage and current ratings of electrical equipment such as generators, motors, lamps, toasters, and ovens determine how much power the appliance can handle. Since the usefulness of the appliance is determined by its ability to convert energy either to or from the electrical form, the calculation of power is of prime importance.

Our problem is shown graphically in Fig. 11.1, where v and i are steady-state sinusoidal signals. Note that we are using the passive sign convention; therefore the power at any instant of time is given by the expression

$$p = vi. \tag{11.1}$$

The power is measured in watts when the voltage is in volts and the current is in amperes. For the problem at hand,

$$v = V_m \cos(\omega t + \theta_v) \tag{11.2}$$

and

$$i = I_m \cos(\omega t + \theta_i), \tag{11.3}$$

where θ_v is the voltage phase angle and θ_i is the current phase angle.

Since we are operating in the sinusoidal steady-state, we are at liberty to choose any convenient reference for zero time. Engineers who are concerned with operating systems involving the transfer of large blocks of power have found it convenient when making power calculations to choose zero time to correspond to the instant of time when the current is passing through a positive maximum. This reference system requires that we shift both the voltage and current by θ_i. Thus Eqs. (11.2) and (11.3) become

$$v = V_m \cos(\omega t + \theta_v - \theta_i), \tag{11.4}$$

$$i = I_m \cos \omega t. \tag{11.5}$$

When we substitute Eqs. (11.4) and (11.5) into Eq. (11.1), our expression for the

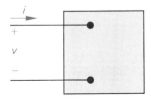

Fig. 11.1 The basic power calculation is to find the average power associated with the voltage and current at a pair of terminals.

instantaneous power becomes

$$p = V_m I_m \cos(\omega t + \theta_v - \theta_i) \cos \omega t, \tag{11.6}$$

which is our starting point for developing the expression for average power.

11.2 REAL AND REACTIVE POWER

The average power associated with sinusoidal signals is the average of the instantaneous power over one period. In equation form, this is written as

$$P = \frac{1}{T} \int_{t_0}^{t_0+T} p \, dt, \tag{11.7}$$

where T is the period of the sinusoidal function. The limits on Eq. (11.7) imply that we can initiate the integration process at any convenient time t_0, but we must terminate the integration exactly one period later. (We could integrate over nT periods, where n is an integer, provided we multiply the integeral by $1/nT$.)

Although we could find the average power by substituting Eq. (11.6) directly into Eq. (11.7) and then performing the integration, we find it more informative to first expand Eq. (11.6) by the trigonometric identity

$$\cos \alpha \cos \beta = \frac{1}{2} \cos(\alpha - \beta) + \frac{1}{2} \cos(\alpha + \beta).$$

Letting $\alpha = (\omega t + \theta_v - \theta_i)$ and $\beta = \omega t$, we can write Eq. (11.6) as

$$p = \frac{V_m I_m}{2} \cos(\theta_v - \theta_i) + \frac{V_m I_m}{2} \cos(2\omega t + \theta_v - \theta_i). \tag{11.8}$$

Now we use the trigonometric identity

$$\cos(\alpha + \beta) = \cos \alpha \cos \beta - \sin \alpha \sin \beta$$

to expand the second term on the right-hand side of Eq. (11.8), to get

$$p = \frac{V_m I_m}{2} \cos(\theta_v - \theta_i) + \frac{V_m I_m}{2} \cos(\theta_v - \theta_i) \cos 2\omega t$$

$$- \frac{V_m I_m}{2} \sin(\theta_v - \theta_i) \sin 2\omega t. \tag{11.9}$$

A careful study of Eq. (11.9) reveals the following characteristics of the instantaneous power.

1. The average value of p is given by the first term on the right-hand side because the integral of either $\cos 2\omega t$ or $\sin 2\omega t$ over one period is zero. Thus the average power is

$$P = \frac{V_m I_m}{2} \cos(\theta_v - \theta_i). \tag{11.10}$$

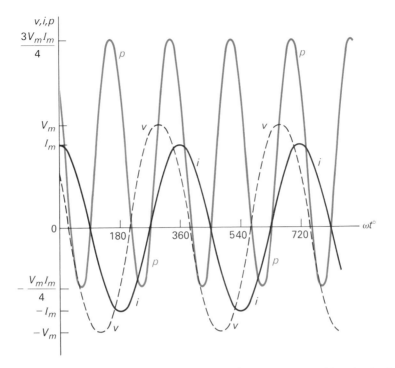

Fig. 11.2   Instantaneous power, voltage, and current versus ωt for steady-state sinusoidal operation.

2. The frequency of the instantaneous power is twice the frequency of the voltage or current. This follows directly from the second two terms on the right-hand side of the equation. A representative relationship between v, i, and p is shown in Fig. 11.2, in which we assume that $\theta_v = 60°$ and $\theta_i = 0°$. In studying Fig. 11.2 note that the instantaneous power goes through two complete cycles for every cycle of either the voltage or the current. Also note that it is possible for the instantaneous power to be negative for a portion of each cycle, even if the network between the terminals is passive. In a completely passive network, negative power implies that energy that has been stored in inductors or capacitors is now being extracted. The fact that the instantaneous power varies with time in the sinusoidal steady-state operation of a circuit explains why some motor-driven appliances (such as a refrigerator) experience vibration, and require resilient mountings to prevent excessive vibration of the appliance itself.

3. If the circuit between the terminals is purely resistive, the voltage and current will be in phase. This means that $\theta_v = \theta_i$ and Eq. (11.9) reduces to

$$p = P + P \cos 2\omega t. \tag{11.11}$$

In writing Eq. (11.11), we have taken advantage of the result given in Eq.

(11.10). The instantaneous power expressed in Eq. (11.11) is referred to as the *instantaneous real power*. Note that this implies that average power (P) is also referred to as "real power," a phrase used to describe power that is transformed from electrical to nonelectrical form. In the case of the purely resistive network, the electrical energy is transformed into thermal energy. Also note from Eq. (11.11) that the instantaneous real power can never be negative. That is, power cannot be extracted from a purely resistive network.

4. If the circuit between the terminals is purely inductive, the voltage and current will be out of phase by precisely 90°. In particular, the current will lag the voltage by 90° (that is, $\theta_i = \theta_v - 90°$); therefore $\theta_v - \theta_i$ will equal $+ 90°$. The expression for the instantaneous power reduces to

$$p = \frac{- V_m I_m}{2} \sin 2\omega t. \qquad (11.12)$$

We observe first that in the case of a purely inductive circuit the average power is zero. Therefore in a purely inductive circuit there is no transformation of energy from electrical to nonelectrical form. The instantaneous power at the terminals at a purely inductive circuit oscillates between the circuit and the source driving the circuit. When p is positive, energy is being stored in the magnetic fields associated with the inductive elements, and when p is negative, energy is being extracted from the magnetic fields of the inductive elements.

5. If the circuit between the terminals is purely capacitive, the voltage and current will also be precisely 90° out of phase. In this case, the current will lead the voltage by 90° (that is, $\theta_i = \theta_v + 90°$); thus $\theta_v - \theta_i$ will equal $- 90°$. The expression for the instantaneous power becomes

$$p = \frac{V_m I_m}{2} \sin 2\omega t. \qquad (11.13)$$

Again, the average power is zero so there is no transformation of energy from electrical to nonelectrical form. In the purely capacitive circuit, the power oscillates between the source driving the circuit and the electric field associated with the capacitive elements.

The power associated with purely inductive or capacitive circuits is referred to as *reactive power*. Inductors and capacitors are referred to as reactive elements in steady-state sinusoidal analysis since their impedances are characterized as inductive reactance and capacitive reactance, respectively. In terms of the general expression for instantaneous power, namely Eq. (11.9), the coefficient of $\sin 2\omega t$ is referred to as the reactive power and is denoted by Q; thus

$$Q = \frac{V_m I_m}{2} \sin (\theta_v - \theta_i). \qquad (11.14)$$

We will find the concept of reactive power very useful. Before pursuing power calculations, let us pause to note that if we use the notation for average and reactive power, we can express Eq. (11.9) as

$$p = P + P \cos 2\omega t - Q \sin 2\omega t. \tag{11.15}$$

Three additional comments regarding Eqs. (11.10) and (11.14) are in order. First, P and Q carry the same dimension. However, in order to distinguish between the real and reactive power, we use the term "var" for the reactive power (var is an acronym for the phrase "volt-amp reactive").

Second, the decision to use the current as the reference leads to Q being positive for inductors (that is, $\theta_v - \theta_i = 90°$) and negative for capacitors (that is, $\theta_v - \theta_i = -90°$). Power engineers recognize this difference in the algebraic sign of Q for inductors and capacitors by saying that inductors demand, or absorb, magnetizing vars and capacitors furnish, or deliver, magnetizing vars. We will have more to say about this convention later.

Third, the angle $(\theta_v - \theta_i)$ is referred to as the *power factor angle*. The cosine of this angle is called the *power factor,* abbreviated pf, and the sine of this angle is called the *reactive factor,* abbreviated rf. Thus

$$\mathrm{pf} = \cos (\theta_v - \theta_i) \tag{11.16}$$

and

$$\mathrm{rf} = \sin (\theta_v - \theta_i). \tag{11.17}$$

We should make note of the fact that once we know the magnitude of the power factor, we know the magnitude of the reactive factor. However, there is an ambiguity regarding the algebraic sign of the reactive factor since $\cos (\theta_v - \theta_i) = \cos (\theta_i - \theta_v)$ and $\sin (\theta_v - \theta_i) = -\sin (\theta_i - \theta_v)$. To completely describe the reactive factor through the power factor, we use the descriptive phrases "lagging power factor" and "leading power factor." Lagging power factor implies current lagging voltage, hence an inductive load. It follows that leading power factor implies current leading voltage, hence a capacitive load.

We will find both the power factor and the reactive factor convenient quantities to use in describing electrical loads.

Example 11.1 illustrates the interpretation of P and Q on the basis of a numerical calculation.

Example 11.1 a) Calculate the average power and the reactive power at the terminals of the network shown in Fig. 11.1 if

$$v = 100 \cos (\omega t + 15°) \text{ V}$$

and

$$i = 4 \sin (\omega t - 15°) \text{ A}.$$

b) State whether the network inside the box is absorbing or delivering average power.

c) State whether the network inside the box is absorbing or supplying magnetizing vars.

Solution   a) Since i is expressed in terms of the sine function, the first step in the calculation for P and Q is to rewrite i as a cosine function. We have

$$i = 4 \cos (\omega t - 105°) \text{ A}.$$

Now we can calculate P and Q directly from Eqs. (11.10) and (11.14). Thus

$$P = \frac{1}{2} (100)(4) \cos [15 - (-105)] = -100 \text{ W}$$

and

$$Q = \frac{1}{2} 100(4) \sin [15 - (-105)] = 173.21 \text{ VAR}.$$

b) We note from Fig. 11.1 that the passive sign convention is used. Therefore the negative value of -100 W means that the network inside the box is delivering average power to the terminals.

c) The passive sign convention means that since Q is positive, the network inside the box is absorbing magnetizing vars at its terminals.   ∎

DRILL EXERCISE
11.1
For each of the following sets of voltage and current, calculate the real and reactive power in the line between networks A and B. In each case, state whether the power flow is from A to B or vice versa. Also state whether magnetizing vars are being transferred from A to B or vice versa.

a) $v = 250 \cos (\omega t + 45°)$; $i = 12 \cos (\omega t - 15°)$

b) $v = 250 \cos (\omega t + 45°)$; $i = 12 \cos (\omega t + 165°)$

c) $v = 250 \cos (\omega t + 45°)$; $i = 12 \cos (\omega t + 105°)$

d) $v = 250 \cos \omega t$; $i = 12 \cos (\omega t - 120°)$

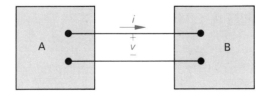

Ans. (a) $P = 750$ W (A to B), $Q = 1299.04$ VAR (A to B); (b) $P = -750$ W (B to A), $Q = -1299.04$ VAR (B to A); (c) $P = 750$ W (A to B), $Q = -1299.04$ VAR (B to A); (d) $P = -750$ W (B to A), $Q = 1299.04$ VAR (A to B).

11.3 THE EFFECTIVE (rms) VALUE OF A SINUSOIDAL SIGNAL

The effective, or rms, value of a sinusoidal signal is important in calculating real and reactive power. The concept of "effective" value comes from a desire to compare the ability of a sinusoidally varying voltage (or current) to deliver energy to a resistor with the ability of a constant (dc) voltage (or current) to deliver energy to a resistor. To make this comparison, we assume that we have a sinusoidal voltage impressed across the terminals of a resistor as shown in Fig. 11.3. Now we ask, "What sinusoidal voltage applied to the resistor R will deliver the same amount of electrical energy to the resistance in T seconds that a constant (dc) voltage would?"

We can answer this as follows. When the sinusoidal voltage is applied to the resistor, the energy delivered to the resistor in T seconds is

$$W_s = \int_{t_0}^{t_0+T} \frac{v^2}{R} \, dt. \tag{11.18}$$

If V_0 denotes the dc voltage applied to the resistor R, then the energy delivered to the resistor in T seconds is

$$W_0 = \frac{V_0^2}{R} T. \tag{11.19}$$

In order for the two energies to be the same, we must have

$$\frac{V_0^2}{R} T = \int_{t_0}^{t_0+T} \frac{v^2}{R} \, dt$$

or

$$V_0 = \sqrt{\frac{1}{T} \int_{t_0}^{t_0+T} v^2 \, dt}. \tag{11.20}$$

The right-hand side of Eq. (11.20) is the effective, or rms, value of the sinusoidal voltage:

$$V_{\text{eff}} = \sqrt{\frac{1}{T} \int_{t_0}^{t_0+T} v^2 \, dt}. \tag{11.21}$$

Equation (11.20) tells us that a constant, or dc, voltage that is equal to the ef-

$V_m \cos (\omega t + \theta_v)$ R

Fig. 11.3 A sinusoidal voltage applied to the terminals of a resistor.

fective value of the sinusoidal voltage delivers the same amount of electrical energy to a resistor every T seconds as does the sinusoidal voltage.

We can use Eq. (11.21) to explain the abbreviation "rms." We see that finding the effective value involves taking the square *root* of the *mean* value of the *square* of the function. Thus the procedure is described as finding the "root-mean-square" value of the function.

Before using Eq. (11.21) to find the effective value of a sinusoidal function, we should note that the concept of an rms value is not limited to sinusoidal functions. In general, the rms value of any periodic function $f(t)$ with a period of T is

$$F_{\mathrm{rms}} = \sqrt{\frac{1}{T} \int_{t_0}^{t_0+T} f^2(t)\ dt}, \tag{11.22}$$

where F_{rms} is used to signify the rms value of $f(t)$.

To find the effective value of a sinusoidal voltage, we can simplify the task by first expanding the squared cosine function via the trigonometric identity

$$\cos^2 \alpha = \frac{1}{2} + \frac{1}{2} \cos 2\alpha.$$

Thus we have

$$V_m^2 \cos^2 (\omega t + \theta_v) = \frac{V_m^2}{2} + \frac{V_m^2}{2} \cos (2\omega t + 2\theta_v) \tag{11.23}$$

When we substitute Eq. (11.23) into Eq. (11.21) the second term will integrate to zero; thus the effective value of the sinusoidal voltage reduces to

$$V_{\mathrm{eff}} = \sqrt{V_m^2/2} = V_m/\sqrt{2} \cong 0.707 V_m, \tag{11.24}$$

from which we see that the effective value of a sinusoidal signal is simply its maximum value divided by $\sqrt{2}$. We have also indicated in Eq. (11.24) that the reciprocal of $\sqrt{2}$ is approximately 0.707.

To see how the effective value relates to the calculation of the average power delivered to the resistor shown in Fig. 11.3, we use the definition of average power to write

$$P = \frac{1}{T} \int_{t_0}^{t_0+T} \frac{v^2}{R}\ dt. \tag{11.25}$$

We can now substitute Eq. (11.21) into Eq. (11.25) to get

$$P = \frac{V_{\mathrm{eff}}^2}{R}, \tag{11.26}$$

from which we conclude that the average power delivered to the terminals of a resistor by a sinusoidal voltage impressed across the resistor is simply the effective value of the voltage squared divided by the resistance.

The average power given by Eq. (11.10) and the reactive power given by

Eq. (11.14) can be written in terms of effective values. Thus

$$P = \frac{V_m I_m}{2} \cos(\theta_v - \theta_i)$$

$$= (V_m/\sqrt{2})(I_m/\sqrt{2}) \cos(\theta_v - \theta_i)$$

$$= V_{\text{eff}} I_{\text{eff}} \cos(\theta_v - \theta_i) \tag{11.27}$$

and, by a similar manipulation,

$$Q = V_{\text{eff}} I_{\text{eff}} \sin(\theta_v - \theta_i). \tag{11.28}$$

The effective value of the sinusoidal signal in power calculations is so widely used that voltage and current ratings of circuits and equipment involved in power utilization are given in terms of rms values. For example, the voltage rating of residential electrical wiring is often 240 V/120 V service. These voltage levels are the rms value of the sinusoidal voltages supplied by the utility company. Such appliances as electric lamps, irons, and toasters all carry rms ratings on their nameplates. For example, a 120-V, 100-W lamp has a resistance of $120^2/100$, or 144 Ω, and draws an rms current of 120/144, or 0.833 A. The peak value of the lamp current is $0.833\sqrt{2}$, or 1.18 A.

The phasor transform of a sinusoidal function can also be expressed in terms of the rms value. The magnitude of the rms phasor is equal to the rms value of the sinusoidal function. If a phasor is based on the rms value, we indicate this by either an explicit statement, a parenthetical rms adjacent to the phasor quantity, or the subscript "eff" as in Eq. (11.33).

Although our primary interest at this point in our study is the effective value of a sinusoidal signal, we conclude this section with an example that illustrates very clearly the calculation of the rms value of a periodic triangular wave and in so doing clarifies the meaning of "root-mean-square."

Example 11.2   Calculate the rms value of the periodic triangular current shown in Fig. 11.4. Express the answer in terms of the peak current I_p.

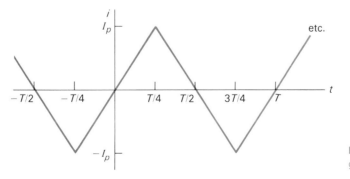

Fig. 11.4   Periodic triangular current.

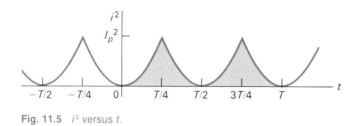

Fig. 11.5 i^2 versus t.

Solution   It follows directly from Eq. (11.22) that the rms value of i will be

$$I_{\text{eff}} = \sqrt{\frac{1}{T}\int_{t_0}^{t_0+T} i^2\,dt}.$$

It is helpful when finding the rms value to interpret the integral under the radical sign as the area under the squared function over an interval of one period. The squared function with the area between 0 and T shaded is shown in Fig. 11.5, which also indicates that for this particular function, the area under the squared current over an interval of one period is equal to four times the area under the squared current over the interval 0 to $T/4$ seconds; that is,

$$\int_{t_0}^{t_0+T} i^2\,dt = 4\int_{0}^{T/4} i^2\,dt.$$

The analytical expression for i in the interval 0 to $T/4$ is

$$i = \frac{4I_p}{T}\,t \qquad (0 < t < T/4).$$

The area under the squared function over one period is

$$\int_{t_0}^{t_0+T} i^2\,dt = 4\int_{0}^{T/4} \frac{16I_p^2}{T^2}\,t^2\,dt = \frac{I_p^2 T}{3}.$$

The mean, or average, value of the function is simply this area over one period divided by the period. Thus

$$i_{\text{mean}} = \left(\frac{1}{T}\right)\left(\frac{I_p^2 T}{3}\right) = \frac{1}{3}I_p^2.$$

The effective, or rms, value of the current is the square root of this mean value, hence

$$I_{\text{eff}} = \frac{I_p}{\sqrt{3}}. \qquad \blacksquare$$

DRILL EXERCISE   11.2   Find the rms value of the half-wave rectified sinusoidal voltage in the figure shown at the top of the following page.

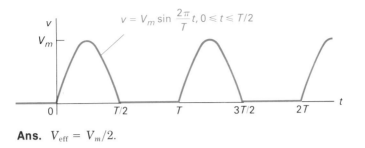

Ans. $V_{\text{eff}} = V_m/2$.

11.4 COMPLEX POWER

Before proceeding to the various methods of calculating the real and reactive power in circuits operating in the sinusoidal steady-state, we pause to introduce the definition of complex, or apparent, power. *Complex power* is the complex sum of the average real power and the reactive power; thus

$$S = P + jQ. \tag{11.29}$$

Dimensionally, complex power is the same as real or reactive power. However, in order to distinguish complex power from either real or reactive power, we use the term "volt amps." Thus we use volt amps for complex power, watts for average real power, and vars for reactive power.

In working with Eq. (11.29) it is convenient to think of P, Q, and $|S|$ as forming the sides of a right triangle, as shown in Fig. 11.6. As we will discover in Section 11.5, the angle θ in the power triangle is the power factor angle $\theta_v - \theta_i$.

11.5 POWER CALCULATIONS

We are now ready to develop equations that can be used to calculate real, reactive, and complex power. We begin by combining Eqs. (11.10), (11.14), and

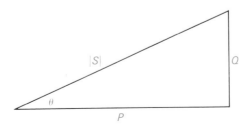

Fig. 11.6 The power triangle.

(11.29) to get

$$S = \frac{V_m I_m}{2} \cos(\theta_v - \theta_i) + j\frac{V_m I_m}{2} \sin(\theta_v - \theta_i)$$

$$= \frac{V_m I_m}{2} [\cos(\theta_v - \theta_i) + j\sin(\theta_v - \theta_i)]$$

$$= \frac{V_m I_m}{2} e^{j(\theta_v - \theta_i)} = \frac{1}{2} V_m I_m \,\underline{/\theta_v - \theta_i}. \tag{11.30}$$

If we use the effective values of the sinusoidal voltage and current, Eq. (11.30) becomes

$$S = V_{\mathrm{eff}} I_{\mathrm{eff}} \,\underline{/\theta_v - \theta_i}. \tag{11.31}$$

Equations (11.30) and (11.31) are very important relationships in power calculations because they show that if the phasor current and voltage are known at a pair of terminals, then the complex power associated with that pair of terminals is either one half of the product of the voltage and the *conjugate* of the current or the product of the rms phasor voltage and the *conjugate* of the rms phasor current. Thus given the phasor voltage and current at a pair of terminals as shown in Fig. 11.7, we find that the complex power is

$$S = \frac{1}{2} \mathbf{V I^*} = P + jQ \tag{11.32}$$

or

$$S = \mathbf{V}_{\mathrm{eff}} \, \mathbf{I}^*_{\mathrm{eff}} = P + jQ. \tag{11.33}$$

Both Eqs. (11.32) and (11.33) assume the passive sign convention. If the current reference is in the direction of the voltage rise across the terminals, we insert a minus sign on the right-hand side of each equation.

As an example of using Eq. (11.32) in a power calculation, let us use the same circuit that we used in Example 11.1. In terms of the phasor representation of the terminal voltage and current, we have

$$\mathbf{V} = 100\,\underline{/15^\circ}$$

and

$$\mathbf{I} = 4\,\underline{/-105^\circ} \text{ A}.$$

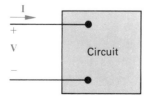

Fig. 11.7 Phasor voltage and current associated with a pair of terminals.

Therefore

$$S = \frac{1}{2}(100\underline{/15°})(4\underline{/+105°})$$

$$= 200\underline{/120°}$$

$$= -100 + j173.21 \text{ VA}.$$

Once we calculate the complex power we can read off both the real and reactive power; thus

$$P = -100 \text{ W}$$

and

$$Q = 173.21 \text{ VAR}.$$

The interpretations of the algebraic signs on P and Q are identical to those given in the solution of Example 11.1.

There are several useful variations of either Eq. (11.33) or (11.32). We will use the rms or effective value form of the equations with the suggestion that you "think" rms when making power calculations.

The first variation of Eq. (11.33) is to replace the voltage by the product of the current times the impedance. That is, we can always represent the circuit inside the box of Fig. 11.7 by an equivalent impedance, as shown in Fig. 11.8, from which it follows directly that

$$\mathbf{V}_{\text{eff}} = \mathbf{Z}\mathbf{I}_{\text{eff}}. \tag{11.34}$$

When we substitute Eq. (11.34) into Eq. (11.33) we get

$$\begin{aligned} S &= \mathbf{Z}\mathbf{I}_{\text{eff}}\mathbf{I}_{\text{eff}}^* \\ &= |\mathbf{I}_{\text{eff}}|^2\mathbf{Z} \\ &= |\mathbf{I}_{\text{eff}}|^2(R + jX) \\ &= |\mathbf{I}_{\text{eff}}|^2 R + j|\mathbf{I}_{\text{eff}}|^2 X = P + jQ, \end{aligned} \tag{11.35}$$

from which we see that

$$P = |\mathbf{I}_{\text{eff}}|^2 R = \frac{1}{2} I_m^2 R \tag{11.36}$$

and

$$Q = |\mathbf{I}_{\text{eff}}|^2 X = \frac{1}{2} I_m^2 X. \tag{11.37}$$

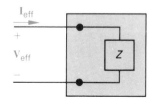

Fig. 11.8 The general circuit in Fig. 11.7 replaced by an equivalent impedance.

In Eq. (11.37), X is the reactance of either the equivalent inductance or equivalent capacitance of the circuit, and will be positive for inductive circuits and negative for capacitive circuits.

A second useful variation of Eq. (11.33) comes from replacing the current by the voltage divided by the impedance. Thus

$$S = \mathbf{V}_{\text{eff}} \left[\frac{\mathbf{V}_{\text{eff}}}{Z}\right]^* = \frac{|\mathbf{V}_{\text{eff}}|^2}{Z^*} = P + jQ. \tag{11.38}$$

It is of interest to note that if Z is a pure resistance element

$$P = \frac{|\mathbf{V}_{\text{eff}}|^2}{R}, \tag{11.39}$$

and if Z is a pure reactive element

$$Q = \frac{|\mathbf{V}_{\text{eff}}|^2}{X} \tag{11.40}$$

In Eq. (11.40), X will be positive for an inductor and negative for a capacitor.

The application of the equations developed in this section are illustrated in the following numerical examples.

11.6 ILLUSTRATIVE EXAMPLES

These examples have been designed to illustrate basic power and reactive power calculations in circuits operating in the sinusoidal steady-state.

Example 11.3 In the circuit shown in Fig. 11.9, a load having an impedance of $39 + j26\ \Omega$ is fed from a voltage source through a line having an impedance of $1 + j4\ \Omega$. The effective, or rms, value of the source voltage is 250 V.

a) Calculate the load current $\mathbf{I}_L$ and voltage $\mathbf{V}_L$.

b) Calculate the average and reactive power delivered to the load.

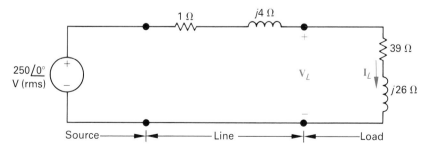

Fig. 11.9 The circuit for Example 11.3.

c) Calculate the average and reactive power delivered to the line.

d) Calculate the average and reactive power supplied by the source.

Solution a) Since the line and load impedances are in series across the voltage source, the load current equals the voltage divided by the total impedance. Thus

$$\mathbf{I}_L = \frac{250\underline{/0^\circ}}{40 + j30} = 4 - j3 = 5\underline{/-36.87^\circ} \text{ A} \quad (\text{rms}).$$

Because the voltage is given in terms of its rms value, the current is also rms. The load voltage is the product of the load current and load impedance:

$$\mathbf{V}_L = (39 + j26)\mathbf{I}_L = 234 - j13 = 234.36\underline{/-3.18^\circ} \text{ V} \quad (\text{rms}).$$

b) The average and reactive power delivered to the load can be computed using Eq. (11.33). Therefore

$$S = \mathbf{V}_L\mathbf{I}_L^* = (234 - j13)(4 + j3)$$
$$= 975 + j650 \text{ VA}.$$

It follows directly that the load is absorbing an average power of 975 watts and a reactive power of 650 magnetizing vars.

c) The average and reactive power delivered to the line are most easily calculated using Eqs. (11.36) and (11.37) since the line current is known. Thus

$$P = (5)^2(1) = 25 \text{ W}$$

and

$$Q = (5^2)(4) = 199 \text{ VAR}.$$

Note that the reactive power associated with the line is positive since the line reactance is inductive.

d) One way to calculate the average and reactive power delivered by the source is to add the complex power delivered to the line to the complex power delivered to the load; thus

$$S_s = 25 + j100 + 975 + j650.$$
$$S_s = (1000 + j750) \text{ VA}.$$

We can also calculate the complex power at the source using Eq. (11.33). Thus

$$S_s = -250 \, \mathbf{I}_L^*.$$

The minus sign is inserted in Eq. (11.33) whenever the current is referenced in the direction of a voltage rise. Thus

$$S_s = -250(4 + j3) = -(1000 + j750) \text{ VA}.$$

The minus sign implies that the source is delivering average power and magnetizing vars. Note this checks our previous calculation for S_s, as it must, since all the average and reactive power absorbed by the line and load must be furnished by the source. ■

Example 11.4   An electrical load operates at 240 V rms. The load absorbs an average power of 8 kW at a lagging power factor of 0.8.

a) Calculate the complex power of the load.
b) Calculate the impedance of the load.

Solution   a) Since the power factor is described as lagging, we know that the load is inductive and the algebraic sign of the reactive power will be positive. From the power triangle shown in Fig. 11.6 it follows that

$$P = |S| \cos \theta$$

and

$$Q = |S| \sin \theta.$$

Now, since $\cos \theta$ is 0.8, $\sin \theta$ is 0.6. Therefore

$$|S| = \frac{P}{\cos \theta} = \frac{8 \text{ kW}}{0.8} = 10 \text{ kVA}$$

and

$$Q = 10 \sin \theta = 6 \text{ kVAR}.$$

Therefore

$$S_L = 8 + j6 \text{ kVA}.$$

b) From Eq. (11.38) it follows that

$$Z^* = \frac{|V|^2}{P + jQ} = \frac{(240)^2}{8000 + j6000}$$
$$= 4.608 - j3.456 \ \Omega$$

Hence

$$Z = 4.608 + j3.456 \ \Omega = 5.76\underline{/36.87°} \ \Omega. \ ■$$

Example 11.5   The three loads in the circuit shown in Fig. 11.10 can be described as follows. Load 1 absorbs an average power of 8 kW at a lagging power factor of 0.8. Load

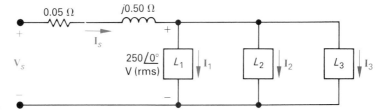

Fig. 11.10   The circuit for Example 11.5.

2 absorbs 20 kVA at leading power factor of 0.6. Load 3 is an impedance of $2.5 + j5.0\ \Omega$. Derive the steady-state expression for $v_s(t)$ if the frequency of the source is 60 Hz.

Solution The currents $\mathbf{I}_1$, $\mathbf{I}_2$, $\mathbf{I}_3$, and $\mathbf{I}_s$ have been placed on the circuit diagram in Fig. 11.10 to facilitate the discussion of the solution. We begin by noting that $\mathbf{I}_s$ is the sum of the three load currents and $\mathbf{V}_s$ is the load voltage plus the drop across the line impedance of $0.05 + j0.5\ \Omega$. Hence, the first step in the solution is to find the three load currents. We can find $\mathbf{I}_1$ and $\mathbf{I}_2$ directly from Eq. (11.33):

$$250\ \mathbf{I}_1^* = 8000 + j6000$$

or

$$\mathbf{I}_1^* = 32 + j24\ \text{A}\quad(\text{rms}).$$

Hence,

$$\mathbf{I}_1 = 32 - j24\ \text{A}\quad(\text{rms}),$$
$$250\ \mathbf{I}_2^* = 12{,}000 - j16{,}000,$$

or

$$\mathbf{I}_2^* = 48 - j64\ \text{A}\quad(\text{rms}).$$

Thus

$$\mathbf{I}_2 = 48 + j64\ \text{A}\quad(\text{rms}).$$

Since load 3 is described in terms of its impedance, $\mathbf{I}_3$ is

$$\mathbf{I}_3 = \frac{250}{2.5 + j5} = 20 - j40\ \text{A}\quad(\text{rms}).$$

Using Kirchhoff's current law we have

$$\mathbf{I}_s = \mathbf{I}_1 + \mathbf{I}_2 + \mathbf{I}_3 = 100 + j0\ \text{A}\quad(\text{rms}).$$

Using Kirchhoff's voltage law gives us

$$\mathbf{V}_s = 250 + (0.05 + j0.50)100$$
$$= 255 + j50 = 259.86\underline{/11.09°}\ \text{V}\quad(\text{rms}).$$

It follows directly that

$$v_s(t) = \sqrt{2}(259.86)\cos(120\pi t + 11.09°)$$
$$= 367.49\cos[377t + 11.09°]\ \text{V}.\quad\blacksquare$$

Example 11.6 a) Calculate the total average and reactive power delivered to each impedance in the circuit in Example 10.11, that is, the circuit in Fig. 10.33.

b) Calculate the average and reactive powers associated with each source in the circuit in Fig. 10.33.

c) Verify that the average power delivered equals the average power absorbed

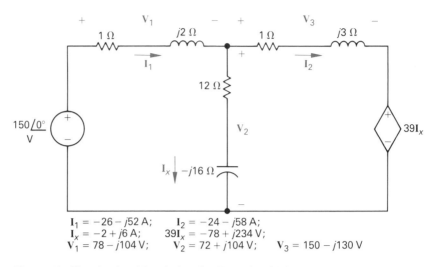

$I_1 = -26 - j52$ A; $I_2 = -24 - j58$ A;
$I_x = -2 + j6$ A; $39I_x = -78 + j234$ V;
$V_1 = 78 - j104$ V; $V_2 = 72 + j104$ V; $V_3 = 150 - j130$ V

Fig. 11.11 The circuit, with solution, for Example 10.11.

and that the magnetizing vars delivered equals the magnetizing vars absorbed.

Solution a) For convenience, the circuit along with the solution of the currents and voltages is repeated in Fig. 11.11. The complex power delivered to the $(1 + j2)$-Ω impedance is

$$S_1 = \frac{1}{2} \mathbf{V}_1 \mathbf{I}_1^* = P_1 + jQ_1$$

$$= \frac{1}{2}(78 - j104)(-26 + j52)$$

$$= \frac{1}{2}(3380 + j6760)$$

$$= 1690 + j3380 \text{ VA.}$$

Thus this impedance is absorbing an average power of 1690 W and 3380 magnetizing vars of reactive power. The complex power delivered to the $(12 - j16)$-Ω impedance is

$$S_2 = \frac{1}{2} \mathbf{V}_2 \mathbf{I}_x^* = P_2 + jQ_2$$

$$= \frac{1}{2}(72 + j104)(-2 - j6)$$

$$= 240 - j320 \text{ VA.}$$

Therefore the impedance in the vertical branch is absorbing 240 W and delivering a reactive power of 320 magnetizing vars. The complex power delivered to the $(1 + j3)$-Ω impedance is

$$S_3 = \frac{1}{2} \mathbf{V}_3 \mathbf{I}_2^* = P_3 + jQ_3$$

$$= \frac{1}{2}(150 - j130)(-24 + j58)$$

$$= 1970 + j5910 \text{ VA}.$$

This impedance is absorbing 1970 W and 5910 magnetizing vars.

b) The complex power associated with the independent voltage source is

$$S_s = -\frac{1}{2} \mathbf{V}_s \mathbf{I}_1^* = P_s + jQ_s$$

$$= -\frac{1}{2}(150)(-26 + j52)$$

$$= 1950 - j3900 \text{ VA}.$$

Note that the independent voltage source is absorbing an average power of 1950 W and is delivering 3900 magnetizing vars. The complex power associated with the current-controlled voltage source is

$$S_x = \frac{1}{2}(39\mathbf{I}_x)(\mathbf{I}_2^*) = P_x + jQ_x$$

$$= \frac{1}{2}(-78 + j234)(-24 + j58)$$

$$= -5850 - j5070 \text{ VA}.$$

The dependent source is delivering both average power and magnetizing vars.

c) The total power absorbed by the passive impedances and the independent voltage source is

$$P_{\text{absorbed}} = P_1 + P_2 + P_3 + P_s = 5850 \text{ W}.$$

The dependent voltage source is the only circuit element delivering average power; thus

$$P_{\text{delivered}} = 5850 \text{ W}.$$

Magnetizing vars are being absorbed by the two horizontal branches; thus

$$Q_{\text{absorbed}} = Q_1 + Q_3 = 9290 \text{ VAR}.$$

Magnetizing vars are being delivered by the independent voltage source, the capacitor in the vertical impedance branch, and the dependent voltage source; therefore

$$Q_{\text{delivered}} = 9290 \text{ VAR}. \quad \blacksquare$$

The apparent power, or volt-amp, requirement of a device that is designed to convert electrical energy to nonelectrical form is more important than the average power requirement. The device must be insulated to withstand the voltage and must be of sufficient size to carry the current even if the power factor is very low. Because the average power represents the useful output of the energy-converting device, it is desirable to operate such devices close to the unity power factor. Many useful appliances (refrigerators, fans, air conditioners, fluorescent lighting fixtures, washing machines) and most industrial loads operate at a lagging power factor. These loads are sometimes corrected either by adding a capacitor to the device itself or by connecting capacitors across the line that is feeding the load. The capacitors are connected at the load end of the line. The use of capacitors across the line is representative of power-factor correction for large industrial loads. Drill Exercise 11.3 and Problems 11.15 and 11.16 give you a chance to make some calculations that show why connecting a capacitor across the terminals of a lagging-power-factor load improves the operation of the circuit.

DRILL EXERCISE
11.3

The load impedance in the circuit in Fig. 11.9 is shunted by a capacitor having a capacitive reactance of -52 Ω. Calculate:

a) the rms phasors $\mathbf{V}_L$ and $\mathbf{I}_L$;

b) the average power and magnetizing vars absorbed by the $(39 + j26)$-Ω load impedance

c) the average power and magnetizing vars absorbed by the $(1 + j4)$-Ω line impedance;

d) the average power and magnetizing vars delivered by the source;

e) the magnetizing vars delivered by the shunting capacitor.

Ans. (a) $252.20\underline{/-4.54°}$ V (rms), $5.38\underline{/-38.23°}$ A (rms); (b) 1129.09 W, 752.73 VAR; (c) 23.52 W, 94.09 VAR; (d) 1152.62 W, -376.36 VAR; (e) 1223.18 VAR.

DRILL EXERCISE
11.4

The rms voltage at the terminals of a load is 440 V. The load is absorbing an average power of 20 kW and a magnetizing reactive power of 10 kVAR. Derive two equivalent impedance models of the load.

Ans. 7.744 Ω in series with 3.872 Ω of inductive reactance; 9.68 Ω in parallel with 19.36 Ω of inductive reactance.

11.7 MAXIMUM POWER TRANSFER

Occasionally when transmitting information through electrical signals it is important to deliver as much power as possible to the load. The efficiency of the transmission may be of secondary importance. For a network operating in the

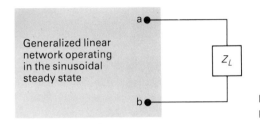

Fig. 11.12 A circuit describing maximum power transfer.

sinusoidal steady-state, maximum average power is delivered to a load when the load impedance is the conjugate of the Thévenin impedance of the network as viewed from the terminals of the load. The problem of maximum power transfer is described in terms of Fig. 11.12. Here we have a linear network operating in the sinusoidal steady-state and we must determine the load impedance Z_L that will result in maximum average power being delivered to the terminals a, b. Any linear network can be viewed from the terminals of the load in terms of a Thévenin equivalent circuit. Thus the problem reduces to finding the value of Z_L that results in maximum average power delivered to Z_L in the circuit in Fig. 11.13.

For maximum average power transfer to the load impedance, Z_L must equal the conjugate of the Thévenin impedance; that is,

$$Z_L = Z_{Th}^*. \qquad (11.41)$$

The result given by Eq. (11.41) can be derived by a straightforward application of elementary calculus. We begin by first expressing Z_{Th} and Z_L in rectangular form; thus

$$Z_{Th} = R_{Th} + jX_{Th} \qquad (11.42)$$

and

$$Z_L = R_L + jX_L. \qquad (11.43)$$

In both Eqs. (11.42) and (11.43), the reactance term carries its own algebraic sign—positive for inductance and negative for capacitance. Since we are making an average power calculation, we will assume that the amplitude of the Thévenin voltage is expressed in terms of its rms value. We will also use

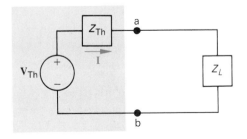

Fig. 11.13 The circuit in Fig. 11.12 with the network replaced by its Thévenin equivalent circuit.

the Thévenin voltage as our reference phasor. It follows directly from Fig. 11.13 that the rms value of the load current $\mathbf{I}$ is

$$\mathbf{I} = \frac{\mathbf{V}_{\mathrm{Th}}\underline{/0^\circ}}{(R_{\mathrm{Th}} + R_L) + j(X_{\mathrm{Th}} + X_L)}. \tag{11.44}$$

The average power delivered to the load is

$$P = |\mathbf{I}|^2 R_L. \tag{11.45}$$

When we substitute Eq. (11.44) into Eq. (11.45), we can write the expression for P as

$$P = \frac{|\mathbf{V}_{\mathrm{Th}}|^2 R_L}{(R_{\mathrm{Th}} + R_L)^2 + (X_{\mathrm{Th}} + X_L)^2}. \tag{11.46}$$

It is important to remember in working with Eq. (11.46) that $\mathbf{V}_{\mathrm{Th}}, R_{\mathrm{Th}}$, and X_{Th} are all fixed quantities, whereas R_L and X_L are independent variables. Therefore, to maximize P, we must find the values of R_L and X_L, where $\partial P/\partial R_L$ and $\partial P/\partial X_L$ are both zero. From Eq. (11.46), we have

$$\frac{\partial P}{\partial X_L} = \frac{-|\mathbf{V}_{\mathrm{Th}}|^2 2R_L(X_L + X_{\mathrm{Th}})}{[(R_L + R_{\mathrm{Th}})^2 + (X_L + X_{\mathrm{Th}})^2]^2}, \tag{11.47}$$

$$\frac{\partial P}{\partial R_L} = \frac{|\mathbf{V}_{\mathrm{Th}}|^2[(R_L + R_{\mathrm{Th}})^2 + (X_L + X_{\mathrm{Th}})^2 - 2R_L(R_L + R_{\mathrm{Th}})]}{[(R_L + R_{\mathrm{Th}})^2 + (X_L + X_{\mathrm{Th}})^2]^2}. \tag{11.48}$$

From Eq. (11.47), we note that $\partial P/\partial X_L$ will be zero when

$$X_L = -X_{\mathrm{Th}}. \tag{11.49}$$

From Eq. (11.48), we note that $\partial P/\partial R_L$ will be zero when

$$R_L = \sqrt{R_{\mathrm{Th}}^2 + (X_L + X_{\mathrm{Th}})^2}. \tag{11.50}$$

When we combine the result expressed in Eq. (11.49) with Eq. (11.50), we see that both derivatives are zero when $Z_L = Z_{\mathrm{Th}}^*$.

The maximum average power that can be delivered to Z_L when it is set equal to the conjugate of Z_{Th} is calculated directly from the circuit in Fig. 11.13. When $Z_L = Z_{\mathrm{Th}}^*$ the rms load current is $\mathbf{V}_{\mathrm{Th}}/2R_L$ and the maximum average power delivered to the load is

$$P_{\max} = \frac{|\mathbf{V}_{\mathrm{Th}}|^2 R_L}{4R_L^2} = \frac{1}{4} \cdot \frac{|\mathbf{V}_{\mathrm{Th}}|^2}{R_L}. \tag{11.51}$$

If the Thévenin voltage is expressed in terms of its maximum amplitude rather than its rms amplitude, Eq. (11.51) becomes

$$P_{\max} = \frac{1}{8} \cdot \frac{V_m^2}{R_L}. \tag{11.52}$$

It is important to note that maximum average power can be delivered to Z_L only if Z_L can be set equal to the conjugate of Z_{Th}. There are situations in

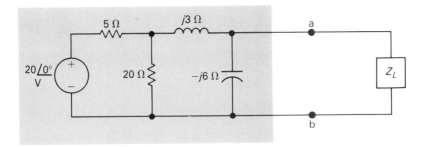

Fig. 11.14 The circuit for Example 11.7.

which this is not possible. First R_L and X_L may be restricted to a limited range of values. In this situation, the optimum condition for R_L and X_L is to adjust X_L as near $-X_{\text{Th}}$ as possible and then adjust R_L as close to $\sqrt{R_{\text{Th}}^2 + (X_L + X_{\text{Th}})^2}$ as possible (see Example 11.8). A second type of restriction occurs when the magnitude of Z_L can be varied but its phase angle cannot. Under this restriction, the greatest amount of power is transferred to the load when the magnitude of Z_L is set equal to the magnitude of Z_{Th}, that is, when

$$|\mathbf{Z}_L| = |\mathbf{Z}_{\text{Th}}|. \tag{11.53}$$

The proof of Eq. (11.53) is left in Problem 11.11 to the reader.

It is worth noting that for purely resistive networks, the condition for maximum power transfer is simply that the load resistance equal the Thévenin resistance. Note that this result was first derived in our introduction to maximum power transfer in Chapter 4.

Examples 11.7, 11.8, and 11.9 illustrate the problem of obtaining maximum power transfer in the basic situations discussed above.

Example 11.7   a) For the circuit shown in Fig. 11.14, determine the impedance Z_L that will result in maximum average power being transferred to Z_L.

b) What is the maximum average power that is transferred to the load impedance determined in part (a)?

Solution   a) We begin by determining the Thévenin equivalent with respect to the load terminals a, b. After two source transformations involving the 20-V source, the 5-Ω resistor, and the 20-Ω resistor, we can simplify the circuit in Fig. 11.14 to that shown in Fig. 11.15, from which it follows that $\mathbf{V}_{\text{Th}}$ is

$$\mathbf{V}_{\text{Th}} = \frac{16\underline{/0^\circ}}{(4 + j3 - j6)}(-j6)$$

$$= 19.2\underline{/-53.13^\circ} = 11.52 - j15.36 \text{ V}$$

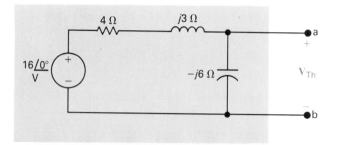

Fig. 11.15   A simplification of·Fig. 11.14 by source transformations.

and

$$Z_{\text{Th}} = \frac{(-j6)(4 + j3)}{4 + j3 - j6} = 5.76 - j1.68 \ \Omega.$$

For maximum average power transfer, the load impedance must be the conjugate of Z_{Th}; therefore

$$Z_L = 5.76 + j1.68 \ \Omega.$$

b) The maximum average power delivered to Z_L can be calculated from the circuit in Fig. 11.16, where the original network has been replaced by its Thévenin equivalent circuit. From the circuit in Fig. 11.16, we see that the rms magnitude of the load current **I** is

$$I_{\text{eff}} = \frac{19.2/\sqrt{2}}{2(5.76)} = 1.1785 \ \text{A}.$$

The average power delivered to the load is

$$P = I_{\text{eff}}^2(5.76) = 8 \ \text{W}. \quad \blacksquare$$

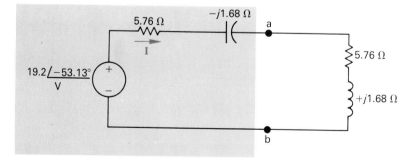

Fig. 11.16   The circuit in Fig. 11.14 with the original network replaced by its Thevenin equivalent circuit.

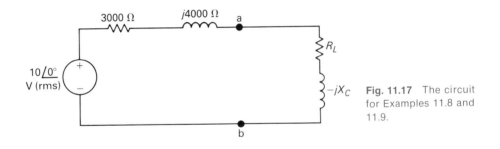

Fig. 11.17 The circuit for Examples 11.8 and 11.9.

Example 11.8 a) For the circuit shown in Fig. 11.17, what value of Z_L will result in maximum average power transfer to Z_L? What is the maximum power in mW?

b) Assume that the load resistance can be varied between 0 and 4000 Ω and that the capacitive reactance of the load can be varied between 0 and -2000 Ω. What settings of R_L and X_L will transfer the most average power to the load? What is the most power that can be transferred to the load under these restrictions?

Solution a) If there are no restrictions on R_L and X_L, the load impedance is set equal to the conjugate of the output or Thévenin impedance. Therefore we set $R_L = 3000$ Ω and $X_C = -4000$ Ω, or

$$Z_L = 3000 - j4000 \ \Omega.$$

Since the source voltage is given in terms of its rms value, the average power delivered to Z_L is

$$P = \frac{1}{4} \frac{(10)^2}{3000} = \frac{25}{3} \text{ mW} = 8.33 \text{ mW}.$$

b) Since R_L and X_L are restricted, we first set X_C as close to -4000 Ω as possible. Thus we set X_C to -2000 Ω. Next, we set R_L as close to $\sqrt{R_{\text{Th}}^2 + (X_L + X_{\text{Th}})^2}$ as possible. Thus

$$R_L = \sqrt{(3000)^2 + (-2000 + 4000)^2}$$
$$= 3605.55 \ \Omega.$$

Now, since R_L can be varied from 0 to 4000 Ω, R_L can be set to 3605.55 Ω. Therefore the load impedance is adjusted to a value of

$$Z_L = 3605.55 - j2000 \ \Omega.$$

With Z_L set at this value, the value of the load current will be

$$\mathbf{I}_{\text{eff}} = \frac{10 \underline{/0°}}{6605.55 + j2000}$$
$$= 1.4489 \underline{/-16.85°} \text{ mA}.$$

The average power delivered to the load is

$$P = (1.4489 \times 10^{-3})^2(3605.55)$$
$$= 7.5694 \text{ mW}.$$

This is the most power that we can deliver to a load given the restrictions on R_L and X_L. Note that this power is less than what can be delivered if there are no restrictions on R_L and X_L, that is, in part (a) we found that we can deliver 8.3333 mW. ■

Example 11.9   A load impedance having a constant phase angle of $-36.87°$ is connected across the load terminals a, b in the circuit in Fig. 11.17. The magnitude of Z_L is varied until the average power delivered is the most possible under the given restriction.

a) Specify Z_L in rectangular form.

b) Calculate the average power delivered to Z_L.

Solution   a) From the result stated in Eq. (11.53), we know that the magnitude of Z_L must equal the magnitude Z_{Th}; therefore

$$|Z_L| = |Z_{Th}| = |3000 + j4000| = 5000 \ \Omega.$$

Now, since we know that the phase angle of Z_L is $-36.87°$, we have

$$Z_L = 5000 \underline{/-36.87°} = 4000 - j3000 \ \Omega.$$

b) With Z_L set equal to $4000 - j3000 \ \Omega$, the load current will be

$$\mathbf{I}_{\text{eff}} = \frac{10}{7000 + j1000} = 1.4142\underline{/-8.13°} \text{ mA}$$

and the average power delivered to the load is

$$P = (1.4142)^2(4) = 8 \text{ mW}.$$

This is the most power that can be delivered by this circuit to a load impedance whose angle is constant at $-36.87°$. Again, we note that this is less than the maximum power that can be delivered if there are no restrictions on Z_L. ■

DRILL EXERCISE
11.5

The source current in the circuit shown is $5 \cos 8000t$ A.

a) What impedance should be connected across the terminals a, b for maximum average power transfer?

b) What is the average power that is transferred to the impedance in part (a)?

c) Assume that the load is restricted to pure resistance. What size resistor connected across a, b will result in the most average power transferred?

d) What is the average power transferred to the resistor in part (c)?

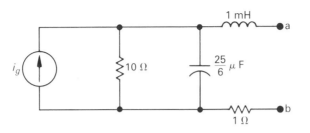

Ans. (a) $10 - j5 \ \Omega$; (b) 28.125 W; (c) 11.18 Ω; (d) 26.56 W.

11.8 SUMMARY

Since power calculations are an important part of steady-state sinusoidal analysis, we have introduced the basic relationships used to make these calculations. The concept of real and reactive power is used by the engineer to distinguish between power that is involved in the transformation between electrical and nonelectrical form (real) and that which always remains in electrical form (reactive). Since an electrical circuit or device must be able to accommodate both real and reactive power simultaneously, we use the concept of apparent power to describe this volt-amp capacity.

In the process of developing the basic power equations, we introduced the "effective" value of a sinusoidal signal. The motivation for defining the effective value stems from the desire to compare the ability of sinusoidal signals to convert electrical energy to nonelectrical form with that of dc, or constant, signals. The usefulness of the effective value is reflected in the fact that all circuits and equipment involved in power utilization are rated in terms of effective voltage and current. We also discussed the use of the acronym "rms" as a substitute for "effective."

In applications in which the efficiency of power transmission is not of prime importance, the conditions necessary for maximum power transfer are of interest. The maximum power transfer theorem was discussed along with the strategies to use when practical constraints are placed on the load impedance.

In Chapter 12, we will introduce three-phase circuits important in the utilization of electrical energy for power applications. Your understanding of three-phase circuits will depend on your grasp of both the phasor method introduced in Chapter 10 and the power computation techniques discussed in this chapter. Hence, the study of Chapter 12 will give you additional experience in using these important analytical tools.

PROBLEMS

11.1 The following sets of values for v and i pertain to the circuit in Fig. 11.1. For each set of values, calculate P and Q and state whether the circuit inside the box is absorbing or delivering (a) average power; and (b) magnetizing vars.

i) $v = 100 \cos (\omega t + 30°)$,
   $i = 10 \cos (\omega t + 60°)$
ii) $v = 100 \cos (\omega t - 45°)$,
   $i = 10 \cos (\omega t - 75°)$
iii) $v = 100 \cos (\omega t + 135°)$,
   $i = 10 \sin (\omega t - 95°)$
iv) $v = 100 \cos (\omega t - 160°)$,
   $i = 10 \sin (\omega t + 45°)$
v) $v = 100 \sin (\omega t + 140°)$,
   $i = 10 \cos (\omega t + 95°)$

11.2   Show that the maximum value of the instantaneous power given by Eq. (11.15) is $P + \sqrt{P^2 + Q^2}$ and the minimum value is $P - \sqrt{P^2 + Q^2}$.

11.3   A load consisting of a 6.25-Ω resistor in parallel with a 5/3-mH inductor is connected across the terminals of a sinusoidal voltage source v_g, where $v_g = 100 \cos 5000t$ V.

a) What is the peak value of the instantaneous power delivered by the source?
b) What is the peak value of the instantaneous power absorbed by the source?
c) What is the average power delivered to the load?
d) What is the reactive power?
e) Does the load absorb or generate magnetizing vars?
f) What is the power factor of the load?
g) What is the reactive factor of the load?

11.4   Find the rms value of the periodic voltage shown in Fig. 11.18.

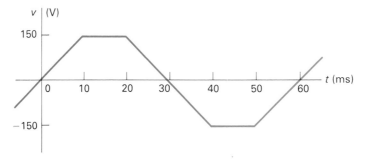

Fig. 11.18   A periodic voltage.

11.5   The periodic voltage in Problem 11.4 is applied across the terminals of a 10-Ω resistor. What is the average power delivered to the resistor?

11.6   The current source in the phasor domain circuit in Fig. 11.19 delivers $26\underline{/0°}$ A (rms).

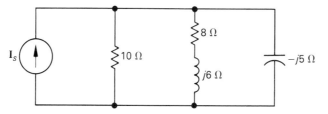

Fig. 11.19   The circuit for Problem 11.6.

a) Find the average and reactive powers at the terminals of the current source.
b) Is the current source absorbing or delivering average power?
c) Is the current source absorbing or delivering magnetizing vars?
d) Find the real (average) and reactive powers associated with each impedance branch in the circuit.
e) Check the balance between delivered and absorbed average power.
f) Check the balance between delivered and absorbed magnetizing vars.

11.7 Three loads are connected in parallel across a 250-V (rms) line as shown in Fig. 11.20. Load 1 absorbs 16 kW and 18 kVAR. Load 2 absorbs 10 kVA at 0.6-pf lead. Load 3 absorbs 8 kW at unity power factor. Find the impedance that is equivalent to the three parallel loads.

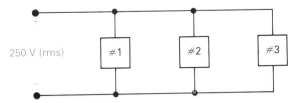

Fig. 11.20 The circuit for Problem 11.7.

11.8 The three loads in Problem 11.7 are fed from a line having a series impedance of $0.01 + j0.08 \ \Omega$ as shown in Fig. 11.21.

a) Calculate the rms value of the voltage ($\mathbf{V}_s$) at the sending end of the line.
b) Calculate the average and reactive powers associated with the line impedance.
c) Calculate the average and reactive powers at the sending end of the line.
d) Calculate the efficiency (η) of the line if the efficiency is defined as

$$\eta = [P_{\text{load}}/P_{\text{sending end}}] \times 100.$$

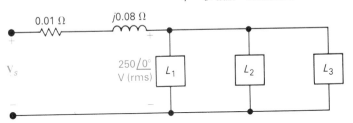

Fig. 11.21 The circuit for Problem 11.8.

11.9 The circuit in Fig. 11.22 represents a residential distribution circuit in which the impedances of the service conductors are negligible. The three loads are described as follows (all voltages are rms):

Load 1: Six 100-W, 120-V lamps in parallel;
Load 2: A 1/3-hp refrigerator motor (that absorbs 300 W and 180 magnetizing vars at 120 V) in parallel with three 60-W, 120-V lamps;
Load 3: A 7.2-kW, 240-V electric oven.

a) Calculate $\mathbf{I}_1$, $\mathbf{I}_2$, and $\mathbf{I}_n$.
b) Calculate the average and reactive powers associated with each voltage source.
c) Assume that the load impedances remain constant when the neutral conductor is inadvertently opened. Calculate the voltage, current, and apparent power associated with each load.

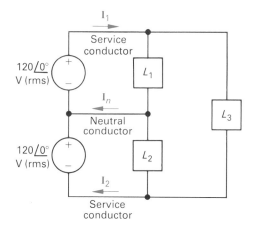

Fig. **11.22** The circuit for Problem 11.9.

11.10 a) Find **V** (rms) and θ for the circuit in Fig. 11.23 if the load absorbs 600 VA at a lagging power factor of 0.8.

b) Construct a phasor diagram of each solution obtained in part (a).

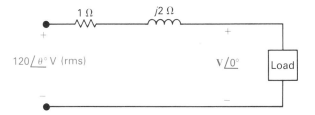

Fig. **11.23** The circuit for Problem 11.10.

11.11 a) Determine the load impedance for the circuit shown in Fig. 11.24 that will result in maximum power being transferred to the load.

b) Determine the maximum power if $v_s = 141.4 \cos 1000t$ V.

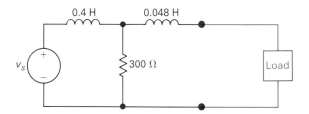

Fig. **11.24** The circuit for Problem 11.11.

11.12 In the circuit shown in Fig. 11.25 the load resistance can be varied from 0 to 4000 Ω. The load capacitor can be varied from 0.1 μF to 0.5 μF.

a) Calculate the average power delivered to a load when $R_L = 2000$ Ω and $C_L = 0.2$ μF.

b) Determine the settings of R_L and C_L that will result in the most average power being transferred to R_L.

c) What is the most average power in part (b)? Is it greater than the power in part (a)?

d) If there are no constraints on R_L and C_L, what is the maximum average power that can be delivered to a load?

e) What are the values of R_L and C_L for the condition of part (d)?

f) Is the average power calculated in part (d) larger than that calculated in part (c)?

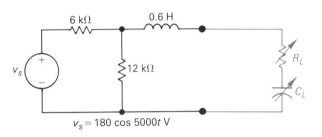

$v_s = 180 \cos 5000t$ V

Fig. 11.25 The circuit for Problem 11.12.

11.13 Prove that if only the magnitude of the load impedance can be varied, the most average power is transferred to the load when $|\mathbf{Z}_L| = |\mathbf{Z}_{Th}|$. (*Hint:* In deriving the expression for the average load power, write the load impedance ($\mathbf{Z}_L$) in the form $\mathbf{Z}_L = |\mathbf{Z}_L| \cos \theta + j|\mathbf{Z}_L| \sin \theta$ and note that only $|\mathbf{Z}_L|$ is variable.)

11.14 Assume that the load consisting of R_L and C_L in the circuit in Fig. 11.25 is replaced by an impedance $\mathbf{Z}_L$ such that $\mathbf{Z}_L = 25m\underline{/-20°}\ \Omega$.

a) What value of m will result in the most average power being transferred to the load?

b) What is the average power for the chosen value of m?

11.15 A factory has an electrical load of 820 kW at a lagging power factor of 0.6. An additional variable power factor load is to be added to the factory. The new load will add 80 kW to the real power load of the factory. The power factor of the added load is to be adjusted so that the overall power factor of the factory is 0.9 lagging.

a) Specify the reactive power associated with the added load.

b) Does the added load absorb or deliver magnetizing vars?

c) What is the power factor of the additional load?

d) Assume that the rms voltage at the input to the factory is 2400 V. What is the rms magnitude of the current into the factory before the variable power factor load is added?

e) What is the rms magnitude of the current into the factory after the variable power factor load has been added?

11.16 The sending end voltage ($\mathbf{V}_s$) in the circuit in Fig. 11.26 is adjusted so that the effective value of the load voltage is maintained at 240 V.

a) Calculate $\mathbf{V}_s$, the power loss in the 0.25-Ω line resistance, and the power factor of the load.

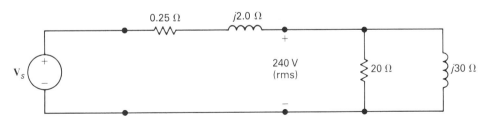

Fig. 11.26 The circuit for Problem 11.16.

b) Add a capacitor, with a capacitive reactance of -30 Ω, in parallel with the load and repeat the calculations of part (a).

c) The addition of the capacitor at the terminals of the load is referred to as a power factor correction procedure. The power factor is improved by making it as close to unity as practical. What are two advantages of operating this circuit at unity power factor?

11.17 The operational amplifier in the circuit in Fig. 11.27 is ideal. Calculate the average power delivered to the 50-Ω resistor when $v_g = \cos 50{,}000t$ V.

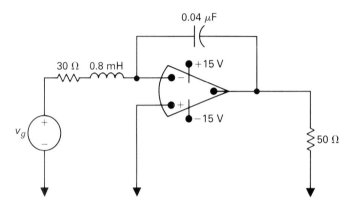

Fig. 11.27 The circuit for Problem 11.17.

12

BALANCED THREE-PHASE CIRCUITS

12.1   INTRODUCTION

The generation, transmission, distribution, and utilization of large blocks of electrical power are accomplished by means of three-phase circuits. The comprehensive analysis of three-phase systems is a field of study in its own right, which we cannot hope to cover in a single chapter. Fortunately, an understanding of only the steady-state sinusoidal behavior of *balanced* three-phase circuits is quite sufficient for engineers who do not specialize in power systems. We will define what we mean by a balanced circuit later in our discussion. For the moment, we note that there are two reasons for our restricting our introduction to balanced operation. First, for economic reasons, three-phase systems are designed to operate in the balanced state. This means that under normal operating conditions the three-phase circuit is so close to being balanced that we are justified in finding the solution that assumes perfect balance. Second, some types of unbalanced operating conditions can be solved by a technique known as the method of symmetrical components, which relies heavily on a thorough understanding of balanced operation. Although we will not discuss the method of symmetrical components, it is worth noting that an understanding of balanced operation is a starting point for an advanced technique used to analyze certain types of unbalanced conditions.

The basic structure of a three-phase system consists of voltage sources connected to loads via transformers† and transmission lines. We can reduce the problem to the analysis of a circuit consisting of a voltage source connected to a load via a line. The omission of the transformer as an element in the system simplifies the discussion without jeopardizing a basic understanding of the calculations involved. The basic circuit is shown in Fig. 12.1. In order to begin analyzing a circuit of this type, we must understand the characteristics of a balanced three-phase set of sinusoidal voltages.

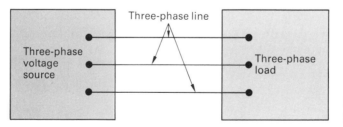

Fig. 12.1 A basic three-phase circuit.

12.2   BALANCED THREE-PHASE VOLTAGES

A set of balanced three-phase voltages consists of three sinusoidal voltages that have identical amplitudes and frequency but are out of phase with each other by exactly 120°. In discussing three-phase circuits, it is standard practice

† Transformers are introduced in Chapter 13.

to refer to the three phases as a, b, and c. Furthermore, the a-phase is almost always used as the reference phase. The three voltages that comprise the three-phase set are referred to as the a-phase voltage, the b-phase voltage, and the c-phase voltage.

Since the phase voltages are out of phase by exactly 120°, there are two possible phase relationships that can exist between the a-phase voltage and the b- and c-phase voltages. One possibility is for the b-phase voltage to lag the a-phase voltage by 120°, in which case the c-phase voltage must lead the a-phase voltage by 120°. This phase relationship is known as the *abc, or positive, phase sequence*. The only other possibility is for the b-phase voltage to lead the a-phase voltage by 120°, in which case the c-phase voltage must lag the a-phase voltage by 120°. This phase relationship is known as the *acb, or negative, phase sequence*. In phasor notation, the two possible sets of balanced phase voltages are

$$\mathbf{V}_a = V_m \,\underline{/0°},$$
$$\mathbf{V}_b = V_m \,\underline{/-120°}, \tag{12.1}$$
$$\mathbf{V}_c = V_m \,\underline{/+120°},$$

and

$$\mathbf{V}_a = V_m \,\underline{/0°},$$
$$\mathbf{V}_b = V_m \,\underline{/+120°}, \tag{12.2}$$
$$\mathbf{V}_c = V_m \,\underline{/-120°}.$$

The phase sequence of the voltages given by Eqs. (12.1) is the abc, or positive, sequence. The phase sequence of the voltages given by Eqs. (12.2) is the acb, or negative, sequence. The phasor diagram representations of the voltage sets given by Eqs. (12.1) and (12.2) are shown in Fig. 12.2, from which we can determine the phase sequence by noting the order of the subscripts as we move clockwise around the diagram. The fact that a three-phase circuit can have one of two possible phase sequences is an important characteristic that must be

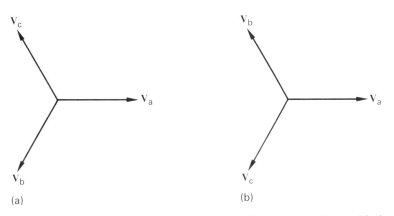

Fig. 12.2 Phasor diagrams of a balanced set of three-phase voltages: (a) abc (positive) sequence; (b) acb (negative) sequence.

taken into account whenever two separate circuits are operated in parallel. The two circuits can operate in parallel only if they have the same phase sequence.

Another important characteristic of a set of balanced three-phase voltages is that the sum of the voltages adds to zero. Thus using either Eqs. (12.1) or Eqs. (12.2) we have

$$\mathbf{V}_a + \mathbf{V}_b + \mathbf{V}_c = 0. \tag{12.3}$$

The fact that the sum of the phasor voltages adds to zero also means that the sum of the instantaneous voltages is zero, that is,

$$v_a + v_b + v_c = 0. \tag{12.4}$$

Another noteworthy observation is that if we know the *phase sequence* and *one voltage in the set, we know the entire set*. Thus if we have a balanced three-phase system, we can focus on determining the voltage (or current) in one phase, because once we know one phase quantity we automatically know the corresponding quantity in the other two phases.

DRILL EXERCISE   What is the phase sequence of each of the following sets of voltages?
12.1
a) $v_a = 208 \cos(\omega t + 76°)$ V,
   $v_b = 208 \cos(\omega t + 316°)$ V,
   $v_c = 208 \cos(\omega t - 164°)$ V

b) $v_a = 4160 \cos(\omega t - 49°)$ V,
   $v_b = 4160 \cos(\omega t - 289°)$ V,
   $v_c = 4160 \cos(\omega t + 191°)$ V

Ans. (a) abc; (b) acb.

12.3   THREE-PHASE VOLTAGE SOURCES

Three-phase voltage sources consist of generators that have three separate windings distributed around the periphery of the stator. Each winding comprises one phase of the generator. The rotor of the generator is an electromagnet driven at synchronous speed by a prime mover such as a steam or gas turbine. As the electromagnet is rotated past the three windings, a sinusoidal voltage is induced in each winding. The phase windings are designed so that the sinusoidal voltages induced in them are equal in amplitude and out of phase with each other by exactly 120°. Since the phase windings are stationary with respect to the rotating electromagnet, the frequency of the voltage induced in each winding is the same.

Normally, the impedance of each phase winding on a three-phase generator is very small compared to the other impedances in the circuit. Therefore, to an approximation, each phase winding can be modeled in an electric circuit by

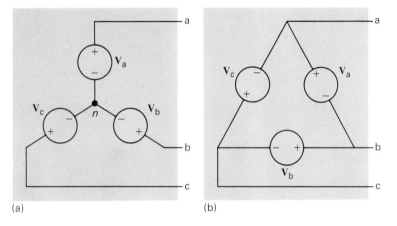

Fig. 12.3 The two basic connections of an ideal three-phase source: (a) Y-connected source; (b) Δ-connected source.

an ideal sinusoidal voltage source. There are two ways of interconnecting the separate phase windings to form a three-phase source. The windings can be connected together in either a wye (Y) or a delta (Δ) configuration. The wye and delta connections are shown in Fig. 12.3, where ideal voltage sources are used to model the phase windings of the three-phase generator. The common terminal in the Y-connected source, labeled "n" in Fig. 12.3(a), is known as the *neutral terminal* of the source. The neutral terminal may or may not be available for external connections.

If the impedance of each phase winding is not negligible, the three-phase source is modeled by placing the winding impedance in series with an ideal sinusoidal voltage source. Since all windings on the machine are of the same construction, the winding impedances are assumed to be identical. The winding impedance of three-phase generators is inductive. The model of a three-phase source including winding impedance is shown in Fig. 12.4, in which R_w is the winding resistance and X_w is the inductive reactance of the winding.

The fact that a three-phase voltage source can be either Y-connected or Δ-connected means that the basic circuit in Fig. 12.1 can take four different configurations, since the three-phase loads can also be either Y-connected or Δ-connected. The four possible arrangements are (1) a Y-connected source and a Y-connected load; (2) a Y-connected source and a Δ-connected load; (3) a Δ-connected source and a Y-connected load; and (4) a Δ-connected source and a Δ-connected load.

We begin our analysis of three-phase circuits with the first arrangement mentioned above. After analyzing the Y–Y circuit, we will show for balanced circuits how the remaining three arrangements can be reduced to a Y–Y equivalent circuit. In other words, the analysis of the Y–Y circuit is the key to solving all balanced three-phase arrangements.

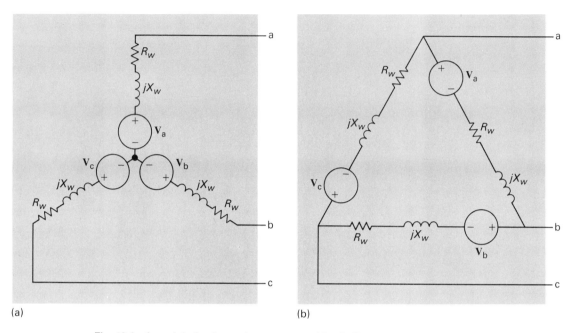

Fig. 12.4 A model of a three-phase source with winding impedance: (a) Y-connected source; (b) Δ-connected source.

12.4 ANALYSIS OF THE WYE–WYE CIRCUIT

We begin our analysis of the Y–Y circuit by assuming that the circuit is *not* balanced! We do this so that we can show what we mean by a balanced three-phase circuit and what the consequences of being balanced are in terms of circuit analysis. The general Y–Y circuit is illustrated in Fig. 12.5, where we have included a fourth conductor that connects the source neutral to the load neutral. The fourth conductor is possible only in the Y–Y arrangement. (More about this later.) We also mention that for convenience in drawing the diagram, we have transformed the Y-connections into "tipped-over tees." In Fig. 12.5, Z_{ga}, Z_{gb}, and Z_{gc} represent the internal impedance associated with each phase winding of the voltage source: Z_{1a}, Z_{1b}, and Z_{1c} represent the impedance of each phase conductor of the line connecting the source to the load; Z_o is the impedance of the neutral conductor that connects the source neutral to the load neutral; and Z_A, Z_B, and Z_C represent the impedance of each phase of the load.

The circuit in Fig. 12.5 can be described by a single node-voltage equation. Using the source neutral as the reference node and letting $\mathbf{V}_N$ denote the node voltage between the nodes N and n, we find that the node-voltage equation is

$$\frac{\mathbf{V}_N}{Z_o} + \frac{\mathbf{V}_N - \mathbf{V}_{a'n}}{Z_A + Z_{1a} + Z_{ga}} + \frac{\mathbf{V}_N - \mathbf{V}_{b'n}}{Z_B + Z_{1b} + Z_{gb}} + \frac{\mathbf{V}_N - \mathbf{V}_{c'n}}{Z_C + Z_{1c} + Z_{gc}} = 0. \quad (12.5)$$

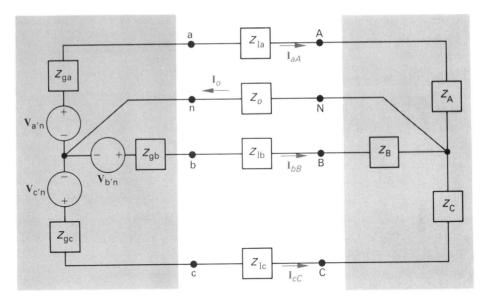

Fig. 12.5 A three-phase Y–Y system.

Before pursuing Eq. (12.5) any further, let us pause to observe that the circuit analysis techniques that we have discussed in the earlier chapters are directly applicable to three-phase circuits. Thus it is not necessary to introduce new analytical techniques in order to analyze three-phase circuits. However, as we will see in the remainder of this chapter, if a three-phase circuit is balanced we can take some significant analytical shortcuts to predict the behavior of the system.

The circuit in Fig. 12.5 is a balanced three-phase circuit if it satisfies *all* of the following criteria:

1. $\mathbf{V}_{a'n}$, $\mathbf{V}_{b'n}$, and $\mathbf{V}_{c'n}$ form a set of balanced three-phase voltages,
2. $Z_{ga} = Z_{gb} = Z_{gc}$,
3. $Z_{1a} = Z_{1b} = Z_{1c}$,
4. $Z_A = Z_B = Z_C$.

There is no restriction on the impedance of the neutral conductor (Z_o); its value has no effect on whether or not the system is balanced.

If the system is balanced, Eq. (12.5) tells us that $\mathbf{V}_N$ must be zero. To see this let

$$Z_\phi = Z_A + Z_{1a} + Z_{ga} \tag{12.6}$$

and then rewrite Eq. (12.5) as

$$\mathbf{V}_N \left(\frac{1}{Z_o} + \frac{3}{Z_\phi} \right) = \frac{\mathbf{V}_{a'n} + \mathbf{V}_{b'n} + \mathbf{V}_{c'n}}{Z_\phi}. \tag{12.7}$$

The right-hand side of Eq. (12.7) is zero because by hypothesis the numerator is a set of balanced three-phase voltages and Z_ϕ is *not* zero. The only value of $\mathbf{V}_N$ that satisfies Eq. (12.7) is zero. Therefore for a balanced three-phase circuit,

$$\mathbf{V}_N = 0. \tag{12.8}$$

Equation (12.8) is an extremely important result. If $\mathbf{V}_N$ is zero, there is no difference in potential between the source neutral (n) and the load neutral (N); consequently, the current in the neutral conductor is zero. These observations tell us that we can either remove the neutral conductor from a balanced Y–Y configuration ($I_o = 0$) or replace it by a perfect short circuit between the nodes n and N ($\mathbf{V}_N = 0$). We find both equivalents convenient to use when modeling balanced three-phase circuits.

Now let us turn our attention to what effect balanced conditions have on the three line currents. It follows directly from Fig. 12.5 that when the system is balanced, the three line currents will be

$$\mathbf{I}_{aA} = \frac{\mathbf{V}_{a'n} - \mathbf{V}_N}{\mathbf{Z}_A + \mathbf{Z}_{la} + \mathbf{Z}_{ga}} = \frac{\mathbf{V}_{a'n}}{\mathbf{Z}_\phi}, \tag{12.9}$$

$$\mathbf{I}_{bB} = \frac{\mathbf{V}_{b'n} - \mathbf{V}_N}{\mathbf{Z}_B + \mathbf{Z}_{lb} + \mathbf{Z}_{gb}} = \frac{\mathbf{V}_{b'n}}{\mathbf{Z}_\phi}, \tag{12.10}$$

$$\mathbf{I}_{cC} = \frac{\mathbf{V}_{c'n} - \mathbf{V}_N}{\mathbf{Z}_c + \mathbf{Z}_{lc} + \mathbf{Z}_{gc}} = \frac{\mathbf{V}_{c'n}}{\mathbf{Z}_\phi}, \tag{12.11}$$

from which we see that in a balanced system the three line currents form a balanced set of three-phase currents. Thus the current in each line will be equal in amplitude and frequency and will be exactly 120° out of phase with the other two line currents. This tells us that if we calculate the current $\mathbf{I}_{aA}$, we can write down the line currents $\mathbf{I}_{bB}$ and $\mathbf{I}_{cC}$ without further computations. We imply by this statement that the phase sequence is known.

We can use Eq. (12.9) to construct a single-phase equivalent circuit of the balanced three-phase Y–Y circuit. It follows from Eq. (12.9) that the current in the a-phase conductor line is simply the voltage generated in the a-phase winding of the generator divided by the total impedance in the a-phase of the circuit. Thus Eq. (12.9) describes the simple circuit in Fig. 12.6, where the neu-

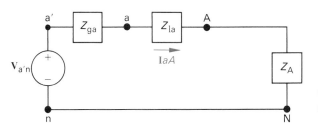

Fig. 12.6 A single-phase equivalent circuit.

tral conductor has been replaced by a perfect short circuit. A word of caution here. The current in the neutral conductor of Fig. 12.6 is *not* the current in the neutral conductor of a balanced three-phase circuit. The current in the neutral conductor is

$$\mathbf{I}_o = \mathbf{I}_{aA} + \mathbf{I}_{bB} + \mathbf{I}_{cC}, \tag{12.12}$$

whereas the current in the neutral conductor in Fig. 12.6 is $\mathbf{I}_{aA}$. Thus the circuit in Fig. 12.6 gives the correct value of the line current but only the a-phase component of the neutral current. Whenever the single-phase equivalent circuit in Fig. 12.6 is applicable, the line currents form a balanced three-phase set and the right-hand side of Eq. (12.12) sums to zero.

Once we know the line currents in the circuit in Fig. 12.5, it is a relatively simple task to calculate any voltages that are of interest. In particular, we are interested in the relationship between the line-to-line voltages and the line-to-neutral voltages. We will establish this relationship at the load terminals. The observations we make will also apply at the source terminals. The line-to-line voltages at the terminals of the load in terms of the line-to-neutral voltages at the load are

$$\mathbf{V}_{AB} = \mathbf{V}_{AN} - \mathbf{V}_{BN}, \tag{12.13}$$

$$\mathbf{V}_{BC} = \mathbf{V}_{BN} - \mathbf{V}_{CN}, \tag{12.14}$$

and

$$\mathbf{V}_{CA} = \mathbf{V}_{CN} - \mathbf{V}_{AN}. \tag{12.15}$$

The double-subscript notation in voltage equations indicates that the voltage is a drop from the first subscript to the second subscript. The relationships given by Eqs. (12.13)–(12.15) are shown in Fig. 12.7. Because we are interested in the balanced state, we have omitted the neutral conductor from the figure.

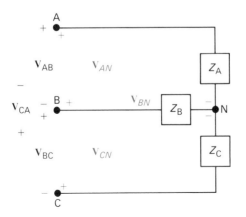

Fig. 12.7 Line-to-line and line-to-neutral voltages.

To show the relationship between the line-to-line voltages and the line-to-neutral voltages, we assume a positive, or abc, sequence. We arbitrarily choose the line-to-neutral voltage of the a-phase as the reference. Thus,

$$\mathbf{V}_{AN} = V_\phi \underline{/0^\circ}, \tag{12.16}$$

$$\mathbf{V}_{BN} = V_\phi \underline{/-120^\circ}, \tag{12.17}$$

and

$$\mathbf{V}_{CN} = V_\phi \underline{/+120^\circ}, \tag{12.18}$$

where V_ϕ represents the magnitude of the line-to-neutral voltage. When we substitute Eqs. (12.16)–(12.18) into Eqs. (12.13)–(12.15), respectively, we get

$$\mathbf{V}_{AB} = V_\phi - V_\phi \underline{/-120^\circ} = \sqrt{3}\, V_\phi \underline{/30^\circ}, \tag{12.19}$$

$$\mathbf{V}_{BC} = V_\phi \underline{/-120^\circ} - V_\phi \underline{/120^\circ} = \sqrt{3}\, V_\phi \underline{/-90^\circ}, \tag{12.20}$$

and

$$\mathbf{V}_{CA} = V_\phi \underline{/120^\circ} - V_\phi \underline{/0^\circ} = \sqrt{3}\, V_\phi \underline{/150^\circ}. \tag{12.21}$$

From Eqs. (12.19)–(12.21), we see that (1) the magnitude of the line-to-line voltage is $\sqrt{3}$ times the magnitude of the line-to-neutral voltage; (2) the line-to-line voltages form a balanced three-phase set of voltages; and (3) the set of line-to-line voltages lead the set of line-to-neutral voltages by 30°. We will leave it to the reader to show that for a negative, or acb, sequence the only change is that the set of line-to-line voltages lags the set of line-to-neutral voltages by 30°. These observations are summarized in the phasor diagrams of Fig. 12.8. We now observe that, in a balanced system, if the line-to-neutral voltage is known at some point in the circuit, the line-to-line voltages at the same point are also known, and vice versa.

Before illustrating balanced three-phase calculations with a numerical example, we must make some additional comments on terminology. In the Y–Y

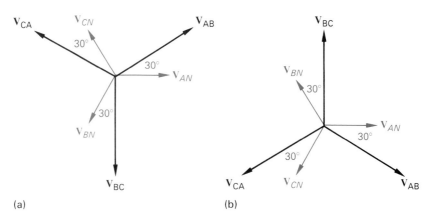

(a)   (b)

Fig. 12.8 Phasor diagrams showing the relationship between line-to-line and line-to-neutral voltages in a balanced system: (a) abc sequence; (b) acb sequence.

system, the line-to-neutral voltage is also called the *phase voltage*. For brevity, the line-to-line voltage is also called the *line voltage*. The *phase current* is defined as the current in each phase of the load or, at the source end of the circuit, the current in each phase of the generator. The *line current* is defined as the current in each phase of the line. For the Y–Y arrangement, the phase current and line current are identical. Because three-phase systems are designed to handle large blocks of electrical power, all voltage and current specifications and calculations are given in terms of rms values. Thus when a three-phase transmission line is rated at 345 kV, this means that the nominal value of the rms line-to-line voltage is 345,000 V. *All voltages and currents in this chapter are understood to be rms values.* Finally, the Greek letter phi (ϕ) is widely used in the literature to denote a per-phase quantity. Thus $\mathbf{V}_\phi$, $\mathbf{I}_\phi$, $\mathbf{Z}_\phi$, P_ϕ, and Q_ϕ are interpreted as voltage/phase, current/phase, impedance/phase, power/phase, and reactive power/phase, respectively.

Example 12.1 is designed to show how we can use the observations made thus far to solve a balanced three-phase Y–Y circuit.

Example 12.1   A three-phase, positive-sequence, Y-connected generator has an impedance of $0.2 + j0.5\ \Omega/\phi$. The internal phase voltage of the generator is 120 V. The generator feeds a balanced three-phase Y-connected load having an impedance of $39 + j28\ \Omega/\phi$. The impedance of the line connecting the generator to the load is $0.8 + j1.5\ \Omega/\phi$. The a-phase internal voltage of the generator is specified as the reference phasor.

a) Construct a single-phase equivalent circuit of the three-phase system.

b) Calculate the three line currents $\mathbf{I}_{aA}$, $\mathbf{I}_{bB}$, and $\mathbf{I}_{cC}$.

c) Calculate the three line-to-neutral voltages at the load: $\mathbf{V}_{AN}$, $\mathbf{V}_{BN}$, $\mathbf{V}_{CN}$.

d) Calculate the line voltages $\mathbf{V}_{AB}$, $\mathbf{V}_{BC}$, and $\mathbf{V}_{CA}$ at the terminals of the load.

e) Calculate the line-to-neutral voltages at the terminals of the generator $\mathbf{V}_{an}$, $\mathbf{V}_{bn}$, and $\mathbf{V}_{cn}$.

f) Calculate the line voltages $\mathbf{V}_{ab}$, $\mathbf{V}_{bc}$, and $\mathbf{V}_{ca}$ at the terminals of the generator.

g) Repeat parts (a) through (f) given that the phase sequence is negative.

Solution   a) The single-phase equivalent circuit is shown in Fig. 12.9.

b) The a-phase line current is

$$\mathbf{I}_{aA} = \frac{120\ \underline{/0^\circ}}{(0.2 + 0.8 + 39) + j(0.5 + 1.5 + 28)}$$

$$= \frac{120\ \underline{/0^\circ}}{40 + j30} = 2.4\ \underline{/-36.87^\circ}\ \text{A}.$$

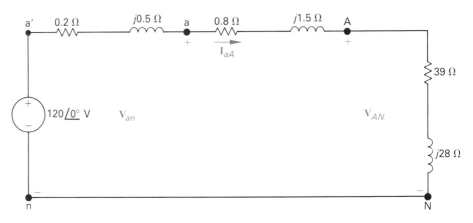

Fig. 12.9 The single-phase equivalent circuit for Example 12.1.

For a positive phase sequence,

$$\mathbf{I}_{bB} = 2.4\,\underline{/-156.87°}\ \text{A},$$
$$\mathbf{I}_{cC} = 2.4\,\underline{/83.13°}\ \text{A}.$$

c) The line-to-neutral voltage at the A terminal of the load is

$$\mathbf{V}_{AN} = (39 + j28)(2.4\,\underline{/-36.87°})$$
$$= 115.22\,\underline{/-1.19°}\ \text{V}.$$

For a positive phase sequence,

$$\mathbf{V}_{BN} = 115.22\,\underline{/-121.19°}\ \text{V},$$
$$\mathbf{V}_{CN} = 115.22\,\underline{/+118.81°}\ \text{V}.$$

d) For a positive phase sequence, the line-to-line voltages lead the line-to-neutral voltages by 30°; thus

$$\mathbf{V}_{AB} = (\sqrt{3}\ \underline{/30°})(\mathbf{V}_{AN})$$
$$= 199.58\,\underline{/28.81°}\ \text{V},$$
$$\mathbf{V}_{BC} = 199.58\,\underline{/-91.19°}\ \text{V},$$
$$\mathbf{V}_{CA} = 199.58\,\underline{/148.81°}\ \text{V}.$$

e) The line-to-neutral voltage at the a-terminal of the source is

$$\mathbf{V}_{an} = 120 - (0.2 + j0.5)(2.4\,\underline{/-36.87°})$$
$$= 120 - 1.29\,\underline{/31.33°}$$
$$= 118.90 - j0.67$$
$$= 118.90\,\underline{/-0.32°}\ \text{V}.$$

For a positive phase sequence,

$$\mathbf{V}_{bn} = 118.90\,\underline{/-120.32°},$$
$$\mathbf{V}_{cn} = 118.90\,\underline{/119.68°}\ \text{V}.$$

f) The line-to-line voltages at the source terminals are

$$\mathbf{V}_{ab} = (\sqrt{3}\;\underline{/30°})\mathbf{V}_{an}$$
$$= 205.94\,\underline{/29.68°}\;\text{V},$$
$$\mathbf{V}_{bc} = 205.94\,\underline{/-90.32°}\;\text{V},$$
$$\mathbf{V}_{ca} = 205.94\,\underline{/149.68°}\;\text{V}.$$

g) Changing the phase sequence has no effect on the single-phase equivalent circuit. The three line currents are

$$\mathbf{I}_{aA} = 2.4\,\underline{/-36.87°}\;\text{A},$$
$$\mathbf{I}_{bB} = 2.4\,\underline{/83.13°}\;\text{A},$$
$$\mathbf{I}_{cC} = 2.4\,\underline{/-156.87°}\;\text{A}.$$

The line-to-neutral voltages at the load are

$$\mathbf{V}_{AN} = 115.22\,\underline{/-1.19°}\;\text{V},$$
$$\mathbf{V}_{BN} = 115.22\,\underline{/+118.81°}\;\text{V},$$
$$\mathbf{V}_{CN} = 115.22\,\underline{/-121.19°}\;\text{V}.$$

For a negative phase sequence, the line-to-line voltages lag the line-to-neutral voltages by 30°:

$$\mathbf{V}_{AB} = (\sqrt{3}\;\underline{/-30°})\mathbf{V}_{AN}$$
$$= 199.58\,\underline{/-31.19°}\;\text{V},$$
$$\mathbf{V}_{BC} = 199.58\,\underline{/88.81°}\;\text{V},$$
$$\mathbf{V}_{CA} = 199.58\,\underline{/-151.19°}\;\text{V}.$$

The line-to-neutral voltages at the terminals of the generator are

$$\mathbf{V}_{an} = 118.90\,\underline{/-0.32°}\;\text{V},$$
$$\mathbf{V}_{bn} = 118.90\,\underline{/119.68°}\;\text{V},$$
$$\mathbf{V}_{cn} = 118.90\,\underline{/-120.32°}\;\text{V}.$$

The line-to-line voltages at the terminals of the generator are

$$\mathbf{V}_{ab} = [\sqrt{3}\;\underline{/-30°}]\mathbf{V}_{an}$$
$$= 205.94\,\underline{/-30.32°}\;\text{V},$$
$$\mathbf{V}_{bc} = 205.94\,\underline{/89.68°}\;\text{V},$$
$$\mathbf{V}_{ca} = 205.94\,\underline{/-150.32°}\;\text{V}. \quad ■$$

In studying Example 12.1, it is important to note that once the a-phase quantity is calculated, the corresponding b- and c-phase values can be tabulated by simply shifting the a-phase value by 120°. For a positive phase sequence, the b-phase lags the a-phase by 120°, whereas the c-phase leads the a-phase by 120°. For a negative phase sequence, the b-phase leads the a-phase by 120° and the c-phase lags the a-phase by 120°. We also call your attention to how easy it is to calculate line-to-line voltages once we know the line-to-neutral voltages.

The line-to-neutral voltage at the terminals of a balanced, three-phase, wye-connected load is 2400 V. The load has an impedance of $16 + j12$ Ω/ϕ and is fed from a line having an impedance of $0.10 + j0.80$ Ω/ϕ. The wye-connected source at the sending end of the line has a phase sequence of acb and an internal impedance of $0.02 + j0.16$ Ω/ϕ. Use the a-phase line-to-neutral voltage at the load at the reference and calculate (a) the line currents $\mathbf{I}_{aA}$, $\mathbf{I}_{bB}$, $\mathbf{I}_{cC}$; (b) the line-to-line voltages as the source $\mathbf{V}_{ab}$, $\mathbf{V}_{bc}$, $\mathbf{V}_{ca}$; and (c) the internal phase-to-neutral voltages at the source $\mathbf{V}_{a'n}$, $\mathbf{V}_{b'n}$, $\mathbf{V}_{c'n}$.

Ans. (a) $\mathbf{I}_{aA} = 120 \underline{/-36.87°}$ A, $\mathbf{I}_{bB} = 120 \underline{/83.13°}$ A, $\mathbf{I}_{cC} = 120 \underline{/-156.87°}$ A; (b) $\mathbf{V}_{ab} = 4275.02 \underline{/-28.38°}$ V, $\mathbf{V}_{bc} = 4275.02 \underline{/91.62°}$ V, $\mathbf{V}_{ca} = 4275.02 \underline{/-148.38°}$ V; (c) $\mathbf{V}_{a'n} = 2482.05 \underline{/1.93°}$ V, $\mathbf{V}_{b'n} = 2482.05 \underline{/121.93°}$ V, $\mathbf{V}_{c'n} = 2482.05 \underline{/-118.07°}$ V.

12.5 ANALYSIS OF THE WYE–DELTA CIRCUIT

If the load in a three-phase circuit is connected in a delta, it can be transformed into a wye by using the delta-to-wye transformation discussed in Section 10.7. When the load is balanced, the impedance of each leg of the wye is one-third the impedance of each leg of the delta; thus

$$Z_Y = \frac{Z_\Delta}{3}, \tag{12.22}$$

which follows directly from Eqs. (10.42)–(10.44). Once the Δ-load has been replaced by its Y-equivalent, the Y-source, Δ-load, three-phase circuit can be modeled by the single-phase equivalent circuit in Fig. 12.6.

After we have used the single-phase equivalent circuit to calculate the line currents, we can find the current in each leg of the original Δ-load by simply dividing the line currents by $\sqrt{3}$ and shifting them through 30°. This relationship between the line currents and phase currents in the delta can be derived using the circuit in Fig. 12.10.

When a load, or source, is connected in a delta, the current in each leg of the delta is the phase current and the voltage across each leg is the phase voltage. We can see from Fig. 12.10 that in the Δ-configuration, the phase voltage is identical to the line voltage.

To see the relationship between the phase currents and line currents, we will assume a positive phase sequence and let I_ϕ represent the magnitude of the phase current. It follows, then, that

$$\mathbf{I}_{AB} = I_\phi \underline{/0°}, \tag{12.23}$$

$$\mathbf{I}_{BC} = I_\phi \underline{/-120°}, \tag{12.24}$$

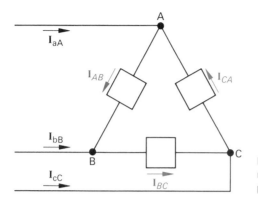

Fig. 12.10 A circuit used to establish the relationship between line currents and phase currents in a balanced delta load.

and

$$\mathbf{I}_{CA} = I_\phi \,\underline{/+120°}. \tag{12.25}$$

In writing these equations, we have arbitrarily selected $\mathbf{I}_{AB}$ as the reference phasor.

We can write the line currents in terms of the phase currents by direct application of Kirchhoff's current law; thus

$$\begin{aligned}
\mathbf{I}_{aA} = \mathbf{I}_{AB} - \mathbf{I}_{CA} &= I_\phi \,\underline{/0°} - I_\phi \,\underline{/120°} \\
&= \sqrt{3}\, I_\phi \,\underline{/-30°},
\end{aligned} \tag{12.26}$$

$$\begin{aligned}
\mathbf{I}_{bB} = \mathbf{I}_{BC} - \mathbf{I}_{AB} &= I_\phi \,\underline{/-120°} - I_\phi \,\underline{/0°} \\
&= \sqrt{3}\, I_\phi \,\underline{/-150°},
\end{aligned} \tag{12.27}$$

$$\begin{aligned}
\mathbf{I}_{cC} = \mathbf{I}_{CA} - \mathbf{I}_{BC} &= I_\phi \,\underline{/120°} - I_\phi \,\underline{/-120°} \\
&= \sqrt{3}\, I_\phi \,\underline{/90°}.
\end{aligned} \tag{12.28}$$

When we compare Eqs. (12.26)–(12.28) with Eqs. (12.23) − (12.25), we see that the magnitude of the line currents is $\sqrt{3}$ times the magnitude of the phase currents and the set of line currents *lags* the set of phase currents by 30°.

We will leave it to the reader to verify that for a negative phase sequence, the line currents are $\sqrt{3}$ times larger than the phase currents and *lead* the phase currents by 30°.

The relationship between the line currents and the phase currents of a Δ-connected load are summarized in Fig. 12.11.

Example 12.2 illustrates the calculations involved in analyzing a balanced three-phase circuit involving a Y-connected source and a Δ-connected load.

Example 12.2   The Y-connected source in Example 12.1 feeds a Δ-connected load through a distribution line having an impedance of $0.3 + j0.9\ \Omega/\phi$. The load impedance

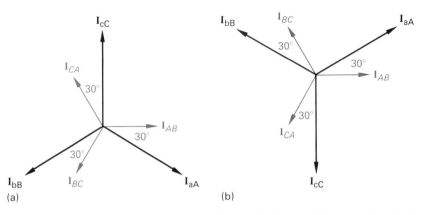

Fig. 12.11   Phasor diagrams showing the relationship between line currents and phase currents in a Δ-connected load: (a) positive sequence; (b) negative sequence.

is $118.5 + j85.8 \ \Omega/\phi$. Use the a-phase internal voltage of the generator as the reference.

a) Construct a single-phase equivalent circuit of the three-phase system.

b) Calculate the line currents $\mathbf{I}_{aA}$, $\mathbf{I}_{bB}$, and $\mathbf{I}_{cC}$.

c) Calculate the phase voltages at the terminals of the load.

d) Calculate the phase currents of the load.

e) Calculate the line voltages at the terminals of the source.

Solution   a) The single-phase equivalent circuit is shown in Fig. 12.12. The load impedance of the Y-equivalent is $(1/3)(118.5 + j85.8)$, or $39.5 + j28.6 \ \Omega/\phi$.

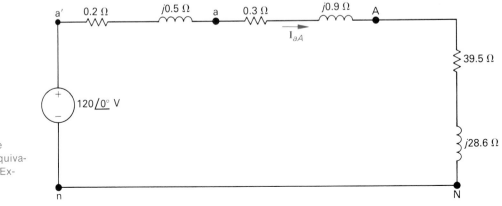

Fig. 12.12   The single-phase equivalent circuit for Example 12.2.

b) The a-phase line current is

$$\mathbf{I}_{aA} = \frac{120\,\underline{/0°}}{(0.2 + 0.3 + 34.5) + j(0.5 + 0.9 + 28.6)}$$

$$= \frac{120\underline{/0°}}{40 + j30} = 2.4\underline{/-36.87°}\ \mathrm{A}.$$

It follows directly that

$$\mathbf{I}_{bB} = 2.4\,\underline{/-156.87°}\ \mathrm{A}$$

and

$$\mathbf{I}_{cC} = 2.4\,\underline{/83.13°}\ \mathrm{A}.$$

c) Since the load is Δ-connected, the phase voltages are the same as the line voltages. To calculate the line voltages, we first calculate $\mathbf{V}_{AN}$:

$$\mathbf{V}_{AN} = [39.5 + j28.6](2.4\,\underline{/-36.87°})$$

$$= 117.04\,\underline{/-0.96°}\ \mathrm{V}.$$

Since we have a positive phase sequence, the line voltage $\mathbf{V}_{AB}$ is

$$\mathbf{V}_{AB} = \sqrt{3}\,\underline{/30°}\ \mathbf{V}_{AN}$$

$$= 202.72\underline{/29.04°}\ \mathrm{V}.$$

Therefore

$$\mathbf{V}_{BC} = 202.72\,\underline{/-90.96°}\ \mathrm{V}$$

and

$$\mathbf{V}_{CA} = 202.72\,\underline{/149.04°}\ \mathrm{V}.$$

d) The phase currents of the load can be calculated directly from the line currents. Then

$$\mathbf{I}_{AB} = \frac{1}{\sqrt{3}}\,\underline{/30°}\ \mathbf{I}_{aA}$$

$$= 1.39\,\underline{/-6.87°}\ \mathrm{A}.$$

Once we know $\mathbf{I}_{AB}$, we also know the other load phase currents:

$$\mathbf{I}_{BC} = 1.39\,\underline{/-126.87°}\ \mathrm{A}$$

and

$$\mathbf{I}_{CA} = 1.39\,\underline{/113.13°}\ \mathrm{A}.$$

Note that we can check our calculation of $\mathbf{I}_{AB}$ using the previously calculated $\mathbf{V}_{AB}$ and the impedance of the Δ-connected load. That is,

$$\mathbf{I}_{AB} = \frac{\mathbf{V}_{AB}}{\mathbf{Z}_\phi} = \frac{202.72\underline{/29.04°}}{118.5 + j85.8}$$

$$= 1.39\,\underline{/-6.87°}\ \mathrm{A}.$$

(Alternative methods of calculation are very helpful in eliminating errors and are highly recommended in all work involving analysis and design.)

e) To calculate the line voltage at the terminals of the source, we first calculate $\mathbf{V}_{an}$. We see from Fig. 12.12 that $\mathbf{V}_{an}$ is the voltage drop across the line impedance plus the load impedance. Thus

$$\mathbf{V}_{an} = [39.8 + j29.5]2.4\underline{/-36.87°}$$
$$= 118.90\underline{/-0.32°} \text{ V.}$$

The line voltage $\mathbf{V}_{ab}$ is

$$\mathbf{V}_{ab} = \sqrt{3}\underline{/30°}\ \mathbf{V}_{an}$$

or

$$\mathbf{V}_{ab} = 205.94\ \underline{/29.68°} \text{ V.}$$

Therefore

$$\mathbf{V}_{bc} = 205.94\ \underline{/-90.32°} \text{ V,}$$
$$\mathbf{V}_{ca} = 205.94\ \underline{/+149.68°} \text{ V.} \quad\blacksquare$$

DRILL EXERCISE
12.3

The line-to-line voltage $\mathbf{V}_{AB}$ at the terminals of a balanced, three-phase, delta-connected load is $4160\ \underline{/0°}$ V. The line current $\mathbf{I}_{aA}$ is $69.28\ \underline{/-10°}$ A.

a) Calculate the per-phase impedance of the load if the phase sequence is positive.

b) Repeat part (a) if the phase sequence is negative.

Ans. (a) $104\ \underline{/-20°}\ \Omega$; (b) $104\ \underline{/+40°}\ \Omega$.

DRILL EXERCISE
12.4

The line voltage at the terminals of a balanced, delta-connected load is 208 V. Each phase of the load consists of a 5.2-Ω resistor in parallel with a 6.933-Ω inductor. What is the magnitude of the current in the line feeding the load?

Ans. 86.60 A.

12.6   ANALYSIS OF THE DELTA–WYE CIRCUIT

In the Δ–Y three-phase circuit, the source is Δ-connected and the load is Y-connected. We can obtain the single-phase equivalent circuit by replacing the balanced Δ-connected source by a Y-equivalent. We can obtain the Y-equivalent of the source by dividing the internal phase voltages of the Δ-source by $\sqrt{3}$ and shifting this set of three-phase voltages by $-30°$ if the phase sequence is positive and by $+30°$ if the phase sequence is negative. The internal impedance of the Y-equivalent is one-third the internal impedance of the Δ-source. The Y-equivalent circuit of a positive-sequence Δ-connected source is illustrated in Fig. 12.13.

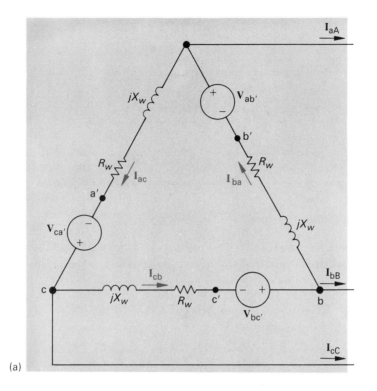

(a)

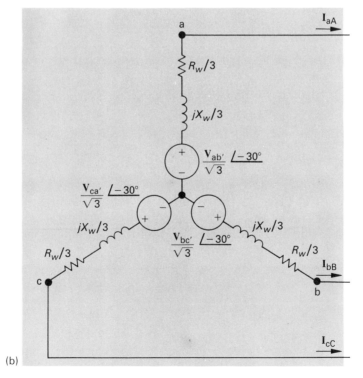

(b)

Fig. 12.13 The Y-equivalent of a balanced three-phase Δ-connected source (positive sequence); (a) Δ source; (b) Y-equivalent.

For a positive phase sequence, the set of Δ-source phase currents ($\mathbf{I}_{ba}$, $\mathbf{I}_{cb}$, and $\mathbf{I}_{ac}$ in Fig. 12.13) lead the set of line currents $\mathbf{I}_{aA}$, $\mathbf{I}_{bB}$, and $\mathbf{I}_{cC}$ by 30°. The magnitude of the phase currents is $1/\sqrt{3}$ times the magnitude of the line currents. For a negative phase sequence, the phase currents in the source lag the line currents.

To show that the Y-source of Fig. 12.13(b) is equivalent to the Δ-source of Fig. 12.13(a), it is necessary to show only that the two circuits produce the same terminal conditions for any balanced *external* connections applied to the terminals a, b, and c. The two test conditions that are easiest to prove are open circuit and short circuit. For open-circuit conditions, the three line currents are zero and the two circuits are equivalent if they deliver the same voltages between the terminals a, b, and c. For an external short circuit connecting the terminals a, b, and c, the line voltages are zero and the two circuits are equivalent if they deliver the same line currents. We will leave it to the reader (in Problem 12.10) to verify that these two circuits are equivalent.

The numerical analysis of a Δ–Y three-phase circuit is illustrated in the following example.

Example 12.3   A balanced, negative-sequence, Δ-connected source has an internal impedance of $0.018 + j0.162\ \Omega/\phi$. At no load, the terminal voltage of the source has a magnitude of 600 V. The source is connected to a Y-connected load, having an impedance of $7.92 - j6.35\ \Omega/\phi$, through a distribution line having an impedance of $0.074 + j0.296\ \Omega/\phi$.

a) Construct a single-phase equivalent circuit of the system using $\mathbf{V}_{ab'}$ as the reference.

b) Calculate the magnitude of the line voltage at the terminals of the load.

c) Calculate the three line currents $\mathbf{I}_{aA}$, $\mathbf{I}_{bB}$, and $\mathbf{I}_{cC}$.

d) Calculate the phase currents $\mathbf{I}_{ba}$, $\mathbf{I}_{cb}$, and $\mathbf{I}_{ac}$ of the source.

e) Calculate the magnitude of the line voltage at the terminals of the source.

Solution   a) At no load, the terminal voltage equals the internal voltage of the source. Therefore the internal voltage of the Δ-source has a magnitude of 600 V. Using $\mathbf{V}_{ab'}$ as a reference, we find that the internal a-phase voltage of the Y-equivalent source is

$$\mathbf{V}_{a'n} = \frac{\mathbf{V}_{ab'}}{\sqrt{3}}\ \underline{/30°} = \frac{600}{\sqrt{3}}\ \underline{/30°}$$

$$\cong 346.41\ \underline{/30°}\ \text{V}.$$

The internal impedance of the equivalent Y-generator is $(1/3)(0.018 + j0.162)$, or $0.006 + j0.054\ \Omega/\phi$. Therefore the single-phase equivalent circuit is as shown in Fig. 12.14.

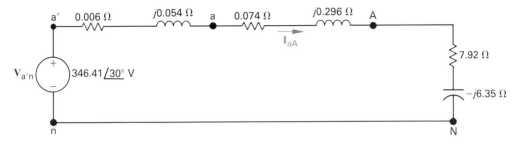

Fig. 12.14   The single-phase equivalent circuit for Example 12.3.

b) It follows directly from the circuit in Fig. 12.14 that

$$\mathbf{I}_{aA} = \frac{346.41 \,/\underline{30°}}{8.00 - j6.00} = 34.64 \,/\underline{66.87°} \text{ A}$$

and

$$\mathbf{V}_{AN} = (7.92 - j6.35)(34.64 \,/\underline{66.87°})$$
$$= 351.65 \,/\underline{28.15°} \text{ V}.$$

The magnitude of the line voltage at the load is

$$|\mathbf{V}_{AB}| = \sqrt{3} \,|\mathbf{V}_{AN}| = 609.08 \text{ V}.$$

c) Using the results of part (b), we find that the three line currents are

$$\mathbf{I}_{aA} = 34.64 \,/\underline{66.87°} \text{ A},$$
$$\mathbf{I}_{bB} = 34.64 \,/\underline{186.87°} \text{ A},$$
$$\mathbf{I}_{cC} = 34.64 \,/\underline{-53.13°} \text{ A}.$$

d) The phase currents of the generator can be calculated directly from the line currents. Since the phase sequence is negative, we have

$$\mathbf{I}_{ba} = \frac{1}{\sqrt{3}} \,/\underline{-30°} \, \mathbf{I}_{aA}$$
$$= 20 \,/\underline{36.87°} \text{ A},$$
$$\mathbf{I}_{cb} = 20 \,/\underline{156.87°} \text{ A},$$
$$\mathbf{I}_{ac} = 20 \,/\underline{-83.13°} \text{ A}.$$

e) From the circuit in Fig. 12.14, we have

$$\mathbf{V}_{an} = (7.994 - j6.054)\mathbf{I}_{aA}$$
$$= 34.64(7.994 - j6.054)/\underline{66.87°}$$
$$= 347.37 \,/\underline{29.73°} \text{ V}.$$

The magnitude of the line voltage at the source will be

$$|\mathbf{V}_{ab}| = \sqrt{3}|\mathbf{V}_{an}| = 601.66 \text{ V}. \quad \blacksquare$$

DRILL EXERCISE
12.5 A balanced, positive-sequence, Δ-connected source has an internal impedance of 0.09 + $j0.81$ Ω/ϕ. The source is feeding a balanced load via a balanced line. The b-phase line current $\mathbf{I}_{bB}$ is $6 \underline{/-120°}$ A and the line voltage $\mathbf{V}_{ab}$ is $480 \underline{/60°}$ V. Calculate the internal source voltage $\mathbf{V}_{ab'}$.

Ans. $481.68 \underline{/60.27°}$ V.

12.7 ANALYSIS OF THE DELTA–DELTA CIRCUIT

In the Δ–Δ circuit, both the source and the load are Δ-connected. The single-phase equivalent circuit of a balanced Δ–Δ system is obtained by replacing both the source and the load with their Y-equivalents. As before, the Y-equivalent circuit is used to solve for line currents and line-to-neutral voltages. Once we know the line currents, we can find the phase currents in both the load and the source using the techniques described in Sections 12.5 and 12.6. The line-to-neutral voltages can be converted to line-to-line voltages as described in Section 12.4. All these techniques have been illustrated in Examples 12.1, 12.2, and 12.3. You can gain additional experience with these types of calculations by solving Problems 12.6–12.13.

12.8 POWER CALCULATIONS IN BALANCED THREE-PHASE CIRCUITS

Thus far, our analysis of balanced three-phase circuits has been limited to the determination of currents and voltages in a given circuit. We are now ready to discuss three-phase power calculations. We begin by discussing the average power delivered to a balanced Y-connected load.

A Y-connected load, along with pertinent currents and voltages, is shown in Fig. 12.15. The average power associated with any one phase can be calculated using the techniques introduced in Chapter 11. Using Eq. (11.27) as a starting point, we find that we can express the average power associated with the a-phase of the load as

$$P_a = |\mathbf{V}_{AN}||\mathbf{I}_{aA}| \cos(\theta_{vA} - \theta_{iA}), \qquad (12.29)$$

where θ_{vA} and θ_{iA} denote the phase angles of $\mathbf{V}_{AN}$ and $\mathbf{I}_{aA}$, respectively. Using the notation introduced in Eq. (12.29), we find that the power associated with the b- and c-phases are

$$P_b = |\mathbf{V}_{BN}||\mathbf{I}_{bB}| \cos(\theta_{vB} - \theta_{iB}) \qquad (12.30)$$

and

$$P_c = |\mathbf{V}_{CN}||\mathbf{I}_{cC}| \cos(\theta_{vC} - \theta_{iC}). \qquad (12.31)$$

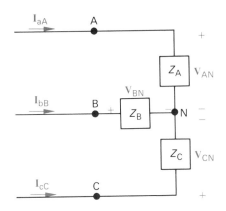

Fig. 12.15 A balanced Y load used to introduce average power calculations in three-phase circuits.

In Eqs. (12.29)–(12.31), all phasor currents and voltages are written in terms of the rms value of the sinusoidal function that they represent.

In a balanced three-phase system, the magnitude of each line-to-neutral voltage is the same, as is the magnitude of each phase current. The argument of the cosine functions is also the same for all three phases. To emphasize these observations, we introduce the following notation to facilitate further discussion of power calculations in balanced three-phase circuits:

$$V_\phi = |\mathbf{V}_{\text{AN}}| = |\mathbf{V}_{\text{BN}}| = |\mathbf{V}_{\text{CN}}|, \tag{12.32}$$

$$I_\phi = |\mathbf{I}_{\text{aA}}| = |\mathbf{I}_{\text{bB}}| = |\mathbf{I}_{\text{cC}}|, \tag{12.33}$$

and

$$\theta_\phi = (\theta_{v\text{A}} - \theta_{i\text{A}}) = (\theta_{v\text{B}} - \theta_{i\text{B}}) = (\theta_{v\text{C}} - \theta_{i\text{C}}). \tag{12.34}$$

We also note from the above observations that for a balanced system the power delivered to each phase of the load is the same; thus

$$P_\text{a} = P_\text{b} = P_\text{c} = P_\phi = V_\phi I_\phi \cos \theta_\phi, \tag{12.35}$$

where P_ϕ stands for average power per phase.

The total average power delivered to the balanced Y-connected load is simply three times the power per phase; thus

$$P_T = 3P_\phi = 3V_\phi I_\phi \cos \theta_\phi. \tag{12.36}$$

It is also desirable to express the total power in terms of the *rms magnitude* of the line voltage and the *rms magnitude* of the line current. If we let V_L represent the rms magnitude of the line voltage and I_L represent the rms magnitude of the line current, then we can modify Eq. (12.36) as follows:

$$P_T = 3 \left(\frac{V_L}{\sqrt{3}} \right) I_L \cos \theta_\phi$$

$$= \sqrt{3} V_L I_L \cos \theta_\phi. \tag{12.37}$$

In deriving Eq. (12.37), we have used the fact that for a balanced Y-connected load, the magnitude of the phase voltage is the magnitude of the line voltage divided by $\sqrt{3}$ and the magnitude of the line current is equal to the magnitude of the phase current. In using Eq. (12.37) to calculate the total power delivered to the load, it is important to remember that θ_ϕ is the phase angle between the *phase voltage* and the *phase current*.

The reactive power and complex power associated with any one phase of a Y-connected load can also be calculated using the techniques introduced in Chapter 11. For the balanced load the expressions for the reactive power are

$$Q_\phi = V_\phi I_\phi \sin \theta_\phi \tag{12.38}$$

and

$$Q_T = 3Q_\phi = \sqrt{3} \, V_L I_L \sin \theta_\phi. \tag{12.39}$$

Equation (11.33) is the basis for expressing the complex power associated with any phase. For a balanced load we have

$$S = \mathbf{V}_{AN} \mathbf{I}_{aA}^* = \mathbf{V}_{BN} \mathbf{I}_{bB}^* = \mathbf{V}_{CN} \mathbf{I}_{cC}^* = \mathbf{V}_\phi \mathbf{I}_\phi^*, \tag{12.40}$$

where $\mathbf{V}_\phi$ and $\mathbf{I}_\phi$ are used to represent a phase voltage and current taken from the same phase. Thus, in general,

$$S_\phi = P_\phi + jQ_\phi = \mathbf{V}_\phi \mathbf{I}_\phi^* \tag{12.41}$$

and

$$S_T = 3S_\phi = \sqrt{3} \, V_L I_L \underline{/\theta_\phi}. \tag{12.42}$$

If the load is Δ-connected, the calculation of power—reactive power or complex power—is basically the same as that for the Y-connected load. A Δ-connected load, along with the pertinent currents and voltages, is shown in Fig. 12.16, from which it follows that the power associated with each phase is

$$P_a = |\mathbf{V}_{AB}||\mathbf{I}_{AB}| \cos (\theta_{vA} - \theta_{iA}), \tag{12.43}$$

$$P_b = |\mathbf{V}_{BC}||\mathbf{I}_{BC}| \cos (\theta_{vB} - \theta_{iB}), \tag{12.44}$$

$$P_c = |\mathbf{V}_{CA}||\mathbf{I}_{CA}| \cos (\theta_{vC} - \theta_{iC}). \tag{12.45}$$

For a balanced load,

$$|\mathbf{V}_{AB}| = |\mathbf{V}_{BC}| = |\mathbf{V}_{CA}| = V_\phi, \tag{12.46}$$

$$|\mathbf{I}_{AB}| = |\mathbf{I}_{BC}| = |\mathbf{I}_{CA}| = I_\phi, \tag{12.47}$$

$$(\theta_{vA} - \theta_{iA}) = (\theta_{vB} - \theta_{iB}) = (\theta_{vC} - \theta_{iC}) = \theta_\phi, \tag{12.48}$$

and

$$P_a = P_b = P_c = P_\phi = V_\phi I_\phi \cos \theta_\phi. \tag{12.49}$$

It is worth noting that Eq. (12.49) is the same as Eq. (12.35). This is equivalent to saying, "In a balanced load the average power per phase is equal to the product of the rms magnitude of the phase voltage, the rms magnitude of the phase

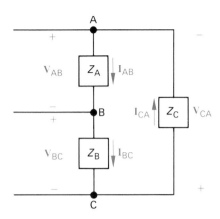

Fig. 12.16   A Δ-connected load used to discuss power calculations.

current, and the cosine of the angle between the phase voltage and phase current."

The total power delivered to a balanced Δ-connected load is

$$P_T = 3P_\phi = 3V_\phi I_\phi \cos \theta_\phi$$

$$= 3 V_L \left(\frac{I_L}{\sqrt{3}}\right) \cos \theta_\phi$$

$$= \sqrt{3} V_L I_L \cos \theta_\phi. \tag{12.50}$$

Note that Eq. (12.50) is the same as Eq. (12.37).

The expressions for reactive power and complex power also have the same form as those developed for the Y-load:

$$Q_\phi = V_\phi I_\phi \sin \theta_\phi, \tag{12.51}$$

$$Q_T = 3Q_\phi = 3V_\phi I_\phi \sin \theta_\phi, \tag{12.52}$$

$$S_\phi = P_\phi + jQ_\phi = V_\phi I_\phi^* \tag{12.53}$$

$$S_T = 3S_\phi = \sqrt{3} V_L I_L \underline{/\theta_\phi}. \tag{12.54}$$

The following examples illustrate power calculations in balanced three-phase circuits.

Example 12.4   a) Calculate the average power per phase delivered to the Y-connected load of Example 12.1.

b) Calculate the total average power delivered to the load.

c) Calculate the total average power lost in the line.

d) Calculate the total average power lost in the generator.

e) Calculate the total number of magnetizing vars absorbed by the load.

f) Calculate the total complex power delivered by the source.

Solution   a) From Example 12.1, we have $V_\phi = 115.22$ V, $I_\phi = 2.4$ A, and $\theta_\phi = -1.19 - (-36.87) = 35.68°$. Therefore

$$P_\phi = (115.22)(2.4) \cos 35.68°$$
$$= 224.64 \text{ W.}$$

We also note that the power per phase can be calculated from $I_\phi^2 R_\phi$, or

$$P_\phi = (2.4)^2(39) = 224.64 \text{ W.}$$

b) The total average power delivered to the load is $P_T = 3P_\phi = 673.92$ W. Since the line voltage was calculated in Example 12.1, we could also use Eq. (12.37); thus

$$P_T = \sqrt{3}\,(199.58)(2.4) \cos 35.68°$$
$$= 673.92 \text{ W.}$$

c) The total power lost in the line is

$$P_\text{line} = (3)(2.4)^2(0.8) = 13.824 \text{ W.}$$

d) The total internal power loss in the generator is

$$P_\text{gen} = 3(2.4)^2(0.2) = 3.456 \text{ W.}$$

e) The total number of magnetizing vars absorbed by the load is

$$Q_T = \sqrt{3}\,(199.58)(2.4) \sin 35.68$$
$$= 483.84 \text{ VAR.}$$

f) The total complex power associated with the source is

$$S_T = 3S_\phi = -3(120)(2.4)\,\underline{/36.87°}$$
$$S_T = -691.20 - j518.40 \text{ VA.}$$

The minus sign tells us that the internal power and magnetizing reactive power are being delivered to the circuit. We can check this result by calculating the total power and reactive power absorbed by the circuit. Thus

$$P = 673.92 + 13.824 + 3.456$$
$$= 691.20 \text{ W} \quad \text{(check);}$$
$$Q = 483.84 + 3(2.4)^2(1.5) + 3(2.4)^2(0.5)$$
$$= 483.84 + 25.92 + 8.64$$
$$= 518.40 \text{ VAR} \quad \text{(check).} \quad \blacksquare$$

Example 12.5   a) Calculate the total complex power delivered to the Δ-connected load of Example 12.2.

b) What percentage of the average power at the sending of the line is delivered to the load?

Solution a) Using the a-phase values from the solution of Example 12.2, we have

$$\mathbf{V}_\phi = \mathbf{V}_{AB} = 202.72\,\underline{/29.04^\circ}\ \text{V},$$
$$\mathbf{I}_\phi = \mathbf{I}_{AB} = 1.39\,\underline{/-6.87^\circ}\ \text{A}.$$

Using Eqs. (12.53) and (12.54) we have

$$S_T = 3[202.72\,\underline{/29.04^\circ}][1.39\,\underline{/+6.87^\circ}]$$
$$= 682.56 + j494.208\ \text{VA}.$$

b) The total power at the sending end of the distribution line will be equal to the total power delivered to the load plus the total power lost in the line; therefore

$$P_{\text{input}} = 682.56 + 3(2.4)^2(0.3)$$
$$= 687.744\ \text{W}.$$

The percentage of the average power at the input of the line reaching the load is 682.56/687.744, or 99.25%. ■

Example 12.6 A balanced three-phase load requires 480 kW at a lagging power factor of 0.8. The load is fed from a line having an impedance of $0.005 + j0.025\ \Omega/\phi$. The line voltage at the terminals of the load is 600 V.

a) Construct a single-phase equivalent circuit of the system.

b) Calculate the magnitude of the line current.

c) Calculate the magnitude of the line voltage at the sending end of the line.

d) Calculate the power factor at the sending end of the line.

Solution a) The single-phase equivalent circuit is shown in Fig. 12.17. We have arbitrarily selected the line-to-neutral voltage at the load as the reference.

b) The line current $\mathbf{I}_{aA}$ is

$$\left(\frac{600}{\sqrt{3}}\right)\mathbf{I}_{aA}^* = (160 + j120)10^3$$

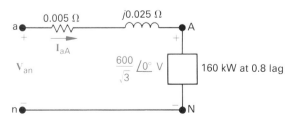

Fig. 12.17 The single-phase equivalent circuit for Example 12.6.

or

$$\mathbf{I}^*_{aA} = 577.35 \underline{/36.87°} \text{ A.}$$

Therefore, $\mathbf{I}_{aA} = 577.35 \underline{/-36.87°}$ A. The magnitude of the line current is the magnitude of $\mathbf{I}_{aA}$:

$$I_L = 577.35 \text{ A.}$$

An alternative solution for I_L is obtained from the expression

$$P_T = \sqrt{3}\, V_L I_L \cos \theta_p$$
$$= \sqrt{3}\,(600)I_L(0.8) = 480{,}000 \text{ W;}$$
$$I_L = \frac{480{,}000}{(\sqrt{3})(600)(0.8)} = \frac{1000}{\sqrt{3}} = 577.35 \text{ A.}$$

c) To calculate the magnitude of the line voltage at the sending end of the line, we first calculate $\mathbf{V}_{an}$. It follows directly from Fig. 12.17 that

$$\mathbf{V}_{an} = \mathbf{V}_{AN} + Z_l \mathbf{I}_{aA}$$
$$= \frac{600}{\sqrt{3}} + (0.005 + j0.025)(577.35 \underline{/-36.87°})$$
$$= 357.51 \underline{/1.57°} \text{ V.}$$

Thus

$$V_L = \sqrt{3}\, |\mathbf{V}_{an}| = 619.23 \text{ V.}$$

d) The power factor at the sending end of the line is the cosine of the phase angle between $\mathbf{V}_{an}$ and $\mathbf{I}_{aA}$:

$$\text{pf} = \cos [1.57° - (-36.87°)]$$
$$= \cos (38.44°) = 0.783.$$

An alternative method for calculating the power factor is to first calculate the complex power at the sending end of the line:

$$S_\phi = (160 + j120)10^3 + (577.35)^2(0.005 + j0.025)$$
$$= 161.67 + j128.33 \text{ kVA}$$
$$= 206.41 \underline{/38.44°} \text{ kVA.}$$

The power factor is

$$\text{pf} = \cos (38.44°) = 0.783.$$

Finally, note that if we calculate the total complex power at the sending end of the line, after first calculating the magnitude of the line current, we can use this value to calculate V_L. That is,

$$\sqrt{3}\, V_L I_L = 3(206.41) \times 10^3$$
$$V_L = \frac{3(206.41) \times 10^3}{\sqrt{3}(577.35)} = 619.23 \text{ V.} \quad \blacksquare$$

Although we are primarily interested in average, reactive, and complex power calculations, the computation of the total *instantaneous* power is also important. The total instantaneous power in a balanced three-phase circuit has an interesting property: It is invariant with time!

This can be shown as follows. Let the instantaneous line-to-neutral voltage v_{AN} be the reference and, as before, θ_ϕ be the phase angle $\theta_{vA} - \theta_{iA}$. Then, for a positive phase sequence, the instantaneous power in each phase is

$$p_a = v_{AN} i_{aA} = V_\phi I_\phi \cos \omega t \cos (\omega t - \theta_\phi),$$

$$p_b = v_{BN} i_{bB} = V_\phi I_\phi \cos (\omega t - 120°) \cos (\omega t - \theta_\phi - 120°),$$

and

$$p_c = v_{CN} i_{cC} = V_\phi I_\phi \cos (\omega t + 120°) \cos (\omega t - \theta_\phi + 120°),$$

where V_ϕ and I_ϕ represent the maximum values of the line-to-neutral voltage and line current, respectively. The instantaneous total power is the sum of the instantaneous phase powers and this sum can be shown to reduce to $1.5 V_\phi I_\phi \cos \theta_\phi$, that is,

$$p_T = p_a + p_b + p_c = 1.5 V_\phi I_\phi \cos \theta_\phi.$$

We will leave this reduction for the reader (see Problem 12.25).

The fact that the total instantaneous power in a three-phase circuit is constant is an important property of three-phase circuits. It means that the torque developed at the shaft of a three-phase motor is constant and this in turn means less vibration in machinery powered by three-phase motors.

DRILL EXERCISE
12.6

The complex power associated with each phase of a balanced load is $384 + j288$ kVA. The line voltage at the terminals of the load is 4160 V.

a) What is the magnitude of the line current feeding the load?

b) If the load is connected in delta and if the impedance of each phase consists of a resistance in parallel with a reactance, calculate R and X.

c) If the load is wye-connected and if the impedance of each phase consists of a resistance in series with a reactance, calculate R and X.

Ans. (a) 199.85 A; (b) $R = 45.07 \ \Omega$, $X = 60.09 \ \Omega$; (c) $R = 9.61 \ \Omega$, $X = 7.21 \ \Omega$.

DRILL EXERCISE
12.7

A balanced bank of delta-connected capacitors is connected in parallel with the load described in Drill Exercise 12.6. The line voltage at the terminals of the load remains at 4160 V. The circuit is operating at a frequency of 60 Hz. The capacitors are adjusted so that the magnitude of the line current feeding the parallel combination of the load and capacitor bank is at its minimum.

a) What is the size of each capacitor in μF?

b) Repeat part (a) given that the capacitors are connected in a wye.

c) What is the magnitude of the line current?

Ans. (a) 44.14 μF; (b) 132.42 μF; (c) 159.88 A.

12.9   MEASUREMENT OF AVERAGE POWER IN THREE-PHASE CIRCUITS

The basic instrument used to measure power in three-phase circuits is the electrodynamometer wattmeter. It contains two coils. One coil, called the *current coil*, is stationary and is designed to carry a current that is proportional to the load current. The second coil, called the *potential coil,* is movable and carries a current that is proportional to the load voltage. The average deflection of the pointer that is attached to the movable coil is proportional to the product of the effective value of the current in the current coil, the effective value of the voltage impressed on the potential coil, and the cosine of the phase angle between this current and voltage. The direction in which the pointer deflects depends on the instantaneous polarity of the current-coil current and the potential-coil voltage. Therefore each coil has one terminal with a polarity mark—usually a plus sign—but sometimes the double polarity mark ± is used. The wattmeter deflects up-scale when (1) the polarity-marked terminal of the current coil is toward the source and (2) the polarity-marked terminal of the potential coil is connected to the same line in which the current coil has been inserted. The important features of the wattmeter are shown in Fig. 12.18.

Before showing how we can use two electrodynamometer wattmeters to measure the total power delivered to a three-phase load, let us make an observation that clearly shows that only two wattmeters are needed. Consider a general network inside a "box" that is being energized by *n* conductors. The system is shown in Fig. 12.19. If we wish to measure, or calculate, the total power at the terminals of the box, we need to know $n - 1$ currents and volt-

Fig. 12.18   The important features of the electrodynamometer wattmeter.

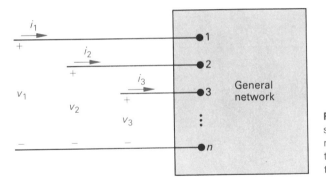

Fig. 12.19 A circuit used to show that only $n - 1$ watt-meters are needed to measure the total power being delivered to the network.

ages. This follows because if we choose one terminal as a reference, then there are only $n - 1$ independent voltages. Likewise, only $n - 1$ independent currents can exist in the n conductors entering the box. Thus the total power involves summing the products of $n - 1$ terms, that is, $p = v_1 i_1 + v_2 i_2 + \cdots + v_{n-1} i_{n-1}$.

Applying this general observation to a three-phase circuit, we can see that for a three-conductor circuit, whether balanced or not, we need only two watt-meters to measure the total power. For a four-conductor circuit, we need three wattmeters if the three-phase circuit is unbalanced, but only two wattmeters if it is balanced. We can conclude then that for any balanced three-phase circuit, we need only two wattmeters to measure the total power.

The two-wattmeter method of measuring the total power in a balanced three-phase circuit reduces to determining the magnitude and algebraic sign of the average power indicated by each wattmeter. We can describe the basic problem in terms of the circuit shown in Fig. 12.20, where the two wattmeters are indicated by the shaded boxes and labeled W_1 and W_2. The coil notations cc and pc stand for current coil and potential coil, respectively. We have elected to insert arbitrarily the current coils of the wattmeters in lines aA and cC. As a consequence, line bB is the reference line for the two potential coils. The load in Fig. 12.20 is connected as a wye and the per-phase load impedance is designated as $Z_\phi = |Z| \underline{/\theta}$. This can be considered a perfectly general representation since any Δ-connected load can be represented by its Y-equivalent and, furthermore, for the balanced case the impedance angle θ is unaffected by the Δ-to-Y transformation.

In the discussion that follows, the current drawn by the potential coil of the wattmeter is considered negligible compared with the line current measured by the current coil of the wattmeter. We assume that the loads can be modeled by passive circuit elements so that the phase angle of the load impedance (θ in Fig. 12.20) lies between $-90°$ (pure capacitance) and $+90°$ (pure inductance). We also assume a positive phase sequence.

On the basis of our introductory discussion of the average deflection of the wattmeter, we can see that wattmeter 1 will respond to the product of $|\mathbf{V}_{AB}|$, $|\mathbf{I}_{aA}|$, and the cosine of the angle between $\mathbf{V}_{AB}$ and $\mathbf{I}_{aA}$. If we denote this watt-

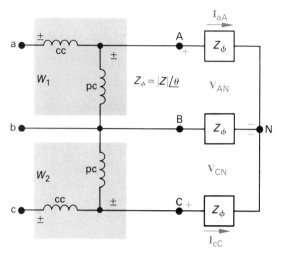

Fig. 12.20 A circuit used to analyze the two-wattmeter method of measuring average power delivered to a balanced load.

meter reading as W_1, we can write

$$W_1 = |\mathbf{V}_{AB}||\mathbf{I}_{aA}| \cos \theta_1$$
$$= V_L I_L \cos \theta_1. \tag{12.55}$$

It follows that

$$W_2 = |\mathbf{V}_{CB}||\mathbf{I}_{cC}| \cos \theta_2$$
$$= V_L I_L \cos \theta_2. \tag{12.56}$$

In Eq. (12.55), θ_1 is the phase angle between $\mathbf{V}_{AB}$ and $\mathbf{I}_{aA}$ and in Eq. (12.56), θ_2 is the phase angle between $\mathbf{V}_{CB}$ and $\mathbf{I}_{cC}$.

To calculate W_1 and W_2, we express θ_1 and θ_2 in terms of the impedance angle θ, which is also the same as the phase angle between the phase voltage and phase current. For a positive phase sequence,

$$\theta_1 = \theta + 30° = \theta_\phi + 30° \tag{12.57}$$

and

$$\theta_2 = \theta - 30° = \theta_\phi - 30°. \tag{12.58}$$

The derivation of Eqs. 12.57 and 12.58 is left as an exercise (see Problem 12.20). When we substitute Eqs. (12.57) and (12.58) into Eqs. (12.55) and (12.56) we get

$$W_1 = V_L I_L \cos (\theta + 30°) \tag{12.59}$$

and

$$W_2 = V_L I_L \cos (\theta - 30°). \tag{12.60}$$

To find the total power, we add W_1 and W_2; thus

$$P_T = W_1 + W_2 = 2V_L I_L \cos \theta_\phi \cos 30° = \sqrt{3} \, V_L I_L \cos \theta_\phi, \tag{12.61}$$

which is the expression for the total power in a three-phase circuit. Therefore we have confirmed that the sum of the two wattmeter readings yields the total power.

A study of Eqs. (12.59) and (12.60) reveals the following characteristics of the two-wattmeter method of measuring three-phase power in a balanced circuit.

1. If the power factor is greater than 0.5, both wattmeters read positive.
2. If the power factor equals 0.5, one wattmeter reads zero.
3. If the power factor is less than 0.5, one wattmeter reads negative.
4. Reversing the phase sequence will interchange the readings on the two wattmeters.

These observations are brought out in the following numerical example and in Problems 12.21, 12.22, and 12.23.

Example 12.7 Calculate the reading of each wattmeter in the circuit of Fig. 12.20 if the line-to-neutral voltage at the load is 120 V and (a) $Z_\phi = 8 + j6 \ \Omega$; (b) $Z_\phi = 8 - j6\Omega$; (c) $Z_\phi = 5 + j5\sqrt{3} \ \Omega$; and (d) $Z_\phi = 10\underline{/-75°}$. (e) Verify for parts (a)–(d) that the sum of the wattmeter readings equals the total power delivered to the load.

Solution a) $Z_\phi = 10 \underline{/36.87°} \ \Omega, \ V_L = 120\sqrt{3} \ \text{V}, \ I_L = 120/10 = 12 \ \text{A}.$

$$W_1 = (120\sqrt{3})(12) \cos (36.87° + 30°)$$
$$= 979.75 \ \text{W},$$
$$W_2 = (120\sqrt{3})(12) \cos (36.87° - 30°)$$
$$= 2476.25 \ \text{W}$$

b) $Z_\phi = 10\underline{/-36.87°} \ \Omega, \ V_L = 120\sqrt{3} \ \text{V}, \ I_L = 12 \ \text{A}.$

$$W_1 = (120\sqrt{3})(12) \cos (-36.87° + 30°)$$
$$= 2476.25 \ \text{W},$$
$$W_2 = (120\sqrt{3})(12) \cos (-36.87° - 30°)$$
$$= 979.75 \ \text{W}$$

c) $Z_\phi = 5(1 + j\sqrt{3}) = 10\underline{/60°} \ \Omega, \ V_L = 120\sqrt{3} \ \text{V}, \ I_L = 12 \ \text{A}.$

$$W_1 = (120\sqrt{3})(12) \cos (60° + 30°) = 0,$$
$$W_2 = (120\sqrt{3})(12) \cos (60° + 30°) = 2160 \ \text{W}$$

d) $Z_\phi = 10\underline{/-75°} \ \Omega, \ V_L = 120\sqrt{3} \ \text{V}, \ I_L = 12 \ \text{A}$

$$W_1 = (120\sqrt{3})(12) \cos (-75° + 30°) = 1763.63 \ \text{W},$$
$$W_2 = (120\sqrt{3})(12) \cos (-75° - 30°) = -645.53 \ \text{W}$$

e)
$$P_T(a) = 3(12)^2(8) = 3456 \text{ W},$$
$$W_1 + W_2 = 979.75 + 2476.25$$
$$= 3456 \text{ W};$$

$$P_T(b) = P_T(a) = 3456 \text{ W},$$
$$W_1 + W_2 = 2476.25 + 979.75$$
$$= 3456 \text{ W};$$

$$P_T(c) = 3(12)^2(5) = 2160 \text{ W},$$
$$W_1 + W_2 = 0 + 2160$$
$$= 2160 \text{ W};$$

$$P_T(d) = 3(12)^2(2.5882) = 1118.10 \text{ W},$$
$$W_1 + W_2 = 1763.63 - 645.53$$
$$= 1118.10 \text{ W.} \quad \blacksquare$$

DRILL EXERCISE
12.8

The two wattmeters in Fig. 12.20 can also be used to compute the total reactive power of the load.

a) Prove this statement by showing that $\sqrt{3}(W_2 - W_1) = \sqrt{3}\, V_L I_L \sin \theta_\phi$.

b) Compute the total reactive power from the wattmeter readings for each of the loads in Example 12.7. Check your computations by calculating the total reactive power directly from the given voltage and impedance.

Ans. (b) 2592 VAR, -2592 VAR, 3741.23 VAR, -4172.80 VAR.

12.10 SUMMARY

Our purpose has been to introduce you to the steady-state sinusoidal behavior of balanced three-phase circuits. When a three-phase circuit operates in a balanced mode, there are significant analytical shortcuts that can be used to calculate currents and voltages of interest. The key to these shortcuts is to reduce a given system to a single-phase equivalent circuit, a technique that relies on being able to make Δ-to-Y transformations at both the source end and load end of the circuit. Once the single-phase equivalent circuit has been derived, it is used to calculate the line current and the line-to-neutral voltages of interest. The current and the voltages obtained from the single-phase equivalent circuit can be translated into any other system current or voltage that is of interest. The translation from the single-phase equivalent circuit values to any other current or voltage in the circuit is based on the following observations.

1. In a balanced system, b- and c-phase currents and voltages are identical to the corresponding a-phase current and voltage except for a 120° shift in phase. In a positive sequence circuit, the b-phase quantity will lag the

a-phase quantity by 120° and the c-phase quantity will lead the a-phase quantity by 120°. For a negative sequence circuit, phases b and c are interchanged with respect to phase a.

2. The set of line voltages is out of phase with the set of line-to-neutral voltages by $\pm 30°$. The plus and minus sign corresponds to positive and negative sequence, respectively.

3. The magnitude of a line voltage is $\sqrt{3}$ times as large as the magnitude of a line-to-neutral voltage.

4. The set of line currents is out of phase with the set of phase currents in Δ-connected sources and loads by $\mp 30°$. The minus and plus sign corresponds to positive and negative sequence, respectively.

5. The magnitude of a line current is $\sqrt{3}$ times as large as the magnitude of a phase current in the Δ-connected source or load.

Real, reactive, or complex power calculations can be made on either a per-phase basis or a total three-phase basis. The techniques for calculating real, reactive, or complex power on a per-phase basis are the same as those introduced in Chapter 11. The calculation of the total real, reactive, or complex power is based on using line current and line voltage, as expressed in Eqs. (12.37), (12.39), and (12.42).

PROBLEMS

All voltages are stated in terms of the rms value.

12.1 For each set of voltages given below, state whether or not the voltages form a balanced three-phase set. If the set is a balanced set, state whether the phase sequence is positive or negative. If the set is not balanced, explain why.

a) $v_a = 180 \cos 377t$ V,
   $v_b = 180 \cos (377t - 120°)$ V,
   $v_c = 180 \cos (377t - 240°)$ V,

b) $v_a = 180 \sin 377t$ V,
   $v_b = 180 \sin (377t + 120°)$ V,
   $v_c = 180 \sin (377t - 120°)$ V

c) $v_a = -400 \sin 377t$ V,
   $v_b = 400 \cos (377t + 210°)$ V,
   $v_c = 400 \cos (377t - 30°)$ V

d) $v_a = 200 \cos (377t + 30°)$ V,
   $v_b = 100 \sqrt{3} \cos (377t + 150°)$ V,
   $v_c = 200 \cos (377t + 270°)$ V

e) $v_a = 208 \cos (377t + 42°)$ V,
   $v_b = 208 \cos (377t - 78°)$ V,
   $v_c = 208 \cos (377t - 201°)$ V

f) $v_a = 240 \cos 377t$ V,
   $v_b = 240 \cos (377t - 120°)$ V,
   $v_c = 240 \cos (397t + 120°)$ V

12.2 Verify that Eq. (12.3) is true for either Eq. (12.1) or Eq. (12.2).

12.3 The time domain expressions for three line-to-neutral voltages at the terminals of a Y-connected load are

$$v_{AN} = 169.71 \cos{(\omega t + 26°)} \text{ V},$$
$$v_{BN} = 169.71 \cos{(\omega t - 94°)} \text{ V},$$
$$v_{CN} = 169.71 \cos{(\omega t + 146°)} \text{ V}.$$

What are the time domain expressions for the three line-to-line voltages v_{AB}, v_{BC}, and v_{CA}?

12.4 Refer to the circuit in Fig. 12.4(b). Assume that there are no external connections to the terminals a, b, c. Assume further that the three windings are from a balanced three-phase generator. How much current will circulate in the Δ-connected generator?

12.5 a) Is the circuit in Fig. 12.21 a balanced or unbalanced three-phase system? Explain.

b) Find $\mathbf{I}_o$.

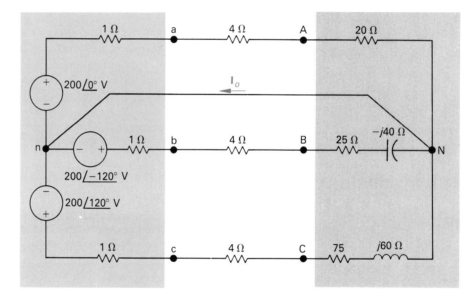

Fig. 12.21 The circuit for Problem 12.5.

12.6 The magnitude of the phase voltage of an ideal balanced three-phase, Y-connected source is 120 V. The source is connected to a balanced Y-connected load by a distribution line that has an impedance of $0.2 + j0.8 \ \Omega/\phi$. The load impedance is $15.8 + j11.2 \ \Omega/\phi$. The phase sequence of the source is abc. Use the a-phase voltage of the source as the reference. Specify the magnitude and phase angle of the following quantities: (a) the three line currents, (b) the three line voltages at the source, (c) the three phase voltages at the load, and (d) the three line voltages at the load.

12.7 The magnitude of the line voltage at the terminals of a balanced Y-connected load is 440 V. The load impedance is $3.387 \ \Omega/\phi$. The load is fed from a line that has an impedance of $0.0693 + j0.04 \ \Omega/\phi$.

a) What is the magnitude of the line current?

b) What is the magnitude of the line voltage at the source?

12.8 A balanced Δ-connected load has an impedance of 60 + j45 Ω/φ. The load is fed through a line having an impedance of 0.8 + j0.6 Ω/φ. The phase voltage at the terminals of the load is 480 V. The phase sequence is negative. Use $\mathbf{V}_{AB}$ as the reference.

a) Calculate the three phase currents of the load.
b) Calculate the three line currents.
c) Calculate the three line voltages at the sending end of the line.

12.9 A balanced Δ-connected load having an impedance of 72 + j54 Ω/φ is connected in parallel with a balanced Y-connected load having an impedance of 50 /0° Ω/φ. The paralleled loads are fed from a line having an impedance of 0.5 + j1.5 Ω/φ. The magnitude of the line-to-neutral voltage of the Y-load is 1200 V.

a) Calculate the magnitude of the current in the line feeding the loads.
b) Calculate the magnitude of the phase current in the Δ-connected load.
c) Calculate the magnitude of the phase current in the Y-connected load.
d) Calculate the magnitude of the line voltage at the sending end of the line.

12.10 A balanced three-phase Δ-connected source is shown in Fig. 12.22.

a) Find the Y-connected equivalent circuit.
b) Show that the Y-connected equivalent circuit delivers the same open-circuit voltage as the original Δ-connected source.
c) Apply an external short circuit to the terminals A, B, and C. Use the Δ-connected source to find the three line currents I_{aA}, I_{bB}, and I_{cC}.
d) Repeat part (c) but use the Y-equivalent source to find the three line currents.

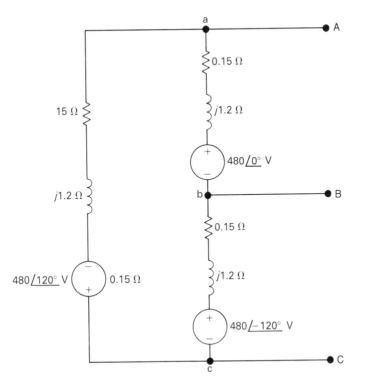

Fig. 12.22 The circuit for Problem 12.10.

12.11 The Δ-connected source of Problem 12.10 is connected to a Y-connected load by means of a balanced three-phase distribution line. The load impedance is $59.5 - j49$ Ω/ϕ and the line impedance is $0.45 + j3.60$ Ω/ϕ.

   a) Construct a single-phase equivalent circuit of the system.
   b) Determine the magnitude of the line voltage at the terminals of the load.
   c) Determine the magnitude of the phase current in the Δ-source.
   d) Determine the magnitude of the line voltage at the terminals of the source.

12.12 In Example 12.3, the solution for $\mathbf{V}_{an}$ is $347.37\,\underline{/29.73°}$ V, where $\mathbf{V}_{an}$ is the line-to-neutral voltage of the equivalent Y-source.

   a) What are the magnitude and phase angle of the line voltage $\mathbf{V}_{ab}$?
   b) Calculate the line voltage using the appropriate phase current, internal voltage, and internal impedance of the Δ-source.
   c) Do these two alternative calculations for $\mathbf{V}_{ab}$ check?

12.13 A three-phase, Δ-connected generator has an internal impedance of $0.009 + j0.09$ Ω/ϕ. When the load is removed from the generator, the magnitude of the terminal voltage is 13,800 V. The generator feeds a Δ-connected load through a transmission line with an impedance of $0.02 + j0.18$ Ω/ϕ. The per-phase impedance of the load is $7.056 + j3.417$ Ω.

   a) Construct a single-phase equivalent circuit.
   b) Calculate the magnitude of the line current.
   c) Calculate the magnitude of the line voltage at the terminals of the load.
   d) Calculate the magnitude of the line voltage at the terminals of the source.
   e) Calculate the magnitude of the phase current in the load.
   f) Calculate the magnitude of the phase current in the source.

12.14 A three-phase line has an impedance of $1 + j3$ Ω/ϕ. The line feeds two balanced three-phase loads that are connected in parallel. The first load is absorbing a total of 7.2 kW and 5.4 kVAR magnetizing vars. The second load is Δ-connected and has an impedance of $11.52 - j8.64$ Ω/ϕ. The line-to-neutral voltage at the load end of the line is 120 V. What is the magnitude of the line voltage at the source end of the line?

12.15 Three balanced three-phase loads are connected in parallel. Load 1 is Y-connected with an impedance of $400 + j300$ Ω/ϕ; load 2 is Δ-connected with an impedance of $2400 - j1800$ Ω/ϕ; and load 3 is $172.8 + j2203.2$ kVA. The loads are fed from a distribution line with an impedance of $2 + j16$ Ω/ϕ. The magnitude of the line voltage at the load end of the line is $24\sqrt{3}$ kV. Calculate the total complex power at the sending end of the line.

12.16 The output of the balanced, positive-sequence, three-phase source in Fig. 12.23 is 41.6 kVA at a lagging power factor of 0.707. The line voltage at the source is 240 V.

   a) Find the line voltage at the load.
   b) Find the total complex power at the terminals of the load.

12.17 The total power delivered to a balanced three-phase load is 360 kW at a lagging power factor of 0.9. The line voltage at the load is 2400 V. The impedance of the distribution line supplying the load is $0.35 + j2.1$ Ω/ϕ. What is the line voltage at the sending end of the line?

12.18 The drop in voltage between the sending end and the load end of the line in Problem 12.17 is considered excessive. In order to reduce the voltage drop along the line, three capacitors are connected across the terminals of the load. The total reactive power associated with this Δ-connected bank of capacitors is 250 kVAR at 2400 V.

   a) What is the line voltage at the sending end of the line if the line voltage at the load is 2400 V?

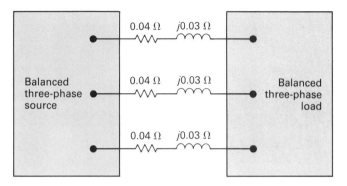

Fig. 12.23 The circuit for Problem 12.16.

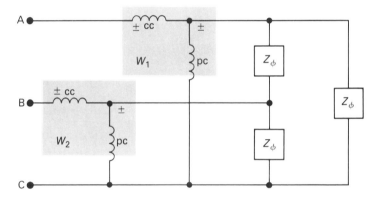

Fig. 12.24 The circuit for Problem 12.21.

    b) What is the rating of the capacitors in μF if the frequency of the source is 60 Hz?

12.19 A balanced three-phase load absorbs 48 kW at a lagging power factor of 0.8 when the line voltage at the terminals of the load is 460 V. Find four equivalent circuits that can be used to model this load.

12.20 Derive Eqs. (12.57) and (12.48). *Hint:* Use $\mathbf{V}_{AN}$ as the reference and compute θ_1 and θ_2 by subtracting the phase angle of the appropriate current from the phase angle of the appropriate voltage. For example, $\theta_1 = (\theta_{AB} - \theta_{aA})$.

12.21 a) Determine the reading of each wattmeter in the circuit in Fig. 12.24 if the phase sequence is negative, the magnitude of the line voltage is 4160 V, and the load impedance is $208\,/-45°\ \Omega/\phi$.
    b) Repeat part (a) given that the phase sequence is positive.

12.22 The wattmeters in the circuit in Fig. 12.20 read as follows: $W_1 = 40{,}823.09$ W and $W_2 = 103{,}176.91$ W. The magnitude of the line voltage is 2400 V. The phase sequence is positive. Find Z_ϕ.

12.23 In the balanced three-phase circuit shown in Fig. 12.25, the current coil of the wattmeter is connected in line aA and the potential coil of the wattmeter is connected

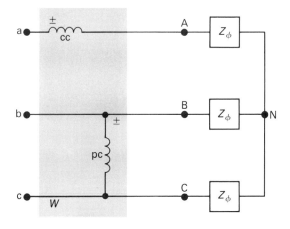

Fig. 12.25 The circuit for Problems 12.23 and 12.24.

across lines b and c. Show that the wattmeter reading multiplied by $\sqrt{3}$ equals the total reactive power associated with the load. The phase sequence is positive.

12.24 The line voltage in the circuit in Fig. 12.25 is 600 V, the phase sequence is positive, and the load impedance is $8 + j6 \ \Omega/\phi$.

    a) Calculate the wattmeter reading.

    b) Calculate the total reactive power associated with the load.

12.25 Show that the total instantaneous power in a balanced three-phase circuit is constant and equal to $1.5 V_\phi I_\phi \cos \theta_\phi$, where V_ϕ and I_ϕ represent the maximum amplitudes of the phase voltage and phase current, respectively.

13

MUTUAL INDUCTANCE

13.1 INTRODUCTION

When we introduced the inductor in Chapter 5, we pointed out that it is a circuit element based on phenomena associated with magnetic fields. In particular, we noted that the source of the magnetic field is current and that a time-varying current will produce a time-varying field. The time-varying field will, in turn, induce a voltage in any conductor linked by the field. We expressed the relationship between the time-varying current and the resulting voltage induced by the time-varying field in terms of the inductance parameter L, namely,

$$v = L\frac{di}{dt}.$$

We now wish to consider the situation in which a time-varying current in one circuit produces a time-varying magnetic field that links a second circuit, causing a voltage to be induced in the second circuit. The voltage induced in the second circuit can be related to the time-varying current in the first circuit by means of an inductance parameter known as *mutual inductance*. Mutual inductance (also measured in henries) is denoted by the letter M. Since the inductance parameter L relates a voltage induced in a circuit to a time-varying current in the same circuit, it is also called *self-inductance*.

When two circuits are linked by a magnetic field, they are described as being magnetically coupled. Magnetic coupling is an important physical phenomenon that is used effectively in both communication and power circuits. The transformer is a device designed entirely around the concept of magnetic coupling. In Section 13.6 we will discuss the transformer as a circuit element.

In order to completely understand the concept of mutual inductance we must expand on the concept of a magnetic field and flux linkage beyond the verbal description given earlier. In Section 13.2, we will review the concept of self-inductance using a more quantitative approach than we used in Chapter 5. Following this review, we will discuss the concept of mutual inductance in Section 13.3.

13.2 A REVIEW OF SELF-INDUCTANCE

The concept of inductance can be traced back to Michael Faraday, who did pioneering work in this area in the early 1800s. Faraday postulated that the magnetic field consists of lines of force surrounding the current-carrying conductor. These lines of force are visualized as energy-storing elastic bands that close upon themselves. As the current increases and decreases, the elastic bands spread out and collapse about the conductor. The voltage induced in the conductor is proportional to the number of lines that collapse into, or cut, the conductor. This image of induced voltage is expressed by what is known today

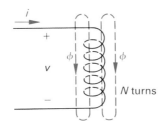

Fig. 13.1   Representation of a magnetic field linking an
N-turn coil.

as *Faraday's law,* that is,

$$v = \frac{d\lambda}{dt}, \tag{13.1}$$

where λ is referred to as the flux linkage and is measured in weber-turns.

To see how Eq. (13.1) leads to the concept of self-inductance, consider the single N-turn coil shown in Fig. 13.1. The magnetic lines of force are shown graphically by the lines threading the N turns and labeled ϕ. The strength of the magnetic field depends on the strength of the current and the spatial orientation of the field depends on the direction of the current. The orientation of the field is related to the direction of the current by the right-hand rule, stated as follows: When the fingers of the right hand are wrapped around the coil such that the fingers point in the direction of the current, the thumb will point in the direction of the magnetic field *inside* the coil. The flux linkage is the product of the magnetic field (ϕ), measured in webers (Wb), and the number of turns linked by the field (N); thus

$$\lambda = N\phi. \tag{13.2}$$

A more detailed picture of flux linkage would introduce partial linkages, that is, portions of the magnetic field linking a fraction of the total turns. However, these details are not necessary for our work if we understand that the flux ϕ in Eq. (13.2) is an equivalent flux that yields the same flux linkage as would be obtained by including the partial linkages in the determination of the total flux linkage λ.

The magnitude of the flux, ϕ, depends on the magnitude of the coil current, the number of turns on the coil, and the magnetic properties of the space occupied by the flux. The relationship between ϕ and i can be written as

$$\phi = \mathscr{P}Ni, \tag{13.3}$$

where $\mathscr{P}$ is the permeance of the space occupied by the field. The permeance is a function of the physical dimensions and the permeability of the space occupied by the flux. When the space is nonmagnetic, the permeance is constant and we have a linear relationship between ϕ and i. When the space is made up of magnetic materials (such as iron, nickel, and cobalt), the permeance varies with the flux and a nonlinear relationship exists between ϕ and i. In our work here, we assume that the core material is nonmagnetic. We will not delve further into the factors that determine the permeance of the space occupied by the magnetic flux. We simply accept permeance as a positive constant of proportionality that relates the strength of the magnetic field to the current that creates it. It is important to note from Eq. (13.3) that the flux is also proportional to the number of turns on the coil.

When we substitute Eqs. (13.2) and (13.3) into Eq. (13.1), we get

$$v = \frac{d\lambda}{dt} = \frac{d(N\phi)}{dt} = N\frac{d\phi}{dt} = N\frac{d}{dt}\,(\mathscr{P}Ni) = N^2\mathscr{P}\frac{di}{dt} = L\frac{di}{dt}, \qquad (13.4)$$

from which we see that the self-inductance is proportional to the square of the number of turns on the coil. We will make use of this observation later in our work.

We now observe that the polarity of the induced voltage in the circuit in Fig. 13.1 reflects the reaction of the field against the current creating the field. For example, when i is increasing, di/dt will be positive and v is positive. Thus energy is required to establish the magnetic field. The rate at which energy is stored in the field is obtained from the product vi. When the field collapses, di/dt is negative and again the polarity of the induced voltage is in opposition to the change. As the field collapses about the coil, energy is returned to the circuit.

With this further insight into the concept of self-inductance in mind, let us turn our attention to the concept of mutual inductance.

13.3   THE CONCEPT OF MUTUAL INDUCTANCE

Mutual inductance is the circuit parameter used to relate the voltage induced in one circuit to a time-varying current in another circuit. This situation arises whenever two or more circuits are linked by a common magnetic field. We will restrict our discussion to the case in which only two circuits are magnetically coupled.

Figure 13.2 shows two coils that are magnetically coupled. The number of turns on each coil are N_1 and N_2, respectively. Coil 1 is energized by a time-varying current source that establishes the current i_1 in the N_1 turns. Coil 2 is not energized and is open. The coils are wound on a nonmagnetic core. The flux produced by the current i_1 can be divided into two components, labeled ϕ_{11} and ϕ_{21} in Fig. 13.2. The flux component ϕ_{11} is the flux produced by i_1, which links only the N_1 turns, whereas the component ϕ_{21} is the flux produced by i_1, which links the N_2 turns as well as the N_1 turns. The first digit in the subscript to the flux gives the coil number and the second digit refers to the coil current. Thus ϕ_{11} is a flux linking coil 1 produced by a current in coil 1, whereas ϕ_{21} is a flux linking coil 2 produced by a current in coil 1.

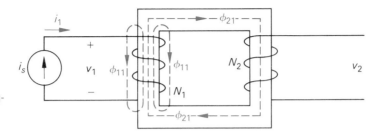

Fig. 13.2 Two magnetically coupled coils.

The total flux linking coil 1 is denoted by ϕ_1 and is the sum of ϕ_{11} and ϕ_{21}, that is,

$$\phi_1 = \phi_{11} + \phi_{21}. \tag{13.5}$$

The flux ϕ_1, as well as its components ϕ_{11} and ϕ_{21}, can be related to the coil current i_1 as follows:

$$\phi_1 = \mathscr{P}_1 N_1 i_1, \tag{13.6}$$

$$\phi_{11} = \mathscr{P}_{11} N_1 i_1, \tag{13.7}$$

and

$$\phi_{21} = \mathscr{P}_{21} N_1 i_1, \tag{13.8}$$

where $\mathscr{P}_1$ is the permeance of the space occupied by the flux ϕ_1, $\mathscr{P}_{11}$ is the permeance of the space occupied by the flux ϕ_{11}, and $\mathscr{P}_{21}$ is the permeance of the space occupied by the flux ϕ_{21}. If we substitute Eqs. (13.6), (13.7), and (13.8) into Eq. (13.5), we get the relationship between the permeance of the space occupied by the total flux ϕ_1 and the permeances of the spaces occupied by its components ϕ_{11} and ϕ_{21}:

$$\mathscr{P}_1 = \mathscr{P}_{11} + \mathscr{P}_{21}. \tag{13.9}$$

We can use Faraday's law to derive expressions for v_1 and v_2. Thus

$$v_1 = \frac{d\lambda_1}{dt} = \frac{d(N_1\phi_1)}{dt} = N_1 \frac{d}{dt}(\phi_{11} + \phi_{21})$$

$$= N_1^2(\mathscr{P}_{11} + \mathscr{P}_{21})\frac{di_1}{dt}$$

$$= N_1^2 \mathscr{P}_1 \frac{di_1}{dt} = L_1 \frac{di_1}{dt} \tag{13.10}$$

and

$$v_2 = \frac{d\lambda_2}{dt} = \frac{d}{dt}(N_2\phi_{21}) = N_2 \frac{d}{dt}(\mathscr{P}_{21}N_1 i_1)$$

$$= N_2 N_1 \mathscr{P}_{21} \frac{di_1}{dt}. \tag{13.11}$$

The coefficient of di_1/dt in Eq. (13.10) is the self-inductance of coil 1 and the coefficient of di_1/dt in Eq. (13.11) is the mutual inductance between coils 1 and 2; thus

$$M_{21} = N_2 N_1 \mathscr{P}_{21}. \tag{13.12}$$

We read the subscripts on M as an inductance that relates the voltage induced in coil 2 to the current in coil 1.

Using the coefficient of mutual inductance, we have

$$v_2 = M_{21} \frac{di_1}{dt}. \tag{13.13}$$

Note that no polarity references have been assigned to v_2 in Fig. 13.2. We will discuss determining the polarity of mutually induced voltages in Section 13.4.

Let us now return to the coupled coils in Fig. 13.2 and excite coil 2 from a time-varying current source (i_2) and leave coil 1 open. This circuit arrangement is shown in Fig. 13.3. In this case, no polarity references are assigned to v_1.

The total flux linking coil 2 is

$$\phi_2 = \phi_{22} + \phi_{12}. \tag{13.14}$$

The flux ϕ_2, as well as its components ϕ_{22} and ϕ_{12}, can be related to the coil current i_2 as follows:

$$\phi_2 = \mathscr{P}_2 N_2 i_2, \tag{13.15}$$

$$\phi_{22} = \mathscr{P}_{22} N_2 i_2, \tag{13.16}$$

and

$$\phi_{12} = \mathscr{P}_{12} N_2 i_2. \tag{13.17}$$

The voltages v_2 and v_1 are

$$v_2 = \frac{d\lambda_2}{dt} = \mathscr{P}_2 N_2^2 \frac{di_2}{dt} = L_2 \frac{di_2}{dt} \tag{13.18}$$

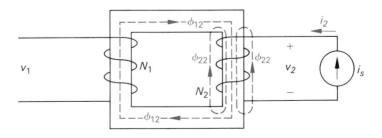

Fig. 13.3   The magnetically coupled coils in Fig. 13.2 with coil 2 excited and coil 1 open.

and

$$v_1 = \frac{d\lambda_{12}}{dt} = \frac{d}{dt}(N_2\phi_{12})$$

$$= N_1 N_2 \mathscr{P}_{12}\frac{di_2}{dt}. \qquad \textbf{(13.19)}$$

The coefficient of mutual inductance that relates the voltage induced in coil 1 to the time-varying current in coil 2 is the coefficient of di_2/dt in Eq. (13.19); thus

$$M_{12} = N_1 N_2 \mathscr{P}_{12}. \qquad \textbf{(13.20)}$$

For nonmagnetic materials, the permeances $\mathscr{P}_{12}$ and $\mathscr{P}_{21}$ will be equal and therefore

$$M_{12} = M_{21} = M. \qquad \textbf{(13.21)}$$

We see then that for linear circuits with just two magnetically coupled coils, it is not necessary to attach subscripts to the coefficient of mutual inductance.

It is also important to recognize that the coefficient of mutual inductance is a function of the self-inductances. We can derive this relationship as follows. From Eqs. (13.10) and (13.18), respectively, we have

$$L_1 = N_1^2 \mathscr{P}_1 \qquad \textbf{(13.22)}$$

and

$$L_2 = N_2^2 \mathscr{P}_2. \qquad \textbf{(13.23)}$$

It follows from Eqs. (13.22) and (13.23) that

$$L_1 L_2 = N_1^2 N_2^2 \mathscr{P}_1 \mathscr{P}_2. \qquad \textbf{(13.24)}$$

Now we use Eq. (13.9) along with corresponding expression for $\mathscr{P}_2$ to write

$$L_1 L_2 = N_1^2 N_2^2 (\mathscr{P}_{11} + \mathscr{P}_{21})(\mathscr{P}_{22} + \mathscr{P}_{12}). \qquad \textbf{(13.25)}$$

But for linear system $\mathscr{P}_{21} = \mathscr{P}_{12}$; thus we can write Eq. (13.25) as

$$L_1 L_2 = (N_1 N_2 \mathscr{P}_{12})^2 \left(1 + \frac{\mathscr{P}_{11}}{\mathscr{P}_{12}}\right)\left(1 + \frac{\mathscr{P}_{22}}{\mathscr{P}_{12}}\right)$$

$$L_1 L_2 = M^2 \left(1 + \frac{\mathscr{P}_{11}}{\mathscr{P}_{12}}\right)\left(1 + \frac{\mathscr{P}_{22}}{\mathscr{P}_{12}}\right). \qquad \textbf{(13.26)}$$

We can express Eq. (13.26) in a more meaningful form by replacing the two terms involving permeances by a single constant; thus

$$\frac{1}{k^2} = \left(1 + \frac{\mathscr{P}_{11}}{\mathscr{P}_{12}}\right)\left(1 + \frac{\mathscr{P}_{22}}{\mathscr{P}_{12}}\right). \qquad \textbf{(13.27)}$$

Equation (13.27) allows us to write Eq. (13.26) as

$$M^2 = k^2 L_1 L_2$$

or

$$M = k\sqrt{L_1 L_2}, \qquad (13.28)$$

where the constant k is known as the coefficient of coupling. We can see from Eq. (13.27) that $1/k^2$ must be greater than 1, which means that k must be less than 1. In fact, the coefficient of coupling must lie between 0 and 1, that is,

$$0 \le k \le 1. \qquad (13.29)$$

The coefficient of coupling equals zero when there is no flux common to the two coils, that is, when $\phi_{12} = \phi_{21} = 0$. This implies that $\mathscr{P}_{12}$ is zero and Eq. (13.27) tells us that $1/k^2$ is infinite or $k = 0$. If there is no flux linkage between the coils, M is obviously zero.

The coefficient of coupling is equal to one when ϕ_{11} and ϕ_{22} are zero. This implies that all the flux that links coil 1 also links coil 2. In terms of Eq. (13.27), $\mathscr{P}_{11} = \mathscr{P}_{22} = 0$. Obviously this represents an ideal state since it is physically impossible to wind two coils so that they share precisely the same flux.

We mention in passing that magnetic materials (such as alloys of iron, cobalt, and nickel) are important because they create a space with very high permeance and are used to establish coefficients of coupling that approach unity. (More about this later.)

In this section, we have seen how the concept of inductance is extended to relate the voltage induced in one circuit to a time-varying current in another circuit. We have also shown how inductance relates to the number of turns on the magnetically coupled coils and a constant, known as permeance, which characterizes the magnetic properties of the space occupied by the flux. For linearly coupled coils, the important results for circuit analysis are contained in Eqs. (13.21), (13.28), and (13.29). We are now ready to discuss the important problem of determining the polarity of the mutually induced voltages.

DRILL EXERCISE
13.1

Two magnetically coupled coils are wound on a nonmagnetic core. The self-inductance of coil 1 is 6 H, the mutual inductance is 9.6 H, the coefficient of coupling is 0.8, and the physical structure of the coils is such that the permeances $\mathscr{P}_{11}$ and $\mathscr{P}_{22}$ are equal.

a) Find L_2 and the turns ratio N_1/N_2.

b) If $N_1 = 800$, what is the value of $\mathscr{P}_1$ and $\mathscr{P}_2$?

Ans. (a) 24 H, 0.5; (b) 9.375×10^{-6} Wb/A.

13.4 POLARITY OF MUTUALLY INDUCED VOLTAGES (THE DOT CONVENTION)

The polarity of a mutually induced voltage also reflects a reaction against the time-varying flux that creates the voltage. For example, in the circuit in Fig. 13.2 when i_1 is increasing in the reference direction, the polarity of v_2 is positive at the upper terminal of coil 2. This follows from the observation that the polarity of v_2 is such that it would establish a current out of the upper terminal of the N_2 coil to produce a flux in opposition to ϕ_{21}. Thus when v_1 is positive at the upper terminal of coil 1, v_2 is positive at the upper terminal of coil 2, and vice versa. (Note that if we excite coil 2 as in Fig. 13.3, we reach the same conclusion concerning the relative polarities of v_1 and v_2.)

Having deduced the polarity of v_2 in relation to the current i_1, we can now express v_2 as a function of i_1 with the proper algebraic sign. If the reference for v_2 is positive at the top terminal of coil 2, then

$$v_2 = M \frac{di_1}{dt}. \tag{13.30}$$

On the other hand, if the reference for v_2 is positive at the lower terminal of coil 2, then

$$v_2 = -M \frac{di_1}{dt}. \tag{13.31}$$

Similarly, for the circuit in Fig. 13.3 where coil 2 is energized by i_2 and coil 1 is open,

$$v_1 = \pm M \frac{di_2}{dt}, \tag{13.32}$$

where we use the positive sign if the reference for v_1 is positive at the top terminal of coil 1, and the minus sign otherwise.

In general, it is too cumbersome to show the details of mutually coupled windings, so that we do not use the "reaction" method for determining polarity signs in writing circuit equations. The polarities of the mutually induced voltages are kept track of by a method known as the *dot convention,* in which a dot is placed on one terminal of each winding. These dots carry the sign information regarding the mutually induced voltages. For example, for the coils in Fig. 13.2 either the two upper terminals or the two lower terminals are given a dot marking. The two possible markings are shown in Fig. 13.4. Note that when the dot markings are used, the coils can be drawn schematically rather than to show how the coils wrap around a core structure.

There are two types of problems associated with polarity dots. One is to determine a proper set of dot markings when the physical arrangement of the coupled coils is given. The other is to determine how the dots are used in writing the circuit equations that describe magnetically coupled coils. We begin our discussion of using the dot convention by describing a systematic method for determining the dot markings.

We assume that we are given a set of two magnetically coupled coils such that the physical arrangement of the two coils and the mode of each winding is known. As an example, consider the coils shown in Fig. 13.5. We follow six steps to determine a set of dot markings.

1. We arbitrarily select one terminal of one coil and give it a dot. We have chosen the D terminal in Fig. 13.5.

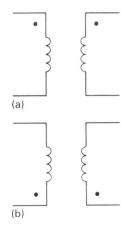

(a)

(b)

Fig. 13.4 Dot markings for the coils in Fig. 13.2.

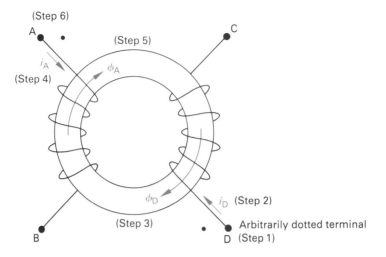

Fig. 13.5   A set of coils showing a method for determining a set of dot markings.

2. We assign a current *into* the arbitrarily selected dotted terminal. This current is labeled i_D in Fig. 13.5.

3. We use the right-hand rule to determine the direction of the magnetic field established by i_D *inside* the coupled coils. This field is labeled ϕ_D in Fig. 13.5.

4. We arbitrarily pick one terminal of the second coil and assign a current *into* this terminal. We have chosen terminal A and show the current as i_A in Fig. 13.5.

5. We use the right-hand rule to determine the direction of the flux established by i_A *inside* the coupled coils. This flux is marked ϕ_A in Fig. 13.5.

6. We compare the directions of the two fluxes ϕ_D and ϕ_A. If the fluxes are additive, we place a dot on the terminal of the second coil where the test current (i_A) enters. (In Fig. 13.5, the fluxes ϕ_D and ϕ_A are additive and therefore a dot is placed on terminal A.) If the fluxes are subtractive, we place a dot on the terminal of the second coil where the test current leaves.

The relative polarities of magnetically coupled coils can also be determined experimentally. This is important because in some situations it is impossible to determine how the coils are wound on the core. One way to experimentally ascertain the polarities of coupled coils is to connect a dc voltage source, a resistor, a switch, and a dc voltmeter to the pair of coils, as shown in Fig. 13.6. The coils in Fig. 13.6 are drawn inside a box to imply that it is not possible to inspect the coils physically in order to determine the polarity markings. The resistor R is used to limit the magnitude of the current supplied by the dc voltage source. The procedure for determining the relative polarity marks is as follows. The coil terminal connected to the positive terminal of the dc

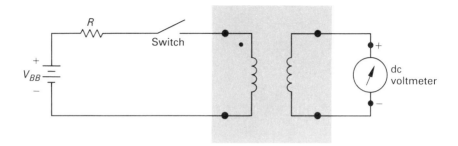

Fig. 13.6 An experimental setup for determining polarity marks.

source via the switch and limiting resistor is given a polarity mark, as shown in Fig. 13.6. At the time when the switch is *closed,* the deflection of the voltmeter is observed. If the momentary deflection is *up-scale,* the coil terminal connected to the positive terminal of the voltmeter is given the polarity mark. If the momentary deflection is *downscale,* the coil terminal connected to the negative terminal of the voltmeter is given the polarity mark.

DRILL EXERCISE 13.2 Assume that the magnetic flux is confined to the core material in the structure shown. Which terminal of coil 2 should be given a dot marking?

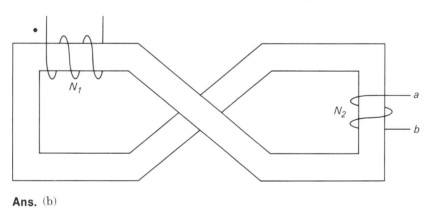

Ans. (b)

Let us now turn our attention to the second type of problem, namely, learning how to use the dot markings when writing circuit equations. We can most easily analyze circuits containing mutual inductance using mesh currents. We can also use the node-voltage method but it is somewhat clumsy to implement. The analysis that follows will be done in terms of mesh currents, using Fig. 13.7 to illustrate the procedure.

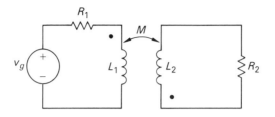

Fig. 13.7 A circuit used to show how the dot markings are used in writing circuit equations.

The circuit in Fig. 13.7 shows the schematic representation of two magnetically coupled coils. The dot markings are given as shown. The self-inductance of each coil is labeled L_1 and L_2 and the mutual inductance is labeled M. The double-headed arrow adjacent to M indicates the pair of coils with this value of mutual inductance. This notation is particularly needed in circuits containing more than one pair of magnetically coupled coils. Our problem is to write the circuit equations that describe the circuit in terms of the coil currents.

At this point in our analysis of the circuit in Fig. 13.7, we are at liberty to choose the reference direction for each coil current. Our arbitrarily selected reference currents are shown in Fig. 13.8.

Once we have chosen the reference directions for i_1 and i_2, we can sum the voltages around each closed path. Because of the mutual inductance M, there will be two voltages across each coil, namely, a self-induced voltage and a mutually induced voltage. The self-induced voltage is a voltage drop in the direction of the current producing the voltage. The polarity of the mutually induced voltage is determined as follows. When the reference direction for a current *enters* the dotted terminal of a coil, the reference polarity of the voltage that it induces in the *other* coil is positive at its dotted terminal. Alternatively, when the reference direction for a current leaves the dotted terminal of a coil, the reference polarity of the voltage that it induces in the other coil is negative at its dotted terminal.

Using the rule stated above, we find that the reference polarity for the voltage induced in coil 1 by the current i_2 will be negative at the dotted termi-

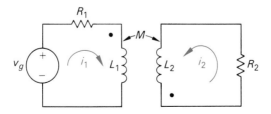

Fig. 13.8 Coil currents i_1 and i_2 used to describe the circuit in Fig. 13.7.

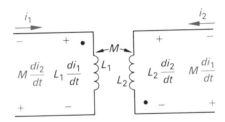

Fig. 13.9 The self- and mutually-induced voltages appearing across the coils in Fig. 13.8.

nal of 1. This voltage ($M\, di_2/dt$) is a voltage rise with respect to i_1. The voltage induced in coil 2 by the current i_1 is $M\, di_1/dt$ and its reference polarity is positive at the dotted terminal of coil 2. This voltage is a voltage rise in the direction of i_2. The self- and mutually induced voltages across coils 1 and 2, along with their polarity marks, are shown in Fig. 13.9.

We can now sum the voltages around each closed loop. In Eqs. (13.33) and (13.34), voltage rises in the reference direction of a current are negative:

$$-v_g + i_1 R_1 + L_1 \frac{di_1}{dt} - M \frac{di_2}{dt} = 0, \tag{13.33}$$

$$i_2 R_2 + L_2 \frac{di_2}{dt} - M \frac{di_1}{dt} = 0. \tag{13.34}$$

Example 13.1 shows how the dot markings are used to formulate a set of circuit equations.

Example 13.1 a) Write a set of mesh-current equations that describe the circuit in Fig. 13.10 in terms of the currents i_1, i_2, and i_3.

b) What is the coefficient of coupling of the magnetically coupled coils?

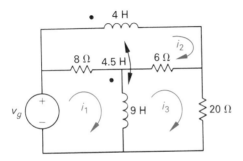

Fig. 13.10 The circuit for Example 13.1.

Solution   a) (In the following set of mesh-current equations, voltage drops appear as positive quantities on the right-hand side of each equation.) Summing the voltages around the first mesh yields

$$v_g = 8(i_1 - i_2) + 9\frac{d}{dt}(i_1 - i_3) + 4.5\frac{di_2}{dt}.$$

The second mesh equation is

$$0 = 4\frac{di_2}{dt} + 4.5\frac{d}{dt}(i_1 - i_3) + 6(i_2 - i_3) + 8(i_2 - i_1).$$

The third mesh equation is

$$0 = 9\frac{d}{dt}(i_3 - i_1) - 4.5\frac{di_2}{dt} + 6(i_3 - i_2) + 20i_3.$$

In studying these equations, note that the voltage induced across the 4-H coil due to the current $(i_1 - i_3)$, that is, $4.5\, d(i_1 - i_3)/dt$, is a voltage drop in the reference direction of i_2. The voltage induced in the 9-H coil due to the current i_2, that is, $4.5\, di_2/dt$, is a voltage drop in the reference direction of i_1 but a voltage rise in the reference direction of i_3.

b) $k = \dfrac{M}{\sqrt{L_1 L_2}} = \dfrac{4.5}{\sqrt{(9)(4)}} = \dfrac{4.5}{6} = 0.75$ ∎

Thus far in our examples, we have written the equations that describe circuits containing mutual inductance in differential form. We will not discuss the solutions of these equations at this time. The transient response of circuits containing mutual inductance will be illustrated after we have introduced the Laplace transform method of solving differential equations (see Chapter 16). For now we will limit our analysis of circuits containing mutual inductance to the sinusoidal steady-state response. Before pursuing the sinusoidal analysis, however, we will discuss the basic energy relationships that pertain to magnetically coupled coils.

DRILL EXERCISE   Write a set of mesh-current equations for the circuit in Example 13.1 if the dot on the
13.3   9-H inductor is at the lower terminal and the reference direction of i_3 is reversed.

Ans. $v_g = 8(i_1 - i_2) + 9\dfrac{d}{dt}(i_1 + i_3) - 4.5\dfrac{di_2}{dt}$,

$0 = 4\dfrac{di_2}{dt} - 4.5\dfrac{d}{dt}(i_1 + i_3) + 6(i_2 + i_3) + 8(i_2 - i_1)$,

$0 = 20i_3 + 6(i_3 + i_2) + 9\dfrac{d}{dt}(i_3 + i_1) - 4.5\dfrac{di_2}{dt}$.

13.5 ENERGY CALCULATIONS

There are three important reasons for our discussion of the total energy stored in magnetically coupled coils. First, we want to show how the energy stored in the pair of coils relates to the coil currents, the self-inductances, and the mutual inductance. Second, we want to show that for linear magnetic coupling the computation of stored energy predicts $M_{12} = M_{21}$. Our third and final reason is that we can use the energy calculation to show that the mutual inductance cannot exceed the square root of the product of the self-inductances, thus supporting our earlier contention that $M = k\sqrt{L_1 L_2}$, where $0 \leq k \leq 1$.

We can use the circuit shown in Fig. 13.11 to derive the expression for the total energy stored in the magnetic fields associated with a pair of coupled coils. We begin by assuming that the currents i_1 and i_2 are zero and that this zero-current state corresponds to zero energy stored in the coils. Now we let i_1 increase from zero to some arbitrary value I_1 and compute the energy stored at the time $i_1 = I_1$. Since i_2 is zero, the total power input into the air of coils is $v_1 i_1$ and the energy stored is

$$\int_0^{W_1} dw = L_1 \int_0^{I_1} i_1 \, di_1$$

$$W_1 = \frac{1}{2} L_1 I_1^2. \tag{13.35}$$

Now we hold i_1 constant at I_1 and increase i_2 from zero to some arbitrary value I_2. During this time interval, the voltage induced in coil 2 due to i_1 is zero because I_1 is constant. The voltage induced in coil 1 due to i_2 is $M_{12} \, di_2/dt$. Therefore the power input to the pair of coils is

$$p = I_1 M_{12} \frac{di_2}{dt} + i_2 v_2.$$

The total energy stored in the pair of coils at the time when i_2 equals I_2 will be

$$\int_{W_1}^{W} dw = \int_0^{I_2} I_1 M_{12} \, di_2 + \int_0^{I_2} L_2 i_2 \, di_2$$

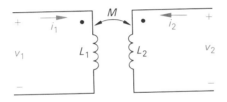

Fig. 13.11 The circuit used to derive the basic energy relationships.

or

$$W = W_1 + I_1 I_2 M_{12} + \frac{1}{2} L_2 I_2^2$$

$$= \frac{1}{2} L_1 I_1^2 + \frac{1}{2} L_2 I_2^2 + I_1 I_2 M_{12}. \tag{13.36}$$

If we reverse the procedure, that is, if we first increase i_2 from zero to I_2 and then increase i_1 from zero to I_1, the total energy stored will be

$$W = \frac{1}{2} L_1 I_1^2 + \frac{1}{2} L_2 I_2^2 + I_1 I_2 M_{21}. \tag{13.37}$$

Equations (13.36) and (13.37) express the total energy stored in a pair of magnetically coupled coils as a function of the coil currents, the self-inductances, and the mutual inductance. Note that the only difference in these equations is the coefficient of the current product $I_1 I_2$. We use Eq. (13.36) if I_1 is established first and Eq. (13.37) if I_2 is established first.

We now argue that if the coupling medium is linear—that is, non-magnetic—the total energy stored will be the same regardless of the order used to establish I_1 and I_2. Therefore for linear coupling, $M_{12} = M_{21}$. We also note that since I_1 and I_2 are arbitrary values of i_1 and i_2, respectively, we can represent the coil currents by their instantaneous values i_1 and i_2; thus at any instant of time, the total energy stored in the coupled coils is

$$w(t) = \frac{1}{2} L_1 i_1^2 + \frac{1}{2} L_2 i_2^2 + M i_1 i_2. \tag{13.38}$$

We derived Eq. (13.38) assuming that both coil currents entered polarity-marked terminals. We leave it to the reader to verify that if one current enters a polarity-marked terminal while the other leaves a polarity-marked terminal, the algebraic sign of the term $M i_1 i_2$ reverses. Thus, in general,

$$w(t) = \frac{1}{2} L_1 i_1^2 + \frac{1}{2} L_2 i_2^2 \pm M i_1 i_2. \tag{13.39}$$

We can use Eq. (13.39) to show that M cannot exceed $\sqrt{L_1 L_2}$. Since the magnetically coupled coils are passive elements, the total energy stored can never be negative. If $w(t)$ can never be negative, it follows from Eq. (13.39) that the quantity

$$\frac{1}{2} L_1 i_1^2 + \frac{1}{2} L_2 i_2^2 - M i_1 i_2$$

must be greater than or equal to zero when i_1 and i_2 are either both positive or both negative. The limiting value of M corresponds to setting the quantity equal to zero, that is,

$$\frac{1}{2} L_1 i_1^2 + \frac{1}{2} L_2 i_2^2 - M i_1 i_2 = 0. \tag{13.40}$$

To find the limiting value of M we add and subtract the term $i_1 i_2 \sqrt{L_1 L_2}$ to the left-hand side of Eq. (13.40). This will generate a term that is a perfect square; thus

$$\left\{ \sqrt{\frac{L_1}{2}} \, i_1 - \sqrt{\frac{L_2}{2}} \, i_2 \right\}^2 + i_1 i_2 (\sqrt{L_1 L_2} - M) = 0. \qquad (13.41)$$

The squared term in Eq. (13.41) can never be negative, but it can be zero. Therefore we can guarantee that $w(t)$ will be greater than or equal to zero only if

$$\sqrt{L_1 L_2} \geq M, \qquad (13.42)$$

which is another way of saying

$$M = k\sqrt{L_1 L_2} \qquad (0 \leq k \leq 1).$$

We derived Eq. (13.42) assuming that i_1 and i_2 are either both positive or both negative. However, we get the same result if i_1 and i_2 are of opposite sign because in this case we obtain the limiting value of M by selecting the plus sign in Eq. (13.39).

DRILL EXERCISE
13.4

The self-inductances of the coils in Fig. 13.11 are $L_1 = 5$ mH and $L_2 = 33.8$ mH. If the coefficient of coupling is 0.96, calculate the energy stored in the system in mJ when (a) $i_1 = 10$ A, $i_2 = 5$ A; (b) $i_1 = -10$ A, $i_2 = -5$ A; (c) $i_1 = -10$ A, $i_2 = 5$ A; (d) $i_1 = 10$ A, $i_2 = -5$ A.

Ans. (a) 1296.50 mJ; (b) 1296.50 mJ; (c) 48.50 mJ; (d) 48.50 mJ.

DRILL EXERCISE
13.5

The coefficient of coupling in Drill Exercise 13.4 is increased to 1.0

a) If i_1 equals 10 A, what value of i_2 will result in zero stored energy?

b) Is there any physically realizable value of i_2 that can make the stored energy negative?

Ans. (a) -3.846 A; (b) no.

13.6 THE LINEAR TRANSFORMER

As we mentioned earlier, the transformer is a device that uses magnetic coupling as the basis for its operation. We are now ready to analyze the sinusoidal steady-state response of the linear transformer. The term "linear" calls attention to the fact that the windings are wound on a nonmagnetic core. The presence of a nonmagnetic core means that there is a linear relationship between the magnetic flux and the winding currents.

The basic circuit using the transformer as a coupling device is shown in Fig. 13.12, where the transformer is used to connect the load to the source. Our

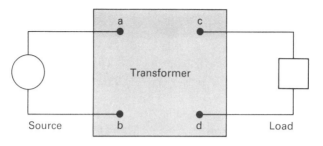

Fig. 13.12 The basic arrangement of a transformer used to connect a load to a source.

problem is to determine how the transformer affects the relationship between the load and the source. We can make one observation immediately, namely, that the transformer will prevent any dc components of current (or voltage) in the source from reaching the load. Thus one reason for using transformers is to isolate portions of a circuit from dc currents and voltages. To see how the transformer affects steady-state sinusoidal currents and voltages, we construct a phasor domain circuit model of the source, transformer, and load.

The phasor domain circuit model of the system in Fig. 13.12 is shown in Fig. 13.13. In discussing this circuit, it is convenient to refer to the transformer winding connected to the source as the *primary* winding and the winding connected to the load as the *secondary* winding. Using this terminology, the transformer circuit parameters are:

R_1 = the resistance of the primary winding,

R_2 = the resistance of the secondary winding,

L_1 = the self-inductance of the primary winding,

L_2 = the self-inductance of the secondary winding, and

M = the mutual inductance.

The internal voltage of the sinusoidal source is $\mathbf{V}_s$ and the internal impedance of the source is Z_s. The load connected to the secondary winding of the

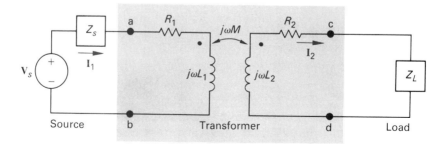

Fig. 13.13 The phasor domain circuit model for the system in Fig. 13.12.

transformer is represented by the impedance Z_L. The phasor currents $\mathbf{I}_1$ and $\mathbf{I}_2$ represent the primary and secondary currents of the transformer, respectively.

The analysis of the circuit in Fig. 13.13 consists of finding $\mathbf{I}_1$ and $\mathbf{I}_2$ as functions of the circuit parameters $\mathbf{V}_s$, Z_s, R_1, L_1, R_2, L_2, M, Z_L, and ω. We are also interested in finding the impedance seen looking into the terminals a, b. To find $\mathbf{I}_1$ and $\mathbf{I}_2$ we first write the two mesh-current equations that describe the circuit:

$$\mathbf{V}_s = (Z_s + R_1 + j\omega L_1)\mathbf{I}_1 - j\omega M\mathbf{I}_2 \tag{13.43}$$

$$0 = -j\omega M\mathbf{I}_1 + (R_2 + j\omega L_2 + Z_L)\mathbf{I}_2. \tag{13.44}$$

To facilitate the algebraic manipulation of Eqs. (13.43) and (13.44), we let

$$Z_{11} = Z_s + R_1 + j\omega L_1 \tag{13.45}$$

and

$$Z_{22} = R_2 + j\omega L_2 + Z_L, \tag{13.46}$$

where we note that Z_{11} is the total self-impedance of the mesh containing the primary winding of the transformer and Z_{22} is the total self-impedance of the mesh containing the secondary winding of the transformer. If we use the notation introduced in Eqs. (13.45) and (13.46), the solutions for $\mathbf{I}_1$ and $\mathbf{I}_2$ from Eqs. (13.43) and (13.44) are

$$\mathbf{I}_1 = \frac{Z_{22}}{Z_{11}Z_{22} + \omega^2 M^2} \cdot \mathbf{V}_s \tag{13.47}$$

and

$$\mathbf{I}_2 = \frac{j\omega M}{Z_{11}Z_{22} + \omega^2 M^2} \cdot \mathbf{V}_s = \frac{j\omega M}{Z_{22}}\mathbf{I}_1. \tag{13.48}$$

The impedance seen by the internal source voltages $\mathbf{V}_s$ is $\mathbf{V}_s/\mathbf{I}_1$, or

$$\frac{\mathbf{V}_s}{\mathbf{I}_1} = Z_{\text{int}} = \frac{Z_{11}Z_{22} + \omega^2 M^2}{Z_{22}} = Z_{11} + \frac{\omega^2 M^2}{Z_{22}}. \tag{13.49}$$

The impedance seen at the terminals of the source is $Z_{\text{int}} - Z_s$; thus

$$Z_{ab} = Z_{11} + \frac{\omega^2 M^2}{Z_{22}} - Z_s$$

$$= R_1 + j\omega L_1 + \frac{\omega^2 M^2}{(R_2 + j\omega L_2 + Z_L)}. \tag{13.50}$$

The impedance given by Eq. (13.50) is of particular interest because it tells us how the transformer modifies the load as seen by the source. In other words, without the transformer the source sees Z_L but with the transformer the source sees a modified version of Z_L.

We begin our study of the impedance Z_{ab} by noting that it is independent of the magnetic polarity of the transformer. This follows because the mutual inductance appears in Eq. (13.50) as a squared quantity. We also observe that the last term in Eq. (13.50) represents the impedance reflected into the primary side of the transformer. That is, if the two windings are decoupled, M becomes zero and Z_{ab} reduces to the self-impedance of the primary winding.

To study the nature of the reflected impedance in more detail, we first express the load impedance in rectangular form:

$$Z_L = R_L + jX_L, \tag{13.51}$$

where the load reactance X_L carries its own algebraic sign. In other words, X_L is a positive number if the load is inductive and a negative number if the load is capacitive. We now use Eq. (13.51) to write the reflected impedance (Z_r), the last term on the right in Eq. (13.50), in rectangular form; thus

$$
\begin{aligned}
Z_r &= \frac{\omega^2 M^2}{R_2 + R_L + j(\omega L_2 + X_L)} \\
&= \frac{\omega^2 M^2 [(R_2 + R_L) - j(\omega L_2 + X_L)]}{(R_2 + R_L)^2 + (\omega L_2 + X_L)^2} \\
&= \frac{\omega^2 M^2}{|Z_{22}|^2} \{(R_2 + R_L) - j(\omega L_2 + X_L)\}.
\end{aligned}
\tag{13.52}
$$

In deriving Eq. (13.52), we have taken advantage of the fact that when Z_L is written in rectangular form, the self-impedance of the mesh containing the secondary winding is

$$Z_{22} = R_2 + R_L + j(\omega L_2 + X_L). \tag{13.53}$$

Now observe from Eq. (13.52) that the self-impedance of the secondary circuit is reflected into the primary circuit by a scaling factor of $(\omega M/|Z_{22}|)^2$ and, furthermore, the sign of the reactive component $(\omega L_2 + X_L)$ is reversed. Thus the linear transformer reflects the *conjugate* of the self-impedance of the secondary circuit (Z_{22}^*) into the primary winding by a scalar multiplier. The multiplier is equal to the square of the ratio of the mutual reactance to the magnitude of the secondary-circuit self-impedance ($[\omega M/|Z_{22}|]^2$). Before we describe Eq. (13.52) any further, let us pause and illustrate the results we have discussed thus far with a numerical example.

Example 13.2   The parameters of a certain linear transformer are $R_1 = 200\ \Omega$, $R_2 = 100\ \Omega$, $L_1 = 9$ H, $L_2 = 4$ H, and $k = 0.5$. The transformer is used to couple an impedance consisting of an 800-Ω resistor in series with a 1-μF capacitor to a sinusoidal voltage source. The 300-V (rms) source has an internal impedance of $500 + j100\ \Omega$ and a frequency of 400 rad/s.

a) Construct a phasor domain equivalent circuit of the system.

b) Calculate the self-impedance of the primary circuit.

c) Calculate the self-impedance of the secondary circuit.

d) Calculate the impedance reflected into the primary winding.

e) Calculate the scaling factor for the reflected impedance.

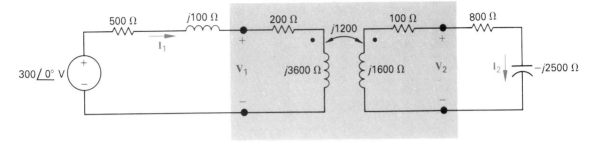

Fig. 13.14 The phasor domain equivalent circuit for Example 13.2(a).

f) Calculate the impedance seen looking into the primary terminals of the transformer.

g) Calculate the rms value of the primary and secondary current.

h) Calculate the rms value of the voltage at the terminals of the load and source.

i) Calculate the average power delivered to the 800-Ω resistor.

j) What percentage of the average power delivered to the transformer is delivered to the load?

Solution   a) The phasor domain equivalent circuit as shown in Fig. 13.14. Note that we have selected the internal voltage of the source as the reference phasor and chosen $\mathbf{V}_1$ and $\mathbf{V}_2$ to represent the terminal voltages of the transformer. In constructing the circuit in Fig. 13.13, we made the following calculations:

$$j\omega L_1 = j(400)(9) = j3600 \ \Omega,$$
$$j\omega L_2 = j(400)(4) = j1600 \ \Omega,$$
$$M = 0.5\sqrt{(9)(4)} = 3 \ \text{H},$$
$$j\omega M = j(400)(3) = j1200 \ \Omega,$$
$$\frac{1}{j\omega C} = \frac{10^6}{j400} = -j\,2500 \ \Omega.$$

b) The self-impedance of the primary circuit is

$$Z_{11} = 500 + j100 + 200 + j3600 \ \Omega = 700 + j3700 \ \Omega.$$

c) The self-impedance of the secondary circuit is

$$Z_{22} = 100 + j1600 + 800 - j2500 = 900 - j900 \ \Omega.$$

d) The impedance reflected into the primary winding is

$$Z_r = \left[\frac{1200}{|900 - j900|}\right]^2 (900 + j900)$$

$$= \frac{8}{9}(900 + j900) = 800 + j800 \ \Omega.$$

e) The scaling factor by which Z_{22}^* is reflected is $8/9$.

f) The impedance seen looking into the primary terminals of the transformer is the impedance of the primary winding plus the reflected impedance; thus

$$Z_{ab} = 200 + j3600 + 800 + j800 = 1000 + j4400 \ \Omega.$$

g) Once we know the input impedance to the transformer, we can easily calculate $\mathbf{I}_1$ as follows:

$$\mathbf{I}_1 = \frac{\mathbf{V}_s}{Z_s + Z_{ab}} = \frac{300 \ \underline{/0^\circ}}{1500 + j4500}$$
$$= 20 - j60 = 63.25 \ \underline{/-71.57^\circ} \ \text{mA rms.}$$

We can calculate $\mathbf{I}_2$ from Eq. (13.48):

$$\mathbf{I}_2 = \frac{j(1200)}{(900 - j900)} \ \mathbf{I}_1 = 59.63 \ \underline{/63.43^\circ} \ \text{mA rms.}$$

h) The voltages at the terminals of the transformer are

$$\mathbf{V}_2 = (800 - j2500) \ \mathbf{I}_2 = 156.52 \underline{/-8.82^\circ} \ \text{V rms,}$$

and

$$\mathbf{V}_1 = Z_{ab} \mathbf{I}_1$$
$$= (1000 + j4400)63.25 \underline{/-71.57^\circ}$$
$$= 285.38 \ \underline{/5.63^\circ} \ \text{V rms.}$$

i) The average power delivered to the load is

$$P = (800)|\mathbf{I}_2|^2$$
$$= 2.84 \ \text{W.}$$

j) The average power delivered to the transformer is

$$P_{ab} = (1000)|\mathbf{I}_1|^2 = 4.00 \ \text{W;}$$

therefore,

$$\eta = \frac{2.84}{4.00} \times 100 = 71.11\%. \quad \blacksquare$$

We are now interested in investigating what happens to the input impedance Z_{ab} as L_1 and L_2 each become infinitely large and, at the same time, the coefficient of coupling approaches its limiting value of unity. We are interested in this behavior because we can approach this condition with transformers wound on ferromagnetic cores. Even though magnetic-core transformers are nonlinear, we can obtain some useful information about such devices by constructing an ideal model that ignores the nonlinearities.

To show how Z_{ab} changes when $k = 1$ and L_1 and L_2 approach infinity, we first introduce the notation

$$Z_{22} = R_2 + R_L + j(\omega L_2 + X_L) = R_{22} + jX_{22}$$

and then rearrange Eq. (13.50) as follows:

$$Z_{ab} = R_1 + \frac{\omega^2 M^2 R_{22}}{R_{22}^2 + X_{22}^2} + j\left[\omega L_1 - \frac{\omega^2 M^2 X_{22}}{R_{22}^2 + X_{22}^2}\right] = R_{ab} + jX_{ab}. \quad (13.54)$$

At this point, we must be careful with the coefficient of j in Eq. (13.54) because as L_1 and L_2 approach infinity, this coefficient ends up being the difference between two very large quantities. Thus before letting L_1 and L_2 increase, we write the coefficient as

$$X_{ab} = \omega L_1 - \frac{(\omega L_1)(\omega L_2) X_{22}}{(R_{22}^2 + X_{22}^2)} = \omega L_1\left(1 - \frac{\omega L_2 X_{22}}{R_{22}^2 + X_{22}^2}\right), \quad (13.55)$$

where we have used the fact that when $k = 1$, $M^2 = L_1 L_2$. When we put the term inside the bracket in Eq. (13.55) over a common denominator, we get

$$X_{ab} = \omega L_1\left(\frac{R_{22}^2 + \omega L_2 X_L + X_L^2}{R_{22}^2 + X_{22}^2}\right). \quad (13.56)$$

We now factor ωL_2 out of the numerator and denominator of Eq. (13.56) to get

$$X_{ab} = \frac{L_1}{L_2}\left[\frac{X_L + (R_{22}^2 + X_L^2)/\omega L_2}{(R_{22}/\omega L_2)^2 + [1 + (X_L/\omega L_2)]^2}\right]. \quad (13.57)$$

Now we make the important observation that as $k \to 1.0$, the ratio L_1/L_2 approaches the constant value of $(N_1/N_2)^2$. This follows from Eqs. (13.22) and (13.33) because as the coupling becomes extremely tight the two permeances $\mathscr{P}_1$ and $\mathscr{P}_2$ become equal. It follows, then, that Eq. (13.57) reduces to

$$X_{ab} = \left(\frac{N_1}{N_2}\right)^2 X_L \quad (13.58)$$

as $L_1 \to \infty$, $L_2 \to \infty$, and $k \to 1.0$.

The same reasoning leads us to conclude that the reflected resistance in Eq. (13.54) simplifies to

$$\frac{\omega^2 M^2 R_{22}}{R_{22}^2 + X_{22}^2} = \frac{L_1}{L_2} R_{22} = \left(\frac{N_1}{N_2}\right)^2 R_{22} \quad (13.59)$$

Applying the results given by Eqs. (13.58) and (13.59) to Eq. (13.54), we get

$$Z_{ab} = R_1 + \left(\frac{N_1}{N_2}\right)^2 (R_2) + \left(\frac{N_1}{N_2}\right)^2 (R_L + jX_L). \quad (13.60)$$

Equation (13.60) is an important result. It tells us that when the coefficient of coupling approaches unity and the self-inductances of the coupled coils approach infinity, the transformer transfers the secondary winding resistance and the load impedance to the primary side by a scaling factor equal to the turns ratio (N_1/N_2) squared. If we think of lumping the winding resistance R_2 with the load, we can regard this special type of transformer as one that can be used to raise or lower the impedance level of a load. In Section 13.7, we expand on this desirable feature by defining the ideal transformer.

A linear transformer is used to couple a load consisting of a 360-Ω resistor in series with a 0.25-H inductor to a sinusoidal voltage source, as shown in Fig. 13.13. The voltage source has an internal impedance of $184 + j0$ Ω and a maximum voltage of 245.20 V, and is operating at 800 rad/s. The transformer parameters are $R_1 = 100$ Ω, $L_1 = 0.5$ H, $R_2 = 40$ Ω, $L_2 = 0.125$ H, and $k = 0.4$. Calculate (a) the reflected impedance; (b) the primary current; (c) the secondary current; and (d) the average power delivered to the primary terminals of the transformer.

Ans. (a) $10.24 - j7.68$ Ω; (b) $0.5 \cos (800t - 53.13°)$ A; (c) $0.08 \cos 800t$; (d) 13.78 W.

13.7 THE IDEAL TRANSFORMER

The ideal transformer consists of two magnetically coupled coils, having N_1 and N_2 turns, respectively, that exhibit the following three properties:

1. The coefficient of coupling is unity ($k = 1$).
2. The self-inductance of each coil is infinite ($L_1 = L_2 = \infty$),
3. The coil losses are negligible.

Because of these three properties, we can describe the terminal behavior of the ideal transformer in terms of two characteristics. First, the magnitude of the volts per turn will be the same for each coil,

$$\left| \frac{v_1}{N_1} \right| = \left| \frac{v_2}{N_2} \right|, \tag{13.61}$$

and second, the magnitude of the ampere turns will be the same for each coil,

$$|i_1 N_1| = |i_2 N_2|. \tag{13.62}$$

We are forced to use magnitude signs in Eqs. (13.61) and (13.62) because we have not yet established reference polarities for the currents and voltages. We will discuss the removal of the magnitude signs momentarily. Figure 13.15 is used to validate Eqs. (13.61) and (13.62). We show two lossless ($R_1 = R_2 = 0$) magnetically coupled coils. In part (a) coil 2 is open, and in part (b) coil 2 is shorted. Although the analysis that follows is carried out in terms of sinusoidal steady-state operation, the results also apply to instantaneous values of v and i.

Referring to Fig. 13.15(a) we note that the voltage at the terminals of the open-circuit coil will be due entirely to the current in coil 1; therefore

$$\mathbf{V}_2 = j\omega M \mathbf{I}_1. \tag{13.63}$$

The current in coil 1 is

$$\mathbf{I}_1 = \frac{\mathbf{V}_1}{j\omega L_1}. \tag{13.64}$$

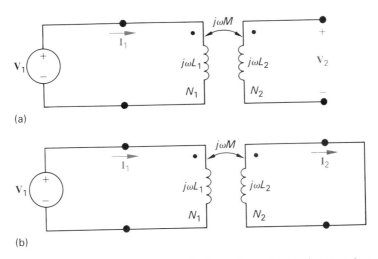

(a)

(b)

Fig. 13.15 The circuits used to verify the volts-per-turn and ampere-turn relationships for the ideal transformer.

It follows directly from Eqs. (13.63) and (13.64) that

$$\mathbf{V}_2 = \frac{M}{L_1} \mathbf{V}_1. \tag{13.65}$$

For unity coupling the mutual inductance equals $\sqrt{L_1 L_2}$ so Eq. (13.65) becomes

$$\mathbf{V}_2 = \sqrt{\frac{L_2}{L_1}} \mathbf{V}_1. \tag{13.66}$$

Now we also note that for unity coupling, the flux linking winding 1 is the same as the flux linking winding 2 so that we need only one permeance to describe the self-inductance of each winding. Thus we can write Eq. (13.66) as

$$\mathbf{V}_2 = \sqrt{\frac{N_2^2 \mathcal{P}}{N_1^2 \mathcal{P}}} \mathbf{V}_1 = \frac{N_2}{N_1} \mathbf{V}_1 \tag{13.67}$$

or

$$\frac{\mathbf{V}_1}{N_1} = \frac{\mathbf{V}_2}{N_2}. \tag{13.68}$$

When we sum the voltages around the shorted coil of Fig. 13.15(b), we get the expression

$$0 = -j\omega M \mathbf{I}_1 + j\omega L_2 \mathbf{I}_2, \tag{13.69}$$

from which we have for $k = 1$

$$\frac{\mathbf{I}_1}{\mathbf{I}_2} = \frac{L_2}{M} = \frac{L_2}{\sqrt{L_1 L_2}} = \sqrt{\frac{L_2}{L_1}} = \frac{N_2}{N_1}. \tag{13.70}$$

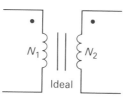

Fig. 13.16 The graphical symbol for an ideal transformer.

Equation (13.70) is equivalent to writing

$$\mathbf{I}_1 N_1 = \mathbf{I}_2 N_2. \tag{13.71}$$

The graphical symbol for the ideal transformer is shown in Fig. 13.16. The vertical bars are used to imply that an ideal transformer can be approximated by coils wound on ferromagnetic cores. These cores frequently consist of laminated sheets of magnetic material and the vertical bars symbolize the laminated construction.

There are several reasons why coils wrapped on a ferromagnetic core can be approximated reasonably well by an ideal transformer. The ferromagnetic material creates a space with very high permeance. This means that most of the magnetic flux is trapped inside the core material; consequently very tight magnetic coupling is established between coils that share the same core. The high permeance also means very high self-inductance, since $L = N^2 \mathscr{P}$. Finally, ferromagnetically coupled coils are very efficient in terms of power transfer from one coil to the other. Efficiencies in excess of 95% are common, so neglecting losses is not a crippling approximation for many applications.

Let us now turn our attention to the removal of the magnitude signs from Eqs. (13.61) and (13.62). Note that magnitude signs did not show up in the derivation of Eq. (13.68) or Eq. (13.71). They were not needed there because we had established reference polarities for voltages and reference directions for currents and, in addition, we knew the magnetic polarity dots of the two coupled coils. The rules for assigning the proper algebraic sign to Eqs. (13.61) and (13.62) are as follows.

1. If the coil voltages v_1 and v_2 are *both* positive or negative at the dot-marked terminal, use a plus sign in Eq. (13.61). Otherwise, use a negative sign.

2. If the coil currents i_1 and i_2 are *both* directed into or out of the dot-marked terminal, use a minus sign in Eq. (13.62). Otherwise, use a plus sign.

These rules are illustrated by the four circuits shown in Fig. 13.17.

It should be apparent at this point in our discussion of the ideal transformer that the ratio of turns on the two windings is an important parameter of the transformer. The turns ratio can be defined as the ratio N_1/N_2 or N_2/N_1. Both ratios appear in the literature. In this text, we will use "a" to denote the ratio N_2/N_1; thus

$$a = N_2/N_1. \tag{13.72}$$

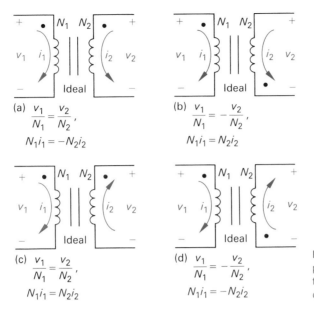

(a) $\dfrac{v_1}{N_1} = \dfrac{v_2}{N_2}$,

$N_1 i_1 = -N_2 i_2$

(b) $\dfrac{v_1}{N_1} = -\dfrac{v_2}{N_2}$,

$N_1 i_1 = N_2 i_2$

(c) $\dfrac{v_1}{N_1} = \dfrac{v_2}{N_2}$,

$N_1 i_1 = N_2 i_2$

(d) $\dfrac{v_1}{N_1} = -\dfrac{v_2}{N_2}$,

$N_1 i_1 = -N_2 i_2$

Fig. 13.17 Circuits that show the proper algebraic signs for relating the terminal voltages and terminal currents of the ideal transformer.

The turns ratio of an ideal transformer can be given on the circuit symbol in several ways, three of which are shown in Fig. 13.18. In part (a), the number of turns in each coil is given explicitly. In part (b), we are told that the ratio N_2/N_1 is 5 to 1 and in part (c) N_2/N_1 is 1 to $\frac{1}{5}$.

Example 13.3 illustrates the analysis of a circuit containing an ideal transformer.

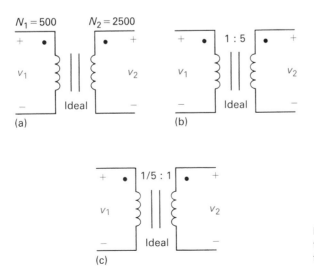

Fig. 13.18 Three ways to show that the turns ratio of an ideal transformer is equal to 5.

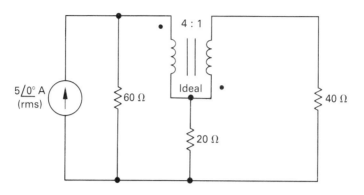

Fig. 13.19 The circuit for Example 13.3.

Example 13.3 a) Find the average power delivered by the sinusoidal current source in the circuit in Fig. 13.19.

b) Find the average power delivered to the 20-Ω resistor.

Solution a) The presence of the ideal transformer encourages the use of the mesh-current method of analysis. Therefore we begin by transforming the current source to an equivalent voltage source. The transformation is shown in Fig. 13.20, where we have also defined the phasor mesh currents and transformer terminal voltages that are used in the solution. Summing the voltages around meshes 1 and 2 generates the following set of equations:

$$300 = 60\mathbf{I}_1 + \mathbf{V}_1 + 20(\mathbf{I}_1 - \mathbf{I}_2),$$
$$0 = 20(\mathbf{I}_2 - \mathbf{I}_1) + \mathbf{V}_2 + 40\mathbf{I}_2.$$

In addition to these two mesh-current equations, we need the constraint equations imposed by the ideal transformer. From the circuit diagram we have $N_2/N_1 = 1/4$. In terms of the currents and voltages defined at the ter-

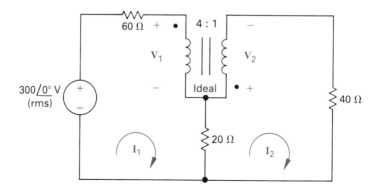

Fig. 13.20 The circuit for the solution of Example 13.3.

minals of the ideal transformer, we have

$$V_2 = \frac{1}{4} V_1$$

and

$$I_2 = -4I_1.$$

We now have four equations and four unknowns. The solutions for V_1, V_2, I_1, and I_2 are as follows:

$$V_1 = 260 \text{ V rms}, \qquad I_1 = 0.25 \text{ A rms},$$
$$V_2 = 65 \text{ V rms}, \qquad I_2 = -1.0 \text{ A rms}.$$

The voltage across the 5-A current source in the circuit in Fig. 13.19 is

$$V_{5A} = V_1 + 20(I_1 - I_2)$$
$$= 260 + 20[0.25 - (-1)]$$
$$= 285 \text{ V rms}.$$

Note that V_{5A} is a voltage rise in the direction of the source current. The average power associated with the current source is

$$P = -(285)(5) = -1425 \text{ W}.$$

The minus sign tells us that the source is delivering power to the circuit.

b) To find the average power delivered to the 20-Ω resistor, we first calculate the current in the resistor. It follows from the circuit shown in Fig. 13.20 that the current oriented down through the 20-Ω resistor is

$$I_{20} = I_1 - I_2 = 0.25 - (-1)$$
$$= 1.25 \text{ A rms}.$$

Therefore the average power dissipated in the resistor is

$$P_{20\,\Omega} = (1.25)^2(20) = 31.25 \text{ W}. \quad ■$$

We implied at the end of Section 13.6 that the ideal transformer can be used to raise or lower the impedance level of a load. The circuit shown in Fig. 13.21 confirms this fact. The impedance seen by the voltage source V_s is V_1/I_1. The voltage and current at the terminals of the load impedance (V_2 and I_2) are

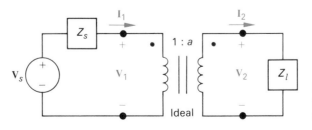

Fig. 13.21 Using an ideal transformer to couple a load to a source.

related to $\mathbf{V}_1$ and $\mathbf{I}_1$ by the transformer turns ratio; thus

$$\mathbf{V}_1 = \mathbf{V}_2/a \tag{13.73}$$

and

$$\mathbf{I}_1 = a\mathbf{I}_2. \tag{13.74}$$

Therefore the impedance seen by the source is

$$Z_{\text{IN}} = \frac{\mathbf{V}_1}{\mathbf{I}_1} = \frac{1}{a^2}\frac{\mathbf{V}_2}{\mathbf{I}_2}. \tag{13.75}$$

But the ratio $\mathbf{V}_2/\mathbf{I}_2$ is the load impedance Z_l; hence Eq. (13.75) becomes

$$Z_{\text{IN}} = \frac{1}{a^2}Z_l. \tag{13.76}$$

Note that the ideal transformer changes the magnitude of Z_l but does not affect its phase angle. Whether Z_{IN} is greater or less than Z_l depends on the turns ratio a.

The ideal transformer—or its practical counterpart, the ferromagnetic core transformer—can be used to match the magnitude of Z_l to the magnitude of Z_s, thus improving the amount of average power transferred from the source to the load.

DRILL EXERCISE 13.7

Make the following changes in the ideal transformer in the circuit of Fig. 13.19: (1) Place the dot on the right-hand side of the transformer at the upper terminal, and (2) change the turns ratio from 4:1 to 3:1. Calculate the average power delivered to the 20-Ω resistor.

Ans. 28.8 W.

DRILL EXERCISE 13.8

Find the average power delivered to the 4-kΩ resistor in the circuit shown.

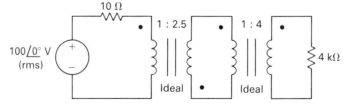

Ans. 160 W.

DRILL EXERCISE 13.9

The ideal transformer connected to the 4-kΩ load in Drill Exercise 13.8 is replaced with an ideal transformer that has a turns ratio of $1:a$.

a) What value of a will result in maximum average power being delivered to the 4-kΩ resistor?

b) What is the maximum average power?

Ans. (a) 8; (b) 250 W.

13.8 EQUIVALENT CIRCUITS FOR MAGNETICALLY COUPLED COILS

There are times when it is convenient to model magnetically coupled coils with an equivalent circuit that does not involve magnetic coupling. Consider the two magnetically coupled coils shown in Fig. 13.22. The resistances R_1 and R_2 represent the winding resistance of each coil. Our goal is to replace the magnetically coupled coils inside the shaded area with a set of inductors that are not magnetically coupled. Before deriving the equivalent circuits, we must point out an important restriction, namely, that the voltage between terminals b and d must be zero. In other words, if terminals b and d can be shorted together without disturbing the voltages and currents in the original circuit in which the magnetically coupled coils are embedded, then the equivalent circuits that are derived in the material that follows can be used to model the coils (see Problem 13.23).

We begin the development of the circuit models by writing the two equations that relate the terminal voltages v_1 and v_2 to the terminal currents i_1 and i_2. For the given references and polarity dots, we have

$$v_1 = L_1 \frac{di_1}{dt} + M \frac{di_2}{dt} \tag{13.77}$$

and

$$v_2 = M \frac{di_1}{dt} + L_2 \frac{di_2}{dt}. \tag{13.78}$$

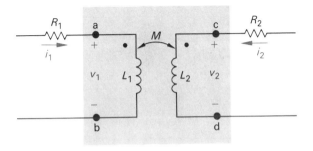

Fig. 13.22   The circuit used to develop an equivalent circuit for magnetically coupled coils.

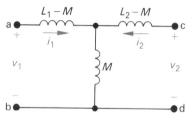

Fig. 13.23 The T-equivalent circuit for the magnetically coupled coils in Fig. 13.22.

In order to arrive at an equivalent circuit for these two magnetically coupled coils, we seek an arrangement of inductors that can be described by a set of
equations equivalent to Eqs. (13.77) and (13.78). The key to finding the
arrangement is to regard Eqs. (13.77) and (13.78) as mesh-current equations
with i_1 and i_2 as the mesh variables. From this vantage point, we see that we
need one mesh with a total inductance of L_1 henries and a second mesh with a
total inductance of L_2 henries. Furthermore, the two meshes must have a common inductance of M henries. The T-arrangement of coils shown in Fig. 13.23
satisfies these requirements. We leave it to the reader to verify that the equations that relate v_1 and v_2 to i_1 and i_2 in the circuit in Fig. 13.23 reduce to Eqs.
(13.77) and (13.78). It is important to note that there is no magnetic coupling
between the inductors in the circuit in Fig. 13.23 and that the voltage between
b and d is zero.

When we use the T-equivalent circuit in Fig. 13.23 to model the magnetically coupled coils in Fig. 13.22, the equivalent circuit that includes the
winding resistances R_1 and R_2 appears as shown in Fig. 13.24.

It is possible to derive a π-equivalent circuit for the magnetically coupled
coils in Fig. 13.22. The derivation of the π-equivalent circuit is based on
solving Eqs. (13.77) and (13.78) for the derivatives di_1/dt and di_2/dt and then
regarding the resulting expressions as a pair of node-voltage equations. Using
Cramer's method for solving simultaneous equations, we find that the expressions for di_1/dt and di_2/dt are

$$\frac{di_1}{dt} = \frac{\begin{vmatrix} v_1 & M \\ v_2 & L_2 \end{vmatrix}}{\begin{vmatrix} L_1 & M \\ M & L_2 \end{vmatrix}} = \frac{L_2}{L_1 L_2 - M^2}\, v_1 - \frac{M}{L_1 L_2 - M^2}\, v_2, \tag{13.79}$$

$$\frac{di_2}{dt} = \frac{\begin{vmatrix} L_1 & v_1 \\ M & v_2 \end{vmatrix}}{L_1 L_2 - M^2} = \frac{-M}{L_1 L_2 - M^2}\, v_1 + \frac{L_1}{L_1 L_2 - M^2}\, v_2. \tag{13.80}$$

Now we solve for i_1 and i_2 by multiplying both sides of Eqs. (13.79) and (13.80)
by dt and then integrating. We get

$$i_1 = i_1(0) + \frac{L_2}{L_1 L_2 - M^2} \int_0^t v_1 \, d\tau - \frac{M}{L_1 L_2 - M^2} \int_0^t v_2 \, d\tau \tag{13.81}$$

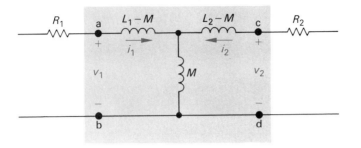

Fig. 13.24   An equivalent circuit that includes the winding resistances R_1 and R_2.

and

$$i_2 = i_2(0) - \frac{M}{L_1 L_2 - M^2} \int_0^t v_1 \, d\tau + \frac{L_1}{L_1 L_2 - M^2} \int_0^t v_2 \, d\tau. \qquad \textbf{(13.82)}$$

If we regard v_1 and v_2 as node voltages, Eqs. (13.81) and (13.82) describe a circuit of the form shown in Fig. 13.25.

All that remains to be done at this point in the derivation of the π-equivalent circuit is to find L_A, L_B, and L_C as functions of L_1, L_2, and M. This is easily done by writing the equations for i_1 and i_2 in the circuit in Fig. 13.25 and then comparing them with Eqs. (13.81) and (13.82). Thus

$$i_1 = i_1(0) + \frac{1}{L_A} \int_0^t v_1 \, d\tau + \frac{1}{L_B} \int_0^t (v_1 - v_2) \, d\tau$$

$$= i_1(0) + \left(\frac{1}{L_A} + \frac{1}{L_B} \right) \int_0^t v_1 \, d\tau - \frac{1}{L_B} \int_0^t v_2 \, d\tau, \qquad \textbf{(13.83)}$$

$$i_2 = i_2(0) + \frac{1}{L_C} \int_0^t v_2 \, d\tau + \frac{1}{L_B} \int_0^t (v_2 - v_1) \, d\tau$$

$$= i_2(0) - \frac{1}{L_B} \int_0^t v_1 \, d\tau + \left(\frac{1}{L_B} + \frac{1}{L_C} \right) \int_0^t v_2 \, d\tau. \qquad \textbf{(13.84)}$$

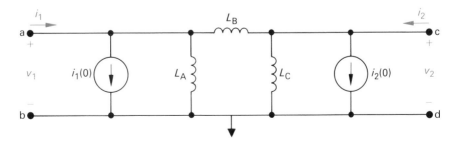

Fig. 13.25   The circuit used to derive the π-equivalent circuit for magnetically coupled coils.

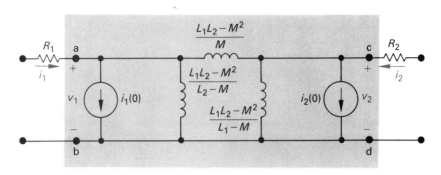

Fig. 13.26   The π-equivalent circuit for the magnetically coupled coils in Fig. 13.22.

If follows directly that

$$\frac{1}{L_{\mathrm{B}}} = \frac{M}{L_1 L_2 - M^2},$$ (13.85)

$$\frac{1}{L_{\mathrm{A}}} = \frac{L_2 - M}{L_1 L_2 - M^2},$$ (13.86)

$$\frac{1}{L_{\mathrm{C}}} = \frac{L_1 - M}{L_1 L_2 - M^2}.$$ (13.87)

When we incorporate Eqs. (13.85)–(13.87) into the circuit in Fig. 13.25, the π-equivalent circuit for the magnetically coupled coils in Fig. 13.22 is as shown in Fig. 13.26.

It is of interest to note that the initial values of i_1 and i_2 appear explicitly in the π-equivalent circuit but are implicit in the T-equivalent circuit. Since in this chapter we are focusing on the sinusoidal steady-state behavior of circuits containing mutual inductance, we can assume that the initial values of i_1 and i_2 are zero and thus eliminate the current sources in the π-equivalent circuit. Thus for sinusoidal steady-state analysis, the circuit in Fig. 13.26 simplifies to that shown in Fig. 13.27.

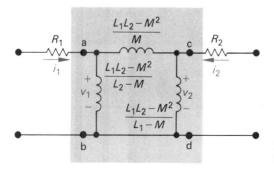

Fig. 13.27   The π-equivalent circuit used for sinusoidal steady-state analysis.

The mutual inductance carries its own algebraic sign in the T- and π-equivalent circuits. That is, if the magnetic polarity of the coupled coils is reversed from that given in Fig. 13.22, then the algebraic sign of M reverses. A reversal in magnetic polarity requires moving one polarity dot without changing the reference polarities of the terminal currents and voltages. The following numerical example illustrates the application of the T-equivalent circuit.

Example 13.4   a) Use the T-equivalent circuit for the magnetically coupled coils in Example 13.2 to find the phasor currents $\mathbf{I}_1$ and $\mathbf{I}_2$. The phasor currents $\mathbf{I}_1$ and $\mathbf{I}_2$ are defined in the circuit in Fig. 13.14.

b) Repeat part (a) given that the polarity dot on the secondary winding is moved to the lower terminal.

Solution   a) For the given polarity dots in Fig. 13.14, M carries a value of $+3$ H in the T-equivalent circuit. Therefore the three inductances in the equivalent circuit are

$$L_1 - M = 9 - 3 = 6 \text{ H},$$
$$L_2 - M = 4 - 3 = 1 \text{ H},$$
$$M = 3 \text{ H}.$$

Thus the T-equivalent circuit is as shown in Fig. 13.28. The phasor domain equivalent circuit at a frequency of 400 rad/s is shown in Fig. 13.29. The phasor domain circuit for the original system is shown in Fig. 13.30. Here the magnetically coupled coils are modeled by the circuit in Fig. 13.29. To find the phasor currents $\mathbf{I}_1$ and $\mathbf{I}_2$, we first find the node voltage across the

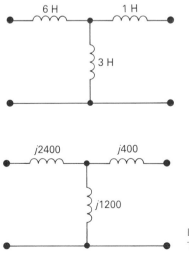

Fig. 13.28   The T-equivalent circuit for the magnetically coupled coils in Example 13.2.

Fig. 13.29   The phasor domain model of the T-equivalent circuit at 400 rad/s.

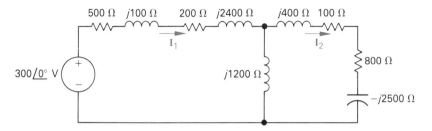

Fig. 13.30   The circuit in Fig. 13.14 with the magnetically coupled coils replaced by their T-equivalent circuit.

1200-Ω inductive reactance. If we use the lower node as the reference, the single node-voltage equation is

$$\frac{V - 300}{700 + j2500} + \frac{V}{j1200} + \frac{V}{900 - j2100} = 0.$$

Solving for V yields

$$V = 136 - j8 = 136.24 \underline{/-3.37^\circ} \text{ V rms.}$$

It follows directly that

$$\mathbf{I}_1 = \frac{300 - (136 - j8)}{700 + j2500} = 63.25 \underline{/-71.57^\circ} \text{ mA rms}$$

and

$$\mathbf{I}_2 = \frac{136 - j8}{900 - j2100} = 59.63 \underline{/63.43^\circ} \text{ mA rms.}$$

When these solutions are checked against those obtained in Example 13.2, we see that they are identical.

b) When the polarity dot is moved to the lower terminal of the secondary coil, M carries a value of -3 H in the T-equivalent circuit. Before carrying out the solution with the new T-equivalent circuit, let us pause long enough to note from Example 13.2 that reversing the algebraic sign of M will have no effect on the solution for $\mathbf{I}_1$ and will shift $\mathbf{I}_2$ by precisely 180°. Therefore we anticipate that the solutions for $\mathbf{I}_1$ and $\mathbf{I}_2$ will be

$$\mathbf{I}_1 = 63.25 \underline{/-71.57^\circ} \text{ mA rms,}$$
$$\mathbf{I}_2 = 59.63 \underline{/-116.57^\circ} \text{ mA rms.}$$

Let us now proceed to find these solutions using the new T-equivalent circuit. With M equal to -3 H, the three inductances in the equivalent circuit are

$$L_1 - M = 9 - (-3) = 12 \text{ H,}$$
$$L_2 - M = 4 - (-3) = 7 \text{ H,}$$
$$M = -3 \text{ H.}$$

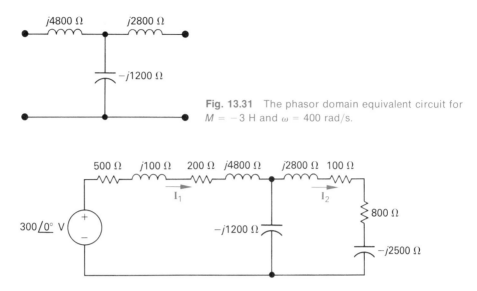

Fig. 13.31 The phasor domain equivalent circuit for $M = -3$ H and $\omega = 400$ rad/s.

Fig. 13.32 The phasor domain equivalent circuit for Example 13.4(b).

At an operating frequency of 400 rad/s, the phasor domain equivalent circuit requires two inductors and a capacitor. The circuit is given in Fig. 13.31. The resulting phasor domain circuit for the original system is shown in Fig. 13.32. As before, we first find the node voltage across the center branch, which in this case is a capacitive reactance of $-j1200$ Ω. If we use the lower node as reference, the node-voltage equation is

$$\frac{\mathbf{V} - 300}{700 + j4900} + \frac{\mathbf{V}}{-j1200} + \frac{\mathbf{V}}{900 + j300} = 0.$$

Solving for $\mathbf{V}$ gives us

$$\mathbf{V} = -8 - j56 = 56.57 \underline{/-98.13°} \text{ V rms.}$$

It follows that

$$\mathbf{I}_1 = \frac{300 - (-8 - j56)}{700 + j4900} = 63.25 \underline{/-71.57°} \text{ mA rms}$$

and

$$\mathbf{I}_2 = \frac{-8 - j56}{900 + j300} = 59.63 \underline{/-116.57°} \text{ mA rms.} \blacksquare$$

DRILL EXERCISE
13.10

a) Show that if the reference direction for i_2 is reversed in both the circuits in Fig. 13.22 and 13.23 the T-equivalent circuit in Fig. 13.23 is still valid.

b) Show that if the dot on the L_2 coil in the circuit in Fig. 13.22 is moved to the lower terminal, the three inductors in the circuit in Fig. 13.23 are $L_1 + M$, $L_2 + M$, and $-M$.

c) Will the circuit derived in part (b) be valid if the reference direction for i_2 is reversed in both circuits?

Ans. (a) Show; (b) show; (c) yes.

DRILL EXERCISE 13.11 Use the T-equivalent circuit for the linear transformer in Drill Exercise 13.6 to find $\mathbf{I}_1$ and $\mathbf{I}_2$. The phasor currents $\mathbf{I}_1$ and $\mathbf{I}_2$ are defined in Fig. 13.13.

Ans. $\mathbf{I}_1 = 500 \,\underline{/-53.13°}$ mA; $\mathbf{I}_2 = 80 \,\underline{/0°}$ mA.

13.9 THE NEED FOR IDEAL TRANSFORMERS IN THE EQUIVALENT CIRCUITS

It is possible for the inductors in the T- and π-equivalent circuits of magnetically coupled coils to have negative values. For example, if $L_1 = 3$ mH, $L_2 = 12$ mH, and $M = 5$ mH, the T-equivalent circuit requires an inductor of -2 mH and the π-equivalent circuit requires an inductor of -4.5 mH. These negative values of inductance are not troublesome when using the equivalent circuits in computations. However, if the equivalent circuits are to be built using circuit components, the negative inductors can be irksome. The aggravation comes from the fact that whenever the frequency of the sinusoidal source changes, the capacitor used to simulate the negative reactance must be changed. For example, at a frequency of 50 krad/s, a -2-mH inductor has an impedance of $-j100 \ \Omega$. This impedance can be modeled with a capacitor having a capacitance of 0.2 μF. If the frequency changes to 25 krad/s, the -2-mH inductor impedance changes to $-j50 \ \Omega$. At 25 krad/s this requires a capacitor with a capacitance of 0.8 μF. Obviously in a situation where the frequency is varied continuously, the use of a capacitor to simulate negative inductance is practically worthless.

The problem of dealing with negative inductances can be circumvented by the introduction of an ideal transformer into the equivalent circuit. This doesn't mean that the modeling problem has been completely solved because ideal transformers can only be approximated. However, in some situations the approximation to an ideal transformer is good enough to warrant a discussion of using the ideal transformer in the T- and π-equivalent circuits of magnetically coupled coils.

There are two different ways to use an ideal transformer in either the T-equivalent or the π-equivalent circuit for magnetically coupled coils. The two possible arrangements for each type of equivalent circuit are shown in Fig. 13.33.

To verify any of the equivalent circuits in Fig. 13.33, we need only show that for any circuit the equations relating v_1 and v_2 to di_1/dt and di_2/dt are

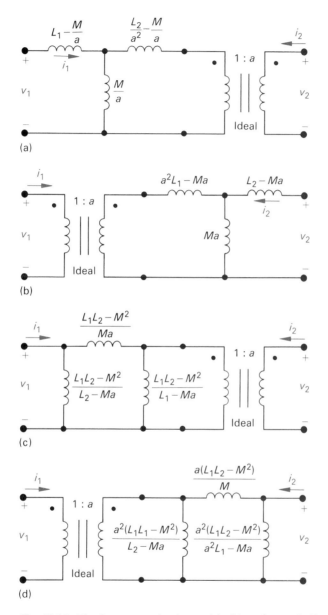

Fig. 13.33   The four ways of using an ideal transformer in the T- and π-equivalent circuit for magnetically coupled coils.

identical to Eqs. (13.77) and (13.78). We will verify the validity of the circuit shown in Fig. 13.33(a) and leave the circuits in Fig. 13.33(b), (c), and (d) as exercises. To facilitate the discussion, we have redrawn the circuit in Fig. 13.33(a) in Fig. 13.34 and added the variables i_0 and v_0. It follows directly from

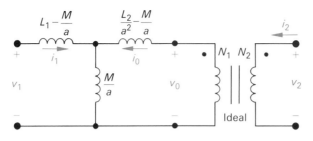

Fig. 13.34 The circuit in Fig. 13.33 with i_0 and v_0 defined.

this circuit that

$$v_1 = \left(L_1 - \frac{M}{a}\right)\frac{di_1}{dt} + \frac{M}{a}\frac{d}{dt}(i_1 + i_0), \tag{13.88}$$

$$v_0 = \left(\frac{L_2}{a^2} - \frac{M}{a}\right)\frac{di_0}{dt} + \frac{M}{a}\frac{d}{dt}(i_0 + i_1). \tag{13.89}$$

The ideal transformer imposes the following constraints on v_0 and i_0:

$$v_0 = v_2/a, \tag{13.90}$$

and

$$i_0 = ai_2. \tag{13.91}$$

When we substitute Eqs. (13.90) and (13.91) into Eqs. (13.88) and (13.89), we get

$$v_1 = L_1\frac{di_1}{dt} + \frac{M}{a}\frac{d}{dt}(ai_2), \tag{13.92}$$

$$\frac{v_2}{a} = \frac{L_2}{a^2}\frac{d}{dt}(ai_2) + \frac{M}{a}\frac{di_1}{dt}. \tag{13.93}$$

It follows from Eqs. (13.92) and (13.93) that

$$v_1 = L_1\frac{di_1}{dt} + M\frac{di_2}{dt}, \tag{13.94}$$

$$v_2 = M\frac{di_1}{dt} + L_2\frac{di_2}{dt}. \tag{13.95}$$

Equations (13.94) and (13.95) are identical to Eqs. (13.77) and (13.78); thus insofar as terminal behavior is concerned, the circuit in Fig. 13.34 is equivalent to the magnetically coupled coils inside the box of Fig. 13.22.

In showing that the circuit in Fig. 13.34 is equivalent to the magnetically coupled coils in Fig. 13.22, we placed *no restrictions* on the turns ratio a. Therefore there is an infinite number of equivalent circuits possible. Furthermore, we can always find a turns ratio that will make all the inductances positive.

Three values of a are of particular interest:

$$a = M/L_1,$$
(13.96)

$$a = L_2/M,$$
(13.97)

and

$$a = \sqrt{L_2/L_1}.$$
(13.98)

The value of a given by Eq. (13.96) will eliminate the inductances $L_1 - M/a$ and $a^2L_1 - aM$ from the T-equivalent circuits and the inductances $[(L_1L_2 - M^2)/(a^2L_1 - aM)]$ and $[a^2(L_1L_2 - M^2)/(a^2L_1 - aM)]$ from the π-equivalent circuits.

The value of a given by Eq. (13.97) will eliminate the inductances $[(L_2/a^2) - (M/a)]$ and $(L_2 - aM)$ from the T-equivalent circuits and the inductances $[(L_1L_2 - M^2)/(L_2 - aM)]$ and $[a^2(L_1L_2 - M^2)/(L_2 - aM)]$ from the π-equivalent circuits.

Also note that when a is chosen to equal M/L_1, the circuits in Fig. 13.33(a) and (c) become identical, and when a is chosen to equal L_2/M, the circuits in Fig. 13.33(b) and (d) become identical. These observations are summarized in Figs. 13.35 and 13.36, respectively. In deriving the expressions for the inductances there, we have made use of the relationship $M = k\sqrt{L_1L_2}$. By expressing the inductances as functions of the self-inductances L_1 and L_2 and the

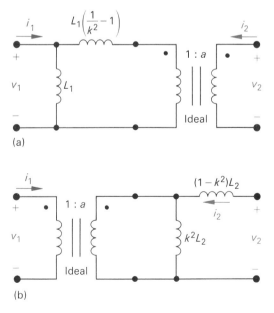

Fig. 13.35   Two equivalent circuits when $a = M/L_1$.

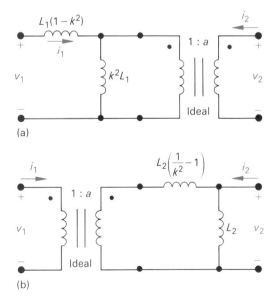

Fig. 13.36   Two equivalent circuits when $a = L_2/M$.

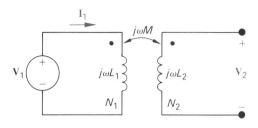

Fig. 13.37 Experimental determination of the ratio M/L_1.

coefficient of coupling k, we see that the values of a given by Eqs. (13.96) and (13.97) not only reduce the number of inductances needed in the equivalent circuit, but also guarantee that all the inductances will be positive. We will leave it to the reader to investigate the consequences of choosing the value of a given by Eq. (13.98).

The values of a given by Eqs. (13.96), (13.97), and (13.98) can be determined experimentally. To obtain the ratio M/L_1, the coil designated as having N_1 turns is driven by a sinusoidal voltage source. The source frequency is set high enough so that $\omega L_1 \gg R_1$, and the N_2 coil is left open. The arrangement is shown in Fig. 13.37.

With the N_2 coil open, we have

$$\mathbf{V}_2 = j\omega M \mathbf{I}_1. \tag{13.99}$$

Now since $j\omega L_1 \gg R_1$ the current $\mathbf{I}_1$ is

$$\mathbf{I}_1 = \mathbf{V}_1/j\omega L_1. \tag{13.100}$$

When we substitute Eq. (13.100) into Eq. (13.99), we obtain

$$\left(\frac{\mathbf{V}_2}{\mathbf{V}_1}\right)_{I_2=0} = \frac{M}{L_1}, \tag{13.101}$$

in which we note that the ratio M/L_1 is the terminal voltage ratio corresponding to coil 2 being open, that is, $\mathbf{I}_2 = 0$.

The ratio L_2/M can be obtained by reversing the procedure. That is, coil 2 is energized and coil 1 is left open. Then

$$\frac{L_2}{M} = \left(\frac{\mathbf{V}_2}{\mathbf{V}_1}\right)_{I_1=0}. \tag{13.102}$$

Finally we observe that the value of a given by Eq. (13.98) is the geometric mean of these two voltage ratios; thus

$$\sqrt{\left(\frac{\mathbf{V}_2}{\mathbf{V}_1}\right)_{I_2=0} \cdot \left(\frac{\mathbf{V}_2}{\mathbf{V}_1}\right)_{I_1=0}} = \sqrt{\frac{M}{L_1} \cdot \frac{L_2}{M}} = \sqrt{\frac{L_2}{L_1}}. \tag{13.103}$$

This is a good place to point out that for coils wound on nonmagnetic cores the voltage ratio is *not* the same as the turns ratio, as it very nearly is for coils wound on ferromagnetic cores. Using the fact that the self-inductances vary as the square of the number of turns, we can see from Eq. (13.103) that the turns ratio is approximately equal to the geometric mean of the two voltage ratios:

$$\sqrt{\frac{L_2}{L_1}} = \frac{N_2}{N_1} = \sqrt{\left(\frac{\mathbf{V}_2}{\mathbf{V}_1}\right)_{I_2=0} \cdot \left(\frac{\mathbf{V}_2}{\mathbf{V}_1}\right)_{I_1=0}}.$$ (13.104)

DRILL EXERCISE
13.12

The circuit shown is the equivalent circuit of a lossless linear transformer. Compute (a) L_1, (b) L_2, (c) M, and (d) k.

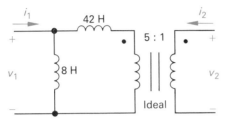

Ans. (a) 8 H; (b) 2 H; (c) 1.6 H; (d) 0.4.

13.10 SUMMARY

Mutual inductance is the circuit parameter that relates the voltage induced in one winding to a changing current in another winding when the two windings are linked by a common magnetic field. The degree of magnetic coupling is described quantitatively by the coefficient of coupling k. The coefficient of coupling is zero for coils that are not linked by a common magnetic flux and is unity for coils that are both completely linked by the same magnetic flux. The mutual inductance is related to the self-inductances (L_1 and L_2) of the two windings through the coefficient of coupling, namely, $M = k\sqrt{L_1 L_2}$. Thus the mutual inductance can never be greater than the geometric mean of the self-inductances.

Magnetic coupling is used advantageously in engineering applications by both nonferromagnetic and ferromagnetic structures. Circuit models for coils wound on nonferromagnetic cores use the parameters of mutual inductance and self-inductance to predict the terminal behavior of the coupled coils. The linear transformer is a good example of a device using a nonferromagnetic core.

Circuit models for coils wound on ferromagnetic cores use the turns ratio along with the volts per turn and the ampere turn balance imposed by the ideal magnetic coupling. Thus mutual inductance does not appear explicitly in the description of the ideal transformer.

T- and π-equivalent circuits for magnetically coupled coils, with and without ideal transformers, enable us to represent magnetically coupled coils by coils with no magnetic coupling. When ideal transformers are used in the T-

and π-equivalent circuits, there is an infinite number of equivalent circuits possible. The ideal transformer can also be used to eliminate the need for negative inductors that are troublesome in laboratory circuit models in which the frequency is to be varied.

PROBLEMS

13.1 Two magnetically coupled coils have self-inductances of 27 mH and 3 mH, respectively. The mutual inductance between the coils is 7.2 mH.

a) What is the coefficient of coupling?
b) For these two coils, what is the largest value that M can have?
c) Assume that the physical structure of these coupled coils is such that $\mathscr{P}_1 = \mathscr{P}_2$. What is the turns ratio N_1/N_2 if N_1 is the number of turns on the 27-mH coil?

13.2 The physical construction of four pairs of magnetically coupled coils is shown in Fig. 13.38. Assume that the magnetic flux is confined to the core material in each structure. Show two possible locations for the dot markings on each pair of coils.

13.3 The polarity markings on two coils is to be determined experimentally. The experimental setup is shown in Fig. 13.39. Assume that the terminal connected to the negative terminal of the battery has been given a polarity mark as shown. When the switch is *opened,* the dc voltmeter "kicks" up-scale. Where should the polarity mark be placed on the coil connected to the voltmeter?

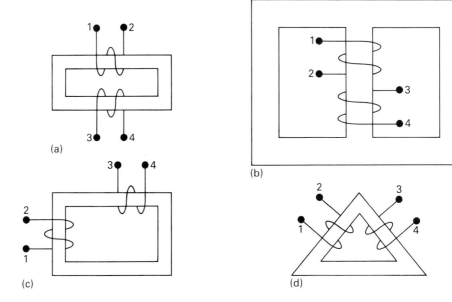

Fig. 13.38 The coupled coils for Problem 13.2.

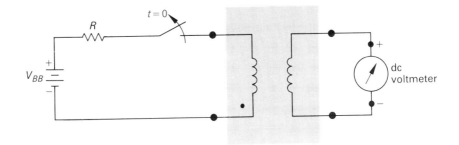

Fig. 13.39 The circuit for Problem 13.3.

13.4 a) Starting with Eq. (13.27), show that the coefficient of coupling can also be expressed as

$$k = \sqrt{\left(\frac{\phi_{21}}{\phi_1}\right)\left(\frac{\phi_{12}}{\phi_2}\right)}.$$

b) On the basis of the fractions (ϕ_{21}/ϕ_1) and (ϕ_{12}/ϕ_2), explain why k is less than 1.0.

13.5 a) Show that the two coupled coils in Fig. 13.40 can be replaced by a single coil having an inductance of $L_{ab} = L_1 + L_2 + 2M$. (*Hint:* Express v_{ab} as a function of i_{ab}.)

b) Show that if the connections to the terminals of the coil labeled L_2 are reversed, $L_{ab} = L_1 + L_2 - 2M$.

13.6 a) Show that the two magnetically coupled coils in Fig. 13.41 can be replaced by a single coil having an inductance of

$$L_{ab} = \frac{L_1 L_2 - M^2}{L_1 + L_2 - 2M}.$$

(*Hint:* Let i_1 and i_2 be clockwise mesh currents in the left and right "windows" of Fig. 13.41, respectively. Sum the voltages around the two meshes. In mesh 1 let v_{ab} be the unspecified applied voltage. Solve for di_1/dt as a function of v_{ab}.)

b) Show that if the magnetic polarity of coil 2 is reversed, then

$$L_{ab} = \frac{L_1 L_2 - M^2}{L_1 + L_2 + 2M}.$$

Fig. 13.40 The coupled coils for Problem 13.5.

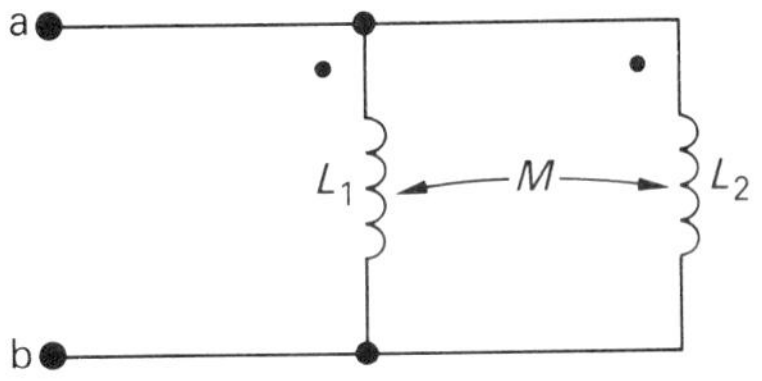

Fig. 13.41 The coupled coils for Problem 13.6.

13.7 A 3000-Ω resistor is connected to a sinusoidal voltage source by a linear transformer. The internal impedance of the source is $10.44 + j13.92 \ \Omega$ when the source is operating at a frequency of 1600 rad/s. The rms voltage at the terminals of the source is 15 V when the source is not loaded. The parameters of the linear transformer are primary winding $R_1 = 13.12 \ \Omega$; $L_1 = 30$ mH; secondary winding $R_2 = 600 \ \Omega$; $L_2 = 3$ H; mutual inductance of the transformer $= 75$ mH.

  a) What is the numerical value of k?
  b) What is the value of the impedance reflected into the primary?
  c) What is the value of the impedance seen by the source?
  d) What is the rms magnitude of the voltage across the 3-kΩ resistor?
  e) What percentage of the average power developed by the source is delivered to the input of the transformer?
  f) What percentage of the average power developed by the source is delivered to the 3-kΩ load resistor?

13.8 The sinusoidal voltage source in the circuit in Fig. 13.42 is operating at a frequency of 75 krad/s. The coefficient of coupling is adjusted until the amplitude of i_1 is maximum.

  a) What is the value of k?
  b) What is the maximum value of i_1 if the rms value of the internal source voltage is 202 V?

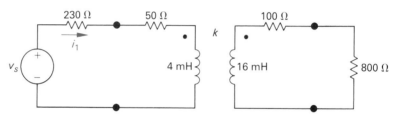

Fig. 13.42   The circuit for Problem 13.8.

13.9 In the circuit in Fig. 13.43 the frequency of the sinusoidal source is 4000 rad/s and the rms value of the internal voltage is 45 V.

  a) Find the Thévenin equivalent circuit with respect to the terminals c, d.
  b) What load impedance would result in maximum average power being transferred to the terminals c, d?
  c) What is the maximum average power that can be transferred?
  d) What percentage of the average power delivered to the transformer is delivered to the load in part (c)?

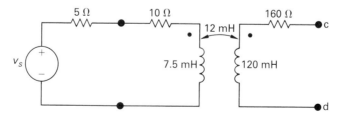

Fig. 13.43   The circuit for Problem 13.9.

13.10   What percentage of the average power delivered to the x, y terminals in the circuit in Fig. 13.44 reaches the 90-Ω resistor?

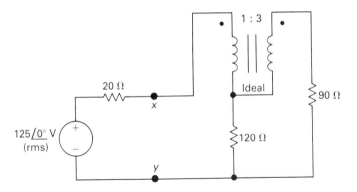

Fig. 13.44 The circuit for Problem 13.10.

13.11 Find the rms value of $\mathbf{I}_a$, $\mathbf{I}_b$, and $\mathbf{I}_c$ in the circuit in Fig. 13.45.

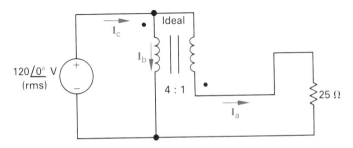

Fig. 13.45 The circuit for Problem 13.11.

13.12 Find the average power delivered to the 50-Ω load in the circuit in Fig. 13.46.

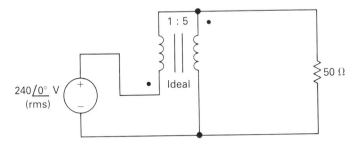

Fig. 13.46 The circuit for Problem 13.12.

13.13 Find the impedance Z_{ab} in the circuit in Fig. 13.47 if $Z_l = 0.125\underline{/45°}$ Ω.

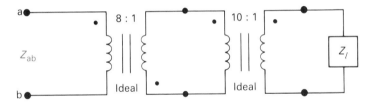

Fig. 13.47 The circuit for Problem 13.13.

13.14   Find the average power delivered in the 1-Ω resistor in the circuit in Fig. 13.48.

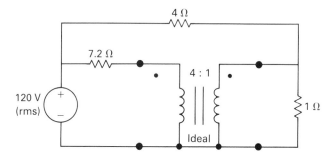

Fig. 13.48   The circuit for Problem 13.14.

13.15   a) Determine the turns ratio a of the ideal transformer in the circuit in Fig. 13.49 so that maximum average power is transferred to the 200-Ω resistor.
b) What is the maximum power that can be transferred?
c) What is the rms voltage across the 200-Ω resistor when maximum power is being delivered?

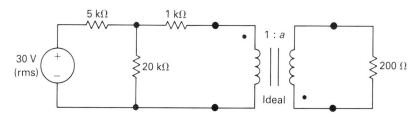

Fig. 13.49   The circuit for Problem 13.15.

13.16   The 8-Ω load in the circuit in Fig. 13.50 is to receive half as much average power as the 4-Ω load. The two loads are to be matched to the sinusoidal voltage source, which has an internal impedance of $600\ \underline{/0°}\ \Omega$. Specify the numerical values of a_1 and a_2.

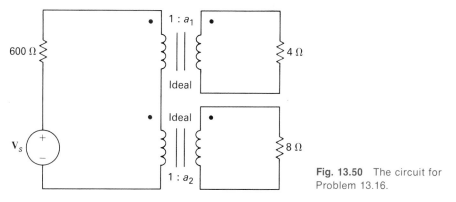

Fig. 13.50   The circuit for Problem 13.16.

13.17   a) Find the T-equivalent circuit for the magnetically coupled coils shown in Fig. 13.51.
b) Find the π-equivalent circuit for the same set of coils. Assume that $i_1(0) = i_2(0) = 0$.

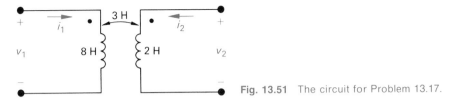

Fig. 13.51 The circuit for Problem 13.17.

13.18 Show that the phasor domain π-equivalent circuit for the magnetically coupled coils in Fig. 13.22 can be derived from the T-equivalent circuit by simply making a Y-to-Δ transformation.

13.19 A sinusoidal voltage source with an internal resistance of 10 Ω is coupled to a load by means of a lossless linear transformer, as shown in Fig. 13.52. The transformer inductances are $L_1 = 2$ mH and $L_2 = 8$ mH. The transformer coefficient of coupling is 0.25.

    a) Specify the phasor domain T-equivalent circuit for the lossless transformer when the source frequency is 10 krad/s.

    b) Use the T-equivalent circuit of part (a) to find the steady-state expression for v_L.

    c) Repeat part (b) given that the polarity dot on the primary side of the transformer is shifted to the lower terminal.

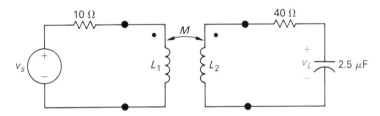

Fig. 13.52 The circuit for Problem 13.19.

13.20 The equivalent circuit of a lossless linear transformer is shown in Fig. 13.53.

    a) Find L_1, L_2, and M.

    b) Find the coefficient of coupling k.

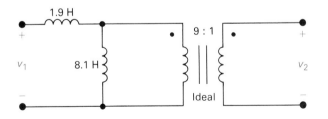

Fig. 13.53 The circuit for Problem 13.20.

13.21 The following measurements were made on a lossless linear transformer.

    With the low-voltage side open, the inductance measured looking into the high-voltage side is 23 H.

    With a 100-V rms sinusoidal voltage applied to the high-voltage winding, the open-circuit voltage measured on the low-voltage winding is 20 V rms.

With a 10-V rms sinusoidal voltage applied to the low-voltage side, the open-circuit voltage measured on the high-voltage side is 46 V.

a) Specify the numerical values of L_1, L_2, M, and k for the transformer.

b) Calculate the turns ratio of the transformer.

c) Construct two possible equivalent circuits for the linear transformer if a is chosen to equal M/L_1.

d) Repeat part (c) given that a is chosen equal to L_2/M.

13.22   a) Use one (your choice) of the equivalent circuits derived in Problem 13.21 to calculate the rms voltage at the terminals of the high-voltage winding when the transformer is used in the circuit in Fig. 13.54. The 50-V (rms) sinusoidal voltage source is operating at a frequency of 400 rad/s.

b) Verify your calculation in part (a) by finding the same voltage without using an equivalent circuit for the linear transformer.

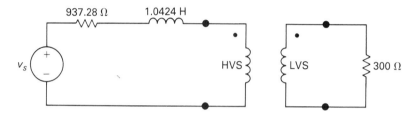

Fig. 13.54   The circuit for Problem 13.22.

13.23   The purpose of this problem is to illustrate a circuit structure in which the T- or π-equivalent circuits derived in Section 13.8 cannot be used because the voltage v_{bd} in Fig. 13.22 is not zero. With this in mind, calculate the voltage v_{bd} in the circuit in Fig. 13.55.

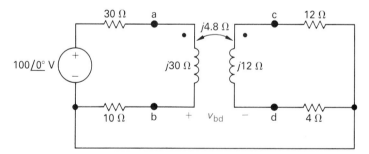

Fig. 13.55   The circuit for Problem 13.23.

14

SERIES AND PARALLEL RESONANCE

14.1 INTRODUCTION

Thus far in our analysis of circuits that are excited by sinusoidal sources, we have concentrated on finding the steady-state currents and voltages under the condition that the frequency of the source, or sources, remain constant. Another important aspect of sinusoidal circuit analysis, which we now wish to consider, is the determination of what happens to the amplitude and phase of the steady-state currents and voltages as the frequency of the source is varied. When we investigate the effect of varying the frequency of the sinusoidal source, we say that we are determining the *frequency response* of the circuit.

There are two primary reasons for our interest in the frequency response of a circuit. First, we are interested in circuits that exhibit discriminatory characteristics insofar as frequency is concerned. That is, we are interested in circuits that will transmit signals at some frequencies noticeably better than at other frequencies, because we can use this characteristic to filter out, or eliminate, the signals in an unwanted frequency range. The ability to design circuits that are frequency-selective is what makes radio, telephone, and television communication possible.

Second, if we know the frequency response we can predict the response of the circuit to any other input. We will have more to say about this in Chapter 17; however, several comments on the significance of this observation are in order here. The fact that the frequency response is related to other responses means that the engineer can carry out a design in terms of frequency specifications knowing that the constraints placed on the frequency response also exert control over the response to nonsinusoidal inputs. Another important point is that we can measure the frequency response in the laboratory and from these data, we can formulate a model of the circuit, or device. We can then use this laboratory-derived model to predict the response of the device to other types of inputs. Thus the study of frequency response is an important part of circuit analysis and design.

We introduce the general topic of frequency response in this chapter by studying the frequency–selectivity characteristics of two specific circuit structures. The first is referred to as the parallel-resonant circuit and the second as the series-resonant circuit.

14.2 INTRODUCTION TO PARALLEL RESONANCE

We begin our study of circuit resonance with the parallel structure shown in Fig. 14.1. We are interested in how the steady-state amplitude and phase of the output voltage v_o vary as the frequency of the sinusoidal current source varies.

Before proceeding with a quantitative analysis of the circuit, let us first get a feel for the behavior of the voltage by analyzing the circuit qualitatively. At very low frequencies, the inductive reactance of the inductor will be small and so the inductor will appear as a very small impedance across the output.

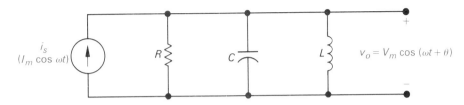

Fig. 14.1 The parallel-resonant circuit.

Thus the output voltage will be very small at very low frequencies. At very high frequencies, the capacitive reactance of the capacitor will be very small and, therefore, will appear as a very small impedance across the output. Consequently, the output voltage will be very small at very high frequencies. At intermediate values of frequency, the impedance of the three parallel branches will have a nonzero value. The output voltage, v_o, will have a nonzero value whenever the impedance has a nonzero value. With these thoughts in mind, we would expect a sketch of the amplitude of v_o vs. the frequency, ω, to have the general shape shown in Fig. 14.2.

Having reasoned that the amplitude of v_o would behave as shown in Fig. 14.2, we face several important questions. At what frequency will V_m be maximum? What is the maximum value of V_m? How sharp is the peak in the neighborhood of the maximum value? In order to answer these questions, and more, we must turn to a quantitative description of V_m as a function of ω.

To find the analytical expression for V_m as a function of ω, we resort to a phasor domain description of v_o. The phasor domain equivalent circuit of the circuit given in Fig. 14.1 is drawn in Fig. 14.3. Because the three branches of the parallel-resonant circuit are in parallel, it is easier to relate $\mathbf{V}_o$ to $\mathbf{I}_s$ through an admittance function, specifically,

$$\mathbf{V}_o = \frac{\mathbf{I}_s}{Y} = \frac{\mathbf{I}_s}{\left(\dfrac{1}{R} + \dfrac{1}{j\omega L} + j\omega C\right)} = \frac{\mathbf{I}_s}{\left[\dfrac{1}{R} + j\left(\omega C - \dfrac{1}{\omega L}\right)\right]}. \tag{14.1}$$

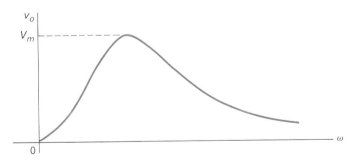

Fig. 14.2 The amplitude of v_o in the circuit in Fig. 14.1 as a function of frequency.

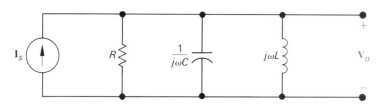

Fig. 14.3   The phasor domain equivalent circuit for the circuit in Fig. 14.1.

Using the current source as the reference, we can express Eq. (14.1) as

$$\mathbf{V}_o = \frac{I_m\,\underline{/0^\circ}}{\sqrt{\dfrac{1}{R^2} + \left(\omega C - \dfrac{1}{\omega L}\right)^2}\,\underline{/\phi}}, \tag{14.2}$$

from which we can see that the amplitude of v_o is

$$V_m = \frac{I_m}{\sqrt{(1/R^2) + [\omega C - (1/\omega L)]^2}} \tag{14.3}$$

and the phase angle between i_s and v_o is $-\phi$, where

$$\tan \phi = (\omega C - 1/\omega L)R. \tag{14.4}$$

For a given source and circuit I_m, R, L, and C are fixed and Eqs. (14.3) and (14.4) tell us how the amplitude and phase of v_o change as ω changes.

The *resonant frequency* for the circuit in Fig. 14.1 is defined as the frequency that makes the impedance seen by the current source purely resistive. This frequency makes the corresponding admittance purely conductive, since by definition $Y = 1/Z$.

If we denote the resonant frequency as ω_0, it is clear from Eq. (14.1) that at resonance

$$\omega_0 C = \frac{1}{\omega_0 L},$$

or

$$\omega_0 = \frac{1}{\sqrt{LC}}. \tag{14.5}$$

Now observe from Eq. (14.3) that the amplitude of v_o—that is, V_m—will be maximum when the magnitude of the admittance is a minimum. Note further that this will occur at the resonant frequency ω_0. We can also see from Eq. (14.3) that when ω equals ω_0 the maximum value of V_m is $I_m R$. When these quantitative results are added to the sketch in Fig. 14.2, we get the curve shown in Fig. 14.4.

If we think of i_s as an input signal and v_0 as an output signal, it is apparent from Fig. 14.4 that the parallel circuit structure in Fig. 14.1 transmits a signal whose frequency is close to ω_0 much better than a signal whose frequency is

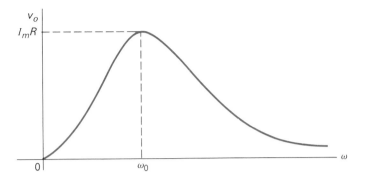

Fig. 14.4 Amplitude versus frequency for the circuit in Fig. 14.1.

much lower than or greater than ω_0. This observation raises the question, How close to ω_0 must the signal frequency be in order to get an acceptable output voltage? In order to answer this question quantitatively, we need to introduce what we mean by the bandwidth of the circuit and its relationship to the resonant frequency. These concepts are introduced in the following section.

14.3 BANDWIDTH AND QUALITY FACTOR

The *bandwidth* for the parallel-resonant circuit in Fig. 14.1 is defined as the range of frequencies in which the amplitude of the output voltage is equal to or greater than the maximum value divided by $\sqrt{2}$. Thus an acceptable output voltage for this circuit is defined as one whose amplitude is at least $1/\sqrt{2}$ times the maximum amplitude that can be transmitted by the circuit. The bandwidth is defined on the amplitude vs. frequency plot shown in Fig. 14.5. We have chosen the factor $1/\sqrt{2}$ because when the amplitude of v_o is reduced by this factor, the average power delivered to the resistor is $\frac{1}{2}$ its maximum value.

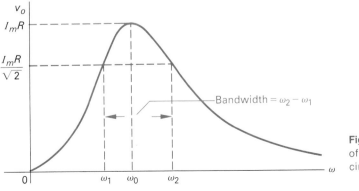

Fig. 14.5 Bandwidth of the parallel resonant circuit.

Therefore the circuit is described as transmitting a signal if the average power delivered to R is at least $\frac{1}{2}$ of the maximum possible value.

The limiting frequencies ω_1 and ω_2, which define the bandwidth, can be found by noting from Eq. (14.3) that they correspond to the values of ω that make

$$\sqrt{\frac{1}{R^2} + \left(\omega C - \frac{1}{\omega L}\right)^2} = \frac{\sqrt{2}}{R}$$

or

$$\left(\omega C - \frac{1}{\omega L}\right)^2 = \frac{1}{R^2}. \qquad (14.6)$$

Since Eq. (14.6) is a fourth-order polynomial in ω, there are four values of ω that yield an output voltage with an amplitude equal to $I_m R / \sqrt{2}$. Two of the four values are negative. These negative values are simply the mirror images of the positive values and have no physical significance. The two positive values of interest are

$$\omega_1 = -\frac{1}{2RC} + \sqrt{\left(\frac{1}{2RC}\right)^2 + \frac{1}{LC}} \qquad (14.7)$$

and

$$\omega_2 = \frac{1}{2RC} + \sqrt{\left(\frac{1}{2RC}\right)^2 + \frac{1}{LC}}, \qquad (14.8)$$

from which it follows that the bandwidth of the circuit is

$$\beta = \omega_2 - \omega_1 = \frac{1}{RC}. \qquad (14.9)$$

With Eqs. (14.5) and (14.9) in mind, we see that we can use the inductor and capacitor to fix the resonant frequency and then the value of R to adjust the bandwidth. Large values of R correspond to a narrow bandwidth, which in turn implies that the circuit is very selective as to which frequencies it transmits. For a circuit that is intended to be frequency-selective, the sharpness of the selectivity is a measure of the quality of the circuit. The quality of the frequency response is described quantitatively in terms of the ratio of the resonant frequency to the bandwidth. This ratio is called the **Q** of the circuit; thus

$$Q = \frac{\omega_0}{\beta}. \qquad (14.10)$$

For the parallel-resonant structure in Fig. 14.1, the quality factor can be expressed in several ways, namely,

$$Q = \omega_0 RC = \frac{R}{\omega_0 L} = R\sqrt{\frac{C}{L}}. \qquad (14.11)$$

The definition of **Q** given by Eq. (14.10) lends itself to laboratory measurements because it is possible to measure both the resonant frequency (ω_0) and the bandwidth (β). For applications outside the realm of circuits, we can also define Q in terms of an energy ratio, namely,

$$Q = 2\pi \frac{\text{maximum energy stored}}{\text{total energy lost per period}}. \qquad (14.12)$$

For the circuit in Fig. 14.1, Eq. (14.12) leads to the same result as is shown by the following computation.

To find the Q of the parallel circuit in Fig. 14.1 using Eq. (14.12), we first compute the total energy stored in the inductor and capacitor at the resonant frequency ω_0. Since the output voltage at resonance is

$$v_o = I_m R \cos \omega_0 t, \qquad (14.13)$$

the current in the inductor is

$$i_L = \frac{I_m R}{\omega_0 L} \cos (\omega_0 t - 90°) = \frac{I_m R}{\omega_0 L} \sin \omega_0 t. \qquad (14.14)$$

It follows from Eqs. (14.13) and (14.14) that

$$w_C(t) = \frac{1}{2} C I_m^2 R^2 \cos^2 \omega_0 t \qquad (14.15)$$

and

$$w_L(t) = \frac{1}{2} L \frac{I_m^2 R^2}{\omega_0^2 L^2} \sin^2 \omega_0 t. \qquad (14.16)$$

At resonance, $C = 1/\omega_0^2 L$; therefore we can write Eq. (14.16) as

$$w_L(t) = \frac{1}{2} C I_m^2 R^2 \sin^2 \omega_0 t. \qquad (14.17)$$

The total energy stored in the circuit is found by adding Eqs. (14.15) and (14.17) to get

$$w_T = w_C(t) + w_L(t) = \frac{1}{2} C I_m^2 R^2, \qquad (14.18)$$

which tells us that the total energy stored in the circuit at resonance is a constant and this constant value must also be the maximum value of the energy stored.

The total energy lost per period can be computed by multiplying the average power dissipated in R by the period. At resonance this will be

$$\text{Total energy lost per period} = \frac{I_m^2}{2} R \left(\frac{1}{f_0} \right). \qquad (14.19)$$

When we substitute Eqs. (14.18) and (14.19) into Eq. (14.12), we get

$$Q = 2\pi \frac{C I_m^2 R^2 / 2}{I_m^2 R / 2 f_0} = 2\pi f_0 R C = \frac{\omega_0}{\beta}. \qquad (14.20)$$

We will use the definition of Q expressed by Eq. (14.10) for our work here. The alternative definition given in Eq. (14.12) is not a useful definition for our study of circuits. (See Problem 14.25.)

Example 14.1   The amplitude of the sinusoidal current source in the circuit in Fig. 14.1 is 50 mA. The circuit parameters are $R = 2 \text{ k}\Omega$, $L = 40 \text{ mH}$, and $C = 0.25 \text{ }\mu\text{F}$.

a) Find $\omega_0, Q, \omega_1, \omega_2$, and the amplitude of the output voltage at ω_0, ω_1, and ω_2.

b) What value of R will produce a bandwidth of 500 rad/s?

c) What is the value of Q in part (b)?

Solution   a)

$$\omega_0 = \sqrt{\frac{1}{LC}} = \sqrt{\frac{10^9}{40(0.25)}} = \sqrt{10^8} = 10^4 \text{ rad/s};$$

$$Q = \omega_0 RC = (10^4)(2000)(0.25 \times 10^{-6}) = 5;$$

$$\beta = \omega_0/Q = 10^4/5 = 2000 \text{ rad/s};$$

$$\omega_1 = -\frac{1}{2RC} + \sqrt{\left(\frac{1}{2RC}\right)^2 + \frac{1}{LC}}$$

$$= -1000 + \sqrt{10^6 + 10^8}$$

$$= -1000 + 1000\sqrt{101}$$

$$= -1000 + 10{,}049.88 = 9049.88 \text{ rad/s};$$

$$\omega_2 = \frac{1}{2RC} + \sqrt{\left(\frac{1}{2RC}\right)^2 + \frac{1}{LC}}$$

$$= 1000 + 10{,}049.88 = 11{,}049.88 \text{ rad/s};$$

$$V_m(\omega_0) = I_m R = 50 \times 10^{-3}(2000) = 100 \text{ V};$$

$$V_m(\omega_1) = V_m(\omega_2) = V_m(\omega_0)/\sqrt{2} = 70.7 \text{ V}$$

b) $R = \dfrac{1}{\beta C} = \dfrac{10^6}{(500)(0.25)} = 8000 \text{ }\Omega$

c) $Q = \dfrac{\omega_0}{\beta} = \dfrac{10^4}{500} = \dfrac{100}{5} = 20$ ■

Note in this numerical example that the resonant frequency is not in the center of the passband. That is, the frequency range from ω_1 to ω_0 is not the same as the range from ω_0 to ω_2. Specifically, $\omega_0 - \omega_1 = 10{,}000 - 9049.88 = 950.12$ rad/s, whereas $\omega_2 - \omega_0 = 11{,}049.88 - 10{,}000 = 1049.88$ rad/s. As Q gets larger ω_0 approaches the center of the passband. You can verify this by calculating ω_1 and ω_2 in Example 14.1 when $R = 8000 \text{ }\Omega$. We will verify this observation analytically in Section 14.4. The exact relationship among ω_1, ω_2, and ω_0 can be obtained from Eqs. (14.7) and (14.8). When these two equations are multiplied together, we get

$$\omega_1 \omega_2 = \omega_0^2;$$

therefore the *resonant frequency is the geometric mean of half-power fre-quencies* ω_1 *and* ω_2, that is,

$$\omega_0 = \sqrt{\omega_1 \omega_2}. \tag{14.21}$$

The frequencies ω_1 and ω_2 are called the half-power frequencies because at either of these frequencies the power dissipated in R is exactly one half the power dissipated in R at the resonant frequency.

**DRILL EXERCISE
14.1** The resonant frequency of the circuit in Fig. 14.1 is 10 Mrad/s. The bandwidth of the circuit is 100 krad/s. If R is 100 kΩ, calculate (a) Q; (b) L; (c) C; (d) ω_1; and (e) ω_2.

Ans. (a) 100; (b) 100 μH; (c) 100 pF; (d) 9.95 Mrad/s; (e) 10.05 Mrad/s.

**DRILL EXERICSE
14.2** The following components are available to construct a parallel-resonant circuit: $L_1 = 7.5$ mH, $L_2 = 15$ mH, $C_1 = 6$ μF, $C_2 = 3$ μF, and $R = 2$ kΩ. Design a circuit that will have the highest possible resonant frequency. Specify (a) ω_0; (b) β; (c) Q; (d)ω_1; and (e) ω_2.

Ans. (a) 10^4 rad/s; (b) 250 rad/s; (c) 40; (d) 9875 rad/s; (e) 10,125 rad/s.

**DRILL EXERCISE
14.3** Use the components in Drill Exercise 14.2 to design a parallel-resonant circuit with the lowest possible resonant frequency. Specify (a) ω_0; (b) β; (c) Q; (d) ω_1; and (e) ω_2.

Ans. (a) 2222.22 rad/s; (b) 55.56 rad/s; (c) 40; (d) 2194.62 rad/s; (e) 2250.17 rad/s.

14.4 FURTHER ANALYSIS OF PARALLEL RESONANCE

Since the purpose of the parallel-resonant structure is to be frequency-selective, it is convenient to describe the behavior of the circuit in terms of ω_0, β, and Q. With this in mind, we return to Eqs. (14.7) and (14.8) and note that

$$\frac{1}{2RC} = \frac{\beta}{2} = \frac{\omega_0}{2Q}. \tag{14.22}$$

By substituting Eq. (14.22) into Eqs. (14.7) and (14.8), we can express the half-power frequencies as functions of the resonant frequency and the Q of the circuit. The expressions become

$$\omega_1 = -\frac{\omega_0}{2Q} + \sqrt{\left(\frac{\omega_0}{2Q}\right)^2 + \omega_0^2}$$

$$= \omega_0 \left\{ -\frac{1}{2Q} + \sqrt{\left(\frac{1}{2Q}\right)^2 + 1} \right\} \tag{14.23}$$

and

$$\omega_2 = \omega_0 \left\{ \frac{1}{2Q} + \sqrt{\left(\frac{1}{2Q}\right)^2 + 1} \right\}. \tag{14.24}$$

We can see from Eqs. (14.23) and (14.24) that when Q is 10 or greater

$$\omega_1 \cong -\frac{1}{2}\frac{\omega_0}{Q} + \omega_0 = \omega_0 - \frac{\beta}{2} \tag{14.25}$$

and

$$\omega_2 \cong \frac{1}{2}\frac{\omega_0}{Q} + \omega_0 = \omega_0 + \frac{\beta}{2}. \tag{14.26}$$

Thus for high-Q circuits the resonant frequency ω_0 lies close to the center of the passband. We can use Eqs. (14.25) and (14.26) to obtain the half-power frequencies with less than 1% error whenever Q is equal to or greater than 5. For example, when Q equals 5, the exact expressions (Eqs. 14.23 and 14.24) predict that $\omega_1/\omega_0 = 0.905$ and $\omega_2/\omega_0 = 1.105$, whereas the approximate expressions (Eqs. 14.25 and 14.26) predict that $\omega_1/\omega_0 = 0.900$ and $\omega_2/\omega_0 = 1.100$.

The behavior of the output voltage near the resonant frequency can be highlighted by expressing the admittance in terms of the resonant frequency ω_0 and the bandwidth β. From Eq. (14.1) we have

$$Y = \frac{1}{R} + j\left(\omega C - \frac{1}{\omega L}\right). \tag{14.27}$$

We now carry out the following series of algebraic manipulations:

$$Y = \frac{1}{R}\left[1 + j\left(\omega RC - \frac{R}{\omega L}\right)\right]$$

$$= \frac{1}{R}\left[1 + j\left(\frac{\omega\omega_0 RC}{\omega_0} - \frac{\omega_0 R}{\omega\omega_0 L}\right)\right]$$

$$= \frac{1}{R}\left[1 + jQ\left(\frac{\omega}{\omega_0} - \frac{\omega_0}{\omega}\right)\right]$$

$$Y = \frac{1}{R}\left[1 + j\frac{\omega_0}{\beta}\left(\frac{\omega^2 - \omega_0^2}{\omega\omega_0}\right)\right], \tag{14.28}$$

which can be further simplified to

$$Y = \frac{1}{R}\left[1 + j\frac{(\omega + \omega_0)(\omega - \omega_0)}{\beta\omega}\right]. \tag{14.29}$$

Equation (14.29) is an exact expression for the admittance for all values of ω since its derivation involves only legitimate algebraic manipulations of the original expression given by Eq. (14.27). However, it has the advantage of showing more clearly how Y varies in the neighborhood of ω_0. As ω approaches ω_0 the ratio $(\omega + \omega_0)/\omega$ approaches 2 and the expression for Y can be approximated as

$$Y \cong \frac{1}{R} \left[1 + j \frac{(\omega - \omega_0)}{\beta/2} \right]. \tag{14.30}$$

The error in using Eq. (14.30) will be less than 5% in both magnitude and phase, provided ω lies within the passband, that is, $\omega_1 \leq \omega \leq \omega_2$.

Example 14.2

a) For the circuit in Example 14.1(a) find $\mathbf{V}_o$ at 10,500 rad/s and 9500 rad/s using the exact formula for Y.

b) Repeat part (a) using the approximate expression for Y.

c) Find the percent error in the magnitude of $\mathbf{V}_o$ that results from using the approximate expression for Y.

Solution

a) At 10,500 rad/s we have

$$Y = \frac{1}{2000} \left[1 + j \frac{(20,500)(500)}{2000(10,500)} \right]$$

$$= \frac{1.1128 \underline{/26.02°}}{2000} = 0.5564 \underline{/26.02°} \text{ m℧};$$

$$\mathbf{V}_o = \frac{50 \times 10^{-3}(2000)}{1.1128 \underline{/26.02°}} = 89.87 \underline{/-26.02°} \text{ V}.$$

At 9500 rad/s we have

$$Y = \frac{1}{2000} \left[1 + j \frac{19,500(-500)}{2000(9500)} \right]$$

$$= \frac{1.1240 \underline{/-27.16°}}{2000} = 0.5620 \underline{/-27.16°} \text{ m℧};$$

$$\mathbf{V}_o = \frac{50 \times 10^{-3}(2000)}{1.1240 \underline{/-27.16°}} = 88.97 \underline{/27.16°} \text{ V}.$$

b) Using the approximate formula for Y we get the following results. At $\omega = 10,500$ rad/s,

$$Y \cong \frac{1}{2000} \left[1 + j \frac{(10,500 - 10,000)}{(2000/2)} \right]$$

$$= \frac{1 + j0.5}{2000} = \frac{1.1180 \underline{/26.57°}}{2000}$$

$$= 0.5590 \underline{/26.57°} \text{ m℧};$$

$$\mathbf{V}_o = \frac{50 \times 10^{-3}(2000)}{1.1180 \underline{/26.57°}} = 89.44 \underline{/-26.57°} \text{ V}.$$

At $\omega = 9500$ rad/s,

$$Y = \frac{1}{2000}\left[1 + j\frac{(9500 - 10{,}000)}{1000}\right]$$

$$= \frac{1 - j0.5}{2000} = 0.5590\underline{/-26.57^\circ} \text{ m℧};$$

$$\mathbf{V}_o = 89.44\underline{/26.57^\circ} \text{ V}.$$

c) The percent error in the magnitude of $\mathbf{V}_o$ at 10,500 rad/s is

$$\% \text{ Error} = \frac{(89.44 - 89.87)}{89.87} \times 100 = -0.48\%.$$

The percent error at 9500 rad/s is

$$\% \text{ Error} = \frac{89.44 - 88.97}{88.97} \times 100 = 0.53\%. \quad \blacksquare$$

Although our primary interest in studying the parallel-resonant circuit is to establish the relationship between the output voltage and the source current, it is of interest to note that at resonance the magnitude of the current in either the inductive or capacitive branch is Q times the magnitude of the source current. We leave the verification of this statement to the reader via Problem 14.6. These large internal currents must be taken into account when selecting components.

The variation with frequency of the phase angle between the output signal and the input signal is also of interest. For the parallel circuit in Fig. 14.1 the phase angle of v_o with respect to i_s is the negative to the phase angle of the admittance Y; thus

$$\theta = -\phi = -\tan^{-1}\frac{(\omega + \omega_0)(\omega - \omega_0)}{\beta\omega}. \tag{14.31}$$

The phase angle θ can also be expressed in terms of Q since $\beta = \omega_0/Q$; thus

$$\theta = -\tan^{-1}\frac{(\omega + \omega_0)(\omega - \omega_0)Q}{\omega_0\omega}, \tag{14.32}$$

from which we see that when $\omega < \omega_0$, the phase angle is positive and as ω approaches zero θ approaches $+90^\circ$. For $\omega > \omega_0$, θ will be negative and as ω approaches infinity ϕ approaches -90°. We know that $\theta = 0$ at $\omega = \omega_0$, $+45^\circ$ at $\omega = \omega_1$, and -45° at $\omega = \omega_2$. These characteristics of θ with respect to ω are shown in Fig. 14.6.

The effect of Q on the phase-angle characteristic is most easily shown by investigating the slope of the θ vs. ω curve at ω_0. Thus by letting

$$\theta = -\tan^{-1}\frac{(\omega^2 - \omega_0^2)Q}{\omega\omega_0} = -\tan^{-1} u,$$

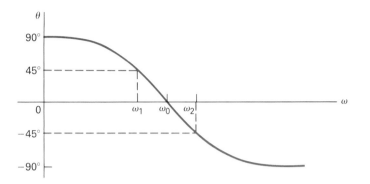

Fig. 14.6   The phase angle θ versus ω for the parallel resonant structure in Fig. 14.1.

we find that

$$\frac{d\theta}{d\omega} = \frac{d\theta}{du} \cdot \frac{du}{d\omega} = -\frac{1}{1+u^2}\frac{du}{d\omega}. \tag{14.33}$$

But

$$\frac{du}{d\omega} = \frac{\omega\omega_0(2\omega)Q - (\omega^2 - \omega_0^2)Q\omega_0}{\omega^2\omega_0^2}$$

$$= \frac{\omega_0 Q(\omega^2 + \omega_0^2)}{\omega^2\omega_0^2}. \tag{14.34}$$

Substituting Eq. (14.34) into Eq. (14.33) gives

$$\frac{d\theta}{d\omega} = -\frac{\omega_0 Q(\omega^2 + \omega_0^2)}{\omega^2\omega_0^2 + (\omega^2 - \omega_0^2)^2 Q^2}. \tag{14.35}$$

At $\omega = \omega_0$, this reduces to

$$\frac{d\theta}{d\omega} = -\frac{2Q}{\omega_0} = -2RC. \tag{14.36}$$

Thus as Q is increased by increasing R, the slope at ω_0 increases. Note that if R is made infinitely large, the phase angle snaps from $+90°$ to $-90°$ at the resonant frequency. The effect of Q on the phase shift is shown in Fig. 14.7.

DRILL EXERCISE
14.4

The components in the circuit in Fig. 14.1 are $R = 40$ kΩ, $L = 0.5$ mH, and $C = 500$ pF.

a) At what frequency will the amplitude of v_o be maximum?

b) At what frequencies will the amplitude of v_o be 92.5% of its maximum value?

Ans.   (a) 318.31 kHz; (b) 316.68 kHz, 319.95 kHz.

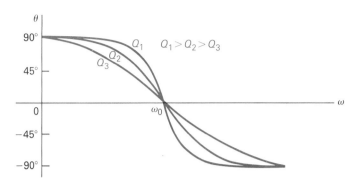

Fig. 14.7   The effect of Q on the phase angle θ versus ω.

14.5   THE FREQUENCY RESPONSE VS. THE NATURAL RESPONSE OF THE PARALLEL *RLC* CIRCUIT

We mentioned in Section 14.1 that we are interested in the frequency response of a circuit because it can be used to predict the time domain behavior of the circuit. Now that we have studied the frequency response of the parallel *RLC* circuit, let us show how attributes of the frequency-response behavior relate to attributes of the natural time domain response. The natural response of the parallel *RLC* circuit was investigated in Chapter 8 and for the underdamped case is given by Eq. (8.29), which is repeated here for convenience:

$$v(t) = B_1 e^{-\alpha t} \cos \omega_d t + B_2 e^{-\alpha t} \sin \omega_d t, \tag{14.37}$$

$$\alpha = \frac{1}{2RC} \tag{14.38}$$

and

$$\omega_d = \sqrt{\omega_0^2 - \alpha^2}. \tag{14.39}$$

The coefficients B_1 and B_2 are determined by the initial conditions and are not germane to the discussion at hand.

We can relate the damping coefficient α and the damped frequency of oscillation ω_d to the Q of the circuit; thus

$$\alpha = \frac{\beta}{2} = \frac{\omega_0}{2Q} \tag{14.40}$$

and

$$\omega_d = \omega_0 \sqrt{1 - 1/4Q^2}. \tag{14.41}$$

We also note that the transition from the overdamped to the underdamped response occurs when $\omega_0^2 = \alpha^2$, or when $Q = 1/2$. Therefore, the natural response will be underdamped whenever the Q of the circuit is greater than

1/2 or, alternatively, whenever the bandwidth is less than $2\omega_0$. From Eq. (14.40) we see that in high-Q circuits the natural response oscillations persist longer than in low-Q circuits. Equation (14.41) tells us that the frequency of oscillation in the underdamped response approaches the resonant frequency ω_0 as Q increases.

We can summarize our observations by saying that if the frequency response of the parallel RLC structure shows a sharp resonant peak, the natural response will be underdamped, the frequency of oscillation will approximate the resonant frequency, and the oscillations will persist over a relatively long interval of time.

DRILL EXERCISE
14.5

The natural response of the circuit in Fig. 14.1 is known to be

$$v_o = 100e^{-100t} \sin 100\sqrt{99}\,t \text{ V}$$

when $C = 1\ \mu$F. Calculate (a) ω_0; (b) β; (c) Q; (d) R; and (e) L.

Ans. (a) 1000 rad/s; (b) 200 rad/s; (c) 5; (d) 5 kΩ; (e) 1H.

14.6 SERIES RESONANCE

The basic series-resonant circuit is shown in Fig. 14.8. Here we are interested in how the amplitude and phase angle of the current vary with the frequency of the sinusoidal voltage source. As the frequency of the source changes, the maximum amplitude of source voltage (V_m) is held constant.

The analysis of the series-resonant circuit follows very closely the analysis of the parallel-resonant structure shown in Fig. 14.1. First we note that the amplitude of the current will approach zero at both very small and very large values of ω. The capacitor blocks the passage of current at very low frequencies, whereas the inductor blocks the passage of current at very high frequencies. The current amplitude will peak at V_m/R when the inductive reactance exactly cancels the capacitive reactance. The frequency at which

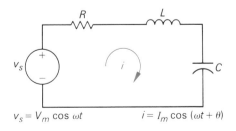

$v_s = V_m \cos \omega t$ $i = I_m \cos (\omega t + \theta)$

Fig. 14.8 The series-resonant circuit.

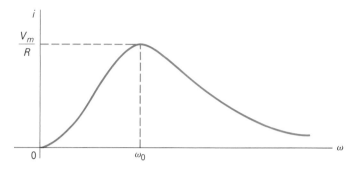

Fig. 14.9   Current amplitude versus frequency for the circuit in Fig. 14.8.

the reactive impedances cancel is the resonant frequency of the circuit; thus $\omega_0 L = 1/\omega_0 C$, or

$$\omega_0 = \frac{1}{\sqrt{LC}}. \tag{14.42}$$

The expression for the resonant frequency of the circuit in Fig. 14.8 is the same as that for the circuit in Fig. 14.1, namely, Eq. (14.15).

The current amplitude (I_m) vs. the frequency (ω) plot has the form shown in Fig. 14.9. We can obtain a more detailed analysis of the curve by expressing the amplitude of i as a function of ω. As before, we use the phasor domain method of analysis; thus

$$\mathbf{I} = I_m\underline{/\theta} = \frac{\mathbf{V}_s}{\mathbf{Z}} = \frac{V_m\underline{/0°}}{|Z|\underline{/\phi}}. \tag{14.43}$$

The series impedance of the circuit in Fig. 14.8 is

$$Z = R + j\left(\omega L - \frac{1}{\omega C}\right).$$

The magnitude and angle of the impedance are, respectively,

$$|Z| = \sqrt{R^2 + [\omega L - (1/\omega C)]^2} \tag{14.44}$$

and

$$\phi = \tan^{-1}\frac{(\omega L - 1/\omega C)}{R}. \tag{14.45}$$

It follows from Eqs. (14.43), (14.44), and (14.45) that

$$I_m = \frac{V_m}{\sqrt{R^2 + [\omega L - (1/\omega C)]^2}}, \tag{14.46}$$

$$\theta = -\phi = -\tan^{-1}\frac{[\omega L - (1/\omega C)]}{R}. \tag{14.47}$$

The bandwidth of the series circuit is defined as the range of frequencies in which the amplitude of the current is equal to or greater than $1/\sqrt{2}$ times the maximum amplitude V_m/R. Thus the frequencies at the edge of the passband are the frequencies where the magnitude of Z equals $\sqrt{2}R$. Setting

$$\sqrt{R^2 + [\omega L - (1/\omega C)]^2} = \sqrt{2}R$$

and solving for the two positive values yields

$$\omega_1 = -\frac{R}{2L} + \sqrt{\left(\frac{R}{2L}\right)^2 + \frac{1}{LC}}, \tag{14.48}$$

$$\omega_2 = \frac{R}{2L} + \sqrt{\left(\frac{R}{2L}\right)^2 + \frac{1}{LC}}. \tag{14.49}$$

It follows from Eqs. (14.48) and (14.49) that the resonant frequency is the geometric mean of the half-power frequencies $(\omega_0 = \sqrt{\omega_1 \omega_2})$ and the bandwidth of the series circuit is

$$\beta = \omega_2 - \omega_1 = \frac{R}{L}. \tag{14.50}$$

These characteristics of the amplitude response of the series circuit in Fig. 14.8 are shown in Fig. 14.10.

Since by definition Q is the ratio of the resonant frequency to the bandwidth, the Q of the series circuit is

$$Q = \frac{\omega_0}{\beta} = \frac{\omega_0 L}{R} = \frac{1}{\omega_0 CR} = \frac{1}{R}\sqrt{\frac{L}{C}}. \tag{14.51}$$

We can express the half-power frequencies and the impedance of the series circuit in terms of the resonant frequency (ω_0) and the Q of the circuit. The ex-

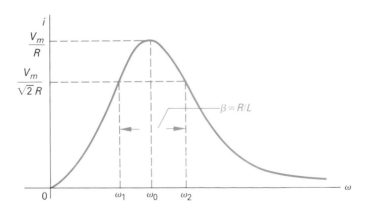

Fig. 14.10 The amplitude response of the series-resonant circuit.

pressions for the half-power frequencies are

$$
\begin{aligned}
\omega_1 &= -\frac{R}{2L} + \sqrt{\left(\frac{R}{2L}\right)^2 + \frac{1}{LC}} \\
&= -\frac{R\omega_0}{2\omega_0 L} + \sqrt{\left(\frac{R}{2\omega_0 L}\right)^2 \omega_0^2 + \omega_0^2} \\
&= \omega_0 \left[-\frac{1}{2Q} + \sqrt{1 + \left(\frac{1}{2Q}\right)^2} \right]
\end{aligned}
\tag{14.52}
$$

and

$$
\omega_2 = \omega_0 \left[\frac{1}{2Q} + \sqrt{1 + \left(\frac{1}{2Q}\right)^2} \right].
\tag{14.53}
$$

The expression for the impedance is

$$
\begin{aligned}
Z &= R + j\left(\omega L - \frac{1}{\omega C}\right) \\
&= R\left[1 + j\left(\frac{\omega L}{R} - \frac{1}{\omega R C}\right)\right] \\
&= R\left[1 + j\,\frac{\omega_0 L}{R}\,\frac{\omega}{\omega_0} - \frac{1}{\omega_0 R C}\,\frac{\omega_0}{\omega}\right] \\
&= R\left[1 + jQ\left(\frac{\omega}{\omega_0} - \frac{\omega_0}{\omega}\right)\right] \\
&= R\left[1 + jQ\,\frac{(\omega - \omega_0)(\omega + \omega_0)}{\omega\omega_0}\right] \\
&= R\left[1 + j\,\frac{(\omega - \omega_0)(\omega + \omega_0)}{\beta\omega}\right].
\end{aligned}
\tag{14.54}
$$

For high-Q circuits, Eqs. (14.52) and (14.53) reduce to

$$
\omega_1 \approx \omega_0 - \frac{\omega_0}{2Q} = \omega_0 - \frac{\beta}{2}
\tag{14.55}
$$

and

$$
\omega_2 \approx \omega_0 + \frac{\omega_0}{2Q} = \omega_0 + \frac{\beta}{2}.
\tag{14.56}
$$

Thus for high-Q circuits the resonant frequency approaches the center of the passband.

For values of ω near resonance, Eq. (14.54) can be simplified to

$$
Z \approx R\left[1 + j\,\frac{(\omega - \omega_0)}{\beta/2}\right].
\tag{14.57}
$$

The phase angle between the source voltage and the current is given by Eq. (14.47) and can be expressed in terms of ω_2 and Q by using Eq. (14.54);

thus

$$\theta = -\tan^{-1} \frac{(\omega - \omega_0)(\omega + \omega_0)}{\beta\omega}. \tag{14.58}$$

When we compare Eq. (14.58) to Eq. (14.31), we see immediately that the behavior of the phase angle between the output current i and the input voltage v_s in the series circuit is identical to the behavior of the phase angle between the output voltage v_o and the input current i_s in the parallel circuit. Thus the phase angle vs. ω characteristic curves shown in Fig. 14.7 also apply to the series circuit.

The amplitude of the voltage across either the inductor or the capacitor at the resonant frequency ω_0 is Q times the amplitude of the source voltage. This can be verified by noting that

$$|V_L| = I_m \omega_0 L = \frac{V_m}{R} \omega_0 L = QV_m \tag{14.59}$$

and

$$|V_c| = \frac{I_m}{\omega_0 C} = \frac{V_m}{\omega_0 CR} = QV_m. \tag{14.60}$$

In high-Q circuits, these large voltages must be taken into account when selecting components.

Problems 14.11–14.14 are designed to give you some practice in analyzing the series-resonant circuit.

DRILL EXERCISE 14.6   The series-resonant circuit in Fig. 14.8 is designed to have a resonant frequency of 5 Mrad/s and a lower half-power frequency of 4.5 Mrad/s. If the capacitor in the circuit equals 0.01 μF, calculate (a) Q; (b) β; (c) L; and (d) R.

Ans. (a) 4.74; (b) 1.0556 Mrad/s; (c) 4 μH; (d) 4.22 Ω.

DRILL EXERCISE 14.7   The components in the circuit in Fig. 14.8 are $R = 10$ Ω, $L = 0.2$ mH, and $C = 5$ nF. Calculate (a) ω_0; (b) Q; (c) β; and (d) the frequencies where the amplitude of the current is 80% of its maximum value.

Ans. (a) 1 Mrad/s; (b) 20; (c) 50 krad/s; (d) 1.019 Mrad/s, 0.981 Mrad/s.

14.7 MORE ON PARALLEL RESONANCE

Thus far we have limited our discussion of parallel resonance to a simple structure where the maximum amplitude of the response signal occurred at the resonant frequency ω_0. We now bring to your attention a more realistic structure

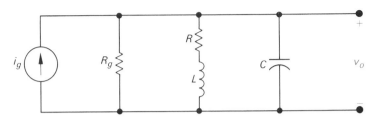

Fig. 14.11 A parallel-resonant circuit containing a low-loss inductor.

where the maximum amplitude of the response does not coincide with unity power-factor resonance. The parallel-resonant structure in Fig. 14.11 is a more practical model because it accounts for the losses in the inductor. The inductor losses are represented by the resistor R.

We can determine the unity power-factor resonant frequency ω_0 by finding the frequency that makes the admittance of the circuit conductive. The expression for the admittance is

$$Y = \frac{1}{R_g} + \frac{1}{R + j\omega L} + j\omega C. \tag{14.61}$$

To find the unity power-factor frequency, we first rewrite Eq. (14.61) in rectangular form; thus

$$Y = \frac{1}{R_g} + \frac{R}{R^2 + \omega^2 L^2} - \frac{j\omega L}{R^2 + \omega^2 L^2} + j\omega C, \tag{14.62}$$

from which we see that the admittance will be purely conductive when the j-terms cancel. The frequency at which this occurs is

$$\omega_0 = \sqrt{\frac{1}{LC} - \left(\frac{R}{L}\right)^2}. \tag{14.63}$$

Note that as the resistance of the inductor approaches zero, the unity power-factor resonant frequency approaches $1/\sqrt{LC}$.

The value of the admittance at ω_0 can be found by substituting Eq. (14.63) into Eq. (14.62). This substitution is greatly facilitated by noting that at ω_0 we have

$$R^2 + \omega_0^2 L^2 = L/C \tag{14.64}$$

and, at the same time, the j-terms vanish; thus

$$Y(\omega_0) = \frac{1}{R_g} + \frac{RC}{L} = \frac{L + R_g RC}{LR_g}, \tag{14.65}$$

from which it follows that

$$Z(\omega_0) = \frac{1}{Y(\omega_0)} = \frac{R_g L}{R_g RC + L}. \qquad (14.66)$$

The amplitude of the output voltage at unity power-factor resonance will be

$$|\mathbf{V}_o| = \frac{|\mathbf{I}_g| R_g L}{R_g RC + L}. \qquad (14.67)$$

An analysis of the amplitude of the output voltage as a function of frequency will reveal that the amplitude is not maximum at ω_0. The analysis requires us to express the amplitude of $\mathbf{V}_o$ as a function of ω, differentiate this expression with respect to ω, and then find the value of ω that makes the derivative zero. Since the algebra involved becomes very unwieldy, we will simply state the result and then illustrate the calculations involved. The frequency corresponding to the maximum amplitude of $\mathbf{V}_o$ is given by

$$\omega_m = \sqrt{\sqrt{\left(\frac{1}{LC}\right)^2 \left(1 + \frac{2R}{R_g}\right) + \left(\frac{R}{L}\right)^2 \left(\frac{2}{LC}\right)} - \left(\frac{R}{L}\right)^2}. \qquad (14.68)$$

It is worth noting that ω_m approaches ω_0 as R approaches zero. This can be seen by comparing Eq. (14.68) with Eq. (14.63).

Example 14.3 The circuit in Fig. 14.11 is driven by a sinusoidal current source. The maximum amplitude of the current is 4 mA; the internal resistance of the source is 10 kΩ; the 25-mH inductor has a resistance of 800 Ω; and the losses associated with the 10-nF capacitor are negligible.

a) Calculate the resonant frequency of the circuit.

b) Calculate the amplitude of the output voltage at resonance.

c) Calculate the frequency at which the amplitude of the output voltage is maximum.

d) Calculate the maximum amplitude of v_o.

Solution a) $\omega_0 = \sqrt{\dfrac{1}{LC} - \left(\dfrac{R}{L}\right)^2} = \sqrt{40 \times 10^8 - \left(\dfrac{800}{0.025}\right)^2}$

$\qquad = 10^4 \sqrt{40 - 10.24} = 54{,}552.73 \text{ rad/s}$

b) $V_m = |\mathbf{V}_o| = \dfrac{I_m R_g L}{R_g RC + L}$

$\qquad = \dfrac{(4 \times 10^{-3})(10^4)(25 \times 10^{-3})}{(10^4)(800)(0.01) \times 10^{-6} + 25 \times 10^{-3}}$

$\qquad = \dfrac{(250)(4)10^{-3}}{0.105} = 9.52 \text{ V}$

c) Before substituting into Eq. (14.68), we make the following preliminary calculations:

$$\left(\frac{1}{LC}\right)^2 = 16 \times 10^{18};$$

$$1 + \frac{2R}{R_g} = 1.16;$$

$$\left(\frac{R}{L}\right)^2 = 10.24 \times 10^8;$$

$$\left(\frac{2}{LC}\right) = 80 \times 10^8;$$

$$\omega_m = \sqrt{\sqrt{16 \times 10^{18}(1.16) + (10.24)(80)10^{16}} - 10.24 \times 10^8}$$

$$= \sqrt{51.70 \times 10^8 - 10.24 \times 10^8}$$

$$= 64{,}406.78 \text{ rad/s}.$$

d) To find the amplitude of v_o at this frequency, we first calculate the admittance; thus

$$Y(\omega_m) = \frac{1}{10^4} + \frac{1}{800 + j\omega_m 25 \times 10^{-3}} + j\omega_m 10 \times 10^{-9}$$

$$= 100 + 247.48 - j498.10 + j644.07 \ \mu\text{U}$$

$$= 376.89\underline{/22.79^\circ} \ \mu\text{U}.$$

Now we can calculate V_m from

$$V_m = |\mathbf{V}_o| = \frac{I_m}{|Y(\omega_m)|}$$

$$= \frac{4 \times 10^{-3}}{376.89} \times 10^6 = 10.61 \text{ V}. \quad \blacksquare$$

Example 14.4   Repeat the calculations of Example 14.3 given that the resistance of the 25-mH inductor is 80 Ω.

Solution   a) $\omega_0 = \sqrt{40 \times 10^8 - 10.24 \times 10^6}$

$$= 1000\sqrt{4000 - 10.24}$$

$$= 63{,}164.55 \text{ rad/s}$$

b) $V_m = \dfrac{(4 \times 10^{-3})(10^4)(25 \times 10^{-3})}{(10^4)(80)(10) \times 10^{-9} + 25 \times 10^{-3}}$

$$= \frac{1}{8 \times 10^{-3} + 25 \times 10^{-3}} = \frac{1000}{33} = 30.30 \text{ V}$$

c) $\omega_m = \sqrt{\sqrt{16 \times 10^{18}(1.016) + (80 \times 10^8)(10.24 \times 10^6)} - 10.24 \times 10^6}$

$= \sqrt{40.42 \times 10^8 - 0.1024 \times 10^8}$

$= 63,496.29 \text{ rad/s}$

d) $Y(\omega_m) = \dfrac{1}{10^4} + \dfrac{1}{80 + j\omega_m 25 \times 10^{-3}} + j\omega_m \times 10^{-8}$

$= (100 + 31.67 - j628.36 + j634.96) \ \mu\mho$

$= 131.83 \underline{/2.87°} \ \mu\mho;$

$V_m = \dfrac{4 \times 10^{-3}}{131.83 \times 10^{-6}} = 30.34 \text{ V} \quad \blacksquare$

By comparing the results obtained in Example 14.4 with those obtained in Example 14.3, we see that as the coil resistance decreases, the unity power-factor frequency (ω_0) approaches the frequency at which the amplitude of v_o is maximum (ω_m). We also note that the amplitude of v_o at ω_0 is very nearly the same as the amplitude of v_o at ω_m. As the coil resistance approaches zero, the maximum amplitude approaches its largest possible value. The largest possible amplitude of v_o is $I_m R_g$. For the circuit in Examples 14.3 and 14.4, this is $(4 \times 10^{-3})(10 \times 10^3)$, or 40 V.

Manufacturers of coils intended for frequency-selective circuits specify the Q of the coil at specific frequencies. The Q of the coil in Example 14.3 at unity power-factor resonance is

$$Q_{\text{coil}} = \frac{\omega_0 L}{R} \cong \frac{(54.55)(25)}{800} = 1.70,$$

while the Q of the coil in Example 14.4 at ω_0 is

$$Q_{\text{coil}} \cong \frac{(63.16)(25)}{80} = 19.74.$$

Thus the higher Q coil results in a larger peak output voltage. For high-Q coils the bandwidth of the circuit in Fig. 14.11 can be approximated by

$$\beta \cong \frac{R}{L} + \frac{1}{R_g C}. \tag{14.69}$$

The derivation of Eq. (14.69) is left to the reader in Problem 14.17.

We can use Eq. (14.69) to estimate the bandwidth of the circuit in Example 14.4 because the Q of the coil is greater than 5. For the circuit in Example 14.4,

$$\beta = \frac{80}{25} \times 10^3 + \frac{10^9}{10^4(10)} = 3200 + 10^4 = 13,200 \text{ rad/s}.$$

The Q for the circuit can now be calculated from the defining equation

$$Q = \frac{\omega_0}{\beta} = \frac{63,164.55}{13,200} = 4.79.$$

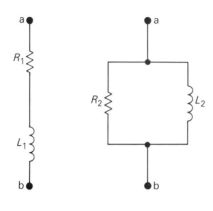

Fig. 14.12 Series–parallel *RL* equivalent circuits.

We see in this instance that even though the Q of the coil is relatively high, the Q of the circuit is significantly reduced by the relatively low internal resistance of the source R_g.

There is a second method for estimating the behavior of the circuit in Fig. 14.11 when the Q of the coil is 5 or greater. This second approach makes use of the equivalent circuit developed in Problem 10.16. The circuit is repeated in Fig. 14.12. The equations that relate R_2 and L_2 to R_1 and L_1 are given as part of Problem 10.16 and repeated here:

$$R_2 = \frac{R_1^2 + \omega^2 L_1^2}{R_1}, \tag{14.70}$$

$$L_2 = \frac{R_1^2 + \omega^2 L_1^2}{\omega^2 L_1}. \tag{14.71}$$

Before showing how we can use the parallel equivalent circuit in Fig. 14.12 to analyze the frequency behavior of the circuit in Fig. 14.11, some preliminary comments are in order. First, the two circuits are exactly equivalent at a single frequency. If ω changes, R_2 and L_2 must be changed in order that the parallel circuit properly represents the original series circuit at the new frequency. However, if we restrict ω to values close to ω_0, we can assume that R_2 and L_2 are essentially constant. In calculating the parameters R_2 and L_2 at the resonant frequency ω_0, it is convenient to express them in terms of the Q of the series circuit. Thus if we let

$$Q_s = \frac{\omega_0 L_1}{R_1}, \tag{14.72}$$

then

$$R_2 = R_1(1 + Q_s^2) \tag{14.73}$$

and

$$L_2 = L_1[1 + (1/Q_s^2)]. \tag{14.74}$$

Example 14.5 illustrates how the parallel equivalent circuit in Fig. 14.12, along with Eqs. (14.73) and (14.74), can be used to estimate the bandwidth of the circuit in Example 14.4.

Example 14.5   a) Find the parallel equivalent circuit for the inductor in the circuit in Example 14.4.

b) Estimate the bandwidth of the circuit using the circuit derived in part (a).

c) Estimate the value of the half-power frequencies ω_1 and ω_2.

d) Calculate $|V_o(\omega_1)|$ and $|V_o(\omega_2)|$. Compare these magnitudes to $|V_o(\omega_0)|/\sqrt{2}$.

Solution   a) From Example 14.4 we have $R = 80 \ \Omega$, $L = 25$ mH, and $\omega_0 = 63{,}496.29$ rad/s. The Q of series-connected coil is calculated in the discussion following Example 14.4 and is found to be 19.74. Therefore, the resistance in the parallel equivalent circuit is

$$R_2 = 80[1 + (19.74)^2] = 31{,}253.41 \ \Omega$$

and the inductance is

$$L_2 = 25[1 + (1/19.74)^2] = 25.06 \cong 25 \text{ mH}.$$

b) The bandwidth of the circuit is

$$\beta = \frac{1}{R_{eq}C} = \frac{10^8}{R_{eq}},$$

where R_{eq} is the parallel combination of the source resistance R_g and R_2. For the circuit at hand,

$$R_{eq} = \frac{(31{,}253.41)(10{,}000)}{41{,}253.41} = 7575.96 \ \Omega.$$

Therefore

$$\beta = \frac{10^8}{7575.96} = 13{,}200 \text{ rad/s}.$$

Note that this value for β is the same as the one we obtained earlier using Eq. (14.69) [see the discussion following Eq. (14.69) and Problem 14.17].

c) Since we have a high-Q coil in the circuit, we can approximate the half-power frequencies by assuming that ω_0 lies in the center of the passband. Thus

$$\omega_1 = \omega_0 - \frac{\beta}{2} = 63{,}164.55 - 6600 = 56{,}564.55 \text{ rad/s},$$

$$\omega_2 = \omega_0 + \frac{\beta}{2} = 63{,}164.55 + 6600 = 69{,}764.55 \text{ rad/s}.$$

d) The magnitude of the output voltage can be calculated from the relationship

$$|V_o(\omega)| = \frac{I_m}{|Y(\omega)|}$$

Thus

$$Y(\omega_1) = \frac{1}{10^4} + \frac{1}{80 + j1414.11} + j5.656 \times 10^{-4} = 1.97\underline{/-44.88°} \times 10^{-4} \, \mho;$$

$$|V_o(\omega_1)| = \frac{4 \times 10^{-3} \times 10^4}{1.97} = 20.26 \, \text{V};$$

$$Y(\omega_2) = \frac{1}{10^4} + \frac{1}{80 + j1744.11} + j697.65 \times 10^{-6} = 1.78\underline{/44.83°} \times 10^{-4} \, \mho;$$

$$|V_o(\omega_2)| = \frac{4 \times 10^{-3} \times 10^4}{1.78} = 22.47 \, \text{V}.$$

From Example 14.4, $|V_o(\omega_0)| = 30.30$. Therefore

$$\frac{|V_o(\omega_0)|}{\sqrt{2}} = \frac{30.30}{\sqrt{2}} = 21.43 \, \text{V}.$$

Since the magnitude of $|V_o(\omega_1)|$ and $|V_o(\omega_2)|$ are very close to $|V_o(\omega_0)|/\sqrt{2}$, it follows that our estimate of the half-power frequencies is very close. ■

Equations (14.73) and (14.74) give the parameters of the parallel equivalent circuit in terms of the series circuit parameters. It is possible to reverse the process, that is, to express R_1 and L_1 as functions of R_2 and L_2. The basic relationships are stated in Problem 10.15 and repeated here for convenience. Thus

$$R_1 = \frac{\omega^2 L_2^2 R_2}{R_2^2 + \omega^2 L_2^2}, \tag{14.75}$$

$$L_1 = \frac{R_2^2 L_2}{R_2^2 + \omega^2 L_2^2}. \tag{14.76}$$

It is more convenient to rewrite Eqs. (14.75) and (14.76) in terms of the Q of the original parallel circuit. The Q of the parallel combination of R_2 and L_2 is contained in Eq. (14.11); thus

$$Q_p = \frac{R_2}{\omega_0 L_2}. \tag{14.77}$$

When we substitute Eq. (14.77) into Eqs. (14.75) and (14.76), the results are

$$R_1 = \frac{R_2}{1 + Q_p^2}, \tag{14.78}$$

$$L_1 = \frac{Q_p^2 L_2}{1 + Q_p^2}. \tag{14.79}$$

We also observe from Eqs. (14.75) and (14.76) that if we calculate the ratio $\omega_0 L_1 / R_1$ we get

$$\frac{\omega_0 L_1}{R_1} = \frac{\omega_0 R_2^2 L_2}{R_2^2 + \omega_0^2 L_2^2} \cdot \frac{R_2^2 + \omega_0^2 L_2^2}{\omega_0^2 L_2^2 R_2} = \frac{R_2}{\omega_0 L_2},$$

or

$$Q_s = Q_p. \tag{14.80}$$

In other words, we can calculate the Q of the equivalent circuits from either the series or parallel equivalent.

The series–parallel RC circuit of Problem 10.13 and 10.14 can also be expressed in terms of the Q of the circuits. The circuit is redrawn in Fig. 14.13 and the appropriate relationships are summarized as follows:

$$Q_p = \omega_0 R_2 C_2, \tag{14.81}$$

$$Q_s = \frac{1}{\omega_0 R_1 C_1}, \tag{14.82}$$

$$R_1 = \frac{R_2}{1 + Q_p^2}, \tag{14.83}$$

$$C_1 = \frac{1 + Q_p^2}{Q_p^2} C_2, \tag{14.84}$$

$$R_2 = (1 + Q_s^2) R_1, \tag{14.85}$$

$$C_2 = \frac{Q_s^2 C_1}{1 + Q_s^2}. \tag{14.86}$$

As in the case of the RL circuits, $Q_s = Q_p$, hence the Q can be calculated from either equivalent circuit.

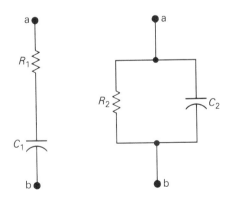

Fig. 14.13   Series–parallel RC equivalent circuits.

The internal resistance of the current source in the circuit in Fig. 14.11 is infinite, that is, $R_g = \infty$. The 16-mH coil has a resistance of 100 Ω. The lossless capacitor is 100 nF. The peak amplitude of the current source is 5 mA at all frequencies. Calculate (a) ω_m; (b) ω_0; (c) $V_m(\omega_0)$; and (d) $V_m(\omega_m)$.

Ans. (a) 24,976.99 rad/s; (b) 24,206.15; (c) 8 V; (d) 8.25 V.

A 40-mH coil has a resistance of 3.2 kΩ. The coil is shunted by a 6.25-pF lossless capacitor. This parallel-resonant circuit is driven by a sinusoidal current source with an internal impedance of $8 + j0$ MΩ. Calculate (a) ω_0; (b) f_0; (c) β; (d) the Q of the circuit; and (e) the Q of the coil.

Ans. (a) 199.84×10^4 rad/s; (b) 318.06 kHz; (c) 100 krad/s; (d) 19.98; (e) 24.98.

14.8 MORE ON SERIES RESONANCE

In the series-resonant circuit shown in Fig. 14.14 the output voltage is picked off the capacitor in order to take advantage of the fact that at resonance the amplitude of the voltage across the capacitor is Q times the amplitude of the source voltage [Eq. (14.60)]. We now point out that the maximum voltage across the capacitor does not occur at the resonant frequency ω_0. To verify this, we express the amplitude of v_o as a function of ω and then proceed to find the value of ω that causes the amplitude to be a maximum. Using phasor circuit analysis, we have

$$\mathbf{V}_o = \frac{(1/j\omega C)}{R + j(\omega L - 1/\omega C)}\,\mathbf{V}_g$$

$$= \frac{\mathbf{V}_g}{1 - \omega^2 LC + j\omega RC}. \tag{14.87}$$

The magnitude of the output voltage is

$$|\mathbf{V}_o| = \frac{V_m}{\sqrt{(1 - \omega^2 LC)^2 + \omega^2 R^2 C^2}}, \tag{14.88}$$

where V_m is the amplitude of the sinusoidal input voltage v_g.

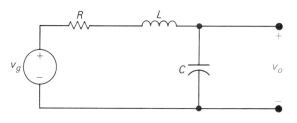

Fig. 14.14 A series-resonant circuit that uses the capacitor voltage as the output signal.

To find the frequency at which the magnitude of v_o is maximum, we differentiate Eq. (14.88) with respect to ω and then find the value of ω that makes the derivative zero. Thus

$$\frac{d|\mathbf{V}_o|}{d\omega} = -\frac{V_m(1/2)[2(1 - \omega^2LC)(-2\omega LC) + 2\omega R^2C^2]}{\{(1 - \omega^2LC)^2 + \omega^2R^2C^2\}^{3/2}}. \tag{14.89}$$

The value of ω greater than zero that makes the derivative equal to zero is

$$\omega_m = \sqrt{\frac{1}{LC} - \frac{1}{2}\left(\frac{R}{L}\right)^2}. \tag{14.90}$$

When we substitute Eq. (14.90) into Eq. (14.88), we find the maximum value of $|\mathbf{V}_o|$; thus

$$|\mathbf{V}_o|(\omega_m) = \frac{V_m}{\sqrt{\frac{R^2C}{L}\left(1 - \frac{R^2C}{4L}\right)}}. \tag{14.91}$$

It is of interest to express Eq. (14.91) in terms of the Q of the circuit. From Eq. (14.51) we see that

$$\frac{R^2C}{L} = \frac{1}{Q^2}, \tag{14.92}$$

from which it follows that

$$|\mathbf{V}_o|(\omega_m) = \frac{QV_m}{\sqrt{1 - (1/2Q)^2}} \tag{14.93}$$

Equations (14.90) and (14.93) tell us that for high-Q circuits the frequency at which the maximum occurs (ω_m) approaches the resonant frequency (ω_0) and the maximum value of $|\mathbf{V}_o|$ approaches QV_m.

Example 14.6   The sinusoidal voltage source in the circuit in Fig. 14.14 delivers a voltage of 10 cos ωt V to a 50-Ω resistor, a 5-mH inductor, and a 0.5-μF capacitor.

a) Calculate ω_0 and Q.
b) What is the peak amplitude of v_o at ω_0?
c) At what frequency will the peak amplitude of v_o be maximum?
d) What is the peak amplitude of v_o at the frequency given in part (c)?
e) Sketch $|\mathbf{V}_o|$ vs. ω.
f) Repeat parts (a) through (e) given that R is decreased to 10 Ω.

Solution   a) $\omega_0 = 1/\sqrt{LC} = 20{,}000$ rad/s; $Q = \omega_0L/R = 2$
b) $|\mathbf{V}_o| = QV_m = 2(10) = 20$ V
c) $\omega_m = \sqrt{4 \times 10^8 - 0.5 \times 10^8} = 18{,}708.29$ rad/s
d) $|\mathbf{V}_o| = 2(10)/\sqrt{1 - 1/16} = 20.66$ V

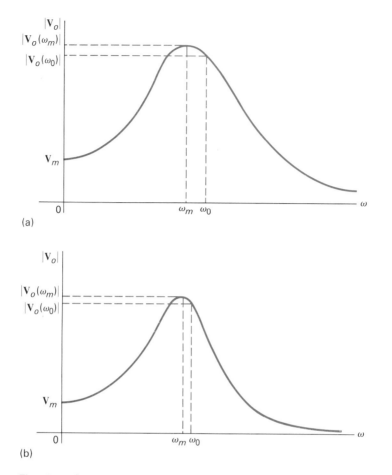

Fig. 14.15 $|\mathbf{V}_o(\omega)|$ versus ω for Example 14.6: (a) $R = 50\ \Omega$; (b) $R = 10\ \Omega$.

e) A sketch of $|\mathbf{V}_o|$ vs. ω is shown in Fig. 14.15(a). The sketch shows the relationship between the magnitude of $\mathbf{V}_o$ at ω_m and ω_0.

f) $\omega_0 = 20{,}000$ rad/s; $Q = \omega_0 L/R = 10$; $|\mathbf{V}_o(\omega_0)| = (10)(10) = 100$ V; $\omega_m = \sqrt{4 \times 10^8 - 10^8/50} = 19{,}949.94$; $|\mathbf{V}_o(\omega_m)| = 100\sqrt{1 - 1/400} = 100.13$ V. ∎

The effect of "loading" the capacitor in the circuit in Fig. 14.14 is of practical importance. That is, the circuit, or device, driven from the capacitor will shunt the capacitor with its input impedance. This impedance is typically resistive in nature and represents a load across the output terminals of the series-resonant circuit. The loading effect is indicated by R_L in the circuit in Fig. 14.16.

The analysis of the circuit in Fig. 14.16 at, or near, the resonant frequency is facilitated by the RC equivalent circuit shown in Fig. 14.13. The parallel

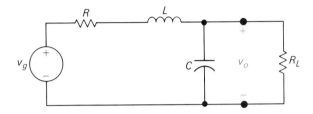

Fig. 14.16 The series-resonant circuit in Fig. 14.14 with a resistive load.

combination of R_L and C can be replaced by a series combination of R_1 and C_1 by means of Eqs. (14.83) and (14.84). So long as the frequency does not deviate greatly from ω_0, the equivalent resistance (R_1) and capacitance (C_1) can be assumed constant. We can see from Eq. (14.84) that if the Q of the parallel combination is greater than 5, $C_1 \cong C$ and the resonant frequency ω_0 is not greatly disturbed by R_L.

The effect of R_L on the magnitude of the output voltage (assuming $\omega_0 R_L C > 5$) can be found as follows. If we use the series-equivalent circuit for R_L and C, the phasor domain equivalent circuit is as shown in Fig. 14.17. Since $C_1 \cong C$ the phasor current at resonance is very nearly $\mathbf{V}_g/(R + R_1)$ and the output voltage is

$$\mathbf{V}_o(\omega_0) \cong \frac{\mathbf{V}_g}{R + R_1} \cdot \left[R_1 - j\frac{1}{\omega_0 C} \right]. \tag{14.94}$$

For $Q_p > 5$, R_1 can be replaced by R_L/Q_p^2.

The coefficient of j in Eq. (14.94) is equivalent to $R/\omega_0 RC$ or QR, where Q is the quality factor of the series-resonant circuit *before* loading. Therefore, we can write Eq. (14.94) as

$$\mathbf{V}_o(\omega_0) = \frac{\mathbf{V}_g Q_p^2}{RQ_p^2 + R_L} \left[\frac{R_L}{Q_p^2} - jQR \right] = \frac{\mathbf{V}_g}{RQ_p^2 + R_L} (R_L - jQRQ_p^2). \tag{14.95}$$

The magnitude of the output voltage at resonance is

$$|\mathbf{V}_o(\omega_0)| = \frac{|\mathbf{V}_g|}{RQ_p^2 + R_L} \sqrt{R_L^2 + Q^2 R^2 Q_p^4}$$

$$= \frac{|\mathbf{V}_g| QRQ_p^2 \sqrt{1 + (R_L^2/Q^2 R^2 Q_p^4)}}{RQ_p^2 [1 + (R_L/RQ_p^2)]}$$

$$= Q|\mathbf{V}_g| \frac{\sqrt{1 + (R_L/Q_p^2 QR)^2}}{[1 + (R_L/RQ_p^2)]}. \tag{14.96}$$

Now since the magnitude of $\mathbf{V}_o$ at resonance before loading is $Q|\mathbf{V}_g|$, we can use Eq. (14.96) to assess the effect that the load resistor has on the magnitude of the output voltage at ω_0. Calculations showing the analysis of the circuit in Fig. 14.17 are illustrated in Example 14.7.

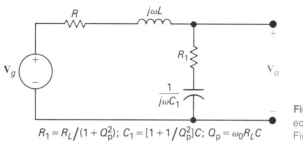

$R_1 = R_L/(1 + Q_p^2); \quad C_1 = [1 + 1/Q_p^2]C; \quad Q_p = \omega_0 R_L C$

Fig. 14.17 A phasor domain equivalent circuit for the circuit in Fig. 14.16.

Example 14.7 Assume that the circuit in Example 14.6 (with $R = 10\ \Omega$) is loaded with a resistance of 10 kΩ.

a) Calculate the Q of the parallel RC portion of the circuit.

b) Calculate the equivalent circuit parameters R_1 and C_1.

c) Calculate the peak amplitude of v_o at ω_0.

d) Repeat parts (a), (b), and (c) given that R_L is increased to 100 kΩ.

Solution a) Assume that Q_p is greater than 5 so that R_L has little effect on ω_0. [If we get a contradiction to this assumption, we would have to calculate ω_0 from the definition of unity power-factor resonance (see Problem 14.19) and the analytical techniques that are based on a highly frequency-selective circuit would have to be abandoned.] We have

$$Q_p = \omega_0 R_L C = (2 \times 10^4)(10^4)(0.5 \times 10^{-6}) = 100.$$

b) The equivalent circuit parameters are

$$R_1 = \frac{R_L}{1 + Q_p^2} \cong \frac{10^4}{10^4} = 1\ \Omega,$$
$$C_1 = (1 + 1/Q_p^2)C \cong 0.5\ \mu\text{F}.$$

c) In anticipation of using Eq. (14.96) to calculate $|\mathbf{V}_o(\omega_0)|$, we make the following preliminary calculations:

$$Q = 10; \quad \text{(See Example 14.6f.)}$$
$$\left[\frac{R_L}{Q_p^2 QR}\right]^2 = \left[\frac{10^4}{(10^4)(10)(10)}\right]^2 = [10^{-2}]^2 = 10^{-4};$$
$$\frac{R_L}{RQ_p^2} = \frac{10^4}{(10)(10^4)} = 10^{-1};$$
$$|\mathbf{V}_o(\omega_0)| = \frac{(10)(10)\sqrt{1 + 10^{-4}}}{(1 + 0.1)}$$
$$= \frac{100\sqrt{1.0001}}{1.1} = 90.91\ \text{V}.$$

It is of interest to note that although R_L has an almost negligible effect on the resonant frequency, it does cause about a 9% drop in the peak amplitude of v_o.

d) When $R_L = 100 \text{ k}\Omega$ we get $Q_p = 1000$, $R_1 = 0.1 \ \Omega$, $C_1 = 0.5 \ \mu\text{F}$, and $|\mathbf{V}_o(\omega_0)| = (10)(10)/1.01 = 99.01 \text{ V}$. ■

The circuit in Fig. 14.16 is also amenable to exact analysis. The interested reader can pursue the exact analysis via Problem 14.20.

DRILL EXERCISE

14.10

The circuit elements in Fig. 14.14 are $R = 80 \ \Omega$, $L = 50 \text{ mH}$, and $C = 0.2 \ \mu\text{F}$.

a) If the circuit is excited from a sinusoidal voltage source $20 \cos \omega_0 t$, what is the rms value of $v_o(t)$?

b) The rms value of v_o is measured with a digital voltmeter that has a resistance of 500 kΩ. What is the rms reading of the voltmeter?

Ans. (a) 88.39 V; (b) 87.84 V.

14.9 SCALING

It is convenient in the design and analysis of frequency-selective circuits to work with element values such as 1 Ω, 1 H, and 1 F. Although these values are unrealistic in terms of practical components, they have the advantage of greatly simplifying computations. Once we have made the computations (using the unrealistic values of R, L, and C), we can transform the element values into realistic values by the process of *scaling*.

There are two types of scaling: *magnitude* and *frequency*. A circuit is scaled in magnitude by multiplying the impedance at a given frequency by the scaling factor k_m. Thus a given circuit is scaled in magnitude by multiplying all resistors and inductors by k_m and all capacitors by $1/k_m$. If we let the subscripts 1 and 2 represent the initial and final values of the parameters, respectively, we have

$$R_2 = k_m R_1,$$
$$L_2 = k_m L_1, \tag{14.97}$$
$$C_2 = C_1/k_m,$$

where k_m is by definition a positive real number. It can be either less than or greater than one.

In frequency scaling, the circuit parameters are changed so that at the new frequency the impedance of each element is the same as it was at the original frequency. Since resistance values are assumed to be independent of frequency, resistors are unaffected by frequency scaling. If we let k_f denote the frequency scaling factor, then both inductors and capacitors are multiplied by $1/k_f$. Thus for frequency scaling,

$$R_2 = R_1,$$
$$L_2 = L_1/k_f,$$
$$C_2 = C_1/k_f.$$

(14.98)

The frequency scaling factor k_f is also a positive real number that can be less than or greater than unity.

A circuit can be simultaneously scaled in both magnitude and frequency. The "new" values (2) in terms of the "old" (1) values are

$$R_2 = k_m R_1,$$
$$L_2 = \frac{k_m}{k_f} L_1,$$
$$C_2 = \frac{1}{k_m k_f} C_1.$$

(14.99)

The following example illustrates the application of the scaling process.

Example 14.8 Scale the circuit in Example 14.1 so that the resonant frequency is shifted to 20 krad/s and the impedance at resonance is 10 kΩ.

Solution From Example 14.1 we know that the original resonant frequency is 10 krad/s and the impedance at this resonant frequency is 2 kΩ. It follows, then, that

$$k_m = \frac{10 \times 10^3}{2 \times 10^3} = 5$$

and

$$k_f = \frac{20 \times 10^3}{10 \times 10^3} = 2.$$

The new values of R, L, and C are

$$R = 5(2) = 10 \text{ k}\Omega,$$
$$L = \frac{5}{2}(40) = 100 \text{ mH},$$

and

$$C = \frac{0.25}{(5)(2)} = 0.025 \text{ } \mu\text{F}. \quad \blacksquare$$

To use the concept of scaling in the design of a frequency-selective circuit, we select the resonant frequency as 1 rad/s and the magnitude of the impedance at resonance as 1 Ω. We then calculate the values of L and C on the basis of the desired Q. Once the values of L and C are computed on the basis of 1 Ω, 1 rad/s, and Q, the circuit is scaled to give the desired resonant frequency and impedance level. The circuit with a resonant frequency of 1 rad/s and an impedance of 1 Ω at resonance is referred to as a *universal* resonant circuit.

For the universal parallel-resonant circuit we have

$$Q = \frac{R}{\omega_0 L} = \frac{1}{L}, \tag{14.100}$$

or

$$Q = \omega_0 RC = C. \tag{14.101}$$

Therefore once Q is specified, L and C can be computed. Observe that since ω_0 is unity, $Q = 1/\beta$ and thus either Q or β can be used to define L or C.

For the universal series-resonant circuit

$$Q = \frac{\omega_0 L}{R} = L, \tag{14.102}$$

or

$$Q = \frac{1}{\omega_0 CR} = \frac{1}{C}. \tag{14.103}$$

Example 14.9   a) Compute the values of L and C for a universal parallel-resonant circuit having a Q of 20.

b) Scale the circuit in part (a) so that the resonant frequency is 1.04 MHz and the impedance at resonance is 25 kΩ.

Solution   a) From Eqs. (14.100) and (14.101) we have $L = 1/20 = 0.05$ H and $C = 20$ F.

b) The magnitude scaling factor is 25×10^3 and the frequency scaling factor is $2.08\pi \times 10^6$. Therefore the circuit values are

$$R = 25 \text{ k}\Omega,$$

$$L = \frac{25 \times 10^3}{2.08\pi \times 10^6}(0.05) = 191.29 \ \mu\text{H},$$

$$C = \frac{20}{(25 \times 10^3)(2.08\pi \times 10^6)} = 122.43 \text{ pF}. \quad \blacksquare$$

DRILL EXERICSE
14.11

Scale the circuit in Drill Exercise 14.1 so that the resonant frequency is shifted to 8 Mrad/s and the impedance at resonance is 400 kΩ. Specify the values of (a) k_m; (b) k_f; (c) R_2; (d) L_2; (e) C_2; (f) Q; and (g) β.

Ans. (a) 4; (b) 0.8; (c) 400 kΩ; (d) 500 μH; (e) 31.25 pF; (f) 100; (g) 80 krad/s.

DRILL EXERCISE   A universal parallel-resonant circuit has a bandwidth of 0.0625 rad/s.

14.12   a) Compute the values of L and C.

b) Scale the circuit in part (a) so that the resonant frequency is 1.35 MHz and the impedance at resonance is 10 kΩ.

Ans. (a) 62.5 mH, 16 F; (b) 10 kΩ, 73.68 μH, 188.63 pF.

DRILL EXERCISE   The circuit shown is to be scaled so that $\omega_0 = 7.5$ Mrad/s and $C = 10$ pF. Specify (a) k_f;

14.13   (b) k_m; (c) R; (d) L; (e) Q; (f) β.

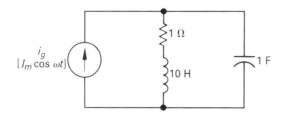

Ans. (a) 25×10^6; (b) 4000; (c) 4000 Ω; (d) 1.6 mH; (e) 3; (f) 2.5 Mrad/s.

14.10   SUMMARY

Our purpose in this chapter has been to study the frequency response of several simple RLC circuits. By frequency response, we mean how the amplitude and phase angle of an output signal are determined as the frequency of the input signal is increased from zero to infinity. While the frequency of the input signal is increased, its amplitude and phase angle are held constant.

The series and parallel RLC circuits discussed in this chapter are of interest because they transmit sinusoidal signals at some frequencies much more effectively than at other frequencies. In order to quantify the frequency-selectivity properties of these circuits, we introduced the concepts of resonance, bandwidth, and quality factor. The resonant frequency (ω_0) is the frequency at which the output is in phase with the input. Thus at resonance, the circuit is operating at unity power factor. The peak amplitude of the output signal either coincides with the resonant frequency (ω_0) or occurs at a frequency very close to resonance (ω_m). The higher the quality factor (Q), the closer the coincidence.

The bandwidth (β) is defined as the range of frequencies for which the peak amplitude of the response is at least $(1/\sqrt{2})$ times the maximum peak amplitude. Circuits that are intended to be very frequency-selective have a narrow bandwidth. How frequency-selective a circuit is depends on both the bandwidth and the resonant frequency. Thus a circuit with a bandwidth of 10 Hz relative to a resonant frequency of 1000 Hz is much more selective than a circuit with a bandwidth of 10 Hz relative to a resonant frequency of 100 Hz. The quality factor (Q) of the resonant circuit recognizes this attribute of frequency selectivity since it is defined as the ratio of the resonant frequency to the bandwidth ($Q = \omega_0/\beta$).

Once we have established the resonant frequency and quality factor for a circuit, we can describe the behavior of the circuit near resonance in terms of ω_0 and Q. The behavior of a circuit where the maximum response occurs at ω_m rather than ω_0 is also related to ω_0 and Q.

The frequency response of a circuit is also important because it reveals something about the natural or step response of the circuit. For the parallel and series RLC circuits discussed in this chapter, we see that a sharp resonant peak predicts a slowly decaying oscillatory response.

PROBLEMS

14.1 Design a parallel-resonant structure like that shown in Fig. 14.1 to meet the following specifications: $C = 400$ pF, $\omega_0 = 10^6$ rad/s, and $Q = 16$.

    a) Specify the numerical values of R and L.
    b) Calculate the numerical values of ω_1 and ω_2.

14.2 The parallel-resonant circuit in Fig. 14.1 has a resonant frequency of 800 krad/s. The upper half-power frequency is 900 krad/s. Find (a) the lower half-power frequency; (b) the bandwidth in kHz; and (c) the Q of the circuit.

14.3 A sinusoidal current source delivers a current having an amplitude of 2 mA. The internal impedance of the source is $100,000 + j0$ Ω. This current source is used with a 25-mH inductor and a 400-kΩ resistor to form a parallel-resonant circuit. The resonant frequency is 500 krad/s.

    a) Specify the numerical value of the circuit capacitor.
    b) What is the Q of the circuit without the source?
    c) What is the Q of the circuit with the source?
    d) What is the bandwidth of the circuit with and without the source?
    e) What is the maximum output voltage that the source can deliver?
    f) What is the maximum value of the capacitor current at resonance?

14.4 The frequency of the sinusoidal voltage source in the circuit in Fig. 14.18 is adjusted until the amplitude of the sinusoidal output voltage is maximum. The maximum amplitude of the source voltage is 400 V.

    a) What is the frequency of v_s in hertz?
    b) What is the amplitude of v_o at the frequency given in part (a)?
    c) What is the bandwidth of the circuit?

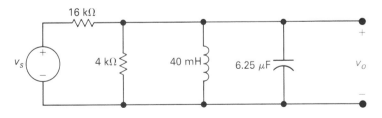

Fig. 14.18 The circuit for Problem 14.4.

d) What is the Q of the circuit?

e) At what frequencies will the amplitude of v_o be $1/\sqrt{2}$ times its maximum value?

f) If the 16-kΩ resistor represents the internal resistance of the source, how much does this source resistance lower the Q of the circuit?

14.5 a) For the circuit in Example 14.1(b), calculate the magnitude of $\mathbf{V}_o$ at 10,500 rad/s and at 9500 rad/s using the exact formula for Y.

b) Repeat part (a) using the approximate formula for Y.

c) Find the percent error in the magnitude of $\mathbf{V}_o$ that results from using the approximate expression for Y.

14.6 Show that when $\omega = \omega_0$ in the circuit in Fig. 14.1 the magnitude of the current in both the inductive and capacitive branches is QI_m.

14.7 Select the circuit values for the parallel-resonant structure in Fig. 14.1 so that $\omega_0 = 500{,}000$ rad/s, $\beta = 20{,}000$ rad/s, $V_m(\omega_0) = 200$ V, and $|I_C(\omega_0)| = 20$ A. Give the values of R, L, C, and I_m.

14.8 Show that Eq. (14.29) predicts $\theta = +45°$ when $\omega = \omega_1$ and $-45°$ when $\omega = \omega_2$. (*Hint:* Use the fact that $\omega_0 = \sqrt{\omega_1 \omega_2}$.)

14.9 A parallel-resonant structure as shown in Fig. 14.1 is to satisfy the following specifications: $f_0 = 159.155$ kHz, $Z(f_0) = 10$ kΩ, $|Z(f_a)| = 5000$ Ω, and $f_a = 150$ kHz.

a) Specify R, L, and C.

b) What is the Q of the circuit?

c) What is the bandwidth of the circuit?

d) If this circuit is driven by a sinusoidal current source having an internal impedance of $40 + j0$ kΩ, what is the Q of the energized circuit?

e) What is the bandwidth of the energized circuit?

14.10 A parallel RLC circuit has energy stored in the 0.4-H inductor but not the capacitor. When the stored energy is released, the voltage appearing across the circuit is

$$v = -10e^{-200t} \sin 2400t \text{ V}, \qquad t > 0.$$

a) Compute the bandwidth, Q, and resonant frequency of the circuit.

b) Compute the half-power frequencies ω_1 and ω_2.

c) If the parallel RLC circuit is driven from a sinusoidal current source having an internal resistance of 34.2 kΩ, what is the Q, bandwidth, and resonant frequency?

14.11 The values of the circuit parameters for the circuit in Fig. 14.8 are $R = 200$ Ω, $L = 12.5$ mH, and $C = 8$ nF.

a) Calculate Q, β, ω_0, ω_1, and ω_2.

b) Assume that the circuit is driven from an ideal sinusoidal voltage source having a maximum amplitude of 100 V. Calculate the amplitude of the voltage across the capacitor at resonance.

c) Repeat parts (a) and (b) given that the sinusoidal voltage source has an internal resistance of 50 Ω.

14.12 For the circuit in Problem 14.11, calculate the magnitude and phase angle of the impedance seen by the voltage source at ω_0 and at frequencies 5000 rad/s larger and smaller than ω_0.

14.13 A coil having a resistance of 15 Ω and an inductance of 25 μH is used in the circuit in Fig. 14.19. The resonant frequency of the circuit in 10 MHz and the bandwidth is 1 MHz. The internal resistance of the sinusoidal voltage source is 50 Ω.

a) Specify the numerical values of R and C.

b) What is the Q of the coil at the resonant frequency ω_0? ($Q_{coil} = \omega_0 L / R_{coil}$)

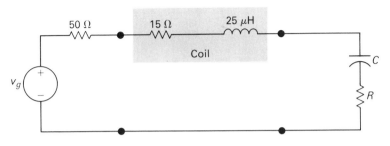

Fig. 14.19 The circuit for Problem 14.13.

14.14 The sinusoidal voltage source in the circuit in Fig. 14.20 has a maximum amplitude of 50 mV. The internal impedance of the source is negligible.

   a) At what frequency is the amplitude of v_o maximum?
   b) What is the maximum amplitude of v_o?
   c) Over what range of frequencies will the amplitude of v_o be equal to or greater than 0.90 of its maximum value?

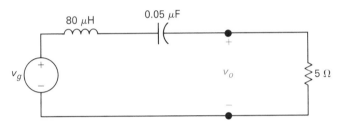

Fig. 14.20 The circuit for Problem 14.14.

14.15 In the circuit in Fig. 14.11, $R_g = 50$ kΩ, $R = 50$ Ω, $L = 100$ μH, and $C = 100$ pF. The peak amplitude of the sinusoidal current source is 1.4 mA.

   a) Calculate the resonant frequency ω_0.
   b) Calculate the amplitude of the output voltage at resonance.
   c) Calculate the frequency at which the amplitude of v_o is maximum.
   d) Calculate the maximum amplitude of v_o.
   e) Estimate the bandwidth of the circuit.
   f) What is the Q of the coil at ω_0?
   g) What is the Q of the circuit at ω_0?

14.16 A 0.4-mH inductor has a Q of 40 at 500 krad/s. This inductor is to be used in the parallel-resonant circuit in Fig. 14.11. The resonant frequency of the circuit is to be 500 krad/s.

   a) Specify the value of the capacitor C. (*Hint:* Use the parallel equivalent circuit for the inductor.)
   b) Specify the value of R_g so that the bandwidth of the circuit is 15,625 rad/s.
   c) What is the Q of the circuit?

14.17 Derive Eq. (14.69). (*Hint:* For a parallel-resonant circuit, $\beta = 1/RC$. For the circuit in Fig. 14.11, R takes the value of R_g in parallel with the equivalent resistor $R_1(1 + Q_s^2)$.)

14.18 The internal impedance of the sinusoidal voltage source in Fig. 14.21 is 40 + $j0$ Ω. The frequency of the source is set at the unity power-factor resonant frequency of the circuit. The amplitude of the source voltage is set to yield an rms value of 500 mV.

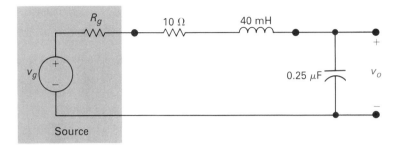

Fig. 14.21 The circuit for Problem 14.18.

   a) What is the rms value of v_o at resonance?

   b) If v_o is measured by an rms voltmeter having a resistance of 10 kΩ, what will the voltmeter read?

   c) Repeat part (b) given that the meter resistance is 100 kΩ.

14.19 a) Show that the impedance seen by the voltage source in the circuit in Fig. 14.16 is

$$Z_{\text{in}} = R + j\omega L + \frac{R_L}{1 + j\omega R_L C}.$$

   b) Show that the impedance found in part (a) is purely resistive when

$$\omega_0 = \sqrt{\frac{1}{LC} - \frac{1}{R_L^2 C^2}}.$$

   c) Show that

$$Z_{\text{in}}(\omega_0) = R + \frac{L}{R_L C} + j0 \ \Omega.$$

   d) Show that

$$|\mathbf{V}_o(\omega_0)| = \frac{|\mathbf{V}_g| R_L \sqrt{LC}}{L + R_L RC}.$$

   (*Hint:* Use the fact that $\omega_0^2 R_L^2 C^2 + 1 = R_L^2 C/L$.)

   e) Use the relationship given in part (d) to calculate $|\mathbf{V}_o(\omega_0)|$ for the circuit in Example 14.7(c) and compare this value to that obtained in Example 14.7(c).

14.20 a) For the circuit in Fig. 14.16 show that the magnitude of $\mathbf{V}_o$ can be expressed as

$$|\mathbf{V}_o(\omega)| = \frac{V_m R_L}{\sqrt{(R + R_L - \omega^2 LCR_L)^2 + \omega^2 (L + RR_L C)^2}},$$

   where V_m is the peak amplitude of the voltage source.

   b) Show that $|\mathbf{V}_o(\omega)|$ is maximum when

$$\omega = \omega_m = \sqrt{\frac{(R + R_L)}{R_L LC} - \frac{(L + RR_L C)^2}{2(LCR_L)^2}}$$

$$= \sqrt{\frac{1}{LC}\left\{1 - \frac{1}{2LC}\left[R^2 C^2 + \left(\frac{L}{R_L}\right)^2\right]\right\}}.$$

c) Show that the expression for $|\mathbf{V}_o(\omega_m)|$ is

$$|\mathbf{V}_o(\omega_m)| = \frac{V_m R_L \sqrt{LC}}{(L + RR_L C) \sqrt{1 - \frac{1}{LC} \left(\frac{L}{2R_L} - \frac{RC}{2}\right)^2}}.$$

(*Hint:* Use the fact that $R + R_L - \omega_m^2 LCR_L = (L + RR_L C)^2 / 2\, LCR_L$.)

d) Use the results of parts (b) and (c) to calculate ω_m and $|\mathbf{V}_o(\omega_m)|$ for the circuit used in Example 14.7.

14.21 In the circuit in Fig. 14.22, the peak amplitude and frequency of the sinusoidal voltage source are constant. The capacitor is adjusted until the peak amplitude of v_o has its maximum value.

  a) What is the expression for C in terms of R, L, and ω?
  b) What is the value of $|\mathbf{V}_o|_{max}$?
  c) What value of C will put the circuit into series resonance?
  d) How do the answers for parts (a) and (c) compare for a high-Q circuit?
  e) How does $|\mathbf{V}_o|_{max}$ compare to $|\mathbf{V}_o(\omega_0)|$ for a high-Q circuit?

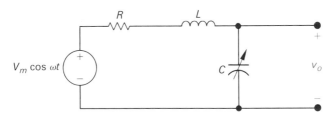

Fig. 14.22 The circuit for Problem 14.21.

14.22 In the circuit shown in Fig. 14.23, the voltage appearing at the input terminals is

$$v_i = 10 \cos 20{,}000\pi t + 10 \cos 10{,}000 \pi t.$$

The circuit capacitors C_1 and C_2 are to be chosen so that the circuit effectively transmits the 5-kHz signal and at the same time effectively blocks the 10-kHz signal.

  a) Specify the numerical values of C_1 and C_2.
  b) Specify the output voltage v_o.

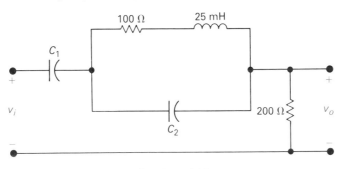

Fig. 14.23 The circuit for Problem 14.22.

14.23 In the circuit in Fig. 14.24, v_g is a sinusoidal signal of fixed peak amplitude and frequency. The capacitance is varied until the peak amplitude of v_o is maximum.

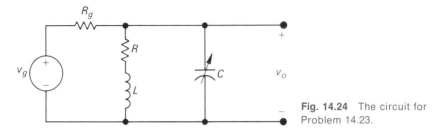

Fig. 14.24 The circuit for Problem 14.23.

a) Derive the expression for C.

b) If $R_g = 10$ kΩ, $R = 800$ Ω, $L = 25$ mH, and $\omega = 64.5$ krad/s, what value of C will make the peak amplitude of v_o maximum?

c) For the numerical values of part (b), what is the ratio of the peak amplitude of v_o to the peak amplitude of v_g?

d) What is the power factor of the circuit when C is adjusted so that v_o has its maximum peak amplitude?

14.24   a) Show that if $R_L = R_C = \sqrt{L/C}$, the circuit shown in Fig. 14.25 will be in resonance for all values of ω.

b) What is the impedance seen by the current source if $R_L = R_C = \sqrt{L/C}$?

c) If $R_L = R_C = \sqrt{L/C}$ and $L = 2.5$ H, $C = 0.1$ μF, $R_g = 20$ kΩ, and $i_g = 5 \cos \omega t$ mA, what is the steady-state expression for v_o?

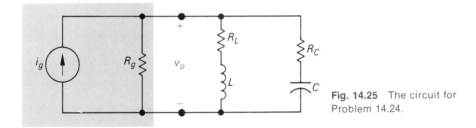

Fig. 14.25 The circuit for Problem 14.24.

14.25   The definition of Q given in Eq. (14.12) is not generally useful in defining the frequency selectivity of a circuit. It is true that Eq. (14.12) gave the same result as Eq. (14.10) for the RLC circuit in Fig. 14.1. However, in general, the two definitions do not lead to the same result. The circuit in Fig. 14.26 illustrates this fact.

a) Calculate the Q of the circuit using Eq. (14.10).

b) Calculate the Q of the circuit using Eq. (14.12). [*Hint:* Note that at resonance the voltage across R is $[V_m R/(R_1 + R)] \cos \omega_0 t$ and the voltage across R_1 is $[V_m R_1/(R + R_1)] \cos \omega_0 t$.]

c) Let $R_1 \to 0$ and comment on the ability of each Q to predict the frequency selectivity of the circuit.

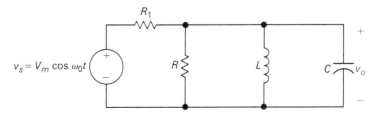

Fig. 14.26 The circuit for Problem 14.25.

14.26 A universal series-resonant circuit has a Q of 40.

a) Find L and C.

b) Scale the circuit in part (a) so that $\omega_0 = 5$ Mrad/s and the impedance at resonance is 50 Ω. Specify the values of R, L, and C.

14.27 a) Calculate ω_0, Q, and β for the circuit shown in Fig. 14.27.

b) Specify the values of k_m, k_f, R, L, and C so that the resonant frequency is changed to 500 krad/s and the impedance at resonance is 100 kΩ.

c) Calculate Q and β for the circuit in part (b).

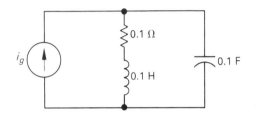

Fig. 14.27   The circuit for Problem 14.27.

14.28 A circuit is scaled in both magnitude and frequency.

a) Show that the Q of the scaled circuit is the same as the Q of the original circuit.

b) Show that the bandwidth of the scaled circuit is k_f times the bandwidth of the original circuit.

14.29 Assume that the ideal op amp in the circuit in Fig. 14.28 is operating in its linear range. The sinusoidal voltage source v_g is generating the signal 200 cos ωt mV. Let the expression for v_o be written as $V_m \cos (\omega t + \theta)$.

a) At what frequency will v_o be 180° out of phase with v_g?

b) What is the value of V_m at the frequency found in part (a)?

c) At what frequencies will V_m be reduced by a factor of $1/\sqrt{2}$ from its value in part (b)?

d) What is the bandwidth of the circuit in rad/s?

e) What is the Q of the circuit?

f) Scale the circuit so that $R_2 = 100$ kΩ and the frequency found in part (a) is shifted to 100 krad/s. What are the values of R_1 and C?

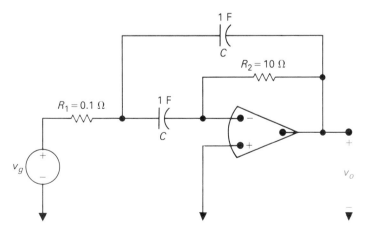

Fig. 14.28   The circuit for Problem 14.29.

15

INTRODUCTION TO THE LAPLACE TRANSFORM

15.1 INTRODUCTION

We are now ready to introduce a very powerful analytical technique that is widely used to study the behavior of linear, lumped-parameter circuits. The method is based on the Laplace transform, which we will define in Section 15.2. Before doing so, we need to explain why another analytical technique is needed. In our earlier study of the transient behavior of circuits (Chapters 6, 7, and 8), we restricted our study to simple circuit structures that could be described by either a single first-order or second-order differential equation. We would now like to expand our capability to include sets of simultaneous, ordinary, differential equations. This will allow us to study the transient behavior of circuits whose description requires more than a single node-voltage differential equation or a single mesh-current differential equation. In Chapters 6, 7, and 8, we also restricted our sources to simple step changes. That is, the sources were allowed to change abruptly only from one dc level to another. The Laplace transform approach will allow us to find the transient response when the signal sources that are switched in or out of a circuit vary with time in ways more involved than simple level jumps.

A second reason for introducing the use of the Laplace transform in circuit analysis is that it gives us a very systematic way for relating the time domain behavior of a circuit to its frequency domain behavior. The ability to work simultaneously in both the time and frequency domains will be greatly enhanced.

As we begin our study of the Laplace transform, it will be helpful if you keep in mind that you have already used the concept of a mathematical transformation to simplify the solution of a problem. For example, the mathematical operations of multiplication and division are transformed into the simpler operations of addition and subtraction by means of the logarithm transform. Thus, we can use the logarithm to transform

$$A = BC \tag{15.1}$$

into an addition problem. Specifically,

$$\log A = \log BC = \log B + \log C. \tag{15.2}$$

To find A, we must be able to carry out the inverse logarithm or antilogarithm operation:

$$A = \text{antilog} \left[\log B + \log C \right]. \tag{15.3}$$

The phasor method introduced in Chapter 10 is a second example of using a mathematical transformation to simplify the solution of a problem. There we reduced the problem of finding the steady-state sinusoidal response of a circuit to the algebraic manipulation of complex numbers. The phasor transform converted the sinusoidal signal to a complex number, which carried the information about the amplitude and phase angle of the signal. Once we determined the phasor value of a signal, we could transform it back to its time domain expression.

Both of these examples point out the essential features of a mathematical transformation. The transformation is designed to create a new domain where the mathematical manipulations are easier to carry out. Once the unknown is found in the new domain, it can be inverse-transformed back to the original domain. In circuit analysis, the Laplace transform is used to transform a set of integrodifferential equations from time domain to a set of algebraic equations in the frequency domain. The solution for an unknown quantity is therefore reduced to the manipulation of algebraic equations. Once the frequency domain expression for the unknown is obtained, it can be inverse-transformed back to the time domain.

In this chapter, we will introduce the Laplace transform, discuss its pertinent characteristics, and present a systematic method for finding inverse transforms. In Chapters 16 and 17, we will show how the Laplace transform is used in circuit analysis. We begin with the definition of the transform.

15.2 DEFINITION OF THE LAPLACE TRANSFORM

The Laplace transform of a function is given by the expression

$$\mathcal{L}\{f(t)\} = \int_0^\infty f(t)e^{-st}\,dt, \tag{15.4}$$

where the symbol $\mathcal{L}\{f(t)\}$ is read as "the Laplace transform of $f(t)$."

The Laplace transform of $f(t)$ is also denoted by the notation of $F(s)$, that is,

$$F(s) = \mathcal{L}\{f(t)\}. \tag{15.5}$$

This notation emphasizes that once the integral in Eq. (15.4) has been evaluated, the resulting expression is a function of s. In our applications, t represents the time domain and since the exponent of e in the integral of Eq. (15.4) must be dimensionless, s must have the dimension of reciprocal time, or frequency. In circuit applications, the Laplace transform transforms the problem from the time domain to the frequency domain.

Before we illustrate some of the important properties of the Laplace transform, some general comments are in order. First note that the integral in Eq. (15.4) is improper because the upper limit is infinite. Thus we are immediately confronted with the question of whether or not the integral converges. That is, does a given $f(t)$ have a Laplace transform? Obviously, the functions of primary interest in engineering analysis have Laplace transforms; otherwise we would not be interested in the transform. We will excite our circuits with sources that have Laplace transforms. Excitation functions such as t^t or e^{t^2}, which do not have Laplace transforms, are of no interest in linear circuit analysis.

Our second general comment about the integral in Eq. (15.4) concerns the lower limit. Because the lower limit on the integral is zero, the Laplace transform ignores $f(t)$ for negative values of t. Or to put it another way, $F(s)$ is determined only by the behavior of $f(t)$ for positive values of t. To call attention to the fact that the lower limit is zero, Eq. (15.4) is frequently referred to as the *one-sided, or unilateral,* Laplace transform. In the two-sided, or bilateral, Laplace transform, the lower limit is minus infinity. We will not use the bilateral form in our work here; hence $F(s)$ is understood to be the one-sided transform. Another point to be made regarding the lower limit concerns the situation when $f(t)$ has a discontinuity at the origin. If $f(t)$ is continuous at the origin, as, for example, in Fig. 15.1(a), there is no ambiguity in the meaning of $f(0)$. On the other hand, if $f(t)$ has a finite discontinuity at the origin, as, for example, in Fig. 15.1(b), the question arises as to whether the Laplace transform integral should include or exclude the discontinuity. That is, should we make the lower limit 0^- and include the discontinuity or should we exclude the discontinuity by making the lower limit 0^+? (The notation 0^- and 0^+ is used to denote values of t just to the left and right of the origin, respectively.) It turns out that we can choose either so long as we are consistent. For reasons to be explained later, we will choose 0^- as the lower limit.

Since we are using 0^- as the lower limit on the Laplace transform integral we need to note right away that the integration from 0^- to 0^+ will be zero *except when there is an impulse function at the origin.* (We will introduce the concept of an impulse function in Section 15.4, and discuss its significance in circuit analysis in Chapter 16.) The important thing to observe now is that the two functions shown in Fig. 15.1 will have the same unilateral Laplace transform since there is no impulse function at the origin.

The one-sided Laplace transform ignores $f(t)$ for $t < 0^-$. What happens prior to 0^- is accounted for by the initial conditions. Thus we will use the Laplace transform to predict the response to a disturbance that occurs after the initial conditions have been established.

With these general observations in mind, we are ready to demonstrate the usefulness of Eq. (15.4). We begin by noting that Laplace transforms can be di-

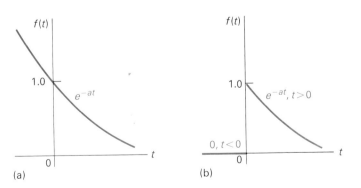

Fig. 15.1   A continuous and a discontinuous function at the origin: (a) $f(t)$ is continuous at the origin; (b) $f(t)$ is discontinuous at the origin.

vided into two types: *functional transforms* and *operational transforms*. The first involves finding the Laplace transform of a specific function, like $\sin \omega t$, t, e^{-at}, etc. The second is concerned with transforms involving mathematical operations on $f(t)$, such as finding the transform of the derivative of $f(t)$. These two types of transforms are discussed in Sections 15.4 and 15.6, respectively. Before introducing functional and operational transforms, we need to pause and introduce the step and impulse functions.

15.3 THE STEP FUNCTION

We noted in our introduction to the unilateral Laplace transform that we would encounter functions that have a discontinuity, or jump, at the origin. We also know from our earlier discussion of transient behavior (Chapters 6, 7, and 8) that switching operations create abrupt changes in currents and voltages. These discontinuities are accommodated mathematically by introducing the step and impulse functions.

The step function is illustrated in Fig. 15.2. The function is zero for $t < 0$ and has a constant value of K for $t > 0$. The mathematical symbol for the step function is $Ku(t)$. Thus

$$
\begin{aligned}
Ku(t) &= 0, \quad t < 0, \\
Ku(t) &= K, \quad t > 0.
\end{aligned}
\tag{15.6}
$$

If K is 1 the function defined by Eq. (15.6) is the unit step.

The step function is not defined at $t = 0$. In situations where we need to define the transition between 0^- and 0^+, we will assume that it is linear and that

$$
Ku(0) = 0.5K.
\tag{15.7}
$$

As before, 0^- and 0^+ represent symmetrical points arbitrarily close to the left and right of the origin. The linear transition from 0^- to 0^+ is illustrated in Fig. 15.3.

A step that occurs at $t = a$ is expressed as $Ku(t - a)$. Thus

$$
\begin{aligned}
Ku(t - a) &= 0, \quad t < a, \\
Ku(t - a) &= K, \quad t > a.
\end{aligned}
\tag{15.8}
$$

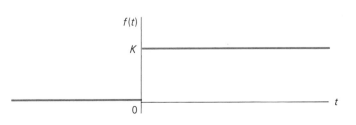

Fig. 15.2 The step function.

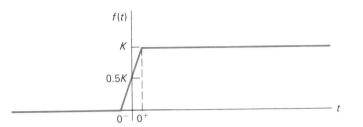

Fig. 15.3 The linear approximation to the step function.

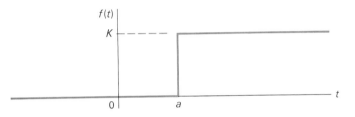

Fig. 15.4 A step function occurring at $t = a$ when $a > 0$.

If $a > 0$ the step occurs to the right of the origin and if $a < 0$ the step occurs to the left of the origin. Equation (15.8) is illustrated in Fig. 15.4. Observe that the step function is 0 when the argument $(t - a)$ is negative and is K when the argument is positive.

A step function that is equal to K for $t < a$ and 0 for $t > a$ is written as $Ku(a - t)$. Thus

$$Ku(a - t) = K, \quad t < a,$$
$$Ku(a - t) = 0, \quad t > a. \tag{15.9}$$

The discontinuity will be to the left of the origin when $a < 0$ and to the right of the origin when $a > 0$. Equation (15.9) is shown in Fig. 15.5.

One application of the step function is to use it to write the mathematical expression for a function that is of finite duration. This is illustrated in Example 15.1. Other applications, in relation to the Laplace transform method, are introduced in the work that follows.

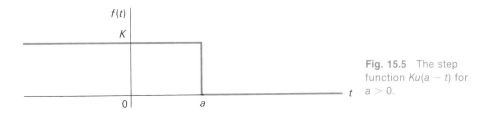

Fig. 15.5 The step function $Ku(a - t)$ for $a > 0$.

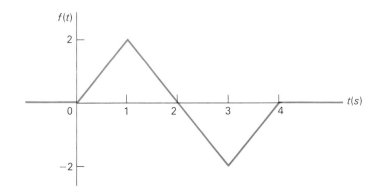

Fig. 15.6   The function for Example 15.1.

Example 15.1   Use step functions to write an expression for the function illustrated in Fig. 15.6.

Solution   The function shown in Fig. 15.6 is made up of linear segments with break points at 0, 1, 3, and 4 seconds. To construct this function we must add and subtract linear functions of the proper slope. We use the step function to initiate these linear segments at the proper time. The breakdown of the function in Fig. 15.6 into its linear segments is shown in Fig. 15.7. We use the step function to initiate a straight line with the following slopes: $+2$ at $t = 0$; -4 at $t = 1$; $+4$ at $t = 3$; and -2 at $t = 4$. Note that when a straight line with a slope of -4 is added to a straight line with a slope of $+2$, the resulting straight line will have a slope of -2. The step function is used to start this addition at $t = 1$. At 3 seconds we must add back a straight line with a slope of $+4$ to give a straight line with a slope of $+2$. Finally the function is made identically zero

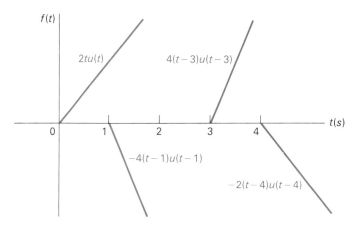

Fig. 15.7   Decomposition of the function shown in Fig. 15.6.

for all $t \geq 4$ seconds by adding in a straight line with a slope of -2 at $t = 4$ seconds. The expression for $f(t)$ is

$$f(t) = 2tu(t) - 4(t - 1)u(t - 1) + 4(t - 3)u(t - 3) - 2(t - 4)u(t - 4). ■$$

DRILL EXERCISE
15.1

Use step functions to write the expression for each of the functions shown.

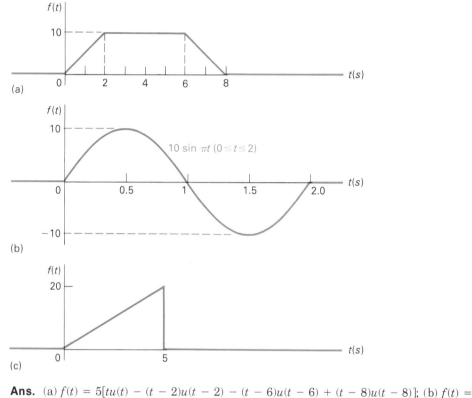

(a)

(b)

(c)

Ans. (a) $f(t) = 5[tu(t) - (t - 2)u(t - 2) - (t - 6)u(t - 6) + (t - 8)u(t - 8)]$; (b) $f(t) = 10 \sin \pi t[u(t) - u(t - 2)]$; (c) $f(t) = 4tu(t) - 4(t - 5)u(t - 5) - 20u(t - 5)$.

15.4 THE IMPULSE FUNCTION

When we have a finite discontinuity in a function, such as that illustrated in Fig. 15.1(b), the derivative of the function is not defined at the point of the discontinuity. The concept of an impulse function enables us to define the derivative at a discontinuity. This means that we can define the Laplace transform of the derivative of a function that has a discontinuity. We can also define the transform of the higher-order derivatives of the function. As we will see later, we will find the impulse function a useful concept in circuit analysis.

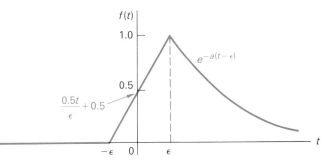

Fig. 15.8 A magnified view of the discontinuity in Fig. 15.1(b), assuming a linear transition between $-\epsilon$ and $+\epsilon$.

In order to define the derivative of a function at a discontinuity, let us assume that the function varies linearly across the discontinuity as shown in Fig. 15.8, where we observe that as $\epsilon \to 0$ we get an abrupt discontinuity at the origin. When we differentiate the function shown in Fig. 15.8, we see that the derivative between $-\epsilon$ and $+\epsilon$ is constant at a value of $\frac{1}{2}\epsilon$. For $t < -\epsilon$ the derivative is zero and for $t > \epsilon$ the derivative is $-ae^{-a(t-\epsilon)}$. These observations are shown graphically in Fig. 15.9.

As ϵ approaches zero, the value of $f'(t)$ between $\pm\epsilon$ approaches infinity. At the same time that the value of $f'(t)$ is becoming infinite, the duration of this large value is approaching zero. Furthermore, the area under $f'(t)$ between $\pm\epsilon$ remains constant as $\epsilon \to 0$. In this case, the area is unity. As ϵ approaches zero, we say that the function between $\pm\epsilon$ approaches a unit impulse function. The impulse function is represented symbolically as $\delta(t)$. Thus for our example we would say that the derivative of $f(t)$ at the origin approaches a unit impulse function as ϵ approaches zero. Symbolically we would write this as

$$f'(0) \to \delta(t) \quad \text{as} \quad \epsilon \to 0.$$

The impulse function is referred to as a *unit* impulse in this case because the *area* under the impulse-generating curve is unity. The strength of the

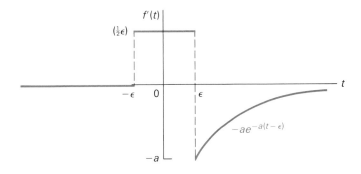

Fig. 15.9 The derivative of the function shown in Fig. 15.8.

impulse function is determined by this area. An impulse of strength K is denoted as $K \delta(t)$, where K is the area under the impulse-generating function. We also note that the impulse function is known as the *Dirac delta function*.

We can summarize the creation of an impulse function as follows. An impulse function is created by defining a function in terms of a variable parameter and then allowing this parameter to approach zero. The variable-parameter function generates an impulse if it exhibits the following three characteristics as the parameter approaches zero:

1. the amplitude approaches infinity,

2. the duration of the function approaches zero, and

3. the area under the variable-parameter function is constant as the parameter changes.

As an example of another function that generates an impulse function, consider the following exponential function:

$$f(t) = \frac{K}{2\epsilon}\, e^{-|t|/\epsilon}. \tag{15.10}$$

As ϵ approaches zero, the function becomes infinite at the origin and at the same time decays to zero in an infinitesimal length of time. The character of $f(t)$ as $\epsilon \to 0$ is illustrated in Fig. 15.10. To show that an impulse function is created as $\epsilon \to 0$, we must also show that the area under the function is independent of ϵ. Thus

$$\begin{aligned}
\text{Area} &= \int_{-\infty}^{0} \frac{K}{2\epsilon} e^{t/\epsilon}\, dt + \int_{0}^{\infty} \frac{K}{2\epsilon} e^{-t/\epsilon}\, dt \\
&= \frac{K}{2\epsilon} \cdot \frac{e^{t/\epsilon}}{(1/\epsilon)}\Big|_{-\infty}^{0} + \frac{K}{2\epsilon} \cdot \frac{e^{-t/\epsilon}}{(-1/\epsilon)}\Big|_{0}^{\infty} \\
&= \frac{K}{2} + \frac{K}{2} = K,
\end{aligned} \tag{15.11}$$

which tells us that the area under the curve is constant and equal to K units. Therefore, as ϵ approaches zero, $f(t)$ approaches $K\,\delta(t)$.

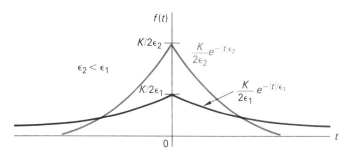

Fig. 15.10 An impulse-generating function.

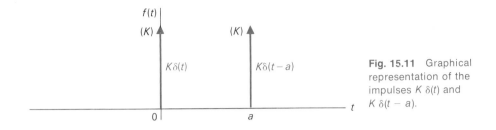

Fig. 15.11 Graphical representation of the impulses $K\,\delta(t)$ and $K\,\delta(t - a)$.

The concept of the impulse function is described mathematically as follows:

$$\int_{-\infty}^{\infty} K\,\delta(t)\,dt = K, \tag{15.12}$$

$$\delta(t) = 0, \qquad t \neq 0. \tag{15.13}$$

Equation (15.12) states that the area under the impulse function is constant. This area represents the strength of the impulse. Equation (15.13) states that the impulse is zero everywhere except at $t = 0$. An impulse that occurs at $t = a$ is denoted as $K\,\delta(t - a)$.

The graphical symbol for the impulse function is simply an arrow. The strength of the impulse is given parenthetically next to the head of the arrow. The impulses $K\,\delta(t)$ and $K\,\delta(t - a)$ are shown in Fig. 15.11.

An important property of the impulse function is known as the *sifting property,* which is expressed as

$$\int_{-\infty}^{\infty} f(t)\,\delta(t - a)\,dt = f(a), \tag{15.14}$$

where the function $f(t)$ is assumed to be continuous at $t = a$, that is, at the location of the impulse. We see from Eq. (15.14) that the impulse function sifts out everything except the value of $f(t)$ at $t = a$. Equation (15.14) follows directly if we note that $\delta(t - a)$ is zero everywhere except at $t = a$. Hence the integral can be written

$$I = \int_{-\infty}^{\infty} f(t)\,\delta(t - a)\,dt = \int_{t_0 - \epsilon}^{t_0 + \epsilon} f(t)\,\delta(t - a)\,dt. \tag{15.15}$$

But since $f(t)$ is continuous at a it takes on the value $f(a)$ as t approaches a and we have

$$I = \int_{a-\epsilon}^{a+\epsilon} f(a)\,\delta(t - a)\,dt = f(a)\int_{a-\epsilon}^{a+\epsilon} \delta(t - a)\,dt = f(a). \tag{15.16}$$

We can use the sifting property of the impulse function to find its Laplace transform; thus

$$\mathcal{L}\{\delta(t)\} = \int_{0^-}^{\infty} \delta(t)e^{-st}\,dt = \int_{0^-}^{\infty} \delta(t)\,dt = 1, \tag{15.17}$$

is an important Laplace transform pair that we will make good use of in our application of the transform to circuit analysis.

It is also possible to define the derivatives of the impulse function and the Laplace transform of these derivatives. We will discuss the first derivative, along with its transform, in the material that follows. We will then state the result for the higher-order derivatives.

The function illustrated in Fig. 15.12(a) generates an impulse function as ϵ approaches zero. The derivative of this impulse-generating function is shown in Fig. 15.12(b) and this function is defined as the derivative of the impulse $[\delta'(t)]$ as ϵ approaches zero. The derivative of the impulse function is sometimes referred to as a *moment function*, or *unit doublet*.

To find the Laplace transform of $\delta'(t)$ we simply apply the defining integral to the function shown in Fig. 15.12(b) and, after integrating, let $\epsilon \to 0$. We have then

$$
\begin{aligned}
\mathcal{L}\{\delta'(t)\} &= \lim_{\epsilon \to 0} \left[\int_{-\epsilon}^{0^-} \frac{1}{\epsilon^2} e^{-st}\, dt + \int_{0^+}^{\epsilon} \left(-\frac{1}{\epsilon^2} \right) e^{-st}\, dt \right] \\
&= \lim_{\epsilon \to 0} \left[\frac{e^{s\epsilon} + e^{-s\epsilon} - 2}{s\epsilon^2} \right] \\
&= \lim_{\epsilon \to 0} \left[\frac{se^{s\epsilon} - se^{-s\epsilon}}{2\epsilon s} \right] \\
&= \lim_{\epsilon \to 0} \left[\frac{s^2 e^{s\epsilon} + s^2 e^{-s\epsilon}}{2s} \right] \\
&= s.
\end{aligned}
\tag{15.18}
$$

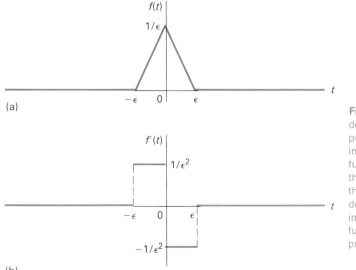

(a)

(b)

Fig. 15.12 The first derivative of the impulse function: (a) impulse-generating function used to define the first derivative of the impulse; (b) first derivative of the impulse-generating function that approaches $\delta'(t)$ as $\epsilon \to 0$.

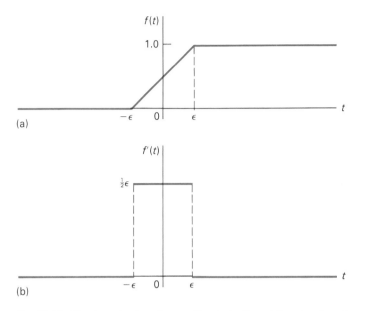

Fig. 15.13 The impulse function as the derivative of the step function: (a) $f(t) \to u(t)$ as $\epsilon \to 0$; (b) $f'(t) \to \delta(t)$ as $\epsilon \to 0$.

Note that in deriving Eq. (15.18), we had to use L'Hôpital's rule twice to evaluate the indeterminate form $0/0$.

Higher-order derivatives can be generated in a manner similar to that used to generate the first derivative (see Problem 15.4) and the defining integral can then be used to find its Laplace transform. For the nth derivative of the impulse function, we find that its Laplace transform is simply s^n, that is,

$$\mathcal{L}\{\delta^{(n)}(t)\} = s^n. \tag{15.19}$$

The last observation we wish to make in this introduction is that an impulse function can be thought of as a derivative of a step function, that is,

$$\delta(t) = \frac{du(t)}{dt}. \tag{15.20}$$

The graphical interpretation of Eq. (15.20) is illustrated in Fig. 15.13. The function shown in Fig. 15.13(a) approaches a unit step function as $\epsilon \to 0$. The function shown in Fig. 15.13(b), which is the derivative of the function illustrated in Fig. 15.13(a), approaches a unit impulse as $\epsilon \to 0$.

We will find the impulse function an extremely useful concept in circuit analysis and we will have more to say about it in the following chapters. We have introduced the concept at this point in our discussion so that we can include discontinuities at the origin in our definition of the Laplace transform.

DRILL EXERCISE
15.2

a) Find the area under the function shown in Fig. 15.12(a).
b) What is the duration of the function when $\epsilon = 0$?
c) What is the magnitude of $f(t)$ when $\epsilon = 0$?

Ans. (a) 1; (b) 0; (c) ∞.

DRILL EXERCISE
15.3

Evaluate the following integrals:

a) $I = \int_{-2}^{4} (t^3 + 4)[\delta(t) + 4\delta(t - 2)]\,dt$;

b) $I = \int_{-3}^{4} t^2[\delta(t) + \delta(t + 2.5) + \delta(t - 5)]\,dt$.

Ans. (a) 52; (b) 6.25.

DRILL EXERCISE
15.4

Find $f(t)$ if

$$f(t) = \frac{1}{2\pi} \int_{-\infty}^{\infty} F(\omega)e^{jt\omega}\,d\omega$$

and

$$F(\omega) = \frac{(3 + j\omega)}{(4 + j\omega)}\,\pi\,\delta(\omega).$$

Ans. $3/8$.

15.5 FUNCTIONAL TRANSFORMS

A functional transform is simply the Laplace transform of a specified function of t. Since we are limiting our introduction to the unilateral, or one-sided, Laplace transform, we will define all of our functions to be zero for $t < 0^-$. This means that $f(t)$ is uniquely defined by stating how $f(t)$ varies for $t > 0^-$.

We derived one functional transform pair in Section 15.4, where we showed that the Laplace transform of the unit impulse function equals one, that is, Eq. (15.17). As a second example, consider the unit step function of Fig. 15.13(a), where we have

$$\mathcal{L}\{u(t)\} = \int_{0^-}^{\infty} f(t)e^{-st}\,dt = \int_{0^+}^{\infty} 1e^{-st}\,dt$$

$$= \frac{e^{-st}}{-s}\Big|_{0^+}^{\infty} = \frac{1}{s}. \tag{15.21}$$

From Eq. (15.21) we see that the Laplace transform of the unit step function is $1/s$.

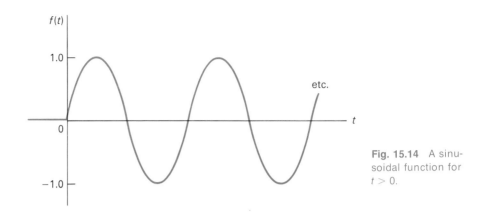

Fig. 15.14 A sinusoidal function for $t > 0$.

The Laplace transform of the decaying exponential function shown in Fig. 15.1(b) is

$$\mathscr{L}\{e^{-at}\} = \int_{0^+}^{\infty} e^{-at}e^{-st}\, dt = \int_{0^+}^{\infty} e^{-(a+s)t}\, dt = \frac{1}{s+a}. \qquad (15.22)$$

In deriving Eqs. (15.21) and (15.22) we have used the fact that integration across the discontinuity at the origin is zero.

Our last example of finding a functional transform is the sinusoidal function shown in Fig. 15.14. The expression for $f(t)$ for $t > 0^-$ is $\sin \omega t$; hence the Laplace transform is

$$\begin{aligned}
\mathscr{L}\{\sin \omega t\} &= \int_{0}^{\infty} [\sin \omega t] e^{-st}\, dt \\
&= \int_{0^-}^{\infty} \left[\frac{e^{j\omega t} - e^{-j\omega t}}{2j} \right] e^{-st}\, dt \\
&= \int_{0^-}^{\infty} \frac{[e^{-(s-j\omega)t} - e^{-(s+j\omega)t}]}{2j}\, dt \\
&= \frac{1}{2j} \left[\frac{1}{s-j\omega} - \frac{1}{s+j\omega} \right] \\
&= \frac{\omega}{s^2 + \omega^2}. \qquad (15.23)
\end{aligned}$$

An abbreviated table of Laplace transform pairs that includes the functions of most interest in an introductory course on circuit applications is given in Table 15.1.

DRILL EXERCISE   Use the defining integral to find the Laplace transform of (a) t and (b) $\cosh \beta t$.
15.5
Ans. (a) $1/s^2$; (b) $s/(s^2 - \beta^2)$.

TABLE 15.1
AN ABBREVIATED TABLE OF LAPLACE TRANSFORM PAIRS

$f(t)$, $(t > 0^-)$		$F(s)$
$\delta(t)$	(impulse)	1
$u(t)$	(step)	$\dfrac{1}{s}$
t	(ramp)	$\dfrac{1}{s^2}$
e^{-at}	(exponential)	$\dfrac{1}{s + a}$
$\sin \omega t$	(sine)	$\dfrac{\omega}{s^2 + \omega^2}$
$\cos \omega t$	(cosine)	$\dfrac{s}{s^2 + \omega^2}$
te^{-at}	(damped ramp)	$\dfrac{1}{(s + a)^2}$
$e^{-at} \sin \omega t$	(damped sine)	$\dfrac{\omega}{(s + a)^2 + \omega^2}$
$e^{-at} \cos \omega t$	(damped cosine)	$\dfrac{s + a}{(s + a)^2 + \omega^2}$

15.6 OPERATIONAL TRANSFORMS

Operational transforms tell us how mathematical operations performed on either $f(t)$ or $F(s)$ are translated into the opposite domain. The operations of primary interest are (a) multiplication by a constant; (b) addition (subtraction); (c) differentiation; (d) integration; (e) translation in the time domain; (f) translation in the frequency domain; and (g) scale changing.

15.6(a) Multiplication by a Constant It follows directly from the defining integral that if

$$\mathscr{L}\{f(t)\} = F(s),$$

then

$$\mathscr{L}\{Kf(t)\} = KF(s). \tag{15.24}$$

Thus the multiplication of $f(t)$ by a constant corresponds to multiplying $F(s)$ by the same constant.

15.6(b) Addition (Subtraction) Addition (subtraction) in the time domain translates into addition (subtraction) in the frequency domain. Thus if

$$\mathscr{L}\{f_1(t)\} = F_1(s),$$
$$\mathscr{L}\{f_2(t)\} = F_2(s),$$

and

$$\mathscr{L}\{f_3(t)\} = F_3(s),$$

then

$$\mathscr{L}\{f_1(t) + f_2(t) - f_3(t)\} = F_1(s) + F_2(s) - F_3(s), \tag{15.25}$$

which is derived by simply substituting the algebraic sum of time domain functions into the defining integral.

15.6(c) Differentiation   Differentiation in the time domain corresponds to multiplying $F(s)$ by s and then subtracting the initial value of $f(t)$, that is, $f(0^-)$, from this product. Thus

$$\mathscr{L}\left\{\frac{df(t)}{dt}\right\} = sF(s) - f(0^-), \tag{15.26}$$

which is obtained directly from the definition of the Laplace transform. Thus

$$\mathscr{L}\left\{\frac{df(t)}{dt}\right\} = \int_{0^-}^{\infty}\left(\frac{df(t)}{dt}\right)e^{-st}\,dt. \tag{15.27}$$

We can evaluate the integral in Eq. (15.27) by integrating by parts. Letting $u = e^{-st}$ and $dv = [df(t)/dt]dt$, we obtain

$$\mathscr{L}\left\{\frac{df(t)}{dt}\right\} = e^{-st}f(t)\,\Big|_{0^-}^{\infty} - \int_{0^-}^{\infty} f(t)[-se^{-st}\,dt]. \tag{15.28}$$

Because we are assuming that $f(t)$ is Laplace transformable, the evaluation of $e^{-st}f(t)$ at $t = \infty$ is zero. Therefore, the right-hand side of Eq. (15.28) reduces to

$$-f(0^-) + s\int_{0^-}^{\infty} f(t)e^{-st}\,dt$$

or

$$sF(s) - f(0^-).$$

This observation completes the derivation of Eq. (15.26), an important result because it tells us that differentiation in the time domain reduces to an algebraic operation in the s-domain.

The Laplace transform of higher-order derivatives can be found using Eq. (15.26) as the starting point. For example, to find the Laplace transform of the second derivative of $f(t)$, we proceed as follows. Let

$$g(t) = \frac{df(t)}{dt}. \tag{15.29}$$

Now we can use Eq. (15.26) to write

$$G(s) = sF(s) - f(0^-). \tag{15.30}$$

But since

$$\frac{dg(t)}{dt} = \frac{d^2f(t)}{dt^2},$$

we can write

$$\mathcal{L}\left\{\frac{dg(t)}{dt}\right\} = \mathcal{L}\left\{\frac{d^2f(t)}{dt^2}\right\} = sG(s) - g(0^-). \qquad \textbf{(15.31)}$$

When we substitute Eq. (15.30) into the right-hand side of Eq. (15.31), we obtain

$$\mathcal{L}\left\{\frac{d^2f(t)}{dt^2}\right\} = s^2F(s) - sf(0^-) - \frac{df(0^-)}{dt}. \qquad \textbf{(15.32)}$$

In writing Eq. (15.32), we have also used Eq. (15.29) to express the value of $g(0^-)$.

The Laplace transform of the nth derivative can be found by the simple expedient of successive application of the above thought process, which leads to the general result

$$\mathcal{L}\left\{\frac{d^nf(t)}{dt^n}\right\} = s^nF(s) - s^{n-1}f(0^-)$$

$$- s^{n-2}\frac{df(0^-)}{dt} - s^{n-3}\frac{d^2f(0^-)}{dt^2} - \cdots - \frac{d^{n-1}f(0^-)}{dt^{n-1}}. \qquad \textbf{(15.33)}$$

15.6(d) Integration Integration in the time domain corresponds to dividing by s in the s-domain. As before, we establish the relationship via the defining integral:

$$\mathcal{L}\left\{\int_{0^-}^{t} f(x)\,dx\right\} = \int_{0^-}^{\infty}\left[\int_{0^-}^{t} f(x)\,dx\right]e^{-st}\,dt. \qquad \textbf{(15.34)}$$

The integral on the right-hand side of Eq. (15.34) can be evaluated by integrating by parts. Thus we let

$$u = \int_{0^-}^{t} f(x)\,dx \quad \text{and} \quad dv = e^{-st}\,dt;$$

then

$$du = f(t)\,dt \quad \text{and} \quad v = -\frac{e^{-st}}{s}.$$

Using the integration-by-parts formula, we get

$$\mathcal{L}\left\{\int_{0^-}^{t} f(x)\,dx\right\} = uv\,\bigg|_{0^-}^{\infty} - \int_{0^-}^{\infty} v\,du$$

$$= -\frac{e^{-st}}{s}\int_{0^-}^{t} f(x)\,dx\,\bigg|_{0^-}^{\infty} + \int_{0^-}^{\infty}\frac{e^{-st}}{s} f(t)\,dt. \qquad \textbf{(15.35)}$$

The first term on the right-hand side of Eq. (15.35) is zero at both the upper and lower limits. The evaluation at the lower limit is obviously zero, whereas the evaluation at the upper limit is zero because we are assuming that $f(t)$ has a Laplace transform. The second term on the right-hand side of Eq. (15.35) is $F(s)/s$; therefore we have

$$\mathscr{L}\left\{\int_{0^-}^{t} f(x)\,dx\right\} = \frac{F(s)}{s}, \tag{15.36}$$

from which we see that the operation of integration in the time domain is transformed to the algebraic operation of multiplying by $1/s$ in the s-domain.

Equations (15.33) and (15.36) form the basis of the earlier statement that the Laplace transform translates a set of integrodifferential equations into a set of algebraic equations.

15.6(e) Translation in the Time Domain   Translation in the time domain corresponds to multiplication by an exponential in the frequency domain. Thus

$$\mathscr{L}\{f(t - a)u(t - a)\} = e^{-as}F(s), \qquad a > 0. \tag{15.37}$$

For example, since we know that

$$\mathscr{L}\{t\} = \frac{1}{s^2}$$

Eq. (15.37) permits us to write down the Laplace transform of $(t - a)u(t - a)$ directly; thus

$$\mathscr{L}\{(t - a)u(t - a)\} = \frac{e^{-as}}{s^2}.$$

The proof of Eq. (15.37) follows from the defining integral. Thus

$$\mathscr{L}\{f(t - a)u(t - a)\} = \int_{0^-}^{\infty} u(t - a)f(t - a)e^{-st}\,dt$$

$$= \int_{a}^{\infty} f(t - a)e^{-st}\,dt. \tag{15.38}$$

In writing Eq. (15.38), we have taken advantage of the fact that $u(t - a)$ equals unity for $t > a$. Now we make a change in the variable of integration. Specifically, we let $x = t - a$, then $x = 0$ when $t = a$ and $x = \infty$ when $t = \infty$. We also note that $dx = dt$. Thus we can write the integral in Eq. (15.38) as

$$\mathscr{L}\{f(t - a)u(t - a) = \int_{0}^{\infty} f(x)e^{-s(x+a)}\,dx$$

$$= e^{-sa}\int_{0}^{\infty} f(x)e^{-sx}\,dx$$

$$= e^{-as}F(s),$$

which is what we set out to prove.

15.6(f) Translation in the Frequency Domain Translation in the frequency domain corresponds to multiplication by an exponential in the time domain. Symbolically, then,

$$\mathcal{L}\{e^{-at}f(t)\} = F(s + a), \tag{15.39}$$

which follows from the defining integral, and its derivation is left as an exercise (see Problem 15.11).

The relationship given by Eq. (15.39) can be used to derive new transform pairs. Thus knowing that

$$\mathcal{L}\{\cos \omega t\} = \frac{s}{s^2 + \omega^2},$$

we can use Eq. (15.36) to deduce that

$$\mathcal{L}\{e^{-at}\cos \omega t\} = \frac{s + a}{(s + a)^2 + \omega^2}.$$

15.6(g) Scale Changing The scale-change property gives the relationship between $f(t)$ and $F(s)$, when the time variable is multiplied by a positive constant. Thus

$$\mathcal{L}\{f(at)\} = \frac{1}{a} F\left(\frac{s}{a}\right), \qquad a > 0, \tag{15.40}$$

the derivation of which is left as an exercise (see Problem 15.15).

The scale-change property is particularly useful in experimental work especially where time scale changes are made to facilitate building a model of a system.

We can use Eq. (15.40) to formulate new transform pairs. Thus knowing that

$$\mathcal{L}\{\cos t\} = \frac{s}{s^2 + 1},$$

it follows from Eq. (15.37) that

$$\mathcal{L}\{\cos \omega t\} = \frac{1}{\omega} \cdot \frac{s/\omega}{(s/\omega)^2 + 1}$$

$$= \frac{s}{s^2 + \omega^2}.$$

An abbreviated table of operational transforms is given in Table 15.2. You will note some entries in the table that were not discussed in this section; however, you will become more familiar with these additional entries by working Problems 15.16 and 15.17.

TABLE 15.2
AN ABBREVIATED TABLE OF OPERATIONAL TRANSFORMS

f(t)	F(s)
$Kf(t)$	$KF(s)$
$f_1(t) + f_2(t) - f_3(t) + \cdots$	$F_1(s) + F_2(s) - F_3(s) + \cdots$
$\dfrac{df(t)}{dt}$	$sF(s) - f(0^-)$
$\dfrac{d^2f(t)}{dt^2}$	$s^2F(s) - sf(0^-) - \dfrac{df(0^-)}{dt}$
$\dfrac{d^nf(t)}{dt^n}$	$s^nF(s) - s^{n-1}f(0^-) - s^{n-2}\dfrac{df(0^-)}{dt}$
	$\quad - s^{n-3}\dfrac{d^2f(0^-)}{dt} - \cdots - \dfrac{d^{n-1}f(0^-)}{dt}$
$\displaystyle\int_0^t f(x)\,dx$	$\dfrac{F(s)}{s}$
$f(t-a)u(t-a),\ a > 0$	$e^{-as}F(s)$
$e^{-at}f(t)$	$F(s+a)$
$f(at),\ a > 0$	$\dfrac{1}{a}F\left(\dfrac{s}{a}\right)$
$tf(t)$	$-\dfrac{dF(s)}{ds}$
$t^nf(t)$	$(-1)^n\dfrac{d^nF(s)}{ds^n}$
$\dfrac{f(t)}{t}$	$\displaystyle\int_s^\infty F(u)\,du$

DRILL EXERCISE 15.6   Use the appropriate operational transform from Table 15.2 to find the Laplace transform of each of the following functions:

(a) t^2e^{-at};   (b) $\dfrac{d}{dt}\left[e^{-at}\cosh \beta t\right]$;   (c) $t \cos \omega t$.

Ans. (a) $2/(s+a)^3$; (b) $\dfrac{\beta^2 - a(s+a)}{(s+a)^2 - \beta^2}$; (c) $\dfrac{s^2 - \omega^2}{(s^2 + \omega^2)^2}$.

15.7   An Illustrative Example

We are now in a position to illustrate how the Laplace transform can be used to solve the ordinary integrodifferential equations that describe the behavior of lumped-parameter circuits. Consider the circuit shown in Fig. 15.15. We will

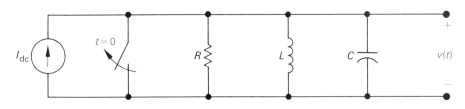

Fig. 15.15   A parallel *RLC* circuit.

assume that there is no initial energy stored in the circuit at the instant when the switch, which is shorting the dc current source, is opened. Our problem is to find the time domain expression for $v(t)$ when $t \geq 0$.

We begin by writing the integrodifferential equation that $v(t)$ must satisfy. Only a single node-voltage equation is needed to describe the circuit. Summing the currents away from the top node in the circuit generates the following equation:

$$\frac{v(t)}{R} + \frac{1}{L} \int_0^t v(x)\, dx + C\frac{dv(t)}{dt} = I_{dc}u(t). \tag{15.41}$$

Note that in writing Eq. (15.41) we indicated the opening of the switch in the step jump of the source current from zero to I_{dc} amperes.

Once the integrodifferential equations (in this example, just one) has been derived, the second step in the Laplace transform method is to transform the equations to the *s*-domain. In transforming Eq. (15.41), we use three operational transforms and one functional transform; thus

$$\frac{V(s)}{R} + \frac{1}{L}\frac{V(s)}{s} + C[sV(s) - v(0^-)] = I_{dc}\left(\frac{1}{s}\right), \tag{15.42}$$

an algebraic equation where $V(s)$ is the unknown variable. We are assuming that the circuit parameters R, L, and C, as well as the source current I_{dc}, are known. (The initial voltage on the capacitor $v(0^-)$ is zero since the initial energy stored in the circuit is zero.) Thus our analytical problem has been reduced to solving an algebraic equation. In general, of course, the problem is reduced to solving a set of algebraic equations.

The third step in the Laplace transform approach is to solve the algebraic equations for the unknowns. In our example, this means solving Eq. (15.42) for $V(s)$. We get

$$V(s)\left[\frac{1}{R} + \frac{1}{sL} + sC\right] = \frac{I_{dc}}{s}$$

$$V(s) = \frac{I_{dc}/C}{s^2 + \dfrac{1}{RC}s + \dfrac{1}{LC}}. \tag{15.43}$$

To find $v(t)$ we must inverse-transform the expression for $V(s)$. We denote this inverse operation as

$$v(t) = \mathcal{L}^{-1}\{V(s)\}. \qquad \textbf{(15.44)}$$

Thus the fourth step in the Laplace transform method is to find the inverse transform of the s-domain unknowns. Finding the inverse transform of s-domain expressions like that given by Eq. (15.43) is the subject of Section 15.8. However, before proceeding with the problem of finding inverse transforms, we need to mention a fifth step in our procedure.

This fifth and last step is not unique to the Laplace transform method. It is a step that conscientious and prudent engineers always incorporate into their analytical thinking, namely, to test any derived solution and make sure that it makes sense in terms of the known behavior of the system being analyzed. This facet of analysis will be demonstrated in the following material.

We have reached a point in our discussion of the Laplace transform where it will be to our advantage to simplify the notation. We do this by dropping the parenthetical t in time domain expressions and the parenthetical s in frequency domain expressions. The time domain will be implied by using lower-case letters for all time domain variables and the corresponding s-domain variables will be represented by upper-case letters. Thus

$$\mathcal{L}\{v\} = V \quad \text{or} \quad v = \mathcal{L}^{-1}\{V\},$$
$$\mathcal{L}\{i\} = I \quad \text{or} \quad i = \mathcal{L}^{-1}\{I\},$$
$$\mathcal{L}\{f\} = F \quad \text{or} \quad f = \mathcal{L}^{-1}\{F\},$$

and so on.

DRILL EXERCISE
15.7

In the circuit in Fig. 15.15, the dc current source is replaced with a sinusoidal source that delivers a current of $1.2 \cos t$ A. The circuit components are $R = 1\ \Omega, C = 0.625$ F, and $L = 1.6$ H. Find the numerical expression for V.

Ans. $V = \dfrac{1.92s^2}{(s^2 + 1.6s + 1)(s^2 + 1)}.$

15.8 INVERSE TRANSFORMS

The expression for $V(s)$ given by Eq. (15.43) is a *rational* function of s, that is, one that can be expressed in the form of a ratio of two polynomials in s such that no nonintegral powers of s appear in the polynomials. This is an important observation to make about Eq. (15.43) because for linear, lumped-parameter, time-invariant circuits, the s-domain expressions for the unknown voltages and currents will always be rational functions of s. (You can support this observation further by working Problems 15.18, 15.19, and 15.20.) Therefore, if we can learn how to inverse-transform rational functions of s, we know that we can solve for the time domain expressions for the voltages and currents. The purpose of this section is to present a very straightforward and systematic technique for finding the inverse transform of a rational function.

In general, we are confronted with the problem of finding the inverse transform of a quotient of two polynomials in s, that is, a function that has the form

$$F(s) = \frac{N(s)}{D(s)} = \frac{a_n s^n + a_{n-1} s^{n-1} + \cdots + a_1 s + a_0}{b_m s^m + b_{m-1} s^{m-1} + \cdots + b_1 s + b_0}. \qquad (15.45)$$

The a and b will be real constants and the m and n positive integers. The ratio of $N(s)$ and $D(s)$ is called a proper rational function if $m > n$ and an improper rational function if $m \leq n$. Only a proper rational function can be expanded as a sum of partial fractions. This restriction poses no problem, as we will see at the end of this section.

A proper rational function is expanded into a sum of partial fractions by writing a term or a series of terms for each root of $D(s)$. This means that $D(s)$ must be in factored form before we can make a partial-fraction expansion. For each distant root of $D(s)$, a single term appears in the sum of partial fractions. For each multiple root of $D(s)$ of multiplicity r, there will be r terms in the expansion. For example, in the rational function

$$\frac{(s + 6)}{s(s + 3)(s + 1)^2},$$

the denominator has four roots. Two of these roots are distinct, namely, at $s = 0$ and $s = -3$; and there is a multiple root of multiplicity 2 at $s = -1$. Thus the partial-fraction expansion of this function will be of the form

$$\frac{s + 6}{s(s + 3)(s + 1)^2} \equiv \frac{K_1}{s} + \frac{K_2}{s + 3} + \frac{K_3}{(s + 1)^2} + \frac{K_4}{(s + 1)}, \qquad (15.46)$$

from which we see that the key to the partial-fraction technique for finding inverse transforms lies in recognizing the $f(t)$ corresponding to each term in the sum of partial fractions. Referring to Table 15.1, you should be able to verify that

$$\mathscr{L}^{-1} \left\{ \frac{s + 6}{s(s + 3)(s + 1)^2} \right\} = [K_1 + K_2 e^{-3t} + K_3 t e^{-t} + K_4 e^{-t}] u(t). \qquad (15.47)$$

All that remains to be done is to establish a technique for determining the coefficients (K_1, K_2, K_3, etc.) that are generated by making a partial-fraction expansion. We will break this problem into three steps. First, we consider the situation where the roots of $D(s)$ are real and distinct. Second, we consider the situation where some of the roots of $D(s)$ are complex and distinct. Finally, we consider the case where $D(s)$ has repeated roots. Before we do so, however, some general comments are in order.

We have used the identity sign $\equiv$ in Eq. (15.46) to emphasize that the operation of expanding a rational function into a sum of partial fractions establishes an identical equation. Thus both sides of the equation must be the same for all values of the variable s. The identity relationship must hold when both sides are subjected to the same mathematical operation. Therefore, the identity is preserved if both sides are differentiated or if both sides are multiplied by the same quantity. These characteristics of an identical equation are pertinent to determining the coefficients. A second point to be made at this time is to caution you to check that the rational function is proper. This is important because there is nothing in the procedure for finding the various K's that will alert you that the results are nonsense if the rational function is improper. We will point out a procedure for checking the K but wasted effort can be avoided by forming the habit of asking yourself, "Is $F(s)$ a proper rational function?"

Now let us consider the problem of determining the coefficients in a partial-fraction expansion when all the roots of $D(s)$ are real and distinct. To find a K associated with a term that arises due to a distinct root of $D(s)$, we multiply both sides of the identity by a factor equal to the denominator beneath the desired K. Now when both sides of the identity are evaluated at the root corresponding to the multiplying factor, the right-hand side will always be the desired K and the left-hand side will always be its numerical value. As an example, consider the following partial-fraction expansion:

$$F(s) = \frac{96(s + 5)(s + 12)}{s(s + 8)(s + 6)} \equiv \frac{K_1}{s} + \frac{K_2}{(s + 8)} + \frac{K_3}{(s + 6)}. \tag{15.48}$$

To find the value of K_1, we multiply both sides by s and then evaluate both sides at $s = 0$. Thus

$$\left.\frac{96(s + 5)(s + 12)}{(s + 8)(s + 6)}\right|_{s=0} \equiv K_1 + \left.\frac{K_2 s}{(s + 8)}\right|_{s=0} + \left.\frac{K_3 s}{(s + 6)}\right|_{s=0}$$

or

$$\frac{96(5)(12)}{(8)(6)} \equiv K_1 = 120. \tag{15.49}$$

To find the value of K_2 we multiply both sides by $(s + 8)$ and then evaluate both sides at $s = -8$. Thus

$$\left.\frac{96(s + 5)(s + 12)}{s(s + 6)}\right|_{s=-8} \equiv \left.\frac{K_1(s + 8)}{s}\right|_{s=-8} + K_2 + \left.\frac{K_3(s + 8)}{(s + 6)}\right|_{s=-8}$$

or

$$\frac{96(-3)(4)}{(-8)(-2)} = K_2 = -72. \tag{15.50}$$

It follows directly that K_3 will be

$$\left.\frac{96(s + 5)(s + 12)}{s(s + 8)}\right|_{s=-6} = K_3 = 48. \tag{15.51}$$

We have seen that

$$\frac{96(s + 5)(s + 12)}{s(s + 8)(s + 6)} \equiv \frac{120}{s} + \frac{48}{s + 6} - \frac{72}{s + 8}. \tag{15.52}$$

At this point in the generation of a partial-fraction expansion, it is a good idea to test the result as a protection against making computational errors. As we already mentioned, a partial-fraction expansion creates an identity; thus both sides of Eq. (15.52) must be the same for all values of s. The choice of test values is completely open and hence values that are easy to verify are chosen. For example, in Eq. (15.52) testing at either -5 or -12 is attractive because in both cases the left-hand side reduces to zero. Choosing -5 yields

$$\frac{120}{-5} + \frac{48}{1} - \frac{72}{3} = -24 + 48 - 24 = 0,$$

while testing at -12 gives

$$\frac{120}{-12} + \frac{48}{-6} - \frac{72}{-4} = -10 - 8 + 18 = 0.$$

Now that we are confident that the numerical values of the various K's are correct, we can proceed to find the inverse transform:

$$\mathcal{L}^{-1} \left\{ \frac{96(s + 5)(s + 12)}{s(s + 8)(s + 6)} \right\} = [120 + 48e^{-6t} - 72e^{-8t}]u(t). \tag{15.53}$$

DRILL EXERCISE
15.8

Find $f(t)$ if $F(s) = (2s + 12)/(s + 1)(s^2 + 5s + 6)$.

Ans. $f(t) = [5e^{-t} - 8e^{-2t} + 3e^{-3t}]u(t).$

The procedure for finding the coefficients associated with distinct complex roots is the same as that for finding those for distinct real roots. The only difference is that the algebra involves the manipulation of complex numbers. We illustrate by expanding the rational function,

$$F(s) = \frac{100(s + 3)}{(s + 6)(s^2 + 6s + 25)}. \tag{15.54}$$

We begin by noting that $F(s)$ is a proper rational function. Next, we must find the roots of the quadratic term $s^2 + 6s + 25$. Thus

$$s^2 + 6s + 25 = (s + 3 - j4)(s + 3 + j4). \tag{15.55}$$

Once we have the denominator in factored form, we proceed as before to get

$$\frac{100(s + 3)}{(s + 6)(s^2 + 6s + 25)} \equiv \frac{K_1}{s + 6} + \frac{K_2}{s + 3 - j4} + \frac{K_3}{s + 3 + j4}. \tag{15.56}$$

To find K_1, K_2, and K_3 we use the same thought process that was demonstrated in the previous example. We get

$$K_1 = \frac{100(s + 3)}{s^2 + 6s + 25}\bigg|_{s=-6} = \frac{100(-3)}{25} = -12; \tag{15.57}$$

$$K_2 = \frac{100(s + 3)}{(s + 6)(s + 3 + j4)}\bigg|_{s=-3+j4} = \frac{100(j4)}{(3 + j4)(j8)}$$

$$= 6 - j8 = 10e^{-j53.13°}; \tag{15.58}$$

$$K_3 = \frac{100(s + 3)}{(s + 6)(s + 3 - j4)}\bigg|_{s=-3-j4} = \frac{100(-j4)}{(3 - j4)(-j8)}$$

$$= 6 + j8 = 10e^{j53.13°}. \tag{15.59}$$

It follows, then, that

$$\frac{100(s + 3)}{(s + 6)(s^2 + 6s + 25)} = \frac{-12}{s + 6} + \frac{10\underline{/-53.13°}}{s + 3 - j4} + \frac{10\underline{/53.13°}}{s + 3 + j4}. \tag{15.60}$$

Once again we need to pause and make some observations. First, in physically realizable circuits, complex roots will always appear in conjugate pairs. Second, the coefficients associated with these conjugate pairs are themselves conjugates. Note, for example, that in the above example K_3 is the conjugate of K_2. Thus for complex conjugate roots, only half the coefficients need actually be calculated.

Before inverse-transforming Eq. (15.60), we make a numerical check of the partial-fraction expansion. Testing at -3 is attractive, since the left-hand side reduces to zero at this value; we get

$$F(s) = \frac{-12}{3} + \frac{10\underline{/-53.13°}}{-j4} + \frac{10\underline{/53.13°}}{j4}$$

$$= -4 + 2.5\underline{/36.87°} + 2.5\underline{/-36.87°}$$

$$= -4 + 2.0 + j1.5 + 2.0 - j1.5 = 0.$$

We now proceed to inverse-transform Eq. (15.60). Thus

$$\mathscr{L}^{-1}\left\{\frac{100(s + 3)}{(s + 6)(s^2 + 6s + 25)}\right\}$$

$$= [-12e^{-6t} + 10e^{-j53.13°}\,e^{-(3-j4)t} + 10e^{j53.13°}\,e^{-(3+j4)t}]u(t). \tag{15.61}$$

In general, it is undesirable to have the function in the time domain contain imaginary components. This is easily taken care of, since the terms involving imaginary components always come in conjugate pairs. Hence the imaginary components are eliminated by simply adding these pairs together. Thus

$$10e^{-j53.13°}\,e^{-(3-j4)t} + 10e^{j53.13°}\,e^{-(3+j4)t} = 10e^{-3t}[e^{j(4t-53.13°)} + e^{-j(4t-53.13°)}]$$

$$= 20e^{-3t}\cos(4t - 53.13°), \tag{15.62}$$

which enables us to simplify Eq. (15.60) to read

$$\mathscr{L}^{-1}\left\{\frac{100(s + 3)}{(s + 6)(s^2 + 6s + 25)}\right\} = [-12e^{-6t} + 20e^{-3t}\cos(4t - 53.13°)]u(t). \quad \textbf{(15.63)}$$

Since distinct complex roots appear rather frequently in lumped-parameter, linear circuit analysis, it is worth our time to pause here and summarize the results of this last example of a new transform pair. Whenever $D(s)$ contains distinct complex roots, that is, factors of the form $(s + \alpha - j\beta)(s + \alpha + j\beta)$, there will be a pair of terms in the partial-fraction expansion of the form

$$\frac{K}{s + \alpha - j\beta} + \frac{K^*}{s + \alpha + j\beta}, \quad \textbf{(15.64)}$$

where the partial-fraction coefficient is, in general, a complex number. In polar form we write

$$K = |K|e^{j\theta} = |K|\underline{/\theta}, \quad \textbf{(15.65)}$$

where the symbol $|K|$ is used to denote the magnitude of the complex coefficient. It follows that

$$K^* = |K|e^{-j\theta} = |K|\underline{/-\theta}. \quad \textbf{(15.66)}$$

The complex conjugate pair denoted by Eq. (15.64) will always inverse-transform as

$$\mathscr{L}^{-1}\left\{\frac{K}{s + \alpha - j\beta} + \frac{K^*}{s + \alpha + j\beta}\right\} = 2|K|e^{-\alpha t}\cos(\beta t + \theta). \quad \textbf{(15.67)}$$

In applying Eq. (15.67) it is important to note that K is defined as the coefficient associated with the denominator term $(s + \alpha - j\beta)$.

DRILL EXERCISE
15.9

Find $v(t)$ in Drill Exercise 15.7.

Ans. $v(t) = [2e^{-0.8t}\cos(0.6t + 233.13°) + 1.2\cos t]u(t)$.

To find the coefficients associated with the terms generated by a multiple root of multiplicity r, we multiply both sides of the identity by the multiple root raised to its rth power. The K appearing over the factor raised to the rth power is found by evaluating both sides of the identity at the multiple root. To find the remaining $(r - 1)$ coefficients associated with the multiple root, we differentiate both sides of the identity $(r - 1)$ times. At the end of each differentiation, we evaluate both sides of the identity at the multiple root. The right-hand side of the identity will always be one of the desired K, and the left-hand side will always be its numerical value. The following example illustrates the technique:

$$\frac{180(s + 30)}{s(s + 5)(s + 3)^2} = \frac{K_1}{s} + \frac{K_2}{(s + 5)} + \frac{K_3}{(s + 3)^2} + \frac{K_4}{(s + 3)}. \tag{15.68}$$

We find K_1 and K_2 as previously described, that is,

$$K_1 = \frac{180(s + 30)}{(s + 5)(s + 3)^2}\bigg|_{s=0} = \frac{180(30)}{(5)(9)} = 120 \tag{15.69}$$

and

$$K_2 = \frac{(180)(s + 30)}{s(s + 3)^2}\bigg|_{s=-5} = \frac{180(25)}{(-5)(4)} = -225. \tag{15.70}$$

To find K_3 we multiply both sides by $(s + 3)^2$, and then evaluate both sides at -3; thus

$$\frac{180(s + 30)}{s(s + 5)}\bigg|_{s=-3} = \frac{K_1(s + 3)^2}{s}\bigg|_{s=-3} + \frac{K_2(s + 3)^2}{(s + 5)}\bigg|_{s=-3} + K_3 + K_4(s + 3)\bigg|_{s=-3} \tag{15.71}$$

$$\frac{180(27)}{(-3)(2)} = K_1 \times 0 + K_2 \times 0 + K_3 + K_4 \times 0 = K_3 = -810. \tag{15.72}$$

To find K_4 we first multiply both sides of Eq. (15.68) by $(s + 3)^2$. Next we differentiate both sides once with respect to s and then evaluate at $s = -3$. Thus

$$\frac{d}{ds}\left[\frac{(180)(s + 30)}{s(s + 5)}\right]_{s=-3} = \frac{d}{ds}\left[\frac{K_1(s + 3)^2}{s}\right]_{s=-3} + \frac{d}{ds}\left[\frac{K_2(s + 3)^2}{(s + 5)}\right]_{s=-3}$$

$$+ \frac{d}{ds}[K_3]\bigg|_{s=-3} + \frac{d}{ds}[K_4(s + 3)]\bigg|_{s=-3} \tag{15.73}$$

$$180\left[\frac{s(s + 5) - (s + 30)(2s + 5)}{s^2(s + 5)^2}\right]_{s=-3} = K_4 \tag{15.74}$$

or

$$180\left[\frac{(-3)(2) - (27)(-1)}{(9)(4)}\right] = K_4 = 105. \tag{15.75}$$

We have then

$$\frac{180(s + 30)}{s(s + 5)(s + 3)^2} = \frac{120}{s} - \frac{225}{s + 5} - \frac{810}{(s + 3)^2} + \frac{105}{s + 3} \tag{15.76}$$

and therefore

$$\mathcal{L}^{-1}\left[\frac{180(s + 30)}{s(s + 5)(s + 3)^2}\right] = [120 - 225e^{-5t} - 810te^{-3t} + 105e^{-3t}]u(t). \tag{15.77}$$

DRILL EXERCISE
15.10

Find $f(t)$ if $F(s) = (4s^2 + 7s + 1)/s(s + 1)^2$.

Ans. $f(t) = [1 + 2te^{-t} + 3e^{-t}]u(t)$.

We handle repeated complex roots in the same way that we did repeated real roots. The only difference is that the algebra involves complex numbers. It should be reiterated that complex roots always appear in conjugate pairs, and the coefficients associated with a conjugate pair are also conjugates, *so that only half of the K's need be evaluated.* As an example, consider the expansion of the function

$$F(s) = \frac{768}{(s^2 + 6s + 25)^2}. \tag{15.78}$$

After factoring the denominator polynomial we can write

$$F(s) = \frac{768}{(s + 3 - j4)^2(s + 3 + j4)^2}$$
$$= \frac{K_1}{(s + 3 - j4)^2} + \frac{K_2}{(s + 3 - j4)} + \frac{K_1^*}{(s + 3 + j4)^2} + \frac{K_2^*}{(s + 3 + j4)}. \tag{15.79}$$

Now we need to evaluate only K_1 and K_2 since K_1^* and K_2^* are conjugate values. The value for K_1 is

$$K_1 = \frac{768}{(s + 3 + j\,4)^2}\bigg|_{s=-3+j4} = \frac{768}{(j8)^2} = -12. \tag{15.80}$$

The evaluation of K_2 is

$$K_2 = \frac{d}{ds}\left\{\frac{768}{(s + 3 + j4)^2}\right\}\bigg|_{s=-3+j4}$$
$$= -\frac{2(768)}{(s + 3 + j4)^3}\bigg|_{s=-3+j4}$$
$$= -\frac{2(768)}{(j8)^3} = -j3 = 3\underline{/-90°}. \tag{15.81}$$

It follows from Eqs. (15.80) and (15.81) that

$$K_1^* = -12 \tag{15.82}$$

and

$$K_2^* = j3 = 3\underline{/90°}. \tag{15.83}$$

The partial-fraction expansion is now grouped by conjugate terms to give

$$F(s) = \left[\frac{-12}{(s + 3 - j4)^2} + \frac{-12}{(s + 3 + j4)^2}\right] + \left[\frac{3\underline{/-90°}}{(s + 3 - j4)} + \frac{3\underline{/90°}}{s + 3 + j4}\right]. \tag{15.84}$$

We are now able to write down the inverse transform of $F(s)$:

$$f(t) = [-24te^{-3t}\cos 4t + 6e^{-3t}\cos(4t - 90°)]u(t). \tag{15.85}$$

It should be apparent from the examples cited in this section that if $F(s)$ has a real root a of multiplicity r in its denominator, the term in a partial-

TABLE 15.3
FOUR USEFUL TRANSFORM PAIRS

Pair Number†	$F(s)$	$f(t)$
(1)	$\dfrac{K}{s + a}$	$Ke^{-at}u(t)$
(2)	$\dfrac{K}{(s + a)^2}$	$Kte^{-at}u(t)$
(3)	$\dfrac{K}{s + \alpha - j\beta} + \dfrac{K^*}{s + \alpha + j\beta}$	$2\|K\|e^{-\alpha t} \cos (\beta t + \theta)u(t)$
(4)	$\dfrac{K}{(s + \alpha - j\beta)^2} + \dfrac{K^*}{(s + \alpha + j\beta)^2}$	$2t\|K\|e^{-\alpha t} \cos (\beta t + \theta)u(t)$

† *Note:* In pairs (1) and (2), K is a real quantity, whereas in pairs (3) and (4) K is the complex quantity $|K|\underline{/\theta}$.

fraction expansion is of the form

$$\frac{K}{(s + a)^r}.$$

The inverse transform of this term is

$$\mathcal{L}^{-1}\left\{\frac{K}{(s + a)^r}\right\} = \left[\frac{Kt^{r-1}e^{-at}}{(r - 1)!}\right]u(t). \qquad (15.86)$$

If $F(s)$ has a complex root of $\alpha + j\beta$ of multiplicity r in its denominator, the term in partial-fraction expansion will be the conjugate pair

$$\frac{K}{(s + \alpha - j\beta)^r} + \frac{K^*}{(s + \alpha + j\beta)^r}.$$

The inverse transform of this pair is

$$\mathcal{L}^{-1}\left\{\frac{K}{(s + \alpha - j\beta)^r} + \frac{K^*}{(s + \alpha + j\beta)^r}\right\} = \left[\frac{2|K|t^{r-1}}{(r - 1)!}e^{-\alpha t}\cos(\beta t + \theta)\right]u(t). \quad (15.87)$$

Equations (15.86) and (15.87) are the key to being able to inverse-transform any partial-fraction expansion by inspection. One further note regarding these two equations. In most circuit analysis problems, r is seldom greater than 2. Therefore, for most problems, the inverse transform of a rational function can be handled with four transform pairs. These transform pairs are listed in Table 15.3

DRILL EXERCISE
15.11

Find $f(t)$ if $F(s) = 40/(s^2 + 4s + 5)^2$.

Ans. $f(t) = [-20te^{-2t}\cos t + 20e^{-2t}\sin t]u(t)$.

We conclude our discussion of partial-fraction expansions by returning to an observation made at the beginning of this section, namely, that improper rational functions pose no serious problem in finding inverse transforms. An improper rational function can always be expanded into a polynomial plus a proper rational function. The polynomial portion of the expansion will inverse-transform into impulse functions and derivatives of impulse functions. The proper rational function portion of the expansion is inverse-transformed by the techniques outlined in this section. To illustrate the procedure, consider the function

$$F(s) = \frac{s^4 + 13s^3 + 66s^2 + 200s + 300}{s^2 + 9s + 20}. \tag{15.88}$$

We divide the denominator into the numerator until the remainder is a proper rational function. Thus

$$F(s) = s^2 + 4s + 10 + \frac{30s + 100}{s^2 + 9s + 20}. \tag{15.89}$$

Next we expand the proper rational function into a sum of partial fractions to get

$$\frac{30s + 100}{s^2 + 9s + 20} = \frac{30s + 100}{(s + 4)(s + 5)} = \frac{-20}{s + 4} + \frac{50}{s + 5}. \tag{15.90}$$

Substituting Eq. (15.90) into Eq. (15.89) yields

$$F(s) = s^2 + 4s + 10 - \frac{20}{s + 4} + \frac{50}{s + 5}. \tag{15.91}$$

Now Eq. (15.91) can be inverse-transformed by inspection; hence

$$f(t) = \frac{d^2\delta}{dt^2} + 4\frac{d\delta}{dt} + 10\delta - [20e^{-4t} - 50e^{-5t}]u(t). \tag{15.92}$$

DRILL EXERCISE
15.12

Find $f(t)$ if $F(s) = s^2/(s + 1)(s + 2)$.

Ans. $f(t) = \delta(t) + [e^{-t} - 4e^{-2t}]u(t)$.

15.9   POLES AND ZEROS OF $F(s)$

The rational function of Eq. (15.45) can also be expressed as the ratio of two factored polynomials; that is, we can write $F(s)$ as

$$F(s) = \frac{K(s + z_1)(s + z_2) \cdots (s + z_n)}{(s + p_1)(s + p_2) \cdots (s + p_m)}, \tag{15.93}$$

where K is the constant a_n/b_m. For example, the function

$$F(s) = \frac{8s^2 + 120s + 400}{2s^4 + 20s^3 + 70s^2 + 100s + 48}$$

can also be written as

$$F(s) = \frac{8(s^2 + 15s + 50)}{2(s^4 + 10s^3 + 35s^2 + 50s + 24)}$$

$$= \frac{4(s + 5)(s + 10)}{(s + 1)(s + 2)(s + 3)(s + 4)}. \qquad \textbf{(15.94)}$$

The roots of the denominator polynomial, that is, $-p_1$, $-p_2$, $-p_3$, . . . , $-p_m$, are called the *poles of $F(s)$* because at these values of s, $F(s)$ becomes infinitely large. In the function described by Eq. (15.94), the poles of $F(s)$ are -1, -2, -3, and -4.

The roots of the numerator polynomial, that is, $-z_1$, $-z_2$, $-z_3$, . . . , $-z_n$, are called the *zeros of $F(s)$* because at these values of s, $F(s)$ becomes zero. In the function described by Eq. (15.94), the zeros of $F(s)$ are -5 and -10.

We will find in the following work that it is sometimes convenient to be able to visualize the poles and zeros of $F(s)$ as points on a complex s-plane. We need a complex plane because as we have already seen in our discussion of partial fractions, the roots of the polynomials may be complex. In the complex s-plane, we use the horizontal axis to plot the real values of s and the vertical axis to plot the imaginary values of s.

As an example of plotting the poles and zeros of $F(s)$, consider the function

$$F(s) = \frac{10(s + 5)(s + 3 - j4)(s + 3 + j4)}{s(s + 10)(s + 6 - j8)(s + 6 + j8)}. \qquad \textbf{(15.95)}$$

The poles of $F(s)$ are at 0, -10, $-6 + j8$, and $-6 - j8$. The zeros are at -5, $-3 + j4$, and $-3 - j4$. The poles and zeros are plotted on the s-plane as shown in Fig. 15.16 using X's to represent poles and O's to represent zeros.

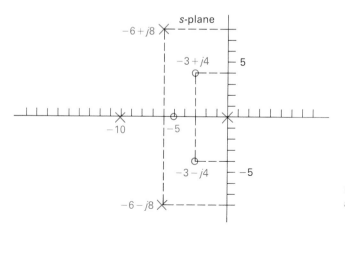

Fig. 15.16 Plotting poles and zeros on the s-plane.

Observe that the poles and zeros designated by $-p_1, -p_2, \ldots, -p_m$ and $-z_1, -z_2, \ldots, -z_n$, respectively, are located in the finite s-plane. It is also possible for $F(s)$ to have either an rth-order pole or an rth-order zero at infinity. For example, the function described by Eq. (15.94) has a second-order zero at infinity because for very large values of s the function reduces to $4/s^2$ and it is apparent that $F(s) = 0$ when $s = \infty$. In our work here we are interested in the poles and zeros that are located in the finite s-plane. Therefore when we refer to the poles and zeros of a rational function of s, we are referring to the finite poles and zeros.

15.10   INITIAL- AND FINAL-VALUE THEOREMS

The initial- and final-value theorems are useful because they enable us to determine from $F(s)$ the behavior of $f(t)$ at 0 and ∞. This means that the initial and final values of $f(t)$ can be checked before actually finding the inverse transform of $F(s)$.

The initial-value theorem states that

$$\lim_{t \to 0^+} [f(t)] = \lim_{s \to \infty} [sF(s)] \tag{15.96}$$

and the final-value theorem states that

$$\lim_{t \to \infty} [f(t)] = \lim_{s \to 0} [sF(s)]. \tag{15.97}$$

The initial-value theorem assumes that $f(t)$ contains no impulse functions. In Eq. (15.97), we must add the restriction that the theorem is valid only if the poles of $F(s)$, except for a first-order pole at the origin, lie in the left half of the s-plane.

To prove Eq. (15.96), we start with the operational transform of the first derivative; thus

$$\mathscr{L}\left\{\frac{df}{dt}\right\} = sF(s) - f(0^-) = \int_{0^-}^{\infty} \frac{df}{dt} e^{-st}\, dt. \tag{15.98}$$

Now we take the limit as $s \to \infty$; thus

$$\lim_{s \to \infty} [sF(s) - f(0^-)] = \lim_{s \to \infty} \int_{0^-}^{\infty} \frac{df}{dt} e^{-st}\, dt. \tag{15.99}$$

Observe that the right-hand side of Eq. (15.99) can be written as

$$\lim_{s \to \infty} \left[\int_{0^-}^{0^+} \frac{df}{dt} e^0\, dt + \int_{0^+}^{\infty} \frac{df}{dt} e^{-st}\, dt \right].$$

As $s \to \infty$, $(df/dt)\, e^{-st} \to 0$; hence the second integral vanishes in the limit. The first integral reduces to $f(0^+) - f(0^-)$, which is independent of s. Thus the right-hand side of Eq. (15.99) becomes

$$\lim_{s \to \infty} \int_{0^-}^{\infty} \frac{df}{dt} e^{-st}\, dt = f(0^+) - f(0^-). \tag{15.100}$$

Since $f(0^-)$ is independent of s, the left-hand side of Eq. (15.99) can be written

$$\lim_{s\to\infty} [sF(s) - f(0^-)] = \lim_{s\to\infty} [sF(s)] - f(0^-). \qquad (15.101)$$

It follows directly from Eqs. (15.100) and (15.101) that

$$\lim_{s\to\infty} [sF(s)] = f(0^+)$$

$$= \lim_{t\to 0^+} [f(t)],$$

which completes the proof of the initial-value theorem.

The proof of the final-value theorem also starts with Eq. (15.98). Here we take the limit as $s \to 0$; thus

$$\lim_{s\to 0} [sF(s) - f(0^-)] = \lim_{s\to 0} \left[\int_{0^-}^{\infty} \frac{df}{dt} e^{-st}\, dt \right]. \qquad (15.102)$$

Since the integration is with respect to t and the limit operation is with respect to s, the right-hand side of Eq. (15.102) reduces to

$$\lim_{s\to 0} \left[\int_{0^-}^{\infty} \frac{df}{dt} e^{-st}\, dt \right] = \int_{0^-}^{\infty} \frac{df}{dt}\, dt. \qquad (15.103)$$

Because the upper limit on the integral is infinite, this integral can also be written as a limit process. Thus

$$\int_{0^-}^{\infty} \frac{df}{dt}\, dt = \lim_{t\to\infty} \int_{0^-}^{t} \frac{df}{dy}\, dy, \qquad (15.104)$$

where we have introduced y as the symbol of integration to avoid confusion with the upper limit on the integral. Carrying out the integration process yields

$$\lim_{t\to\infty} [f(t) - f(0^-)] = \lim_{t\to\infty} [f(t)] - f(0^-). \qquad (15.105)$$

When we substitute Eq. (15.105) into Eq. (15.102), we can write

$$\lim_{s\to 0} [sF(s)] - f(0^-) = \lim_{t\to\infty} [f(t)] - f(0^-). \qquad (15.106)$$

Now since $f(0^-)$ cancels out, Eq. (15.106) reduces to the final-value theorem, namely,

$$\lim_{s\to 0} [sF(s)] = \lim_{t\to\infty} [f(t)].$$

The final-value theorem is useful only if $f(\infty)$ is finite. This will be true only if all the poles of $F(s)$, except for a simple pole at the origin, lie in the left half of the s-plane.

To illustrate the application of the initial- and final-value theorems, we will apply them to some of the functions we used to illustrate partial-fraction expansions. Since we already have the time domain expressions, we can easily test the theorems. We begin by applying the initial-value theorem to the transform pair of Eq. (15.53). We have

$$\lim_{s \to \infty} [sF(s)] = \lim_{s \to \infty} \left[\frac{96(s + 5)(s + 12)}{(s + 8)(s + 6)} \right]$$

$$= \lim_{s \to \infty} \left[\frac{96[1 + (5/s)][1 + (12/s)]}{[1 + (8/s)][1 + (6/s)]} \right] = 96;$$

$$\lim_{t \to 0^+} [f(t)] = \lim_{t \to 0^+} [120 + 48e^{-6t} - 72e^{-8t}]u(t)$$

$$= [120 + 48 - 72](1) = 96.$$

In applying the final-value theorem, note that all the poles of $sF(s)$ lie in the left half of the s-plane. Thus

$$\lim_{s \to 0} [sF(s)] = \frac{(96)(5)(12)}{(8)(6)} = 120,$$

$$\lim_{t \to \infty} [f(t)] = [120 + 0 - 0](1) = 120.$$

As a second example, consider the transform pair given by Eq. (15.63). First we have the initial-value theorem:

$$\lim_{s \to \infty} [sF(s)] = \lim_{s \to \infty} \left[\frac{100s^2[1 + (3/s)]}{s^3[1 + (6/s)][1 + (6/s) + (25/s^2)]} \right] = 0.$$

$$\lim_{t \to 0^+} [f(t)] = [-12 + 20 \cos (-53.13°)](1) = -12 + 12 = 0.$$

The final-value theorem gives

$$\lim_{s \to 0} [sF(s)] = \lim_{s \to 0} \left[\frac{100s(s + 3)}{(s + 6)(s^2 + 6s + 25)} \right] = 0,$$

$$\lim_{t \to \infty} [f(t)] = \lim_{t \to \infty} [-12e^{-6t} + 20e^{-3t} \cos (4t - 53.13°)]u(t) = 0.$$

As a final example of illustrating these theorems, consider the expression for $V(s)$ given by Eq. (15.43). Although we cannot calculate $v(t)$ until the circuit parameters are specified, we can check to see if $V(s)$ predicts the correct values of $v(0^+)$ and $v(\infty)$. We know from the statement of the problem that generated $V(s)$ that $v(0^+)$ is zero. We also know that $v(\infty)$ must be zero since the ideal inductor will be a perfect short circuit across the dc current source. Finally, we know that the poles of $V(s)$ must lie in the left half of the s-plane since $R, L,$ and C are positive constants. It follows that the poles of $sV(s)$ also lie in the left half of the s-plane.

Applying the initial-value theorem yields

$$\lim_{s \to \infty} [sV(s)] = \lim_{s \to \infty} \left[\frac{s(I_{dc}/C)}{s^2 \left[1 + \dfrac{1}{RCs} + \dfrac{1}{LCs^2} \right]} \right] = 0.$$

Applying the final-value theorem gives us

$$\lim_{s \to 0} [sV(s)] = \lim_{s \to 0} \left[\frac{s(I_{dc}/C)}{s^2 + \dfrac{1}{RC} s + \dfrac{1}{LC}} \right] = 0.$$

We see that our derived expression for $V(s)$ correctly predicts the initial and final values of $v(t)$. The initial- and final-value theorems give us one method of testing the s-domain expressions for the unknown variables before working out the inverse transform.

DRILL EXERCISE
15.13

Use the initial- and final-value theorems to find the initial and final values of $f(t)$ in Drill Exercises 15.8, 15.10, and 15.11.

Ans. 0, 0; 4, 1; and 0, 0.

DRILL EXERCISE
15.14

a) Use the initial-value theorem to find the initial value of v in Drill Exercise 15.7.
b) Can the final-value theorem be used to find the steady-state value of v? Why?

Ans. (a) 0; (b) no, because V has a pair of poles on the imaginary axis.

15.11 SUMMARY

We have introduced the Laplace transform as a technique for solving constant-coefficient integrodifferential equations. In studying and using the Laplace transform, we want to keep in mind that it is particularly well suited for solving ordinary integrodifferential equations with constant coefficients where the system boundary conditions involve the initial conditions of the dependent variables, that is, initial currents and voltages in circuit problems.

The most important feature of the Laplace transform method is that it transforms an integrodifferential equation into an algebraic equation, or a set of integrodifferential equations into a set of algebraic equations. Thus all the techniques for handling algebraic equations are available to us once we transfer the problem from the t-domain to the s-domain.

The application of the Laplace transform to circuit analysis involves both functional transforms and operational transforms. We must keep in mind that although not all possible functions and operations have Laplace transforms, enough functions and operations that are important in circuit analysis do, which makes the Laplace transform a meaningful analytical tool.

In general, the problem of returning to the t-domain from the s-domain involves finding the inverse transform of a rational function of s. This operation is greatly expedited by expanding the rational function of s into a sum of partial fractions.

The Laplace transform method of circuit analysis can be divided into five fundamental steps:

1. writing the integrodifferential time domain equations that describe the circuit;
2. transforming the integrodifferential equations into s-domain algebraic equations;
3. solving the s-domain algebraic equations for the unknowns of interest;
4. inverse-transforming the s-domain variables back to the time domain; and
5. testing the derived time domain expressions to make sure they make sense in terms of the known physical behavior of the circuit.

In Chapter 16, we will discuss the application of the Laplace transform to circuit problems. We will begin by showing how we can eliminate step 1 from the procedure enumerated above!

PROBLEMS

15.1 Use step functions to write the expression for each of the functions shown in Fig. 15.17.

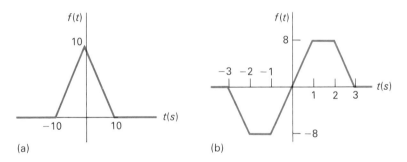

Fig. 15.17 The functions for Problem 15.1.

15.2 Explain why the following function generates an impulse function as $\epsilon \to 0$:

$$f(t) = \frac{\epsilon/\pi}{\epsilon^2 + t^2}, \qquad -\infty \le t \le \infty.$$

15.3 In Section 15.2, we used the sifting property of the impulse function to show that $\mathcal{L}\{\delta(t)\} = 1$. Show that we can obtain the same result by finding the Laplace transform of the rectangular pulse that exists between $\pm \epsilon$ in Fig. 15.9, and then finding the limit of this transform as $\epsilon \to 0$.

15.4 The triangular pulses shown in Fig. 15.18 are equivalent to the rectangular pulses in Fig. 15.12(b), because they both enclose the same area $(1/\epsilon)$ and they both approach infinity proportional to $1/\epsilon^2$ as $\epsilon \to 0$. Use this triangular pulse representation for $\delta'(t)$ to find the Laplace transform of $\delta''(t)$.

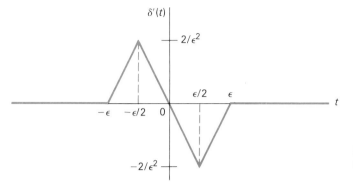

Fig. 15.18 Graphical representation of $\delta'(t)$.

15.5 a) Show that

$$\int_{-\infty}^{\infty} f(t)\delta'(t - a)\, dt = -f'(a).$$

(*Hint:* Integrate by parts.)

b) Use the formula in part (a) to show that

$$\mathscr{L}\{\delta'(t)\} = s.$$

15.6 Find the Laplace transform of each of the following functions:

a) $f(t) = te^{-at}$;
b) $f(t) = \cos \omega t$;
c) $f(t) = \cos (\omega t + \theta)$;
d) $f(t) = \sinh t$.

15.7 Find the Laplace transform, when $\epsilon \to 0$, of the derivative of the exponential function illustrated in Fig. 15.8 by each of the following two methods.

a) First differentiate the function and then find the transform of the resulting function.

b) Use the operational transform given by Eq. (15.26).

15.8 Show that

$$\mathscr{L}\{\delta^{(n)}(t)\} = s^n.$$

15.9 Find each of the following:

a) $\mathscr{L}\left\{\dfrac{d}{dt} (\sin \omega t)\right\}$;

b) $\mathscr{L}\left\{\dfrac{d}{dt} (\cos \omega t)\right\}$;

c) $\mathscr{L}\left\{\dfrac{d^3}{dt^3} (t^2)\right\}$.

d) Check the results of parts (a), (b), and (c) by first differentiating and then transforming.

15.10 Find each of the following.

a) $\mathscr{L}\left\{\displaystyle\int_{0^-}^{t} e^{-ax}\, dx\right\}$;

b) $\mathscr{L}\left\{\displaystyle\int_{0^-}^{t} y\, dy\right\}$.

c) Check the results of parts (a) and (b) by first integrating and then transforming.

15.11 Show that

$$\mathcal{L}\{e^{-at}f(t)\} = F(s + a).$$

15.12 a) Sketch each of the following functions:

$$f(t) = 40e^{-(t-10)}u(t - 10);$$
$$f(t) = (t - 4)u(t - 4) - (t - 8)u(t - 8) - (t - 16)u(t - 16) + (t - 20)u(t - 20).$$

b) Find the Laplace transform of each function.

15.13 a) Find the Laplace transform of the function illustrated in Fig. 15.19.
  b) Find the Laplace transform of the first derivative of the function illustrated in Fig. 15.19.

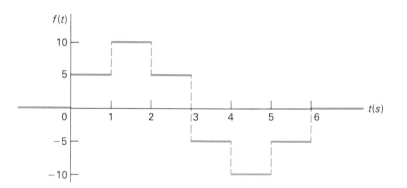

Fig. 15.19 The function for Problem 15.13.

15.14 Find the Laplace transform of each of the following functions:

a) $f(t) = \dfrac{d}{dt}[e^{-at} \sin \omega t];$

b) $f(t) = \displaystyle\int_{0^-}^{t} e^{-ax} \cos \omega x \, dx.$

c) Verify the results obtained in parts (a) and (b) by first carrying out the indicated mathematical operation and then finding the Laplace transform.

15.15 a) Show that

$$\mathcal{L}\{f(at)\} = \frac{1}{a}F\left(\frac{s}{a}\right).$$

b) Use the result of part (a) along with the answers derived in Problem 15.6(d) to find

$$\mathcal{L}\{\sinh \beta t\}.$$

15.16 a) Given that $F(s) = \mathcal{L}\{f(t)\}$, show that

$$-\frac{dF(s)}{ds} = \mathcal{L}\{tf(t)\}.$$

b) Show that

$$(-1)^n \frac{d^n F(s)}{ds^n} = \mathcal{L}\{t^n f(t)\}.$$

c) Use the result of part (b) to find $\mathcal{L}\{t^5\}$, $\mathcal{L}\{t \sin \beta t\}$, and $\mathcal{L}\{te^{-at} \cos \beta t\}$.

15.17 a) Show that if $F(s) = \mathcal{L}\{f(t)\}$ and $[f(t)/t]$ is Laplace transformable, then

$$\int_s^\infty F(u) \, du = \mathcal{L}\left\{\frac{f(t)}{t}\right\}.$$

Hint: Use the defining integral to write

$$\int_s^\infty F(u) \, du = \int_s^\infty \left(\int_0^\infty f(t)e^{-ut} \, dt \right) du$$

and then reverse the order of integration.

b) Start with the result obtained in Problem 15.16(c) for $\mathcal{L}\{t \sin \beta t\}$ and use the operational transform given in Problem 15.17(a) to find $\mathcal{L}\{\sin \beta t\}$.

15.18   There is no energy stored in the circuit shown in Fig. 15.20 at the time when the switch is closed.

a) Derive the integrodifferential equation that governs the behavior of the current i.

b) Show that

$$I(s) = \frac{V_{dc}}{L} \cdot \frac{1}{s^2 + \dfrac{R}{L} s + \dfrac{1}{LC}}.$$

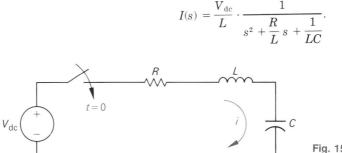

Fig. 15.20   The circuit for Problem 15.18.

15.19   The current in the inductor and the voltage across the capacitor are zero at the time when the switch is closed in the circuit shown in Fig. 15.21.

a) Derive the integrodifferential equation that governs the behavior of the voltage v.

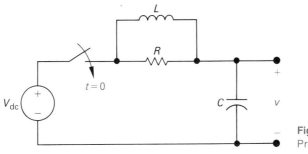

Fig. 15.21   The circuit for Problem 15.19.

b) Show that

$$V(s) = \frac{V_{dc}}{RC} \cdot \frac{s + (R/L)}{s\left(s^2 + \dfrac{1}{RC}s + \dfrac{1}{LC}\right)}.$$

15.20 There is no energy stored in the circuit shown in Fig. 15.22 at the time when the switch is closed.

    a) Derive the integrodifferential equations that govern the behavior of the mesh currents i_1 and i_2.

    b) Show that

$$I_1 = \frac{(s^2L_2C + R_2Cs + 1)V_g}{s^3L_1L_2C + s^2C(R_1L_2 + R_2L_1) + s(R_1R_2C + L_1 + L_2) + R_1 + R_2}.$$

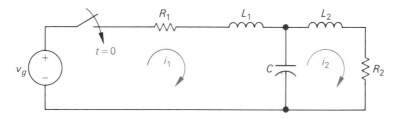

Fig. 15.22 The circuit for Problem 15.20.

15.21 Find $f(t)$ for each of the following functions:

    a) $F(s) = \dfrac{36(s + 4)}{s(s + 2)(s + 8)};$

    b) $F(s) = \dfrac{30(s + 1)(s + 2)}{s(s + 3)(s^2 + 9s + 20)};$

    c) $F(s) = \dfrac{500(s + 6)}{s(s^2 + 12s + 100)};$

    d) $F(s) = \dfrac{200(s + 10)}{s(s + 5)^2};$

    e) $F(s) = \dfrac{800s}{(s^2 + 10s + 125)^2};$

    f) $F(s) = \dfrac{10(s^2 + 9s + 18)}{s(s + 5)}.$

15.22 Derive the transform pair given by Eq. (15.67).

15.23 a) Derive the transform pair given by Eq. (15.86).
    b) Derive the transform pair given by Eq. (15.87).

15.24 Apply the initial- and final-value theorems to each transform pair in Problem 15.21.

15.25 Use the initial- and final-value theorems to check the initial and final values of the current in the circuit in Problem 15.18.

15.26 Use the initial- and final-value theorems to check the initial and final values of the voltage in the circuit in Problem 15.19.

16

THE LAPLACE TRANSFORM IN CIRCUIT ANALYSIS

16.1   INTRODUCTION

The Laplace transform has two characteristics that make it an attractive tool in circuit analysis. First, it transforms a set of linear, constant-coefficient, differential equations into a set of linear polynomial equations. Second, it automatically introduces into the polynomial equations the initial values of the current and voltage variables. Thus initial conditions are an inherent part of the transform process. This contrasts with the classical approach to the solution of a differential equation in which initial conditions enter at the time when the unknown coefficients are evaluated.

Our purpose in this chapter is to develop a systematic method for using the Laplace transform to find the transient behavior of a circuit. The five-step procedure we listed in Section 15.6 is the basis of our current discussion. The first step in making efficient use of the transform method is to eliminate the necessity of writing the integrodifferential equations that describe the circuit. We can do this by developing *s*-domain equivalent circuits for the circuit elements. This will allow us to construct the circuit to be analyzed directly in the *s*-domain. Once we have formulated the *s*-domain circuit, we use the analytical tools already developed (such as node voltages, mesh currents, and circuit reductions) to describe the circuit by algebraic equations. The solution of these *s*-domain equations will yield the unknown currents and voltages as rational functions of *s*, which are then inverse-transformed by partial fraction expansions. Finally, we test the time domain expressions to ensure that the solutions make sense in terms of given initial conditions and known final values.

In Section 16.2, we will develop the *s*-domain equivalent circuits for the circuit elements. As we start our analysis of *s*-domain circuits, it is important to keep in mind that the dimension of a transformed voltage is volt-seconds and the dimension of a transformed current is ampere-seconds. A voltage-to-current ratio in the *s*-domain carries the dimension of volts/amperes and therefore an impedance in the *s*-domain is measured in ohms and an admittance is measured in siemens, or mhos.

16.2   CIRCUIT ELEMENTS IN THE *s*-DOMAIN

The procedure for developing the *s*-domain equivalent circuit for each circuit element is a simple one. First, we write the time domain equation that relates the terminal voltage to the terminal current. Next, we take the Laplace transform of the time domain equation. This will generate an algebraic relationship between the *s*-domain current and voltage. Finally, we construct a circuit model that satisfies the relationship between the *s*-domain current and voltage. We will use the passive sign convention in all our derivations.

We begin with the resistance element. From Ohm's law, we have

$$v = Ri. \tag{16.1}$$

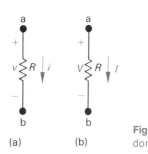

Fig. 16.1   The resistance element: (a) time domain; (b) frequency domain.

(a)   (b)

Since R is a constant, the Laplace transform of Eq. (16.1) is

$$V = RI, \qquad (16.2)$$

where

$$V = \mathcal{L}\{v\} \quad \text{and} \quad I = \mathcal{L}\{i\}.$$

Equation (16.2) tells us that the s-domain equivalent circuit of a resistor is simply a resistance of R ohms that carries a current of I ampere-seconds and has a terminal voltage of V volt-seconds.

The time and frequency domain circuits of the resistor are shown in Fig. 16.1. Note that the resistance element is unchanged in going from the time domain to the frequency domain.

An inductor carrying an initial current of I_0 amperes is shown in Fig. 16.2. The time domain equation that relates the terminal voltage to the terminal current is

$$v = L \frac{di}{dt}. \qquad (16.3)$$

When we find the Laplace transform of Eq. (16.3), we get

$$V = L[sI - i(0^-)]$$
$$= sLI - LI_0. \qquad (16.4)$$

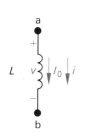

Fig. 16.2   An inductor of L henries carrying an initial current of I_0 amperes.

Equation 16-4 can be satisfied by two different circuit configurations. The first equivalent circuit consists of an impedance of sL ohms in series with an independent voltage source of LI_0 volt-seconds. This circuit is shown in Fig. 16.3. In studying the s-domain equivalent circuit of Fig. 16.3, note that the polarity marks on the voltage source LI_0 are in agreement with the minus sign in Eq. (16.4). It is also important to note that I_0 carries its own algebraic sign. That is, if the initial value of i is opposite to the reference direction for i, then I_0 will have a negative value.

The second equivalent circuit that satisfies Eq. (16.4) consists of an impedance of sL ohms in parallel with an independent current source of I_0/s ampere-seconds. This alternative equivalent circuit is shown in Fig. 16.4.

There are several ways to derive the alternative equivalent circuit shown in Fig. 16.4. One way is to simply solve Eq. (16.4) for the current I and then

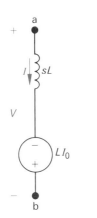

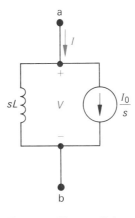

Fig. 16.3 The series equivalent circuit for an inductor of L henries carrying an initial current of I_0 amperes.

Fig. 16.4 The parallel equivalent circuit for an inductor of L henries carrying an initial current of I_0 amperes.

Fig. 16.5 The s-domain circuit for an inductor when the initial current is zero.

construct the circuit to satisfy the resulting equation. Thus

$$I = \frac{V + LI_0}{sL} = \frac{V}{sL} + \frac{I_0}{s}. \tag{16.5}$$

It is easy to see that the circuit shown in Fig. 16.4 satisfies Eq. (16.5). Two other ways to derive the circuit of Fig. 16.4 are (1) to find the Norton equivalent of the circuit shown in Fig. 16.3, and (2) to start with the inductor current as a function of the inductor voltage and then find the Laplace transform of the resulting integral equation. These two approaches are left as exercises in Problems 16.1 and 16.2.

It is worth noting that if the initial energy stored in the inductor is zero, that is, if $I_0 = 0$, then the s-domain equivalent circuit of the inductor reduces to an inductor with an impedance of sL ohms. This circuit is shown in Fig. 16.5.

There are also two s-domain equivalent circuits for an initially charged capacitor. A capacitor initially charged to V_0 volts is shown in Fig. 16.6. The terminal current is

$$i = C \frac{dv}{dt}. \tag{16.6}$$

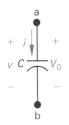

Fig. 16.6 A capacitor of C farads initially charged to V_0 volts.

Transforming Eq. (16.6) yields

$$I = C[sV - v(0^-)]$$

or

$$I = sCV - CV_0, \tag{16.7}$$

from which we see that the s-domain current I is the sum of two branch currents. One branch consists of an admittance of sC mhos and the second branch

TABLE 16.1
SUMMARY OF THE s-DOMAIN EQUIVALENT CIRCUITS

Time domain	Frequency domain

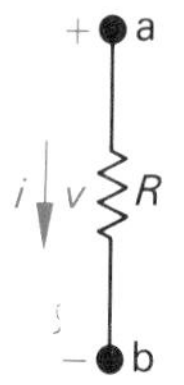

$v = Ri$

$V = RI$

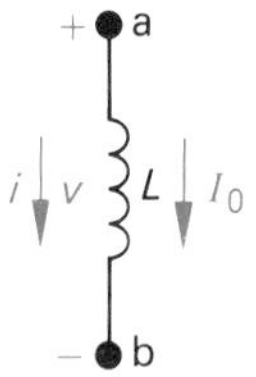

$v = L \, di/dt,$

$i = \dfrac{1}{L} \int_{0^-}^{t} v \, dx + I_0$

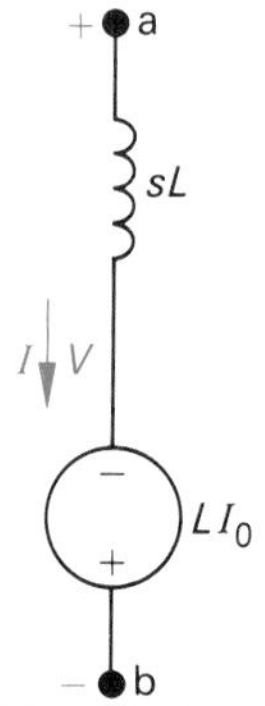

$V = sLI - LI_0$

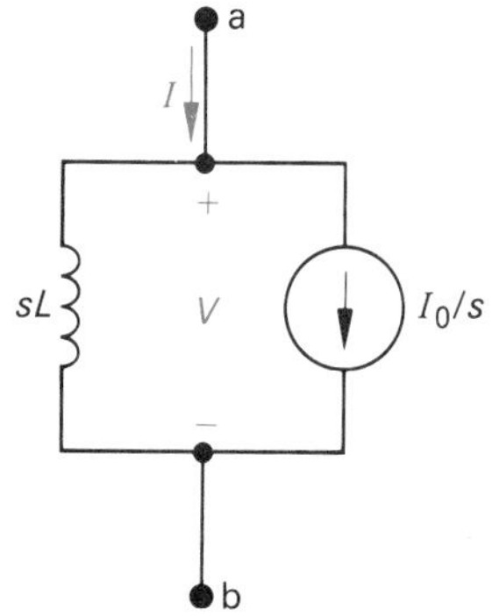

$I = \dfrac{V}{sL} + \dfrac{I_0}{s}$

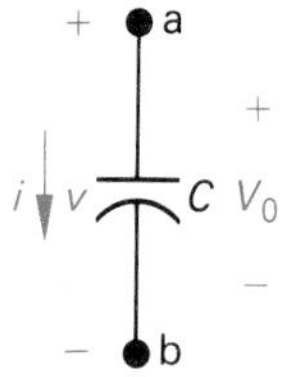

$i = C \, dv/dt,$

$v = \dfrac{1}{C} \int_{0^-}^{t} i \, dx + V_0$

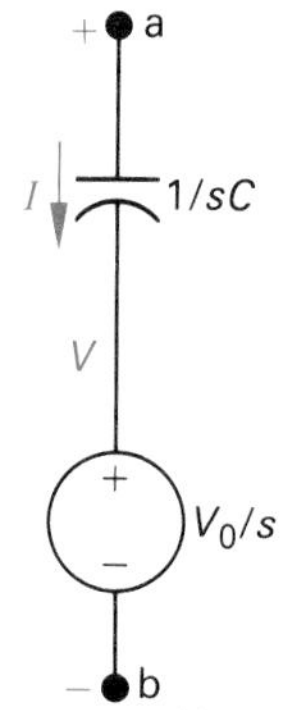

$V = \dfrac{I}{sC} + \dfrac{V_0}{s}$

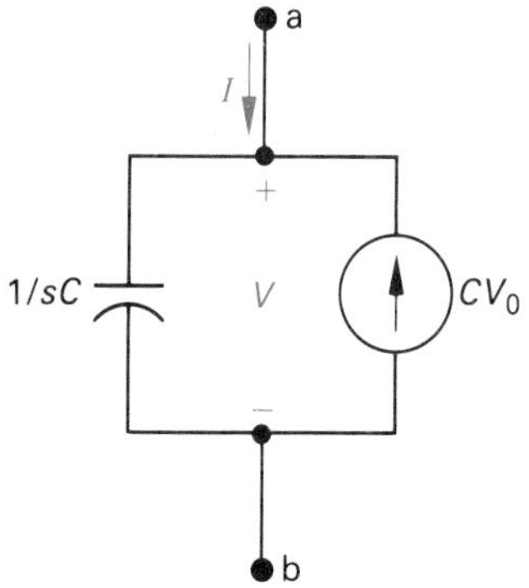

$I = sCV - CV_0$

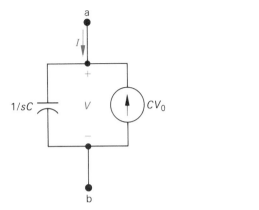

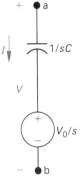

Fig. 16.7 The parallel equivalent circuit for a capacitor initially charged to V_0 volts.

Fig. 16.8 The series equivalent circuit for a capacitor initially charged to V_0 volts.

Fig. 16.9 The *s*-domain circuit for a capacitor when the initial voltage is zero.

consists of an independent current source of CV_0 ampere-seconds. This equivalent circuit is shown in Fig. 16.7.

The series equivalent circuit for the charged capacitor can be derived by solving Eq. (16.7) for V; thus

$$V = \left(\frac{1}{sC}\right) I + \frac{V_0}{s}. \tag{16.8}$$

The circuit that satisfies Eq. (16.8) is shown in Fig. 16.8.

In the equivalent circuits shown in Figs. 16.7 and 16.8, V_0 carries its own algebraic sign. That is, if the polarity of V_0 is opposite to the reference polarity for v, V_0 will be a negative quantity. If the initial voltage on the capacitor is zero, both equivalent circuits reduce to an impedance of $1/sC$ ohms, as shown in Fig. 16.9.

The *s*-domain circuits developed in this section are summarized in Table 16.1. We illustrate the application of these circuits in Section 16.4.

16.3 CIRCUIT ANALYSIS IN THE *s*-DOMAIN

Before illustrating how the *s*-domain equivalent circuits are used in analysis, we need to make some observations that are the basis for all the work that follows.

First we know that if there is no energy stored in the inductor or capacitor, the relationship between the terminal voltage and current for each passive element takes the form

$$V = ZI, \tag{16.9}$$

where Z refers to the *s*-domain impedance of the element. Thus a resistor has an impedance of R ohms, an inductor has an impedance of sL ohms, and a capacitor has an impedance of $1/sC$ ohms. The observation contained in Eq. (16.9) is also contained in Figs. 16.1(b), 16.5, and 16.9. Equation (16.9) is sometimes referred to as Ohm's law for the s-domain.

The reciprocal of the impedance is admittance; therefore the s-domain admittance of a resistor is $1/R$ mhos, an inductor has an admittance of $1/sL$ mhos, and a capacitor has an admittance of sC mhos.

The rules for combining impedances and admittances in the s-domain are precisely the same as those for combining impedances and admittances in phasor domain circuits. Thus series–parallel simplifications and Δ-to-Y conversions are applicable to s-domain analysis.

The second important observation to make at this time is that Kirchhoff's laws are applicable to s-domain currents and voltages. This follows from the operational transform, which states that the Laplace transform of a sum of time domain functions is the sum of the transforms of the individual functions (see Table 15.2). Thus given that the algebraic sum of the currents at a junction is zero in the time domain, it follows that the algebraic sum of the transformed currents is also zero. A similar statement holds for the algebraic sum of the transformed voltages around a closed path. The s-domain version of Kirchhoff's laws is given as follows:

$$\mathrm{alg} \sum I\text{s} = 0, \tag{16.10}$$

$$\mathrm{alg} \sum V\text{'s} = 0. \tag{16.11}$$

The third observation requires an understanding of the implications of the first two observations. The fact that the voltage and current at the terminals of a passive element are related by an algebraic equation and that Kirchhoff's laws still hold means that all the techniques of circuit analysis developed for pure resistive networks can be used in s-domain analysis. Thus node voltages, mesh currents, source transformations, and Thévenin–Norton equivalents are all valid techniques, even when there is energy initially stored in the inductors and capacitors. Initially stored energy requires that we modify Eq. (16.9), but the modifications simply require the addition of independent sources either in series or parallel with the element impedances and the addition of these sources is governed by Kirchhoff's laws.

DRILL EXERCISE
16.1

A 125-Ω resistor, a 4-mH inductor, and a 0.1-μF capacitor are connected in parallel.

a) Express the admittance of this parallel combination of elements as a rational function of s.

b) Compute the numerical values of the zeros and poles.

Ans. (a) $10^{-7}(s^2 + 80{,}000s + 25 \times 10^8)/s$; (b) $z_1 = 40{,}000 + j30{,}000$; $z_2 = 40{,}000 - j30{,}000$, $p_1 = 0$.

DRILL EXERCISE
16.2

The parallel circuit in Drill Exercise 16.1 is placed in series with a 500-Ω resistor.

a) Express the impedance of this series combination as a rational function of s.

b) Compute the numerical values of the zeros and poles.

Ans. (a) $500(s + 50,000)^2/(s^2 + 80,000s + 25 \times 10^8)$; (b) $z_1 = z_2 = 50,000$;
$p_1 = 40,000 + j30,000$, $p_2 = 40,000 - j30,000$.

16.4   ILLUSTRATIVE EXAMPLES

In order to illustrate how we can use the Laplace transform to determine the transient behavior of linear, lumped-parameter circuits, we begin with some circuit structures that we analyzed in Chapters 6, 7, and 8. We start by analyzing familiar circuits because we will gain confidence in our ability to use the Laplace transform approach when we see that it yields the same results we obtained in our earlier work.

The first circuit we will analyze is the RC circuit shown in Fig. 16.10. The capacitor is initially charged to V_0 volts and we are interested in the time domain expressions for i and v. Since this circuit was first analyzed in Chapter 6, you may wish to review Section 6.3 before proceeding with the present analysis.

Let us start by first finding i. In transferring the circuit in Fig. 16.10 to the s-domain we have a choice of two equivalent circuits for the charged capacitor. Since we are interested in the current, the series equivalent circuit is more attractive because it will result in a single mesh circuit in the frequency domain. Thus we construct the s-domain circuit shown in Fig. 16.11.

Summing the voltages around the mesh in Fig. 16.11 generates the expression

$$\frac{V_0}{s} = \frac{1}{sC}I + RI. \tag{16.12}$$

Solving Eq. (16.12) for I yields

$$I = \frac{CV_0}{(RCs + 1)} = \frac{V_0/R}{[s + (1/RC)]}. \tag{16.13}$$

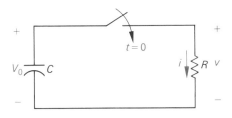

Fig. 16.10   The capacitor discharge circuit.

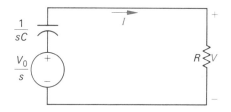

Fig. 16.11 An s-domain equivalent circuit for the circuit in Fig. 16.10.

Note that the expression for I is a proper rational function of s and can be inverse-transformed by inspection. We get

$$i = \frac{V_0}{R} e^{-t/RC} u(t),$$ (16.14)

which is equivalent to the expression for the current derived by the classical methods discussed in Chapter 6 since $\tau = RC$ (see Eq. 6.21).

Now let us consider the problem of finding the voltage v. Once we have found i, the easiest way to determine v is to simply apply Ohm's law, that is, we see from the circuit that

$$v = Ri = V_0 e^{-t/RC} u(t).$$ (16.15)

We now illustrate a way to find v from the circuit without first finding the current i. In this alternative approach, we return to the original circuit of Fig. 16.10 and transfer it to the s-domain using the parallel equivalent circuit for the charged capacitor. The parallel equivalent circuit is attractive now because its use makes it possible to describe the resulting circuit in terms of a single node voltage. The new s-domain equivalent circuit is shown in Fig. 16.12.

The node-voltage equation that describes the circuit in Fig. 16.12 is

$$\frac{V}{R} + sCV = CV_0.$$ (16.16)

Solving Eq. (16.16) for V gives

$$V = \frac{V_0}{s + (1/RC)}.$$ (16.17)

Inverse-transforming Eq. (16.17) leads to the same expression for v given by

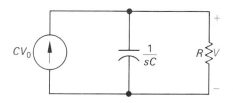

Fig. 16.12 An s-domain equivalent circuit for the circuit in Fig. 16.10.

Eq. (16.12), namely,

$$v = V_0 e^{-t/RC} = V_0 e^{-t/\tau}\, u(t). \tag{16.18}$$

Our purpose in deriving v by the direct use of the transform method is to show that the choice of which s-domain equivalent circuit to use is influenced by what response signal is of interest.

DRILL EXERCISE
16.3

The switch in the circuit shown has been in position a for a long time. At $t = 0$ the switch is thrown to position b.

a) Find I, V_1, and V_2 as rational functions of s.

b) Find the time domain expressions for i, v_1, and v_2.

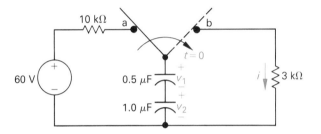

Ans. (a) $I = 0.02/(s + 1000)$, $V_1 = 40/(s + 1000)$, $V_2 = 20/(s + 1000)$;
(b) $i = 20e^{-1000t}u(t)$ mA, $v_1 = 40e^{-1000t}u(t)$ V, $v_2 = 20e^{-1000t}u(t)$ V.

The second circuit we will analyze is the parallel RLC circuit we first analyzed in Example 8.6. The circuit is redrawn in Fig. 16.13. The problem is to find the expression for i_L after the constant current source is switched across the paralleled elements. The initial energy stored in the circuit is zero.

As before, we begin our analysis by constructing the s-domain equivalent circuit, which is shown in Fig. 16.14. Note how easy it is to transform an independent source from the time domain to the frequency domain. The source is transferred to the s-domain by simply determining the Laplace transform of the time domain function of the source. For the case at hand, the opening of the

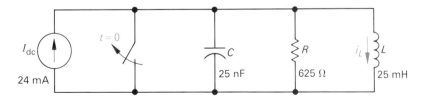

Fig. 16.13   Step response of a parallel RLC circuit.

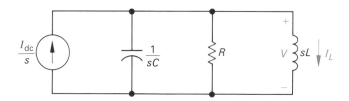

Fig. 16.14   The s-domain equivalent circuit for the circuit in Fig. 16.13.

switch results in a step change in the current applied to the circuit. Therefore, the s-domain current source is $\mathcal{L}\{I_{\mathrm{dc}}u(t)\}$, or I_{dc}/s. To find I_L, we first solve for V and then use the fact that

$$I_L = \frac{V}{sL} \tag{16.19}$$

to establish the s-domain expression for I_L. Summing the currents away from the top node in the circuit in Fig. 16.14 generates the expression

$$sCV + \frac{V}{R} + \frac{V}{sL} = \frac{I_{\mathrm{dc}}}{s}. \tag{16.20}$$

When we solve Eq. (16.20) for V we obtain

$$V = \frac{I_{\mathrm{dc}}/C}{s^2 + \dfrac{1}{RC}s + \dfrac{1}{LC}}. \tag{16.21}$$

Substituting this expression for V in Eq. (16.19) gives

$$I_L = \frac{I_{\mathrm{dc}}/LC}{s\left(s^2 + \dfrac{1}{RC}s + \dfrac{1}{LC}\right)}. \tag{16.22}$$

When the numerical values of R, L, C, and I_{dc} are substituted into Eq. (16.22), we get

$$I_L = \frac{384 \times 10^5}{s(s^2 + 64{,}000s + 16 \times 10^8)}. \tag{16.23}$$

Before expanding Eq. (16.23) into a sum of partial fractions, we factor the quadratic term in the denominator. Thus

$$I_L = \frac{384 \times 10^5}{s(s + 32{,}000 - j24{,}000)(s + 32{,}000 + j24{,}000)}. \tag{16.24}$$

At this point in our analysis, we can test the s-domain expression for I_L by checking to see if the final-value theorem predicts the correct value for i_L at $t = \infty$. First, let us note that all the poles of I_L, except for the first-order pole at the origin, lie in the left half of the s-plane and therefore the theorem is applicable. We know from the behavior of the circuit that after the switch has been open for a long time the inductor will short-circuit the current source. Therefore, the final value of i_L must be 24 mA. The limit of sI_L as $s \to 0$ is

$$\lim_{s \to 0} [sI_L] = \frac{384 \times 10^5}{16 \times 10^8} = 24 \text{ mA.} \tag{16.25}$$

(Since currents in the s-domain carry the dimension of ampere-second, the dimension of sI_L will be amperes.)

We now proceed with the partial-fraction expansion of Eq. (16.24). Thus

$$I_L = \frac{K_1}{s} + \frac{K_2}{s + 32,000 - j24,000} + \frac{K_2^*}{s + 32,000 + j24,000}. \tag{16.26}$$

The partial-fraction coefficients are

$$K_1 = \frac{384 \times 10^5}{16 \times 10^8} = 24 \times 10^{-3}; \tag{16.27}$$

$$K_2 = \frac{384 \times 10^5}{(-32,000 + j24,000)(j48,000)}$$

$$= 20 \times 10^{-3} \underline{/126.87°}. \tag{16.28}$$

When we substitute the numerical values of K_1 and K_2 into Eq. (16.26) and inverse-transform the resulting expression, we get

$$i_L = [24 + 40e^{-32,000t} \cos (24,000t + 126.87°)]u(t) \text{ mA.} \tag{16.29}$$

To verify that the answer given by Eq. (16.29) is equivalent to the answer given for Example 8.6, we observe that

$$40 \cos (24,000t + 126.87°) = -24 \cos 24,000t - 32 \sin 24,000t.$$

If we were not using a previous solution to check the validity of our work, we would test Eq. (16.29) to make sure that $i_L(0)$ satisfied the given initial conditions and $i_L(\infty)$ satisfied the known behavior of the circuit.

DRILL EXERCISE
16.4

The energy stored in the circuit shown is zero at the time when the switch is closed.

a) Find the s-domain expression for I.
b) Find the time domain expression for i when $t > 0$.
c) Find the s-domain expression for V.
d) Find the time domain expression for v when $t > 0$.

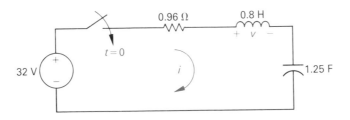

Ans. (a) $I = 40/(s^2 + 1.2s + 1)$; (b) $i = [50e^{-0.6t} \sin 0.8t]u(t)$ A;
(c) $V = 32s/(s^2 + 1.2s + 1)$; (d) $v = [40e^{-0.6t} \cos (0.8t + 36.87°)]u(t)$ V.

As a third example of using the Laplace transform to find the transient behavior of a circuit, we return to the circuit in Fig. 16.13 and replace the dc current source with a sinusoidal source. The current source is stipulated to be

$$i_g = I_m \cos \omega t \text{ A,} \tag{16.30}$$

where $I_m = 24$ mA and $\omega = 40,000$ rad/s. As before, we assume that the initial energy stored in the circuit is zero.

The s-domain expression for the source current is

$$I_g = \frac{sI_m}{s^2 + \omega^2}. \tag{16.31}$$

The voltage across the parallel elements is

$$V = \frac{(I_g/C)s}{(s^2 + s)/(RC + 1/LC)}. \tag{16.32}$$

When we substitute Eq. (16.31) into Eq. (16.32), there results

$$V = \frac{(I_m/C)s^2}{(s^2 + \omega^2)\left(s^2 + \dfrac{s}{RC} + \dfrac{1}{LC}\right)}, \tag{16.33}$$

from which it follows that

$$I_L = \frac{V}{sL} = \frac{(I_m/LC)s}{(s^2 + \omega^2)\left(s^2 + \dfrac{s}{RC} + \dfrac{1}{LC}\right)}. \tag{16.34}$$

When we substitute the numerical values of I_m, ω, R, L, and C into Eq. (16.34) we get

$$I_L = \frac{384 \times 10^5 s}{(s^2 + 16 \times 10^8)(s^2 + 64,000s + 16 \times 10^8)}. \tag{16.35}$$

We now write the denominator in factored form to give

$$I_L = \frac{384 \times 10^5 s}{(s - ja)(s + ja)(s + \alpha - j\beta)(s + \alpha + j\beta)} \tag{16.36}$$

where $a = 40,000$, $\alpha = 32,000$, and $\beta = 24,000$.

We cannot test the final value of i_L via the final-value theorem since I_L has a pair of poles on the imaginary axis, that is, poles at $\pm j4 \times 10^4$. Thus we must first find i_L and then check the validity of our expression from known circuit behavior.

If we expand Eq. (16.36) into a sum of partial fractions we generate the following equation:

$$I_L = \frac{K_1}{s - j40{,}000} + \frac{K_1^*}{s + j40{,}000}$$

$$+ \frac{K_2}{s + 32{,}000 - j24{,}000} + \frac{K_2^*}{s + 32{,}000 + j24{,}000}. \tag{16.37}$$

The numerical values of the coefficients K_1 and K_2 are

$$K_1 = \frac{384 \times 10^5 (j40{,}000)}{(j80{,}000)(32{,}000 + j16{,}000)(32{,}000 + j64{,}000)}$$

$$= 7.5 \times 10^{-3} \underline{/-90°}; \tag{16.38}$$

$$K_2 = \frac{384 \times 10^5 (-32{,}000 + j24{,}000)}{(-32{,}000 - j16{,}000)(-32{,}000 + j64{,}000)j48{,}000}$$

$$= 12.5 \times 10^{-3} \underline{/90°}. \tag{16.39}$$

When we substitute the numerical values given by Eqs. (16.38) and (16.39) into Eq. (16.37) and inverse-transform the result, we get

$$i_L = [15 \cos (40{,}000t - 90°) + 25e^{-32{,}000t} \cos (24{,}000t + 90°)] \text{ mA}$$

$$= [15 \sin 40{,}000t - 25e^{-32{,}000t} \sin 24{,}000t]u(t) \text{ mA}. \tag{16.40}$$

We now test Eq. (16.40) to see whether it makes sense in terms of the given initial conditions and the known circuit behavior after the switch has been open for a long time. For $t = 0$, Eq. (16.40) predicts zero initial current, which agrees with the fact that the initial energy in the circuit is zero. Equation (16.40) predicts a steady-state current of

$$i_{L_{ss}} = 15 \sin 40{,}000t \text{ mA}, \tag{16.41}$$

which can be verified by the phasor method discussed in Chapter 10 and is left as an exercise.

Our fourth example is shown in Fig. 16.15. The problem is to find the branch currents i_1 and i_2 that arise when the 336-V dc voltage source is suddenly applied to the circuit. The initial energy stored in the circuit is zero. We were not able to solve this type of transient response by the analytical technique presented in Chapter 7 because it involves the solution of two simulta-

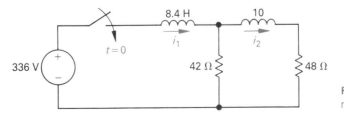

Fig. 16.15 A multiple-mesh *RL* circuit.

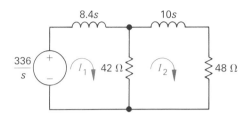

Fig. 16.16 The s-domain equivalent circuit for the circuit in Fig. 16.15.

neous first-order differential equations. Although there are techniques for solving simultaneous differential equations in the time domain, we have chosen not to introduce them in this text. However, since the Laplace transform method changes the problem to the solution of algebraic equations, we will now analyze multiple mesh and node circuits.

The s-domain circuit of the circuit in Fig. 16.15 is shown in Fig. 16.16. The two mesh-current equations for that circuit are

$$\frac{336}{s} = (42 + 8.4s)I_1 - 42I_2, \tag{16.42}$$

$$0 = -42I_1 + (90 + 10s)I_2. \tag{16.43}$$

Using Cramer's method to solve for I_1 and I_2 we have

$$\triangle = \begin{vmatrix} 42 + 8.4s & -42 \\ -42 & 90 + 10s \end{vmatrix}$$

$$= 84(s^2 + 14s + 24)$$

$$= 84(s + 2)(s + 12); \tag{16.44}$$

$$N_1 = \begin{vmatrix} \dfrac{336}{s} & -42 \\ 0 & 90 + 10s \end{vmatrix}$$

$$= \frac{3360(s + 9)}{s}; \tag{16.45}$$

and

$$N_2 = \begin{vmatrix} 8.4s + 42 & \dfrac{336}{s} \\ -42 & 0 \end{vmatrix}$$

$$= \frac{14{,}112}{s}. \tag{16.46}$$

It follows from Eqs. (16.44), (16.45), and (16.46) that

$$I_1 = \frac{N_1}{\triangle} = \frac{40(s + 9)}{s(s + 2)(s + 12)} \tag{16.47}$$

and

$$I_2 = \frac{N_2}{\triangle} = \frac{168}{s(s + 2)(s + 12)}. \tag{16.48}$$

Expanding I_1 and I_2 into a sum of partial fractions gives

$$I_1 = \frac{15}{s} - \frac{14}{s + 2} - \frac{1}{s + 12}, \tag{16.49}$$

$$I_2 = \frac{7}{s} - \frac{8.4}{s + 2} + \frac{1.4}{s + 12}. \tag{16.50}$$

Now the expressions for i_1 and i_2 are obtained by inverse-transforming Eqs. (16.49) and (16.50), respectively; thus

$$i_1 = [15 - 14e^{-2t} - e^{-12t}]u(t) \text{ A} \tag{16.51}$$

and

$$i_2 = [7 - 8.4e^{-2t} + 1.4e^{-12t}]u(t) \text{ A}. \tag{16.52}$$

Next we test our solutions to see whether they make sense in terms of the given circuit. Since there is no energy stored in the circuit at the instant when the switch is closed, it follows that both $i_1(0^-)$ and $i_2(0^-)$ must be zero. Our solutions are in agreement with these initial values. After the switch has been closed for a long period, the two inductors will appear as short circuits. Therefore the final values of i_1 and i_2 will be

$$i_1(\infty) = \frac{336(90)}{(42)(48)} = 15 \text{ A} \tag{16.53}$$

and

$$i_2(\infty) = \frac{15(42)}{90} = 7 \text{ A}. \tag{16.54}$$

Our solutions are also in agreement with these final values.

One final test, which involves the numerical values of the exponents, is to calculate the voltage drop across the 42-Ω resistor by three different methods. From the circuit we see that the voltage across the 42-Ω resistor (positive at the top of the resistor) is

$$v = 42(i_1 - i_2) = 336 - 8.4\frac{di_1}{dt} = 48i_2 + 10\frac{di_2}{dt}. \tag{16.55}$$

We leave it to the reader to verify that regardless of which form of Eq. (16.55) we use, the voltage is

$$v = [336 - 235.2e^{-2t} - 100.80e^{-12t}]u(t) \text{ V}.$$

At this point, we are confident that our solutions for i_1 and i_2 are correct.

DRILL EXERCISE   The dc current and voltage sources are applied simultaneously to the circuit shown.
16.5   There is no energy stored in the circuit at the instant of application.

   a) Derive the s-domain expressions for V_1 and V_2.
   b) Derive, for $t > 0$, the time domain expressions for v_1 and v_2.
   c) Calculate $v_1(0^+)$ and $v_2(0^+)$.
   d) Compute the steady-state values of v_1 and v_2.

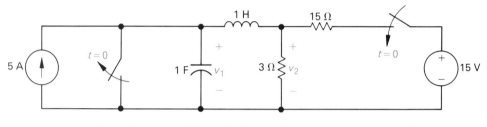

Ans. (a) $V_1 = 5(s + 3)/s(s + 0.5)(s + 2)$, $V_2 = 2.5(s^2 + 6)/s(s + 0.5)(s + 2)$;
(b) $v_1 = [15 - \frac{50}{3} e^{-0.5t} + \frac{5}{3} e^{-2t}]u(t)$ V, $v_2 = [15 - \frac{125}{6} e^{-0.5t} + \frac{25}{3} e^{-2t}]u(t)$V;
(c) $v_1(0^+) = 0$, $v_2(0^+) = 2.5$ V; (d) $v_1 = v_2 = 15$ V.

   In our next example, we will show how Thévenin's equivalent can be used in
the s-domain. The circuit to be analyzed is illustrated in Fig. 16.17. Our problem
is to find the capacitor current that results from closing the switch. The energy
stored in the circuit prior to closing the switch is zero.
   To find i_C we will first construct the s-domain equivalent circuit of the circuit shown in Fig. 16.17 and then find the Thévenin equivalent of the s-domain
circuit with respect to the terminals of the capacitor. The s-domain circuit is illustrated in Fig. 16.18.
   The Thévenin voltage is the open-circuit voltage at the terminals a, b.
Under open-circuit conditions, there is no voltage across the 60-Ω resistor;
hence

$$V_t = \frac{(480/s)(0.002s)}{20 + 0.002s} = \frac{480}{s + 10^4}.$$ (16.56)

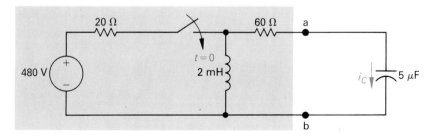

Fig. 16.17   A circuit showing the use of Thévenin's equivalent in the s-domain.

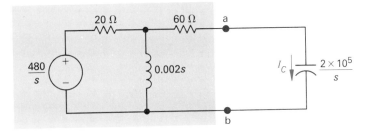

Fig. 16.18   The s-domain model of the circuit shown in Fig. 16.17.

The Thévenin impedance seen from the terminals a, b is equal to the 60-Ω resistor in series with the parallel combination of the 20-Ω resistor and the 2-mH inductor; thus

$$Z_t = 60 + \frac{0.002s(20)}{20 + 0.002s} = \frac{80(s + 7500)}{s + 10^4}. \tag{16.57}$$

Using the Thévenin equivalent, we reduce the circuit in Fig. 16.18 to that shown in Fig. 16.19, from which we see that the capacitor current I_C equals the Thévenin voltage divided by the total series impedance. Thus

$$I_C = \frac{480/(s + 10^4)}{[80(s + 7500)/(s + 10^4)] + [(2 \times 10^5)/s]}. \tag{16.58}$$

We can simplify Eq. (16.58) to

$$I_C = \frac{6s}{s^2 + 10,000s + 25 \times 10^6}$$

$$= \frac{6s}{(s + 5000)^2}. \tag{16.59}$$

A partial-fraction expansion of Eq. (16.59) generates

$$I_C = \frac{-30,000}{(s + 5000)^2} + \frac{6}{(s + 5000)}, \tag{16.60}$$

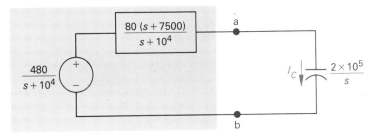

Fig. 16.19   A simplified version of the circuit in Fig. 16.18 using a Thévenin equivalent.

the inverse transform of which is

$$i_C = [-30,000te^{-5000t} + 6e^{-5000t}]u(t) \text{ A.} \qquad \text{(16.61)}$$

We now test Eq. (16.61) to see whether it makes sense in terms of known circuit behavior. From Eq. (16.61) we have

$$i_C(0) = 6 \text{ A.} \qquad \text{(16.62)}$$

This agrees with the initial current in the capacitor as calculated from the circuit in Fig. 16.17. Since the initial inductor current is zero and the initial capacitor voltage is zero, the initial capacitor current will be 480/80, or 6 A. The final value of the current is zero, which is also in agreement with Eq. (16.61).

Also note from Eq. (16.61) that the current reverses sign when t exceeds 6/30,000, or 200 μs. The fact that i_C reverses sign makes sense because when the switch first closes, the capacitor begins to charge. Eventually this charge is reduced to zero since the inductor is a short circuit at $t = \infty$. The charging and discharging of the capacitor is reflected in the sign reversal of i_C.

Assume that the voltage drop across the capacitor v_{ab} is also of interest. Once we know i_C, we can find v_{ab} by integration in the time domain, that is,

$$v_C = 2 \times 10^5 \int_{0^-}^{t} (6 - 30,000x)e^{-5000x} \, dx. \qquad \text{(16.63)}$$

Although the integration called for in Eq. (16.63) is not difficult, it can be avoided altogether by first finding the s-domain expression for V_C and then finding v_C via an inverse transform. Thus

$$V_C = \frac{1}{sC} I_C = \frac{(2 \times 10^5)}{s} \cdot \frac{6s}{(s + 5000)^2}$$

$$= \frac{12 \times 10^5}{(s + 5000)^2}, \qquad \text{(16.64)}$$

from which it follows directly that

$$v_C = 12 \times 10^5 te^{-5000t} u(t). \qquad \text{(16.65)}$$

We leave it to the reader to verify that Eq. (16.65) is consistent with Eq. (16.61) and that it also supports the observations made with regard to the behavior of i_C (see Problem 16.17.)

DRILL EXERCISE
16.6

The initial charge on the capacitor in the circuit shown is zero.

a) Find the s-domain Thévenin equivalent circuit with respect to the terminals a, b.
b) Find the s-domain expression for the current that the circuit delivers to a load consisting of a 0.4-H inductor in series with a 1-Ω resistor.

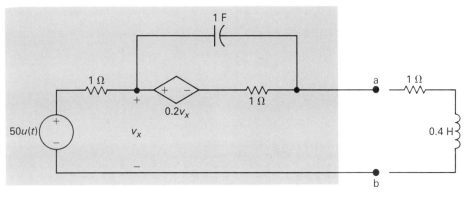

Ans. (a) $V_t = V_{ab} = 50(s + 0.8)/s(s + 1)$, $Z_t = (s + 1.8)/(s + 1)$;
(b) $I_{ab} = [125(s + 0.8)]/[s(s^2 + 6s + 7)]$.

In our sixth and last introductory example, we will illustrate how the transient response of a circuit that contains mutual inductance is analyzed by the Laplace transform. The circuit is shown in Fig. 16.20. The make-before-break switch has been in position a for a long time. At $t = 0$ the switch moves instantaneously to position b. Our problem is to derive the time domain expression for i_2.

We begin our analysis by redrawing the circuit in Fig. 16.20 with the switch in position b and the magnetically coupled coils replaced by a T-equivalent circuit. The circuit is shown in Fig. 16.21.

We now transfer the circuit shown in Fig. 16.21 to the s-domain. In so doing, we make note of the fact that

$$i_1(0^-) = \frac{60}{12} = 5 \text{ A} \tag{16.66}$$

and

$$i_2(0^-) = 0. \tag{16.67}$$

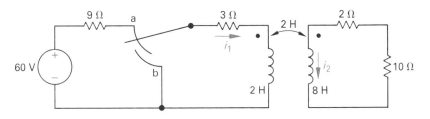

Fig. 16.20   A circuit containing magnetically coupled coils.

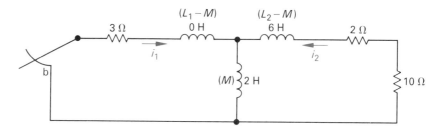

Fig. 16.21   The circuit in Fig. 16.20 with the magnetically coupled coils replaced by a T-equivalent circuit.

Since we plan to use mesh analysis in the s-domain, we use the series equivalent circuit for an inductor carrying an initial current. The s-domain circuit is shown in Fig. 16.22, where we see that only one independent voltage source appears in the circuit. This source appears in the vertical leg of the tee to account for the fact that the initial value of the current in the 2-H inductor is $i_1(0^-) + i_2(0^-)$, or 5 A. There is no voltage source in the branch carrying i_1 because $L_1 - M = 0$.

The two s-domain mesh equations that describe the circuit in Fig. 16.22 are

$$(3 + 2s)I_1 + 2sI_2 = 10 \tag{16.68}$$

and

$$2sI_1 + (12 + 8s)I_2 = 10. \tag{16.69}$$

Solving for I_2 yields

$$I_2 = \frac{2.5}{(s + 1)(s + 3)}. \tag{16.70}$$

Expanding Eq. (16.70) into a sum of partial fractions generates

$$I_2 = \frac{1.25}{s + 1} - \frac{1.25}{s + 3}. \tag{16.71}$$

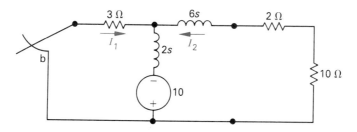

Fig. 16.22   The s-domain equivalent circuit for the circuit in Fig. 16.21.

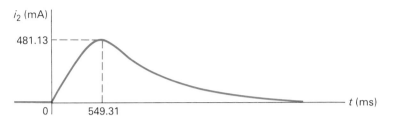

Fig. 16.23   The plot of i_2 versus t for the circuit in Fig. 16.20.

It follows directly from Eq. (16.71) that

$$i_2 = [1.25e^{-t} - 1.25e^{-3t}]u(t) \text{ A}. \tag{16.72}$$

A study of Eq. (16.72) reveals that i_2 increases from zero to a peak value of 481.13 mA 549.31 ms after the switch is moved to position b. Thereafter, i_2 decreases exponentially toward zero. A plot of i_2 vs. t is shown in Fig. 16.23. This response makes sense in terms of the known physical behavior of the magnetically coupled coils. A current can exist in the L_2 inductor only if there is a time-varying current in the L_1 inductor. As i_1 decreases from its initial value of 5 A, i_2 increases from zero and then approaches zero as i_1 approaches zero. (See Problem 16.18.)

DRILL EXERCISE
16.7

a) Verify from Eq. (16.72) that i_2 reaches a peak value of 481.13 mA at $t = 549.31$ ms.
b) Find i_1, for $t > 0$, for the circuit in Fig. 16.20.
c) Compute di_1/dt when i_2 is at its peak.
d) Express i_2 as a function of di_1/dt when i_2 is at its peak value.
e) Use (c) and (d) to calculate the peak value of i_2.

Ans. (a) $di_2/dt = 0$ when $t = \frac{1}{2} \ln 3$ (s); (b) $i_1 = 2.5[e^{-t} + e^{-3t}]u(t)$ A; (c) -2.89A/s; (d) $i_2 = -(M di_1/dt)/12$; (e) 481.13 mA.

16.5   THE IMPULSE FUNCTION IN CIRCUIT ANALYSIS

Impulse functions occur in circuit analysis either because of a switching operation or because the circuit is excited by an impulsive source. We begin our discussion by first showing how an impulse function can be created by a switching operation.

In the circuit of Fig. 16.24, the capacitor C_1 is charged to an initial voltage of V_0 at the time the switch is closed. The initial charge on C_2 is zero. Our problem is to find the expression for $i(t)$ as R approaches zero. The s-domain equivalent circuit is shown in Fig. 16.25.

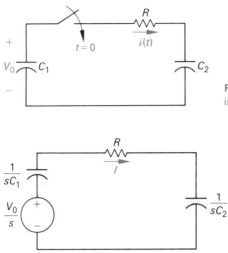

Fig. 16.24　A circuit showing the creation of an impulsive current.

Fig. 16.25　The s-domain equivalent circuit for the circuit in Fig. 16.24.

It follows directly from Fig. 16.25 that

$$I = \frac{V_0/s}{[R + (1/sC_1) + (1/sC_2)]}$$

$$= \frac{V_0/R}{s + (1/RC_e)} \tag{16.73}$$

where we have introduced C_e for the equivalent capacitance $C_1 C_2/(C_1 + C_2)$. Equation (16.73) can be inverse-transformed by inspection; thus

$$i = \left[\frac{V_0}{R} e^{-t/RC_e} \right] u(t), \tag{16.74}$$

which tells us that as R decreases the initial current (V_0/R) increases and the time constant (RC_e) decreases. Thus as R gets smaller, the current starts from a larger initial value and then drops off more rapidly. These characteristics of i are shown in Fig. 16.26.

At this point in our analysis, we begin to suspect that i is approaching an impulse function as R approaches zero, because we see the initial value of i approaching infinity and the duration of i approaching zero. It remains for us to check whether the area under the current function is independent of R. Physically the total area under the i vs. t curve represents the total charge transferred to C_2 after the switch is closed; thus

$$\text{Area} = q = \int_{0^-}^{\infty} \frac{V_0}{R} e^{-t/RC_e} \, dt = V_0 C_e, \tag{16.75}$$

which tells us that the total charge transferred to C_2 is independent of R and

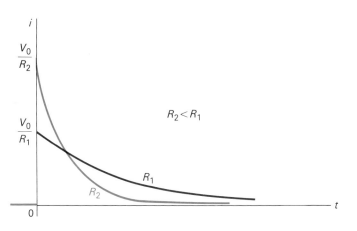

Fig. 16.26   The plot of $i(t)$ versus t for two different values of R.

equals $V_0 C_e$ coulombs. Thus as R approaches zero, the current approaches an impulse of strength $V_0 C_e$; that is,

$$i \to V_0 C_e \, \delta(t). \tag{16.76}$$

The physical interpretation of Eq. (16.76) is that when $R = 0$ a finite amount of charge is transferred to C_2 instantaneously. If we return to Fig. 16.24 and make R zero, we can understand why we get an instantaneous transfer of charge. With R equal to zero, we create a contradiction when we close the switch. That is, we apply a voltage across a capacitor that has a zero initial voltage. The only way we can have an instantaneous change in the capacitor voltage is to have an instantaneous transfer of charge. When the switch is closed, the voltage across C_2 does not jump to V_0 but to its final value of

$$v_2 = \frac{C_1 V_0}{C_1 + C_2}. \tag{16.77}$$

We leave the derivation of Eq. (16.77) to the reader via Problem 16.21.

It is important to point out that if we set R equal to zero at the very outset, the Laplace transform analysis will predict the impulsive current response. Thus

$$I = \frac{V_0/s}{(1/sC_1) + (1/sC_2)} = \frac{C_1 C_2 V_0}{C_1 + C_2} = C_e V_0. \tag{16.78}$$

In writing Eq. (16.78), we use the capacitor voltages that exist at $t = 0^-$. The inverse transform of a constant is the constant times the impulse function; therefore from Eq. (16.78) we have

$$i = C_e V_0 \, \delta(t). \tag{16.79}$$

The ability of the Laplace transform to predict correctly the occurrence of an

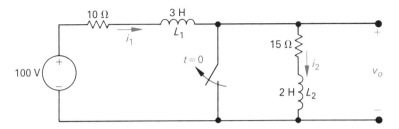

Fig. 16.27   A circuit showing the creation of an impulsive voltage.

impulsive response is one reason why the transform is widely used to analyze the transient behavior of linear, lumped-parameter, time-invariant circuits.

As a second example of a switching operation producing an impulsive response, consider the circuit illustrated in Fig. 16.27. Our problem is to find the time domain expression for v_o after the switch has been opened. Note that the opening of the switch forces an instantaneous change in the current of L_2, which causes v_o to contain an impulsive component.

The s-domain equivalent with the switch open is shown in Fig. 16.28. In deriving the circuit shown in Fig. 16.28, we have used the fact that the current in the 3-H inductor at $t = 0^-$ is 10 A and the current in the 2-H inductor at 0^- is zero. Using the initial conditions at $t = 0^-$ is a direct consequence of our using 0^- as the lower limit on the defining integral of the Laplace transform.

The expression for V_o can be derived from a single node-voltage equation. Summing the currents away from the node between the 15-Ω resistor and the 30-V source gives

$$\frac{V_o}{(2s + 15)} + \frac{V_o - [(100/s) + 30]}{3s + 10} = 0. \tag{16.80}$$

Solving for V_o yields

$$V_o = \frac{40(s + 7.5)}{s(s + 5)} + \frac{12(s + 7.5)}{(s + 5)}. \tag{16.81}$$

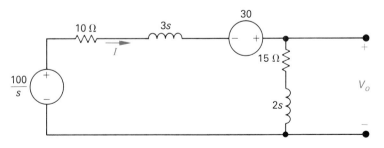

Fig. 16.28   The s-domain equivalent circuit for the circuit in Fig. 16.27.

At this point in our derivation, we see that v_o will contain an impulse term because the second term on the right-hand side of Eq. (16.81) is an improper rational function. This improper fraction can be expressed as a constant plus a rational function by simply dividing the denominator into the numerator, that is,

$$\frac{12(s + 7.5)}{s + 5} = 12 + \frac{30}{s + 5}. \tag{16.82}$$

If we combine Eq. (16.82) with the partial-fraction expansion of the first term on the right-hand side of Eq. (16.81), we have

$$
\begin{aligned}
V_o &= \frac{60}{s} - \frac{20}{s + 5} + 12 + \frac{30}{s + 5} \\
&= 12 + \frac{60}{s} + \frac{10}{s + 5}, \tag{16.83}
\end{aligned}
$$

from which it follows that

$$v_o = 12\delta(t) + [60 + 10e^{-5t}]u(t) \text{ V}. \tag{16.84}$$

Does our solution make sense? Before answering that question, let us first derive the expression for the current when $t > 0^-$. After the switch has been opened, the current in L_1 is the same as the current in L_2. If we reference the current in a clockwise direction around the mesh, the s-domain expression is

$$
\begin{aligned}
I &= \frac{(100/s) + 30}{5s + 15} = \frac{20}{s(s + 5)} + \frac{6}{s + 5} \\
&= \frac{4}{s} - \frac{4}{s + 5} + \frac{6}{s + 5} \\
&= \frac{4}{s} + \frac{2}{s + 5}. \tag{16.85}
\end{aligned}
$$

Inverse-transforming Eq. (16.85) gives

$$i = [4 + 2e^{-5t}]u(t) \text{ A}. \tag{16.86}$$

We now make the following observations: (1) Before the switch is opened the current in L_1 is 10 A and the current in L_2 is 0 A, and (2) from Eq. (16.86) we know that at $t = 0^+$ the current in L_1 and L_2 is 6 A. It follows, then, that the current in L_1 changes instantaneously from 10 A to 6 A while the current in L_2 changes instantaneously from 0 A to 6 A. From this value of 6 A the current decreases exponentially to a final value of 4 A. This final value is easily verified by noting from the circuit that it should equal 100/25, or 4 A. These characteristics of i_1 and i_2 are shown graphically in Fig. 16.29.

How can we check to see whether these instantaneous jumps in the inductor current make sense in terms of the physical behavior of the circuit? First, we note that the switching operation places the two inductors in series. Any impulsive voltage appearing across the 3-H inductor must be exactly bal-

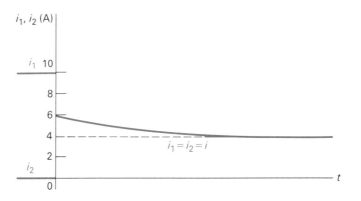

Fig. 16.29   The inductor currents versus t for the circuit in Fig. 16.27.

anced by an impulsive voltage across the 2-H inductor because the sum of the impulsive voltages around a closed path must equal zero. Faraday's law tells us that the induced voltage is proportional to the change in flux linkage ($v = d\lambda/dt$). Therefore in this series circuit the *change* in flux linkage must sum to zero. In other words, the total flux linkage immediately after switching has occurred is the same as that before the switching took place. For the circuit under investigation, the flux linkage before switching is

$$\lambda = L_1 i_1 + L_2 i_2 = 3(10) + 2(0) = 30 \text{ Wb-turns.} \tag{16.87}$$

Immediately after switching,

$$\lambda = (L_1 + L_2)i(0^+) = 5i(0^+). \tag{16.88}$$

Combining Eqs. (16.87) and (16.88) gives

$$i(0^+) = 30/5 = 6 \text{ A.} \tag{16.89}$$

Thus our solution for i (Eq. 16.86) is in agreement with the principle of the conservation of flux linkage.

We can now test the validity of Eq. (16.84). First we check the impulsive term $12\delta(t)$. The instantaneous jump of i_2 from 0 to 6 A at $t = 0$ gives rise to an impulse of strength $6\,\delta(t)$ in the derivative of i_2. This impulse in the derivative of i_2 gives rise to the $12\,\delta(t)$ in the voltage across the 2-H inductor. For $t > 0^+$, di_2/dt is $-10e^{-5t}$ A/s; therefore the voltage v_o is

$$v_o = 15(4 + 2e^{-5t}) + 2(-10e^{-5t})$$
$$= [60 + 10e^{-5t}]u(t) \text{ V.} \tag{16.90}$$

We observe that Eq. (16.90) agrees with the last two terms on the right-hand side of Eq. (16.84); thus we have confirmed that Eq. (16.84) does make sense in terms of known circuit behavior.

We can also check the instantaneous drop from 10 A to 6 A in the current i_1. This drop gives rise to an impulse of $-4\,\delta(t)$ in the derivative of i_1. Therefore

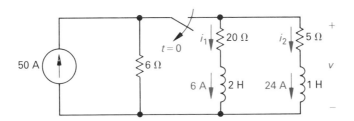

Fig. 16.30   A circuit showing an impulsive response.

the voltage across L_1 contains an impulse of $-12\,\delta(t)$ at the origin. This impulse exactly balances the impulse across L_2; that is, the sum of the impulsive voltages around a closed path equals zero.

A third example of an impulsive response that results from a switching operation is illustrated by the circuit shown in Fig. 16.30. The switch has been closed for a long time prior to opening. The initial currents in the 2-H and 1-H inductors are 6 A and 24 A, as shown in Fig. 16.30. The first step in solving for v, i_1, and i_2 is to construct the s-domain equivalent circuit, which is shown in Fig. 16.31.

The solution for V is found by writing a single node-voltage equation; thus

$$\frac{V + 12}{2s + 20} + \frac{V + 24}{s + 5} = 0. \tag{16.91}$$

Solving Eq. (16.91) for V gives

$$V = -\frac{20(s + 9)}{[s + (25/3)]}. \tag{16.92}$$

Since Eq. (16.92) is an improper rational function, we divide the denominator

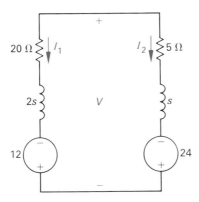

Fig. 16.31   The s-domain equivalent circuit for the circuit in Fig. 16.30.

into the numerator to obtain

$$V = -20 - \frac{40/3}{(s + 25/3)}. \tag{16.93}$$

We can now inverse-transform Eq. (16.93) to get

$$v(t) = -20\ \delta(t) - \frac{40}{3}\ e^{-(25/3)t}\ u(t). \tag{16.94}$$

The solutions for I_1 and I_2 are

$$I_1 = \frac{V + 12}{2s + 20} = \frac{-10(s + 9)}{(s + 10)[s + (25/3)]} + \frac{6}{s + 10} = -\frac{4}{s + (25/3)} \tag{16.95}$$

and

$$I_2 = \frac{V + 24}{(s + 5)} = \frac{-20(s + 9)}{(s + 5)[s + (25/3)]} + \frac{24}{(s + 5)} = \frac{4}{s + (25/3)}. \tag{16.96}$$

The expressions for i_1 and i_2 are

$$i_1 = -4e^{-(25/3)t}\ u(t)\ \text{A} \tag{16.97}$$

and

$$i_2 = 4e^{-(25/3)t}\ u(t)\ \text{A}. \tag{16.98}$$

We are now at that point in our analysis when we must check to see whether our solutions make sense. We begin by noting that $i_1 = -i_2$ for $t > 0$. This result is in agreement with Kirchhoff's current law. Next, we note that when the switch is open i_1 jumps from $+6$ A to -4 A and i_2 jumps from $+24$ A to $+4$ A. We can verify that these abrupt changes are correct by noting the following. Since the two inductors are in parallel, the change in flux linkage must be the same in each branch; thus

$$\Delta i_1 L_1 = \Delta i_2 L_2. \tag{16.99}$$

Kirchhoff's current law requires that

$$i_1(0^-) + \Delta i_1 + i_2(0^-) + \Delta i_2 = 0. \tag{16.100}$$

When we substitute the numerical values $L_1 = 2$ H, $L_2 = 1$ H, $i_1(0^-) = 6$ A, and $i_2(0^-) = 24$ A into Eqs. (16.99) and (16.100), we find that the solutions for Δi_1 and Δi_2 are

$$\Delta i_1 = -10\ \text{A}, \tag{16.101}$$

$$\Delta i_2 = -20\ \text{A}, \tag{16.102}$$

which are in agreement with the initial jumps in i_1 and i_2, respectively.

Finally we note that the sudden drop in i_1 from $+6$ A to -4 A generates an impulse in the derivative of i_1 equal to $-10\ \delta(t)$. Likewise, the abrupt change in i_2 from $+24$ A to $+4$ A generates an impulse in the derivative of i_2 equal to $-20\,\delta(t)$. Thus

$$L_1 \frac{di_1}{dt} = L_2 \frac{di_2}{dt} = -20\,\delta(t) \qquad (16.103)$$

at the instant of switching. The result given by Eq. (16.103) is in agreement with our solution for v, namely, Eq. (16.94).

The second type of circuit problem involving impulse functions is the use of impulsive driving sources. When a circuit is driven by an impulsive source, the effect is to impart a finite amount of energy into the system instantaneously. A mechanical analogy is the striking of a bell with an impulsive blow of the clapper. Once the energy has been transferred to the bell, the natural response of the bell determines the metallic tone that is emitted and the duration of the tone.

In the circuit shown in Fig. 16.32, an impulsive voltage source having a strength of V_0 volt-seconds is applied to a series connection of a resistor and an inductor. At the time when the voltage source is applied, the initial energy in the inductor is zero; therefore the initial current is zero. This means that there is no voltage drop across R and so the impulsive voltage source appears directly across L. An impulsive voltage at the terminals of an inductor will establish an instantaneous current. The value of the current can be found directly from the integral equation

$$i = \frac{1}{L} \int_{0^-}^{t} V_0\,\delta(x)\,dx. \qquad (16.104)$$

Using the sifting property of the impulse function, we find that Eq. (16.104) yields

$$i(0^+) = \frac{V_0}{L}\ \text{A}. \qquad (16.105)$$

Thus, in an infinitesimal moment, the impulsive voltage source has stored

$$w = \frac{1}{2} L \left(\frac{V_0}{L}\right)^2 = \frac{1}{2}\frac{V_0^2}{L} \qquad (16.106)$$

joules in the inductor.

The current V_0/L now decays to zero in accordance with the natural

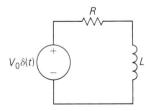

Fig. 16.32   An *RL* circuit excited by an impulsive voltage source.

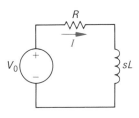

Fig. 16.33   The s-domain equivalent circuit for the circuit in Fig. 16.32.

response of the circuit, that is,

$$i = \frac{V_0}{L} e^{-t/\tau} u(t),$$ (16.107)

where $\tau = L/R$. When a circuit is driven by only an impulsive source, the response is always the natural response of the circuit.

We can also obtain Eq. (16.107) by a direct application of the Laplace transform method. The s-domain equivalent of the circuit in Fig. 16.32 is shown in Fig. 16.33, from which we can write

$$I = \frac{V_0}{R + sL} = \frac{V_0/L}{s + R/L}.$$ (16.108)

It follows that

$$i = \frac{V_0}{L} e^{-(R/L)t} = \frac{V_0}{L} e^{-t/\tau} u(t).$$ (16.109)

We see that the Laplace transform method gives the correct solution for $t \geq 0^+$.

Finally let us consider the case in which internally generated impulses and externally applied impulses occur simultaneously. The Laplace transform approach automatically ensures the correct solution for $t \geq 0^+$ if inductor currents and capacitor voltages at $t = 0^-$ are used in constructing the s-domain equivalent circuit and externally applied impulses are represented by their transforms. To illustrate, we return to the circuit in Fig. 16.27 and add an impulsive voltage source of $50 \, \delta(t)$ in series with the 100-V source. The new arrangement is shown in Fig. 16.34.

At $t = 0^-$ we have $i_1(0^-) = 10$ A and $i_2(0^-) = 0$ A. The Laplace transform of $50 \, \delta(t)$ equals 50. If we use these values, the s-domain equivalent circuit is as shown in Fig. 16.35.

The expression for I is

$$\begin{aligned} I &= \frac{50 + (100/s) + 30}{25 + 5s} = \frac{16}{s + 5} + \frac{20}{s(s + 5)} \\ &= \frac{16}{s + 5} + \frac{4}{s} - \frac{4}{s + 5} \\ &= \frac{12}{s + 5} + \frac{4}{s}, \end{aligned}$$ (16.110)

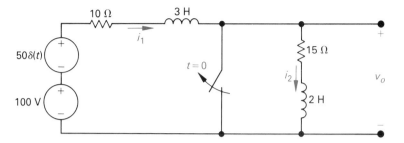

Fig. 16.34 The circuit in Fig. 16.27 with an impulsive voltage source added in series with the 100-V source.

from which we have

$$i(t) = [12e^{-5t} + 4]u(t) \text{ A}. \tag{16.111}$$

The expression for V_o is

$$V_o = (15 + 2s)I = \frac{32(s + 7.5)}{s + 5} + \frac{40(s + 7.5)}{s(s + 5)}$$

$$= 32\left[1 + \frac{2.5}{s + 5}\right] + \frac{60}{s} - \frac{20}{s + 5}$$

$$= 32 + \frac{60}{s + 5} + \frac{60}{s}, \tag{16.112}$$

from which we have

$$v_o = 32\,\delta(t) + [60e^{-5t} + 60]u(t) \text{ V}. \tag{16.113}$$

Now let us test our results to see whether they make sense. From Eq. (16.111) we see that the current in L_1 and L_2 is 16 A at $t = 0^+$. As in the previous case, the switch operation will cause i_1 to decrease instantaneously from

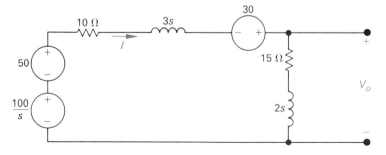

Fig. 16.35 The s-domain equivalent circuit for the circuit in Fig. 16.34.

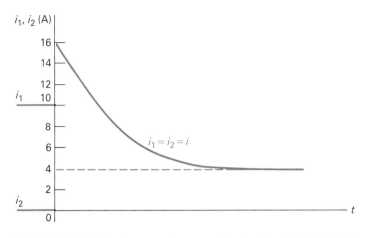

Fig. 16.36 The inductor currents versus t for the circuit in Fig. 16.34.

10 A to 6 A and, at the same time, cause i_2 to increase from 0 to 6 A. Superimposed on this is the establishment of 10 A in L_1 and L_2 by the impulsive voltage source. That is,

$$i = \frac{1}{(3 + 2)} \int_{0^-}^{t} 50 \, \delta(x) \, dx = 10 \text{ A}. \tag{16.114}$$

Therefore, i_1 increases suddenly from 10 A to 16 A, while i_2 increases suddenly from 0 to 16 A. The final value of i is 4 A. A sketch of i_1, i_2, and i is shown in Fig. 16.36.

It is also of interest to note that the abrupt changes in i_1 and i_2 can be found without using superposition. In this circuit, the sum of the impulsive voltages across L_1 (3 H) and L_2 (2 H) equals $50 \, \delta(t)$. Thus the change in flux linkage must sum to 50, that is,

$$\Delta\lambda_1 + \Delta\lambda_2 = 50. \tag{16.115}$$

Since $\lambda = Li$, we can express Eq. (16.115) as

$$3 \, \Delta i_1 + 2 \, \Delta i_2 = 50. \tag{16.116}$$

But since i_1 and i_2 must be equal after the switching takes place,

$$i_1(0^-) + \Delta i_1 = i_2(0^-) + \Delta i_2. \tag{16.117}$$

It follows that

$$10 + \Delta i_1 = 0 + \Delta i_2. \tag{16.118}$$

Solving Eqs. (16.116) and (16.118) for Δi_1 and Δi_2 yields the equations

$$\Delta i_1 = 6 \text{ A} \tag{16.119}$$

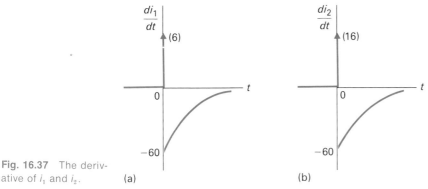

Fig. 16.37 The derivative of i_1 and i_2.

(a)   (b)

and

$$\Delta i_2 = 16 \text{ A} \tag{16.120}$$

which are in agreement with our previous check.

    We can also see from Fig. 16.36 that the derivatives of i_1 and i_2 will contain an impulse at $t = 0$. Specifically, the derivative of i_1 will have an impulse of $6\,\delta(t)$ and i_2 will have an impulse of $16\,\delta(t)$. The derivatives of i_1 and i_2 are illustrated in Figs. 16.37(a) and (b), respectively.

    Now let us turn our attention to Eq. (16.113). The impulsive component $32\,\delta(t)$ is in agreement with the fact that di_2/dt has an impulse of $16\,\delta(t)$ at the origin. The terms $60e^{-5t} + 60$ are in agreement with the fact that for $t > 0^+$,

$$v_o = 15i + 2\frac{di}{dt}.$$

    The impulsive component of di_1/dt can be tested by noting that it will produce an impulsive voltage of $(3)6\,\delta(t)$, or $18\,\delta(t)$ across L_1. This voltage, along with $32\,\delta(t)$ across L_2, adds to $50\,\delta(t)$. Thus the algebraic sum of the impulsive voltages around the mesh adds to zero.

    We can summarize our discussion of impulse functions in circuit analysis as follows. The Laplace transform will correctly predict the creation of impulsive currents and voltages that arise from switching, provided the s-domain equivalent circuits are based on initial conditions at $t = 0^-$, that is, on the initial conditions that exist prior to the disturbance caused by the switching. The Laplace transform will correctly predict the response to impulsive driving sources by simply representing these sources in the s-domain by their correct transforms.

DRILL EXERCISE
16.8

The switch in the circuit shown has been in position a for a long time. At $t = 0$ the switch moves to position b. Compute (a) $v_1(0^-)$; (b) $v_2(0^-)$; (c) $v_3(0^-)$; (d) $i(t)$; (e) $v_1(0^+)$; (f) $v_2(0^+)$; (g) $v_3(0^+)$.

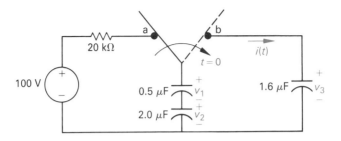

Ans. (a) 80 V; (b) 20 V; (c) 0 V; (d) 32 $\delta(t)$ μA; (e) 16 V; (f) 4 V; (g) 20 V.

DRILL EXERCISE
16.9

The switch in the circuit shown has been closed for a long time. The switch opens at $t =$ 0. Compute (a) $i_1(0^-)$; (b) $i_1(0^+)$; (c) $i_2(0^-)$; (d) $i_2(0^+)$; (e) $i_1(t)$; (f) $i_2(t)$; and (g) $v(t)$.

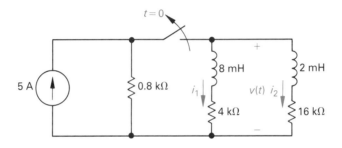

Ans. (a) 0.8 A; (b) 0.6 A; (c) 0.2 A; (d) -0.6 A; (e) $0.6e^{-2\times10^6 t}u(t)$ A; (f) $-0.6e^{-2\times10^6 t}u(t)$ A; (g) -1.6×10^{-3} $\delta(t) - 7200e^{-2\times10^6 t}u(t)$ V.

16.6 SUMMARY

Our purpose has been to show, largely by example, how we can use the Laplace transform to find the transient response of a linear, lumped-parameter circuit. The transform approach is greatly facilitated by transferring the circuit to be analyzed directly to the s-domain. The s-domain equivalent circuits contain independent voltage or current sources, which account for the initial voltage on a capacitor or the initial current in an inductor. Kirchhoff's laws hold for s-domain currents and voltages. The voltage and current at the terminals of a passive element can always be related by means of an impedance and an independent source to account for initial conditions. If initial conditions are zero, the relationship reduces to Ohm's law in the s-domain, that is, $V = Z(s)I$. As a consequence of the fact that these fundamental laws are applicable in the s-domain, all the techniques of circuit analysis that are applicable to resistive circuits, or phasor domain circuits, can be used.

Impulse functions can be created in a circuit by a switching operation and impulsive sources can be used to excite a circuit. The Laplace transform method correctly predicts the creation of an impulsive response and correctly predicts the response to an impulsive driving source. This ability to accommodate impulse functions is a distinct advantage of the Laplace approach. In testing whether an impulse response resulting from a switching operation makes sense in terms of circuit behavior, the principle of conservation of charge is useful in capacitive circuits and the principle of conservation of flux linkage is useful in inductive circuits.

PROBLEMS

16.1 Find the Norton equivalent of the circuit shown in Fig. 16.3.

16.2 Derive the s-domain equivalent circuit shown in Fig. 16.4 by first expressing the inductor current i as a function of the terminal voltage v and then finding the Laplace transform of this time domain integral equation.

16.3 Find the Thévenin equivalent of the circuit shown in Fig. 16.7.

16.4 An 820-Ω resistor, a 40-mH inductor, and a 0.25-μF capacitor are in series.

    a) Express the s-domain impedance of this series combination as a rational function.
    b) Give the numerical values of the poles and zeros of the impedance.

16.5 A 480-Ω resistor is in series with an 80-mH inductor. This series combination is in parallel with a 0.5-μF capacitor.

    a) Express the equivalent s-domain impedance of these paralleled branches as a rational function.
    b) Determine the numerical values of the poles and zeros.

16.6 Find the poles and zeros of the admittance seen looking into the terminals a, b of the circuit shown in Fig. 16.38.

16.7 Find the poles and zeros of the impedance seen looking into the terminals a, b of the circuit shown in Fig. 16.39.

16.8 The switch in the circuit shown in Fig. 16.40 has been closed for a long time. At $t = 0$ the switch is opened.

    a) Express I and V as rational functions of s.
    b) Find i and v for $t > 0$.

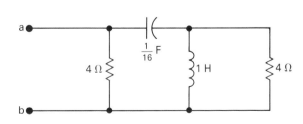

Fig. 16.38 The circuit for Problem 16.6.

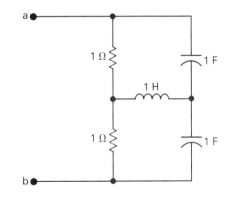

Fig. 16.39 The circuit for Problem 16.7.

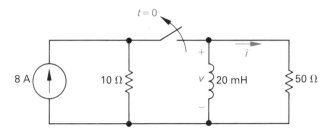

Fig. 16.40 The circuit for Problem 16.8.

16.9 The switch in the circuit in Fig. 16.41 has been closed for a long time. At $t = 0$ the switch is opened.

    a) Find $V, I_1,$ and I_2.
    b) Find $v, i_1,$ and i_2.

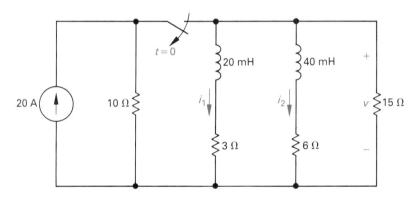

Fig. 16.41 The circuit for Problem 16.9.

16.10 The initial energy stored in the circuit in Fig. 16.42 is zero. Time is measured from the instant the switch is opened.

    a) Express V_o as a rational function of s.

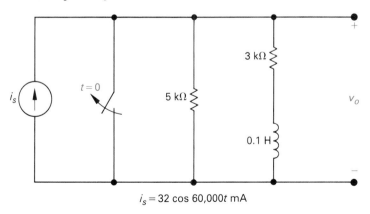

$i_s = 32 \cos 60,000t$ mA

Fig. 16.42 The circuit for Problem 16.10.

b) Find the time domain expression for v_o.
c) Use phasor domain analysis to check the steady-state component of v_o.

16.11 Find V_o and v_o in the circuit shown in Fig. 16.43 if the initial energy is zero and the switch is closed at $t = 0$.

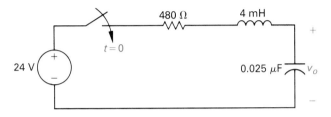

Fig. 16.43 The circuit for Problem 16.11.

16.12 The energy stored in the circuit shown in Fig. 16.44 is zero at the time when the switch is closed.

a) Find v_C.
b) Find i_C.

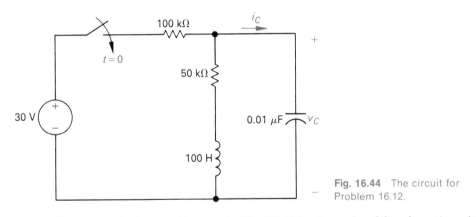

Fig. 16.44 The circuit for Problem 16.12.

16.13 The switch in the circuit shown in Fig. 16.45 has been closed for a long time. At $t = 0$ the switch is opened.

a) Find v.
b) Find i.

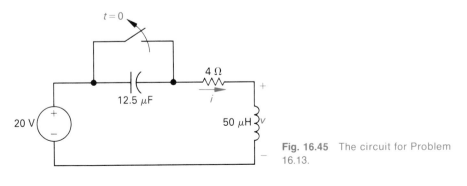

Fig. 16.45 The circuit for Problem 16.13.

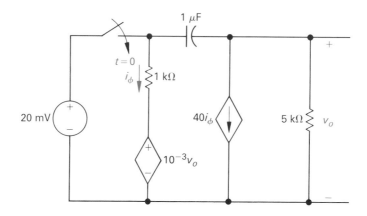

Fig. 16.46 The circuit for Problem 16.14.

16.14 There is no charge on the capacitor in the circuit shown in Fig. 16.46 when the switch is closed.

   a) Find the numerical expression for v_o when $t > 0$.
   b) Find the charge on the capacitor (in μC) after the switch has been closed for a long time.

16.15 The switch in the circuit shown in Fig. 16.47 has been open for a long time. The voltage of the sinusoidal source is $v_g = V_m \sin(\omega t + \phi)$. The switch closes at $t = 0$. Note that the angle ϕ in the voltage expression determines the value of the voltage at the moment when the switch closes, that is, $v_g(0) = V_m \sin \phi$.

   a) Use the Laplace transform method to find i for $t > 0$.
   b) Using the expression derived in part (a), write the expression for the current after the switch has been closed for a long time.
   c) Using the expression derived in part (a), write the expression for the transient component of i.
   d) Find the steady-state expression for i using the phasor method. Verify that your expression is equivalent to that obtained in part (b).
   e) Specify the value of ϕ so that the circuit passes directly into steady-state operation when the switch is closed.

16.16 a) Find the s-domain Thévenin equivalent with respect to the terminals a, b for the circuit shown in Fig. 16.48.
   b) Use the Thévenin equivalent found in part (a) to derive the time domain expression for v_{ab}.

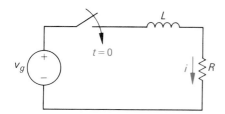

Fig. 16.47 The circuit for Problem 16.15.

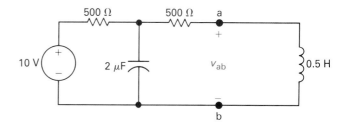

Fig. 16.48 The circuit for Problem 16.16.

16.17 Beginning with Eq. (16.65), show that the capacitor current in the circuit in Fig. 16.17 is positive for $0 < t < 200\ \mu s$ and negative for $t > 200\ \mu s$. Also show that at 200 μs the current is zero and this corresponds to when dv_C/dt is zero.

16.18 The make-before-break switch in the circuit shown in Fig. 16.49 has been in position a for a long time. At $t = 0$ the switch moves to position b.

    a) Construct the s-domain equivalent circuit that applies when the switch is in position b.

    b) Show that

$$I_2 = \frac{MV_{dc}}{(R_1 + sL_1)(R_2 + sL_2) - s^2M^2}.$$

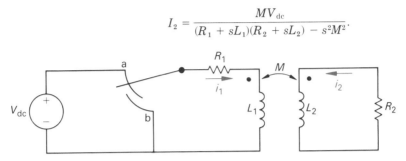

Fig. 16.49 The circuit for Problem 16.18.

16.19 The switch in the circuit shown in Fig. 16.50 has been closed for a long time. The switch opens at $t = 0$. Find (a) i_1; (b) i_2; (c) i_0; and (d) v.

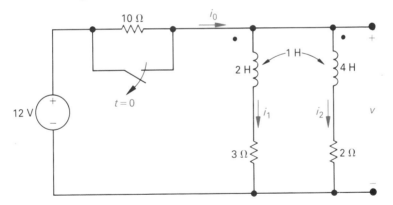

Fig. 16.50 The circuit for Problem 16.19.

16.20 Show that after V_0C_e coulombs are transferred from C_1 to C_2 in the circuit shown in Fig. 16.24, the voltage across each capcitor is $C_1V_0/(C_1 + C_2)$. (*Hint:* Use the conservation-of-charge principle.)

16.21 The inductor L_1 in the circuit shown in Fig. 16.51 is carrying an initial current of ρ amperes at the instant the switch opens. Find (a) $v(t)$; (b) $i_1(t)$; (c) $i_2(t)$; and (d) $\lambda(t)$, where $\lambda(t)$ is the total flux linkage in the circuit.

16.22 a) Let $R \rightarrow \infty$ in the circuit shown in Fig. 16.51 and use the solutions derived in Problem 16.21 to find $v(t)$, $i_1(t)$, and $i_2(t)$.

    b) Let $R = \infty$ in the circuit shown in Fig. 16.51 and use the Laplace transform method to find $v(t)$, $i_1(t)$, and $i_2(t)$.

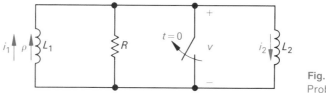

Fig. 16.51 The circuit for Problems 16.21 and 16.22.

16.23 The switch in the circuit shown in Fig. 16.52 has been open for a long time before closing at $t = 0$.

    a) Find v and i for $t \geq 0$.

    b) Test your solutions and make sure they are in agreement with known circuit behavior.

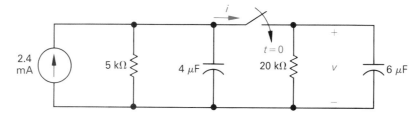

Fig. 16.52 The circuit for Problem 16.23.

16.24 The parallel combination of R_2 and C_2 in the circuit shown in Fig. 16.53 represents the input circuit to a cathode-ray oscilloscope. The parallel combination of R_1 and C_1 is a circuit model of a compensating lead that is used to connect the CRO to the source. There is no energy stored in C_1 or C_2 at the time when the 10-V source is connected to the CRO via the compensating lead. The circuit values are $C_1 = 5$ pF, $C_2 = 20$ pF, $R_1 = 1$ MΩ, and $R_2 = 4$ MΩ. Find (a) v_o and (b) i.

    c) Repeat parts (a) and (b) if C_1 is changed to 80 pF.

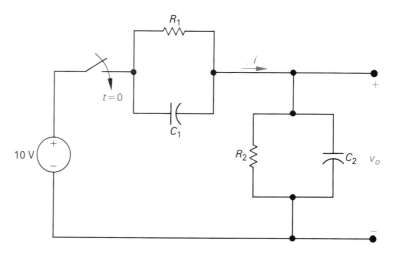

Fig. 16.53 The circuit for Problems 16.24 and 16.25.

16.25 Show that if $R_1C_1 = R_2C_2$ in the circuit shown in Fig. 16.53 v_o will be a scaled replica of the source voltage.

16.26 The ideal operational amplifier in the circuit shown in Fig. 16.54 is operating in its linear range. Find $v_o(t)$ for $t > 0$ if the switch closes at $t = 0$.

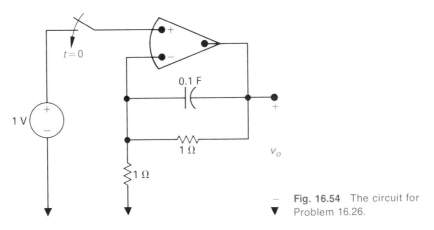

Fig. 16.54 The circuit for Problem 16.26.

16.27 The voltage source in the circuit shown in Fig. 16.55 is the unit ramp function $tu(t)$. How long after the ramp is turned on will the ideal op amp saturate?

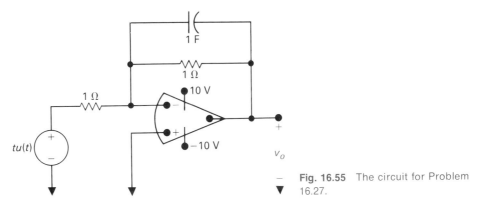

Fig. 16.55 The circuit for Problem 16.27.

16.28 Find $v_o(t)$ in the circuit shown in Fig. 16.56 if the ideal op amp operates within its linear range and $v_g = 4.8u(t)$.

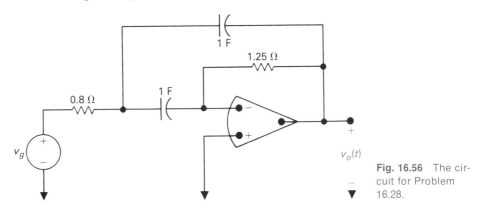

Fig. 16.56 The circuit for Problem 16.28.

17

THE TRANSFER FUNCTION

17.1   INTRODUCTION

Our purpose in this chapter is to introduce the concept of the transfer function and show how it can be used in circuit analysis. After defining the transfer function and discussing its properties, we will divide our discussion of its applications into three parts. First we will discuss the use of the transfer function in a partial-fraction expansion. Next, we will show how the transfer function can be used in conjunction with the convolution integral to find the response of a circuit. Finally, we discuss how the transfer function can be used to find the sinusoidal steady-state response of a circuit.

We use the transfer function to find the response of a circuit to externally applied excitation sources. The contribution of initially stored energy to the response is not included in the transfer function solution. Therefore, we pause and illustrate how we use the principle of superposition to divide the response function into two parts, one due to excitation sources and one due to initial conditions.

17.2   AN s-DOMAIN APPLICATION OF SUPERPOSITION

Since we are analyzing linear, lumped-parameter circuits, we can use superposition to divide the response into components that can be identified with particular driving sources and initial conditions. Our purpose here is simply to confirm this observation with a specific example so that as you study the following material, you will keep the concept of the transfer function in its proper perspective.

Our illustrative circuit is shown in Fig. 17.1. We assume that at the instant when the two sources are applied to the circuit the inductor is carrying an initial current of ρ amperes and the capacitor is carrying an initial voltage of γ volts. The desired response of the circuit is the voltage across the resistor R_2, labeled v_2 in Fig. 17.1.

The s-domain equivalent circuit is shown in Fig. 17.2. We have opted for the parallel equivalents for L and C since we anticipated solving for V_2 via the node-voltage method.

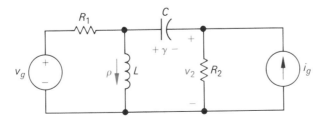

Fig. 17.1   A circuit showing the use of superposition in s-domain analysis.

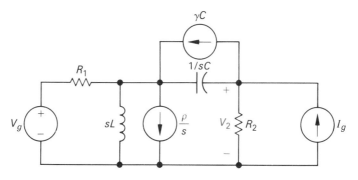

Fig. 17.2   The s-domain equivalent for the circuit in Fig. 17.1.

To find V_2 by superposition, we find the component of V_2 due to each source acting alone and then sum the components. We begin with V_g acting alone. The three current sources are deactivated by opening each source. The resulting circuit is shown in Fig. 17.3. We have added the node voltage V_1' in Fig. 17.3 to facilitate the analysis. The prime marks on V_1 and V_2 indicate that they are the components of V_1 and V_2 due to V_g acting alone. The two equations that describe the circuit in Fig. 17.3 are

$$\left[\frac{1}{R_1} + \frac{1}{sL} + sC \right] V_1' - sCV_2' = \frac{V_g}{R_1}, \tag{17.1}$$

$$- sCV_1' + \left[\frac{1}{R_2} + sC \right] V_2' = 0 \tag{17.2}$$

For convenience we introduce the following notation:

$$Y_{11} = \frac{1}{R_1} + \frac{1}{sL} + sC, \tag{17.3}$$

$$Y_{12} = - sC, \tag{17.4}$$

$$Y_{22} = \frac{1}{R_2} + sC. \tag{17.5}$$

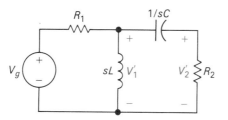

Fig. 17.3   The circuit in Fig. 17.2 with V_g acting alone.

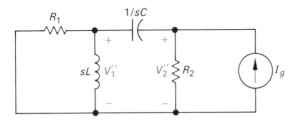

Fig. 17.4   The circuit in Fig. 17.2 with I_g acting alone.

Substituting Eqs. (17.3), (17.4), and (17.5) into Eqs. (17.1) and (17.2) gives

$$Y_{11}V_1' + Y_{12}V_2' = V_g/R_1, \tag{17.6}$$

$$Y_{12}V_1' + Y_{22}V_2' = 0. \tag{17.7}$$

Solving Eqs. (17.6) and (17.7) for V_2' gives

$$V_2' = \frac{-Y_{12}/R_1}{(Y_{11}Y_{22} - Y_{12}^2)} \cdot V_g \tag{17.8}$$

With the current source I_g acting alone, the circuit in Fig. 17.2 reduces to that shown in Fig. 17.4. Here V_1'' and V_2'' are the components of V_1 and V_2 due to I_g. If we use the notation introduced in Eqs. (17.3), (17.4), and (17.5), the two node-voltage equations that describe the circuit in Fig. 17.4 are

$$Y_{11}V_1'' + Y_{12}V_2'' = 0, \tag{17.9}$$

$$Y_{12}V_1'' + Y_{22}V_2'' = I_g. \tag{17.10}$$

Solving Eqs. (17.9) and (17.10) for V_2'' gives

$$V_2'' = \frac{Y_{11}}{Y_{11}Y_{22} - Y_{12}^2} \cdot I_g. \tag{17.11}$$

To find the component of V_2 due to the initial energy stored in the inductor, that is, V_2''', we must solve the circuit shown in Fig. 17.5, where we see

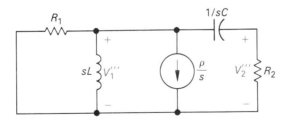

Fig. 17.5   The circuit in Fig. 17.2 with the energized inductor acting alone.

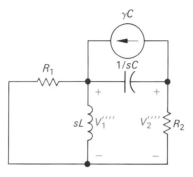

Fig. 17.6 The circuit in Fig. 17.2 with the energized capacitor acting alone.

that

$$Y_{11}V_1''' + Y_{12}V_2''' = -\rho/s, \tag{17.12}$$

$$Y_{12}V_1''' + Y_{22}V_2''' = 0; \tag{17.13}$$

thus

$$V_2''' = \frac{(Y_{12}/s)}{Y_{11}Y_{22} - Y_{12}^2} \cdot \rho. \tag{17.14}$$

The component of V_2 (V_2'''') due to the initial energy stored in the capacitor is found from the circuit shown in Fig. 17.6. The node-voltage equations describing this circuit are

$$Y_{11}V_1'''' + Y_{12}V_2'''' = \gamma C, \tag{17.15}$$

$$Y_{12}V_1'''' + Y_{22}V_2'''' = -\gamma C. \tag{17.16}$$

Solving for V_2'''' we get

$$V_2'''' = \frac{-(Y_{11} + Y_{12})C}{Y_{11}Y_{22} - Y_{12}^2} \cdot \gamma. \tag{17.17}$$

The expression for V_2 is

$$\begin{aligned}
V_2 &= V_2' + V_2'' + V_2''' + V_2'''' \\
&= \frac{-(Y_{12}/R_1)}{Y_{11}Y_{22} - Y_{12}^2} \cdot V_g + \frac{Y_{11}}{Y_{11}Y_{22} - Y_{12}^2} \cdot I_g \\
&\quad + \frac{Y_{12}/s}{Y_{11}Y_{22} - Y_{12}^2} \cdot \rho + \frac{-C(Y_{11} + Y_{12})}{Y_{11}Y_{22} - Y_{12}^2} \cdot \gamma.
\end{aligned} \tag{17.18}$$

We can find V_2 without using superposition by solving the two node-voltage equations that describe the circuit in Fig. 17.2. Thus

$$Y_{11}V_1 + Y_{12}V_2 = \frac{V_g}{R_1} + \gamma C - \frac{\rho}{s}, \tag{17.19}$$

$$Y_{12}V_1 + Y_{22}V_2 = I_g - \gamma C. \tag{17.20}$$

We leave it to the reader (Problem 17.2) to verify that the solution of Eqs. (17.19) and (17.20) for V_2 gives the same result as Eq. (17.18).

The fact that we can subdivide the response of a linear circuit into components that can be associated with specific sources of energy enables us to develop a technique that focuses on the relationship between an output and a source. This relationship is known as the transfer function and is introduced in the following section.

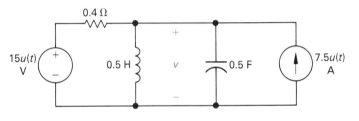

17.3   THE TRANSFER FUNCTION

The transfer function is defined as the s-domain ratio of the output (response) to the input (source). In finding this ratio, we set all initial conditions equal to zero. Furthermore, the ratio applies to a single source. If more than one source exists in the circuit, we define a transfer function for each source and then use superposition to find the total response. In our work here we will focus our attention on circuits in which the initial energy stored is zero and the excitation is from a single source.

The transfer function is defined mathematically as

$$H(s) = \frac{Y(s)}{X(s)},\tag{17.21}$$

where $Y(s)$ is the Laplace transform of the output signal and $X(s)$ is the Laplace transform of the input signal. Note that the transfer function depends on what is defined as the output signal. Thus a given circuit can generate many transfer functions. Consider, for example, the series circuit shown in Fig. 17.7.

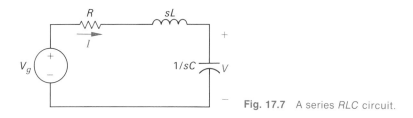

Fig. 17.7   A series RLC circuit.

If the current is defined as the response signal of the circuit, then

$$H(s) = \frac{I}{V_g} = \frac{1}{R + sL + 1/sC} = \frac{sC}{s^2LC + RCs + 1}. \qquad (17.22)$$

In deriving Eq. (17.22), we recognize that I corresponds to the output $Y(s)$ and V_g corresponds to the input $X(s)$.

If the voltage across the capacitor is defined as the output signal of the circuit shown in Fig. 17.7, then the transfer function is

$$H(s) = \frac{V}{V_g} = \frac{1/sC}{R + sL + 1/sC} = \frac{1}{s^2LC + RCs + 1}. \qquad (17.23)$$

Example 17.1 illustrates the computation of a transfer function when the numerical values of R, L, and C are known.

Example 17.1   The circuit in Fig. 17.8 is driven by the voltage source v_g. The reponse signal is the voltage across the capacitor, that is, v_o.

a) Calculate the numerical expression for the transfer function.

b) Calculate the numerical values for the poles and zeroes of the transfer function.

Solution   a) The first step in finding the transfer function is to construct the s-domain equivalent circuit, shown in Fig. 17.9. By definition, the transfer function is

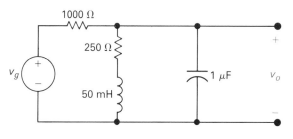

Fig. 17.8   The circuit for Example 17.1.

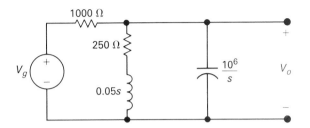

Fig. 17.9 The s-domain equivalent circuit for the circuit in Fig. 17.8

the ratio of V_o/V_g, which can be computed from a single node-voltage equation. Summing the currents away from the upper node generates the following expression:

$$\frac{V_o - V_g}{1000} + \frac{V_o}{250 + 0.05s} + \frac{V_o s}{10^6} = 0.$$

Solving for V_o yields

$$V_o = \frac{1000(s + 5000)V_g}{s^2 + 6000s + 25 \times 10^6}.$$

It follows that the transfer function is

$$H(s) = \frac{V_o}{V_g} = \frac{1000(s + 5000)}{s^2 + 6000s + 25 \times 10^6}.$$

b) The poles of $H(s)$ are the roots of the denominator polynomial. Therefore

$$-p_1 = -3000 - j4000$$

and

$$-p_2 = -3000 + j4000$$

The zeros of $H(s)$ are the roots of the numerator polynomial; thus $H(s)$ has a zero at

$$-z_1 = -5000. \quad \blacksquare$$

For linear, lumped-parameter circuits $H(s)$ is always a rational function of s. Complex poles and zeros always appear in conjugate pairs. The poles of $H(s)$ must lie in the left half of the s-plane if the response to a bounded driving source is to be bounded. The zeros of $H(s)$ may lie in either the right half or the left half of the s-plane. If the zeros of $H(s)$ lie only in the left half of the s-plane, then $H(s)$ is said to be minimum phase.

With these general characteristics of $H(s)$ in mind, we are ready to discuss the role that $H(s)$ plays in the determination of the response function. We begin with the partial-fraction expansion technique for finding $y(t)$.

DRILL EXERCISE
17.2
a) Derive the numerical expression for the transfer function V_o/I_g for the circuit shown.

b) Give the numerical value of each pole and zero of $H(s)$.

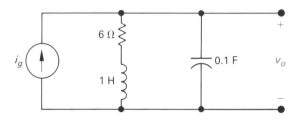

Ans. (a) $H(s) = 10(s + 6)/(s^2 + 6s + 10)$; (b) $-p_1 = -3 + j1$, $-p_2 = -3 - j1$, $-z = -6$.

17.4   THE TRANSFER FUNCTION IN PARTIAL-FRACTION EXPANSIONS

It follows from Eq. (17.21) that the circuit output can be written as the product of the transfer function times the driving function, that is,

$$Y(s) = H(s)X(s). \tag{17.24}$$

We have already noted that $H(s)$ is a rational function of s and reference to Table 15.1 will show that $X(s)$ is also a rational function of s for the excitation functions of most interest in circuit analysis.

When we expand the right-hand side of Eq. (17.24) into a sum of partial fractions, we see that there will be a term for each pole of $H(s)$ and a term for each pole of $X(s)$. The terms generated by the poles of $H(s)$ give rise to the transient component of the total response, whereas the terms generated by the poles of $X(s)$ give rise to the steady-state component of the response. By steady-state response, we mean the response that exists after the transient components have become negligible. Let us illustrate these general observations with a specific example.

Example 17.2
The circuit in Example 17.1 (Fig. 17.8) is driven by a voltage source whose voltage increases linearly with time, namely, $v_g = 50tu(t)$.

a) Use the transfer function to find v_o.

b) Identify the transient component of the response.

c) Identify the steady-state component of the response.

d) Sketch v_o vs. t for $0 \le t \le 1.5$ ms.

Solution   a) From Example 17.1 we have

$$H(s) = \frac{1000(s + 5000)}{s^2 + 6000s + 25 \times 10^6}.$$

The transform of the driving voltage is $50/s^2$; therefore the s-domain expression for the output voltage is

$$V_o = \frac{1000(s + 5000)}{s^2 + 6000s + 25 \times 10^6} \cdot \frac{50}{s^2}.$$

The partial-fraction expansion of V_o is

$$V_o = \frac{K_1}{s + 3000 - j4000} + \frac{K_1^*}{s + 3000 + j4000} + \frac{K_2}{s^2} + \frac{K_3}{s}.$$

The coefficients K_1, K_2, and K_3 are evaluated using the techniques described in Section 15.7. The results are

$$K_1 = 5\sqrt{5} \times 10^{-4}\underline{/79.90°},$$
$$K_1^* = 5\sqrt{5} \times 10^{-4}\underline{/-79.70°},$$
$$K_2 = 10,$$
$$K_3 = -4 \times 10^{-4}.$$

The time domain expression for v_o is

$$v_o = [10\sqrt{5} \times 10^{-4}e^{-3000t} \cos (4000t + 79.70°) + 10t - 4 \times 10^{-4}]u(t) \text{ V}.$$

b) The transient component of v_o is the term

$$10\sqrt{5} \times 10^{-4}e^{-3000t} \cos (4000t + 79.70°).$$

Note that this term is generated by the poles $(-3000 + j4000)$ and $(-3000 - j4000)$ of the transfer function.

c) The steady-state component of the response is

$$[10t - 4 \times 10^{-4}]u(t).$$

These two terms are generated by the second-order pole (at the origin) of the driving voltage.

d) A sketch of v_o vs. t is shown in Fig. 17.10. Note that the deviation from the steady-state solution $(10,000t - 0.4)$ mV is imperceptible after approximately 1 ms. ■

Example 17.2 shows very clearly how the transfer function $H(s)$ relates to the response of a circuit through a partial-fraction expansion. However, the example does raise questions about the practicality of driving a circuit with an ever-increasing voltage that generates a response that also increases without limit. We know that eventually the circuit components will fail under the

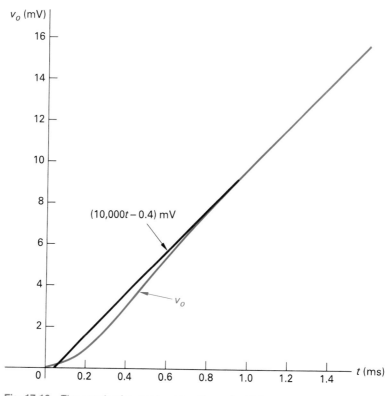

Fig. 17.10   The graph of v_o versus t for Example 17.2.

stress of excessive voltage and when this happens our linear model is no longer valid. The ramp response is of interest in practical applications where the ramp function increases to a maximum value over a finite time interval. If the run-up time is long compared to the time constants of the circuit, then the solution assuming an unbounded ramp is valid for this run-up time interval.

We must make two additional observations regarding Eq. (17.24). First, if the circuit is driven by a unit impulse source, then the response of the circuit is equal to the inverse transform of the transfer function. Thus if

$$x(t) = \delta(t),$$

then

$$X(s) = 1$$

and

$$Y(s) = H(s). \tag{17.25}$$

It follows from Eq. (17.25) that

$$y(t) = h(t), \tag{17.26}$$

where we see that the inverse transform of the transfer function is equal to the unit impulse response of the circuit. Note that this is also the natural response of the circuit since the application of an impulsive source is equivalent to instantaneously storing energy in the circuit. The subsequent release of this energy gives rise to the natural response. (See Problem 17.4.)

The second observation relates to the response of the circuit due to a delayed input. If the input is delayed by a seconds, then

$$X(s) = \mathcal{L}\{x(t - a)u(t - a)\} = e^{-as}X(s),$$

and from Eq. (17.24) we see that the response becomes

$$Y(s) = H(s)X(s)e^{-as}. \tag{17.27}$$

If $y(t) = \mathcal{L}^{-1}[H(s)X(s)]$, then from Eq. (17.27) we have

$$y(t - a)u(t - a) = \mathcal{L}^{-1}[H(s)X(s)e^{-as}]. \tag{17.28}$$

Therefore the consequences of delaying the input by a seconds is simply to delay the response function by a seconds. A circuit that exhibits this characteristic is said to be time-invariant.

We are now ready to discuss how the transfer function is used in conjunction with the convolution integral.

DRILL EXERCISE
17.3

Find (a) the unit step and (b) the unit impulse response of the circuit in Drill Exercise 17.2.

Ans. (a) $[6 + 10e^{-3t} \cos (t + 126.87°)]u(t)$ V; (b) $31.62e^{-3t} \cos (t - 71.57°)u(t)$ V.

17.5 THE TRANSFER FUNCTION AND THE CONVOLUTION INTEGRAL

The convolution integral relates the output $[y(t)]$ of a linear, time-invariant circuit to the input of the circuit $[x(t)]$ and the circuit's impulse response $[h(t)]$. The integral relationship can be expressed in two ways; thus

$$y(t) = \int_{-\infty}^{\infty} h(\lambda)x(t - \lambda) \, d\lambda = \int_{-\infty}^{\infty} h(t - \lambda)x(\lambda) \, d\lambda. \tag{17.29}$$

We are interested in the convolution integral for several reasons. First of all, it allows us to work entirely in the time domain in situations where $x(t)$ and $h(t)$ may be known only through experimental data. Where experimental data are the basis for computations, the transform method may be awkward or even impossible. Second, the convolution integral introduces the concepts of memory and weighting function into analysis. We will show how the concept of memory enables us to look at the impulse response (or weighting function) $h(t)$

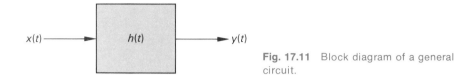

Fig. 17.11 Block diagram of a general circuit.

and predict, to some degree, how closely the output wave form will replicate the input waveform. Finally, the convolution integral provides a formal procedure for finding the inverse transform of products of Laplace transforms.

The derivation of Eq. (17.29) is based on the assumption that the circuit is linear and time-invariant. Because the circuit is linear, the principle of superposition is valid and because the circuit is time-invariant, the response to a delayed input is delayed by exactly the amount of the input delay. With these thoughts in mind, consider Fig. 17.11, where the block containing $h(t)$ represents any linear, time-invariant circuit whose impulse response is known, $x(t)$ represents the excitation signal, and $y(t)$ represents the desired output signal.

Assume that $x(t)$ is the general excitation signal shown in Fig. 17.12(a). For convenience we assume that $x(t)$ is zero for $t < 0^-$. (Once we see the derivation of the convolution integral assuming $x(t)$ is zero for $t < 0^-$, the extension of the integral to include excitation functions that exist over all time will be apparent.) Note that we permit a discontinuity in $x(t)$ at the origin, that is, a jump between 0^- and 0^+.

Now we approximate $x(t)$ by a series of rectangular pulses of uniform width $\Delta\lambda$, as shown in Fig. 17.12(b). Thus

$$x(t) = x_0(t) + x_1(t) + \cdots + x_i(t) + \cdots , \qquad (17.30)$$

where $x_i(t)$ is a rectangular pulse that equals $x(\lambda_i)$ between λ_i and λ_{i+1} and is zero elsewhere. (Note that the ith pulse can be expressed in terms of step functions, that is, $x_i(t) = x(\lambda_i)\{u(t - \lambda_i) - u[t - (\lambda_i + \Delta\lambda)]\}$.)

The next step in the approximation of $x(t)$ is to make $\Delta\lambda$ sufficiently small so that the ith component can be approximated by an impulse function of strength $x(\lambda_i) \Delta\lambda$. The impulse representation is shown in Fig. 17.12(c). The strength of each impulse is shown in brackets beside each arrow. The impulse representation of $x(t)$ is

$$x(t) = x(\lambda_0) \Delta\lambda\delta(t - \lambda_0) + x(\lambda_1) \Delta\lambda\delta(t - \lambda_1) + \cdots$$
$$+ x(\lambda_i) \Delta\lambda\delta(t - \lambda_i) + \cdots . \qquad (17.31)$$

Now when $x(t)$ is represented by a series of impulse functions (which occur at equally spaced intervals of time, that is, at λ_0, λ_1, λ_2, etc.), the response function $y(t)$ will consist of the sum of a series of uniformly delayed impulse responses. The strength of each impulse response depends on the strength of the impulse driving the circuit. For example, assume that the unit impulse response of the circuit contained in the box in Fig. 17.11 is the exponential decay function shown in Fig. 17.13(a). Then the approximation of $y(t)$ is the sum of the impulse responses shown in Fig. 17.13(b).

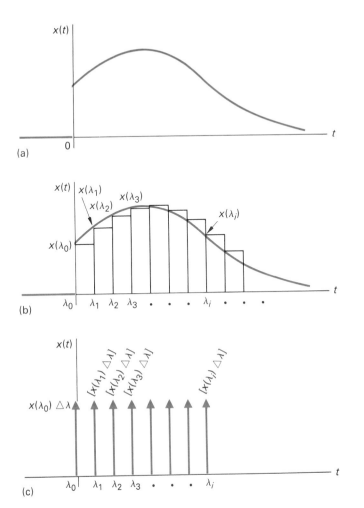

Fig. 17.12 The excitation signal $x(t)$: (a) a general excitation signal; (b) approximating $x(t)$ with a series of pulses; (c) approximating $x(t)$ with a series of impulses.

Analytically the expression for $y(t)$ is

$$y(t) = x(\lambda_0) \, \Delta\lambda h(t - \lambda_0) + x(\lambda_1) \, \Delta\lambda h(t - \lambda_1) + x(\lambda_2) \, \Delta\lambda h(t - \lambda_2) + \cdots$$
$$+ x(\lambda_i) \, \Delta\lambda h(t - \lambda_i) + \cdots . \quad \textbf{(17.32)}$$

As $\Delta\lambda$ approaches zero, the summation in Eq. (17.32) approaches a continuous integration, that is,

$$\sum_{i=0}^{\infty} x(\lambda_i) h(t - \lambda_i) \, \Delta\lambda \rightarrow \int_0^{\infty} x(\lambda) h(t - \lambda) \, d\lambda. \quad \textbf{(17.33)}$$

Therefore we can write

$$y(t) = \int_0^{\infty} x(\lambda) h(t - \lambda) \, d\lambda. \quad \textbf{(17.34)}$$

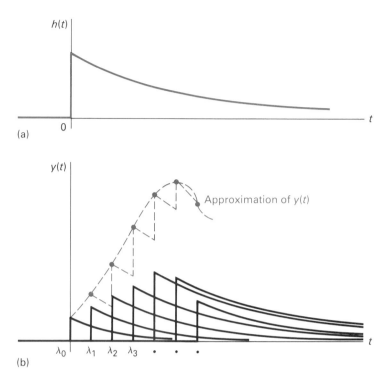

Fig. 17.13 The approximation of $y(t)$: (a) impulse response of the box in Fig. 17.11; (b) summing the impulse responses.

If $x(t)$ exists over all time, then the lower limit on Eq. (17.34) becomes $-\infty$; thus in general we write

$$y(t) = \int_{-\infty}^{\infty} x(\lambda)h(t - \lambda) \, d\lambda, \qquad (17.35)$$

which is the second form of the convolution integral given in Eq. (17.29). The first form of the integral can be derived from Eq. (17.35) by making a change in the variable of integration. Thus we let $u = t - \lambda$ and then note that $du = -d\lambda$, $u = -\infty$ when $\lambda = \infty$, and $u = +\infty$ when $\lambda = -\infty$. With these observations in mind, we can write Eq. (17.35) as

$$y(t) = \int_{\infty}^{-\infty} x(t - u)h(u)(-du)$$

or

$$y(t) = \int_{-\infty}^{\infty} x(t - u)h(u) \, du. \qquad (17.36)$$

But since u is just a symbol of integration, Eq. (17.36) is equivalent to the first form of the convolution integral given by Eq. (17.29).

The integral relationship between $y(t)$, $h(t)$, and $x(t)$, expressed by Eq. (17.29), is often written in a shorthand notation; thus

$$y(t) = h(t) * x(t) = x(t) * h(t), \qquad \textbf{(17.37)}$$

where the asterisk is used to signify the integral relationship between $h(t)$ and $x(t)$. Thus $h(t) * x(t)$ is read as "$h(t)$ is convolved with $x(t)$" and implies that

$$h(t) * x(t) = \int_{-\infty}^{\infty} h(\lambda)x(t - \lambda) \, d\lambda,$$

whereas $x(t) * h(t)$ is read as "$x(t)$ is convolved with $h(t)$" and implies that

$$x(t) * (t) = \int_{-\infty}^{\infty} x(\lambda)h(t - \lambda) \, d\lambda.$$

The integrals in Eq. (17.29) give the most general relationship for the convolution of two functions. However, in our applications of the convolution integral we can change the lower limit to zero and the upper limit to t. Thus we can write Eq. (17.29) as

$$y(t) = \int_0^t h(\lambda)x(t - \lambda) \, d\lambda = \int_0^t x(\lambda)h(t - \lambda) \, d\lambda. \qquad \textbf{(17.38)}$$

We can change the limits for two reasons. First, for physically realizable circuits $h(t)$ will be zero for $t < 0$. In other words, there can be no impulse response before an impulse is applied. Second, we start measuring time at the instant the excitation $[x(t)]$ is turned on; therefore $x(t) = 0$ for $t < 0^-$. These observations regarding the limits on the convolution integral are reinforced by the following discussion on the graphical interpretation of the integral.

A graphical interpretation of the convolution integrals contained in Eq. (17.38) is very important when it comes to using the integral as a computational tool. We begin with an interpretation of the first integral. For purposes of discussion, we assume that the impulse response of our circuit is the exponential decay function shown in Fig. 17.14(a) and the excitation function has the waveform shown in Fig. 17.14(b). In each of these plots we have replaced t with λ, the symbol of integration. The graphical interpretation of $x(t - \lambda)$ is illustrated in Fig. 17.14(c) and (d). Replacing λ with $-\lambda$ simply folds the excitation function over the vertical axis and replacing $-\lambda$ with $t - \lambda$ slides the "folded" function to the right. It is this folding operation that gives rise to the term "convolution." At any specified value of t, the response function $y(t)$ is the area under the product function $h(\lambda)x(t - \lambda)$. This is shown in Fig. 17.14(e). It should be apparent from the plot shown there why the lower limit on the convolution integral is zero and the upper limit is t. For $\lambda < 0$ the product $h(\lambda)x(t - \lambda)$ is zero because $h(\lambda)$ is zero. For $\lambda > t$ the product $h(\lambda)x(t - \lambda)$ is zero because $x(t - \lambda)$ is zero.

The graphical interpretation of the second form of the convolution integral is shown in Fig. 17.15. Note that the product function in part (e) of the figure confirms the use of zero for the lower limit and t for the upper limit.

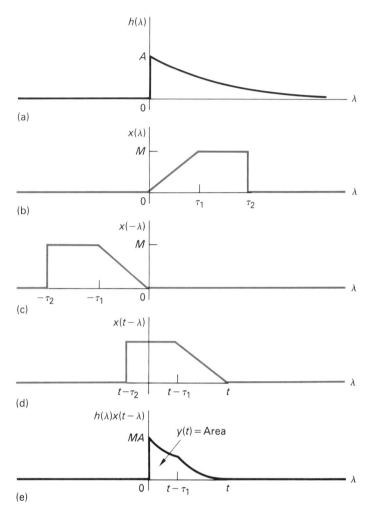

(a)

(b)

(c)

(d)

(e)

Fig. 17.14   Graphical interpretation of the convolution integral $\int_0^t h(\lambda)x(t - \lambda)d\lambda$: (a) impulse response; (b) excitation function; (c) folded excitation function; (d) folded excitation function displaced t units; (e) product $h(\lambda)x(t - \lambda)$.

Example 17.3 illustrates how we can use the convolution integral, in conjunction with the unit impulse response, to find the response of a circuit.

Example 17.3   The excitation voltage v_i for the circuit shown in Fig. 17.16(a) is shown in Fig. 17.16(b).

a) Use the convolution integral to find $v_o(t)$.

b) Plot $v_o(t)$ over the range of $0 \le t \le 15$ s.

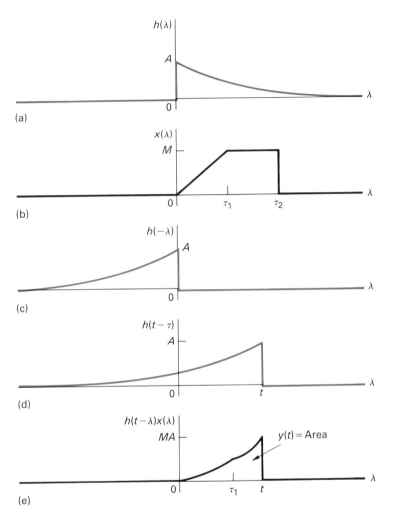

Fig. 17.15   Graphical representation of the convolution integral $\int_0^t h(t - \lambda)x(\lambda)d\lambda$: (a) impulse response; (b) excitation function; (c) folded impulse response; (d) folded impulse response displaced t units; (e) product $h(t - \lambda)x(\lambda)$.

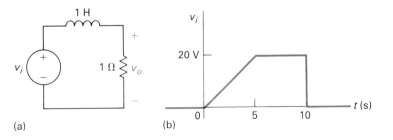

Fig. 17.16   The circuit and excitation voltage for Example 17.3: (a) circuit; (b) excitation voltage.

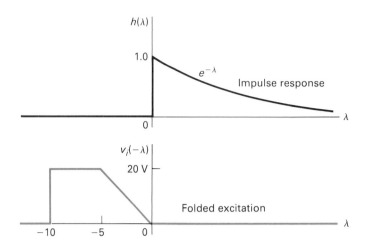

Fig. 17.17   The impulse response and the folded excitation function for Example 17.3.

Solution   a) The first step in using the convolution integral is to find the unit impulse response of the circuit. The expression for V_o is obtained from the s-domain equivalent of the circuit in Fig. 17.16(a). Thus

$$V_o = \frac{V_i}{(s + 1)} \quad (1).$$

When v_i is a unit impulse function $[\delta(t)]$, then

$$v_o = h(t) = e^{-t}u(t),$$

from which it follows that

$$h(\lambda) = e^{-\lambda}u(\lambda).$$

Using the first form of the convolution integral given in Eq. (17.38), we construct the impulse response and folded excitation function as shown in Fig. 17.17, which is very helpful in selecting the limits on the convolution integral. We can see that as we slide the folded excitation function to the right, we will have to break the integration into three intervals, namely, $0 \le t \le 5$, $5 \le t \le 10$, and $10 \le t \le \infty$. These break points are dictated by the breaks in the excitation function at 0, 5, and 10 seconds. The positioning of the folded excitation for each of these intervals is shown in Fig. 17.18. The analytical expression for v_i in the time interval $0 \le t \le 5$ is

$$v_i = 4t \quad (0 \le t \le 5 \text{ s}).$$

It follows that the analytical expression for the folded excitation function in the interval $t - 5 \le \lambda \le t$ is

$$v_i(t - \lambda) = 4(t - \lambda) \quad (t - 5 \le \lambda \le t).$$

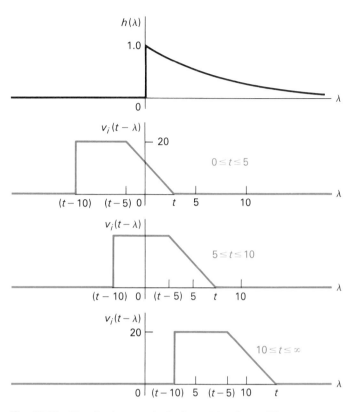

Fig. 17.18   The displacement of $v_i(t - \lambda)$ for three different time intervals.

We can now set up the three integral expressions for $v_o(t)$. We have

for $0 \le t \le 5$ s: $v_o = \displaystyle\int_0^t 4(t - \lambda)e^{-\lambda} \, d\lambda = 4[e^{-t} + t - 1]$ V;

for $5 \le t \le 10$ s: $v_o = \displaystyle\int_0^{t-5} 20e^{-\lambda} \, d\lambda + \int_{t-5}^t 4(t - \lambda)e^{-\lambda} \, d\lambda$

$= 4[5 + e^{-t} - e^{-(t-5)}]$ V;

for 10 s $\le t \le \infty$: $v_o = \displaystyle\int_{t-10}^{t-5} 20e^{-\lambda} \, d\lambda + \int_{t-5}^t 4(t - \lambda)e^{-\lambda} \, d\lambda$

$= 4[e^{-t} - e^{-(t-5)} + 5e^{-(t-10)}]$ V.

b) We have computed $v_o(t)$ for one-second intervals of time using the appropriate equation. The results are tabulated in Table 17.1 and shown graphically in Fig. 17.19.   ■

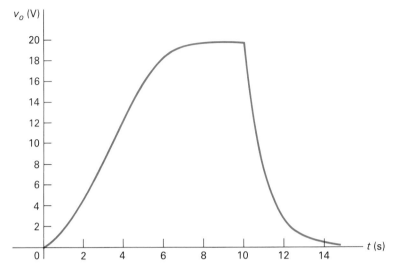

Fig. 17.19   The voltage response versus time for Example 17.3.

We mentioned at the beginning of this section that the convolution integral introduces the concepts of memory and weighting function into circuit analysis. These concepts are most easily introduced through the graphical interpretation of the convolution integral. The concepts of memory and weighting function become evident when the folding and sliding of the excitation function are viewed on a time scale that is characterized as past, present, and future. On such a time scale, the vertical axis, over which the excitation function $x(t)$ is folded, represents the present value, past values of $x(t)$ lie to the right of the vertical axis, and future values of $x(t)$ lie to the left of the vertical axis. This description of $x(t)$ is shown in Fig. 17.20, where for illustrative purposes we have used the excitation function from Example 17.3.

When we combine the past, present, and future viewpoints of $x(t - \tau)$ with the impulse response of the circuit, we see that the impulse response weights $x(t)$ according to present and past values. For example, referring to Fig. 17.18

TABLE 17.1
NUMERICAL VALUES OF $v_o(t)$

t	v_o	t	v_o	t	v_o
1	1.47	6	18.54	11	7.35
2	4.54	7	19.56	12	2.70
3	8.20	8	19.80	13	0.99
4	12.07	9	19.93	14	0.37
5	16.03	10	19.97	15	0.13

Fig. 17.20   The past, present, and future values of the excitation function.

we see that the impulse response in Example 17.3 gives less weight to past values of $x(t)$ than to the present value of $x(t)$. In other words, the circuit remembers less and less about the past values of the input. Note, for example, in Fig. 17.19 how quickly v_o approaches zero when the present value of the input is zero (that is, when $t > 10$ s). In other words, since the present value of the input receives more weight than the past values of the input, the output quickly approaches the present value of the input.

It is this multiplication of $x(t - \lambda)$ by $h(\lambda)$ that gives rise to referring to the impulse response as the circuit *weighting* function. At the same time the weighting function is described in terms of how much memory it has. For example, if the impulse response, or weighting function, is flat, as shown in Fig. 17.21(a), it gives equal weight to all values of $x(t)$, past and present. Such a circuit has a perfect memory. On the other hand, if the impulse response is an impulse function, as shown in Fig. 17.21(b), it gives no weight to past values of $x(t)$. Such a circuit has no memory. Now we make the important observation that the more memory a circuit has, the more distortion there is between the waveform of the excitation function and the waveform of the response function. This is most easily seen by assuming that the circuit has no memory—that is, $h(t) = A\delta(t)$—and then noting from the convolution integral that

$$y(t) = \int_0^t h(\lambda)x(t - \lambda)\, d\lambda$$

$$= \int_0^t A\delta(\lambda)x(t - \lambda)\, d\lambda$$

$$= Ax(t). \tag{17.39}$$

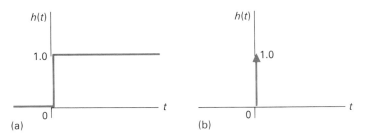

Fig. 17.21   Weighting functions: (a) perfect memory; (b) no memory.

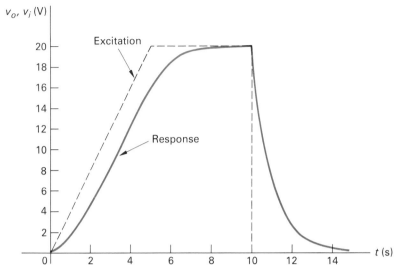

Fig. 17.22 The input and output waveforms for Example 17.3.

We see from Eq. (17.39) that if the circuit has no memory, the output will be a scaled replica of the input.

The distortion between input and output for a circuit that has some memory is illustrated by the circuit in Example 17.3. The distortion is clearly evident when we plot the input and output waveforms on the same graph. This is done in Fig. 17.22.

DRILL EXERCISE
17.4

A rectangular voltage pulse $v_i = [u(t) - u(t - 1)]$ V is applied to the circuit shown. Use the convolution integral to find v_o.

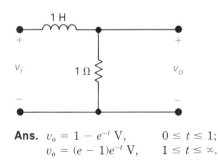

Ans. $v_o = 1 - e^{-t}$ V, $\qquad 0 \leq t \leq 1$;
$v_o = (e - 1)e^{-t}$ V, $\qquad 1 \leq t \leq \infty$.

DRILL EXERCISE
17.5

Interchange the inductor and resistor in Drill Exercise 17.4 and again use the convolution integral to find v_o.

Ans. $v_o = e^{-t}$ V, $\qquad 0 \leq t \leq 1$;
$v_o = (1 - e)e^{-t}$ V, $\qquad 1 \leq t \leq \infty$.

17.6 THE TRANSFER FUNCTION AND THE STEADY-STATE SINUSOIDAL RESPONSE

The transfer function can be used to find the steady-state response of a circuit that is driven by a sinusoidal source. To see how the transfer function relates the steady-state response to the excitation source, we assume that

$$x(t) = A \cos (\omega t + \phi) \tag{17.40}$$

and then use Eq. (17.24) to find the steady-state solution of $y(t)$. To find the Laplace transform of $x(t)$ we first write $x(t)$ as

$$x(t) = A \cos \omega t \cos \phi - A \sin \omega t \sin \phi, \tag{17.41}$$

from which it follows that

$$X(s) = \frac{[A \cos \phi]s}{s^2 + \omega^2} - \frac{(A \sin \phi)\omega}{s^2 + \omega^2} = \frac{A(s \cos \phi - \omega \sin \phi)}{s^2 + \omega^2}. \tag{17.42}$$

Substituting Eq. (17.42) into Eq. (17.24) gives us the s-domain expression for the response; thus

$$Y(s) = H(s) \frac{A(s \cos \phi - \omega \sin \phi)}{s^2 + \omega^2}. \tag{17.43}$$

We now visualize the partial-fraction expansion of Eq. (17.43). The number of terms in the expansion will depend on the number of poles of $H(s)$. Since $H(s)$ is not specified beyond being the transfer function of a physically realizable circuit, we write the expansion of Eq. (17.43) as

$$Y(s) = \frac{K_1}{s - j\omega} + \frac{K_1^*}{s + j\omega} + \sum \text{ terms generated by the poles of } H(s). \tag{17.44}$$

In Eq. (17.44) the first two terms are due to the complex conjugate poles of the driving source, that is, $s^2 + \omega^2 = (s - j\omega)(s + j\omega)$. Now we make the critical observation that the terms generated by the poles of $H(s)$ will not contribute to the steady-state response of $y(t)$. This follows from the fact that all the poles of $H(s)$ lie in the left half of the s-plane and consequently the corresponding time domain terms approach zero as t increases. Thus the first two terms on the right-hand side of Eq. (17.44) determine the steady-state response. Our problem is reduced to finding the partial-fraction coefficient K_1. We have

$$K_1 = \frac{H(s)A(s \cos \phi - \omega \sin \phi)}{s + j\omega} \bigg|_{s=j\omega} = \frac{H(j\omega) \cdot A(j\omega \cos \phi - \omega \sin \phi)}{2j\omega}$$

$$= \frac{H(j\omega)A(\cos \phi + j \sin \phi)}{2} = \frac{1}{2} H(j\omega)Ae^{j\phi}. \tag{17.45}$$

In general, $H(j\omega)$ will be a complex quantity and we signify this by writing it in polar form; thus

$$H(j\omega) = |H(j\omega)|e^{j\theta(\omega)}. \tag{17.46}$$

Note from Eq. (17.46) that both the magnitude $|H(j\omega)|$ and phase angle $[\phi(\omega)]$ of the transfer function vary with the frequency ω. When we substitute Eq. (17.46) into Eq. (17.45), the expression for K_1 becomes

$$K_1 = \frac{A}{2}|H(j\omega)|e^{j[\theta(\omega)+\phi]}. \tag{17.47}$$

The steady-state solution for $y(t)$ is obtained by inverse-transforming Eq. (17.44) and, in the process, ignoring the terms generated by the poles of $H(s)$; thus we have

$$y_{ss}(t) = A|H(j\omega)|\cos[\omega t + \phi + \theta(\omega)], \tag{17.48}$$

which tells us how to use the transfer function to find the steady-state sinusoidal response of a circuit. The amplitude of the response equals the amplitude of the source (A) times the magnitude of the transfer function $|H(j\omega)|$ and the phase angle of the response $[\phi + \theta(\omega)]$ equals the phase angle of the source (ϕ) plus the phase angle of the transfer function $[\theta(\omega)]$. Both $|H(j\omega)|$ and $\theta(\omega)$ are evaluated at the frequency of the source (ω).

Example 17.4 illustrates how we can use the transfer function to find the steady-state sinusoidal response of a circuit.

Example 17.4 The circuit in Example 17.1 is driven by a sinusoidal voltage source. The source voltage is $120\cos(5000t + 30°)$ V. Find the steady-state expression for v_o.

Solution From Example 17.1 we have

$$H(s) = \frac{1000(s + 5000)}{s^2 + 6000s + 25 \times 10^6}.$$

The frequency of the voltage source is 5000 rad/s; hence we evaluate $H(s)$ at $H(j5000)$. Thus

$$\begin{aligned} H(j5000) &= \frac{1000(5000 + j5000)}{-25 \times 10^6 + j5000(6000) + 25 \times 10^6} \\ &= \frac{1+j1}{j6} = \frac{1-j1}{6} = \frac{\sqrt{2}}{6}\underline{/-45°}. \end{aligned}$$

It follows directly from Eq. (17.48) that

$$v_{o_{ss}} = \frac{(120)\sqrt{2}}{6}\cos(5000t + 30° - 45°) = 20\sqrt{2}\cos(5000t - 15°) \text{ V.} \quad\blacksquare$$

The observation that the transfer function can be used to calculate the steady-state sinusoidal response of a circuit is an important one. First of all, it tells us that we can use the transfer function to find the steady-state frequency response of the circuit. That is, since the amplitude and phase of the response depend on the magnitude and phase of $H(j\omega)$, we simply evaluate the magnitude and phase of $H(j\omega)$ over the frequency range of interest. The variation of $|H(j\omega)|$ and $\theta(\omega)$ with ω tells us exactly how the amplitude and phase of the steady-state response signal will vary with ω. Second, we observe that if $H(j\omega)$ is known then, at least theoretically, $H(s)$ is also known. In other words, we can reverse the process. Instead of using $H(s)$ to find $H(j\omega)$, we use $H(j\omega)$ to find $H(s)$. Once we know $H(s)$, we can find the response to other excitation sources. In this application, $H(j\omega)$ is determined experimentally and $H(s)$ is then constructed from the experimental data. From a practical point of view, this experimental approach is not always possible; however, in some cases it does provide a viable method for deriving $H(s)$. From a theoretical point of view, the relationship between $H(s)$ and $H(j\omega)$ provides a link between the time domain and the frequency domain.

DRILL EXERCISE
17.6

The current source in the circuit in Drill Exercise 17.2 is delivering $3\sqrt{2}\cos 2t$ A. Use the transfer function to compute the steady-state expression for v_o.

Ans. $20\cos(2t - 45°)$ V.

DRILL EXERCISE
17.7

a) For the circuit shown, find the steady-state expression for v_o when $v_g = 10\cos 50{,}000t$ V.
b) Replace the 50-kΩ resistor with a variable resistor and compute the value of resistance necessary to cause v_o to lead v_g by 120°.

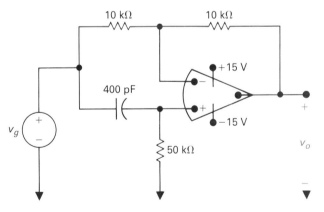

Ans. (a) $10\cos(50{,}000t + 90°)$ V; (b) 28,867.51 Ω.

17.7 BODE DIAGRAMS

The behavior of the transfer function as the frequency of the sinusoidal source is varied is an important characteristic of a circuit. One effective way to describe how the amplitude and phase angle of $H(j\omega)$ vary with frequency is by using graphical plots. The most efficient method for generating and plotting the amplitude and phase data is the digital computer. Although we can rely on the digital computer to give us accurate numerical plots of $|H(j\omega)|$ and $\theta(\omega)$ vs. ω, there are situations where the intelligent use of the computer can be greatly facilitated by first making some preliminary sketches using Bode diagrams. Thus a Bode diagram, or plot, can be thought of as a graphical technique for getting a "feel" for the frequency response of a circuit. As we will soon see, these plots are most useful in circuits where the poles and zeros of $H(s)$ are reasonably well separated.

A Bode diagram consists of two separate plots: One shows how the amplitude of $H(j\omega)$ varies with frequency and the second shows how the phase angle of $H(j\omega)$ varies with frequency. For reasons that will soon be apparent, both plots are made on semilog graph paper. In both the amplitude and phase plots, the frequency is plotted on the horizontal log scale and the amplitude and phase angle are plotted on the linear vertical scale. These amplitude and phase diagrams are called Bode plots in recognition of the pioneering work done in this area by H. W. Bode.[†]

In order to simplify our development of the Bode diagrams, we will assume to start with that all the poles and zeros of $H(s)$ are real and first order. For the purposes of our discussion, it will be helpful to have a specific expression for $H(s)$ in mind. We will base our discussion on the following:

$$H(s) = \frac{K(s + z_1)}{s(s + p_1)}, \tag{17.49}$$

from which it follows directly that

$$H(j\omega) = \frac{K(j\omega + z_1)}{j\omega(j\omega + p_1)}. \tag{17.50}$$

The first step in making the Bode diagrams is to put the expression for $H(j\omega)$ in what we call a *standard form*, which is derived by simply dividing out the poles and zeros. Thus

$$H(j\omega) = \frac{Kz_1(1 + j\omega/z_1)}{p_1(j\omega)(1 + j\omega/p_1)}. \tag{17.51}$$

Next we let the constant quantity Kz_1/p_1 be represented by K_o and at the same time express $H(j\omega)$ in polar form; thus

$$H(j\omega) = \frac{K_o|1 + j\omega/z_1|\underline{/\psi_1}}{|\omega|\underline{/90°}|1 + j\omega/p_1|\underline{/\beta_1}} = \frac{K_o|1 + j\omega/z_1|}{\omega|1 + j\omega/p_1|} \underline{/\psi_1 - 90° - \beta_1}. \tag{17.52}$$

From Eq. (17.52) we have

$$|H(j\omega)| = \frac{K_o|1 + j\omega/z_1|}{\omega|1 + j\omega/p_1|} \tag{17.53}$$

[†] See H. W. Bode, *Network Analysis and Feedback Design*. New York: Van Nostrand, 1945.

and

$$\theta(\omega) = \psi_1 - 90° - \beta_1. \tag{17.54}$$

The phase angles ψ_1 and β_1 are by definition

$$\psi_1 = \tan^{-1} \omega/z_1 \tag{17.55}$$

and

$$\beta_1 = \tan^{-1} \omega/p_1. \tag{17.56}$$

The Bode diagrams consist of plotting Eq. (17.53) (amplitude) and Eq. (17.54) (phase) as functions of ω.

The amplitude plot involves the multiplication and division of factors associated with the poles and zeros of $H(s)$. We can reduce this multiplication and division to addition and subtraction by expressing the amplitude of $H(j\omega)$ in terms of a logarithmic value. The value used is the decibel. The amplitude of $H(j\omega)$ in decibels is

$$A_{dB} = 20 \log_{10} |H(j\omega)|. \tag{17.57}$$

Expressing Eq. (17.53) in terms of decibels gives us

$$A_{dB} = 20 \log_{10} \frac{K_o|1 + j\omega/z_1|}{\omega|1 + j\omega/p_1|}$$
$$= 20 \log_{10} K_o + 20 \log_{10} |1 + j\omega/z_1| - 20 \log_{10} \omega - 20 \log_{10} |1 + j\omega/p_1|. \tag{17.58}$$

The key to plotting Eq. (17.58) is to plot each term in the equation separately and then combine the separate plots graphically. The individual factors are easy to plot because they can be approximated in all cases by straight lines.

The plot of $20 \log_{10} K_o$ is a horizontal straight line since K_o is not a function of frequency. The value of this term is positive for $K_o > 1$, zero for $K_o = 1$, and negative for $K_o < 1$.

The plot of $20 \log_{10} |1 + j\omega/z_1|$ is approximated by two straight lines. For small values of ω the magnitude of $|1 + j\omega/z_1|$ is approximately 1 and therefore

$$20 \log_{10} |1 + j\omega/z_1| \to 0 \quad \text{as } \omega \to 0. \tag{17.59}$$

For large values of ω the magnitude of $|1 + j\omega/z_1|$ is approximately ω/z_1 and therefore

$$20 \log_{10} |1 + j\omega/z_1| \to 20 \log_{10}(\omega/z_1) \quad \text{as } \omega \to \infty. \tag{17.60}$$

On a log frequency scale, $20 \log_{10} (\omega/z_1)$ is a straight line with a slope of 20 dB/decade (a decade is a ten-to-one change in frequency). Furthermore, this straight line intersects the 0-dB axis at $\omega = z_1$. This value of ω is known as the *corner frequency*. Thus on the basis of Eqs. (17.59) and (17.60) we see that the amplitude plot of a first-order zero can be approximated by two straight lines, as shown in Fig. 17.23.

The plot of $-20 \log_{10} \omega$ is a straight line having a slope of -20 dB/decade that intersects the 0-dB axis at $\omega = 1$.

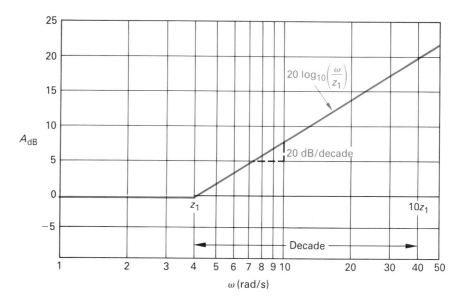

Fig. 17.23 A straight-line approximation of the amplitude plot of a first-order zero.

The plot of $-20 \log_{10} |1 + j\omega/p_1|$ is also approximated by two straight lines. Here the two straight lines intersect on the 0-dB axis at $\omega = p_1$. For large values of ω the straight line $-20 \log_{10} (\omega/p_1)$ has a slope of -20 db/decade. The straight-line approximation of the amplitude plot of a first-order pole is shown in Fig. 17.24.

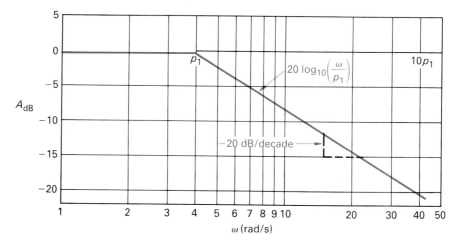

Fig. 17.24 A straight-line approximation of the amplitude plot of a first-order pole.

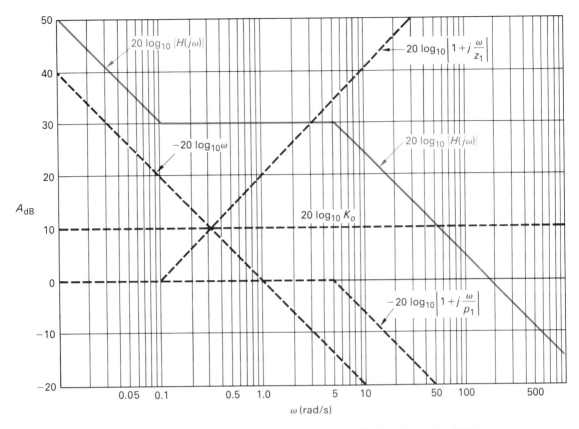

Fig. 17.25 A straight-line approximation of the amplitude plot for Eq. 17.58.

A graphical plot of Eq. (17.58) is shown in Fig. 17.25, assuming $K_o = \sqrt{10}$, $z_1 = 0.1$ rad/s, and $p_1 = 5$ rad/s. Each term in Eq. (17.58) is labeled on Fig. 17.25 so that you can verify how the individual terms add up to create the resultant plot, which is labeled $20 \log_{10} |H(j\omega)|$ in Fig. 17.25.

Example 17.5 also illustrates the construction of a straight-line amplitude plot for a transfer function characterized by first-order poles and zeros.

Example 17.5 The numerical expression for a transfer function is

$$H(s) = \frac{10^4(s + 1)}{(s + 10)(s + 100)}.$$

a) Construct a straight-line approximation of the Bode amplitude plot.
b) Calculate $20 \log_{10} |H(j\omega)|$ at $\omega = 50$ rad/s and $\omega = 1000$ rad/s.
c) Plot the values computed in part (b) on the straight-line graph.

Solution   a) We begin by writing $H(j\omega)$ in standard form; thus

$$H(j\omega) = \frac{10(1 + j\omega)}{[1 + j(\omega/10)][1 + j(\omega/100)]}.$$

The expression for the amplitude of $H(j\omega)$ in dB is

$$A_{dB} = 20 \log_{10} |H(j\omega)| = 20 \log_{10} 10 + 20 \log_{10} |1 + j\omega|$$
$$- 20 \log_{10} \left|1 + j\frac{\omega}{10}\right| - 20 \log_{10} \left|1 + j\frac{\omega}{100}\right|.$$

The straight-line plot is shown in Fig. 17.26, where we note that each term contributing to the overall amplitude is identified.

b) We have

$$H(j50) = \frac{10(1 + j50)}{(1 + j5)(1 + j0.5)} = 87.72\underline{/-16.40°},$$

$$20 \log_{10} |H(j50)| = 20 \log_{10} 87.72 = 38.86 \text{ dB},$$

$$H(j1000) = \frac{10(1 + j1000)}{(1 + j100)(1 + j10)} = 9.95\underline{/-83.77°},$$

$$20 \log_{10} |9.95| = 19.96 \text{ dB}.$$

c) See Fig. 17.26.   ■

The straight-line plots for first-order poles and zeros can be made more accurate by correcting the amplitude values at the corner frequency, one half the corner frequency, and twice the corner frequency. At the corner frequency, the actual value in decibels is

$$A_{dB_c} = \pm 20 \log_{10} |1 + j1| = \pm 20 \log_{10} \sqrt{2} \approx \pm 3 \text{ dB}. \quad \textbf{(17.61)}$$

The actual value of the amplitude in decibels at one half of the corner frequency is

$$A_{dB_{c/2}} = \pm 20 \log_{10} \left|1 + j\frac{1}{2}\right| = \pm 20 \log_{10} \sqrt{5/4} \approx \pm 1 \text{ dB}. \quad \textbf{(17.62)}$$

At twice the corner frequency the actual value in decibels is

$$A_{dB_{2c}} = \pm 20 \log_{10} |1 + j2| = \pm 20 \log_{10} \sqrt{5} \approx \pm 7 \text{ dB}. \quad \textbf{(17.63)}$$

In Eqs. (17.61), (17.62), and (17.63), the plus sign applies to a first-order zero and the minus sign applies to a first-order pole. Since the straight-line approximation of the amplitude plot gives 0 dB at the corner and one half the corner frequencies and ± 6 dB at twice the corner frequency, the corrections are ± 3 dB at the corner frequency and ± 1 dB at both one half the corner frequency and twice the corner frequency. These corrections are summarized in Fig. 17.27.

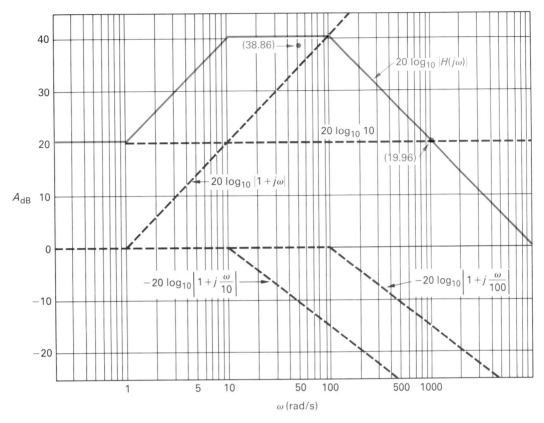

Fig. 17.26 The amplitude plot for Example 17.5.

It is worth noting at this point that a two-to-one change in frequency is called an *octave*. A slope of 20 dB/decade is equivalent to 6.02 dB/octave which, for graphical purposes, is equivalent to 6 dB/octave. Thus the corrections enumerated above correspond to one octave below and one octave above the corner frequency.

If the poles and zeros of $H(s)$ are well separated, it is relatively easy to insert these corrections into the overall amplitude plot and achieve a reasonably accurate curve. However, if the poles and zeros are close together, the overlapping corrections are difficult to evaluate and one is better off using the straight-line plot as a first estimate of the amplitude characteristic and then using the computer to refine the calculations in the frequency range of interest.

Phase-angle plots can also be made using straight-line approximations. In making phase plots, it is worth noting that the phase angle associated with the constant K_o is zero and the phase angle associated with a first-order zero or

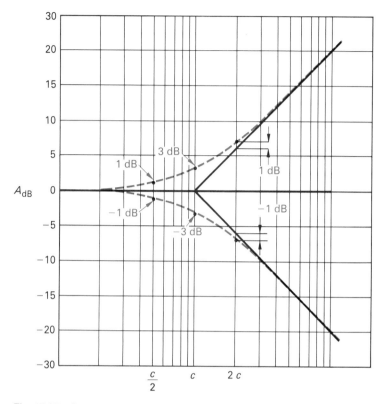

Fig. 17.27 Corrected amplitude plots for a first-order zero and pole.

pole at the origin is a constant $\pm 90°$. For a first-order zero or pole not at the origin, the straight-line approximations are as follows. For frequencies less than one tenth the corner frequency, the phase angle is assumed to be zero. For frequencies greater than ten times the corner frequency, the phase angle is assumed to be $\pm 90°$. Between one tenth the corner frequency and ten times the corner frequency, the phase-angle plot is a straight-line that goes through $0°$ at one tenth the corner frequency, $\pm 45°$ at the corner frequency, and $\pm 90°$ at ten times the corner frequency. In all these statements, the plus sign applies to the first-order zero and the minus sign to the first-order pole. The straight-line approximation for a first-order zero and pole is shown in Fig. 17.28, where we have also shown (via dashed curves) the exact variation of the phase angle as the frequency varies. We have done this so you can see how closely the straight-line plot approximates the actual variation in phase angle. The maximum deviation between the straight-line plot and the actual plot is approximately $6°$.

The straight-line approximation of the phase angle of the transfer function given by Eq. (17.49) is given in Fig. 17.29. The equation for the phase

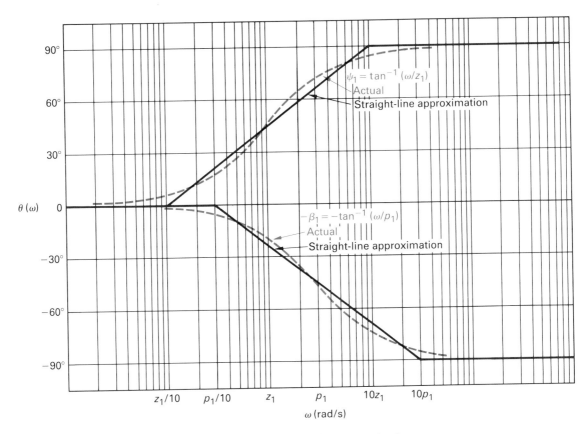

Fig. 17.28 Phase-angle plots for a first-order zero and pole.

angle is given by Eq. (17.54) and the plot corresponds to $z_1 = 0.1$ rad/s and $p_1 = 5$ rad/s.

Example 17.6   a) Make a straight-line phase-angle plot for the transfer function in Example 17.5.

b) Compute the phase angle $\theta(\omega)$ at $\omega = 50, 500,$ and 1000 rad/s.

c) Plot the values of part (b) on the diagram of part (a).

Solution   a) From Example 17.5,

$$H(j\omega) = \frac{10(1 + j\omega)}{[1 + j\,(\omega/10)][1 + j\,(\omega/100)]}$$

$$= \frac{10|1 + j\omega|}{\left|1 + j\,\dfrac{\omega}{10}\right|\left|1 + j\,\dfrac{\omega}{100}\right|} \underline{/\psi_1 - \beta_1 - \beta_2}.$$

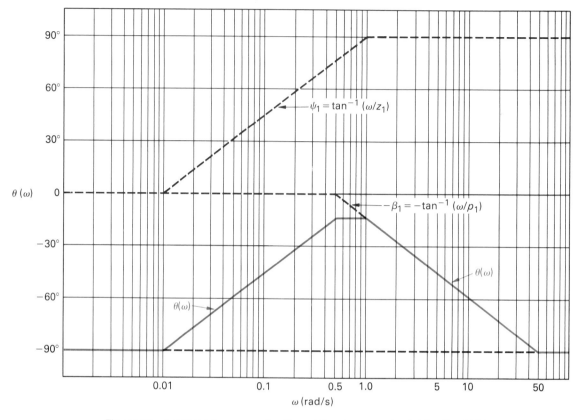

Fig. 17.29 A straight-line approximation of the phase-angle plot for Eq. 17.54.

Therefore

$$\theta(\omega) = \psi_1 - \beta_1 - \beta_2,$$

where $\psi_1 = \tan^{-1} \omega$, $\beta_1 = \tan^{-1} (\omega/10)$, and $\beta_2 = \tan^{-1} (\omega/100)$. The straight-line approximation of $\theta(\omega)$ is shown in Fig. 17.30.

b) We have

$$H(j50) = 87.72\underline{/-16.40°},$$
$$H(j500) = 19.61\underline{/-77.66°},$$
$$H(j1000) = 9.95\underline{/-83.77°}.$$

Thus

$$\theta(50) = -16.40°, \qquad \theta(500) = -77.66°, \quad \text{and} \quad \theta(1000) = -83.77°.$$

c) See Fig. 17.30. ∎

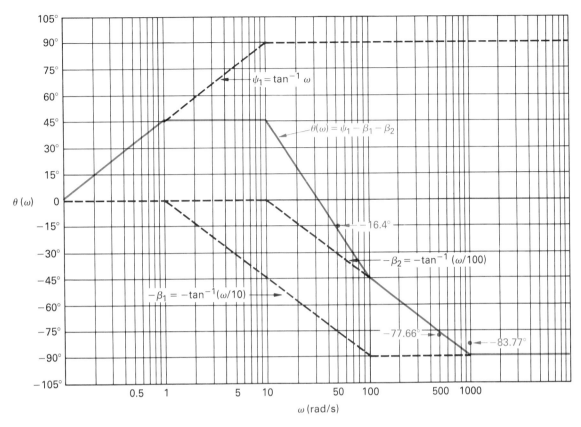

Fig. 17.30 A straight-line approximation of $\theta(\omega)$ for Example 17.6.

DRILL EXERCISE
17.8

The numerical expression for a transfer function is

$$H(s) = \frac{10^5(s + 5)}{(s + 100)(s + 5000)}.$$

On the basis of a straight-line approximation of $|H(j\omega)|$ vs. ω estimate (a) the maximum amplitude of $H(j\omega)$ in decibels, and (b) the value of $\omega > 0$ where the amplitude of $H(j\omega)$ equals unity.

Ans. (a) 26 dB; (b) 98 krad/s.

DRILL EXERCISE
17.9

Approximate the phase angle of the transfer function in Drill Exercise 17.8 by means of a straight-line plot.

a) Use the straight-line plot to predict the phase angle of $H(s)$ at frequencies of 30, 50, 100, and 5000 rad/s.

b) Calculate the actual value of the phase angle at 30, 50, 100, and 5000 rad/s.

Ans. (a) 58.5°, 58.5°, 45°, -45°; (b) 63.49°, 57.15°, 40.99°, -43.91°.

17.8 BODE DIAGRAMS: COMPLEX POLES AND ZEROS

Complex poles or zeros in the expression for $H(s)$ require special attention when we are making amplitude and phase-angle plots. We will focus our attention on the contribution that a pair of complex poles make to the amplitude and phase-angle plots. Once the rules for handling complex poles are understood, their application to a pair of complex zeros will be apparent.

The complex poles and zeros of $H(s)$ always appear in conjugate pairs. The first step in making either an amplitude or a phase-angle plot of a transfer function that contains complex poles is to combine the conjugate pair into a single quadratic term. Thus, given that

$$H(s) = \frac{K}{(s + \alpha - j\beta)(s + \alpha + j\beta)}, \tag{17.64}$$

we first rewrite the product $(s + \alpha - j\beta)(s + \alpha + j\beta)$ as

$$(s + \alpha)^2 + \beta^2 = s^2 + 2\alpha s + \alpha^2 + \beta^2.$$

When making Bode diagrams, we write the quadratic term in the more convenient form of

$$s^2 + 2\alpha s + \alpha^2 + \beta^2 = s^2 + 2\zeta\omega_n s + \omega_n^2. \tag{17.65}$$

A direct comparison of the two forms shows that

$$\omega_n^2 = \alpha^2 + \beta^2 \tag{17.66}$$

and

$$\zeta\omega_n = \alpha. \tag{17.67}$$

As we will soon see, ω_n is the corner frequency of the quadratic factor. Zeta (ζ) is known as the damping coefficient of the quadratic term. The critical value of ζ is one. If ζ is less than one, the roots of the quadratic factor are complex and Eq. (17.65) is used to represent the complex poles. If ζ is equal to or greater than one, the quadratic factor is factored into $(s + p_1)(s + p_2)$ and the amplitude and phase plots are made in accordance with the discussion in Section 17.7. Assuming ζ is less than one, we rewrite Eq. (17.64) as

$$H(s) = \frac{K}{s^2 + 2\zeta\omega_n s + \omega_n^2}. \tag{17.68}$$

We then write Eq. (17.68) in standard form by dividing through by the poles and zeros. For the quadratic term this means we divide through by ω_n^2. Thus

Eq. (17.68) becomes

$$H(s) = \frac{K}{\omega_n^2} \cdot \frac{1}{1 + (s/\omega_n)^2 + 2\zeta(s/\omega_n)}, \qquad (17.69)$$

from which it follows directly that

$$H(j\omega) = \frac{K_o}{[1 - (\omega^2/\omega_n^2) + j(2\zeta\omega/\omega_n)]}, \qquad (17.70)$$

where

$$K_o = (K/\omega_n^2).$$

Before we discuss the amplitude and phase-angle diagrams associated with Eq. (17.70), it is convenient to replace the ratio ω/ω_n by a new variable u; thus

$$H(j\omega) = \frac{K_o}{(1 - u^2 + j2\zeta u)}. \qquad (17.71)$$

Now we write $H(j\omega)$ in polar form:

$$H(j\omega) = \frac{K_o}{|(1 - u^2) + j2\zeta u|\underline{/\beta_1}}, \qquad (17.72)$$

from which we have

$$A_{\text{dB}} = 20 \log_{10} |H(j\omega)| = 20 \log_{10} K_o - 20 \log_{10} |(1 - u^2) + j2\zeta u| \quad (17.73)$$

and

$$\theta(\omega) = -\beta_1 = -\tan^{-1} \frac{2\zeta u}{(1 - u^2)}. \qquad (17.74)$$

The contribution of the quadratic factor to the amplitude of $H(j\omega)$ is wrapped up in the term $-20 \log_{10} |1 - u^2 + j2\zeta u|$. Now since $u = \omega/\omega_n$ we see that $u \to 0$ as $\omega \to 0$ and $u \to \infty$ as $\omega \to \infty$. To see how the term behaves as ω ranges from 0 to ∞, we note that

$$-20 \log_{10} |(1 - u^2) + j2\zeta u| = -20 \log_{10} \sqrt{(1 - u^2)^2 + 4\zeta^2 u^2}$$
$$= -10 \log_{10} [u^4 + 2u^2(2\zeta^2 - 1) + 1]; \quad (17.75)$$

as $u \to 0$,

$$-10 \log_{10} [u^4 + 2u^2(2\zeta^2 - 1) + 1] \to 0; \qquad (17.76)$$

and as $u \to \infty$,

$$-10 \log_{10} [u^4 + 2u^2(2\zeta^2 - 1) + 1] \to -40 \log_{10} u. \qquad (17.77)$$

From Eqs. (17.76) and (17.77) we conclude that the approximate amplitude plot consists of two straight lines. For $\omega < \omega_n$ the straight line lies along the 0-dB axis and for $\omega > \omega_n$ the straight line has a slope of -40 dB/decade. These two straight lines join on the 0-dB axis at $u = 1$ or $\omega = \omega_n$. The straight-line approximation for a quadratic factor with $\zeta < 1$ is shown in Fig. 17.31.

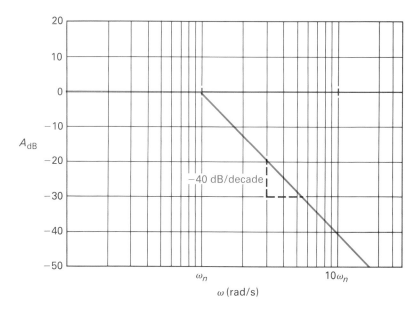

Fig. 17.31   The amplitude plot for a pair of complex poles.

It is not as easy to correct the straight-line amplitude plot for a pair of complex poles as it is for a first-order real pole, because the corrections depend on the damping coefficient ζ. The effect of ζ on the amplitude plot is shown in Fig. 17.32. Note that as ζ becomes very small, there is a large peak in the amplitude in the neighborhood of the corner frequency ω_n ($u = 1$). When ζ is equal to or greater than $1/\sqrt{2}$, the corrected amplitude plot lies entirely below the straight-line approximation. For sketching purposes, the straight-line amplitude plot can be corrected by locating four points on the actual curve. These four points correspond to one half the corner frequency, the frequency at which the amplitude reaches its peak value, the corner frequency, and the frequency at which the amplitude is zero. These four points are illustrated in Fig. 17.33.

At one half the corner frequency (point 1) the actual amplitude is

$$A_{\mathrm{dB}}(\omega_n/2) = -10 \log_{10} (\zeta^2 + 0.5625). \tag{17.78}$$

The amplitude will peak (point 2) at a frequency of

$$\omega_p = \omega_n \sqrt{1 - 2\zeta^2} \tag{17.79}$$

and will have a peak amplitude of

$$A_{\mathrm{dB}}(\omega_p) = -10 \log_{10} [4\zeta^2(1 - \zeta^2)]. \tag{17.80}$$

At the corner frequency (point 3) the actual amplitude is

$$A_{\mathrm{dB}}(\omega_n) = -20 \log_{10} 2\zeta. \tag{17.81}$$

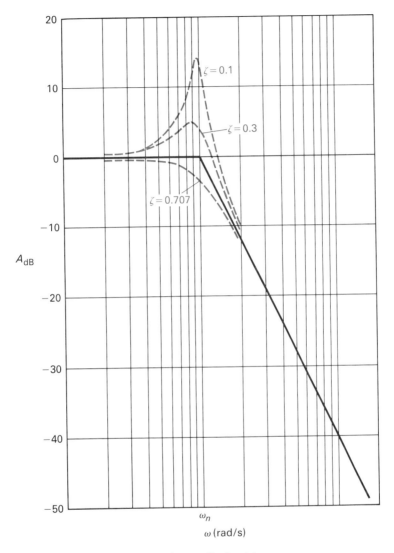

Fig. 17.32   The effect of ζ on the amplitude plot.

The corrected amplitude plot crosses the 0-dB axis (point 4) at

$$\omega_0 = \omega_n \sqrt{2(1 - 2\zeta^2)} = \sqrt{2}\,\omega_p. \tag{17.82}$$

The derivation of Eqs. (17.78), (17.81), and (17.82) follow from Eq. (17.75). Equations (17.78) and (17.81) are obtained by evaluating Eq. (17.75) at $u = 0.5$ and $u = 1.0$, respectively. Equation (17.82) corresponds to finding the value of u that makes $u^4 + 2u^2(2\zeta^2 - 1) + 1 = 1$. The derivation of Eq. (17.79) requires

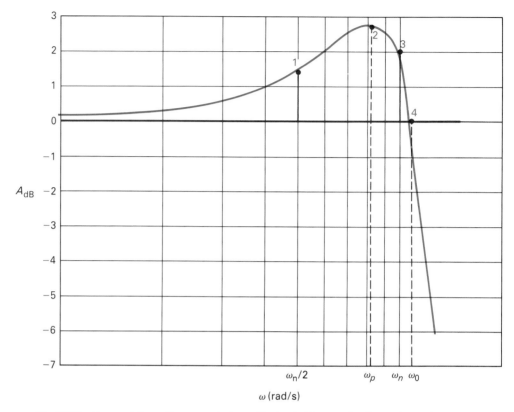

Fig. 17.33 Four points on the corrected amplitude plot for a pair of complex poles.

differentiating Eq. (17.75) with respect to u and finding the value of u where the derivative is zero. Equation (17.80) is the evaluation of Eq. (17.75) at the value of u found in Eq. (17.79).

The amplitude plot for a transfer function with a pair of complex poles is illustrated in the following example.

Example 17.7 The transfer function for a linear circuit is

$$H(s) = \frac{2500}{s^2 + 20s + 2500}.$$

a) What is the value of the corner frequency in rad/s?
b) What is the value of K_o?
c) What is the value of the damping coefficient?
d) Make a straight-line amplitude plot ranging from 10 to 500 rad/s.
e) Calculate the actual amplitude in dB at $\omega_n/2$, ω_p, ω_n, and ω_0.

Solution   a) From the expression for $H(s)$, $\omega_n^2 = 2500$; therefore $\omega_n = 50$ rad/s.

b) By definition K_o is $2500/\omega_n^2$ or 1.

c) The coefficient of s equals $2\zeta\omega_n$; therefore

$$\zeta = \frac{20}{2\omega_n} = 0.20.$$

d) See Fig. 17.34.

e) The actual amplitudes are

$$A_{dB}(\omega_n/2) = -10 \log_{10} (0.6025) = 2.2 \text{ dB},$$
$$\omega_p = 50\sqrt{0.92} = 47.96 \text{ rad/s},$$
$$A_{dB}(\omega_p) = -10 \log_{10} (0.16)(0.96) = 8.14 \text{ dB},$$
$$A_{dB}(\omega_n) = -20 \log_{10} (0.4) = 7.96 \text{ dB},$$
$$\omega_0 = \sqrt{2}\omega_p = 67.82 \text{ rad/s},$$
$$A_{dB}(\omega_0) = 0 \text{ dB}.$$

(See Fig. 17.34 for a sketch of the corrected plot.)   ■

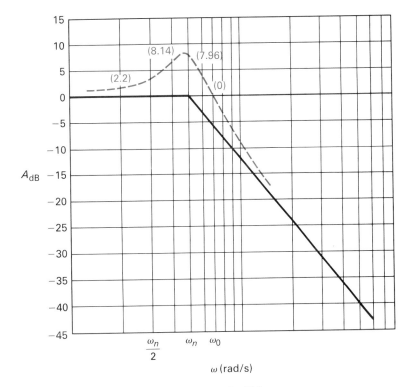

Fig. 17.34   The amplitude plot for Example 17.7.

The phase-angle plot for a pair of complex poles is a plot of Eq. (17.74). A study of Eq. (17.74) shows that the phase angle is zero at zero frequency, is −90° at the corner frequency, and approaches −180° as $\omega(u)$ becomes large. As in the case of the amplitude plot, ζ is very important in determining the exact shape of the phase-angle plot. For small values of ζ the phase angle changes very rapidly in the vicinity of the corner frequency. The effect of ζ on the phase-angle plot is shown in Fig. 17.35.

It is also possible to make a straight-line approximation of the phase-angle plot for a pair of complex poles. We make the straight-line approximation by drawing a line tangent to the phase-angle curve at the corner frequency and extending this line until it intersects with the 0° and −180° lines. The line tangent to the phase-angle curve at −90° has a slope of −2.3/ζ radians per decade (−132/ζ degrees/decade) and intersects the 0° and −180° lines at $u_1 = 4.81^{-\zeta}$ and $u_2 = 4.81^{\zeta}$, respectively. The straight-line approximation for $\zeta = 0.3$ is shown in Fig. 17.36, where we also show the actual phase-angle plot. In comparing the straight-line approximation to the actual curve, you can see that

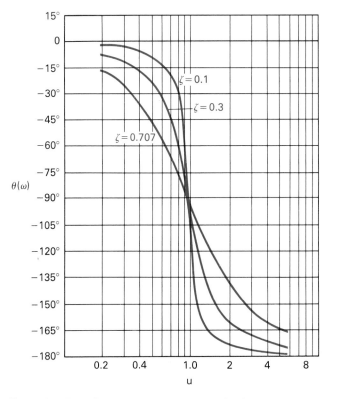

Fig. 17.35 The effect of ζ on the phase-angle plot.

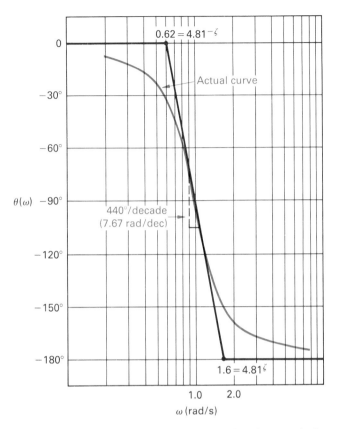

Fig. 17.36   A straight-line approximation of the phase angle for a pair of complex poles.

the approximation is reasonable in the vicinity of the corner frequency. However, in the neighborhood of u_1 and u_2 the error is quite large.

We will summarize our discussion of Bode diagrams with the following example.

Example 17.8   The transfer function for a linear circuit is

$$H(s) = \frac{100(s+1)}{(s^2 + 8s + 100)}.$$

a) Make a straight-line amplitude plot of $20 \log_{10} |H(j\omega)|$.

b) Estimate the maximum amplitude of $H(j\omega)$ in dB using the straight-line plot adjusted for corrections.

c) Use the straight-line amplitude plot to estimate the frequency, greater than zero, where the amplitude of $H(j\omega)$ equals 1.

d) What is the actual amplitude at the frequency of part (c)?

e) Make a straight-line phase-angle plot of $H(j\omega)$.

f) What is the maximum value of $\theta(\omega)$ predicted by the straight-line plot?

g) What is the actual value of $\theta(\omega)$ at the frequency corresponding to the prediction of part (f)?

Solution   a) The first step in making Bode diagrams is to put $H(j\omega)$ in standard form. Since $H(s)$ contains a quadratic factor, we first check the value of ζ. Here we find that $\zeta = 0.4$ and $\omega_n = 10$. Therefore, we write

$$H(s) = \frac{(s + 1)}{1 + (s/10)^2 + 0.8(s/10)},$$

from which it follows that

$$H(j\omega) = \frac{|1 + j\omega|\underline{/\psi_1}}{|1 - (\omega/10)^2 + j0.8(\omega/10)|\underline{/\beta_1}}.$$

Note that for the quadratic factor $u = \omega/10$. The amplitude of $H(j\omega)$ in dB is

$$A_{dB} = 20 \log_{10} |1 + j\omega| - 20 \log_{10}\left[\left|1 - \left(\frac{\omega}{10}\right)^2 + j0.8\left(\frac{\omega}{10}\right)\right|\right]$$

and the phase angle is

$$\theta(\omega) = \psi_1 - \beta_1,$$

where

$$\psi_1 = \tan^{-1}\omega$$

and

$$\beta_1 = \tan^{-1}\frac{0.8(\omega/10)}{1 - (\omega/10)^2}.$$

The amplitude plot is shown in Fig. 17.37.

b) From Fig. 17.37 we see that the straight-line approximation of A_{dB} predicts a maximum amplitude of 20 dB at 10 rad/s. The correction due to the first-order zero is negligible at this frequency. For $\zeta = 0.4$, the correction at 10 rad/s is $-20 \log_{10}(2\zeta)$, or 1.94 dB. The peak correction is $-10 \log_{10}[4\zeta^2(1 - \zeta^2)]$ or 2.70 dB at $\omega_n\sqrt{1 - 2\zeta^2}$, or 8.25 rad/s. Therefore at the corner frequency the estimated value of A_{dB} is $20 + 1.94$, or 21.94 dB. At ω_p the estimated value of A_{dB} is $18.2 + 2.7$, or 20.9 dB. Thus the peak amplitude is estimated at 21.94 dB, which occurs at a frequency of 10 rad/s. We leave it to the reader to verify that the actual value of $20 \log_{10} |H(j10)|$ is 21.98 dB.

c) From Fig. 17.37, A_{dB} is zero at $\omega = 100$ rad/s. Zero dB corresponds to $|H(j\omega)|$ equal to unity. Therefore, the straight-line approximation plot predicts that $|H(j\omega)| = 1.0$ at 100 rad/s.

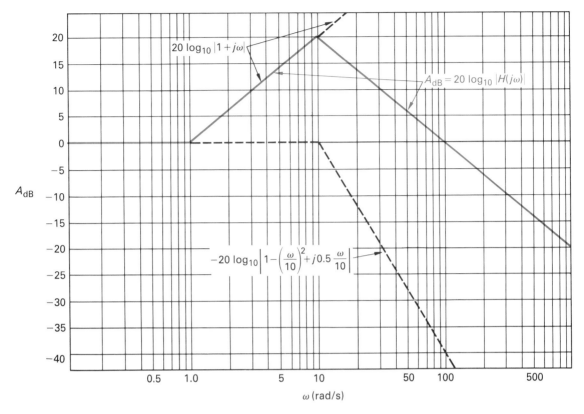

Fig. 17.37   The amplitude plot for Example 17.8.

d) The actual amplitude is

$$H(j100) = \frac{|1 + j100|}{|-99 + j8|} = 1.007.$$

The phase-angle plot is shown in Fig. 17.38. The straight-line approximation of the angle associated with the quadratic factor intercepts the $0°$ axis at $10[4.81^{-0.4}]$, or 18.74 rad/s. (Careful study shows that the straight-line segment of $\theta(\omega)$ between 5.34 and 10 rad/s does not have the same slope as the segment between 10 and 18.74 rad/s.)

f) $\theta(\omega)_{max} \cong 70°$ at $\omega = 5.34$ rad/s.

g) We have $\theta(5.34) = 79.38 - 30.82 = 48.56°$. Note that the large error in the predicted angle is due to the fact that it occurs close to the intercept at u_1. Phase-angle estimates from straight-line plots are, in general, not very satisfactory. The straight-line plot does show the general behavior of the phase-angle variation. In this case, it predicts that the phase angle starts from $0°$, rises to a maximum value, and eventually approaches $-90°$.   ∎

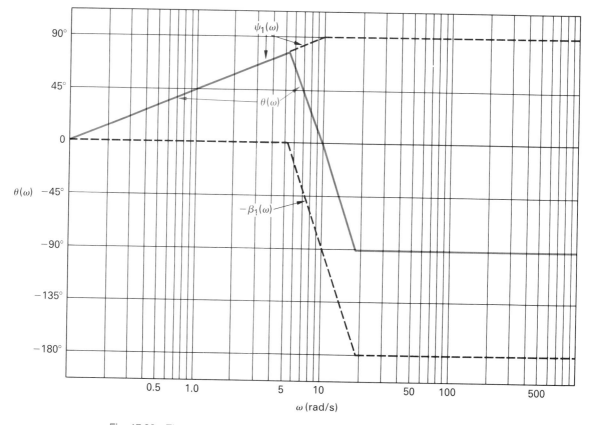

Fig. 17.38   The phase-angle plot for Example 17.8.

DRILL EXERCISE
17.10

The numerical expression for a current transfer function is

$$H(s) = \frac{I_o}{I_i} = \frac{25 \times 10^8}{s^2 + 20{,}000s + 25 \times 10^8}.$$

Compute the following:

a) the corner frequency;

b) the damping coefficient;

c) the frequencies when $H(j\omega)$ is unity;

d) the peak amplitude of $H(j\omega)$ in decibels;

e) the frequency at which the peak occurs;

f) the amplitude of $H(j\omega)$ at one half the corner frequency.

Ans.   (a) 50 krad/s; (b) 0.2; (c) 0, 67.82 krad/s; (d) 8.14 dB; (e) 47.96 krad/s; (f) 2.20 dB.

DRILL EXERCISE
17.11

The numerical expression for a voltage transfer function is

$$H(s) = \frac{V_o}{V_g} = \frac{32 \times 10^5}{s^2 + 400s + 64 \times 10^4}.$$

a) Use a straight-line amplitude plot to find the frequency where the amplitude of $H(j\omega)$ equals unity.
b) What is the actual amplitude at the frequency found in part (a)?
c) What is the peak amplitude of $H(j\omega)$ in decibels?
d) At what frequency does the output voltage reach its peak amplitude?
e) If the amplitude of the source voltage is 10 V, what is the peak amplitude of the output voltage in V?

Ans. (a) 1800 rad/s; (b) 1.19; (c) 20.28 dB; (d) 748.33 rad/s; (e) 103.28 V.

17.9 THE DECIBEL

The amplitude of $H(j\omega)$ in decibels was defined in Eq. (17.57). However, the original definition of the decibel involves power ratios. It is worth tracing back to this definition because it is still a widely accepted use of the term. The decibel was introduced by telephone engineers who were concerned with the power loss across cascaded circuits used to transmit telephone signals. The problem is defined in terms of Fig. 17.39, where p_i is the power input to the system, p_1 is the power output of circuit A, p_2 is the power output of circuit B, and p_o is the power output of the system. The power gain of each circuit is defined as the ratio of the power out to the power in. Thus

$$\sigma_A = \frac{p_1}{p_i}, \qquad \sigma_B = \frac{p_2}{p_1}, \quad \text{and} \quad \sigma_C = \frac{p_o}{p_2}.$$

The overall power gain of the system is simply the product of the individual gains; thus

$$\frac{p_o}{p_i} = \frac{p_1}{p_i} \cdot \frac{p_2}{p_1} \cdot \frac{p_o}{p_2} = \sigma_A\sigma_B\sigma_C. \tag{17.83}$$

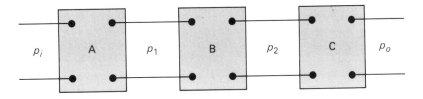

Fig. 17.39 Three cascaded circuits.

The multiplication of power ratios is converted to addition via the logarithm, that is,

$$\log_{10} \frac{p_o}{p_i} = \log_{10} \sigma_A + \log_{10} \sigma_B + \log_{10} \sigma_C. \tag{17.84}$$

This log ratio of the powers was named the *bel* in honor of Alexander Graham Bell. Thus the overall power gain, in bels, can be calculated by simply summing the power gains, in bels, of each segment of the transmission system. It turned out in practice that the bel was an inconveniently large quantity and one tenth of a bel was found to be a more useful measure of power gain. The *decibel* is defined as being equal to one tenth of a bel. It follows, then, that the number of decibels is equal to ten times the number of bels, that is,

$$\text{Number of decibels} = 10 \log_{10} \frac{p_o}{p_i}. \tag{17.85}$$

When we use the decibel as a measure of power ratios, there are situations in which the resistance seen looking into the circuit is equal to the resistance loading the circuit. This is illustrated in Fig. 17.40. When the input resistance equals the load resistance, the power ratio can be converted to either a voltage ratio or a current ratio; thus

$$\frac{p_o}{p_i} = \frac{v_{out}^2/R_L}{v_{in}^2/R_{in}} = \left(\frac{v_{out}}{v_{in}}\right)^2 \tag{17.86}$$

or

$$\frac{p_o}{p_i} = \frac{i_{out}^2 R_L}{i_{in}^2 R_L} = \left(\frac{i_{out}}{i_{in}}\right)^2. \tag{17.87}$$

We see from Eqs. (17.86) and (17.87) that in this case the number of decibels becomes

$$\text{Number of decibels} = 20 \log_{10} \frac{v_o}{v_{in}} = 20 \log_{10} \frac{i_o}{i_{in}}. \tag{17.88}$$

The definition of the decibel used in Bode diagrams, that is, Eq. (17.57), is borrowed from the results expressed by Eq. (17.88). Equation (17.57) goes

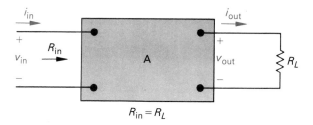

Fig. 17.40   A circuit in which the input resistance equals the load resistance.

TABLE 17.2
A TABLE OF dB-RATIO PAIRS

dB	Ratio	dB	Ratio
0	1.00	30	31.62
3	1.41	40	100.00
6	2.00	60	10^3
10	3.16	80	10^4
15	5.62	100	10^5
20	10.00	120	10^6

beyond Eq. (17.88) because it applies to any transfer function involving a voltage ratio, a current ratio, a voltage-to-current ratio, or a current-to-voltage ratio. We advise you to keep the original definition of the decibel firmly in mind because it is of fundamental importance in many engineering applications.

When working with transfer function amplitudes expressed in decibels, it is helpful to have a table that translates the decibel value to the actual value of the output/input ratio. Some useful pairs are given in Table 17.2. The ratio corresponding to a negative dB value is the reciprocal of the positive ratio. For example, -3 dB corresponds to an output/input ratio of $1/1.41$, or 0.707. It is worth noting that -3 dB corresponds to the half-power frequencies of the parallel- and series-resonant circuits discussed in Chapter 14.

17.10 SUMMARY

The transfer function is the ratio of the Laplace transform of the output to the Laplace transform of the input. In finding the ratio, we assume that all initial conditions are equal to zero. The ratio applies to a single-output/single-input circuit; hence for circuits driven by multiple inputs the transfer function is useful only if the principle of superposition is used to find the total response.

When we use the transfer function in a partial-fraction expansion of the response, it is evident that the poles of $H(s)$ govern the waveform of the transient component of the response. Therefore, if the excitation function is bounded, the response is bounded only if the poles of $H(s)$ lie in the left half of the s-plane.

The inverse transform of the transfer function is the impulse response of the circuit. We can use the impulse response with the excitation function to find the response of the circuit through the convolution integral. The convolution integral allows the analysis to be done entirely in the time domain. The convolution process also leads to the concepts of memory and system-weighting function. By viewing the impulse response as a weighting function, the analyst can get some feel for how much distortion the signal will undergo as it is transmitted through the circuit. The rule of thumb is that the shorter the memory of the circuit, the less distortion.

Finally, we found the transfer function very useful in predicting the steady-state sinusoidal response of the circuit. For circuits where the poles and zeros of the transfer function are well separated, we can obtain a good feel for the frequency response of the circuit by making straight-line Bode diagrams of the magnitude and phase of $H(j\omega)$. These straight-line plots can also highlight the range of frequencies of particular interest and therefore can be used to guide the generation of detailed computer plots.

PROBLEMS

17.1 The switch in the circuit shown in Fig. 17.41 moves from position a to position b at $t = 0$. Use the principle of superposition to find $v_o(t)$ for $t > 0$.

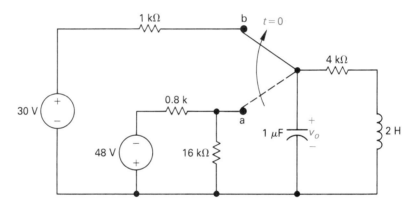

Fig. 17.41 The circuit for Problem 17.1.

17.2 Verify that the solution of Eqs. (17.19) and (17.20) for V_2 yields the same expression as that given by Eq. (17.18).

17.3 a) Find the numerical expression for the transfer function $[V_o/V_i]$ of each circuit in Fig. 17.42.
   b) Give the numerical value of the poles and zeros of each transfer function.

17.4 The voltage source in the circuit in Example 17.1 is changed to a unit impulse, that is, $v_g = \delta(t)$.

   a) How much energy does the impulsive voltage source store in the capacitor?
   b) How much energy does it store in the inductor?
   c) Use the transfer function to find $v_o(t)$.
   d) Show that the response found in part (c) is identical to the response generated by first charging the capacitor to 1000 V and then releasing the charge to the circuit, as shown in Fig. 17.43.

17.5 a) Find the numerical expression for the transfer function $H(s) = V_o/V_i$ for the circuit in Fig. 17.44.
   b) Give the numerical value of each pole and zero of $H(s)$.

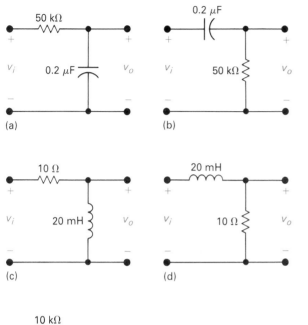

(a)

(b)

(c)

(d)

(e)

Fig. 17.42 The circuits for Problem 17.3.

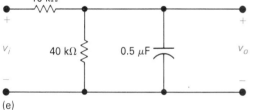

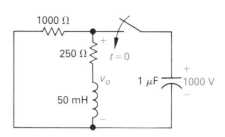

Fig. 17.43 The circuit for Problem 17.4(d).

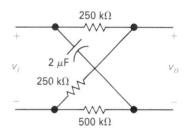

Fig. 17.44 The circuit for Problem 17.5.

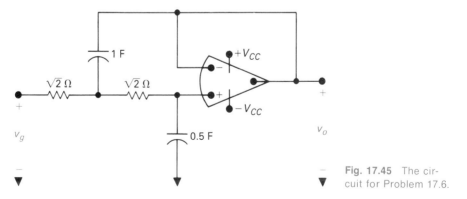

Fig. 17.45 The circuit for Problem 17.6.

17.6 a) Derive the numerical expression of the transfer function $H(s) = V_o/V_g$ for the circuit in Fig. 17.45.

b) Give the numerical value of each pole and zero of $H(s)$

17.7 a) Find $h(t) * x(t)$ when $h(t)$ and $x(t)$ are the rectangular pulses shown in Fig. 17.46(a).

b) Repeat part (a) when $x(t)$ changes to the rectangular pulse shown in Fig. 17.46(b).

c) Repeat part (a) when $h(t)$ changes to the rectangular pulse shown in Fig. 17.46(c).

17.8 Use the convolution integral to evaluate $y(t)$ at 3, 6, 9, 12, 15, and 18 s if

$$y(t) = h(t) * x(t) = x(t) * h(t)$$

and $h(t)$ and $x(t)$ have the waveforms shown in Fig. 17.47.

17.9 a) Use the convolution integral to find $y(t)$ if $h(t)$ and $x(t)$ have the waveforms shown in Fig. 17.48.

b) Sketch the waveform of $y(t)$ for $0 \le t \le 50$ s.

17.10 a) Use the convolution integral to find $y(t)$ if $h(t)$ and $x(t)$ have the waveforms shown in Fig. 17.49.

b) Sketch the waveform of $y(t)$ for $-5 \le t \le 20$ s.

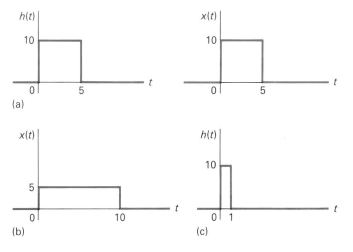

Fig. 17.46 The functions for Problem 17.7.

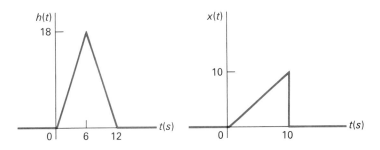

Fig. 17.47 The waveforms for Problem 17.8.

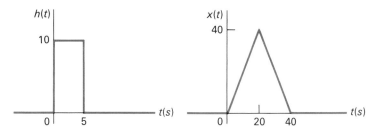

Fig. 17.48   The waveforms for Problem 17.9.

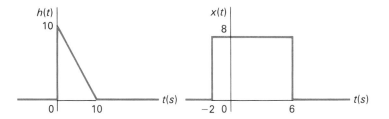

Fig. 17.49   The waveforms for Problem 17.10.

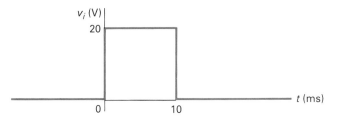

Fig. 17.50   The rectangular voltage pulse for Problem 17.11.

17.11   a) Use the convolution integral to find the output voltage of the circuit in Fig. 17.42(a) if the input voltage is the rectangular pulse shown in Fig. 17.50.

    b) Sketch $v_o(t)$ vs. t for the time interval $0 \le t \le 20$ ms.

17.12   a) Repeat Problem 17.11 given that the resistor in the circuit in Fig. 17.42(a) is reduced to 5 kΩ.

    b) Does decreasing the resistor increase or decrease the "memory" of the circuit?

    c) Which circuit comes closest to transmitting a replica of the input voltage?

17.13   Use the convolution integral to find the inverse transform of the following functions:

    a) $F(s) = \dfrac{a}{s(s + a)}$;

    b) $F(s) = \dfrac{\omega}{(s + a)(s^2 + \omega^2)}$.

17.14 The transfer function for a linear time-invariant circuit is

$$H(s) = \frac{V_o}{I_i} = \frac{625(s + 100)}{s^2 + 100s + 12,500}.$$

If $i_i(t) = 4 \cos (50t + 60°)$, what is the steady-state expression for $v_o(t)$?

17.15 The operational amplifier in the circuit in Fig. 17.51 is ideal and operating within its linear region.

    a) Calculate the transfer function V_o/V_s.
    b) If $v_s = 10 \cos 200t$ V, what is the steady-state expression for v_o?

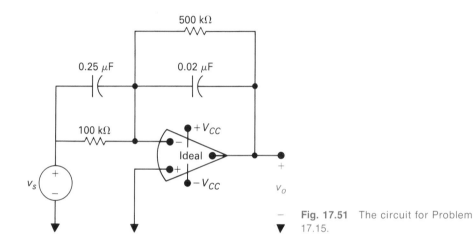

Fig. 17.51 The circuit for Problem 17.15.

17.16 The unit step response of a circuit is known to be

$$v_o = 50e^{-30,000t} \sin 40,000t \text{ V}.$$

    a) What is the transfer function of the circuit?
    b) What is the impulse response of the circuit?
    c) If the input sinusoidal voltage source is

$$v_g = 5 \cos (40,000t + 9°) \text{ mV},$$

what is the steady-state expression for v_o?

17.17 Make straight-line (uncorrected) amplitude and phase-angle plots for each of the transfer functions derived in Problem 17.3.

17.18 Make straight-line amplitude and phase-angle plots for the voltage transfer function derived in Problem 17.5.

17.19 a) Derive the numerical expression of the transfer function V_o/I_g for the circuit in Fig. 17.52.
    b) Make a corrected amplitude plot for the transfer function derived in part (a).
    c) At what frequency is the amplitude maximum?
    d) What is the maximum amplitude in dB?
    e) At what frequencies is the amplitude down 3 dB from the maximum?
    f) What is the bandwidth of the circuit?
    g) Check your graphical results by calculating the actual amplitude in dB at the frequencies read from the plot.

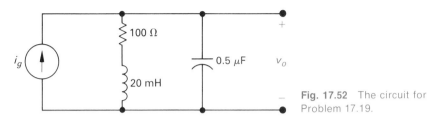

Fig. 17.52 The circuit for Problem 17.19.

17.20   Use Bode diagrams to describe the behavior of the circuit in Problem 10.32 as R_x is varied from zero to infinity.

17.21   Given the following current transfer function:

$$H(s) = \frac{I_o}{I_i} = \frac{10^6}{s^2 + 200s + 10^6}.$$

    a) At what frequencies (rad/s) is the ratio of I_o/I_i equal to unity?
    b) At what frequency is the ratio maximum?
    c) What is the maximum value of the ratio?

17.22   The circuit shown in Fig. 17.53 resembles the interstage coupling network of an amplifier.

    a) Show that

$$H(s) = \frac{V_o}{V_i} = \frac{(1/R_1C_2)s}{s^2 + [(1/R_1C_1) + (1/R_2C_2) + (1/R_1C_2)]s + (1/R_1C_1R_2C_2)}.$$

    b) Find the numerical expression for $H(s)$ if $R_1 = 40\,k\Omega$, $C_1 = 0.1\,\mu F$, $R_2 = 10\,k\Omega$, and $C_2 = 250\,pF$.
    c) Give the numerical values of the poles and zeros of $H(s)$.
    d) Give an approximate numerical expression for $H(s)$ for values of s much less than the highest corner frequency, that is, for values of s close to the lowest corner frequency.
    e) Give an approximate numerical expression for $H(s)$ for values of s much larger than the lowest corner frequency.
    f) Show that the expression derived in part (d) is equivalent to the transfer function of the circuit in Fig. 17.53 when C_2 is neglected at low frequencies.
    g) Show that the expression derived in part (e) is equivalent to the transfer function of the circuit in Fig. 17.53 when C_1 is neglected at high frequencies.

This problem illustrates that when the numerical values of the circuit parameters are known, it is sometimes possible to use different circuit models in different frequency ranges. Quite frequently electronic-amplifier equivalent circuits can be divided into models that apply to low-, mid, and high-frequency ranges.

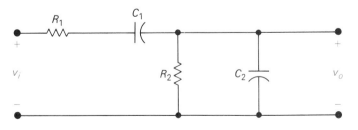

Fig. 17.53 The circuit for Problem 17.22.

17.23 a) Find the resistance seen looking into the terminals a, b of the circuit in Fig. 17.54.

b) Find the power loss through the network, in decibels, when the output power is the power delivered to the 80-Ω resistor.

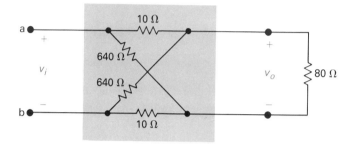

Fig. 17.54 The circuit for Problem 17.23.

17.24 The amplitude plot of a transfer function is shown in Fig. 17.55. What is the numerical expression for $H(s)$?

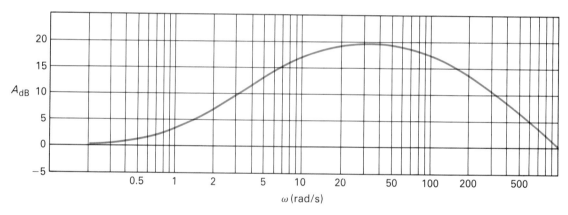

Fig. 17.55 The amplitude plot for Problem 17.24.

FOURIER SERIES

18

18.1 INTRODUCTION

We have devoted a considerable amount of time in the preceding chapters to a discussion of steady-state sinusoidal analysis. One reason for this interest in the sinusoidal excitation function is that it enables us to find the steady-state response to nonsinusoidal, but periodic, excitations. A periodic function is a function that repeats itself every T seconds. For example, the triangular wave illustrated in Fig. 18.1 is a nonsinusoidal, but periodic, waveform.

A periodic function is defined as one that satisfies the relationship

$$f(t) = f(t \pm nT), \tag{18.1}$$

where n is an integer (1, 2, 3, . . .) and T is the period. The period is the smallest interval of time required for $f(t)$ to pass through all its values. The function shown in Fig. 18.1 is periodic because

$$f(t_0) = f(t_0 - T) = f(t_0 + T) = f(t_0 + 2T) = \cdots$$

for any arbitrarily chosen value of t_0.

Why are we interested in periodic functions? One reason is that many electrical sources of practical value generate periodic waveforms. For example, nonfiltered electronic rectifiers driven from a sinusoidal source produce rectified sine waves that are nonsinusoidal but periodic. The waveforms of the full-wave and half-wave sinusoidal rectifiers are shown in Fig. 18.2(a) and (b), respectively.

The sweep generator used to control the electron beam of a cathode-ray oscilloscope produces a periodic triangular wave like that shown in Fig. 18.3.

Electronic oscillators, which are useful in laboratory testing of equipment, are designed to produce nonsinusoidal periodic waveforms. Function generators, which are capable of producing square-wave, triangular-wave, and rectangular-pulse waveforms, are found in most testing laboratories. Typical waveforms are illustrated in Fig. 18.4.

Another practical problem that stimulates our interest in periodic functions is the fact that power generators, although designed to produce a sinusoidal waveform, cannot in practice be made to produce a pure sine wave. The

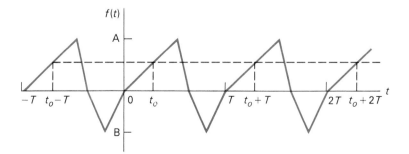

Fig. 18.1 A periodic waveform.

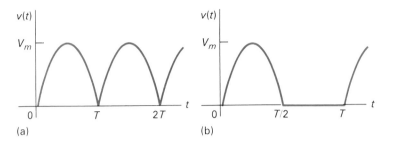

Fig. 18.2 Output waveforms of a nonfiltered sinusoidal rectifier: (a) full-wave rectification; (b) half-wave rectification.

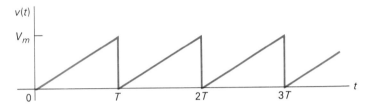

Fig. 18.3 The triangular waveform of a cathode-ray oscilloscope sweep generator.

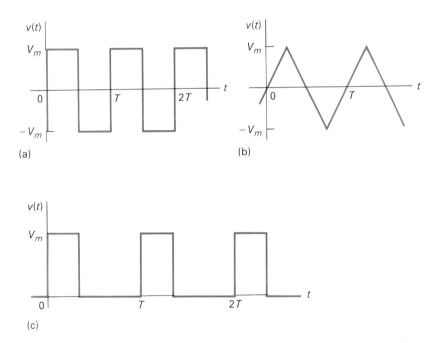

Fig. 18.4 Waveforms produced by function generators used in laboratory testing: (a) square wave; (b) triangular wave; (c) rectangular-pulse waveform.

distorted sinusoidal wave is, however, periodic. Power engineers are naturally interested in ascertaining the consequences of driving their systems with a slightly distorted sinusoidal voltage.

Our interest in periodic functions also stems from the general observation that any nonlinearity in an otherwise linear circuit creates a nonsinusoidal periodic function. The rectifier circuit alluded to earlier is one example of this phenomenon. Magnetic saturation, which occurs in both machines and transformers, is another example of a nonlinearity that generates a nonsinusoidal periodic function. An electronic clipping circuit, which uses transistor saturation, is another illustration of using nonlinearity to produce a periodic function.

We should also point out that nonsinusoidal periodic functions are important in the analysis of nonelectrical systems. Problems involving mechanical vibrations, fluid flow, and heat flow all make use of periodic functions. In fact, it was the study and analysis of heat flow in a metal rod that led the French mathematician Jean Baptiste Joseph Fourier (1768–1830) to the trigonometric series representation of a periodic function. Today this series bears his name and is our starting point for finding the steady-state response to periodic excitations of electrical circuits.

18.2   FOURIER SERIES ANALYSIS: AN OVERVIEW

What Fourier discovered in his investigations of heat flow problems is that a periodic function can be represented by an infinite sum of sine or cosine functions that are harmonically related. That is, the period of any trigonometric term in the infinite series is an integral multiple, or harmonic, of the fundamental period (T) of the periodic function. Thus given that $f(t)$ is periodic, Fourier was able to show that $f(t)$ can be expressed as

$$f(t) = a_v + \sum_{n=1}^{\infty} a_n \cos n\omega_0 t + b_n \sin n\omega_0 t, \qquad \textbf{(18.2)}$$

where n is the integer sequence 1, 2, 3, ...

In Eq. (18.2), a_v, a_n, and b_n are known as the *Fourier coefficients* and are calculated from $f(t)$. The term $\omega_0(2\pi/T)$ represents the fundamental frequency of the periodic function $f(t)$. The integral multiples of ω_0—that is, $2\omega_0$, $3\omega_0$, $4\omega_0$, and so on—are known as the harmonic frequencies of $f(t)$. Thus $2\omega_0$ is the second harmonic, $3\omega_0$ is the third harmonic, and $n\omega_0$ is the nth harmonic of $f(t)$.

We will discuss the determination of the Fourier coefficients in Section 18.3. Before pursuing the analytical details of using a Fourier series in circuit analysis, it will be helpful to first look at the problem in general terms. Our first general observation is that from an application point of view all the periodic functions of interest can be expressed in terms of a Fourier series. Mathematically, the sufficient conditions on a periodic function $f(t)$ that ensures expressing $f(t)$ as a convergent Fourier series are known as *Dirichlet's condi-*

tions. These conditions are:

1. $f(t)$ is single-valued;
2. $f(t)$ has a finite number of discontinuities in the periodic interval;
3. $f(t)$ has a finite number of maxima and minima in the periodic interval; and
4. the integral $\int_{t_0}^{t_0+T} |f(t)| \, dt$ exists.

Any periodic function that is generated by a physically realizable source will satisfy Dirichlet's conditions. It is worth noting that these are *sufficient* conditions and not *necessary* conditions. Thus if $f(t)$ meets these requirements, we know it can be expressed as a Fourier series. However, if $f(t)$ does not meet these requirements, it may still be expressible as a Fourier series. We should point out that the *necessary* conditions on $f(t)$ are not known.

A second observation to make is that once $f(t)$ is known and the Fourier coefficients (a_v, a_n, b_n) have been calculated, the periodic source is resolved into a dc source (a_v) plus a sum of sinusoidal sources (a_n, b_n). Since the periodic source is driving a linear circuit, we can use the principle of superposition to find the steady-state response. In particular, we first calculate the response to each source generated by the Fourier series representation of $f(t)$ and then add the individual responses to find the total response. The steady-state response due to a specific sinusoidal source is most easily found using the phasor method of analysis.

The third general observation to be made at this time is that the procedure for finding the steady-state response to a periodic excitation is straightforward and involves no new techniques of circuit analysis. At the same time it is important to note that this procedure produces the Fourier series representation of the steady-state response; consequently the actual shape of the response is not known. Furthermore, the response waveform can be estimated only by adding a sufficient number of terms together. Even though the Fourier series approach to finding the steady-state response does have some drawbacks, it is important to keep in mind that it introduces a way of thinking about a problem that is as important as introducing a way of getting quantitative results. In fact, the conceptual picture is even more important in some respects than the quantitative one.

18.3 THE FOURIER COEFFICIENTS

Once a periodic function has been defined over its fundamental period, the Fourier coefficients can be found from the following relationships:

$$a_v = \frac{1}{T} \int_{t_0}^{t_0+T} f(t) \, dt, \tag{18.3}$$

$$a_k = \frac{2}{T} \int_{t_0}^{t_0+T} f(t) \cos k\omega_0 t \, dt, \tag{18.4}$$

and

$$b_k = \frac{2}{T} \int_{t_0}^{t_0+T} f(t) \sin k\omega_0 t \, dt. \tag{18.5}$$

In Eqs. (18.4) and (18.5) the subscript k is used to indicate the kth coefficient in the integer sequence 1, 2, 3, ... Note that a_v is the average value of $f(t)$, a_k is twice the average value of $f(t) \cos k\omega_0 t$, and b_k is twice the average value of $f(t) \sin k\omega_0 t$.

We can easily derive Eqs. (18.3), (18.4), and (18.5) from Eq. (18.2) if we remember the following integral relationships, which hold when m and n are integers:

$$\int_{t_0}^{t_0+T} \sin m\omega_0 t \, dt = 0 \quad \text{for all } m, \tag{18.6}$$

$$\int_{t_0}^{t_0+T} \cos m\omega_0 t \, dt = 0 \quad \text{for all } m, \tag{18.7}$$

$$\int_{t_0}^{t_0+T} \cos m\omega_0 t \sin n\omega_0 t \, dt = 0 \quad \text{for all } m, n, \tag{18.8}$$

$$\int_{t_0}^{t_0+T} \sin m\omega_0 t \sin n\omega_0 t \, dt = 0 \qquad \text{for } m \neq n,$$
$$= T/2 \quad \text{for } m = n \tag{18.9}$$

and

$$\int_{t_0}^{t_0+T} \cos m\omega_0 t \cos n\omega_0 t \, dt = 0 \qquad \text{for } m \neq n,$$
$$= T/2 \quad \text{for } m = n. \tag{18.10}$$

We leave the verification of Eqs. (18.6)–(18.10) for the reader to determine in Problem 18.1.

With Eqs. (18.6)–(18.10) in mind, we derive Eqs. (18.3), (18.4), and (18.5) as follows. To derive Eq. (18.3), we simply integrate both sides of Eq. (18.2) over one period; thus

$$\int_{t_0}^{t_0+T} f(t) \, dt = \int_{t_0}^{t_0+T} \left[a_v + \sum_{n=1}^{\infty} a_n \cos n\omega_0 t + b_n \sin n\omega_0 t \right] dt$$

$$= \int_{t_0}^{t_0+T} a_v \, dt + \sum_{n=1}^{\infty} \int_{t_0}^{t_0+T} [a_n \cos n\omega_0 t + b_n \sin n\omega_0 t] \, dt$$

$$= a_v T + 0. \tag{18.11}$$

Equation (18.3) follows directly from Eq. (18.11).

To derive the expression for the kth value of a_n, we must first multiply Eq. (18.2) by $\cos k\omega_0 t$ and then integrate both sides over one period of $f(t)$.

Thus

$$\int_{t_0}^{t_0+T} f(t) \cos k\omega_0 t \, dt = \int_{t_0}^{t_0+T} a_v \cos k\omega_0 t \, dt + \sum_{n=1}^{\infty} \int_{t_0}^{t_0+T} [a_n \cos n\omega_0 t \cos k\omega_0 t$$
$$+ \, b_n \sin n\omega_0 t \cos k\omega_0 t] \, dt$$
$$= 0 + a_k(T/2) + 0. \tag{18.12}$$

Solving Eq. (18.12) for a_k yields the expression given by Eq. (18.4).

We can obtain the expression for the kth value of b_n by first multiplying both sides of Eq. (18.2) by $\sin k\omega_0 t$ and then integrating each side over one period of $f(t)$.

Example 18.1 illustrates how Eqs. (18.3), (18.4), and (18.5) are used to find the Fourier coefficients for a given periodic function.

Example 18.1 Find the Fourier series for the periodic voltage shown in Fig. 18.5.

Solution When using Eqs. (18.3), (18.4), and (18.5) to find a_v, a_k, and b_k, we are at liberty to choose the value of t_0. For the periodic voltage of Fig. 18.5, the best choice for t_0 is zero. Any other choice will make the required integrations more cumbersome. The expression for $v(t)$ between 0 and T is

$$v(t) = \left(\frac{V_m}{T}\right) t.$$

The equation for a_v is

$$a_v = \frac{1}{T} \int_0^T \left(\frac{V_m}{T}\right) t \, dt = \frac{1}{2} V_m.$$

The equation for the kth value of a_n is

$$a_k = \frac{2}{T} \int_0^T \left(\frac{V_m}{T}\right) t \cos k\omega_0 t \, dt$$
$$= \frac{2V_m}{T^2} \left\{ \frac{1}{k^2\omega_0^2} \cos k\omega_0 t + \frac{t}{k\omega_0} \sin k\omega_0 t \Big|_0^T \right\}$$
$$= \frac{2V_m}{T^2} \left\{ \frac{1}{k^2\omega_0^2} (\cos 2\pi k - 1) \right\} = 0 \quad \text{for all } k.$$

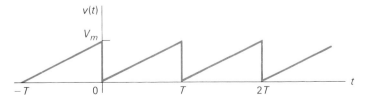

Fig. 18.5 The periodic voltage for Example 18.1.

The equation for the kth value of b_n is

$$b_k = \frac{2}{T} \int_0^T \left(\frac{V_m}{T}\right) t \, \sin k\omega_0 t \, dt$$

$$= \frac{2V_m}{T^2} \left\{ \frac{1}{k^2\omega_0^2} \sin k\omega_0 t - \frac{t}{k\omega_0} \cos k\omega_0 t \Big|_0^T \right\}$$

$$= \frac{2V_m}{T^2} \left\{ 0 - \frac{T}{k\omega_0} \cos 2\pi k \right\}$$

$$= \frac{-V_m}{\pi k}.$$

The Fourier series for $v(t)$ is

$$v(t) = \frac{V_m}{2} - \frac{V_m}{\pi} \sum_{n=1}^{\infty} \frac{1}{n} \sin n\omega_0 t$$

$$= \frac{V_m}{2} - \frac{V_m}{\pi} \sin \omega_0 t - \frac{V_m}{2\pi} \sin 2\omega_0 t - \frac{V_m}{3\pi} \sin 3\omega_0 t - \cdots. \quad \blacksquare$$

Finding the Fourier coefficients is, in general, a tedious chore. Therefore anything we can do to simplify the task is beneficial. It turns out that if a periodic function possesses certain types of symmetry, the amount of work involved in finding the coefficients is greatly reduced. In Section 18.4 we discuss how symmetry affects the coefficients in a Fourier series.

DRILL EXERCISE
18.1

Derive the expressions for a_v, a_k, and b_k for the periodic voltage function shown if $V_m = 60\pi$ V.

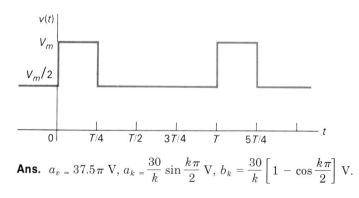

Ans. $a_v = 37.5\pi$ V, $a_k = \dfrac{30}{k} \sin \dfrac{k\pi}{2}$ V, $b_k = \dfrac{30}{k} \left[1 - \cos \dfrac{k\pi}{2} \right]$ V.

DRILL EXERCISE
18.2

Refer to Drill Exercise 18.1.

a) What is the average value of the periodic voltage?

b) Compute the numerical values of a_1 through a_5 and b_1 through b_5.

c) If $T = 628.32$ ms, what is the fundamental frequency in rad/s?

d) What is the frequency of the fifth harmonic in Hz?

e) Write the Fourier series up to and including the fifth harmonic.

Ans. (a) 117.81 V; (b) 30 V, 0 V, − 10 V, 0 V, and 6 V; 30 V, 30 V, 10 V, 0 V, and 6 V; (c) 10 rad/s; (d) 7.96 Hz; (e) $v(t) = 117.81 + 30 \cos 10t + 30 \sin 10t + 30 \sin 20t - 10 \cos 30t + 10 \sin 30t + 6 \cos 50t + 6 \sin 50t$.

18.4   THE EFFECT OF SYMMETRY ON THE FOURIER COEFFICIENTS

There are four types of symmetry that can be used to simplify the task of evaluating the Fourier coefficients:

1. even-function symmetry;

2. odd-function symmetry;

3. half-wave symmetry;

4. quarter-wave symmetry.

The effect of each type of symmetry on the Fourier coefficients is discussed below.

18.4(a) Even-function Symmetry

A function is defined as even if

$$f(t) = f(-t). \tag{18.13}$$

Functions that satisfy Eq. (18.13) are said to be even because polynomial functions with only even exponents possess this characteristic. For even periodic functions, the equations for the Fourier coefficients reduce to

$$a_v = \frac{2}{T} \int_0^{T/2} f(t) \, dt, \tag{18.14}$$

$$a_k = \frac{4}{T} \int_0^{T/2} f(t) \cos k\omega_0 t \, dt, \tag{18.15}$$

and

$$b_k = 0 \quad \text{for all } k. \tag{18.16}$$

Note that all the b-coefficients are zero if the periodic function is even. An even periodic function is illustrated in Fig. 18.6. The derivations of Eqs. (18.14), (18.15), and (18.16) follow directly from Eqs. (18.3), (18.4), and (18.5). In each derivation, we select t_0 to equal $- T/2$ and then break the interval of integration into the range from $- T/2$ to 0 and 0 to $T/2$. Thus

$$a_v = \frac{1}{T} \int_{-T/2}^{T/2} f(t) \, dt$$

$$= \frac{1}{T} \int_{-T/2}^0 f(t) \, dt + \frac{1}{T} \int_0^{T/2} f(t) \, dt. \tag{18.17}$$

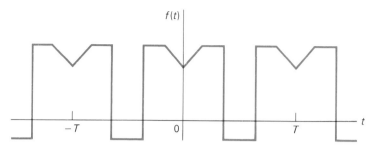

Fig. 18.6 An even periodic function $f(t) = f(-t)$.

Now we change the variable of integration in the first integral on the right-hand side of Eq. (18.17). Specifically, we let $t = -x$ and note that $f(t) = f(-x) = f(x)$ because the function is even. We also observe that $x = T/2$ when $t = -T/2$ and $dt = -dx$. It follows that

$$\int_{-T/2}^{0} f(t)\ dt = \int_{T/2}^{0} f(x)(-dx) = \int_{0}^{T/2} f(x)\ dx, \tag{18.18}$$

which tells us that the integration from $-T/2$ to 0 is identical to the integration from 0 to $T/2$; therefore Eq. (18.17) is the same as Eq. (18.14). The derivation of Eq. (18.15) proceeds along similar lines. Here we have

$$a_k = \frac{2}{T}\int_{-T/2}^{0} f(t)\cos k\omega_0 t\ dt + \frac{2}{T}\int_{0}^{T/2} f(t)\cos k\omega_0 t\ dt. \tag{18.19}$$

But

$$\int_{-T/2}^{0} f(t)\cos k\omega_0 t\ dt = \int_{T/2}^{0} f(x)\cos(-k\omega_0 x)(-dx)$$
$$= \int_{0}^{T/2} f(x)\cos k\omega_0 x\ dx. \tag{18.20}$$

As before we see that the integration from $-T/2$ to 0 is identical to that from 0 to $T/2$. When we combine Eq. (18.20) with Eq. (18.19) the result is Eq. (18.15).

All the b-coefficients are zero when $f(t)$ is an even periodic function because the integration from $-T/2$ to 0 is the exact negative of the integration from 0 to $T/2$. That is,

$$\int_{-T/2}^{0} f(t)\sin k\omega_0 t\ dt = \int_{T/2}^{0} f(x)\sin(-k\omega_0 x)(-dx)$$
$$= -\int_{0}^{T/2} f(x)\sin k\omega_0 x\ dx. \tag{18.21}$$

It is important to note that when we use Eqs. (18.14) and (18.15) to find the Fourier coefficients, the interval of integration must be between 0 and $T/2$.

18.4(b) Odd-function Symmetry   A function is defined as odd if

$$f(t) = -f(-t). \qquad (18.22)$$

Functions that satisfy Eq. (18.22) are said to be odd because polynomial functions with only odd exponents have this characteristic. For odd periodic functions, the Fourier coefficients are given by the equations

$$a_v = 0, \qquad (18.23)$$

$$a_k = 0 \quad \text{for all } k, \qquad (18.24)$$

$$b_k = \frac{4}{T} \int_0^{T/2} f(t) \sin k\omega_0 t \, dt. \qquad (18.25)$$

Note that all the a-coefficients are zero if the periodic function is odd. An odd periodic function is shown in Fig. 18.7.

The derivations of Eqs. (18.23), (18.24), and (18.25) follow the same thought process that was used in deriving Eqs. (18.14), (18.15), and (18.16), and consequently are left in Problem 18.4 to the reader.

The evenness, or oddness, of a periodic function can be destroyed by shifting the function along the time axis. In other words, it may be possible by the judicious choice of where $t = 0$ to give a periodic function even or odd symmetry. For example, the triangular function shown in Fig. 18.8(a) is neither even nor odd. However, the function can be made even as shown in Fig. 18.8(b) or odd as shown in Fig. 18.8(c).

18.4(c) Half-wave Symmetry   A periodic function possesses half-wave symmetry if it satisfies the constraint

$$f(t) = -f(t - T/2). \qquad (18.26)$$

Equation (18.26) tells us that a periodic function has half-wave symmetry if after it is shifted one-half period and inverted it is identical to the original

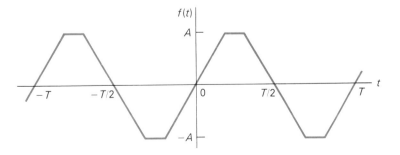

Fig. 18.7   An odd periodic function $f(t) = -f(-t)$.

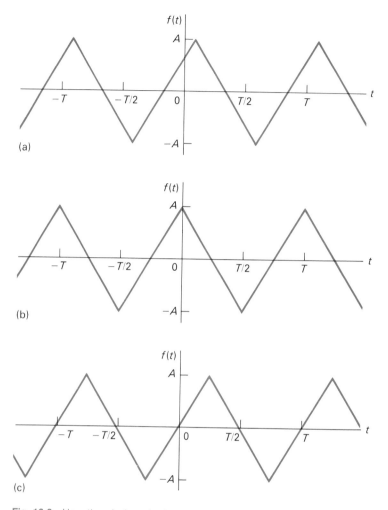

Fig. 18.8   How the choice of where $t = 0$ can make a periodic function even, odd, or neither even nor odd: (a) a periodic triangular wave that is neither even nor odd; (b) the triangular wave of (a) made even by shifting the function along the t-axis; (c) the triangular wave of (a) made odd by shifting the function along the t-axis.

function. For example, the functions shown in Figs. 18.7 and 18.8 have half-wave symmetry whereas those shown in Figs. 18.5 and 18.6 do not. Note that half-wave symmetry is not a function of where $t = 0$.

   If a periodic function has half-wave symmetry, both a_k and b_k are zero for even values of k. We know that a_v is also zero since the average value of a function with half-wave symmetry is zero. The expressions for the Fourier coefficients are

$$a_v = 0, \tag{18.27}$$

$$a_k = 0 \quad \text{for } k \text{ even}, \tag{18.28}$$

$$a_k = \frac{4}{T} \int_0^{T/2} f(t) \cos k\omega_0 t \, dt \quad \text{for } k \text{ odd}, \tag{18.29}$$

$$b_k = 0 \quad \text{for } k \text{ even}, \tag{18.30}$$

$$b_k = \frac{4}{T} \int_0^{T/2} f(t) \sin k\omega_0 t \, dt \quad \text{for } k \text{ odd}. \tag{18.31}$$

The results given by Eqs. (18.27)–(18.31) are derived by starting with Eqs. (18.3), (18.4), and (18.5) and choosing the interval of integration to be from $-T/2$ to $T/2$. This range is then divided into the intervals $-T/2$ to 0 and 0 to $T/2$. For example, the derivation for a_k is as follows:

$$a_k = \frac{2}{T} \int_{t_0}^{t_0+T} f(t) \cos k\omega_0 t \, dt$$

$$= \frac{2}{T} \int_{-T/2}^{T/2} f(t) \cos k\omega_0 t \, dt$$

$$= \frac{2}{T} \int_{-T/2}^{0} f(t) \cos k\omega_0 t \, dt + \frac{2}{T} \int_{0}^{T/2} f(t) \cos k\omega_0 t \, dt. \tag{18.32}$$

Now we make a change in variable in the first integral on the right-hand side of Eq. (18.32). Specifically, we let

$$t = x - T/2;$$

then

$$x = T/2, \quad \text{when } t = 0;$$
$$x = 0, \quad \text{when } t = -T/2;$$

and

$$dt = dx.$$

Therefore, we can rewrite the first integral as

$$\int_{-T/2}^{0} f(t) \cos k\omega_0 t \, dt = \int_{0}^{T/2} f(x - T/2) \cos k\omega_0 (x - T/2) \, dx. \tag{18.33}$$

Now observe that

$$\cos k\omega_0 (x - T/2) = \cos (k\omega_0 x - k\pi) = \cos k\pi \cos k\omega_0 x$$

and, by hypothesis,

$$f(x - T/2) = -f(x).$$

Therefore, we can express Eq. (18.33) as

$$\int_{-T/2}^{0} f(t) \cos k\omega_0 t \, dt = \int_{0}^{T/2} [-f(x)] \cos k\pi \cos k\omega_0 x \, dx \qquad \textbf{(18.34)}$$

When we incorporate results of Eq. (18.34) into Eq. (18.32), we get

$$a_k = \frac{2}{T} (1 - \cos k\pi) \int_{0}^{T/2} f(t) \cos k\omega_0 t \, dt. \qquad \textbf{(18.35)}$$

Now observe that $\cos k\pi$ is 1 when k is even and -1 when k is odd. Therefore, Eq. (18.35) generates Eqs. (18.28) and (18.29).

We leave it to the reader to verify that this same thought process can be used to derive the results given by Eqs. (18.30) and (18.31) (see Problem 18.5).

We can summarize our observations about a periodic function that possesses half-wave symmetry by noting that the Fourier series representation of such a function has zero average, or dc, value and contains only odd harmonics.

18.4(d) Quarter-wave Symmetry The term "quarter-wave symmetry" is used to describe a periodic function that has half-wave symmetry and, in addition, has symmetry about the midpoint of the positive and negative half-cycles. The function illustrated in Fig. 18.9(a) has quarter-wave symmetry about the midpoint of the positive and negative half-cycles. The function in Fig. 18.9(b) does not have quarter-wave symmetry although it does have half-wave symmetry.

A periodic function that has quarter-wave symmetry can always be made either even or odd by the proper choice of the point where t equals zero. For example, the function shown in Fig. 18.9(a) is odd and can be made even by shifting the function $T/4$ units either right or left along the t-axis. On the other hand, the function in Fig. 18.9(b) can never be made either even or odd. To take advantage of quarter-wave symmetry in the calculation of the Fourier coefficients, the point where t equals zero must be chosen to make the function either even or odd.

If the function is made even, then

$$a_v = 0, \quad \text{because of half-wave symmetry;}$$

$$a_k = 0, \quad \text{for } k \text{ even, because of half-wave symmetry;}$$

$$a_k = \frac{8}{T} \int_{0}^{T/4} f(t) \cos k\omega_0 t \, dt, \quad \text{for } k \text{ odd;} \qquad \textbf{(18.36)}$$

$$b_k = 0, \quad \text{for all } k, \text{ because the function is even.}$$

Equations (18.36) result from the fact that the function has quarter-wave symmetry in addition to being even. Recall that quarter-wave symmetry is superimposed on top of half-wave symmetry and so we can eliminate a_v and a_k for k even. When we compare Eq. (18.36) with Eq. (18.29), we see that combining quarter-wave symmetry with evenness allows us to shorten the range of inte-

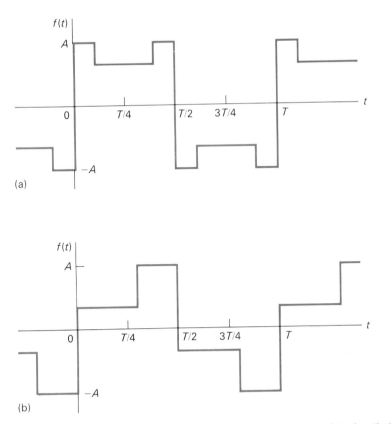

Fig. 18.9   (a) A function that has quarter-wave symmetry; (b) a function that does not have quarter-wave symmetry.

gration from 0 to $T/2$ to 0 to $T/4$. The derivation of Eq. (18.36) is left as an exercise in Problem 18.6.

If the quarter-wave symmetrical function is made odd, then we have

$$a_v = 0, \quad \text{because the function is odd;}$$
$$a_k = 0, \quad \text{for all } k, \text{ because the function is odd;}$$
$$b_k = 0, \quad \text{for } k \text{ even, because of half-wave symmetry;} \qquad \textbf{(18.37)}$$
$$b_k = \frac{8}{T} \int_0^{T/4} f(t) \sin k\omega_0 t \, dt, \quad \text{for } k \text{ odd.}$$

Equations (18.37) are a direct consequence of quarter-wave symmetry and oddness. Again quarter-wave symmetry allows us to shorten the interval of integration from 0 to $T/2$ to 0 to $T/4$. The derivation of Eq. (18.37) is left as an exercise in Problem 18.7.

Example 18.2 shows how symmetry can be used to simplify the task of finding the Fourier coefficients.

Example 18.2   Find the Fourier series representation for the current waveform shown in Fig. 18.10.

Solution   We begin by looking for degrees of symmetry in the waveform. We find that the function is odd and, in addition, has half-wave and quarter-wave symmetry. Because the function is odd, all the a-coefficients are zero, that is, $a_v = 0$ and $a_k = 0$ for all k. Because the function has half-wave symmetry, $b_k = 0$ for even values of k. Because the function has quarter-wave symmetry, the expression for b_k for odd values of k is

$$b_k = \frac{8}{T} \int_0^{T/4} i(t) \sin k\omega_0 t \, dt.$$

In the interval $0 \leq t \leq T/4$, the expression for $i(t)$ is

$$i(t) = \frac{4I_m}{T} t.$$

Thus

$$
\begin{aligned}
b_k &= \frac{8}{T} \int_0^{T/4} \frac{4I_m}{T} t \sin k\omega_0 t \, dt \\
&= \frac{32 I_m}{T^2} \left\{ \frac{\sin k\omega_0 t}{k^2 \omega_0^2} - \frac{t \cos k\omega_0 t}{k\omega_0} \Big|_0^{T/4} \right\} \\
&= \frac{8 I_m}{\pi^2 k^2} \sin \frac{k\pi}{2} \quad (k \text{ is odd}).
\end{aligned}
$$

The Fourier series representation of $i(t)$ is

$$
\begin{aligned}
i(t) &= \frac{8 I_m}{\pi^2} \sum_{n=1,3,5,\ldots}^{\infty} \frac{1}{k^2} \sin \frac{n\pi}{2} \sin n\omega_0 t \\
&= \frac{8 I_m}{\pi^2} \left\{ \sin \omega_0 t - \frac{1}{9} \sin 3\omega_0 t + \frac{1}{25} \sin 5\omega_0 t - \frac{1}{49} \sin 7\omega_0 t + \cdots \right\}. \quad \blacksquare
\end{aligned}
$$

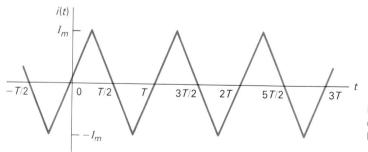

Fig. 18.10 The periodic waveform for Example 18.2.

DRILL EXERCISE
18.3 Derive the Fourier series for the following periodic voltage.

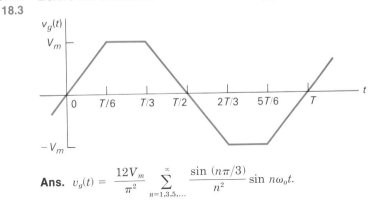

Ans. $v_g(t) = \dfrac{12V_m}{\pi^2} \displaystyle\sum_{n=1,3,5,\ldots}^{\infty} \dfrac{\sin (n\pi/3)}{n^2} \sin n\omega_0 t.$

18.5   AN ALTERNATIVE TRIGONOMETRIC FORM OF THE FOURIER SERIES

In circuit application of the Fourier series, it is more convenient to combine the cosine and sine terms in the series into a single term. This allows us to represent each harmonic of $v(t)$, or $i(t)$, as a single phasor quantity. The cosine and sine terms can be joined together in either a cosine expression or a sine expression. Since we have chosen to use the cosine format in the phasor method of analysis, we will use the cosine expression in the alternative form of the series. Thus the Fourier series given by Eq. (18.2) can also be written as

$$f(t) = a_v + \sum_{n=1}^{\infty} A_n \cos (n\omega_0 t - \theta_n), \tag{18.38}$$

where A_n and θ_n are defined by the complex quantity

$$a_n - jb_n = \sqrt{a_n^2 + b_n^2} \; \underline{/-\theta_n} = A_n \underline{/-\theta_n}. \tag{18.39}$$

We can derive Eqs. 18.38 and 18.39 by using the phasor method to add the cosine and sine terms in Eq. (18.2). We begin by expressing the sine functions as cosine functions, that is, we rewrite Eq. (18.2) as

$$f(t) = a_v + \sum_{n=1}^{\infty} a_n \cos n\omega_0 t + b_n \cos (n\omega_0 t - 90°). \tag{18.40}$$

Now we add the terms under the summation sign using phasors; thus

$$\mathscr{P}\{a_n \cos n\omega_0 t\} = a_n \underline{/0°} \tag{18.41}$$

and

$$\mathscr{P}\{b_n \cos (n\omega_0 t - 90°)\} = b_n \underline{/-90°} = -jb_n. \tag{18.42}$$

It follows that

$$\mathscr{P}\{a_n \cos n\omega_0 t + b_n \cos (n\omega_0 t - 90°)\} = a_n - jb_n$$
$$= \sqrt{a_n^2 + b_n^2}\,\underline{/-\theta_n}$$
$$= A_n\,\underline{/-\theta_n}. \qquad (18.43)$$

When Eq. (18.43) is inverse-transformed, we get

$$a_n \cos n\omega_0 t + b_n \cos (n\omega_0 t - 90°) = \mathscr{P}^{-1}\{A_n\,\underline{/-\theta_n}\}$$
$$= A_n \cos (n\omega_0 t - \theta_n). \qquad (18.44)$$

The substitution of Eq. (18.44) into Eq. (18.40) yields the result given by Eq. (18.38) and Eq. (18.43) corresponds to Eq. (18.39).

It is worth noting that if the periodic function is either even or odd, A_n reduces to either a_n (even) or b_n (odd) and θ_n is either 0° (even) or 90° (odd).

Example 18.3   a) Derive the expressions for a_k and b_k for the periodic function shown in Fig. 18.11.

b) Write the first four terms of the Fourier series representation of $v(t)$ using the format of Eq. (18.38).

Solution   a) The voltage $v(t)$ is neither even or odd nor does it have half-wave symmetry. Therefore, we use Eqs. (18.4) and (18.5) to find a_k and b_k. Choosing t_0 as zero, we have

$$a_k = \frac{2}{T}\left\{ \int_0^{T/4} V_m \cos k\omega_0 t\, dt + \int_{T/4}^{T} [0] \cos k\omega_0 t\, dt \right\}$$
$$= \frac{2V_m}{T} \cdot \frac{\sin k\omega_0 t}{k\omega_0}\bigg|_0^{T/4} = \frac{V_m}{k\pi} \sin \frac{k\pi}{2}$$

and

$$b_k = \frac{2}{T} \int_0^{T/4} V_m \sin k\omega_0 t\, dt$$
$$= \frac{2V_m}{T}\left[\frac{-\cos k\omega_0 t}{k\omega_0}\bigg|_0^{T/4} \right]$$
$$= \frac{V_m}{k\pi}\left[1 - \cos \frac{k\pi}{2} \right].$$

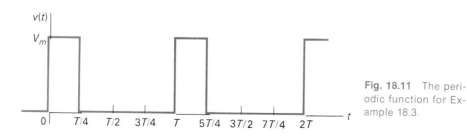

Fig. 18.11   The periodic function for Example 18.3.

b) The average value of $v(t)$ is

$$a_v = \frac{V_m(T/4)}{T} = \frac{V_m}{4}.$$

The values of $a_k - jb_k$ for $k = 1, 2,$ and 3 are

$$a_1 - jb_1 = \frac{V_m}{\pi} - j\frac{V_m}{\pi} = \frac{\sqrt{2}\,V_m}{\pi}\,\underline{/-45°};$$

$$a_2 - jb_2 = 0 - j\frac{V_m}{\pi} = \frac{V_m}{\pi}\,\underline{/-90°};$$

and

$$a_3 - jb_3 = \frac{-V_m}{3\pi} - j\frac{V_m}{3\pi} = \frac{\sqrt{2}\,V_m}{3\pi}\,\underline{/-135°}.$$

Thus the first four terms in the Fourier series representation of $v(t)$ are

$$v(t) = \frac{V_m}{4} + \frac{\sqrt{2}\,V_m}{\pi}\cos(\omega_0 t - 45°) + \frac{V_m}{\pi}\cos(2\omega_0 t - 90°)$$

$$+ \frac{\sqrt{2}\,V_m}{3\pi}\cos(3\omega_0 t - 135°) + \cdots. \quad \blacksquare$$

We are now ready to illustrate how we can use a Fourier series representation of a periodic excitation function to find the steady-state response of a linear circuit.

DRILL EXERCISE
18.4

a) Compute A_1 through A_5 and θ_1 through θ_5 for the periodic function in Drill Exercise 18.1.

b) Write the Fourier series for $v(t)$ up to and including the fifth harmonic using the format of Eq. (18.38).

Ans. (a) 42.43, 30, 14.14, 0, 8.49, $+45°$, $+90°$, $+135°$, not defined, $+45°$; (b) $v(t) = 117.81 + 42.43\cos(10t - 45°) + 30\cos(20t - 90°) + 14.14\cos(30t - 135°) + 8.49\cos(50t - 45°) + \cdots$ V.

18.6 AN ILLUSTRATIVE APPLICATION

The RC circuit shown in Fig. 18.12(a) illustrates how a Fourier series is used in circuit analysis. The circuit is energized with the periodic square-wave voltage shown in Fig. 18.12(b). The voltage across the capacitor is the desired response, or output, signal.

The first step in finding the steady-state response is to represent the periodic excitation source with its Fourier series. After noting that the source has odd, half-wave, and quarter-wave symmetry, we know that the Fourier coeffi-

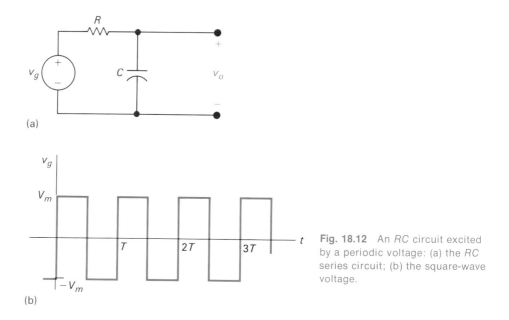

Fig. 18.12 An *RC* circuit excited by a periodic voltage: (a) the *RC* series circuit; (b) the square-wave voltage.

cients reduce to b_k with k restricted to odd integer values. Thus

$$b_k = \frac{8}{T} \int_0^{T/4} V_m \sin k\omega_0 t \; dt$$

$$= \frac{4V_m}{\pi k} \quad (k \text{ odd}). \tag{18.45}$$

It follows directly that the Fourier series representation of v_g is

$$v_g = \frac{4V_m}{\pi} \sum_{n=1,3,5,\ldots}^{\infty} \frac{1}{n} \sin \omega_0 t. \tag{18.46}$$

If we write the series in expanded form, we have

$$v_g = \frac{4V_m}{\pi} \sin \omega_0 t + \frac{4V_m}{3\pi} \sin 3\omega_0 t$$

$$+ \frac{4V_m}{5\pi} \sin 5\omega_0 t + \frac{4V_m}{7\pi} \sin 7\omega_0 t + \cdots. \tag{18.47}$$

Now when we visualize the voltage source expressed by Eq. (18.47), we see that we have the equivalent of an infinite number of series-connected sinusoidal sources, each source having its own amplitude and frequency. To find the contribution of each source to the output voltage, we use the principle of superposition.

For any one of the sinusoidal sources, the phasor domain expression for the output voltage is

$$\mathbf{V}_o = \frac{\mathbf{V}_g}{1 + j\omega RC}.$$ (18.48)

Since all the voltage sources are expressed as sine functions, we will interpret a phasor in terms of the sine instead of the cosine, as we have done in the past. This simply means that when we go from the phasor back to the time domain, we write the time domain expressions as $\sin(\omega t + \theta)$ instead of $\cos(\omega t + \theta)$.

The phasor outptut voltage due to the fundamental-frequency sinusoidal source is

$$\mathbf{V}_{o1} = \frac{(4V_m/\pi)\underline{/0°}}{1 + j\omega_0 RC}.$$ (18.49)

Writing $\mathbf{V}_{o1}$ in polar form gives

$$\mathbf{V}_{o1} = \frac{(4V_m)\underline{/-\beta_1}}{\pi\sqrt{1 + \omega_0^2 R^2 C^2}},$$ (18.50)

where

$$\beta_1 = \tan^{-1}\omega_0 RC.$$ (18.51)

From Eq. (18.50) we can write the time domain expression for the fundamental-frequency component of v_o; thus

$$v_{o1} = \frac{4V_m}{\pi\sqrt{1 + \omega_0^2 R^2 C^2}} \sin(\omega_0 t - \beta_1).$$ (18.52)

The third-harmonic component of the output voltage is derived in a similar fashion. The third-harmonic phasor voltage is

$$\mathbf{V}_{o3} = \frac{[4V_m/3\pi]\underline{/0°}}{1 + j3\omega_0 RC}$$

$$= \frac{4V_m}{3\pi\sqrt{1 + 9\omega_0^2 R^2 C^2}} \underline{/-\beta_3},$$ (18.53)

where

$$\beta_3 = \tan^{-1} 3\omega_0 RC.$$ (18.54)

The time domain expression for the third-harmonic output voltage is

$$v_{o3} = \frac{4V_m}{3\pi\sqrt{1 + 9\omega_0^2 R^2 C^2}} \sin(3\omega_0 t - \beta_3).$$ (18.55)

At this point, it should be apparent that the expression for the kth harmonic component of the output voltage is

$$v_{ok} = \frac{4V_m}{k\pi\sqrt{1 + k^2\omega_0^2 R^2 C^2}} \sin(k\omega_0 t - \beta_k), \quad k \text{ odd},$$ (18.56)

where

$$\beta_k = \tan^{-1} k\omega_0 RC, \quad k \text{ odd.} \tag{18.57}$$

We can now write down the Fourier series representation of the output voltage, as follows:

$$v_o(t) = \frac{4V_m}{\pi} \sum_{n=1,3,5,\dots}^{\infty} \frac{\sin(n\omega_0 t - \beta_n)}{n\sqrt{1 + (n\omega_0 RC)^2}}. \tag{18.58}$$

The derivation of Eq. (18.58) has not been difficult. But, although we have an analytical expression for the steady-state output, it is not immediately apparent from Eq. (18.58) what $v_o(t)$ looks like. As we mentioned earlier, this is a shortcoming of the Fourier series approach. We do not imply by this observation that Eq. (18.58) is useless, since it can be useful in getting some feel for the steady-state waveform of $v_o(t)$ if we focus our thinking on the frequency response of the circuit. For example, if C is very large, then $1/n\omega_0 C$ will be very small for the higher-order harmonics. Thus the capacitor short-circuits the high frequency components of the input waveform. This means that the higher-order harmonics in Eq. (18.58) will be negligible compared to the lower-order harmonics. This is reflected in Eq. (18.58) by noting that for large C,

$$v_o \cong \frac{4V_m}{\pi\omega_0 RC} \sum_{n=1,3,5,\dots}^{\infty} \frac{1}{n^2} \sin(n\omega_0 t - 90°)$$

$$\cong \frac{-4V_m}{\pi\omega_0 RC} \sum_{n=1,3,5,\dots}^{\infty} \frac{1}{n^2} \cos n\omega_0 t. \tag{18.59}$$

We see from Eq. (18.59) that the amplitude of the harmonic in the output is decreasing by $1/n^2$ compared to $1/n$ for the input harmonics. If C is so large that only the fundamental component is significant, then to a first approximation

$$v_o(t) \cong \frac{-4V_m}{\pi\omega_0 RC} \cos \omega_0 t \tag{18.60}$$

and our Fourier analysis tells us that the square-wave input is deformed into a sinusoidal output.

Now let's see what happens as C approaches zero. From the circuit we note that v_o and v_g will be the same when C equals zero since the capacitive branch looks like an open circuit at all frequencies. Equation (18.58) predicts the same result because as $C \to 0$ we get

$$v_o = \frac{4V_m}{\pi} \sum_{n=1,3,5,\dots}^{n} \frac{1}{n} \sin n\omega_0 t. \tag{18.61}$$

But Eq. (18.61) is identical to Eq. (18.46) and therefore $v_o \to v_g$ as $C \to 0$.

Thus Eq. (18.58) has proved to be useful because it has enabled us to predict that the output will be a highly distorted replica of the input waveform if C is large and a reasonable replica of the input waveform if C is small. In Chapter 17, we looked at distortion between the input and output from the point of view of how much memory the system weighting function had. In the frequency domain, we look at distortion between the steady-state input and output in terms of how the amplitude and phase of the harmonics are altered as they are transmitted through the circuit. When the network significantly alters the amplitude and phase relationship among the harmonics at the output relative to their relationship at the input, the output will be a distorted version of the input. Thus in the frequency domain we speak of amplitude distortion and phase distortion.

For the circuit in our illustrative example, amplitude distortion is present because the amplitudes of the input harmonics decrease as $1/n$ while the amplitudes of the output harmonics decrease as

$$\frac{1}{n} \cdot \frac{1}{\sqrt{1 + (n\omega_0 RC)^2}}.$$

Our illustrative circuit also exhibits phase distortion because the phase angle of each input harmonic is zero while the phase angle of the nth harmonic in the output signal is $\tan^{-1} n\omega_0 RC$.

For the simple RC circuit in Fig. 18.12(a) we can derive the expression for the steady-state response without resorting to the Fourier series representation of the excitation function. We will do this extra analysis here because to do so adds to our understanding of the Fourier series approach to obtaining a steady-state solution.

To find the steady-state expression for v_o by straightforward circuit analysis, we reason as follows. The square-wave excitation function alternates between charging the capacitor toward $+V_m$ and $-V_m$. Once the circuit reaches steady-state operation, this alternate charging becomes periodic. Furthermore, we know from our analysis of the single-time constant RC circuit (Chapter 7) that the response to abrupt changes in the driving voltage will be exponential. Thus the steady-state waveform of the voltage across the capacitor in the circuit in Fig. 18.12(a) is as shown in Fig. 18.13.

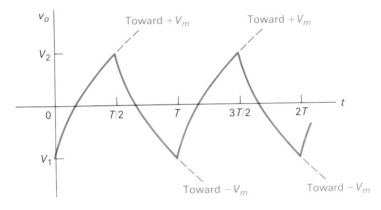

Fig. 18.13   The steady-state waveform of v_o for the circuit in Fig. 18.12(a).

The analytical expressions for $v_o(t)$ in the time intervals $0 \le t \le T/2$ and $T/2 \le t \le T$ are

$$v_o = V_m + (V_1 - V_m)e^{-t/RC}, \qquad 0 \le t \le T/2, \qquad \text{(18.62)}$$

$$v_o = -V_m + (V_2 + V_m)e^{-[t-(T/2)]/RC}, \qquad T/2 \le t \le T. \qquad \text{(18.63)}$$

Equations (18.62) and (18.63) are derived using the methods of Chapter 7 and are summarized by Eq. (7.48). The values of V_1 and V_2 can be found by noting from Eq. (18.62) that

$$V_2 = V_m + (V_1 - V_m)e^{-T/2RC} \qquad \text{(18.64)}$$

and from Eq. (18.63) that

$$V_1 = -V_m + (V_2 + V_m)e^{-T/2RC}. \qquad \text{(18.65)}$$

Solving Eqs. (18.64) and (18.65) for V_1 and V_2 yields

$$V_2 = -V_1 = \frac{V_m(1 - e^{-T/2RC})}{(1 + e^{-T/2RC})}. \qquad \text{(18.66)}$$

When we substitute Eq. (18.66) into Eqs. (18.62) and (18.63), we get

$$v_o = V_m - \frac{2V_m}{(1 + e^{-T/2RC})} e^{-t/RC}, \qquad 0 \le t \le T/2 \qquad \text{(18.67)}$$

and

$$v_o = -V_m + \frac{2V_m}{(1 + e^{-T/2RC})} e^{-[t-(T/2)]/RC}, \qquad T/2 \le t \le T. \qquad \text{(18.68)}$$

It is of interest to note from Eqs. (18.67) and (18.68) that $v_o(t)$ has half-wave symmetry and therefore the average value of v_o is zero. This is in agreement with the Fourier series solution for the steady-state response, namely, that since there is no zero frequency component in the excitation function, there can be no zero frequency component in the response. Equations (18.67) and (18.68) also show the effect of changing the size of the capacitor. If C is very small, the exponential functions quickly vanish and v_o equals V_m between 0 and $T/2$ and $-V_m$ between $T/2$ and T. In other words, v_o approaches v_g as C approaches zero. If C is very large, the output waveform becomes triangular in shape, as shown in Fig. 18.14. This triangular waveform can be seen by noting that for large C the exponential terms $e^{-t/RC}$ and $e^{-[t-(T/2)]/RC}$ can be approximated by the linear terms $1 - (t/RC)$ and $1 - \{[t - (T/2)]/RC\}$, respectively. The Fourier series of this triangular waveform is given by Eq. (18.59). The results are summarized by the sketches in Fig. 18.14.

DRILL EXERCISE
18.5

a) Show that for large values of C Eq. (18.67) can be approximated by the expression

$$v_o(t) \cong \frac{-V_m T}{4RC} + \frac{V_m}{RC} t.$$

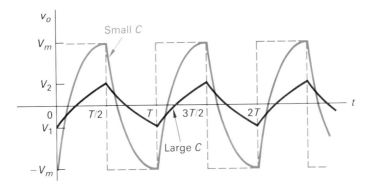

Fig. 18.14   The effect of capacitor size on the steady-state response.

Note that this is the equation of the triangular wave for $0 \le t \le T/2$. (*Hints:* (1) Let $e^{-t/RC} \cong 1 - (t/RC)$ and $e^{-T/2RC} \cong 1 - (T/2RC)$; (2) put the resulting expression over the common denominator $2 - (T/2RC)$; (3) simplify the numerator; and (4) for large C assume that $T/2RC$ is much less than 2.)

b) Substitute the peak value of the triangular wave into the solution for Problem 18.9(b) and show that the result is Eq. (18.59).

Ans.   (a) Derivation; (b) From Problem 18.9(b), $a_k = -8V_p/\pi^2 k^2$, where $k = 1, 3, 5, \ldots$ and V_p is the peak value of the triangular wave. Therefore for the triangular wave described in part (a),

$$a_k = \frac{-2V_m T}{\pi^2 k^2 RC} = \frac{-4V_m}{\pi \omega_0 RC k^2}.$$

Finally, we verify that the steady-state response given by Eqs. (18.67) and (18.68) is equivalent to the Fourier series solution given by Eq. (18.58). To show this we simply derive the Fourier series representation of the periodic function described by Eqs. (18.67) and (18.68). We have already noted that our periodic voltage response has half-wave symmetry. Therefore, the Fourier series will contain only odd harmonics. For k odd, we have

$$a_k = \frac{4}{T} \int_0^{T/2} \left[V_m - \frac{2V_m e^{-t/RC}}{(1 + e^{-T/2RC})} \right] \cos k\omega_0 t \, dt$$

$$= \frac{-8RCV_m}{T[1 + (k\omega_0 RC)^2]}, \quad (k \text{ odd}); \tag{18.69}$$

$$b_k = \frac{4}{T} \int_0^{T/2} \left[V_m - \frac{2V_m e^{-t/RC}}{(1 + e^{-T/2RC})} \right] \sin k\omega_0 t \, dt$$

$$= \frac{4V_m}{k\pi} - \frac{8k\omega_0 V_m R^2 C^2}{T[1 + (k\omega_0 RC)^2]}, \quad (k \text{ odd}). \tag{18.70}$$

To show that the results given by Eqs. (18.69) and (18.70) are consistent with Eq. (18.58) it is necessary to prove that

$$\sqrt{a_k^2 + b_k^2} = \frac{4V_m}{k\pi} \frac{1}{\sqrt{1 + (k\omega_0 RC)^2}} \tag{18.71}$$

and

$$\frac{a_k}{b_k} = -k\omega_0 RC. \tag{18.72}$$

We leave the verification of Eqs. (18.69)–(18.72) as exercises (see Problems 18.16 and 18.17). Equations (18.71) and (18.72) are used with Eqs. (18.38) and (18.39) to derive the Fourier series expression given by Eq. (18.58). The details are left to the reader in Problem 18.18.

We have seen through this illustrative circuit how we can use the Fourier series in conjunction with the principle of superposition to obtain the steady-state response to a periodic driving function. The principal shortcoming of the Fourier series approach is the difficulty in ascertaining the waveform of the response. However, we also observed that by thinking in terms of a circuit's frequency response, we can deduce that a reasonable approximation of the steady-state response is possible by using a finite number of appropriate terms in the Fourier series representation. See, for example, Problems 18.22 and 18.24.

18.7 AVERAGE POWER CALCULATIONS WITH PERIODIC FUNCTIONS

If we are given the Fourier series representation of the voltage and current at a pair of terminals in a linear, lumped-parameter circuit, it is very easy to express the average power at the terminals as a function of the harmonic voltages and currents. Using the trigonometric form of the Fourier series expressed in Eq. (18.38), we can write the periodic voltage and current at the terminals of a network as

$$v = V_{dc} + \sum_{n=1}^{\infty} V_n \cos (n\omega_0 t - \theta_{vn}) \tag{18.73}$$

and

$$i = I_{dc} + \sum_{n=1}^{\infty} I_n \cos (n\omega_0 t - \theta_{in}). \tag{18.74}$$

The notation used in Eqs. (18.73) and (18.74) is defined as follows:

V_{dc} = the amplitude of the dc voltage component;

V_n = the amplitude of the nth harmonic voltage;

θ_{vn} = the phase angle of the nth harmonic voltage;

I_{dc} = the amplitude of the dc current component;

$$I_n = \text{the amplitude of the } n\text{th harmonic current};$$

$$\theta_{in} = \text{the phase angle of the } n\text{th harmonic current}.$$

We assume that the current is in the direction of the voltage drop across the terminals (passive sign convention) so that the instantaneous power at the terminals is vi. The average power is

$$P = \frac{1}{T} \int_{t_0}^{t_0+T} p \, dt = \frac{1}{T} \int_{t_0}^{t_0+T} vi \, dt. \tag{18.75}$$

To find the expression for the average power, we substitute Eqs. (18.73) and (18.74) into Eq. (18.75) and integrate. At first glance this appears to be a formidable task since the product vi requires multiplying two infinite series! However, the only terms to survive the integration process will be products of voltage and current at the same frequency. A review of Eqs. (18.8), (18.9), and (18.10) should convince you of the validity of this observation. Therefore, Eq. (18.75) reduces to

$$P = \frac{1}{T} V_{dc} I_{dc} t \Big|_{t_0}^{t_0+T} + \sum_{n=1}^{\infty} \frac{1}{T} \int_{t_0}^{t_0+T} V_n I_n \cos(n\omega_0 t - \theta_{vn}) \cos(n\omega_0 t - \theta_{in}) \, dt. \tag{18.76}$$

Now using the trigonometric identity

$$\cos \alpha \cos \beta = \frac{1}{2} \cos(\alpha - \beta) + \frac{1}{2} \cos(\alpha + \beta),$$

we can simplify Eq. (18.76) to

$$P = V_{dc} I_{dc} + \frac{1}{T} \sum_{n=1}^{\infty} \frac{V_n I_n}{2} \int_{t_0}^{t_0+T} [\cos(\theta_{vn} - \theta_{in}) + \cos(2n\omega_0 t - \theta_{vn} - \theta_{in})]. \tag{18.77}$$

The second term under the integral sign integrates to zero; therefore we have

$$P = V_{dc} I_{dc} + \sum_{n=1}^{\infty} \frac{V_n I_n}{2} \cos(\theta_{vn} - \theta_{in}). \tag{18.78}$$

Equation (18.78) is an important result. It tells us that in the case of an interaction between a periodic voltage and the corresponding periodic current, the total average power is the sum of the average powers obtained from the interaction of currents and voltages of the same frequency. Currents and voltages of different frequencies do not interact to produce average power. Therefore, in average-power calculations involving periodic functions, the total average power is the superposition of the average powers associated with each harmonic voltage and current.

Example 18.4   Assume that the periodic square-wave voltage in Example 18.3 is applied across the terminals of a 15-Ω resistor. The value of V_m is 60 V and that of T is 5 ms.

a) Write the first five nonzero terms of the Fourier series representation of $v(t)$. Use the trigonometric form given in Eq. (18.38).

b) Calculate the average power associated with each term in part (a).

c) Calculate the total average power delivered to the 60-Ω resistor.

d) What percentage of the total power is delivered by the first five terms of the Fourier series?

Solution   a) The dc component of $v(t)$ is

$$a_v = \frac{(60)(T/4)}{T} = 15 \text{ V.}$$

From Example 18.3 we have

$$A_1 = \sqrt{2}\ 60/\pi = 27.01,$$
$$\theta_1 = 45°,$$
$$A_2 = 60/\pi = 19.10,$$
$$\theta_2 = 90°,$$
$$A_3 = 20\sqrt{2}/\pi = 9.00,$$
$$\theta_3 = 135°,$$
$$A_4 = 0,$$
$$A_5 = 5.40,$$
$$\theta_5 = 45°,$$
$$\omega_0 = \frac{2\pi}{T} = \frac{2\pi(1000)}{5} = 400\pi \text{ rad/s.}$$

Thus the first five nonzero terms of the Fourier series are

$$
\begin{aligned}
v(t) = \ & 15 + 27.01 \cos (400\pi t - 45°) \\
& + 19.10 \cos (800\pi t - 90°) \\
& + 9.00 \cos (1200\ \pi t - 135°) \\
& + 5.40 \cos (2000\pi t - 45°) + \cdots .
\end{aligned}
$$

b) Since the voltage is applied to the terminals of a resistor, we can find the power associated with each term as follows:

$$P_{\text{dc}} = (15)^2/15 = 15 \text{ W,}$$

$$P_1 = \frac{1}{2}\frac{(27.01)^2}{15} = 24.32 \text{ W,}$$

$$P_2 = \frac{1}{2}\frac{(19.10)^2}{15} = 12.16 \text{ W,}$$

$$P_3 = \frac{1}{2}\frac{(9)^2}{15} = 2.70 \text{ W,}$$

$$P_5 = \frac{1}{2}\frac{(5.4)^2}{15} = 0.97 \text{ W.}$$

c) To calculate the total average power delivered to the 15-Ω resistor, we first calculate the rms value of $v(t)$; thus

$$V_{\text{rms}} = \sqrt{\frac{(60)^2(T/4)}{T}} = \sqrt{900} = 30 \text{ V}.$$

Thus the total average power delivered to the 15-Ω resistor is

$$P_T = \frac{(30)^2}{15} = 60 \text{ W}.$$

d) The total power delivered by the first five nonzero terms is

$$P = P_{\text{dc}} + P_1 + P_2 + P_3 + P_5 = 55.15.$$

This is $(55.15/60)(100)$, or 91.92% of the total. ■

DRILL EXERCISE 18.6 The trapezoidal voltage function in Drill Exercise 18.3 is applied to the circuit shown. If $12V_m = 986.96$ V and $T = 6283.19$ ms, estimate the average power delivered to the 1-Ω resistor.

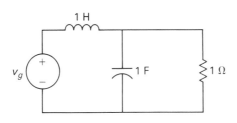

Ans. 3750 W.

18.8 THE rms VALUE OF A PERIODIC FUNCTION

The rms value of a periodic function can be expressed in terms of the Fourier coefficients; by definition,

$$F_{\text{rms}} = \sqrt{\frac{1}{T} \int_{t_0}^{t_0+T} f(t)^2 \, dt}. \tag{18.79}$$

Representing $f(t)$ by its Fourier series, we have

$$F_{\text{rms}} = \sqrt{\frac{1}{T} \int_{t_0}^{t_0+T} \left[a_v + \sum_{n=1}^{\infty} A_n \cos (n\omega_0 t - \theta_n) \right]^2 \, dt}. \tag{18.80}$$

The integral of the squared time function simplifies since the only terms to survive the integration over a period will be the product of the dc term and the harmonic products of the same frequency. All other products will integrate to zero. Therefore, Eq. (18.80) reduces to

$$F_{rms} = \sqrt{\frac{1}{T}\left[a_v^2 T + \sum_{n=1}^{\infty}\frac{T}{2}A_n^2\right]} = \sqrt{a_v^2 + \sum_{n=1}^{\infty}\frac{A_n^2}{2}}$$

$$= \sqrt{a_v^2 + \sum_{n=1}^{\infty}\left(\frac{A_n}{\sqrt{2}}\right)^2}.$$ (18.81)

Equation (18.81) tells us that the rms value of a periodic function is the square root of the sum obtained by adding the square of the rms value of each harmonic to the square of the dc value. For example, assume that a periodic voltage is represented by the finite series

$v = 10 + 30 \cos(\omega_0 t - \theta_1)$
$\qquad + 20 \cos(2\omega_0 t - \theta_2) + 5\cos(3\omega_0 t - \theta_3) + 2\cos(5\omega_0 t - \theta_5).$

The rms value of this voltage is

$$V = \sqrt{(10)^2 + (30/\sqrt{2})^2 + (20/\sqrt{2})^2 + (5/\sqrt{2})^2 + (2/\sqrt{2})^2}$$
$$= \sqrt{764.5} = 27.65 \text{ V}.$$

In the usual case an infinite number of terms are required to represent a periodic function by a Fourier series, and therefore Eq. (18.81) yields an estimate of the true rms value. This is illustrated in the following example.

Example 18.5   Use Eq. (18.81) to estimate the rms value of the voltage in Example 18.4.

Solution   From Example 18.4 we have

$$V_{dc} = 15 \text{ V},$$
$$V_1 = 27.01/\sqrt{2}, \quad \text{the rms value of the fundamental,}$$
$$V_2 = 19.10/\sqrt{2}, \quad \text{the rms value of the second harmonic,}$$
$$V_3 = 9.00/\sqrt{2}, \quad \text{the rms value of the third harmonic,}$$
$$V_5 = 5.40/\sqrt{2}, \quad \text{the rms value of the fifth harmonic.}$$

Therefore,

$$V_{rms} = \sqrt{15^2 + (27.01/\sqrt{2})^2 + (19.10/\sqrt{2})^2 + (9.00/\sqrt{2})^2 + (5.40/\sqrt{2})^2}$$
$$= 28.76 \text{ V}.$$

As we know from Example 18.4, the true rms value is 30 V. We will approach this value as more and more harmonics are included in Eq. (18.81). For example, if the harmonics through $k = 9$ are included, Eq. (18.81) yields a value of 29.32 V. ∎

DRILL EXERCISE
18.7

a) Find the rms value of the voltage in Drill Exercise 18.3 if $V_m = 100$ V.

b) Estimate the rms value of the voltage using the first three terms in the Fourier series representation of $v_g(t)$.

Ans. (a) 74.5536 V; (b) 74.5306 V.

18.9 THE EXPONENTIAL FORM OF THE FOURIER SERIES

The exponential form of the Fourier series is of interest because it allows us to express the series in a very concise form. The exponential form of the series is

$$f(t) = \sum_{n=-\infty}^{\infty} C_n e^{jn\omega_0 t}, \tag{18.82}$$

where

$$C_n = \frac{1}{T} \int_{t_0}^{t_0+T} f(t) e^{-jn\omega_0 t} \, dt. \tag{18.83}$$

To derive Eqs. (18.82) and (18.83), we return to Eq. (18.2) and replace the cosine and sine functions with their exponential equivalents

$$\cos n\omega_0 t = \frac{e^{jn\omega_0 t} + e^{-jn\omega_0 t}}{2} \tag{18.84}$$

and

$$\sin n\omega_0 t = \frac{e^{jn\omega_0 t} - e^{-jn\omega_0 t}}{2j}. \tag{18.85}$$

When we substitute Eqs. (18.84) and (18.85) into Eq. (18.2) we have

$$
\begin{aligned}
f(t) &= a_v + \sum_{n=1}^{\infty} \frac{a_n}{2} (e^{jn\omega_0 t} + e^{-jn\omega_0 t}) + \frac{b_n}{2j} (e^{jn\omega_0 t} - e^{-jn\omega_0 t}) \\
&= a_v + \sum_{n=1}^{\infty} \left(\frac{a_n - jb_n}{2} \right) e^{jn\omega_0 t} + \left(\frac{a_n + jb_n}{2} \right) e^{-jn\omega_0 t}.
\end{aligned} \tag{18.86}
$$

Now we define C_n as

$$C_n = \frac{1}{2}(a_n - jb_n) = \frac{A_n}{2} \underline{/-\theta_n}, \quad n = 1, 2, 3, \ldots \tag{18.87}$$

It follows directly from the definition of C_n that

$$
\begin{aligned}
C_n &= \frac{1}{2} \left\{ \frac{2}{T} \int_{t_0}^{t_0+T} f(t) \cos n\omega_0 t \, dt - j\frac{2}{T} \int_{t_0}^{t_0+T} f(t) \sin n\omega_0 t \, dt \right\} \\
&= \frac{1}{T} \int_{t_0}^{t_0+T} f(t)(\cos n\omega_0 t - j \sin n\omega_0 t) \, dt \\
&= \frac{1}{T} \int_{t_0}^{t_0+T} f(t) e^{-jn\omega_0 t} \, dt,
\end{aligned} \tag{18.88}
$$

which completes the derivation of Eq. (18.83). To complete the derivation of

Eq. (18.82), we continue as follows. First, we observe from Eq. (18.88) that

$$C_0 = \frac{1}{T} \int_{t_0}^{t_0+T} f(t) \, dt = a_v. \tag{18.89}$$

Next we note that

$$C_{-n} = \frac{1}{T} \int_{t_0}^{t_0+T} f(t) e^{jn\omega_0 t} \, dt = C_n^* = \frac{1}{2}(a_n + jb_n). \tag{18.90}$$

When we substitute the observations contained in Eqs. (18.87), (18.89), and (18.90) into Eq. (18.86), we have

$$f(t) = C_0 + \sum_{n=1}^{\infty} C_n e^{jn\omega_0 t} + C_n^* e^{-jn\omega_0 t}$$
$$= \sum_{n=0}^{\infty} C_n e^{jn\omega_0 t} + \sum_{n=1}^{\infty} C_n^* e^{-jn\omega_0 t}. \tag{18.91}$$

Now observe that the second summation on the right-hand side of Eq. (18.91) is equivalent to summing $C_n e^{jn\omega_0 t}$ from -1 to $-\infty$; that is,

$$\sum_{n=1}^{\infty} C_n^* e^{-jn\omega_0 t} = \sum_{n=-1}^{-\infty} C_n e^{jn\omega_0 t}. \tag{18.92}$$

Since the summation from -1 to $-\infty$ is the same as the summation from $-\infty$ to -1, we can use Eq. (18.92) to rewrite Eq. (18.91) as

$$f(t) = \sum_{n=0}^{\infty} C_n e^{jn\omega_0 t} + \sum_{-\infty}^{-1} C_n e^{jn\omega_0 t}$$
$$= \sum_{-\infty}^{\infty} C_n e^{jn\omega_0 t}, \tag{18.93}$$

which completes the derivation of Eq. (18.82).

The process of finding the exponential Fourier series representation of a periodic function is illustrated in the following example.

Example 18.6 Find the exponential Fourier series for the periodic voltage shown in Fig. 18.15.

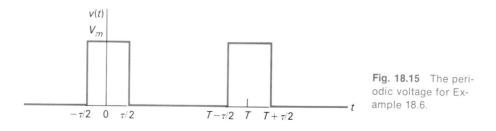

Fig. 18.15 The periodic voltage for Example 18.6.

Solution   Using $-\tau/2$ as the starting point for the integration, we have from Eq. (18.83)

$$
\begin{aligned}
C_n &= \frac{1}{T} \int_{-\tau/2}^{\tau/2} V_m e^{-jn\omega_0 t} \, dt \\
&= \frac{V_m}{T} \left[\frac{e^{-jn\omega_0 t}}{-jn\omega_0} \right]_{-\tau/2}^{\tau/2} \\
&= \frac{jV_m}{n\omega_0 T} \left[e^{-jn\omega_0\tau/2} - e^{jn\omega_0\tau/2} \right] \\
&= \frac{2V_m}{n\omega_0 T} \sin n\omega_0\tau/2.
\end{aligned}
$$

There are a couple of observations we can make about C_n for this particular example. First since $v(t)$ has even symmetry, $b_n = 0$ for all n and hence we should expect C_n to be real. Second, the amplitude of C_n follows a $(\sin x)/x$ distribution. We can see this by rewriting C_n as

$$
C_n = \frac{V_m \tau}{T} \cdot \frac{\sin (n\omega_0\tau/2)}{(n\omega_0\tau/2)}.
$$

We will have more to say about this in Section 18.10. The exponential series representation of $v(t)$ is

$$
\begin{aligned}
v(t) &= \sum_{n=-\infty}^{n=\infty} \left[\frac{V_m \tau}{T} \right] \frac{\sin n\omega_0\tau/2}{(n\omega_0\tau/2)} e^{jn\omega_0 t} \\
&= \left(\frac{V_m \tau}{T} \right) \sum_{n=-\infty}^{n=\infty} \frac{\sin n\omega_0\tau/2}{(n\omega_0\tau/2)} e^{jn\omega_0 t}. \quad \blacksquare
\end{aligned}
$$

The rms value of a periodic function can also be expressed in terms of the complex Fourier coefficients. From Eqs. (18.81), (18.87), and (18.89) we have

$$
F_{\text{rms}} = \sqrt{a_v^2 + \sum_{n=1}^{\infty} \frac{(a_n^2 + b_n^2)}{2}}; \tag{18.94}
$$

$$
|C_n| = \frac{\sqrt{a_n^2 + b_n^2}}{2}; \tag{18.95}
$$

$$
C_0^2 = a_v^2. \tag{18.96}
$$

Substituting Eqs. (18.95) and (18.96) into Eq. (18.94) yields the desired expression; thus

$$
F_{\text{rms}} = \sqrt{C_0^2 + 2 \sum_{n=1}^{\infty} |C_n|^2}. \tag{18.97}
$$

DRILL EXERCISE   Derive the expression for the Fourier coefficients C_n for the periodic function shown.
18.8   (*Hint:* Take advantage of symmetry by using the fact that $C_n = (a_n - jb_n)/2$.)

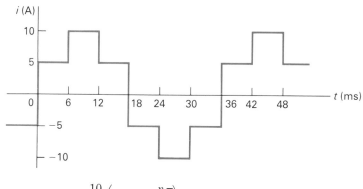

Ans. $C_n = -j\dfrac{10}{\pi n}\left(1 + \cos\dfrac{n\pi}{3}\right)$, n odd.

DRILL EXERCISE
18.9

a) Calculate the rms value of the periodic current in Drill Exercise 18.8

b) Estimate the rms value using C_1 through C_{11}.

c) What is the percentage error in the value found in part (b) based on the true value found in part (a)?

d) For this periodic function, how many terms must be used to estimate the rms value before the error is less than 1%?

Ans. (a) $\sqrt{50}$ A; (b) 6.980 A; (c) -1.28%; (d) $n = 17$; therefore, the first six nonzero harmonic terms of the series are required.

18.10   AMPLITUDE AND PHASE SPECTRA

It should be apparent at this time that a periodic time function is defined by its Fourier coefficients and its period. That is, once a_v, a_n, b_n, and T are known, it is possible, at least theoretically, to construct $f(t)$. Once a_n and b_n are given, we also know the amplitude (A_n) and phase angle (θ_n) of each harmonic. We have already noted that we cannot, in general, visualize what the periodic function looks like in the time domain from a description of the coefficients and phase angles; nevertheless we recognize that the periodic function is completely characterized by these quantities. Thus given sufficient computing time we can synthesize the time domain waveform from the amplitude and phase-angle data. It should also be noted that when a periodic driving function is exciting a circuit that is highly frequency-selective, the Fourier series of the steady-state response will be dominated by just a few terms. Thus the description of the response in terms of amplitude and phase may provide a good understanding of the output waveform.

The description of a periodic function in terms of the amplitude and phase angle of each term in the Fourier series of $f(t)$ can be presented graphically. The plot of the amplitude of each term versus the frequency is called the *amplitude spectrum* of $f(t)$, and the plot of the phase angle versus the frequency is called the *phase spectrum of f(t)*. Because the amplitude and phase-angle data occur at discrete values of the frequency (that is, at ω_0, $2\omega_0$, $3\omega_0$, etc.), these plots are also referred to as *line spectra*.

Amplitude and phase-spectra plots are based on either Eq. (18.38) (A_n and θ_n) or Eq. (18.82) (C_n). We will focus on Eq. (18.82) and leave the plots based on Eq. (18.38) to Problem 18.34. To illustrate the amplitude and phase-angle spectra, which are based on the exponential form of the Fourier series, we will use the periodic voltage of Example 18.6. To facilitate our discussion, we will assume that $V_m = 5$ V and $\tau = T/5$. It follows directly from Example 18.6 that

$$C_n = \frac{V_m \tau}{T} \frac{\sin (n\omega_0 \tau/2)}{n\omega_0 \tau/2}, \tag{18.98}$$

which for the assumed values of V_m and τ reduces to

$$C_n = 1 \cdot \frac{\sin (n\pi/5)}{(n\pi/5)}. \tag{18.99}$$

The plot of the magnitude of C_n from Eq. (18.99) is shown in Fig. 18.16 for values of n ranging from -10 to $+10$. Figure 18.16 clearly shows that the amplitude spectrum is bounded by the envelope of the $|(\sin x)/x|$ function. We have used the order of the harmonic as the frequency scale since the numerical value of T is not specified. Once we know T, we also know ω_0 and the frequency corresponding to each harmonic.

It is of interest to sketch $|(\sin x)/x|$ versus x, where x is given in radians as shown in Fig. 18.17, because the sketch shows that the function goes through zero whenever x is an integral multiple of π. From Eq. (18.98) we see that

$$n\omega_0 \left(\frac{\tau}{2}\right) = \frac{n\pi\tau}{T} = \frac{n\pi}{(T/\tau)}. \tag{18.100}$$

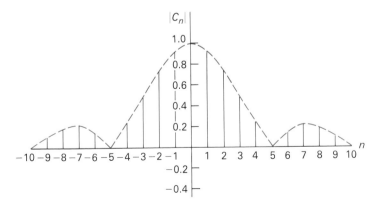

Fig. 18.16 The plot of $|C_n|$ versus n when $\tau = T/5$ for Example 18.6.

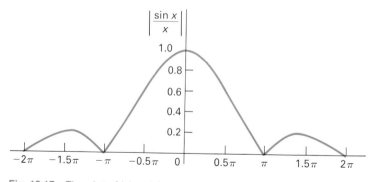

Fig. 18.17 The plot of $|\sin x / x|$ versus x.

From Eq. (18.100) we can deduce that the amplitude spectrum goes through zero whenever $n\tau/T$ is an integer. For example, in our plot τ/T is $\frac{1}{5}$ and therefore the envelope goes through zero at $n = 5, 10, 15$, etc. In other words, the 5th, 10th, 15th (etc.) harmonics are all zero. As the reciprocal of τ/T becomes an increasingly larger and larger integer, the number of harmonics between every π radians increases. If $n\tau/T$ is not an integer, the amplitude spectrum still follows the $|(\sin x)/x|$ envelope. However, the envelope is not zero at an integeral multiple of ω_0.

Since C_n is real for all n, the phase angle associated with C_n is either zero or 180° depending on the algebraic sign of $(\sin n\pi/5)/(n\pi/5)$. For example, the phase angle is zero for $n = 0, \pm 1, \pm 2, \pm 3$, and ± 4. The phase angle is not defined at $n = \pm 5$ since $C_{\pm 5}$ is zero. The phase angle equals 180° at $n = \pm 6, \pm 7, \pm 8$, and ± 9, and is not defined at ± 10. This pattern repeats itself as n takes on larger integer values. The phase angle of C_n given by Eq. (18.99) is sketched in Fig. 18.18.

Now we raise an interesting question. What happens to the amplitude and phase spectra if $f(t)$ is shifted along the time axis? To answer this, we will shift the periodic voltage in Example 18.6 t_0 units to the right. By hypothesis,

$$v(t) = \sum_{n=-\infty}^{\infty} C_n e^{jn\omega_0 t}; \tag{18.101}$$

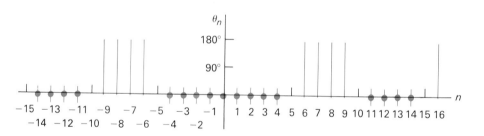

Fig. 18.18 The phase angle of C_n.

therefore

$$v(t - t_0) = \sum_{n=-\infty}^{\infty} C_n e^{jn\omega_0(t-t_0)},$$

$$= \sum_{n=-\infty}^{\infty} C_n e^{-jn\omega_0 t_0} e^{jn\omega_0 t}, \tag{18.102}$$

from which we see that shifting the origin has no effect on the amplitude spectrum since

$$|C_n| = |C_n e^{-jn\omega_0 t_0}|. \tag{18.103}$$

However, it follows from Eq. (18.87) that the phase spectrum has changed to $-(\theta_n + n\omega_0 t_0)$ radians.

As an example, let us shift the periodic voltage in Example 18.6 $\tau/2$ units to the right. As before, we assume that $\tau = T/5$; then the new phase angle θ'_n is

$$\theta'_n = -(\theta_n + n\pi/5). \tag{18.104}$$

We have plotted Eq. (18.104) in Fig. 18.19 for n ranging from -8 to $+8$. Note that no phase angle is associated with a zero amplitude coefficient.

You may wonder why we have devoted so much attention to the amplitude spectrum of the periodic pulse in Example 18.6. It turns out that this particu-

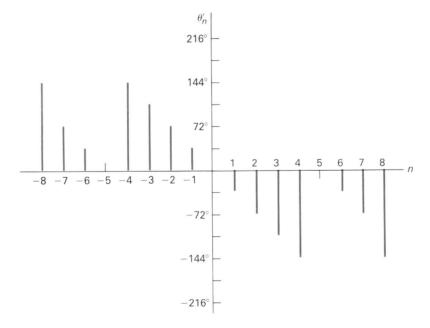

Fig. 18.19 The plot of θ'_n versus n for Eq. (18.104).

lar periodic waveform is an excellent vehicle for illustrating the transition from the Fourier series representation of a periodic function to the Fourier transform representation of a nonperiodic function. We will discuss the Fourier transform in Chapter 19.

Write the exponential Fourier series for the periodic current in Drill Exercise 18.8 if the function is shifted along the time axis 9 ms to the left.

Ans. $i(t) = \dfrac{10}{\pi} \sum\limits_{\substack{n=-\infty \\ \text{(odd)}}}^{n=\infty} \dfrac{1}{n}\left(1 + \cos\dfrac{n\pi}{3}\right) e^{(j\pi/2)(n-1)} e^{jn\omega_0 t}.$

18.11 MEAN-SQUARE ERROR

In practical applications where a Fourier series is used to represent a periodic function, we are forced to truncate the series to a finite number of turns. This means that the periodic function is actually being represented by a partial sum, that is,

$$f(t) \cong S_N(t) = \sum_{n=-N}^{N} C_n e^{jn\omega_0 t}. \tag{18.105}$$

The error in this approximation to $f(t)$ is defined as the difference between $f(t)$ and its representation by the partial sum, that is,

$$\epsilon(t) = f(t) - S_N(t). \tag{18.106}$$

This error function is a direct consequence of having to work with a finite number of terms in the series representation of $f(t)$.

Once we focus our attention on the error function $\epsilon(t)$, we begin to raise questions about its behavior and this in turn raises questions about the series being used to represent $f(t)$. Are there other trigonometric series that can be used to approximate $f(t)$? (A trigonometric series is a Fourier series only if the coefficients are found in accordance with Eqs. 18.3, 18.4, and 18.5 or Eq. 18.83.) The answer to the question is yes but in this text we will not investigate these other types of trigonometric series.† Our purpose in bringing this aspect of series analysis to your attention is to point out the nature of the error function when $f(t)$ is represented by a Fourier series.

The phrase "mean-square error" is used to designate finding the mean (average) value of the error squared. If we let $\overline{\epsilon^2}$ denote the mean-square error,

---

† See James B. Ley, Samuel G. Lutz, and Charles F. Rehberg, *Linear Circuit Analysis,* Chapter 6 (New York: McGraw-Hill, 1959) and E. A. Gullemin, *The Mathematics of Circuit Analysis,* Chapter 7 (New York: John Wiley, 1949).

then by definition

$$\overline{\epsilon^2} = \frac{1}{T} \int_{t_0}^{t_0+T} \epsilon^2(t) \; dt. \tag{18.107}$$

Now we make the observation that the *mean-square error is minimum if the coefficients in the partial-sum approximation to f(t) are Fourier coefficients.*

In order to prove that the Fourier coefficients minimize the integral in Eq. (18.107) we proceed as follows. First we postulate that $f(t)$ is approximated by a finite trigonometric series where the coefficients in the series are unspecified, that is, we let

$$S_N(t) = \sum_{n=-N}^{N} D_n e^{jn\omega_0 t}, \tag{18.108}$$

where the coefficients D_n are not necessarily the Fourier coefficients. The second step in our proof is to solve for the values of D_n in Eq. (18.108) that will minimize the integral in Eq. (18.107).

The first step in the solution for D_n is to express the mean-square error as a function of the partial sum given by Eq. (18.108); thus

$$\overline{\epsilon^2} = \frac{1}{T} \int_{-T/2}^{T/2} \left[f(t) - \sum_{n=-N}^{N} D_n e^{jn\omega_0 t} \right]^2 dt. \tag{18.109}$$

In writing Eq. (18.109) we have, for convenience, chosen t_0 equal to $-T/2$.

If we let D_k represent the kth value of D_n, our problem is reduced to finding the expression for D_k such that

$$\frac{d\overline{\epsilon^2}}{dD_k} = 0 \tag{18.110}$$

for all k.

It follows from Eq. (18.109) that

$$\frac{d\overline{\epsilon^2}}{dD_k} = \frac{d}{dD_k} \left\{ \frac{1}{T} \int_{-T/2}^{T/2} \left[f(t) - \sum_{n=-N}^{N} D_n e^{jn\omega_0 t} \right]^2 dt \right\}. \tag{18.111}$$

The integration is with respect to time and therefore the differentiation can be moved inside the integral; thus Eq. (18.111) leads to

$$\frac{d\overline{\epsilon^2}}{dD_k} = \frac{1}{T} \int_{-T/2}^{T/2} \frac{d}{dD_k} \left[f(t) - \sum_{n=-N}^{N} D_n e^{jn\omega_0 t} \right]^2 dt$$

$$= \frac{1}{T} \int_{-T/2}^{T/2} 2 \left[f(t) - \sum_{n=-N}^{N} D_n e^{jn\omega_0 t} \right] (-e^{jk\omega_0 t}) \; dt$$

$$= -\frac{2}{T} \int_{-T/2}^{T/2} \left[f(t) - \sum_{n=-N}^{N} D_n e^{jn\omega_0 t} \right] e^{jk\omega_0 t} \; dt. \tag{18.112}$$

In order for the right-hand side of Eq. (18.112) to equal zero, the integral must be zero; hence

$$\int_{-T/2}^{T/2} f(t)e^{jk\omega_0 t}\, dt = \int_{-T/2}^{T/2} e^{jk\omega_0 t} \sum_{n=-N}^{N} D_n e^{jn\omega_0 t}\, dt. \tag{18.113}$$

The next step in the derivation of the expression for D_n is to evaluate the right-hand side of Eq. (18.113). We begin the evaluation by noting that the integration is with respect to time and the summation is with respect to N; therefore the operations can be interchanged. Letting I represent the integral of the right-hand side, we have

$$\begin{aligned}
I &= \sum_{n=-N}^{N} D_n \int_{-T/2}^{T/2} e^{j(k+n)\omega_0 t}\, dt \\
&= \sum_{n=-N}^{N} D_n \frac{e^{j(k+n)\omega_0 t}}{j(k+n)\omega_0}\Bigg|_{-T/2}^{T/2} \\
&= \sum_{n=-N}^{N} D_n T \frac{\sin (k+n)\pi}{(k+n)\pi}. \tag{18.114}
\end{aligned}$$

Now we make the observation that

$$\frac{\sin (k+n)\pi}{(k+n)\pi} = 0$$

except when $n = -k$. At $n = -k$ the ratio is 1. Therefore Eq. (18.114) reduces to

$$I = TD_{-k}. \tag{18.115}$$

Since Eq. (18.115) equals the left-hand side of Eq. (18.113), we have

$$D_{-k} = \frac{1}{T}\int_{-T/2}^{T/2} f(t)e^{jk\omega_0 t}\, dt, \tag{18.116}$$

which is equivalent to writing

$$D_k = \frac{1}{T}\int_{-T/2}^{T/2} f(t)e^{-jk\omega_0 t}\, dt. \tag{18.117}$$

Equation (18.117) is the formula for the Fourier coefficients. Therefore, we have shown that the mean-square error is minimized when $f(t)$ is approximated by a partial sum of trigonometric terms with Fourier coefficients.

DRILL EXERCISE
18.11

a) Calculate the mean-square error if the periodic current in Example 18.2 is approximated by the first term in its Fourier series and $I_m = 10$ A.

b) Repeat part (a) given that the periodic current is approximated by $10 \sin \omega_0 t$. (*Hint:* Observe that because of symmetry the computation of $\overline{\epsilon^2}$ can be made by integrating from 0 to $T/4$ and then dividing this result by $T/4$.)

Ans. (a) 0.4822 A^2; (b) 2.2764 A^2.

18.12   SUMMARY

Nonsinusoidal, but periodic, waveforms are frequently encountered in the operation of electrical systems. We have shown how we can use the Fourier series representation of such a periodic driving function to find the steady-state response function. The problem of finding the steady-state response is reduced to three major steps. The first step is to find the Fourier series representation of the periodic excitation. This representation consists of a dc source plus a series of harmonically related sinusoidal sources. The second step is to find the response of the circuit due to the dc source and then each sinusoidal source acting alone. The third and final step is to simply add the individual responses to form the Fourier series representation of the steady-state response. This third step is nothing more than invoking the principle of superposition in linear analysis.

In using the Fourier series approach there are several general observations worth remembering. First, any symmetry that exists in the waveform of a periodic function greatly simplifies the task of finding the Fourier coefficients. Therefore, when possible, always choose the point on the time axis where t equals zero with symmetry in mind. Second, since the Fourier series translates the problem from the time domain to the frequency domain, the frequency response of the circuit being analyzed becomes of primary importance in predicting the waveform of the output. Thus think in terms of the filtering (bandwidth) characteristics of the circuit when trying to ascertain how closely the output waveform will follow the input waveform. Third, only harmonics of the same frequency interact to produce average power. Therefore, a portion of the total average power delivered by a periodic waveform can be associated with each harmonic.

We are now ready to consider the Fourier transform which will allow us to apply some of the ideas introduced by Fourier series analysis to non-periodic time domain functions.

PROBLEMS

18.1 a) Verify Eqs. (18.6) and (18.7).
   b) Verify Eq. (18.8). (*Hint:* Use the trigonometric identity $\cos \alpha \sin \beta = \frac{1}{2} \sin (\alpha + \beta) - \frac{1}{2} \sin (\alpha - \beta)$.)
   c) Verify Eq. (18.9). (*Hint:* Use the trigonometric identity $\sin \alpha \sin \beta = \frac{1}{2} \cos (\alpha - \beta) - \frac{1}{2} \cos (\alpha + \beta)$.)
   d) Verify Eq. (18.10). (*Hint:* Use the trigonometric identity $\cos \alpha \cos \beta = \frac{1}{2} \cos (\alpha - \beta) + \frac{1}{2} \cos (\alpha + \beta)$.)

18.2 Derive Eq. (18.5).

18.3 Find the Fourier series expressions for the periodic voltage functions shown in Fig. 18.20.

18.4 Derive the expressions for the Fourier coefficients of an odd periodic function. (*Hint:* Use the same technique as the one used in the text in deriving Eqs. 18.14, 18.15, and 18.16.)

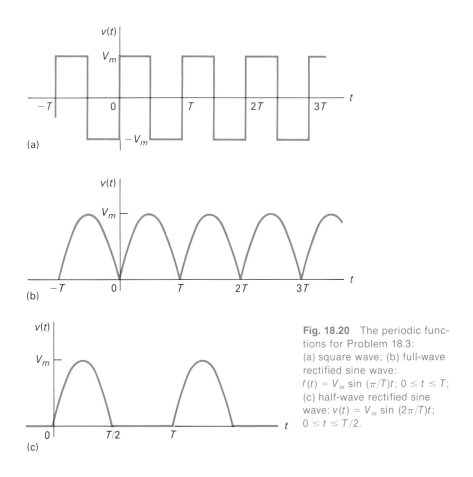

Fig. 18.20 The periodic functions for Problem 18.3: (a) square wave; (b) full-wave rectified sine wave: $f(t) = V_m \sin (\pi/T)t$; $0 \le t \le T$; (c) half-wave rectified sine wave: $v(t) = V_m \sin (2\pi/T)t$; $0 \le t \le T/2$.

18.5 Show that if $f(t) = -f(t - T/2)$, the Fourier coefficients b_k are given by the following expressions:

$$b_k = 0, \qquad\qquad\qquad \text{for } k \text{ even};$$

$$b_k = \frac{4}{T} \int_0^{T/2} f(t) \sin k\omega_0 t \; dt, \quad \text{for } k \text{ odd}.$$

(*Hint:* Use the same technique as the one used in the text to derive Eqs. 18.28 and 18.29.)

18.6 Derive Eq. (18.36). (*Hint:* Start with Eq. (18.29) and divide the interval of integration into 0 to $T/4$ and $T/4$ to $T/2$. Note that because of evenness and quarter-wave symmetry, $f(t) = -f(T/2 - t)$ in the interval $T/4 \le t \le T/2$. Let $x = T/2 - t$ in the second interval and combine the resulting integral with the integration between 0 and $T/4$.)

18.7 Derive Eq. (18.37). Follow the hint given in Problem 18.6 except that because of oddness and quarter-wave symmetry $f(t) = f(T/2 - t)$ in the interval $T/4 \le t \le T/2$.

18.8 For each of the periodic functions in Fig. 18.21 specify (a) ω_0 in rad/s; (b) f_0 in Hz; (c) the value of a_v; and (d) the equations for a_k and b_k. (e) For each function express $v(t)$ as a Fourier series.

18.9 Find the Fourier series of each periodic function shown in Fig. 18.22.

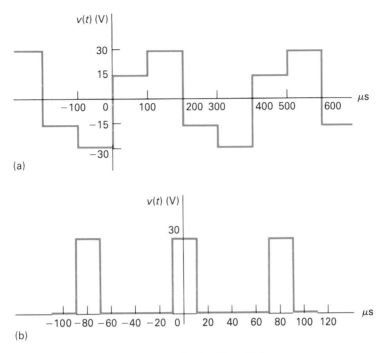

(a)

(b)

Fig. 18.21 The periodic voltages for Problem 18.8.

18.10 Given $f(t) = t^2$ over the interval $-1 < t < 1$.

  a) Construct a periodic function that satisfies this $f(t)$ between -1 and $+1$, has a period of 4 seconds, and has half-wave symmetry.
  b) Is the function even or odd?
  c) Does the function have quarter-wave symmetry?

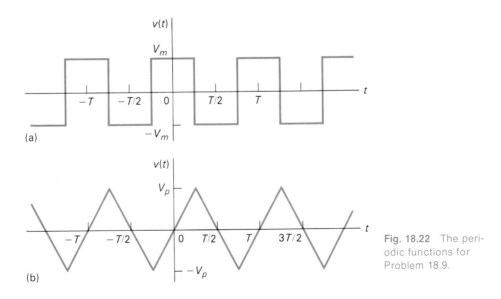

(a)

(b)

Fig. 18.22 The periodic functions for Problem 18.9.

18.11   Repeat Problem 18.10 given that $f(t) = t^3$ over the interval $-1 < t < 1$.

18.12   Given $v(t) = 20 \sin \pi|t|$ V over the interval $-1 \le t \le 1$. The function then repeats itself.

a) What is the fundamental frequency in rad/s?
b) Is the function even?
c) Is the function odd?
d) Does the function have half-wave symmetry?

18.13   One period of a periodic function is described by the following equations:

$$i(t) = 5t, \qquad\qquad -2 \text{ ms} \le t \le 2 \text{ ms};$$
$$i(t) = 10 \text{ mA}, \qquad\quad 2 \text{ ms} \le t \le 6 \text{ ms};$$
$$i(t) = 0.04 - 5t, \qquad 6 \text{ ms} \le t \le 10 \text{ ms};$$
$$i(t) = -10 \text{ mA}, \qquad 10 \text{ ms} \le t \le 14 \text{ ms}.$$

a) What is the fundamental frequency in Hz?
b) Is the function even?
c) Is the function odd?
d) Does the function have half-wave symmetry?
e) Does the function have quarter-wave symmetry?
f) Evaluate the Fourier coefficients a_v, a_1, a_2, b_1, and b_2.

18.14   It is sometimes possible to use symmetry to find the Fourier coefficients even though the original function is not symmetrical! With this thought in mind, consider the function in Drill Exercise 18.1. Observe that $v(t)$ can be divided into the two functions illustrated in Fig. 18.23(a) and (b). Furthermore, we can make $v_2(t)$ an even function by shifting it $T/8$ units to the left. This is illustrated in Fig. 18.23(c). At this point we note that $v(t) = v_1(t) + v_2(t)$ and that the Fourier series of $v_1(t)$ is a single-term series consisting of $V_m/2$. To find the Fourier series of $v_2(t)$ we first find the Fourier series of $v_2(t + T/8)$ and then shift this series $T/8$ units to the right. Use the technique outlined above to verify the Fourier series given as the answer to Drill Exercise 18.2(e).

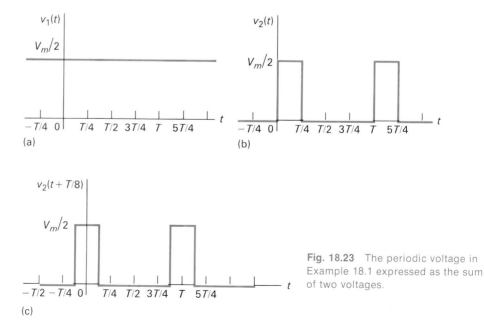

(a)

(b)

(c)

Fig. 18.23   The periodic voltage in Example 18.1 expressed as the sum of two voltages.

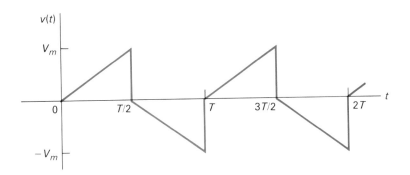

Fig. 18.24 The periodic function for Problem 18.15.

18.15 Derive the Fourier series for the periodic function shown in Fig. 18.24. Write the series in the form of Eq. (18.38).

18.16 Derive Eqs. (18.69) and (18.70).

18.17 a) Derive Eq. (18.71). (*Hint:* Note that $b_k = (4V_m/\pi k) + k\omega_0 RCa_k$. Use this expression for b_k to find $a_k^2 + b_k^2$ in terms of a_k. Now use the expression for a_k to derive Eq. (18.71).)

   b) Derive Eq. (18.72).

18.18 Show that when we combine Eqs. (18.71) and (18.72) with Eqs. (18.38) and (18.39) the result is Eq. (18.58). (*Hint:* Note from the definition of β_k that

$$\frac{a_k}{b_k} = -\tan \beta_k$$

and from the definition of θ_k that

$$\tan \theta_k = -\cot \beta_k.$$

Now use the trigonometric identity

$$\tan x = \cot (90 - x)$$

to show that $\theta_k = (90 + \beta_k).$)

18.19 The square-wave voltage shown in Fig. 18.25(a) is applied to the circuit shown in Fig. 18.25(b).

   a) Find the Fourier series representation of the steady-state current i.

   b) Find the steady-state expression for i by straightforward circuit analysis.

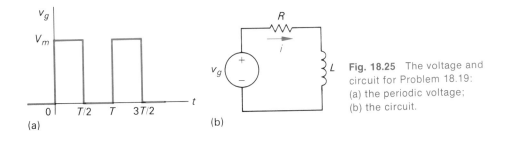

(a)

(b)

Fig. 18.25 The voltage and circuit for Problem 18.19: (a) the periodic voltage; (b) the circuit.

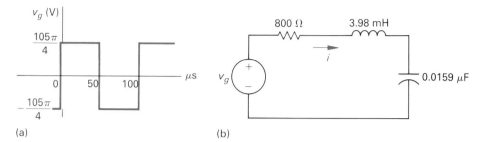

(a)   (b)

Fig. 18.26 The voltage and circuit for Problem 18.20: (a) the periodic voltage ($T = 100\ \mu s$); (b) the circuit.

18.20 The periodic square-wave voltage in Fig. 18.26(a) is applied to the circuit shown in Fig. 18.26(b). Derive the first four terms in the Fourier series that represents the steady-state current.

18.21 The periodic triangular waveform shown in Fig. 18.27(a) is applied to the bridge circuit shown in Fig. 18.27(b).

   a) Determine the first three terms in the Fourier series representation of v_o.
   b) If the bridge circuit is regarded as a filter, what frequency component present in the input voltage is eliminated from the output voltage?

18.22 It is claimed that the circuit shown in Fig. 18.28(a) will convert the square-wave voltage shown in Fig. 18.28(b) to an almost sinusoidal voltage with a frequency of 10 Mrad/s. Do you agree? Explain your answer.

18.23 The periodic sawtooth voltage shown in Fig. 18.29(a) is applied to the circuit shown in Fig. 18.29(b). Derive the first four terms in the Fourier series representation of the steady-state output voltage v_o.

18.24. The full-wave rectified sine-wave voltage shown in Fig. 18.30(a) is applied to the circuit shown in Fig. 18.30(b).

   a) Find the first four terms in the Fourier series representation of v_o.
   b) What is a good approximation for the steady-state expression of v_o?

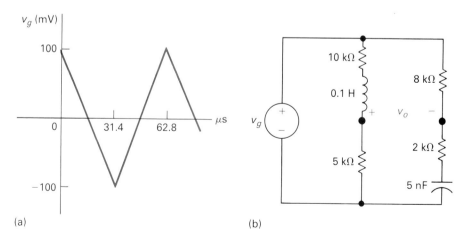

(a)   (b)

Fig. 18.27 The voltage and circuit for Problem 18.21: (a) the periodic voltage ($T = 62.8\ \mu s$); (b) the bridge circuit.

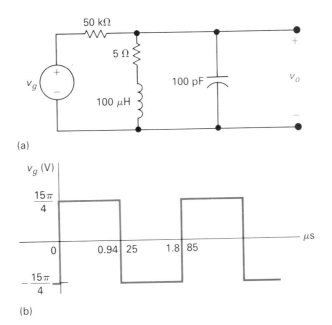

(a)

(b)

Fig. 18.28 The circuit and voltage for Problem 18.22:
(a) the circuit; (b) the periodic voltage.

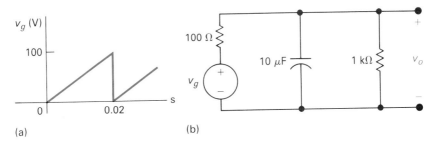

(a)

(b)

Fig. 18.29 The voltage and circuit for Problem 18.23: (a) the sawtooth voltage;
(b) the circuit.

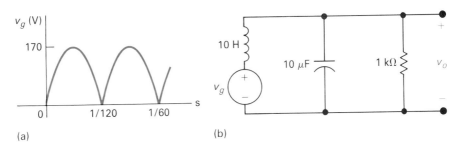

(a)

(b)

Fig. 18.30 The voltage and circuit for Problem 18.24: (a) the full-wave rectified sine wave;
(b) the circuit.

18.25   The Fourier series for the triangular current shown in Fig. 18.31 is

$$i(t) = 5\pi - 10 \sum_{n=1}^{\infty} \frac{1}{n} \sin n100\pi t.$$

a) Estimate the rms value of i using the first three terms of the Fourier series.
b) Repeat part (a) using the first five terms.
c) Repeat part (a) using the first seven terms.
d) Calculate the true rms value of i.

18.26   The voltage and current at the terminals of a network are

$$v = 5 + 300 \cos 1000t + 100 \sin 3000t \text{ V},$$
$$i = 3 + 10 \sin (1000t + 30°) + 6 \cos (3000t - 30°) \text{ A}.$$

The current is in the direction of the voltage drop across the terminals.

a) What is the average power at the terminals?
b) What is the rms value of the voltage?
c) What is the rms value of the current?

18.27   The square-wave voltage shown in Fig. 18.32(a) is applied to the circuit shown in Fig. 18.32(b). Find the average power associated with the fifth harmonic of the source voltage.

18.28   The Fourier series for the half-wave rectified sine wave shown in Fig. 18.33 is

$$v(t) = \frac{A}{\pi} + \frac{A}{2} \sin \omega_0 t - \frac{A}{\pi} \sum_{n=2}^{\infty} \frac{[1 + \cos n\pi]}{(n^2 - 1)} \cos n\omega_0 t.$$

If this voltage is applied to a 10-Ω resistor, what percentage of the total power delivered to the resistor is carried by the second harmonic voltage?

18.29   Derive the expression for C_n in the exponential Fourier series for the periodic waveform shown in Fig. 18.34.

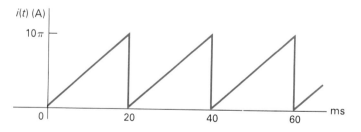

Fig. 18.31   The periodic current for Problem 18.25.

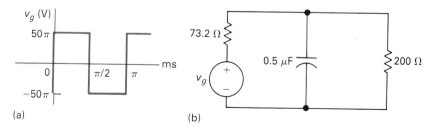

(a)   (b)

Fig. 18.32   The voltage and circuit for Problem 18.27: (a) the square-wave voltage; (b) the circuit.

18.30 a) The periodic voltage in Problem 18.29 is applied across a 2.5-Ω resistor. If V_m is 10 V and $\tau = T/8$ s, what is the average power delivered to the resistor?

b) What percentage of the total average power is associated with the first five harmonics?

18.31 A function is defined to equal $4t$ on the interval $-2 \le t \le 2$. The function is periodic with a period of 8 s and has half-wave symmetry. Choose t equal to zero where $f(t) = 0$ and $f'(t) = 4$. Find the exponential Fourier series for the function.

18.32 The periodic current source in the circuit shown in Fig. 18.35(a) has the waveform shown in Fig. 18.35(b). If the period of i is 628.3185 μs, estimate the average power delivered to the 5-kΩ resistor.

18.33 Use Eq. (18.97) to estimate the rms value of the periodic function described in Problem 18.31.

18.34 a) Make an amplitude and phase plot, based on Eq. (18.38), for the periodic voltage in Example 18.3. Assume V_m is 40 V. Plot both amplitude and phase versus $n\omega_0$ where $n = 0, 1, 2, 3, \ldots$.

b) Repeat part (a) except to base the plots on Eq. (18.82).

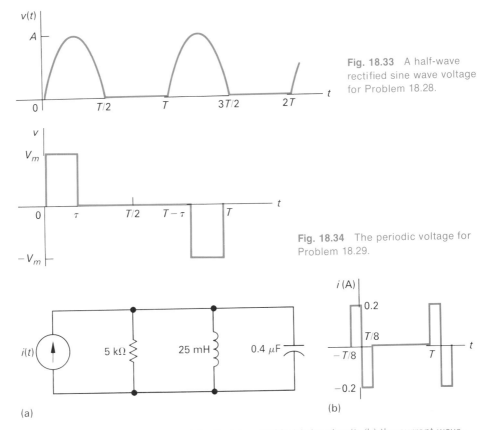

Fig. **18.33** A half-wave rectified sine wave voltage for Problem 18.28.

Fig. **18.34** The periodic voltage for Problem 18.29.

(a)

(b)

Fig. **18.35** The circuit and current for Problem 18.32: (a) the circuit; (b) the current waveform.

18.35 A periodic function is represented by a Fourier series that has a finite number of terms. The amplitude and phase spectra are shown in Fig. 18.36(a) and (b), respectively.

    a) Write the expression for the periodic current using the form given by Eq. (18.38).
    b) Is the current an even or odd function of t?
    c) Does the current have half-wave symmetry?
    d) Calculate the rms value of the current in mA.
    e) Write the exponential form of the Fourier series.
    f) Make the amplitude and phase spectra plots on the basis of the exponential series.

18.36   a) Approximate the periodic square-wave voltage shown in Fig. 18.37 with the first term in its Fourier series representation. Using this partial-sum approximation to $v(t)$, sketch $\epsilon(t)$ versus t over the range $-T/2 < t < T/2$.
    b) Sketch $\epsilon^2(t)$ versus t over the range $-T/2 < t < T/2$.
    c) Calculate $\overline{\epsilon^2}$.
    d) Approximate $v(t)$ with the single term $10 \sin \omega_0 t$. Calculate $\overline{\epsilon^2}$.
    e) Compare the value of $\overline{\epsilon^2}$ in part (c) with that obtained in part (d).

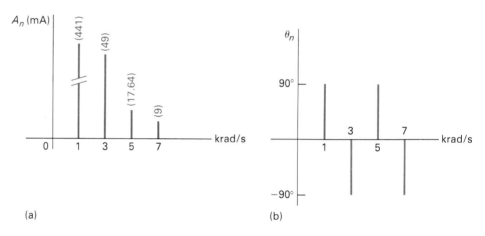

(a)   (b)

Fig. 18.36 The amplitude and phase spectra for Problem 18.35: (a) the amplitude spectrum; (b) the phase-angle spectrum.

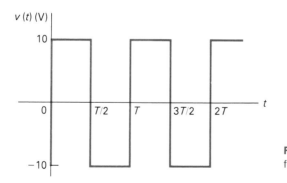

Fig. 18.37 The periodic voltage for Problem 18.36.

THE FOURIER TRANSFORM

19

19.1 INTRODUCTION

In Chapter 18, we discussed the representation of a periodic function by means of a Fourier series. This series representation enables us to describe the periodic function in terms of the frequency domain attributes of amplitude and phase. The Fourier transform enables us to extend this frequency domain description to functions that are not periodic. The idea of transforming an aperiodic function from the time domain to the frequency domain has already been introduced through the Laplace transform. You may wonder, then, why yet another transformation is necessary. Strictly speaking, the Fourier transform is not a new transform. It is a special case of the *bilateral* Laplace transform with the real part of the complex frequency set equal to zero. However, in introducing the idea of the Fourier transform, it is more satisfying in terms of physical interpretation to view it as a limiting case of a Fourier series. We will adopt this point of view in Section 19.2, where we derive the Fourier transform equations.

There are problems in communications theory and signal processing where the Fourier transform is more useful than the Laplace transform. Although we cannot pursue this area of study in this introductory text, it seems appropriate to introduce the Fourier transform at a time when the ideas underlying the Laplace transform and the Fourier series are still fresh in your mind.

19.2 DERIVATION OF THE FOURIER TRANSFORM

The derivation of the Fourier transform as a limiting case of a Fourier series begins with the exponential form of the series; thus

$$f(t) = \sum_{n=-\infty}^{\infty} C_n e^{jn\omega_0 t}, \tag{19.1}$$

where

$$C_n = \frac{1}{T} \int_{-T/2}^{T/2} f(t)e^{-jn\omega_0 t} \, dt. \tag{19.2}$$

In writing Eq. (19.2), we have elected to start the integration at t_0 equal to $-T/2$.

The transition from a periodic to an aperiodic function is accomplished by allowing the fundamental period T to increase without limit. That is, if T becomes infinite, the function never repeats itself and hence is aperiodic. As T increases, the separation between adjacent harmonic frequencies becomes smaller and smaller. In particular,

$$\Delta\omega = (n + 1)\omega_0 - n\omega_0 = \omega_0 = \frac{2\pi}{T}, \tag{19.3}$$

and as T gets larger and larger the incremental separation ($\Delta\omega$) approaches a differential separation $d\omega$. It follows from Eq. (19.3) that

$$\frac{1}{T} \to \frac{d\omega}{2\pi} \quad \text{as} \quad T \to \infty. \tag{19.4}$$

We conclude that as the period increases the frequency moves from being a discrete variable to becoming a continuous variable. We note this transition as

$$n\omega_0 \to \omega \quad \text{as } T \to \infty. \tag{19.5}$$

Now let us turn our attention to Eq. (19.2). As the period increases, the Fourier coefficients C_n get smaller. In the limit C_n approaches zero as T approaches infinity. This makes sense since we would expect the Fourier coefficients to vanish as the function loses its periodicity. However, it is interesting to note with respect to Eq. (19.2) the limiting value of the product C_nT. We see from Eq. (19.2) that

$$C_nT \to \int_{-\infty}^{\infty} f(t)e^{-j\omega t}\, dt \quad \text{as } T \to \infty. \tag{19.6}$$

In writing Eq. (19.6) we have taken advantage of the relationship given in Eq. (19.5). The integral in Eq. (19.6) is the Fourier transform of $f(t)$ and is denoted as

$$F(\omega) = \mathscr{F}\{f(t)\} = \int_{-\infty}^{\infty} f(t)e^{-j\omega t}\, dt. \tag{19.7}$$

An explicit expression for the inverse Fourier transform is obtained by investigating the limiting form of Eq. (19.1) as $T \to \infty$. We begin by first multiplying and dividing by T; thus

$$f(t) = \sum_{n=-\infty}^{\infty} (C_nT)e^{jn\omega_0 t}\left(\frac{1}{T}\right). \tag{19.8}$$

As T approaches infinity, the summation approaches integration, C_nT approaches $F(\omega)$, $n\omega_0$ approaches ω, and $1/T$ approaches $d\omega/2\pi$. Thus, in the limit, Eq. (19.8) becomes

$$f(t) = \frac{1}{2\pi}\int_{-\infty}^{\infty} F(\omega)e^{j\omega t}\, d\omega. \tag{19.9}$$

Equations (19.7) and (19.9) define the Fourier transform. Equation (19.7) transforms the time domain expression $f(t)$ into its corresponding frequency domain expression $F(\omega)$ and Eq. (19.9) defines the inverse operation of transforming $F(\omega)$ into $f(t)$.

It will be informative at this time to derive the Fourier transform of the single voltage pulse shown in Fig. 19.1. Note that this single voltage pulse corresponds to the periodic voltage in Example 18.6 if we let $T \to \infty$. The Fourier

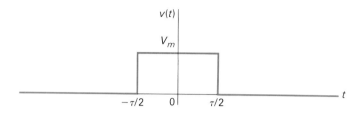

Fig. 19.1 A voltage pulse.

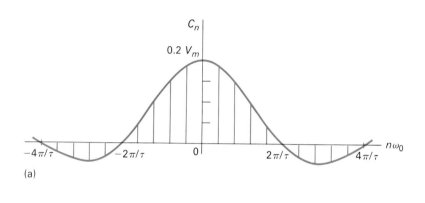

(a)

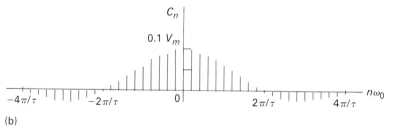

(b)

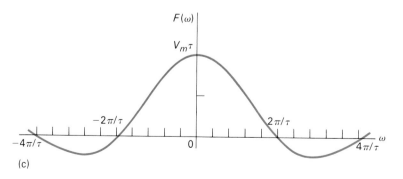

(c)

Fig. 19.2 Transition of the amplitude spectrum as $f(t)$ goes from periodic to aperiodic: (a) C_n versus $n\omega_0$, $(T/\tau) = 5$; (b) $|C_n|$ versus $n\omega_0$, $(T/\tau) = 10$; (c) $F(\omega)$ versus ω.

transform of $v(t)$ is found directly from Eq. (19.7); thus

$$F(\omega) = \int_{-\tau/2}^{\tau/2} V_m e^{-j\omega t} \, dt$$

$$= V_m \left. \frac{e^{-j\omega t}}{(-j\omega)} \right|_{-\tau/2}^{\tau/2}$$

$$= \frac{V_m}{-j\omega} (-2j \sin \omega\tau/2), \tag{19.10}$$

which can be put in the form of $(\sin x)/x$ by multiplying the numerator and denominator by τ. We then have

$$F(\omega) = V_m \tau \cdot \frac{\sin \omega\tau/2}{\omega\tau/2}. \tag{19.11}$$

For the periodic train of voltage pulses in Example 18.6, the expression for the Fourier coefficients is (see Example 18.6)

$$C_n = \frac{V_m \tau}{T} \frac{\sin n\omega_0\tau/2}{(n\omega_0\tau/2)}. \tag{19.12}$$

If we compare Eqs. (19.11) and (19.12) we see very clearly that as the time domain function goes from periodic to aperiodic, the amplitude spectrum goes from a discrete line spectrum to a continuous spectrum. Furthermore, the envelope of the line spectrum has the same shape as the continuous spectrum. Thus as T increases, the spectrum of lines gets denser and the amplitudes get smaller but the envelope does not change in shape. Thus we have the physical interpretation of the Fourier transform $F(\omega)$ as being a measure of the frequency content of $f(t)$. The observations are illustrated graphically in Fig. 19.2. In plotting the amplitude spectrum we assume that τ is constant and T is increasing.

19.3 CONVERGENCE OF THE FOURIER INTEGRAL

A given function of time $[f(t)]$ has a Fourier transform if the integral in Eq. (19.7) converges. If $f(t)$ is a "well-behaved" function that differs from zero over a *finite* interval of time, then there is no problem with convergence. By "well-behaved" we imply that $f(t)$ is single-valued and encloses a finite area over the range of integration. From a practical point of view, all pulses of finite duration in which we are interested are well-behaved functions. The evaluation of the Fourier transform of the rectangular pulse discussed in Section 19.2 illustrates this point.

If $f(t)$ is different from zero over an infinite interval, then the convergence of the Fourier integral depends on the behavior of $f(t)$ as t approaches plus or minus infinity. A single-valued function that is nonzero over an infinite in-

Fig. 19.3   The exponential function $Ke^{-at}u(t)$.

terval will have a Fourier transform if the integral

$$\int_{-\infty}^{\infty} |f(t)|\ dt$$

exists and if any discontinuities in $f(t)$ are finite. An example of this type of function is the exponential decaying function illustrated in Fig. 19.3.

The Fourier transform of $f(t)$ is

$$F(\omega) = \int_{-\infty}^{\infty} f(t)e^{-j\omega t}\ dt = \int_{0}^{\infty} Ke^{-at}e^{-j\omega t}\ dt$$

$$= \frac{Ke^{-(a+j\omega)t}}{-(a+j\omega)}\Big|_{0}^{\infty} = \frac{K}{-(a+j\omega)}\ (0-1)$$

$$= \frac{K}{a+j\omega}. \qquad\qquad (19.13)$$

A third important group of functions are those functions that are of great practical interest but do not in a strict sense have a Fourier transform. For example, the integral in Eq. (19.7) does not converge if $f(t)$ is a constant. The same can be said if $f(t)$ is a sinusoidal function (cos $\omega_0 t$) or a step function $[Ku(t)]$. Since these functions are of great interest in circuit analysis, we do not want to exclude them from Fourier analysis. In order to include these functions, we resort to a little mathematical "subterfuge." We create a function in the time domain that has a Fourier transform and at the same time can be made arbitrarily close to the function of interest. Next we find the Fourier transform of the approximating function and then evaluate the limiting value of $F(\omega)$ as the approximating function approaches $f(t)$. The limiting value of $F(\omega)$ is defined as the Fourier transform of $f(t)$. We will demonstrate this technique by finding the Fourier transform of a constant.

We can approximate a constant with the exponential function

$$f(t) = Ae^{-\epsilon|t|}, \qquad \epsilon > 0. \qquad\qquad (19.14)$$

Observe that as $\epsilon \to 0$, $f(t) \to A$. The approximation is shown graphically in Fig. 19.4.

The Fourier transform of $f(t)$ is

$$F(\omega) = \int_{-\infty}^{0} Ae^{\epsilon t}e^{-j\omega t}\ dt + \int_{0}^{\infty} Ae^{-\epsilon t}e^{-j\omega t}\ dt. \qquad\qquad (19.15)$$

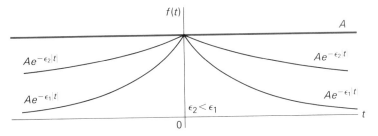

Fig. 19.4   The approximation of a constant with an exponential function.

Carrying out the integration called for in Eq. (19.15) yields

$$F(\omega) = \frac{A}{\epsilon - j\omega} + \frac{A}{\epsilon + j\omega} = \frac{2\epsilon A}{\epsilon^2 + \omega^2}. \tag{19.16}$$

The function given by Eq. (19.16) generates an impulse function at $\omega = 0$ as $\epsilon \to 0$. You can verify this by showing that $F(\omega)$ approaches infinity at $\omega = 0$ as $\epsilon \to 0$; the duration of $F(\omega)$ approaches zero as $\epsilon \to 0$; and the area under $F(\omega)$ is independent of ϵ. The area under $F(\omega)$ is the strength of the impulse and is

$$\int_{-\infty}^{\infty} \frac{2\epsilon A}{\epsilon^2 + \omega^2} \, d\omega = 4\epsilon A \int_{0}^{\infty} \frac{d\omega}{\epsilon^2 + \omega^2} = 2\pi A. \tag{19.17}$$

We conclude that in the limit, $f(t)$ approaches a constant A and $F(\omega)$ approaches an impulse function $2\pi A \delta(\omega)$. Therefore, the Fourier transform of a constant A is defined as $2\pi A \delta(\omega)$; thus

$$\mathcal{F}\{A\} = 2\pi A \delta(\omega). \tag{19.18}$$

In Section 19.5 we will have more to say about Fourier transforms that are defined through a limit process. Before doing so, we take time to point out in the following section how we can take advantage of the Laplace transform to find the Fourier transform of functions for which the Fourier integral converges.

DRILL EXERCISE   Use the defining integral to find the Fourier transform of the following functions.

19.1   a) $f(t) = -A$, $-\tau/2 \le t < 0$
$\quad\quad f(t) = A$, $0 < t \le \tau/2$
$\quad\quad f(t) = 0$,   elsewhere

b) $f(t) = 0$, $t < 0$
$\quad\quad f(t) = te^{-at}$, $t \ge 0, a > 0$

Ans.   (a) $-j(2A/\omega)\left(1 - \cos\dfrac{\omega\tau}{2}\right)$; (b) $1/(a + j\omega)^2$.

The Fourier transform of $f(t)$ is given by the following expressions:

$$F(\omega) = 0, \qquad -\infty \le \omega < -2,$$
$$F(\omega) = 2, \qquad -2 < \omega < -1,$$
$$F(\omega) = 1, \qquad -1 < \omega < 1,$$
$$F(\omega) = 2, \qquad 1 < \omega < 2,$$
$$F(\omega) = 0, \qquad 2 < \omega \le \infty.$$

Find $f(t)$.

Ans. $f(t) = \dfrac{1}{\pi t}(2 \sin 2t - \sin t).$

19.4 USING LAPLACE TRANSFORMS TO FIND FOURIER TRANSFORMS

A table of unilateral, or one-sided, Laplace transform pairs can be used to find the Fourier transform of functions *for which the Fourier integral converges.* The Fourier integral will converge when all the poles of $F(s)$ lie in the left-half of the s-plane. (If $F(s)$ has poles in the right-half plane or along the imaginary axis, then $f(t)$ will not satisfy the constraint that $\int_{-\infty}^{\infty}|f(t)|\ dt$ exists.) The rules for using Laplace transforms to find the Fourier transforms of such functions are described below.

1. If $f(t)$ is zero for $t \le 0^-$, we can obtain the Fourier transform of $f(t)$ from the Laplace transform of $f(t)$ by simply replacing s by $j\omega$. Thus

$$\mathcal{F}\{f(t)\} = \mathcal{L}\{f(t)\}_{s=j\omega}. \qquad (19.19)$$

For example, if

$$f(t) = 0, \qquad\qquad t \le 0^-,$$
$$f(t) = e^{-at} \cos \omega_0 t, \qquad t \ge 0^+,$$

then

$$\mathcal{F}\{f(t)\} = \frac{s + a}{(s + a)^2 + \omega_0^2}\bigg|_{s=j\omega}$$
$$= \frac{j\omega + a}{(j\omega + a)^2 + \omega_0^2}.$$

2. Since the range of integration on the Fourier integral goes from $-\infty$ to $+\infty$, the Fourier transform of a negative-time function exists. A negative-time function is defined as a function that is nonzero for negative values of time and zero for positive values of time. To find the Fourier transform of a negative-time function, we proceed as follows. First, we reflect the negative-time function over to the positive time domain and then find its

one-sided Laplace transform. The Fourier transform of the original time function is obtained by replacing s by $-j\omega$. Therefore, when $f(t) = 0$ for $t \geq 0^+$,

$$\mathscr{F}\{f(t)\} = \mathscr{L}\{f(-t)\}_{s=-j\omega}. \tag{19.20}$$

For example, if

$$f(t) = 0, \qquad\qquad t \geq 0^+,$$
$$f(t) = e^{at} \cos \omega_0 t, \qquad t \leq 0^-,$$

then

$$f(-t) = 0, \qquad t \leq 0^-$$

and

$$f(-t) = e^{-at} \cos \omega_0 t, \qquad t \geq 0^+.$$

Both $f(t)$ and its mirror image are plotted in Fig. 19.5. The Fourier transform of $f(t)$ is

$$\mathscr{F}\{f(t)\} = \mathscr{L}\{f(-t)\}_{s=-j\omega} = \frac{s + a}{(s + a)^2 + \omega_0^2}\bigg|_{s=-j\omega}$$

$$= \frac{-j\omega + a}{(-j\omega + a)^2 + \omega_0^2}.$$

3. Functions that are nonzero over all time can be resolved into positive- and negative-time functions. We can use Eqs. (19.19) and (19.20) to find the Fourier transform of the positive- and negative-time functions, respectively. The Fourier transform of the original function is the sum of the two transforms. Thus if we let

$$f^+(t) = f(t) \quad \text{for } t > 0$$

and

$$f^-(t) = f(t) \quad \text{for } t < 0,$$

then

$$f(t) = f^+(t) + f^-(t)$$

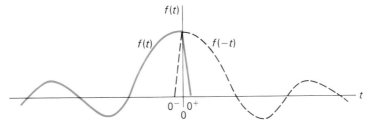

Fig. 19.5   Reflection of a negative-time function over to the positive time domain.

and

$$\mathscr{F}\{f(t)\} = \mathscr{F}\{f^+(t)\} + \mathscr{F}\{f^-(t)\}$$
$$= \mathscr{L}\{f^+(t)\}_{s=j\omega} + \mathscr{L}\{f^-(-t)\}_{s=-j\omega}. \qquad \textbf{(19.21)}$$

As an example of using Eq. (19.21), we will find the Fourier transform of $e^{-a|t|}$. For the given function the positive- and negative-time functions are

$$f^+(t) = e^{-at}$$

and

$$f^-(t) = e^{at}.$$

It follows directly that

$$\mathscr{L}\{f^+(t)\} = \frac{1}{s + a}$$

and

$$\mathscr{L}\{f^-(-t)\} = \frac{1}{s + a}.$$

Therefore, from Eq. (19.21) we have

$$\mathscr{F}\{e^{-a|t|}\} = \frac{1}{s + a}\bigg|_{s=j\omega} + \frac{1}{s + a}\bigg|_{s=-j\omega}$$

$$= \frac{1}{j\omega + a} + \frac{1}{-j\omega + a}$$

$$= \frac{2a}{\omega^2 + a^2}.$$

It is worth noting that if $f(t)$ is even, Eq. (19.21) reduces to

$$\mathscr{F}\{f(t)\} = \mathscr{L}\{f(t)\}_{s=j\omega} + \mathscr{L}\{f(t)\}_{s=-j\omega}. \qquad \textbf{(19.22)}$$

If $f(t)$ is odd, then Eq. (19.21) becomes

$$\mathscr{F}\{f(t)\} = \mathscr{L}\{f(t)\}_{s=j\omega} - \mathscr{L}\{f(t)\}_{s=-j\omega}. \qquad \textbf{(19.23)}$$

DRILL EXERCISE
19.3

Find the Fourier transform of each of the following functions. In all the functions, a is a positive real constant.

a) $f(t) = 0, \qquad\qquad t < 0,$
$ f(t) = e^{-at} \sin \omega_0 t, \qquad t \geq 0$

b) $f(t) = 0, \qquad\qquad t > 0,$
$ f(t) = -te^{at}, \qquad\quad t \leq 0$

c) $f(t) = te^{-at}, \qquad\quad t \geq 0,$
$ f(t) = te^{at}, \qquad\qquad t \leq 0$

Ans. (a) $\dfrac{\omega_0}{(a + j\omega)^2 + \omega_0^2}$; (b) $\dfrac{1}{(a - j\omega)^2}$; (c) $\dfrac{- j4a\omega}{(a^2 + \omega^2)^2}$.

19.5 FOURIER TRANSFORMS IN THE LIMIT

As we pointed out in Section 19.4, there are several practical functions whose Fourier transforms must be defined through a limit process. We now return to these types of functions and develop their transforms.

We have already shown that the Fourier transform of a constant A is $2\pi A\delta(\omega)$ (Eq. 19.18). Our next function of interest is the signum function, which is defined as $+1$ for $t > 0$ and -1 for $t < 0$. The signum function is denoted as sgn (t) and can be expressed in terms of unit step functions; thus

$$\text{sgn } (t) = u(t) - u(-t). \tag{19.24}$$

The function is shown graphically in Fig. 19.6.

To find the Fourier transform of the signum function, we first create a function that approaches the signum function in the limit. Thus

$$\text{sgn } (t) = \lim_{\epsilon \to 0} \left[e^{-\epsilon t} u(t) - e^{\epsilon t} u(-t) \right]. \tag{19.25}$$

The function inside the brackets is plotted in Fig. 19.7. It is important to note that the function shown there has a Fourier transform because the Fourier integral converges. Since $f(t)$ is an odd function, we can use Eq. (19.23) to find its Fourier transform. We have

$$\mathcal{F}\{f(t)\} = \frac{1}{s + \epsilon}\bigg|_{s=j\omega} - \frac{1}{s + \epsilon}\bigg|_{s=-j\omega}$$

$$= \frac{1}{j\omega + \epsilon} - \frac{1}{-j\omega + \epsilon}$$

$$= \frac{- 2j\omega}{\omega^2 + \epsilon^2} \tag{19.26}$$

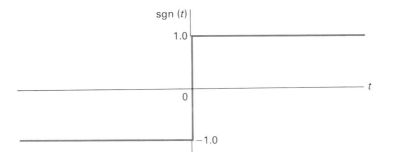

Fig. 19.6 The signum function.

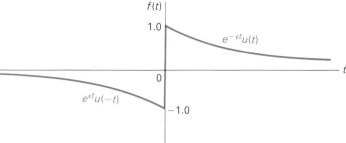

Fig. 19.7 A function that approaches the sgn(t) as ϵ approaches zero.

as $\epsilon \to 0$, $f(t) \to \text{sgn}\ (t)$ and $\mathscr{F}\{f(t)\} \to 2/j\omega$; therefore we have

$$\mathscr{F}\{\text{sgn}\ (t)\} = \frac{2}{j\omega}. \tag{19.27}$$

To find the Fourier transform of the unit step function, we can use the results given by Eqs. (19.18) and (19.27). We do this by recognizing that the unit step function can be expressed as

$$u(t) = \frac{1}{2} + \frac{1}{2}\ \text{sgn}\ (t). \tag{19.28}$$

Thus

$$\mathscr{F}\{u(t)\} = \mathscr{F}\left\{\frac{1}{2}\right\} + \mathscr{F}\left\{\frac{1}{2}\ \text{sgn}\ (t)\right\}$$

$$= \pi\delta(\omega) + \frac{1}{j\omega}. \tag{19.29}$$

To find the Fourier transform of the cos $\omega_0 t$, we return to the inverse-transform integral of Eq. (19.9) and observe that if

$$F(\omega) = 2\pi\delta(\omega - \omega_0), \tag{19.30}$$

then

$$f(t) = \frac{1}{2\pi} \int_{-\infty}^{\infty} [2\pi\delta(\omega - \omega_0]e^{j\omega t}\ d\omega. \tag{19.31}$$

Using the sifting property of the impulse function, we reduce Eq. (19.31) to

$$f(t) = e^{j\omega_0 t}. \tag{19.32}$$

We have, then, from Eqs. (19.30) and (19.32) that

$$\mathscr{F}\{e^{j\omega_0 t}\} = 2\pi\delta(\omega - \omega_0). \tag{19.33}$$

Now we can use Eq. (19.33) to find the Fourier transform of the cos $\omega_0 t$ since

$$\cos \omega_0 t = \frac{e^{j\omega_0 t} + e^{-j\omega_0 t}}{2}. \tag{19.34}$$

TABLE 19.1
FOURIER TRANSFORMS OF ELEMENTARY FUNCTIONS

f(t)	F(ω)
$\delta(t)$ (impulse)	1
A (constant)	$2\pi A\delta(\omega)$
sgn (t) (signum)	$\dfrac{2}{j\omega}$
$u(t)$ (step)	$\pi\delta(\omega) + \dfrac{1}{j\omega}$
$e^{-at}u(t)$ (positive-time exponential)	$\dfrac{1}{a + j\omega}$
$e^{at}u(-t)$ (negative-time exponential)	$\dfrac{1}{a - j\omega}$
$e^{-a|t|}$ (positive- and negative-time exponential)	$\dfrac{2a}{a^2 + \omega^2}$
$e^{j\omega_0 t}$ (complex exponential)	$2\pi\delta(\omega - \omega_0)$
cos $\omega_0 t$ (cosine)	$\pi[\delta(\omega + \omega_0) + \delta(\omega - \omega_0)]$
sin $\omega_0 t$ (sine)	$j\pi[\delta(\omega + \omega_0) - \delta(\omega - \omega_0)]$

Thus

$$\mathscr{F}\{\cos \omega_0 t\} = \frac{1}{2}\left[\mathscr{F}\{e^{j\omega_0 t}\} + \mathscr{F}\{e^{-j\omega_0 t}\}\right]$$

$$= \frac{1}{2}\left[2\pi\delta(\omega - \omega_0) + 2\pi\delta(\omega + \omega_0)\right]$$

$$= \pi\delta(\omega - \omega_0) + \pi\delta(\omega + \omega_0). \tag{19.35}$$

The Fourier transform of the sin $\omega_0 t$ can be found by a similar manipulation and is left as an exercise in Drill Exercise 19.4.

A summary of the transform pairs of the important elementary functions is given in Table 19.1.

Now that we have developed the Fourier transforms of the elementary functions, we need to discuss the properties of the transform that enhance our ability to describe aperiodic time domain behavior in terms of frequency domain behavior. These properties are discussed in the following four sections.

DRILL EXERCISE
19.4

Find $\mathscr{F}\{\sin \omega_0 t\}$.

Ans. $j\pi[\delta(\omega + \omega_0) - \delta(\omega - \omega_0)]$.

19.6 SOME MATHEMATICAL PROPERTIES

The first mathematical property we call to your attention is the fact that $F(\omega)$ is a complex quantity and can be expressed in either rectangular or polar form. Thus from the defining integral we have

$$
\begin{aligned}
F(\omega) &= \int_{-\infty}^{\infty} f(t)e^{-j\omega t}\, dt \\
&= \int_{-\infty}^{\infty} f(t)[\cos \omega t - j \sin \omega t]\, dt \\
&= \int_{-\infty}^{\infty} f(t) \cos \omega t\, dt - j\int_{-\infty}^{\infty} f(t) \sin \omega t\, dt.
\end{aligned} \tag{19.36}
$$

Now we let

$$
A(\omega) = \int_{-\infty}^{\infty} f(t) \cos \omega t\, dt \tag{19.37}
$$

and

$$
B(\omega) = -\int_{-\infty}^{\infty} f(t) \sin \omega t\, dt. \tag{19.38}
$$

Thus using the definitions given by Eqs. (19.37) and (19.38) in Eq. (19.36) we have

$$
F(\omega) = A(\omega) + jB(\omega) = |F(\omega)|e^{j\theta(\omega)}. \tag{19.39}
$$

Now we make the following observations about $F(\omega)$:

1. The real part of $F(\omega)$, that is, $A(\omega)$, is an even function of ω; in other words, $A(\omega) = A(-\omega)$.
2. The imaginary part of $F(\omega)$, that is, $B(\omega)$, is an odd function of ω; in other words, $B(\omega) = -B(-\omega)$.
3. The magnitude of $F(\omega)$, that is, $\sqrt{A^2(\omega) + B^2(\omega)}$, is an even function of ω.
4. The phase angle of $F(\omega)$, that is, $\theta(\omega) = \tan^{-1} B(\omega)/A(\omega)$, is an odd function of ω.
5. Replacing ω by $-\omega$ generates the conjugate of $F(\omega)$; in other words, $F(-\omega) = F^*(\omega)$.

With these properties of $F(\omega)$ in mind, we make the additional observations that if $f(t)$ is an even function, $F(\omega)$ is real, and if $f(t)$ is an odd function, $F(\omega)$ is imaginary. If $f(t)$ is even, we have from Eqs. (19.37) and (19.38)

$$
A(\omega) = 2\int_{0}^{\infty} f(t) \cos \omega t\, dt \tag{19.40}
$$

and

$$
B(\omega) = 0. \tag{19.41}
$$

If $f(t)$ is an odd function, then

$$A(\omega) = 0, \tag{19.42}$$

$$B(\omega) = -2 \int_0^\infty f(t) \sin \omega t \, dt. \tag{19.43}$$

The derivations of Eqs. (19.40)–(19.43) are left as exercises in Problems 19.5 and 19.6.

It is of interest to note that if $f(t)$ is an even function, its Fourier transform is an even function and if $f(t)$ is an odd function, its Fourier transform is an odd function. We can make an additional observation about the relationship between $f(t)$ and its Fourier transform if $f(t)$ is an even function. This observation comes from the inverse Fourier integral. Thus if $f(t)$ is even, then

$$
\begin{aligned}
f(t) &= \frac{1}{2\pi} \int_{-\infty}^{\infty} F(\omega) e^{j\omega t} \, d\omega \\
&= \frac{1}{2\pi} \int_{-\infty}^{\infty} A(\omega) e^{j\omega t} \, d\omega \\
&= \frac{1}{2\pi} \int_{-\infty}^{\infty} A(\omega)[\cos \omega t + j \sin \omega t] \, d\omega \\
&= \frac{1}{2\pi} \int_{-\infty}^{\infty} A(\omega) \cos \omega t \, d\omega + 0 \\
&= \frac{2}{2\pi} \int_{0}^{\infty} A(\omega) \cos \omega t \, d\omega. \tag{19.44}
\end{aligned}
$$

Now compare Eq. (19.44) with Eq. (19.40) and note that except for a factor of $1/2\pi$ these two equations are of the same form. Thus the waveforms of $A(\omega)$ and $f(t)$ become interchangeable if $f(t)$ is an even function. For example, we have already observed that a rectangular pulse in the time domain produces a frequency spectrum of the form $\sin \omega / \omega$. Specifically, the Fourier transform of the voltage pulse shown in Fig. 19.1 is given by Eq. (19.11). It follows then that a rectangular pulse in the frequency domain must be generated by a time domain function of the form $\sin t / t$. Let us illustrate by finding the time domain function $f(t)$ corresponding to the frequency spectrum shown in Fig. 19.8. It

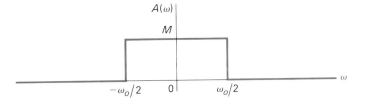

Fig. 19.8 A rectangular frequency spectrum.

follows from Eq. (19.44) that

$$f(t) = \frac{2}{2\pi} \int_0^{\omega_0/2} M \cos \omega t \, d\omega$$

$$= \frac{2M}{2\pi} \left[\frac{\sin \omega t}{t} \right] \Big|_0^{\omega_0/2}$$

$$= \frac{1}{2\pi} \left[2M \frac{\sin \omega_0 t/2}{t} \right]$$

$$= \frac{1}{2\pi} \left[M\omega_0 \frac{\sin (\omega_0 t/2)}{\omega_0 t/2} \right]. \qquad (19.45)$$

We will have more to say about the frequency spectrum of a rectangular pulse in the time domain versus the rectangular frequency spectrum of $(\sin t)/t$ after we introduce Parseval's theorem.

DRILL EXERCISE
19.5

Given that $f(t)$ is a real function of t, show that the inversion integral reduces to

$$f(t) = \frac{1}{2\pi} \int_{-\infty}^{\infty} [A(\omega) \cos \omega t - B(\omega) \sin \omega t] \, d\omega.$$

DRILL EXERCISE
19.6

Given that $f(t)$ is a real, odd function of t, show that the inversion integral reduces to

$$f(t) = -\frac{1}{2\pi} \int_{-\infty}^{\infty} B(\omega) \sin \omega t \, d\omega.$$

19.7 OPERATIONAL TRANSFORMS

Fourier transforms, like Laplace transforms, can be grouped into the major classifications of functional and operational. Up to this point we have concentrated on the functional transforms. We will now discuss some of the important operational transforms. Since these operational transforms are very similar, with regard to the Laplace transform, to those discussed in Chapter 15, we leave their proofs as exercises in Problems 19.7–19.14.

1. *Multiplication by a constant.* If $\mathcal{F}\{f(t)\} = F(\omega)$, then

$$\mathcal{F}\{Kf(t)\} = KF(\omega). \qquad (19.46)$$

2. *Addition (subtraction).* If

$$\mathcal{F}\{f_1(t)\} = F_1(\omega),$$
$$\mathcal{F}\{f_2(t)\} = F_2(\omega),$$

and

$$\mathcal{F}\{f_3(t)\} = F_3(\omega),$$

then
$$\mathscr{F}\{f_1(t) - f_2(t) + f_3(t)\} = F_1(\omega) - F_2(\omega) + F_3(\omega). \tag{19.47}$$

3. *Differentiation.* The Fourier transform of the first derivative of $f(t)$ is
$$\mathscr{F}\left\{\frac{df(t)}{dt}\right\} = j\omega F(\omega). \tag{19.48}$$

For the nth derivative of $f(t)$ we have
$$\mathscr{F}\left\{\frac{d^n f(t)}{dt^n}\right\} = (j\omega)^n F(\omega). \tag{19.49}$$

Equations (19.48) and (19.49) are valid provided $f(t)$ is zero at plus and minus infinity.

4. *Integration.* If $g(t) = \int_{-\infty}^{t} f(x)\,dx$, then
$$\mathscr{F}\{g(t)\} = \frac{F(\omega)}{j\omega}. \tag{19.50}$$

In order for Eq. (19.50) to be valid, it is necessary that
$$\int_{-\infty}^{\infty} f(x)\,dx = 0.$$

5. *Scale change.* Dimensionally, time and frequency are reciprocals of each other. Therefore when time is stretched out, frequency will be compressed and vice versa. These observations are reflected in the functional transform
$$\mathscr{F}\{f(at)\} = \frac{1}{a} F\left(\frac{\omega}{a}\right), \qquad a > 0. \tag{19.51}$$

Note that when $0 < a < 1.0$ time is stretched out, whereas when $a > 1.0$ time is compressed.

6. *Translation in the time domain.* The effect of translating a function in the time domain is to alter the phase spectrum and leave the amplitude spectrum untouched. Thus
$$\mathscr{F}\{f(t - a)\} = e^{-j\omega a} F(\omega). \tag{19.52}$$

if a is positive in Eq. (19.52), the time function is delayed and if a is negative, the time function is advanced.

7. *Translation in the frequency domain.* Translation in the frequency domain corresponds to multiplication by the complex exponential in the time domain:
$$\mathscr{F}\{e^{j\omega_0 t} f(t)\} = F(\omega - \omega_0). \tag{19.53}$$

8. *Modulation.* Amplitude modulation is the process of varying the amplitude of a sinusoidal carrier. If the modulating signal is denoted as $f(t)$,

then the modulated carrier becomes $f(t) \cos \omega_0 t$. The amplitude spectrum of the modulated carrier is one half the amplitude spectrum $f(t)$ centered at $\pm \omega_0$, that is,

$$\mathscr{F}\{f(t) \cos \omega_0 t\} = \frac{1}{2} F(\omega - \omega_0) + \frac{1}{2} F(\omega + \omega_0). \tag{19.54}$$

9. *Convolution in the time domain.* Convolution in the time domain corresponds to multiplication in the frequency domain. That is, given that

$$y(t) = \int_{-\infty}^{\infty} x(\lambda) h(t - \lambda) \, d\lambda,$$

then

$$\mathscr{F}\{y(t)\} = Y(\omega) = X(\omega)H(\omega). \tag{19.55}$$

Equation (19.55) is of major importance in applications of the Fourier transform because it tells us that the transform of the response function $Y(\omega)$ is the product of the input transform $X(\omega)$ and the system function $H(\omega)$. We will have more to say about this in Section 19.8.

10. *Convolution in the frequency domain.* Convolution in the frequency domain corresponds to finding the Fourier transform of the product of two time functions. Thus given that

$$f(t) = f_1(t) f_2(t),$$

then

$$F(\omega) = \frac{1}{2\pi} \int_{-\infty}^{\infty} F_1(u) F_2(\omega - u) \, du. \tag{19.56}$$

These ten operational transforms are summarized in Table 19.2 along with another operational transform that is introduced in Problem 19.14.

DRILL EXERCISE 19.7

a) Find the second derivative of the function described in Problem 19.1(b).

b) Find the Fourier transform of the second derivative.

c) Use the result obtained in part (b) to find the Fourier transform of the function in part (a). (*Hint:* Use the operational transform of differentiation.)

Ans. (a) $\dfrac{d^2 f}{dt^2} = \dfrac{2A}{\tau} \delta\!\left(t + \dfrac{\tau}{2}\right) - \dfrac{4A}{\tau} \delta(t) + \dfrac{2A}{\tau} \delta\!\left(t - \dfrac{\tau}{2}\right);$ (b) $\dfrac{4A}{\tau}\left(\cos \dfrac{\omega \tau}{2} - 1\right);$

(c) $\dfrac{4A}{\omega^2 \tau}\left(1 - \cos \dfrac{\omega \tau}{2}\right).$

DRILL EXERCISE 19.8

The rectangular voltage pulse in Fig. 19.1 can be expressed as the difference between two step voltages, that is,

$$v(t) = V_m u\!\left(t + \frac{\tau}{2}\right) - V_m u\!\left(t - \frac{\tau}{2}\right) \text{ V}.$$

Use the operational transform for translation in the time domain to find the Fourier transform of $v(t)$.

Ans. $V(\omega) = V_m \tau \dfrac{\sin (\omega\tau/2)}{(\omega\tau/2)}$.

TABLE 19.2
OPERATIONAL TRANSFORMS

$f(t)$	$F(\omega)$
$Kf(t)$	$KF(\omega)$
$f_1(t) - f_2(t) + f_3(t)$	$F_1(\omega) - F_2(\omega) + F_3(\omega)$
$\dfrac{d^n f(t)}{dt^n}$	$(j\omega)^n F(\omega)$
$\displaystyle\int_{-\infty}^{t} f(x)\, dx$	$\dfrac{F(\omega)}{j\omega}$
$f(at)$	$\dfrac{1}{a} F\left(\dfrac{\omega}{a}\right), \quad a > 0$
$f(t - a)$	$e^{-j\omega a} F(\omega)$
$e^{j\omega_0 t} f(t)$	$F(\omega - \omega_0)$
$f(t) \cos \omega_0 t$	$\dfrac{1}{2} F(\omega - \omega_0) + \dfrac{1}{2} F(\omega + \omega_0)$
$\displaystyle\int_{-\infty}^{\infty} x(\lambda) h(t - \lambda)\, d\lambda$	$X(\omega)H(\omega)$
$f_1(t) f_2(t)$	$\dfrac{1}{2} \displaystyle\int_{-\infty}^{\infty} F_1(u) F_2(\omega - u)\, du$
$t^n f(t)$	$(j)^n \dfrac{d^n F(\omega)}{d\omega^n}$

19.8 CIRCUIT APPLICATIONS

The Laplace transform is more widely used to find the response of a circuit than is the Fourier transform. There are two fundamental reasons for this choice. First, the Laplace transform integral converges for a wider range of driving functions and second, it accommodates initial conditions. Despite the advantages of the Laplace transform in solving this type of problem, it is informative to illustrate how we can use the Fourier transform to find the response. The fundamental relationship underlying the use of the Fourier transform in transient analysis is Eq. (19.55), which relates the transform of the response $Y(\omega)$ to the transform of the input $X(\omega)$ and the transfer function $H(\omega)$ of the circuit. It is also necessary to recognize that $H(\omega)$ is the familiar $H(s)$ with s replaced by $j\omega$.

Example 19.1 illustrates how the Fourier transform is used to find the response of a circuit.

Example 19.1   Use the Fourier transform to find $i_o(t)$ in the circuit shown in Fig. 19.9. The current source $i_g(t)$ is the signum function 20 sgn t A.

Solution   The Fourier transform of the driving source is

$$I_g(\omega) = \mathscr{F}\{20 \text{ sgn } t\} = 20 \left(\frac{2}{j\omega}\right) = \frac{40}{j\omega}.$$

The transfer function of the circuit is the ratio of $\mathbf{I}_o$ to $\mathbf{I}_g$; thus

$$H(\omega) = \frac{\mathbf{I}_o}{\mathbf{I}_g} = \frac{1}{4 + j\omega}.$$

The Fourier transform of $i_o(t)$ is

$$I_o(\omega) = I_g(\omega)H(\omega) = \frac{40}{j\omega(4 + j\omega)}.$$

Now we expand $I_o(\omega)$ into a sum of partial fractions; thus

$$I_o(\omega) = \frac{K_1}{j\omega} + \frac{K_2}{4 + j\omega}.$$

Evaluating K_1 and K_2 gives

$$K_1 = \frac{40}{4} = 10$$

and

$$K_2 = \frac{40}{-4} = -10.$$

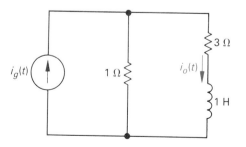

Fig. 19.9   The circuit for Example 19.1.

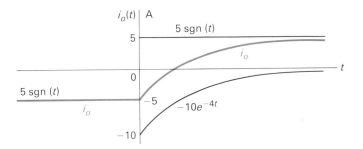

Fig. 19.10 The plot of $i_o(t)$ versus t.

Therefore we have

$$I_o(\omega) = \frac{10}{j\omega} - \frac{10}{4 + j\omega}.$$

The response is

$$i_o(t) = \mathcal{F}^{-1}[I_o(\omega)] = [5 \text{ sgn } t - 10e^{-4t}u(t)].$$

The response is sketched in Fig. 19.10. Does our solution make sense in terms of known circuit behavior? The answer is yes, as we can see from the following observations. The current source delivers -20 A to the circuit between $-\infty$ and 0 seconds. The resistance in each branch governs how the -20 A divides between the two branches. In particular, one fourth of the -20 A appears in the i_o branch; therefore i_o is -5 A for $t < 0$. When the current source jumps from -20 A to $+20$ A at $t = 0$, i_o will approach its final value of $+5$ A exponentially with a time constant of $(\frac{1}{4})$ s. ■

An important characteristic of the Fourier transform is that it directly yields the steady-state response to a sinusoidal driving function. This follows because the Fourier transform of $\cos \omega_0 t$ assumes that the function exists over all time. This feature is illustrated in the following example.

Example 19.2   The current source in the circuit of Example 19.1 (Fig. 19.9) is changed to a sinusoidal source. The expression for the current is

$$i_g(t) = 50 \cos 3t \text{ A}.$$

Use the Fourier transform method to find $i_o(t)$.

Solution   The transform of the driving function is

$$I_g(\omega) = 50\pi[\delta(\omega - 3) + \delta(\omega + 3)].$$

As before, the transfer function of the circuit is

$$H(\omega) = \frac{1}{4 + j\omega}.$$

It follows that the transform of the current response is

$$I_o(\omega) = 50\pi \frac{[\delta(\omega - 3) + \delta(\omega + 3)]}{(4 + j\omega)}.$$

Because of the sifting property of the impulse function, the easiest way to find the inverse transform of $I_o(\omega)$ is by the inversion integral; thus

$$i_o(t) = \mathcal{F}^{-1}\{I_o(\omega)\} = \frac{50\pi}{2\pi} \int_{-\infty}^{\infty} \left[\frac{\delta(\omega - 3) + \delta(\omega + 3)}{(4 + j\omega)} \right] e^{j\omega t}\, d\omega$$

$$= 25 \left[\frac{e^{j3t}}{4 + j3} + \frac{e^{-j3t}}{4 - j3} \right]$$

$$= 25 \left[\frac{e^{j3t} e^{-j36.87°}}{5} + \frac{e^{-j3t} e^{j36.87°}}{5} \right]$$

$$= 5[2 \cos (3t - 36.87°)]$$

$$= 10 \cos (3t - 36.87°).$$

We leave it to the reader to verify that the solution for $i_o(t)$ is identical to that obtained by phasor analysis. ∎

DRILL EXERCISE
19.9

The current source in the circuit shown delivers a current of 10 sgn t A. The response is the voltage across the 1-H inductor. Compute (a) $I_g(\omega)$; (b) $H(j\omega)$; (c) $V_o(\omega)$; (d) $v_o(t)$; (e) $i_1(0^-)$; (f) $i_1(0^+)$; (g) $i_2(0^-)$; (h) $i_2(0^+)$; (i) $v_o(0^-)$; and (j) $v_o(0^+)$.

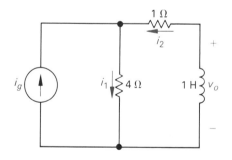

Ans. (a) $20/j\omega$; (b) $4j\omega/(5 + j\omega)$; (c) $80/(5 + j\omega)$; (d) $80e^{-5t}u(t)$ V; (e) -2 A; (f) 18 A; (g) 8 A; (h) 8 A; (i) 0 V; (j) 80 V.

DRILL EXERCISE
19.10

The voltage source in the circuit shown is generating the voltage

$$v_g = e^t u(-t) + u(t) \text{ V}.$$

a) Use the Fourier transform method to find v_a.

b) Compute $v_a(0^-)$, $v_a(0^+)$, and $v_a(\infty)$.

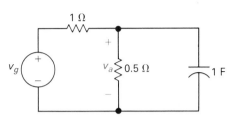

Ans. (a) $v_a = \frac{1}{4}e^t u(-t) - \frac{1}{12}e^{-3t}u(t) + \frac{1}{6} + \frac{1}{6}\,\mathrm{sgn}\,t$ V; (b) $\frac{1}{4}$V, $\frac{1}{4}$V, $\frac{1}{3}$V.

19.9 PARSEVAL'S THEOREM

Parseval's theorem is of interest because it relates the energy associated with a time domain function to the Fourier transform of the function. If we imagine our time domain function $f(t)$ to be either the voltage across, or the current in, a 1-Ω resistor, then the energy associated with this function is

$$W_{1\Omega} = \int_{-\infty}^{\infty} f^2(t)\ dt. \tag{19.57}$$

Parseval's theorem tells us that this same energy can be calculated by an integration in the frequency domain. Specifically, the theorem states that

$$\int_{-\infty}^{\infty} f^2(t)\ dt = \frac{1}{2\pi} \int_{-\infty}^{\infty} |F(\omega)|^2\ d\omega. \tag{19.58}$$

Therefore, the 1-Ω energy associated with $f(t)$ can be calculated either by integrating the square of $f(t)$ over all time or by integrating $1/2\pi$ times the square of the Fourier transform of $f(t)$ over all frequency. Parseval's theorem is valid provided both integrals exist.

We begin the derivation of Eq. (19.58) by rewriting the kernel of the integral on the left-hand side as $f(t)$ times itself and then expressing one $f(t)$ in terms of the inversion integral; thus

$$\int_{-\infty}^{\infty} f^2(t)\ dt = \int_{-\infty}^{\infty} f(t) \cdot f(t)\ dt$$

$$= \int_{-\infty}^{\infty} f(t) \left[\frac{1}{2\pi} \int_{-\infty}^{\infty} F(\omega)e^{j\omega t}\ d\omega \right] dt. \tag{19.59}$$

Now we can move $f(t)$ inside the interior integral since the integration is with respect to ω and we can factor the constant $1/2\pi$ outside both integrations. Thus we write Eq. (19.59) as

$$\int_{-\infty}^{\infty} f^2(t)\ dt = \frac{1}{2\pi} \int_{-\infty}^{\infty} \left[\int_{-\infty}^{\infty} F(\omega)f(t)e^{j\omega t}\ d\omega \right] dt. \tag{19.60}$$

Now we reverse the order of integration and in so doing recognize that $F(\omega)$

can be factored out of the integration with respect to t; thus

$$\int_{-\infty}^{\infty} f^2(t)\, dt = \frac{1}{2\pi} \int_{-\infty}^{\infty} F(\omega) \left[\int_{-\infty}^{\infty} f(t) e^{j\omega t}\, dt \right] d\omega. \tag{19.61}$$

In Eq. (19.61) the interior integral is $F(-\omega)$; thus the equation reduces to

$$\int_{-\infty}^{\infty} f^2(t)\, dt = \frac{1}{2\pi} \int_{-\infty}^{\infty} F(\omega) F(-\omega)\, d\omega. \tag{19.62}$$

In Section 19.7, we noted that $F(-\omega) = F^*(\omega)$; thus the product $F(\omega)F(-\omega)$ is simply the magnitude of $F(\omega)$ squared and Eq. (19.62) is equivalent to Eq. (19.58). We also noted that $|F(\omega)|$ is an even function of ω and therefore Eq. (19.58) can also be written as

$$\int_{-\infty}^{\infty} f^2(t)\, dt = \frac{1}{\pi} \int_{0}^{\infty} |F(\omega)|^2\, d\omega. \tag{19.63}$$

Let us demonstrate the validity of Eq. (19.63) with a specific example. If

$$f(t) = e^{-a|t|},$$

then the left-hand side of Eq. (19.63) becomes

$$\int_{-\infty}^{\infty} e^{-2a|t|}\, dt = \int_{-\infty}^{0} e^{2at}\, dt + \int_{0}^{\infty} e^{-2at}\, dt$$

$$= \frac{e^{2at}}{2a} \bigg|_{-\infty}^{0} + \frac{e^{-2at}}{-2a} \bigg|_{0}^{\infty}$$

$$= \frac{1}{2a} + \frac{1}{2a} = \frac{1}{a}. \tag{19.64}$$

The Fourier transform of $f(t)$ is

$$F(\omega) = \frac{2a}{a^2 + \omega^2}$$

and therefore the right-hand side of Eq. (19.63) becomes

$$\frac{1}{\pi} \int_{0}^{\infty} \frac{4a^2}{[a^2 + \omega^2]^2}\, d\omega = \frac{4a^2}{\pi} \cdot \frac{1}{2a^2} \left[\frac{\omega}{\omega^2 + a^2} + \frac{1}{a} \tan^{-1} \frac{\omega}{a} \right]_{0}^{\infty}$$

$$= \frac{2}{\pi} \left[0 + \frac{\pi}{2a} - 0 - 0 \right]$$

$$= \frac{1}{a}. \tag{19.65}$$

Note that the result given by Eq. (19.65) is the same as that given by Eq. (19.64).

Parseval's theorem gives us the physical interpretation that the magnitude of the Fourier transform squared $[|F(\omega)|^2]$ is an energy density (J/Hz). We can see this by writing the right-hand side of Eq. (19.63) as

$$\frac{1}{\pi} \int_0^\infty |F(2\pi f)|^2 2\pi \ df = 2 \int_0^\infty |F(2\pi f)|^2 \ df, \tag{19.66}$$

where $|F(2\pi f)|^2 \ df$ is the energy in an infinitesimal band of frequencies (df) and the total 1-Ω energy associated with $f(t)$ is the summation (integration) of $|F(2\pi f)|^2 \ df$ over all frequency. Observe that we can associate a portion of the total energy with a specified band of frequencies. That is, the 1-Ω energy in the frequency band from ω_1 to ω_2 is

$$W_{1\Omega} = \frac{1}{\pi} \int_{\omega_1}^{\omega_2} |F(\omega)|^2 \ d\omega. \tag{19.67}$$

Note that if the integration in the frequency domain is expressed as

$$\frac{1}{2\pi} \int_{-\infty}^\infty |F(\omega)|^2 \ d\omega$$

instead of

$$\frac{1}{\pi} \int_0^\infty |F(\omega)|^2 \ d\omega,$$

then Eq. (19.67) is written

$$W_{1\Omega} = \frac{1}{2\pi} \int_{-\omega_2}^{-\omega_1} |F(\omega)|^2 \ d\omega + \frac{1}{2\pi} \int_{\omega_1}^{\omega_2} |F(\omega)|^2 \ d\omega. \tag{19.68}$$

The graphical interpretation of Eq. (19.68) is shown in Fig. 19.11.

Examples 19.3, 19.4, and 19.5 have been designed to illustrate calculations involving Parseval's theorem.

Example 19.3 The current in a 40-Ω resistor is

$$i = 20e^{-2t}u(t) \ \text{A}.$$

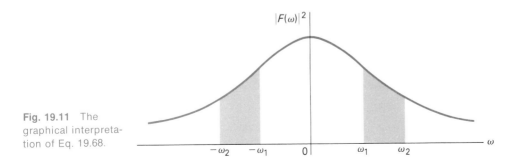

Fig. 19.11 The graphical interpretation of Eq. 19.68.

What percentage of the total energy dissipated in the resistor can be associated with the frequency band $0 \le \omega \le 2\sqrt{3}$ rad/s?

Solution The total energy dissipated in the 40-Ω resistor is

$$
W_{40\Omega} = 40 \int_0^\infty 400 e^{-4t} \, dt
$$

$$
= 16,000 \left. \frac{e^{-4t}}{-4} \right|_0^\infty
$$

$$
= 4000 \text{ J}.
$$

We can check our total energy calculation using Parseval's theorem; thus

$$
F(\omega) = \frac{20}{2 + j\omega}.
$$

Therefore,

$$
|F(\omega)| = \frac{20}{\sqrt{4 + \omega^2}}
$$

and

$$
W_{40\Omega} = \frac{40}{\pi} \int_0^\infty \frac{400 \, d\omega}{4 + \omega^2}
$$

$$
= \frac{16,000}{\pi} \left[\frac{1}{2} \tan^{-1} \left. \frac{\omega}{2} \right|_0^\infty \right]
$$

$$
= \frac{8000}{\pi} \left(\frac{\pi}{2} \right) = 4000 \text{ J}.
$$

The energy associated with the frequency band $0 \le \omega \le 2\sqrt{3}$ rad/s is

$$
W_{40\Omega} = \frac{40}{\pi} \int_0^{2\sqrt{3}} \frac{400 \, d\omega}{4 + \omega^2}
$$

$$
= \frac{16,000}{\pi} \left[\frac{1}{2} \tan^{-1} \left. \frac{\omega}{2} \right|_0^{2\sqrt{3}} \right]
$$

$$
= \frac{8000}{\pi} \cdot \frac{\pi}{3} = \frac{8000}{3} \text{ J}.
$$

Therefore, the percentage of the total energy associated with this range of frequencies is

$$
\eta = \frac{8000/3}{4000} \times 100 = 66.67\%. \quad \blacksquare
$$

Example 19.4 The input voltage to an ideal bandpass filter is

$$
v(t) = 120 e^{-24t} u(t) \text{ V}.
$$

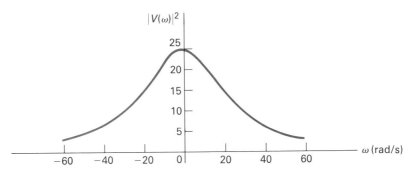

Fig. 19.12 $|V(\omega)|^2$ versus ω for Example 19.4.

The ideal filter passes all frequencies that lie between 24 and 48 rad/s, without attenuation, and completely rejects all frequencies outside this pass band.

a) Sketch $|V(\omega)|^2$ for the filter input voltage.

b) Sketch $|V_o(\omega)|^2$ for the filter output voltage.

c) What percentage of the total energy available at the input of the filter is available at the output of the filter?

Solution   a) The Fourier transform of the filter input voltage is

$$V(\omega) = \frac{120}{24 + j\omega}.$$

Therefore

$$|V(\omega)|^2 = \frac{14,400}{576 + \omega^2}.$$

The sketch of $|V(\omega)|^2$ versus ω is shown in Fig. 19.12.

b) Since the ideal bandpass filter rejects all frequencies outside the pass band, the plot of $|V_o(\omega)|^2$ versus ω appears as shown in Fig. 19.13.

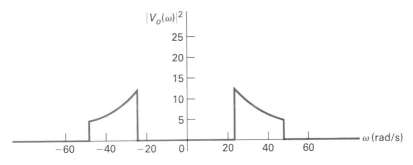

Fig. 19.13 $|V_o(\omega)|^2$ versus ω for Example 19.4.

c) The total 1-Ω energy available at the input to the filter is

$$W_i = \frac{1}{\pi} \int_0^\infty \frac{14{,}400}{576 + \omega^2} \, d\omega$$

$$= \frac{14{,}400}{\pi} \left[\frac{1}{24} \tan^{-1} \frac{\omega}{24} \Big|_0^\infty \right]$$

$$= \frac{600}{\pi} \cdot \frac{\pi}{2} = 300 \text{ J.}$$

The total 1-Ω energy available at the output of the filter is

$$W_o = \frac{1}{\pi} \int_{24}^{48} \frac{14{,}400}{576 + \omega^2} \, d\omega$$

$$= \frac{600}{\pi} \tan^{-1} \frac{\omega}{24} \Big|_{24}^{48}$$

$$= \frac{600}{\pi} [\tan^{-1} 2 - \tan^{-1} 1]$$

$$= \frac{600}{\pi} \left[\frac{\pi}{2.84} - \frac{\pi}{4} \right]$$

$$= 61.45 \text{ J.}$$

Thus the percentage of the input energy available at the output is

$$\eta = \frac{61.45}{300} \times 100 = 20.48\%. \quad \blacksquare$$

It is worth noting that Parseval's theorem makes it possible to calculate the energy available at the output of the filter without knowing the time domain expression for $v_o(t)$.

Example 19.5   The input voltage to the low-pass RC filter circuit of Fig. 19.14 is

$$v_i(t) = 15e^{-5t}u(t) \text{ V.}$$

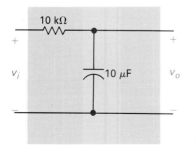

10 kΩ

v_i

10 μF

v_o

Fig. 19.14 The low-pass RC filter for Example 19.5.

a) What percentage of the 1-Ω energy available at the input is available at the output?

b) What percentage of the output energy is associated with the frequency range $0 \leq \omega \leq 10$ rad/s?

Solution   a) The 1-Ω energy at the input to the filter is

$$W_i = \int_0^\infty [15e^{-5t}]^2 \, dt$$

$$= 225 \left. \frac{e^{-10t}}{-10} \right|_0^\infty = 22.5 \text{ J.}$$

The Fourier transform of the output voltage is

$$V_o(\omega) = V_i(\omega)H(\omega),$$

where

$$V_i(\omega) = \frac{15}{5 + j\omega}$$

and

$$H(\omega) = \frac{1/RC}{1/RC + j\omega} = \frac{10}{10 + j\omega}.$$

It follows that

$$V_o(\omega) = \frac{150}{(5 + j\omega)(10 + j\omega)}$$

and

$$|V_o(\omega)|^2 = \frac{22{,}500}{(25 + \omega^2)(100 + \omega^2)}.$$

The 1-Ω energy available at the output of the filter is

$$W_o = \frac{1}{\pi} \int_0^\infty \frac{22{,}500 \, d\omega}{(25 + \omega^2)(100 + \omega^2)}.$$

We can easily evaluate the integral by expanding the kernel into a sum of partial fractions; thus

$$\frac{22{,}500}{(25 + \omega^2)(100 + \omega^2)} = \frac{300}{25 + \omega^2} - \frac{300}{100 + \omega^2}.$$

We have then

$$W_o = \frac{300}{\pi} \left\{ \int_0^\infty \frac{d\omega}{25 + \omega^2} - \int_0^\infty \frac{d\omega}{100 + \omega^2} \right\}$$

$$= \frac{300}{\pi} \left\{ \frac{1}{5} \left(\frac{\pi}{2} \right) - \frac{1}{10} \left(\frac{\pi}{2} \right) \right\}$$

$$= 15 \text{ J.}$$

The energy available at the output is therefore 66.67% of the energy available at the input, that is,

$$\eta = \frac{15}{22.5} (100) = 66.67\%.$$

b) The output energy associated with the frequency range $0 \le \omega \le 10$ rad/s is

$$W_o' = \frac{300}{\pi} \left\{ \int_0^{10} \frac{d\omega}{(25 + \omega^2)} - \int_0^{10} \frac{d\omega}{100 + \omega^2} \right\}$$

$$= \frac{300}{\pi} \left\{ \frac{1}{5} \tan^{-1} \frac{10}{5} - \frac{1}{10} \tan^{-1} \frac{10}{10} \right\} = \frac{30}{\pi} \left\{ \frac{\pi}{2.84} - \frac{\pi}{4} \right\}$$

$$= 13.64 \text{ J}.$$

Since the total 1-Ω energy available at the output is 15 J, the percentage associated with the frequency range 0 to 10 rad/s is 90.97%. ∎

We conclude our discussion of Parseval's theorem by calculating the energy associated with the rectangular voltage pulse. In Section 19.2 we found the Fourier transform of the voltage pulse to be

$$V(\omega) = V_m \tau \cdot \frac{\sin \omega\tau/2}{\omega\tau/2}. \tag{19.69}$$

To facilitate our present discussion, we have redrawn the voltage pulse along with its Fourier transform in Figs. 19.15(a) and (b), respectively. We can see from the figure that as the width of the voltage pulse (τ) becomes smaller, the dominant portion of the amplitude spectrum (that is, the spectrum from $-2\pi/\tau$ to $2\pi/\tau$ spreads out over a wider range of frequencies. This observation is in agreement with our earlier comments regarding the operational transform involving a scale change—in other words, when time is compressed frequency is stretched out and vice versa. In order to transmit a single rectangular pulse with reasonable fidelity, the bandwidth of the system must be at least wide enough to accommodate the dominant portion of the amplitude spectrum. Thus the cutoff frequency should be at least $2\pi/\tau$ rad/s, or $1/\tau$ Hz.

Parseval's theorem can be used to calculate the fraction of the total energy associated with $v(t)$ that lies in the frequency range $0 \le \omega \le 2\pi/\tau$. It follows from Eq. (19.69) that

$$W = \frac{1}{\pi} \int_0^{2\pi/\tau} V_m^2 \tau^2 \frac{\sin^2 \omega\tau/2}{(\omega\tau/2)^2} \, d\omega. \tag{19.70}$$

To carry out the integration called for in Eq. (19.70), we let

$$x = \frac{\omega\tau}{2} \tag{19.71}$$

and note that

$$dx = \frac{\tau}{2} \, d\omega \tag{19.72}$$

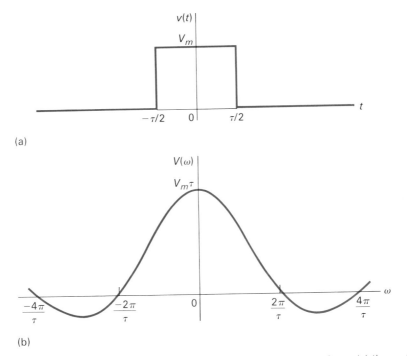

Fig. 19.15   The rectangular voltage pulse and its Fourier transform: (a) the rectangular voltage pulse; (b) the Fourier transform of $v(t)$.

and

$$x = \pi, \quad \text{when } \omega = 2\pi/\tau. \tag{19.73}$$

If we use the results given by Eqs. (19.71), (19.72), and (19.73), Eq. (19.70) becomes

$$W = \frac{2V_m^2 \tau}{\pi} \int_0^\pi \frac{\sin^2 x}{x^2} \, dx. \tag{19.74}$$

The integral in Eq. (19.74) can be integrated by parts. If we let

$$u = \sin^2 x \tag{19.75}$$

and

$$dv = \frac{dx}{x^2}, \tag{19.76}$$

then

$$du = 2 \sin x \cos x \, dx = \sin 2x \, dx \tag{19.77}$$

and

$$v = -\frac{1}{x}. \tag{19.78}$$

It follows then that

$$\int_0^\pi \frac{\sin^2 x}{x^2}\, dx = -\left.\frac{\sin^2 x}{x}\right|_0^\pi - \int_0^\pi -\frac{1}{x}\sin 2x\, dx$$

$$= 0 + \int_0^\pi \frac{\sin 2x\, dx}{x}. \tag{19.79}$$

When we substitute Eq. (19.79) into Eq. (19.74), we have

$$W = \frac{4V_m^2\tau}{\pi}\int_0^\pi \frac{\sin 2x}{(2x)}\, dx. \tag{19.80}$$

To evaluate the integral in Eq. (19.80), it is necessary to first put it in the form of $\sin y/y$. This is easily done by letting $y = 2x$ and noting that $dy = 2\, dx$ and $y = 2\pi$ when $x = \pi$. Thus Eq. (19.80) becomes

$$W = \frac{2V_m^2\tau}{\pi}\int_0^{2\pi} \frac{\sin y}{y}\, dy. \tag{19.81}$$

The value of the integral in Eq. (19.81) can be found in a table of sine integrals.† Its value is 1.41815; thus we have

$$W = \frac{2V_m^2\tau}{\pi}\,(1.41815). \tag{19.82}$$

The total 1-Ω energy associated with $v(t)$ can be calculated either from the time domain integration or the evaluation of Eq. (19.81) with the upper limit equal to infinity. In either case, the total energy is found to be

$$W_t = V_m^2\tau. \tag{19.83}$$

The fraction of the total energy associated with the band of frequencies between 0 and $2\pi/\tau$ is

$$\eta = \frac{W}{W_t} = \frac{2V_m^2\tau(1.41815)}{\pi(V_m^2\tau)}$$

$$= 0.9028. \tag{19.84}$$

Therefore, approximately 90% of the energy associated with $v(t)$ is contained in the dominant portion of the amplitude spectrum.

DRILL EXERCISE
9.11

The voltage across a 50-Ω resistor is

$$v = 4te^{-t}u(t)\ \text{V}.$$

What percentage of the total energy dissipated in the resistor can be associated with the frequency band $0 \le \omega \le \sqrt{3}$ rad/s?

Ans. 94.23%.

---

† M. Abramowitz and I. Stegun, *Handbook of Mathematical Functions*, p. 244 (New York: Dover, 1965).

DRILL EXERCISE
19.12

Assume that the magnitude of the Fourier transform of $v(t)$ is as shown. If this voltage is applied to a 6-kΩ resistor, calculate the total energy delivered to the resistor.

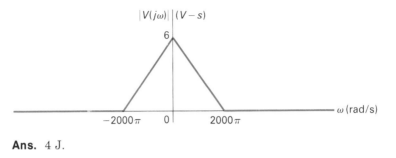

Ans. 4 J.

19.10 **SUMMARY**

The Fourier transform is of primary interest at this introductory level because of the frequency domain description that it gives us of an aperiodic time domain function.

The Fourier transforms of the aperiodic functions that are of most interest in circuit applications can be divided into two categories. In the first category are those transforms that exist because the Fourier integral converges for the given time domain function. Functions that are of finite duration or that approach zero as t approaches plus or minus infinity produce transforms that fall in this category. In the second category are those transforms that exist because they are defined in terms of a limit process. The constant, the signum function, the step function, and the sinusoidal function are examples of functions for which the Fourier integral does not converge and hence have transforms that are defined by a limit process. Whenever the Fourier transform is defined in terms of a limit, it contains an impulse function.

In comparing the Fourier transform to the unilateral Laplace transform, we note several important differences. The Fourier transform will accommodate negative-time functions as well as positive-time functions and therefore is suited to problems that are described in terms of events that start at $t = -\infty$. The unilateral Laplace transform, on the other hand, is suited to problems that are described in terms of initial conditions and events that occur for $t > 0$. For example, the Fourier transform automatically generates the steady-state response of a circuit that is excited by a sinusoidal source because the transform assumes that the sinusoidal excitation has existed over all time. The unilateral Laplace transform produces the response of this same circuit from the moment the sinusoidal source is turned on. Furthermore, the Laplace transform method will predict what happens if the sinusoidal source is switched on to a circuit that already has energy stored in its reactive elements.

The key to using the Fourier transform in circuit analysis is to recognize that the Fourier transform of the response function is obtained by multiplying the transform of the excitation function by the system transfer function. In Fourier analysis, the system transfer function $[H(j\omega) = H(\omega)]$ is derived from either the s-domain or the phasor domain circuit.

Parseval's theorem gives us further insight into the physical interpretation of the Fourier transform. Specifically, it reveals that the magnitude of the transform squared is a measure of the energy density (J/Hz) in the frequency domain. Thus, through Parsevel's theorem, it is possible to associate a fraction of the total energy contained in $f(t)$ with a specific band of frequencies. Knowing how the energy is distributed over the frequency spectrum is of interest in electrical communication systems.

As we mentioned in the introduction to this chapter, the important applications of the Fourier transform in electrical engineering lie in the areas of communication theory and signal processing. We have chosen to introduce the transform here because of its ties with both the Fourier series and the Laplace transform.

PROBLEMS

19.1   Use the defining integral to find the Fourier transform of the following functions.

a) $f(t) = A \sin \dfrac{\pi}{2} t,$ $-2 \le t \le 2,$

$f(t) = 0,$ elsewhere

b) $f(t) = \dfrac{2A}{\tau} t + A,$ $-\dfrac{\tau}{2} \le t \le 0,$

$f(t) = -\dfrac{2A}{\tau} t + A,$ $0 \le t \le \tau/2,$

$f(t) = 0,$ elsewhere

19.2   Find the Fourier transform of each of the following functions. In all of the functions, a is a positive real constant and $-\infty \le t \le \infty$.

a) $f(t) = |t|e^{-a|t|}$
b) $f(t) = t^3 e^{-a|t|}$
c) $f(t) = e^{-a|t|} \cos \omega_0 t$
d) $f(t) = e^{-a|t|} \sin \omega_0 t$
e) $f(t) = \delta(t - t_0)$

19.3   Use the inversion integral (Eq. 19.9) to show that $\mathcal{F}^{-1}\{2/j\omega\} = \operatorname{sgn} t$. (*Hint:* Use Drill Exercise 19.6.)

19.4   Find $\mathcal{F}\{\cos \omega_0 t\}$ by using the approximating function

$$f(t) = e^{-\epsilon|t|} \cos \omega_0 t,$$

where ϵ is a positive real constant.

19.5   Show that if $f(t)$ is an even function,

$$A(\omega) = 2 \int_0^\infty f(t) \cos \omega t \, dt$$

and

$$B(\omega) = 0.$$

19.6 Show that if $f(t)$ is an odd function,

$$A(\omega) = 0$$

and

$$B(\omega) = -2 \int_0^\infty f(t) \sin \omega t \, dt.$$

19.7 a) Show that $\mathcal{F}\{df(t)/dt\} = j\omega F(\omega)$, where $F(\omega) = \mathcal{F}\{f(t)\}$. (*Hint:* Use the defining integral and integrate by parts.)
b) What is the restriction on $f(t)$ if the result given in part (a) is valid?
c) Show that $\mathcal{F}\{d^n f(t)/dt^n\} = (j\omega)^n F(\omega)$, where $F(\omega) = \mathcal{F}\{f(t)\}$.

19.8 a) Show that

$$\mathcal{F}\left\{ \int_{-\infty}^t f(x) \, dx \right\} = \frac{F(\omega)}{j\omega},$$

where $F(\omega) = \mathcal{F}\{f(x)\}$. (*Hint:* Use the defining integral and integrate by parts.)
b) What is the restriction on $f(x)$ if the result given in part (a) is valid?
c) If $f(x) = e^{-ax}u(x)$, can the operational transform in part (a) be used? Explain.

19.9 a) Show that

$$\mathcal{F}\{f(at)\} = \frac{1}{a} F\left(\frac{\omega}{a}\right), \qquad a > 0.$$

b) Given $f(at) = e^{-a|t|}$ for $a > 0$. Sketch $F(\omega) = \mathcal{F}\{f(at)\}$ for $a = 0.5$, 1.0, and 2.0. Do your sketches reflect the observation that "compression" in the time domain correspond to "stretching" in the frequency domain?

19.10 Derive each of the following operational transforms.
a) $\mathcal{F}\{f(t - a)\} = e^{-j\omega a} F(\omega)$
b) $\mathcal{F}\{e^{j\omega_0 t} f(t)\} = F(\omega - \omega_0)$
c) $\mathcal{F}\{f(t) \cos \omega_0 t\} = \frac{1}{2}F(\omega - \omega_0) + \frac{1}{2}F(\omega + \omega_0)$

19.11 Given $y(t) = \int_{-\infty}^\infty x(\lambda)h(t - \lambda) \, d\lambda$. Show that $Y(\omega) = \mathcal{F}\{y(t)\} = X(\omega)H(\omega)$, where $X(\omega) = \mathcal{F}\{x(t)\}$ and $H(\omega) = \mathcal{F}\{h(t)\}$. (*Hint:* Use the defining integral to write $\mathcal{F}\{y(t)\} = \int_{-\infty}^\infty [\int_{-\infty}^\infty x(\lambda)h(t - \lambda \, d\lambda]e^{-j\omega t} \, dt$. Next, reverse the order of integration and then make a change in the variable of integration, that is, let $u = t - \lambda$.)

19.12 Given $f(t) = f_1(t)f_2(t)$. Show that $F(\omega) = (1/2\pi)\int_{-\infty}^\infty F_1(u)F_2(\omega - u) \, du$. (*Hint:* First, use the defining integral to express $F(\omega)$ as

$$F(\omega) = \int_{-\infty}^\infty f_1(t)f_2(t)e^{-j\omega t} \, dt.$$

Second, use the inversion integral to write

$$f_1(t) = \frac{1}{2\pi} \int_{-\infty}^\infty F_1(u)e^{jut} \, du.$$

Third, substitute the expression for $f_1(t)$ into the defining integral and then interchange the order of integration.)

19.13   Given $f(t) = f_1(t)f_2(t)$, where

$$f_1(t) = \cos \omega_0 t \quad \text{and}$$
$$f_2(t) = 1, \quad -\tau/2 < t < \tau/2,$$
$$f_2(t) = 0, \quad \text{elsewhere.}$$

a) Use convolution in the frequency domain to find $F(\omega)$.
b) What happens to $F(\omega)$ as the width of $f_2(t)$ increases so that $f(t)$ includes more and more cycles of $f_1(t)$?

19.14   (a) Show that

$$(j)^n \frac{d^n F(\omega)}{d\omega^n} = \mathscr{F}\{t^n f(t)\}.$$

b) Use the result of part (a) to find each of the following Fourier transforms:
(i) $\mathscr{F}\{te^{-at}u(t)\};$ (ii) $\mathscr{F}\{|t|e^{-a|t|}\};$ (iii) $\mathscr{F}\{te^{-a|t|}\}.$

19.15   a) Use the Fourier transform method to find $i(t)$ in the circuit shown in Fig. 19.16. The initial value of $i(t)$ is zero and the source current is $10u(t)$ A.
b) Sketch $i(t)$ versus t.

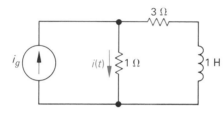

Fig. 19.16   The circuit for Problem 19.15.

19.16   a) Use the Fourier transform method to find $v(t)$ in the circuit shown in Fig. 19.17 if $v_g(t) = 40 \operatorname{sgn} t$.
b) Sketch $v(t)$ versus t.

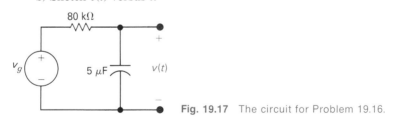

Fig. 19.17   The circuit for Problem 19.16.

19.17   The current source in the circuit in Fig. 19.18 is given by the expression

$$i_g = e^{-2|t|} \text{ A.}$$

a) Find $v_o(t)$.
b) What is the value of $v_o(0^-)$?
c) What is the value of $v_o(0^+)$?
d) Use the Laplace transform method to find $v_o(t)$ for $t > 0^+$.
e) Does the solution obtained in part (d) agree with $v_o(t)$ for $t > 0^+$ from part (a)?

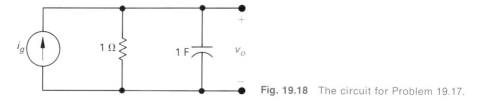

Fig. 19.18   The circuit for Problem 19.17.

19.18 Given that $F(\omega) = e^{\omega} u(-\omega) + e^{-\omega} u(\omega)$.

    a) Find $f(t)$.
    b) Find the 1-Ω energy associated with $f(t)$ via time domain integration.
    c) Repeat part (b) using frequency domain integration.
    d) Find the value of ω_1 if $f(t)$ has 90% of the energy in the frequency band $0 \leq \omega \leq \omega_1$.

19.19 The amplitude spectrum of the input voltage to the high-pass RC filter in Fig. 19.19 is

$$V_i(\omega) = \frac{100}{|\omega|}, \qquad 100 \leq |\omega| \leq 200 \text{ rad/s},$$

$$V_i(\omega) = 0, \qquad \text{elsewhere.}$$

    a) Sketch $|V_i(\omega)|^2$ for $-300 \leq \omega \leq 300$ rad/s.
    b) Sketch $|V_o(\omega)|^2$ for $-300 \leq \omega \leq 300$ rad/s.
    c) Calculate the 1-Ω energy at the input of the filter.
    d) Calculate the 1-Ω energy at the output of the filter.
    e) What percentage of the input energy is available at the output?
    f) Repeat parts (d) and (e) given that R is changed to 5 kΩ.

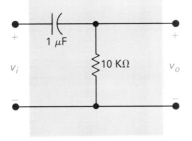

1 μF

V_i

10 KΩ

V_o

Fig. 19.19 The circuit for Problem 19.19.

19.20 The input voltage to the high-pass RC filter circuit in Fig. 19.20 is

$$v_i(t) = Ae^{-at} u(t).$$

Let α denote the corner frequency of the filter, that is, $\alpha = 1/RC$.

    a) What percentage of the energy at the output of the filter is associated with the frequency band $0 \leq \omega \leq \alpha$ if $\alpha = a$?
    b) Repeat part (a) given that $\alpha = \sqrt{3}\,a$.
    c) Repeat part (a) given that $\alpha = a/\sqrt{3}$.

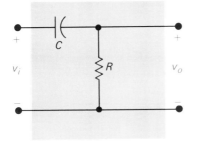

C

V_i

R

V_o

Fig. 19.20 The circuit for Problem 19.20.

TWO-PORT CIRCUITS

20.1 INTRODUCTION

Up to this point in our study, we have frequently focused our attention on the behavior of a circuit at a specified pair of terminals. Recall that we introduced the Thévenin and Norton equivalent circuits solely for the purpose of simplifying the analysis of a circuit relative to a pair of terminals. We also find it convenient, in the analysis of some electrical systems, to focus our attention on two pairs of terminals. We find this point of view convenient in those situations when a signal is fed into one pair of terminals and then after being processed by the system is extracted at a second pair of terminals. Because the terminal pairs represent the points in the system where signals are either fed in or extracted, they are referred to as the *ports* of the system. In this chapter, we will limit our discussion to circuits in which we have one input port and one output port.

The basic two-port building block is illustrated in Fig. 20.1. There are several restrictions in using this building block. First of all, there can be no energy stored within the circuit. Second, there can be no independent sources within the circuit. Dependent sources are permissible. Third, the current into the port must equal the current out of the port, that is, $i_1 = i_1'$ and $i_2 = i_2'$. The fourth, and final, restriction is that all external connections are made to either the input port or the output port. No external connections between ports are allowed. In other words, no external connections can be made between terminals a–c, a–d, b–c, or b–d. These restrictions simply limit the range of circuit problems to which the two-port formulation is amenable.

The fundamental principle underlying two-port modeling of a system is that only the terminal variables, that is, i_1, v_1, i_2, and v_2, are of interest. There is no interest in calculating the currents and voltages that exist inside the circuit. We have already seen this emphasis on terminal behavior stressed in our analysis of operational amplifier circuits. In this chapter we will formalize the approach by introducing the two-port parameters.

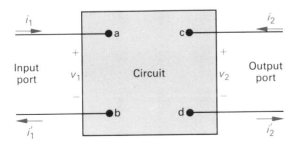

Fig. 20.1   The two-port building block.

20.2   THE TERMINAL EQUATIONS

In viewing a circuit as a two-port network, we are interested in relating the current and voltage at one port to the current and voltage at the second port. The reference polarities of the terminal voltages and the reference directions of the terminal currents are as shown in Fig. 20.1. It is worth noting that the references at each port are symmetrical with respect to each other, that is, at each port the current is directed into the upper terminal and each port voltage is a rise from the lower to the upper terminal. This symmetry makes it easier to generalize the analysis of a network in terms of two-port blocks and is the reason for its nearly universal use in the literature.

The most general description of the two-port network is carried out in the s-domain. For purely resistive networks the analysis reduces to solving resistive circuits. Sinusoidal steady-state problems can be solved either by first finding the appropriate s-domain expressions and then replacing s by $j\omega$ or by direct analysis in the phasor domain. In the work that follows, all equations are written in the s-domain. Thus resistive networks and sinusoidal state-state solutions simply become special cases. The basic building block in terms of the s-domain variables I_1, V_1, I_2, and V_2 is shown in Fig. 20.2.

Of the four terminal variables I_1, V_1, I_2, and V_2, only two are independent. Thus for a given circuit, once we specify two of the variables we can find the two remaining unknowns. For example, given V_1 and V_2 plus the circuit within the box, we can determine I_1 and I_2. It follows that we can describe a two-port network with just two simultaneous equations. Although we need just two simultaneous equations, we have a choice of six different ways in which to combine the four variables. The six sets of equations are

$$V_1 = z_{11}I_1 + z_{12}I_2,$$
$$V_2 = z_{21}I_1 + z_{22}I_2; \tag{20.1}$$

$$I_1 = y_{11}V_1 + y_{12}V_2,$$
$$I_2 = y_{21}V_1 + y_{22}V_2; \tag{20.2}$$

$$V_1 = a_{11}V_2 - a_{12}I_2,$$
$$I_1 = a_{21}V_2 - a_{22}I_2; \tag{20.3}$$

$$V_2 = b_{11}V_1 - b_{12}I_1,$$
$$I_2 = b_{21}V_1 - b_{22}I_1; \tag{20.4}$$

$$V_1 = h_{11}I_1 + h_{12}V_2,$$
$$I_2 = h_{21}I_1 + h_{22}V_2; \tag{20.5}$$

$$I_1 = g_{11}V_1 + g_{12}I_2,$$
$$V_2 = g_{21}V_1 + g_{22}I_2. \tag{20.6}$$

These six sets of equations can also be thought of as three pairs of mutually inverse relations. The first set (Eqs. 20.1) gives the input and output volt-

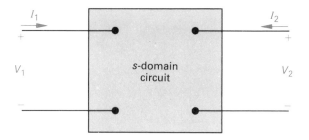

Fig. 20.2   The s-domain two-port basic building block.

ages as functions of the input and output currents, whereas the second set (Eqs. 20.2) gives the inverse relationship, that is, the input and output currents as functions of the input and output voltages. Further study of the remaining four sets will show that Eqs. (20.3) and (20.4) are inverse relations and so too are Eqs. (20.5) and (20.6).

The coefficients of the current and/or voltage variables on the right-hand side of Eqs. (20.1)–(20.6) are called the *parameters* of the two-port circuit. Thus when using Eqs. (20.1), we refer to the *z*-parameters of the circuit. Similarly, we refer to the *y*-parameters, the *a*-parameters, the *b*-parameters, the *h*-parameters, and the *g*-parameters of the network.

20.3   THE TWO-PORT PARAMETERS

The parameters for any given circuit can be found either by computation or by measurement. The computation, or measurement, to be made is determined directly from the parameter equations. For example, suppose our problem is to find the *z*-parameters for a given circuit. It follows directly from Eqs. (20.1) that

$$z_{11} = \left. \frac{V_1}{I_1} \right|_{I_2=0}, \qquad \text{(20.7)}$$

$$z_{12} = \left. \frac{V_1}{I_2} \right|_{I_1=0}, \qquad \text{(20.8)}$$

$$z_{21} = \left. \frac{V_2}{I_1} \right|_{I_2=0}, \qquad \text{(20.9)}$$

and

$$z_{22} = \left. \frac{V_2}{I_2} \right|_{I_1=0} \qquad \text{(20.10)}$$

A study of Eqs. (20.7)–(20.10) reveals that the four z-parameters can be described as follows:

z_{11} is the impedance seen looking into port 1 when port 2 is open.

z_{12} is a transfer impedance. It is the ratio of the port-1 voltage to the port-2 current when port 1 is open.

z_{21} is a transfer impedance. It is the ratio of the port-2 voltage to the port-1 current when port 2 is open.

z_{22} is the impedance seen looking into port 2 when port 1 is open.

We see, then, that the impedance parameters can be either calculated or measured by first opening port 2 and determining the ratios V_1/I_1 and V_2/I_1 and then opening port 1 and determining the ratios V_1/I_2 and V_2/I_2. The determination of the z-parameters for a resistive circuit is illustrated in the following example.

Example 20.1 Find the z-parameters for the circuit shown in Fig. 20.3.

Solution Since the circuit is purely resistive, the s-domain circuit is also purely resistive. With port 2 open, that is, $I_2 = 0$, the resistance seen looking into port 1 will be the 20-Ω resistor in parallel with the series combination of the 5-Ω and 15-Ω resistors. Therefore

$$z_{11} = \frac{V_1}{I_1}\bigg|_{I_2=0} = \frac{(20)(20)}{40} = 10 \ \Omega.$$

When I_2 is zero, V_2 is

$$V_2 = \frac{V_1}{(15 + 5)}(15) = 0.75\,V_1$$

and therefore

$$z_{21} = \frac{V_2}{I_1}\bigg|_{I_2=0} = \frac{0.75\,V_1}{V_1/10} = 7.5 \ \Omega.$$

When I_1 is zero, the resistance seen looking into port 2 will be the 15-Ω resistor

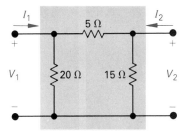

Fig. 20.3 The circuit for Example 20.1.

in parallel with the series combination of the 5-Ω and 20-Ω resistors. Therefore

$$z_{22} = \left.\frac{V_2}{I_2}\right|_{I_1=0} = \frac{(15)(25)}{40} = 9.375 \ \Omega.$$

When port 1 is open, I_1 is zero and the voltage V_1 is

$$V_1 = \frac{V_2}{(5+20)}(20) = 0.8V_2.$$

With port 1 open, the current into port 2 is

$$I_2 = \frac{V_2}{9.375}.$$

It follows that

$$z_{12} = \left.\frac{V_1}{I_2}\right|_{I_1=0} = \frac{0.8V_2}{V_2/9.375} = 7.5 \ \Omega. \quad \blacksquare$$

We see from Eqs. (20.7)–(20.10) and Example 20.1 why the parameters in Eqs. (20.1) are called the z-parameters. Each parameter is the ratio of a voltage to a current and therefore is an impedance with the dimension of ohms.

The remaining port parameters are either calculated, or measured, using the same thought process that we used to find the z-parameters. A given port parameter is obtained from either opening or shorting a port. Furthermore, a given port parameter will be either an impedance, an admittance, or a dimensionless ratio. The dimensionless ratio will be either the ratio of two voltages or the ratio of two currents. These observations are summarized by Eqs. (20.11)–(20.15):

$$y_{11} = \left.\frac{I_1}{V_1}\right|_{V_2=0} \mho,$$

$$y_{12} = \left.\frac{I_1}{V_2}\right|_{V_1=0} \mho, \quad \textbf{(20.11)}$$

$$y_{21} = \left.\frac{I_2}{V_1}\right|_{V_2=0} \mho,$$

$$y_{22} = \left.\frac{I_2}{V_2}\right|_{V_1=0} \mho;$$

$$a_{11} = \left.\frac{V_1}{V_2}\right|_{I_2=0},$$

$$a_{12} = -\left.\frac{V_1}{I_2}\right|_{V_2=0} \Omega, \quad \textbf{(20.12)}$$

$$a_{21} = \left.\frac{I_1}{V_2}\right|_{I_2=0} \mho,$$

$$a_{22} = -\left.\frac{I_1}{I_2}\right|_{V_2=0};$$

$$b_{11} = \left.\frac{V_2}{V_1}\right|_{I_1=0},$$

$$b_{12} = -\left.\frac{V_2}{I_1}\right|_{V_1=0} \Omega, \quad \textbf{(20.13)}$$

$$b_{21} = \left.\frac{I_2}{V_1}\right|_{I_1=0} \mho,$$

$$b_{22} = -\left.\frac{I_2}{I_1}\right|_{V_1=0};$$

$$h_{11} = \left.\frac{V_1}{I_1}\right|_{V_2=0} \Omega,$$

$$h_{12} = \left.\frac{V_1}{V_2}\right|_{I_1=0}, \quad \textbf{(20.14)}$$

$$h_{21} = \left.\frac{I_2}{I_1}\right|_{V_2=0},$$

$$h_{22} = \left.\frac{I_2}{V_2}\right|_{I_1=0} \mho;$$

$$g_{11} = \left.\frac{I_1}{V_1}\right|_{I_2=0} \mho,$$

$$g_{12} = \left.\frac{I_1}{I_2}\right|_{V_1=0}, \quad \textbf{(20.15)}$$

$$g_{21} = \left.\frac{V_2}{V_1}\right|_{I_2=0},$$

$$g_{22} = \left.\frac{V_2}{I_1}\right|_{V_1=0} \Omega.$$

DRILL EXERCISE
20.1
Find the y-parameters for the circuit in Fig. 20.3.

Ans. $y_{11} = 0.25$ ℧; $y_{12} = y_{21} = -0.2$ ℧; $y_{22} = \frac{4}{15}$ ℧.

DRILL EXERCISE
20.2
Find the a- and b-parameters for the circuit in Fig. 20.3.

Ans. $a_{11} = \frac{4}{3}$; $a_{12} = 5$ Ω; $a_{21} = \frac{2}{15}$ ℧; $a_{22} = 1.25$; $b_{11} = 1.25$; $b_{12} = 5$ Ω; $b_{21} = \frac{2}{15}$ ℧; $b_{22} = \frac{4}{3}$.

The two-port parameters are also described in relation to the reciprocal sets of equations. The impedance and admittance parameters are grouped into the *immittance* parameters. The term immittance is used to denote a quantity that is either an impedance or an admittance. The a- and b-parameters are called the *transmission* parameters because they describe the voltage and current at one end of the two-port in terms of the voltage and current at the other end. Thus the a- or b-parameters give a measure of how the voltage and current are transmitted through the network. The immittance and transmission parameters are characterized as being the natural choices for relating the port variables. That is, they relate either voltage variables to current variables or input variables to output variables. The h- and g-parameters, on the other hand, relate cross-variables, that is, an input voltage and output current to an output voltage and input current. Therefore, the h- and g-parameters are called *hybrid* parameters.

Example 20.2

The following measurements pertain to a two-port circuit operating in the sinusoidal steady-state. With port 2 open, a voltage equal to $150 \cos 4000t$ V is applied to port 1. The current into port 1 is $25 \cos (4000t - 45°)$ A and the port-2 voltage is $100 \cos (4000t + 15°)$ V. With port 2 short-circuited, a voltage equal to $30 \cos 4000t$ V is applied to port 1. The current into port 1 is $1.5 \cos (4000t + 30°)$ A and the current into port 2 is $0.25 \cos (4000t + 150°)$ A. Find the a-parameters that can be used to describe the sinusoidal steady-state behavior of the circuit.

Solution

From the first set of measurements, we have

$$\mathbf{V}_1 = 150\underline{/0°} \text{ V}; \qquad \mathbf{I}_1 = 25\underline{/-45°} \text{ A}; \qquad \mathbf{I}_2 = 0 \text{ A}; \qquad \mathbf{V}_2 = 100\underline{/15°} \text{ V}.$$

It follows from Eqs. (20.12) that

$$a_{11} = \left.\frac{\mathbf{V}_1}{\mathbf{V}_2}\right|_{I_2=0} = \frac{150\underline{/0°}}{100\underline{/15°}} = 1.5\underline{/-15°}$$

and

$$a_{21} = \left.\frac{\mathbf{I}_1}{\mathbf{V}_2}\right|_{I_2=0} = \frac{25\underline{/-45°}}{100\underline{/15°}} = 0.25\underline{/-60°} \text{ ℧}.$$

From the second set of measurements we have

$$\mathbf{V}_1 = 30\underline{/0°} \text{ V}; \qquad \mathbf{I}_1 = 1.5\underline{/30°} \text{ A}; \qquad \mathbf{I}_2 = 0.25\underline{/150°} \text{ A}; \qquad \mathbf{V}_2 = 0.$$

Therefore

$$a_{12} = -\left.\frac{\mathbf{V}_1}{\mathbf{I}_2}\right|_{V_2=0} = \frac{-30\underline{/0°}}{0.25\underline{/150°}} = 120\underline{/30°} \text{ } \Omega$$

and

$$a_{22} = -\left.\frac{\mathbf{I}_1}{\mathbf{I}_2}\right|_{V_2=0} = \frac{-1.5\underline{/30°}}{0.25\underline{/150°}} = 6\underline{/60°}. \quad \blacksquare$$

DRILL EXERCISE 20.3 The following measurements were made on a two-port resistive circuit. With 10 mV applied to port 2 and port 1 open, the current into port 2 is 0.25 μA and the voltage across port 1 is 5 μV. With port 2 short-circuited and 50 mV applied to port 1, the current into port 1 is 50 μA and the current into port 2 is 2 mA. Find the h-parameters of the network.

Ans. $h_{11} = 1000 \text{ } \Omega$; $h_{12} = 5 \times 10^{-4}$; $h_{21} = 40$; $h_{22} = 25 \text{ } \mu\text{℧}$.

Since our six sets of equations relate to the same variables, the parameters associated with any pair of equations must be related to the parameters of all the other pairs of equations. In other words, if we know one set of parameters, we can derive all the other sets of parameters from this known set. There is a considerable amount of algebra involved in deriving all the interrelationships and hence we will merely list the results in Table 20.1.

Although we will not derive all the relationships listed in Table 20.1, we will derive the relationships between the z-parameters and y-parameters as well as the relationships between the z-parameters and the a-parameters. These derivations will illustrate the general thought process involved in relating one set of parameters to the other. To find the z-parameters as functions of the y-parameters we solve Eqs. (20.2) for V_1 and V_2 and then compare the coefficients of I_1 and I_2 in the resulting expressions with the coefficients of I_1 and I_2 in Eqs. (20.1). From Eqs. (20.2) we have

$$V_1 = \frac{\begin{vmatrix} I_1 & y_{12} \\ I_2 & y_{22} \end{vmatrix}}{\begin{vmatrix} y_{11} & y_{12} \\ y_{21} & y_{22} \end{vmatrix}} = \frac{y_{22}}{\Delta y} I_1 - \frac{y_{12}}{\Delta y} I_2 \tag{20.16}$$

and

$$V_2 = \frac{\begin{vmatrix} y_{11} & I_1 \\ y_{21} & I_2 \end{vmatrix}}{\Delta y} = -\frac{y_{21} I_1}{\Delta y} + \frac{y_{11}}{\Delta y} I_2. \tag{20.17}$$

TABLE 20.1
PARAMETER CONVERSION TABLE

$$z_{11} = \frac{y_{22}}{\Delta y} = \frac{a_{11}}{a_{21}} = \frac{b_{22}}{b_{21}} = \frac{\Delta h}{h_{22}} = \frac{1}{g_{11}}$$

$$z_{12} = -\frac{y_{12}}{\Delta y} = \frac{\Delta a}{a_{21}} = \frac{1}{b_{21}} = \frac{h_{12}}{h_{22}} = -\frac{g_{12}}{g_{11}}$$

$$z_{21} = -\frac{y_{21}}{\Delta y} = \frac{1}{a_{21}} = \frac{\Delta b}{b_{21}} = -\frac{h_{21}}{h_{22}} = \frac{g_{21}}{g_{11}}$$

$$z_{22} = \frac{y_{11}}{\Delta y} = \frac{a_{22}}{a_{21}} = \frac{b_{11}}{b_{21}} = \frac{1}{h_{22}} = \frac{\Delta g}{g_{11}}$$

$$y_{11} = \frac{z_{22}}{\Delta z} = \frac{a_{22}}{a_{12}} = \frac{b_{11}}{b_{12}} = \frac{1}{h_{11}} = \frac{\Delta g}{g_{22}}$$

$$y_{12} = -\frac{z_{12}}{\Delta z} = -\frac{\Delta a}{a_{12}} = -\frac{1}{b_{12}} = -\frac{h_{12}}{h_{11}} = \frac{g_{12}}{g_{22}}$$

$$y_{21} = -\frac{z_{21}}{\Delta z} = -\frac{1}{a_{12}} = -\frac{\Delta b}{b_{12}} = \frac{h_{21}}{h_{11}} = -\frac{g_{21}}{g_{22}}$$

$$y_{22} = \frac{z_{11}}{\Delta z} = \frac{a_{11}}{a_{12}} = \frac{b_{22}}{b_{12}} = \frac{\Delta h}{h_{11}} = \frac{1}{g_{22}}$$

$$a_{11} = \frac{z_{11}}{z_{21}} = -\frac{y_{22}}{y_{21}} = \frac{b_{22}}{\Delta b} = -\frac{\Delta h}{h_{21}} = \frac{1}{g_{21}}$$

$$a_{12} = \frac{\Delta z}{z_{21}} = -\frac{1}{y_{21}} = \frac{b_{12}}{\Delta b} = \frac{-h_{11}}{h_{21}} = \frac{g_{22}}{g_{21}}$$

$$a_{21} = \frac{1}{z_{21}} = -\frac{\Delta y}{y_{21}} = \frac{b_{21}}{\Delta b} = -\frac{h_{22}}{h_{21}} = \frac{g_{11}}{g_{21}}$$

$$a_{22} = \frac{z_{22}}{z_{21}} = -\frac{y_{11}}{y_{21}} = \frac{b_{11}}{\Delta b} = -\frac{1}{h_{21}} = \frac{\Delta g}{g_{21}}$$

$$b_{11} = \frac{z_{22}}{z_{12}} = -\frac{y_{11}}{y_{12}} = \frac{a_{22}}{\Delta a} = \frac{1}{h_{12}} = -\frac{\Delta g}{g_{12}}$$

$$b_{12} = \frac{\Delta z}{z_{12}} = -\frac{1}{y_{12}} = \frac{a_{12}}{\Delta a} = \frac{h_{11}}{h_{12}} = -\frac{g_{22}}{g_{12}}$$

$$b_{21} = \frac{1}{z_{12}} = -\frac{\Delta y}{y_{12}} = \frac{a_{21}}{\Delta a} = \frac{h_{22}}{h_{12}} = -\frac{g_{11}}{g_{12}}$$

$$b_{22} = \frac{z_{11}}{z_{12}} = -\frac{y_{22}}{y_{12}} = \frac{a_{11}}{\Delta a} = \frac{\Delta h}{h_{12}} = -\frac{1}{g_{12}}$$

$$h_{11} = \frac{\Delta z}{z_{22}} = \frac{1}{y_{11}} = \frac{a_{12}}{a_{22}} = \frac{b_{12}}{b_{11}} = \frac{g_{22}}{\Delta g}$$

$$h_{12} = \frac{z_{12}}{z_{22}} = -\frac{y_{12}}{y_{11}} = \frac{\Delta a}{a_{22}} = \frac{1}{b_{11}} = -\frac{g_{12}}{\Delta g}$$

$$h_{21} = -\frac{z_{21}}{z_{22}} = \frac{y_{21}}{y_{11}} = -\frac{1}{a_{22}} = -\frac{\Delta b}{b_{11}} = -\frac{g_{21}}{\Delta g}$$

$$h_{22} = \frac{1}{z_{22}} = \frac{\Delta y}{y_{11}} = \frac{a_{21}}{a_{22}} = \frac{b_{21}}{b_{11}} = \frac{g_{11}}{\Delta g}$$

$$g_{11} = \frac{1}{z_{11}} = \frac{\Delta y}{y_{22}} = \frac{a_{21}}{a_{11}} = \frac{b_{21}}{b_{22}} = \frac{h_{22}}{\Delta h}$$

$$g_{12} = -\frac{z_{12}}{z_{11}} = \frac{y_{12}}{y_{22}} = -\frac{\Delta a}{a_{11}} = -\frac{1}{b_{22}} = -\frac{h_{12}}{\Delta h}$$

$$g_{21} = \frac{z_{21}}{z_{11}} = -\frac{y_{21}}{y_{22}} = \frac{1}{a_{11}} = \frac{\Delta b}{b_{22}} = -\frac{h_{21}}{\Delta h}$$

$$g_{22} = \frac{\Delta z}{z_{11}} = \frac{1}{y_{22}} = \frac{a_{12}}{a_{11}} = \frac{b_{12}}{b_{22}} = \frac{h_{11}}{\Delta h}$$

$$\Delta z = z_{11}z_{22} - z_{12}z_{21}$$
$$\Delta y = y_{11}y_{22} - y_{12}y_{21}$$
$$\Delta a = a_{11}a_{22} - a_{12}a_{21}$$
$$\Delta b = b_{11}b_{22} - b_{12}b_{21}$$
$$\Delta h = h_{11}h_{22} - h_{12}h_{21}$$
$$\Delta g = g_{11}g_{22} - g_{12}g_{21}$$

By comparing Eqs. (20.16) and (20.17) with Eqs. (20.1), we see that

$$z_{11} = \frac{y_{22}}{\Delta y}, \tag{20.18}$$

$$z_{12} = -\frac{y_{12}}{\Delta y}, \tag{20.19}$$

$$z_{21} = -\frac{y_{21}}{\Delta y}, \tag{20.20}$$

and

$$z_{22} = \frac{y_{11}}{\Delta y}. \tag{20.21}$$

To find the z-parameters as functions of the a-parameters, we rearrange Eqs. (20.3) in the form of Eqs. (20.1) and then compare coefficients. From the second equation in Eqs. (20.3) we have

$$V_2 = \frac{1}{a_{21}} I_1 + \frac{a_{22}}{a_{21}} I_2. \tag{20.22}$$

Therefore when Eq. (20.22) is substituted in the first equation of Eqs. (20.3) the result is

$$V_1 = \frac{a_{11}}{a_{21}} I_1 + \left(\frac{a_{11} a_{22}}{a_{21}} - a_{12}\right) I_2. \tag{20.23}$$

From Eq. (20.23) we have

$$z_{11} = \frac{a_{11}}{a_{21}}, \tag{20.24}$$

$$z_{12} = \frac{\Delta a}{a_{21}}; \tag{20.25}$$

and from Eq. (20.22) we have

$$z_{21} = \frac{1}{a_{21}}, \tag{20.26}$$

$$z_{22} = \frac{a_{22}}{a_{21}}. \tag{20.27}$$

The usefulness of the parameter conversion table is illustrated by the following example.

Example 20.3 Two sets of measurements are made on a two-port resistive circuit. The first set of measurements is made with port 2 open and the second set of measurements is made with port 2 short-circuited. The results are given below.

Port 2 open:	*Port 2 short-circuited:*
$V_1 = 10$ mV,	$V_1 = 24$ mV,
$I_1 = 10$ μA,	$I_1 = 20$ μA,
$V_2 = -40$ V;	$I_2 = 1$ mA.

Find the h-parameters of the circuit.

Solution   We can find h_{11} and h_{21} directly from the short-circuit test. Thus

$$h_{11} = \left.\frac{V_1}{I_1}\right|_{V_2=0} = \frac{24 \times 10^{-3}}{20 \times 10^{-6}} = 1.2 \text{ k}\Omega$$

and

$$h_{21} = \left.\frac{I_2}{I_1}\right|_{V_2=0} = \frac{10^{-3}}{20 \times 10^{-6}} = 50.$$

The parameters h_{12} and h_{22} cannot be found directly from the open-circuit test. However, by checking Eqs. (20.7)–(20.15) we see that the four a-parameters can be derived from the test data. Therefore, h_{12} and h_{22} can be found through the conversion table. Specifically,

$$h_{12} = \frac{\Delta a}{a_{22}} \quad \text{and} \quad h_{22} = \frac{a_{21}}{a_{22}}.$$

The a-parameters are

$$a_{11} = \left.\frac{V_1}{V_2}\right|_{I_2=0} = \frac{10 \times 10^{-3}}{-40} = -0.25 \times 10^{-3},$$

$$a_{21} = \left.\frac{I_1}{V_2}\right|_{I_2=0} = \frac{10 \times 10^{-6}}{-40} = -0.25 \times 10^{-6} \text{ } \mho,$$

$$a_{12} = -\left.\frac{V_1}{I_2}\right|_{V_2=0} = -\frac{24 \times 10^{-3}}{10^{-3}} = -24 \text{ } \Omega,$$

$$a_{22} = -\left.\frac{I_1}{I_2}\right|_{V_2=0} = -\frac{20 \times 10^{-6}}{10^{-3}} = -20 \times 10^{-3}.$$

The numerical value of Δa is

$$\Delta a = a_{11} a_{22} - a_{12} a_{21}$$
$$= 5 \times 10^{-6} - 6 \times 10^{-6} = -10^{-6}.$$

Thus

$$h_{12} = \frac{\Delta a}{a_{22}} = \frac{-10^{-6}}{-20 \times 10^{-3}} = 5 \times 10^{-5}$$

and

$$h_{22} = \frac{a_{21}}{a_{22}} = \frac{-0.25 \times 10^{-6}}{-20 \times 10^{-3}} = 1.25 \text{ } \mu\mho. \quad\blacksquare$$

DRILL EXERCISE   The following measurements were made on a two-port resistive circuit: with port 1
20.4   open, $V_2 = 15$ V, $V_1 = 10$ V, and $I_2 = 30$ A; with port 1 short-circuited, $V_2 = 10$ V, $I_2 = 4$ A, and $I_1 = -5$ A. Calculate the y-parameters.

Ans. $y_{11} = 0.75$ $\mho$; $y_{12} = -0.5$ $\mho$; $y_{21} = 2.4$ $\mho$; $y_{22} = 0.4$ $\mho$.

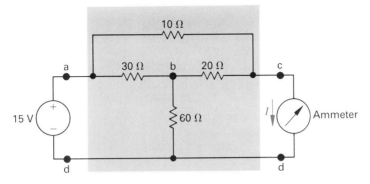

Fig. 20.4   A reciprocal two-port circuit.

If a two-port circuit is reciprocal, the following relationships exist among the port parameters:

$$z_{12} = z_{21}, \tag{20.28}$$

$$y_{12} = y_{21}, \tag{20.29}$$

$$a_{11}a_{22} - a_{12}a_{21} = \Delta a = 1, \tag{20.30}$$

$$b_{11}b_{22} - b_{12}b_{21} = \Delta b = 1, \tag{20.31}$$

$$h_{12} = -h_{21}, \tag{20.32}$$

$$g_{12} = -g_{21}. \tag{20.33}$$

A two-port is reciprocal if the interchange of an ideal voltage source at one port with an ideal ammeter at the second port produces the same ammeter reading. Consider, for example, the resistive circuit shown in Fig. 20.4. When a voltage source of 15 V is applied to the port ad, it produces a current of 1.75 A in the ammeter at port cd. The ammeter current is easily determined once the voltage V_{bd} is known. Thus

$$\frac{V_{bd}}{60} + \frac{V_{bd} - 15}{30} + \frac{V_{bd}}{20} = 0, \tag{20.34}$$

from which it follows that $V_{bd} = 5$ V and therefore

$$I = \frac{5}{20} + \frac{15}{10} = 1.75 \text{ A.} \tag{20.35}$$

If the voltage source and ammeter are interchanged, the ammeter will still read 1.75 A. This result is verified by solving the circuit shown in Fig. 20.5, for which we have

$$\frac{V_{bd}}{60} + \frac{V_{bd}}{30} + \frac{V_{bd} - 15}{20} = 0. \tag{20.36}$$

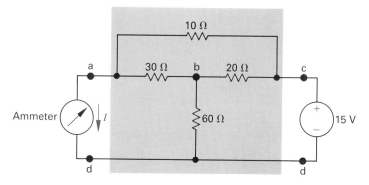

Fig. 20.5 The circuit in Fig. 20.4 with the voltage source and ammeter interchanged.

From Eq. (20.36) we get $V_{bd} = 7.5$ V. The current I_{ad} equals

$$I_{ad} = \frac{7.5}{30} + \frac{15}{10} = 1.75 \text{ A}. \tag{20.37}$$

A two-port is also reciprocal if the interchange of an ideal current source at one port with an ideal voltmeter at the second port produces the same reading of the voltmeter. (See Drill Exercise 20.5.)

DRILL EXERCISE
20.5

a) Calculate the reading of the ideal voltmeter in the circuit shown.

b) Interchange the voltmeter and the ideal current source and calculate the voltmeter reading.

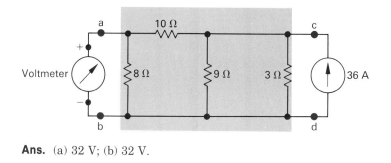

Ans. (a) 32 V; (b) 32 V.

It is worth noting that for a reciprocal two-port circuit only three calculations, or measurements, are needed to determine a set of parameters.

A reciprocal two-port circuit is symmetrical if its ports can be interchanged without disturbing the values of the terminal currents and voltages. Four examples of symmetrical two-port circuits are shown in Fig. 20.6. If a

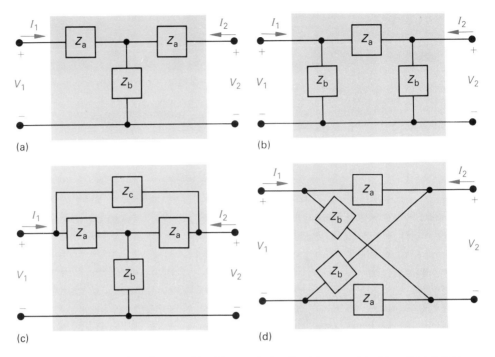

Fig. 20.6 Four examples of symmetrical two-port circuits: (a) symmetrical tee; (b) symmetrical pi; (c) symmetrical bridged-tee; (d) symmetrical lattice.

reciprocal two-port circuit is symmetrical, the following additional relationships exist among the port parameters:

$$z_{11} = z_{22}, \tag{20.38}$$

$$y_{11} = y_{22}, \tag{20.39}$$

$$a_{11} = a_{22}, \tag{20.40}$$

$$b_{11} = b_{22}, \tag{20.41}$$

$$h_{11}h_{22} - h_{12}h_{21} = \Delta h = 1, \tag{20.42}$$

$$g_{11}g_{22} - g_{12}g_{21} = \Delta g = 1. \tag{20.43}$$

For a symmetrical reciprocal network, only two calculations, or measurements, are necessary to find all the two-port parameters.

**DRILL EXERCISE
20.6**
The following measurements were made on a symmetrical, reciprocal, resistive, two-port network: with port 2 open, $V_1 = 95$ V and $I_1 = 5$ A; with a short-circuit across port 2, $V_1 = 11.52$ V and $I_2 = -2.72$ A. Calculate the z-parameters of the two-port network.

Ans. $z_{11} = z_{22} = 19 \ \Omega$, $z_{12} = z_{21} = 17 \ \Omega$.

20.4 ANALYSIS OF THE TERMINATED TWO-PORT CIRCUIT

In the typical application of a two-port model of a network, it is driven at port 1 and loaded at port 2. The s-domain circuit diagram for the typically terminated two-port model is shown in Fig. 20.7, where Z_g represents the internal impedance of the source, V_g is the internal voltage of the source, and Z_L is the load impedance. The analysis of the circuit in Fig. 20.7 involves expressing the terminal currents and voltages as functions of the two-port parameters V_g, Z_g, and Z_L.

There are six characteristics of the terminated two-port circuit that define its terminal behavior:

1. the input impedance $Z_{in} = V_1/I_1$, or admittance $Y_{in} = I_1/V_1$;
2. the output current I_2;
3. the Thévenin voltage and impedance (V_t, Z_t) with respect to port 2;
4. the current gain I_2/I_1;
5. the voltage gain V_2/V_1;
6. the voltage gain V_2/V_g.

To illustrate how these six characteristics are derived, we will develop the expressions using the z-parameters to model the two-port portion of the circuit. The expressions involving the y-, a-, b-, h-, and g-parameters are summarized in Table 20.2. The derivation of any one of the desired expressions involves the algebraic manipulation of the two-port equations along with the two constraint equations imposed by the terminations. If we use the z-parameter equations, the four equations that describe the circuit in Fig. 20.7 are

$$V_1 = z_{11}I_1 + z_{12}I_2, \tag{20.44}$$

$$V_2 = z_{21}I_1 + z_{22}I_2, \tag{20.45}$$

$$V_1 = V_g - I_1 Z_g, \tag{20.46}$$

$$V_2 = -I_2 Z_L. \tag{20.47}$$

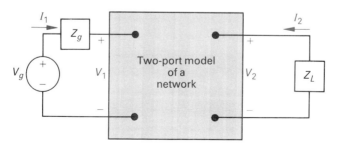

Fig. 20.7 A terminated two-port model.

TABLE 20.2
TERMINATED TWO-PORT EQUATIONS

z-parameters

$$Z_{in} = z_{11} - \frac{z_{12}z_{21}}{z_{22} + Z_L}$$

$$I_2 = \frac{-z_{21}V_g}{(z_{11} + Z_g)(z_{22} + Z_L) - z_{12}z_{21}}$$

$$V_t = \frac{z_{21}}{(z_{11} + Z_g)}V_g$$

$$Z_t = z_{22} - \frac{z_{12}z_{21}}{z_{11} + Z_g}$$

$$\frac{I_2}{I_1} = \frac{-z_{21}}{z_{22} + Z_L}$$

$$\frac{V_2}{V_1} = \frac{z_{21}Z_L}{z_{11}Z_L + \Delta z}$$

$$\frac{V_2}{V_g} = \frac{z_{21}Z_L}{(z_{11} + Z_g)(z_{22} + Z_L) - z_{12}z_{21}}$$

y-parameters

$$Y_{in} = y_{11} - \frac{y_{12}y_{21}Z_L}{1 + y_{22}Z_L}$$

$$I_2 = \frac{y_{21}V_g}{1 + y_{22}Z_L + y_{11}Z_g + \Delta_y Z_g Z_L}$$

$$V_t = \frac{-y_{21}V_g}{y_{22} + \Delta_y Z_g}$$

$$Z_t = \frac{1 + y_{11}Z_g}{y_{22} + \Delta_y Z_g}$$

$$\frac{I_2}{I_1} = \frac{y_{21}}{y_{11} + \Delta_y Z_L}$$

$$\frac{V_2}{V_1} = \frac{-y_{21}Z_L}{1 + y_{22}Z_L}$$

$$\frac{V_2}{V_g} = \frac{y_{21}Z_L}{y_{12}y_{21}Z_g Z_L - (1 + y_{11}Z_g)(1 + y_{22}Z_L)}$$

a-parameters

$$Z_{in} = \frac{a_{11}Z_L + a_{12}}{a_{21}Z_L + a_{22}}$$

$$I_2 = \frac{-V_g}{a_{11}Z_L + a_{12} + a_{21}Z_g Z_L + a_{22}Z_g}$$

$$V_t = \frac{V_g}{a_{11} + a_{21}Z_g}$$

$$Z_t = \frac{a_{12} + a_{22}Z_g}{a_{11} + a_{21}Z_g}$$

$$\frac{I_2}{I_1} = \frac{-1}{a_{21}Z_L + a_{22}}$$

$$\frac{V_2}{V_1} = \frac{Z_L}{a_{11}Z_L + a_{12}}$$

$$\frac{V_2}{V_g} = \frac{Z_L}{(a_{11} + a_{21}Z_g)Z_L + a_{12} + a_{22}Z_g}$$

b-parameters

$$Z_{in} = \frac{b_{22}Z_L + b_{12}}{b_{21}Z_L + b_{11}}$$

$$I_2 = \frac{-V_g\Delta_b}{b_{11}Z_g + b_{21}Z_g Z_L + b_{22}Z_L + b_{12}}$$

$$V_t = \frac{V_g\Delta_b}{b_{22} + b_{21}Z_g}$$

$$Z_t = \frac{b_{11}Z_g + b_{12}}{b_{21}Z_g + b_{22}}$$

$$\frac{I_2}{I_1} = \frac{-\Delta_b}{b_{11} + b_{21}Z_L}$$

$$\frac{V_2}{V_1} = \frac{\Delta_b Z_L}{b_{12} + b_{22}Z_L}$$

$$\frac{V_2}{V_g} = \frac{\Delta_b Z_L}{b_{12} + b_{11}Z_g + b_{22}Z_L + b_{21}Z_g Z_L}$$

h-parameters

$$Z_{in} = h_{11} - \frac{h_{12}h_{21}Z_L}{1 + h_{22}Z_L}$$

$$I_2 = \frac{h_{21}V_g}{(1 + h_{22}Z_L)(h_{11} + Z_g) - h_{12}h_{21}Z_L}$$

$$V_t = \frac{-h_{21}V_g}{h_{22}Z_g + \Delta_h}$$

$$Z_t = \frac{Z_g + h_{11}}{h_{22}Z_g + \Delta_h}$$

$$\frac{I_2}{I_1} = \frac{h_{21}}{1 + h_{22}Z_L}$$

$$\frac{V_2}{V_1} = \frac{-h_{21}Z_L}{\Delta_h Z_L + h_{11}}$$

$$\frac{V_2}{V_g} = \frac{-h_{21}Z_L}{(h_{11} + Z_g)(1 + h_{22}Z_L) - h_{12}h_{21}Z_L}$$

(Continued on page 740)

TABLE 20.2 (Cont.)

g-parameters

$$Y_{in} = g_{11} - \frac{g_{12}g_{21}}{g_{22} + Z_L}$$

$$\frac{I_2}{I_1} = \frac{-g_{21}}{g_{11}Z_L + \Delta_g}$$

$$I_2 = \frac{-g_{21}V_g}{(1 + g_{11}Z_g)(g_{22} + Z_L) - g_{12}g_{21}Z_g}$$

$$\frac{V_2}{V_1} = \frac{g_{21}Z_L}{g_{22} + Z_L}$$

$$V_t = \frac{g_{21}V_g}{1 + g_{11}Z_g}$$

$$\frac{V_2}{V_g} = \frac{g_{21}Z_L}{(1 + g_{11}Z_g)(g_{22} + Z_L) - g_{12}g_{21}Z_g}$$

$$Z_t = g_{22} - \frac{g_{12}g_{21}Z_g}{1 + g_{11}Z_g}$$

Equations (20.46) and (20.47) describe the constraints imposed by the terminations.

To find the impedance seen looking into port 1, that is, $Z_{in} = V_1/I_1$, we proceed as follows. In Eq. (20.45), we replace V_2 with $-I_2Z_L$ and solve the resulting expression for I_2. We get

$$I_2 = \frac{-z_{21}I_1}{z_L + z_{22}}, \tag{20.48}$$

which we then substitute into Eq. (20.44) and solve for Z_{in}; thus

$$Z_{in} = z_{11} - \frac{z_{12}z_{21}}{z_{22} + Z_L}. \tag{20.49}$$

To find the terminal current I_2, we first solve Eq. (20.44) for I_1 after replacing V_1 with the right-hand side of Eq. (20.46). The result is

$$I_1 = \frac{V_g - z_{12}I_2}{z_{11} + Z_g}. \tag{20.50}$$

We now substitute this expression for I_1 into Eq. (20.48) and solve the resulting equation for I_2. We get

$$I_2 = \frac{-z_{21}V_g}{(z_{11} + Z_g)(z_{22} + Z_L) - z_{12}z_{21}}. \tag{20.51}$$

The Thévenin voltage with respect to port 2 is equal to V_2 when $I_2 = 0$. With $I_2 = 0$, Eqs. (20.44) and (20.45) combine to yield

$$V_2 \Big|_{I_2=0} = z_{21}I_1 = z_{21}\frac{V_1}{z_{11}}. \tag{20.52}$$

But $V_1 = V_g - I_1Z_g$ and $I_1 = V_g/(Z_g + z_{11})$; therefore when the results are substituted into Eq. (20.52) the open-circuit value of V_2 becomes

$$V_2 \Big|_{I_2=0} = V_t = \frac{z_{21}}{Z_g + z_{11}} V_g. \tag{20.53}$$

The Thévenin, or output, impedance is the ratio V_2/I_2 when V_g is replaced by a short circuit. When V_g is zero, Eq. (20.46) reduces to

$$V_1 = -I_1 Z_g. \qquad (20.54)$$

When we substitute Eq. (20.54) into Eq. (20.44) we get

$$I_1 = \frac{-z_{12} I_2}{z_{11} + Z_g}. \qquad (20.55)$$

Equation (20.55) is now used to replace I_1 in Eq. (20.45) with the result that

$$\left.\frac{V_2}{I_2}\right|_{V_g=0} = Z_t = z_{22} - \frac{z_{12} z_{21}}{z_{11} + Z_g}. \qquad (20.56)$$

The current gain I_2/I_1 comes directly from Eq. (20.48):

$$\frac{I_2}{I_1} = \frac{-z_{21}}{Z_L + z_{22}}. \qquad (20.57)$$

To derive the expression for the voltage gain V_2/V_1 we start by replacing I_2 in Eq. (20.45) with its value given by Eq. (20.47); thus

$$V_2 = z_{21} I_1 + z_{22} \left(\frac{-V_2}{Z_L}\right). \qquad (20.58)$$

Next we solve Eq. (20.44) for I_1 as a function of V_1 and V_2; thus

$$z_{11} I_1 = V_1 - z_{12} \left(\frac{-V_2}{Z_L}\right)$$

or

$$I_1 = \frac{V_1}{z_{11}} + \frac{z_{12} V_2}{z_{11} Z_L}. \qquad (20.59)$$

Now we replace I_1 in Eq. (20.58) by Eq. (20.59) and solve the resulting expression for V_2/V_1; thus

$$\begin{aligned} \frac{V_2}{V_1} &= \frac{z_{21} Z_L}{z_{11} Z_L + z_{11} z_{22} - z_{12} z_{21}} \\ &= \frac{z_{21} Z_L}{z_{11} Z_L + \Delta z}. \end{aligned} \qquad (20.60)$$

The voltage ratio V_2/V_g is derived by first combining Eqs. (20.44), (20.46), and (20.47) to find I_1 as a function of V_2 and V_g. The result is

$$I_1 = \frac{z_{12} V_2}{Z_L(z_{11} + Z_g)} + \frac{V_g}{z_{11} + Z_g}. \qquad (20.61)$$

We can now use Eqs. (20.61) and (20.47) in conjunction with Eq. (20.45) to derive an expression involving only V_2 and V_g, that is,

$$V_2 = \frac{z_{21} z_{12} V_2}{Z_L(z_{11} + Z_g)} + \frac{z_{21} V_g}{z_{11} + Z_g} - \frac{z_{22}}{Z_L} V_2, \qquad (20.62)$$

which can be manipulated to get the desired voltage ratio:

$$\frac{V_2}{V_g} = \frac{z_{21}Z_L}{(z_{11} + Z_g)(z_{22} + Z_L) - z_{12}z_{21}}. \tag{20.63}$$

The expressions for these six attributes of the terminated two-port circuit are summarized as the first entries in Table 20.2. The corresponding expressions in terms of the y-, a-, b-, h-, and g-parameters are also listed in the table.

Example 20.4 illustrates the usefulness of the relationships listed in Table 20.2.

Example 20.4   The two-port circuit in Fig. 20.8 is described in terms of its b-parameters. The values of the parameters are

$$b_{11} = -20, \qquad b_{12} = -3000 \ \Omega, \qquad b_{21} = -2 \ \text{m℧}, \qquad b_{22} = -0.2.$$

a) Find the phasor voltage V_2.

b) Find the average power delivered to the 5-kΩ load.

c) Find the average power delivered to the input port.

d) Find the load impedance for maximum average power transfer.

e) Find the maximum average power delivered to the load in part (d).

Solution   a) To find V_2 we have two choices from the entries in Table 20.2. We can choose to find I_2 and then find V_2 from the relationship $V_2 = -I_2Z_L$ or we can find the voltage gain V_2/V_g and calculate V_2 from the gain. We will use the latter approach. For the given values of the b-parameters we have

$$\begin{aligned}
\Delta_b &= (-20)(-0.2) - (-3000)(-2 \times 10^{-3}) \\
&= 4 - 6 \\
&= -2.
\end{aligned}$$

From Table 20.2 we have

$$\begin{aligned}
\frac{V_2}{V_g} &= \frac{\Delta_b Z_L}{b_{12} + b_{11}Z_g + b_{22}Z_L + b_{21}Z_gZ_L} \\
&= \frac{(-2)(5000)}{-3000 + (-20)500 + (-0.2)5000 + [-2 \times 10^{-3}(500)(5000)]} \\
&= \frac{10}{19}.
\end{aligned}$$

It follows that

$$V_2 = \left(\frac{10}{19}\right)500 = 263.16\underline{/0°} \ \text{V}.$$

b) The average power delivered to the 5000-Ω load is

$$P_2 = \frac{(263.16)^2}{2(5000)} = 6.93 \ \text{W}.$$

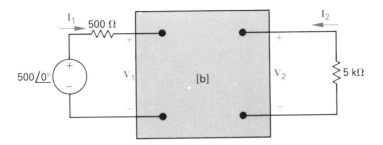

Fig. 20.8 The circuit for Example 20.4.

c) To find the average power delivered to the input port, we first find the input impedance Z_{in}. From Table 20.2 we have

$$Z_{in} = \frac{b_{22}Z_L + b_{12}}{b_{21}Z_L + b_{11}}$$

$$= \frac{(-0.2)(5000) - 3000}{-2 \times 10^{-3}(5000) - 20}$$

$$= \frac{400}{3} = 133.33 \ \Omega.$$

Now I_1 follows directly:

$$I_1 = \frac{500}{500 + 133.33} = 789.47 \ \text{mA}.$$

The average power delivered to the input port is

$$P_1 = \frac{(0.78947)^2}{2} (133.33)$$

$$= 41.55 \ \text{W}.$$

d) The load impedance for maximum power transfer is equal to the conjugate of the Thévenin impedance seen from looking into port 2. From Table 20.2 we have

$$Z_t = \frac{b_{11}Z_g + b_{12}}{b_{21}Z_g + b_{22}}$$

$$= \frac{(-20)(500) - 3000}{(-2 \cdot 10^{-3})(500) - 0.2}$$

$$= \frac{13,000}{1.2} = 10,833.33 \ \Omega.$$

Therefore $Z_L = Z_t^* = 10,833.33 \ \Omega$.

e) To find the maximum average power delivered to Z_L we first find V_2 from the voltage-gain expression V_2/V_g. When Z_L is 10,833.33 Ω we find this gain to be

$$\frac{\mathbf{V}_2}{\mathbf{V}_g} = 0.8333.$$

Thus

$$\mathbf{V}_2 = (0.8333)(500) = 416.67 \text{ V}$$

and

$$P_2(\text{maximum}) = \frac{1}{2} \cdot \frac{(416.67)^2}{10,833.33}$$
$$= 8.01 \text{ W.} \quad \blacksquare$$

DRILL EXERCISE 20.7 The b-parameters of the two-port network in Fig. 20.7 are $b_{11} = 2000/3$, $b_{12} = \frac{2}{3}$ MΩ, $b_{21} = \frac{1}{15}$ ℧, and $b_{22} = -100/3$. The network is driven by a sinusoidal current source having a maximum amplitude of 100 μA and an internal impedance of $1000 + j0$ Ω. The network is terminated in a resistive load of 10 kΩ.

a) Calculate the average power delivered to the load resistor.

b) Calculate the load resistance for maximum average power.

c) Calculate the maximum average power delivered to the resistor in part (b).

Ans. (a) 80 mW; (b) 40 kΩ; (c) 125 mW.

20.5 INTERCONNECTED TWO-PORTS

In the design of large, complex systems, it is usually easier to synthesize the system by first designing subsections of the system. The complete system is then fabricated by interconnecting these simpler, easier-to-design, smaller units. If the subsections are modeled by two-port circuits, then the synthesis of the complete system involves the analysis of interconnected two-ports.

There are five fundamental ways of interconnecting two-port circuits: (1) cascade, (2) series, (3) parallel, (4) series–parallel, and (5) parallel–series interconnections. These five basic interconnections are illustrated in Fig. 20.9.

We will analyze and illustrate only the cascade connection in this section. However, in passing, we note that if these four latter connections meet certain requirements, we can obtain the parameters that describe the interconnected circuits by the simple addition of the parameters of the individual networks. In particular, the z-parameters describe the series connection, the y-parameters describe the parallel connection, the h-parameters the series–parallel connection, and the g-parameters the parallel–series connection.[†]

---

† The interested reader can find a detailed discussion of these four interconnections in Henry Ruston and Joseph Bordogna, *Electric Networks: Functions, Filters, Analysis*, Chapter 4 (New York: McGraw-Hill, 1966).

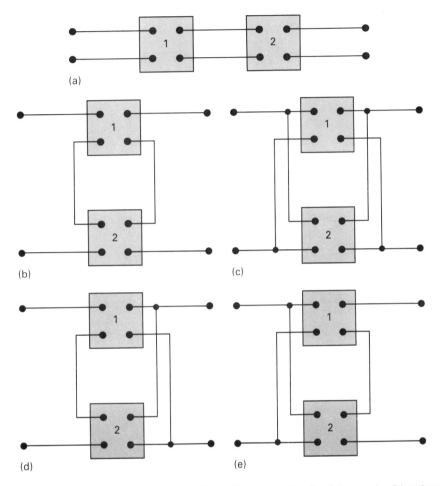

Fig. 20.9 The five basic interconnections of two-port circuits: (a) cascade; (b) series; (c) parallel; (d) series–series; (e) parallel–series.

The cascade connection is important because of its frequent occurrence in the modeling of large systems. Unlike the other four basic interconnections, there are no restrictions in using the parameters of the individual two-port circuits to obtain the parameters of the interconnected circuits. The a-parameters are best suited for describing the cascade connection. We will analyze the cascade connection using the circuit shown in Fig. 20.10, where we have used a single prime to denote the a-parameters of the first circuit and a double prime to denote the a-parameters of the second circuit. The output voltage and current of the first circuit are labeled V_2' and I_2' and the input voltage and current of the second circuit are labeled V_1' and I_1'. Our problem is to derive the a-parameter equations that relate V_2 and I_2 to V_1 and I_1. That is, we seek the

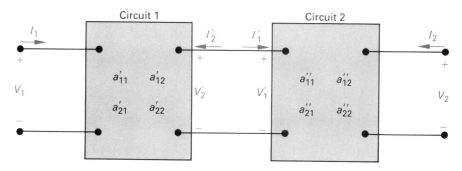

Fig. 20.10 The cascade connection.

pair of equations

$$V_1 = a_{11}V_2 - a_{12}I_2, \tag{20.64}$$

$$I_1 = a_{21}V_2 - a_{22}I_2, \tag{20.65}$$

where the a-parameters are given explicitly in terms of the a-parameters of the individual circuits.

We begin the derivation by noting from Fig. 20.10 that

$$V_1 = a'_{11}V'_2 - a'_{12}I'_2 \tag{20.66}$$

and

$$I_1 = a'_{21}V'_2 - a'_{22}I'_2. \tag{20.67}$$

Next, we observe that the interconnection means that $V'_2 = V'_1$ and $I'_2 = -I'_1$. When these constraints are substituted into Eqs. (20.66) and (20.67) we get

$$V_1 = a'_{11}V'_1 + a'_{12}I'_1 \tag{20.68}$$

and

$$I_1 = a'_{21}V'_1 + a'_{22}I'_1. \tag{20.69}$$

The voltage V'_1 and the current I'_1 are related to V_2 and I_2 through the a-parameters of the second circuit; thus

$$V'_1 = a''_{11}V_2 - a''_{12}I_2 \tag{20.70}$$

and

$$I'_1 = a''_{21}V_2 - a''_{22}I_2. \tag{20.71}$$

We can substitute Eqs. (20.70) and (20.71) into Eqs. (20.68) and (20.69) to generate the sought-after relationships between V_1, I_1 and V_2, I_2. We get

$$V_1 = (a'_{11}a''_{11} + a'_{12}a''_{21})V_2 - (a'_{11}a''_{12} + a'_{12}a''_{22})I_2 \tag{20.72}$$

and

$$I_1 = (a'_{21}a''_{11} + a'_{22}a''_{21})V_2 - (a'_{21}a''_{12} + a'_{22}a''_{22})I_2. \tag{20.73}$$

By comparing Eqs. (20.72) and (20.73) with Eqs. (20.64) and (20.65) we get the desired expressions for the a-parameters of the interconnected networks, namely,

$$a_{11} = a'_{11}a''_{11} + a'_{12}a''_{21}, \qquad (20.74)$$

$$a_{12} = a'_{11}a''_{12} + a'_{12}a''_{22}, \qquad (20.75)$$

$$a_{21} = a'_{21}a''_{11} + a'_{22}a''_{21}, \qquad (20.76)$$

$$a_{22} = a'_{21}a''_{12} + a'_{22}a''_{22}. \qquad (20.77)$$

Example 20.5 illustrates how we can use Eqs. (20.74)–(20.77) to analyze the cascade connection of two amplifier circuits.

Example 20.5 Two identical amplifiers are connected in cascade as shown in Fig. 20.11. Each amplifier is described in terms of its h-parameters. The values are $h_{11} = 1000 \ \Omega$, $h_{12} = 0.0015$, $h_{21} = 100$, and $h_{22} = 100 \ \mu\mho$. Find the voltage gain V_2/V_g.

Solution The first step to finding the overall voltage gain V_2/V_g is to convert from the h-parameters to the a-parameters. Since the amplifiers are identical, one set of a-parameters describes the amplifiers:

$$a'_{11} = \frac{-\Delta_h}{h_{21}} = \frac{+0.05}{100} = 5 \times 10^{-4},$$

$$a'_{12} = \frac{-h_{11}}{h_{21}} = \frac{-1000}{100} = -10 \ \Omega$$

$$a'_{21} = \frac{-h_{22}}{h_{21}} = \frac{-100 \times 10^{-6}}{100} = -10^{-6} \ \mho,$$

$$a'_{22} = \frac{-1}{h_{21}} = \frac{-1}{100} = -10^{-2}.$$

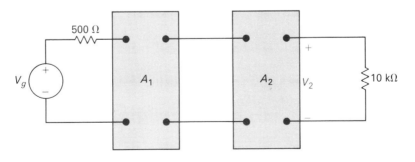

Fig. 20.11 The circuit for Example 20.5.

Next we use Eqs. (20.74)–(20.77) to compute the a-parameters of the cascaded amplifiers:

$$a_{11} = a'_{11}a'_{11} + a'_{12}a'_{21}$$
$$= 25 \times 10^{-8} + (-10)(-10^{-6}) = 10.25 \times 10^{-6},$$
$$a_{12} = a'_{11}a'_{12} + a'_{12}a'_{22}$$
$$= -10(-0.01) + (-10)(5 \times 10^{-4}) = 0.095 \ \Omega,$$
$$a_{21} = a'_{21}a'_{11} + a'_{22}a'_{21}$$
$$= (-10^{-6})(5 \times 10^{-4}) + (-10^{-6})(-0.01) = 0.0095 \times 10^{-6} \ \mho,$$
$$a_{22} = a'_{21}a'_{12} + a'_{22}a'_{22}$$
$$= (-10)(-10^{-6}) + (-10^{-2})^2 = 1.1 \times 10^{-4}.$$

From Table 20.2 we have

$$\frac{V_2}{V_g} = \frac{Z_L}{(a_{11} + a_{21}Z_g)Z_L + a_{12} + a_{22}Z_g}$$

$$= \frac{10^4}{[10.25 \times 10^{-6} + 0.0095 \times 10^{-6}(500)]10^4 + 0.095 + 1.1 \times 10^{-4}(500)}$$

$$= \frac{10^4}{0.15 + 0.095 + 0.055} = \frac{10^5}{3} = 33{,}333.33.$$

Thus an input signal of 150 μV is amplified to an output signal of 5 V. For an alternative approach to finding the voltage gain V_2/V_g see Problem 20.24. ∎

If more than two units are connected in cascade, the a-parameters of the equivalent two-port can be found by successively reducing the original set of two-ports one pair at a time.

DRILL EXERCISE
20.8

Each element in the symmetrical bridged-tee circuit in Fig. 20.6(c) is a 15-Ω resistor. Two of these bridged tees are connected in cascade between a dc voltage source and a resistive load. The dc voltage source has a no-load voltage of 100 V and an internal resistance of 8 Ω. The load resistor is adjusted until maximum power is delivered to the load. Calculate (a) the load resistance; (b) the load voltage; and (c) the load power.

Ans. (a) 14.44 Ω; (b) 16 V; (c) 17.73 W.

20.6 SUMMARY

Two-port models of circuits are useful when it is convenient to describe the performance of the circuit in terms of its behavior at two pairs of terminals. Each pair of terminals is referred to as a port. The port where the signal is fed into the system is called the input and the port where the signal is extracted is called the output. The two-port model is restricted to systems where (1) there are no independent sources in the internal network between the ports; (2) there is no energy stored in the network; and (3) the passive elements (R, L, and C) are linear and time-invariant. It is also understood that no external connections are made between the input and output ports.

The two-port model is described in terms of four variables: the input voltage (V_1) and current (I_1) and the output voltage (V_2) and current (I_2). The two-port model is constructed in the s-domain so that dc circuits and phasor domain circuits are treated as special cases, that is, $s = 0$ for dc models and $s = j\omega$ for sinusoidal steady-state models. Only two of the four variables are independent, which means that there are six possible ways to relate the four variables. The six sets of equations group the variables as follows: (1) the voltages as functions of the currents (z-parameters); (2) the currents as functions of the voltages (y-parameters); (3) the input port variables as functions of the output port variables (a-parameters); (4) the output port variables as functions of the input port variables (b-parameters); (5) the input voltage and output current as functions of the output voltage and input current (h-parameters); and (6) the output voltage and input current as functions of the input voltage and output current (g-parameters). These six sets of equations can be grouped into three sets of reciprocal pairs, and identified by the parameters as (1) the immittance parameters (z and y); (2) the transmission parameters (a and b); and (3) the hybrid parameters (h and g).

Because all six sets of equations involve the same four variables, all the parameters are interrelated. Thus if we know one set of parameters, we can calculate the remaining five sets. The interrelationships are summarized in Table 20.1.

If the two-port circuit satisfies the principle of reciprocity, it is referred to as a reciprocal two-port. For reciprocal two-ports, special relationships exist among the port parameters. In particular, $z_{12} = z_{21}$, $y_{12} = y_{21}$, $\Delta_a = \Delta_b = 1$, $h_{12} = -h_{21}$, and $g_{12} = -g_{21}$.

If a reciprocal network is symmetrical, the input and output ports can be interchanged without disturbing the values of the port variables. For symmetrical two-ports, $z_{11} = z_{22}$, $y_{11} = y_{22}$, $a_{11} = a_{22}$, $b_{11} = b_{22}$, and $\Delta_h = \Delta_g = 1$.

The analysis of the typically terminated two-port (Fig. 20.7) is greatly facilitated by Table 20.2, which lists the explicit relationships for the pertinent characteristics of the terminated two-port as functions of the parameters, the source impedance (Z_g) and the load impedance Z_L. The characteristics of interest are the input impedance V_1/I_1, the output current I_2, the Thévenin equivalent circuit with respect to the output port V_t and Z_t, the current gain I_2/I_1, and the voltage gains V_2/V_1 and V_2/V_g.

Two-port equivalent circuits, as functions of the port parameters, are introduced in Problems 20.24–20.28.

PROBLEMS

20.1 Find the h- and g-parameters for the circuit in Example 20.1.

20.2 Find the z-parameters for the circuit shown in Fig. 20.12.

20.3 Find the b-parameters for the circuit shown in Fig. 20.13.

20.4 Select the values of R_1, R_2, and R_3 in the circuit in Fig. 20.14 so that $h_{11} = 14\ \Omega$, $h_{12} = 0.8$, $h_{21} = -0.8$, and $h_{22} = 0.04\ \mho$.

20.5 Find the a-parameters of the circuit shown in Fig. 20.15.

20.6 The following measurements pertain to a two-port resistive circuit. With port 2 short-circuited and 20 V applied to port 1, the current into port 1 is 2 A and the current into port 2 is -0.8 A. With port 1 short-circuited and 25 V applied to port 2, the current into port 2 is 1.4 A and the current into port 1 is -1.0 A. Find the y-parameters of the two-port.

20.7 Find the phasor domain values of the a-parameters for the two port circuit shown in Fig. 20.16.

20.8 Find the g-parameters for the operational amplifier circuit shown in Fig. 20.17.

20.9 Derive the expressions for the h-parameters as functions of the g-parameters.

20.10 Derive the expressions for the b-parameters as functions of the h-parameters.

20.11 The operational amplifier in the circuit shown in Fig. 20.18 is ideal. Find the h-parameters of the circuit.

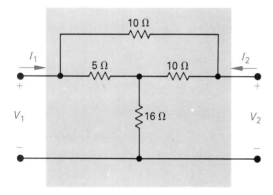

Fig. 20.12 The circuit for Problem 20.2.

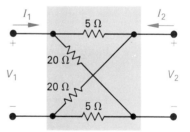

Fig. 20.13 The circuit for Problem 20.3.

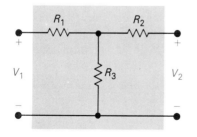

Fig. 20.14 The circuit for Problem 20.4.

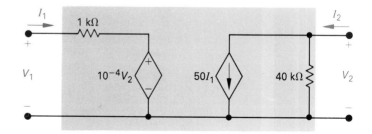

Fig. 20.15 The circuit for Problem 20.5.

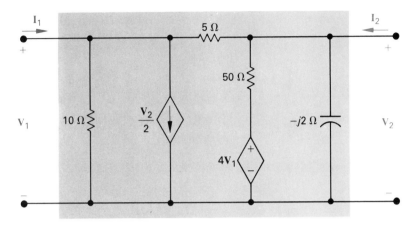

Fig. 20.16   The two-port circuit for Problem 20.7.

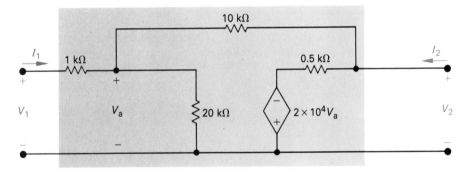

Fig. 20.17   The circuit for Problem 20.8.

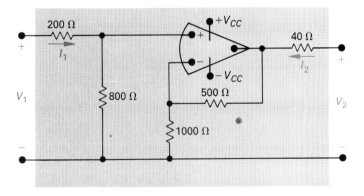

Fig. 20.18   The circuit for Problem 20.11.

20.12   Find the s-domain expressions for the y-parameters of the two-port circuit shown in Fig. 20.19.

20.13   Find the s-domain expressions for the a-parameters of the two-port circuit shown in Fig. 20.20.

20.14   Is the two-port circuit shown in Fig. 20.21 symmetrical? Justify your answer.

20.15   Find the s-domain expressions for the y-parameters of the circuit shown in Fig. 20.22.

20.16   Derive the expression for the input impedance ($Z_{\text{in}} = V_1/I_1$) of the circuit in Fig. 20.7 in terms of the a-parameters.

20.17   Derive the expression for the current gain I_2/I_1 of the circuit in Fig. 20.7 in terms of the h-parameters.

20.18   Derive the expression for the voltage gain V_2/V_1 of the circuit in Fig. 20.7 in terms of the g-parameters.

20.19   Find the Thévenin equivalent circuit with respect to port 2 of the circuit in Fig. 20.7 in terms of the y-parameters.

20.20   In the circuit in Fig. 20.23, the g-parameters of the two-port are

$$g_{11} = -20 \times 10^{-4} \ \mho, \qquad g_{12} = 3 \times 10^{-2}, \qquad g_{21} = 2000, \qquad g_{22} = -20,000 \ \Omega.$$

The internal impedance of the source is $1000 + j0 \ \Omega$ and the load impedance is $10,000 + j0 \ \Omega$.

    a) Find $\mathbf{V}_2$.
    b) Find the average power delivered to Z_L in mW.
    c) Find the load impedance Z_L that will result in maximum average power transfer to Z_L.
    d) Find the maximum average power in part (c).
    e) Find $\mathbf{V}_1$ and $\mathbf{I}_1$ under conditions of maximum power transfer.
    f) How much average power is delivered to port 1 when maximum average power is delivered to Z_L?

20.21   a) Find the s-domain expressions for the g-parameters of the circuit in Fig. 20.24.
    b) Port 2 in Fig. 20.24 is terminated in a resistance of 50 kΩ and port 1 is driven by a step voltage source $v_1(t) = 28u(t)$ V. Find $v_2(t)$ for $t > 0$ if $C = 0.01 \ \mu$F and $L = 2$ H.

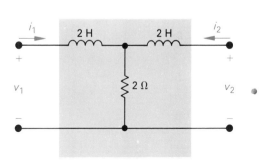

Fig. 20.19   The circuit for Problem 20.12.

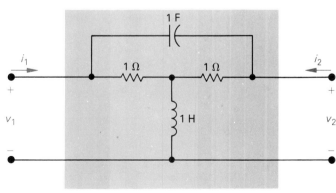

Fig. 20.20   The circuit for Problem 20.13.

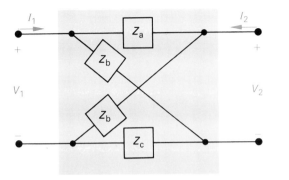

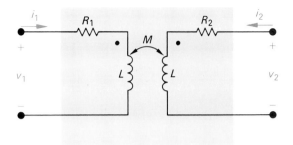

Fig. 20.22 The circuit for Problem 20.15.

Fig. 20.21 The circuit for Problem 20.14.

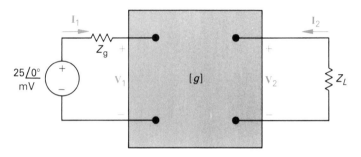

Fig. 20.23 The circuit for Problem 20.20.

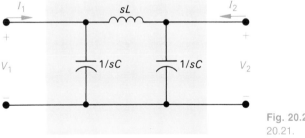

Fig. 20.24 The circuit for Problem 20.21.

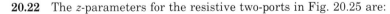

20.22 The z-parameters for the resistive two-ports in Fig. 20.25 are:

$$z'_{11} = 100 \ \Omega, \qquad z''_{11} = 80 \ \Omega,$$
$$z'_{12} = -500 \ \Omega, \qquad z''_{12} = -400 \ \Omega,$$
$$z'_{21} = 1000 \ \Omega, \qquad z''_{21} = 2000 \ \Omega,$$
$$z'_{22} = 10{,}000 \ \Omega, \qquad z''_{22} = 16{,}000 \ \Omega.$$

Find the input resistance $Z_{in} = V_1/I_1$.

20.23 Networks A and B in the circuit in Fig. 20.26 are reciprocal and symmetrical. For network A, it is known that $a'_{11} = 4$ and $a'_{12} = 5 \ \Omega$.

    a) Find the a-parameters of network B.

    b) Find the voltage ratio V_2/V_1 when $I_2 = 0$.

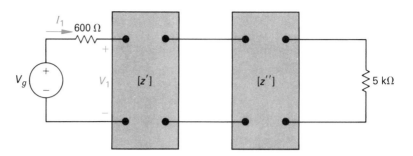

Fig. 20.25   The circuit for Problem 20.22.

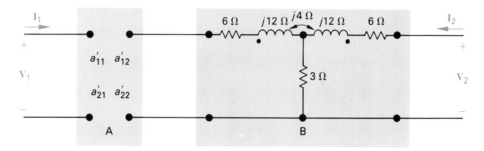

Fig. 20.26   The circuit for Problem 20.23.

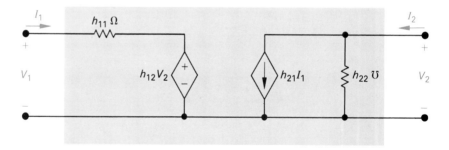

Fig. 20.27   The circuit for Problem 20.24.

20.24   a) Show that the circuit in Fig. 20.27 is an equivalent circuit that is satisfied by the h-parameter equations.
   b) Use the h-parameter equivalent circuit of part (a) to find the voltage gain V_2/V_g in the circuit in Fig. 20.11.

20.25   a) Show that the circuit in Fig. 20.28 is an equivalent circuit that is satisfied by the z-parameter equations.
   b) Assume that the equivalent circuit in Fig. 20.28 is driven by a voltage source having an internal impedance of Z_g ohms. Calculate the Thévenin equivalent circuit with respect to port 2. Check your results against the appropriate entries in Table 20.2.

20.26   a) Show that the circuit in Fig. 20.29 is also an equivalent circuit that is satisfied by the z-parameter equations.
   b) Assume that the equivalent circuit in Fig. 20.29 is terminated in an impedance

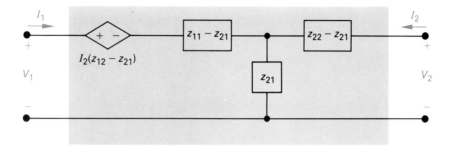

Fig. 20.28   The z-parameter equivalent circuit for Problem 20.25.

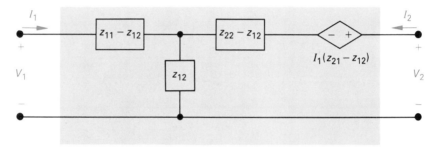

Fig. 20.29   The z-parameter equivalent circuit for Problem 20.26.

of Z_L ohms at port 2. Find the input impedance V_1/I_1. Check your result against the appropriate entry in Table 20.2.

20.27   a) Derive two equivalent circuits that are satisfied by the y-parameters equations. (*Hint:* Start with Eqs. 20.2. Add and subtract $y_{21} V_2$ to the first equation of the set. Construct a circuit that satisfies the resulting set of equations, by thinking in terms of node voltages. Derive an alternative equivalent circuit by first altering the second equation in Eqs. 20.2.)

   b) Assume that port 1 is driven by a voltage source having an internal impedance Z_g and port 2 is loaded with an impedance Z_L. Find the current gain I_2/I_1. Check your result against the appropriate entry in Table 20.2.

20.28   Derive the equivalent circuit that is satisfied by the g-parameter equations.

APPENDIXES

APPENDIX A SOLUTION OF LINEAR SIMULTANEOUS EQUATIONS

A.1 INTRODUCTION

Circuit analysis frequently involves the solution of linear simultaneous equations. Our purpose here is to review the use of determinants to solve such a set of equations. The theory of determinants (with applications) can be found in most intermediate-level algebra texts. (A particularly good reference for engineering students is Chapter 1 of *The Mathematics of Circuit Analysis* by E. A. Guillemin, John Wiley & Sons, Inc., New York, 1949). In our review here we shall limit our discussion to the mechanics of solving simultaneous equations with determinants.

A.2 PRELIMINARY STEPS

The first step in solving a set of simultaneous equations by determinants is to write the equations in a rectangular (square) format. That is, the equations are arranged in a vertical stack such that each variable occupies the same horizontal position in every equation. For example, in Eqs. (A.1), the variables i_1, i_2, and i_3 occupy the first, second, and third position, respectively, on the left-hand side of each equation:

$$\begin{aligned}
21i_1 - 9i_2 - 12i_3 &= -33, \\
-3i_1 + 6i_2 - 2i_3 &= 3, \\
-8i_1 - 4i_2 + 22i_3 &= 50.
\end{aligned} \tag{A.1}$$

Alternatively one can describe this set of equations by saying that i_1 occupies the first column in the array, i_2 the second column, and i_3 the third column.

If one or more variables are missing from a given equation, they can be inserted by simply making their coefficient zero. Thus Eqs. (A.2) can be "squared up" as shown by Eqs. (A.3):

$$\begin{aligned}
2v_1 - v_2 &= 4, \\
4v_2 + 3v_3 &= 16, \\
7v_1 + 2v_3 &= 5;
\end{aligned} \tag{A.2}$$

$$\begin{aligned}
2v_1 - v_2 + 0v_3 &= 4, \\
0v_1 + 4v_2 + 3v_3 &= 16, \\
7v_1 + 0v_2 + 2v_3 &= 5.
\end{aligned} \tag{A.3}$$

A.3 CRAMER'S METHOD

The value of each unknown variable in the set of equations is expressed as the ratio of two determinants. If we let N, with an appropriate subscript, represent the numerator

757

determinant and Δ represent the denominator determinant, then the kth unknown x_k is

$$x_k = \frac{N_k}{\Delta}. \tag{A.4}$$

The denominator determinant Δ is the same for every unknown variable and is called the characteristic determinant of the set of equations. The numerator determinant N_k varies with each unknown. Equation (A.4) is referred to as Cramer's method for solving simultaneous equations.

A.4 THE CHARACTERISTIC DETERMINANT

Once we have organized the set of simultaneous equations into an ordered array as illustrated by Eqs. (A.1) and (A.3), it is a simple matter to form the characteristic determinant. The characteristic determinant is the square array made up from the coefficients of the unknown variables. For example, the characteristic determinants of Eqs. (A.1) and (A.3) are

$$\Delta = \begin{vmatrix} 21 & -9 & -12 \\ -3 & 6 & -2 \\ -8 & -4 & 22 \end{vmatrix} \tag{A.5}$$

and

$$\Delta = \begin{vmatrix} 2 & -1 & 0 \\ 0 & 4 & 3 \\ 7 & 0 & 2 \end{vmatrix}, \tag{A.6}$$

respectively.

A.5 THE NUMERATOR DETERMINANT

The numerator determinant N_k is formed from the characteristic determinant by replacing the kth column in the characteristic determinant by the column of values appearing on the right-hand side of the equations. For example, the numerator determinants for evaluating i_1, i_2, and i_3 in Eqs. (A.1) are

$$N_1 = \begin{vmatrix} -33 & -9 & -12 \\ 3 & 6 & -2 \\ 50 & -4 & 22 \end{vmatrix}, \tag{A.7}$$

$$N_2 = \begin{vmatrix} 21 & -33 & -12 \\ -3 & 3 & -2 \\ -8 & 50 & 22 \end{vmatrix}, \tag{A.8}$$

and

$$N_3 = \begin{vmatrix} 21 & -9 & -33 \\ -3 & 6 & 3 \\ -8 & -4 & 50 \end{vmatrix}. \tag{A.9}$$

The numerator determinants for the evaluation of v_1, v_2, and v_3 in Eqs. (A.3) are

$$N_1 = \begin{vmatrix} 4 & -1 & 0 \\ 16 & 4 & 3 \\ 5 & 0 & 2 \end{vmatrix}, \tag{A.10}$$

$$N_2 = \begin{vmatrix} 2 & 4 & 0 \\ 0 & 16 & 3 \\ 7 & 5 & 2 \end{vmatrix}, \tag{A.11}$$

and

$$N_3 = \begin{vmatrix} 2 & -1 & 4 \\ 0 & 4 & 16 \\ 7 & 0 & 5 \end{vmatrix}. \tag{A.12}$$

A.6 EVALUATION OF A DETERMINANT

The value of a determinant is found by expanding it in terms of its minors. The minor of any element in a determinant is the determinant that remains after the row and column occupied by the element have been deleted. For example, the minor of the element 6 in Eq. (A.7) is

$$\begin{vmatrix} -33 & -12 \\ 50 & 22 \end{vmatrix},$$

while the minor of the element 22 in Eq. (A.7) is

$$\begin{vmatrix} -33 & -9 \\ 3 & 6 \end{vmatrix}.$$

The cofactor of an element is its minor multiplied by the sign-controlling factor

$$-1^{(i+k)},$$

where i and k denote the row and column, respectively, occupied by the element. Thus the cofactor of the element 6 in Eq. (A.7) is

$$-1^{(2+2)} \begin{vmatrix} -33 & -12 \\ 50 & 22 \end{vmatrix}$$

and the cofactor of the element 22 is

$$-1^{(3+3)} \begin{vmatrix} -33 & -9 \\ 3 & 6 \end{vmatrix}.$$

The cofactor of an element is also referred to as its signed minor.

The sign-controlling factor $-1^{(i+k)}$ will equal plus or minus one depending on whether $i + k$ is an even or odd integer. Thus the algebraic sign of a cofactor alternates between ± 1 as we move along a row or column. For a 3×3 determinant, the $+$ and $-$

signs form the checkerboard pattern illustrated below:

$$\begin{vmatrix} + & - & + \\ - & + & - \\ + & - & + \end{vmatrix}$$

A determinant can be expanded along any row or column. Thus the first step in making an expansion is to select a row i or a column k. Once a row or column has been selected, each element in that row or column is multiplied by its signed minor, or cofactor. The value of the determinant is the sum of these products. As an example, let us evaluate the determinant in Eq. (A.5) by expanding it along its first column. Following the rules enumerated above, we write the expansion as

$$\Delta = 21(1)\begin{vmatrix} 6 & -2 \\ -4 & 22 \end{vmatrix} - 3(-1)\begin{vmatrix} -9 & -12 \\ -4 & 22 \end{vmatrix} - 8(1)\begin{vmatrix} -9 & -12 \\ 6 & -2 \end{vmatrix}. \tag{A.13}$$

The 2×2 determinants in Eq. (A.13) can also be expanded by minors. The minor of an element in a 2×2 determinant is a single element. It follows that the expansion reduces to multiplying the upper left element by the lower right element and then subtracting from this product the product of the lower left element times the upper right element. Using this observation, we evaluate Eq. (A.13) to

$$\Delta = 21(132 - 8) + 3(-198 - 48) - 8(18 + 72)$$
$$= 2604 - 738 - 720 = 1146. \tag{A.14}$$

Had we elected to expand the determinant along the second row of elements we would have written

$$\Delta = -3(-1)\begin{vmatrix} -9 & -12 \\ -4 & 22 \end{vmatrix} + 6(+1)\begin{vmatrix} 21 & -12 \\ -8 & 22 \end{vmatrix} - 2(-1)\begin{vmatrix} 21 & -9 \\ -8 & -4 \end{vmatrix}$$
$$= 3(-198 - 48) + 6(462 - 96) + 2(-84 - 72)$$
$$= -738 + 2196 - 312 = 1146. \tag{A.15}$$

The numerical values of the determinants N_1, N_2, and N_3 given by Eqs. (A.7), (A.8), and (A.9) are

$$N_1 = 1146, \tag{A.16}$$

$$N_2 = 2292, \tag{A.17}$$

and

$$N_3 = 3438. \tag{A.18}$$

It follows from Eqs. (A.15)–(A.18) that the solutions for i_1, i_2, and i_3 in Eq. (A.1) are

$$i_1 = \frac{N_1}{\Delta} = 1 \text{ A,}$$

$$i_2 = \frac{N_2}{\Delta} = 2 \text{ A,}$$

and

$$i_3 = \frac{N_3}{\Delta} = 3 \text{ A.} \tag{A.19}$$

We will leave it to the reader to verify the solutions for v_1, v_2, and v_3 in Eqs. (A.3) are

$$v_1 = \frac{49}{-5} = -9.8 \text{ V,}$$

$$v_2 = \frac{118}{-5} = -23.6 \text{ V,} \tag{A.20}$$

and

$$v_3 = \frac{-184}{-5} = 36.8 \text{ V.}$$

APPENDIX  COMPLEX NUMBERS

B.1 INTRODUCTION

Complex numbers were invented and introduced into the number system to permit the extraction of square roots of negative numbers. The invention of complex numbers simplifies the solution of problems that would otherwise be very difficult. The equation $x^2 + 8x + 41 = 0$ has no solution in a number system that excludes complex numbers. We find the concept of a complex number, and the capability of manipulating these numbers algebraically, extremely useful in circuit analysis.

B.2 NOTATION

There are two ways to designate a complex number: the *cartesian,* or *rectangular,* form and the *polar,* or *trigonometric,* form. In the rectangular form, a complex number is written in terms of its real and imaginary components; hence

$$n = a + jb, \tag{B.1}$$

where a is the real component, b is the imaginary component, and j is by definition $\sqrt{-1}$.†

In polar form, a complex number is written in terms of its magnitude, or modulus, and angle, or argument; hence

$$n = ce^{j\theta}, \tag{B.2}$$

where c is the magnitude, θ is the angle, e is the base of the natural logarithm, and, as before, $j = \sqrt{-1}$. In the literature, the symbol $\underline{/\theta}$ is frequently used in place of $e^{j\theta}$, that is, the polar form is written

$$n = c\underline{/\theta} \tag{B.3}$$

Although Eq. (B.3) is more convenient in printing text material, Eq. (B.2) is of primary importance in mathematical operations because the rules for manipulating an exponential quantity are well known. For example, since $(y^x)^n = y^{xn}$, then $(e^{j\theta})^n = e^{jn\theta}$; and since $y^{-x} = 1/y^x$, then $e^{-j\theta} = 1/e^{j\theta}$, and so forth.

Since there are two ways of expressing the same complex number, we need to relate one form to the other. The transition from the polar to the rectangular form makes

---

† You may be more familiar with the notation $i = \sqrt{-1}$. In electrical engineering, i is used as the symbol for current and hence in electrical engineering literature, j is used to denote $\sqrt{-1}$.

use of Euler's identity,

$$e^{\pm j\theta} = \cos\theta \pm j\sin\theta. \tag{B.4}$$

A complex number in polar form can be put in rectangular form by writing

$$
\begin{aligned}
ce^{j\theta} &= c[\cos\theta + j\sin\theta]\\
&= c\cos\theta + jc\sin\theta\\
&= a + jb.
\end{aligned} \tag{B.5}
$$

The transition from rectangular to polar form makes use of the geometry of the right triangle, namely,

$$
\begin{aligned}
a + jb &= [\sqrt{a^2 + b^2}]e^{j\theta}\\
&= ce^{j\theta},
\end{aligned} \tag{B.6}
$$

where

$$\tan\theta = b/a. \tag{B.7}$$

It is not obvious from Eq. (B.7) in which quadrant the angle θ lies. The ambiguity can be resolved by a graphical representation of the complex number.

B.3 GRAPHICAL REPRESENTATION OF A COMPLEX NUMBER

A complex number is represented graphically on a complex number plane, which uses the horizontal axis for plotting the real component and the vertical axis for plotting the imaginary component. The angle of the complex number is measured *counterclockwise* from the *positive real axis*. The graphical plot of the complex number $n = a + jb = c\underline{/\theta}$, if we assume that a and b are both positive, is shown in Fig. B.1. The graphical representation of a complex number makes very clear the relationship between the rectangular and polar forms. Any point in the complex number plane is uniquely defined by giving either its distance from each axis (that is, a and b) or its radial distance from the origin (c) and the angle of the radial measurement θ.

It follows from Fig. B.1 that θ is in the first quadrant when a and b are both positive; in the second quadrant when a is negative and b is positive; in the third quadrant when a and b are both negative; and in the fourth quadrant when a is positive and b is negative. These observations are illustrated in Fig. B.2, where we have plotted $4 + j3$, $-4 + j3$, $-4 - j3$, and $4 - j3$.

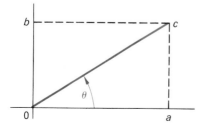

Fig. B.1 The graphical representation of $a + jb$ when a and b are both positive.

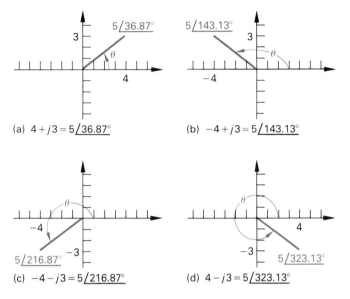

Fig. B.2   The graphical representation of four complex numbers.

It should be pointed out that we can also specify θ as a clockwise angle from the positive real axis. Thus in Fig. B.2(c) we could also designate $-4 - j3$ as $5\underline{/-143.13°}$. In Fig. B.2(d) we observe that $5\underline{/323.13°} = 5\underline{/-36.87°}$. It is customary to express θ in terms of negative values when θ lies in the third or fourth quadrant.

The graphical interpretation of a complex number also shows the relationship between a complex number and its conjugate. *The conjugate of a complex number is formed by reversing the sign of its imaginary component.* Thus the conjugate of $a + jb$ is $a - jb$ and the conjugate of $-a + jb$ is $-a - jb$. When we write a complex number in polar form, we form its conjugate by simply reversing the sign of the angle θ. Therefore, the conjugate of $c\underline{/\theta}$ is $c\underline{/-\theta}$. The conjugate of a complex number is designated with an asterisk. In other words, n^* is understood to be the conjugate of n. When the conjugate of a complex number is plotted on the complex number plane, we see that conjugation simply reflects the complex number about the real axis. This is illustrated in Fig. B.3.

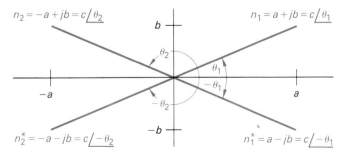

Fig. B.3   The complex numbers n_1 and n_2 and conjugates n_1^* and n_2^*.

B.4   ARITHMETIC OPERATIONS

Addition (Subtraction)

In order to add or subtract complex numbers, we must express the numbers in rectangular form. Complex numbers are added by adding the real parts to form the real part of the sum and by adding the imaginary parts to form the imaginary part of the sum. Thus if we are given

$$n_1 = 8 + j16$$

and

$$n_2 = 12 - j3,$$

then

$$n_1 + n_2 = (8 + 12) + j(16 - 3) = 20 + j13.$$

Subtraction follows the same rule. Thus,

$$n_2 - n_1 = (12 - 8) + j(-3 - 16) = 4 - j19.$$

If the numbers to be added or subtracted are given in polar form, they are first converted to rectangular form. For example, if

$$n_1 = 10\underline{/53.13°}$$

and

$$n_2 = 5\underline{/-135°},$$

then

$$\begin{aligned} n_1 + n_2 &= 6 + j8 - 3.535 - j3.535 \\ &= (6 - 3.535) + j(8 - 3.535) \\ &= 2.465 + j4.465 = 5.10\underline{/61.10°} \end{aligned}$$

and

$$\begin{aligned} n_1 - n_2 &= 6 + j8 - (-3.535 - j3.535) \\ &= 9.535 + j11.535 \\ &= 14.966\underline{/50.42°}. \end{aligned}$$

Multiplication (Division)

Multiplication or division of complex numbers can be carried out with the numbers written in either rectangular or polar form. However, in most cases, the polar form is more convenient. As an example, let us find the product $n_1 n_2$ when $n_1 = 8 + j10$ and $n_2 = 5 - j4$. Using the rectangular form, we have

$$\begin{aligned} n_1 n_2 &= (8 + j10)(5 - j4) = 40 - j32 + j50 + 40 \\ &= 80 + j18 \\ &= 82\underline{/12.68°}. \end{aligned}$$

If we use the polar form, the multiplication $n_1 n_2$ becomes

$$n_1 n_2 = [12.81\underline{/51.34°}][6.40\underline{/-38.66°}]$$
$$= 82\underline{/12.68°} = 80 + j18.$$

The first step in dividing two complex numbers that are to be divided in rectangular form is to multiply the numerator and denominator by the *conjugate of the denominator*. This will reduce the denominator to a real number. We then divide the real number into the new numerator. As an example we will find the value of n_1/n_2, where $n_1 = 6 + j3$ and $n_2 = 3 - j1$. We have

$$\frac{n_1}{n_2} = \frac{6 + j3}{3 - j1} = \frac{(6 + j3)(3 + j1)}{(3 - j1)(3 + j1)} = \frac{18 + j6 + j9 - 3}{9 + 1}$$
$$= \frac{15 + j15}{10} = 1.5 + j1.5 = 2.12\underline{/45°}.$$

In polar form, the division of n_1 by n_2 is

$$\frac{n_1}{n_2} = \frac{6.71\underline{/26.57°}}{3.16\underline{/-18.43°}} = 2.12\underline{/45°} = 1.5 + j1.5.$$

B.5   USEFUL IDENTITIES

In working with complex numbers and quantities, the following identities are very useful:

$$\pm j^2 = \mp 1, \tag{B.8}$$

$$(-j)(j) = 1, \tag{B.9}$$

$$j = \frac{1}{-j}, \tag{B.10}$$

$$e^{\pm j\pi} = -1, \tag{B.11}$$

$$e^{\pm j\pi/2} = \pm j. \tag{B.12}$$

Given that $n = a + jb = c\underline{/\theta}$, it follows that

$$nn^* = a^2 + b^2 = c^2, \tag{B.13}$$

$$n + n^* = 2a, \tag{B.14}$$

$$n - n^* = j2b, \tag{B.15}$$

$$n/n^* = 1\underline{/2\theta}. \tag{B.16}$$

B.6 INTEGER POWER OF A COMPLEX NUMBER

To raise a complex number to an integer power k, it is easier to first write the complex number in polar form. Thus

$$n^k = (a + jb)^k = (ce^{j\theta})^k = c^k e^{jk\theta}$$
$$= c^k(\cos k\theta + j \sin k\theta).$$

For example,

$$[2e^{j12°}]^5 = 2^5 e^{j60°} = 32e^{j60°} = 16 + j27.71$$

and

$$(3 + j4)^4 = [5e^{j53.13°}]^4 = 5^4 e^{j212.52°}$$
$$= 625e^{j212.52°} = -527 - j336.$$

B.7 ROOTS OF A COMPLEX NUMBER

To find the kth root of a complex number, we must recognize that we are solving the equation

$$x^k - ce^{j\theta} = 0, \tag{B.17}$$

which is an equation of the kth degree and therefore has k roots.

To find the k roots, we first note that

$$ce^{j\theta} = ce^{j(\theta+2\pi)} = ce^{j(\theta+4\pi)} = \cdots . \tag{B.18}$$

It follows from Eqs. (B.17) and (B.18) that

$$x_1 = (ce^{j\theta})^{1/k} = c^{1/k} e^{j\theta/k}, \tag{B.19}$$

$$x_2 = [ce^{j(\theta+2\pi)}]^{1/k} = c^{1/k} e^{j(\theta+2\pi)/k}, \tag{B.20}$$

$$x_3 = [ce^{j(\theta+4\pi)}]^{1/k} = c^{1/k} e^{j(\theta+4\pi)/k}, \tag{B.21}$$

$$\vdots$$

We continue the process outlined by Eqs. (B.19), (B.20), and (B.21), until the roots start repeating. This will happen when the multiple of π is equal to $2k$. For example, let us find the four roots of $81e^{j60°}$. We have

$$x_1 = (81)^{1/4} e^{j60/4} = 3e^{j15°},$$
$$x_2 = (81)^{1/4} e^{j(60+360)/4} = 3e^{j105°},$$
$$x_3 = (81)^{1/4} e^{j(60+720)/4} = 3e^{j195°},$$
$$x_4 = (81)^{1/4} e^{j(60+1080)/4} = 3e^{j285°},$$
$$x_5 = (81)^{1/4} e^{j(60+1440)/4} = 3e^{j375°} = 3e^{j15°}$$

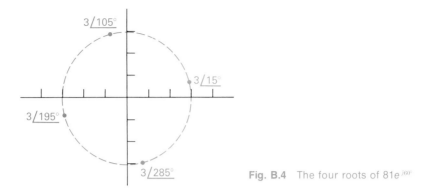

Fig. B.4 The four roots of $81e^{j60°}$

But x_5 is the same as x_1, so the roots have started to repeat. Therefore, we know the four roots of $81e^{j60°}$ are the values given by x_1, x_2, x_3, and x_4.

It is worth noting that the roots of a complex number lie on a circle in the complex number plane. The radius of the circle is $c^{1/k}$. The roots are uniformly distributed around the circle, the angle between adjacent roots being equal to $2\pi/k$ radians, or $360/k$ degrees. The four roots of $81e^{j60°}$ are shown plotted in Fig. B.4.

1. $\sin(\alpha \pm \beta) = \sin \alpha \cos \beta \pm \cos \alpha \sin \beta$

2. $\cos(\alpha \pm \beta) = \cos \alpha \cos \beta \mp \sin \alpha \sin \beta$

3. $\sin \alpha + \sin \beta = 2 \sin \dfrac{(\alpha + \beta)}{2} \cos \dfrac{(\alpha - \beta)}{2}$

4. $\sin \alpha - \sin \beta = 2 \cos \left(\dfrac{\alpha + \beta}{2}\right) \sin \left(\dfrac{\alpha - \beta}{2}\right)$

5. $\cos \alpha + \cos \beta = 2 \cos \left(\dfrac{\alpha + \beta}{2}\right) \cos \left(\dfrac{\alpha - \beta}{2}\right)$

6. $\cos \alpha - \cos \beta = -2 \sin \left(\dfrac{\alpha + \beta}{2}\right) \sin \left(\dfrac{\alpha - \beta}{2}\right)$

7. $2 \sin \alpha \sin \beta = \cos(\alpha - \beta) - \cos(\alpha + \beta)$

8. $2 \cos \alpha \cos \beta = \cos(\alpha - \beta) + \cos(\alpha + \beta)$

9. $2 \sin \alpha \cos \beta = \sin(\alpha + \beta) + \sin(\alpha - \beta)$

10. $\sin 2\alpha = 2 \sin \alpha \cos \alpha$

11. $\cos 2\alpha = 2 \cos^2 \alpha - 1 = 1 - 2 \sin^2 \alpha$

12. $\cos^2 \alpha = \frac{1}{2} + \frac{1}{2} \cos 2\alpha$

13. $\sin^2 \alpha = \frac{1}{2} - \frac{1}{2} \cos 2\alpha$

14. $\tan(\alpha \pm \beta) = \dfrac{\tan \alpha \pm \tan \beta}{1 \mp \tan \alpha \tan \beta}$

15. $\tan 2\alpha = \dfrac{2 \tan \alpha}{1 - \tan^2 \alpha}$

APPENDIX D

AN ABBREVIATED TABLE OF INTEGRALS

1. $\displaystyle\int xe^{ax}\,dx = \frac{e^{ax}}{a^2}\,(ax - 1)$

2. $\displaystyle\int x^2 e^{ax}\,dx = \frac{e^{ax}}{a^3}\,(a^2 x^2 - 2ax + 2)$

3. $\displaystyle\int x \sin ax\,dx = \frac{1}{a^2}\sin ax - \frac{x}{a}\cos ax$

4. $\displaystyle\int x \cos ax\,dx = \frac{1}{a^2}\cos ax + \frac{x}{a}\sin ax$

5. $\displaystyle\int e^{ax} \sin bx\,dx = \frac{e^{ax}}{a^2 + b^2}\,(a \sin bx - b \cos bx)$

6. $\displaystyle\int e^{ax} \cos bx\,dx = \frac{e^{ax}}{a^2 + b^2}\,(a \cos bx + b \sin bx)$

7. $\displaystyle\int \frac{dx}{x^2 + a^2} = \frac{1}{a}\tan^{-1}\frac{x}{a}$

8. $\displaystyle\int \frac{dx}{(x^2 + a^2)^2} = \frac{1}{2a^2}\left[\frac{x}{x^2 + a^2} + \frac{1}{a}\tan^{-1}\frac{x}{a}\right]$

9. $\displaystyle\int \sin ax \sin bx\,dx =$

$$\frac{\sin (a - b)x}{2(a - b)} - \frac{\sin (a + b)x}{2(a + b)},\ a^2 \neq b^2$$

10. $\displaystyle\int \cos ax \cos bx\,dx =$

$$\frac{\sin (a - b)x}{2(a - b)} + \frac{\sin (a + b)x}{2(a + b)},\ a^2 \neq b^2$$

11. $\displaystyle\int \sin ax \cos bx\,dx =$

$$-\frac{\cos (a - b)x}{2(a - b)} - \frac{\cos (a + b)x}{2(a + b)},\ a^2 \neq b^2$$

12. $\displaystyle\int \sin^2 ax\,dx = \frac{x}{2} - \frac{\sin 2ax}{4a}$

13. $\displaystyle\int \cos^2 ax\,dx = \frac{x}{2} + \frac{\sin 2ax}{4a}$

14. $\displaystyle\int_0^{\infty} \frac{a\,dx}{a^2 + x^2} = \frac{\pi}{2},\ a > 0;$

$$= 0,\ a = 0;$$

$$= \frac{-\pi}{2},\ a < 0$$

15. $\displaystyle\int_0^{\infty} \frac{\sin ax}{x}\,dx = \frac{\pi}{2},\ a > 0$

$$= \frac{-\pi}{2},\ a > 0$$

ANSWERS TO
SELECTED PROBLEMS

ANSWERS TO SELECTED PROBLEMS

Chapter 1

1.1 (a) 6.24×10^{13} e/s; (b) 15×10^{12} cm

1.3 156.04 μm/s

1.5 (a) 300 W A to B; (b) 500 W B to A; (c) 200 W B to A;
(d) 400 W A to B

Chapter 2

2.3 (a) 9 A; (b) 72 V; (c) 864 W

2.7 (a) $i_1 = 4$ A, $i_2 = 10$ A, $i_3 = 6$ A, $i_4 = 7$ A, $i_5 = 1$ A;
(b) $v_2 = 60$ V, $v_3 = 12$ V, $v_4 = 28$ V, $v_5 = 72$ V;
(c) $p_{\text{del}} = 1100$ W;
(d) $p_6 = 600$ W, $p_{10} = 160$ W, $p_2 = 72$ W, $p_4 = 196$ W, $p_{72} = 72$W;
(e) yes

2.9 $i_B = 50$ μA, $i_E = 1$ mA, $i_C = 0.95$ mA, $v_{3d} = 0.6$ V, $v_{bd} = 1.2$ V, $i_2 = 30$ μA,
$i_1 = 80$ μA, $v_{ab} = 4.8$ V, $i_{CC} = 1.03$ mA, $v_{13} = 3.5$ V

2.10 $v_1 = -2$ V, $v_g = 0.5$ V

Chapter 3

3.3 (a) 35 kΩ; (b) 10 Ω; (c) 40 Ω

3.5 (a) 5 A, 14 W

3.7 (a) 10 V; (b) $P_1 = 0.40$ W, $P_2 = 0.20$ W; (c) $R_1 = 1600$ Ω, $R_2 = 800$ Ω

3.9 100 A

3.12 (a) 74.63 mA; (b) 79.68 mA; (c) -6.34%

3.14 (a) $R_1 = 1480$ Ω, $R_2 = 13,500$ Ω, $R_3 = 15,000$ Ω;
(b) 487.5 V

3.17 (a) 6500 Ω; 0; 1000 Ω; 9000 Ω; 81 kΩ; ∞

3.19 (a) 20 mA; (b) 150 Ω; (c) 7.01 mA

3.20 (a) 200 Ω; (b) 24 mA; (c) R_x, 80 mW, 500 Ω, 8 mW

Chapter 4

4.1 $i_1 = 1.89$ A; $i_2 = 1.24$ A; $i_3 = 0.65$ A; $i_4 = 2.26$ A; $i_5 = -1.61$ A

4.4 $v_0 = 5$ V

4.6 (a) 32 V; (b) $P_{50\text{ V}} = 310$ W (del) $p_{15i_o} = 110.4$ W (abs)

4.7 $v = 2$ V

4.11 733.5 W (abs)

4.14 276.25 V

4.15 (a) $i_0 = 15$ A, $i_1 = 1.5$ A, $i_2 = 10$ A, $i_3 = 5$ A, $i_4 = 10$ A;
(b) $p_{del} = P_{abs} = 1495$ W

4.17 $i = -0.50$ mA

4.19 $v_t = v_{ab} = -200$ V, $R_t = 54$ kΩ

4.21 $v_t = 0$ V, $R_t = 36$ Ω

4.24 (a) 3.5 kΩ; (b) 87.5 mW; (c) 43.96%

4.26 $i_0(90$ V$) = 18$ A, $i_0(30$ A$) = -12$ A, $i_0(45$ V$) = 6$ A, $i_0 = 12$ A

Chapter 5

5.2 (a) $i = 0$, $t \leq 0$;
$i = 50t$, $0 \leq t \leq 0.005$ s;
$i = 0.5 - 50t$, $0.005 \leq t \leq 0.01$ s;
$i = 0$, $t \geq 0.01$ s;
(b) $v = 0$, $t < 0$;
$v = 1$ V, $0 < t < 0.005$ s;
$v = -1$ V, $0.005 < t < 0.01$ s;
$v = 0$, $t > 0.01$ s;
$p = 0$, $t < 0$;
$p = 50t$, $0 < t < 0.005$ s;
$p = -0.5 + 50t$, $0.005 < t < 0.01$ s;
$p = 0$, $t > 0.01$ s;
$w = 0$, $t \leq 0$;
$w = 25t^2$, $0 \leq t \leq 0.005$ s;
$w = 25t^2 - 0.5t + 2.5 \times 10^{-3}$, $0.005 \leq t \leq 0.01$ s;
$w = 0$, $t \geq 0.01$ s

5.4 154.26 V

5.5 (a) 625 μJ; (b) $A_1 = -10^6$, $A_2 = 100$;
(c) $i = 0.125[-1.2 + 2000t]e^{-2000t}$ A, $t > 0$

5.8 8 H

5.9 (a) $6e^{-t}$ A, $t \geq 0$; (b) $i_1(t) = [4e^{-t} - 2]$ A, $t \geq 0$; (c) $i_2(t) = [2e^{-t} + 2]$ A, $t \geq 0$; (d) 36 J; (e) 54 J; (f) 18 J; (g) yes

5.12 (a) $v = 10e^{-t}$ V, $t \geq 0$;
(b) $v_1(t) = [\frac{20}{3} e^{-t} - \frac{8}{3}]$ V, $t \geq 0$;
(c) $v_2(t) = [\frac{10}{3} e^{-t} + \frac{8}{3}]$ V, $t \geq 0$;
(d) 100 μJ; (e) 132 μJ; (f) 32 μJ; (g) yes

Chapter 6

6.1 (a) 30 Ω; (b) 1.5 H; (c) 50 ms; (d) 48 J; (e) 47.12 J

6.3 (a) $i(t) = [4 - 3e^{-0.5t}]$ A, $t \geq 0$; (b) 810.93 ms

6.5 (a) $v = 30e^{-2000t}$ V, $t \geq 0$;
$i_1 = -1.25[e^{-2000t} + 3]$ A, $t \geq 0$;
$i_2 = 3.75[1 - e^{-2000t}]$ A, $t \geq 0$;
(b) 112.50 mJ, (c) 37.5 mJ

6.7 (a) 400 kΩ; (b) 0.5 μF; (c) 200 ms; (d) 16 μJ; (e) 10.11 μJ

6.10 (a) $i = 4e^{-5t}$ mA, $t > 0$;
$v_1 = 20[e^{-5t} + 4]$ V, $t \geq 0$;
$v_2 = 80[1 - e^{-5t}]$ V, $t \geq 0$;
(b) 200 mJ; (c) 160 mJ, 40 mJ

Chapter 7

7.1 (a) $V_s = 500$ V, $R = 50$ Ω, $L = 0.25$ H; (b) 3.47 ms

7.3 $i(t) = [-2 + 7e^{-500t}]$ A, $t \geq 0$

7.4 $i(t) = [2 + 6e^{-5t}]$ A, $t \geq 0$; $v(t) = -120e^{-5t}$ V, $t > 0$

7.7 (a) 5 mA, 40 kΩ, 0.5 μF, 20 ms; (b) 18.33 ms

7.9 $v_c = [75 - 120e^{-100t}]$ V, $t \geq 0$

7.10 (a) $i(0^-) = 0$, $i(0^+) = 3$ mA; (b) $i(\infty) = 0$, (c) 4 ms;
(d) $i(t) = 3e^{-250t}$ mA, $t > 0$;
(e) $v(t) = [6 - 3.6e^{-250t}]$ V, $t > 0$

7.14 115.13 ms

7.15 $v_c(t) = [25 - 25e^{-100t}]$ V, $t \geq 0$

Chapter 8

8.1 (a) $s_1 = -200$ rad/s, $s_2 = -800$ rad/s; (b) overdamped; (c) 5 H;
(d) 3125 Ω

8.3 $i_L(0) = 0$, $i_R(0) = -30$ mA, $i_c(0) = 30$ mA;
(b) $v(t) = [20e^{-400t} - 80e^{-1600t}]$ V, $t \geq 0$;
(c) $i_L(t) = [-8e^{-400t} + 8e^{-1600t}]$ mA, $t \geq 0$

8.6 (a) $v(t) = [30 \cos 1000t - 48 \sin 1000t]$ V, $t \geq 0$;
(b) $f = 159.15$ Hz; (c) 56.60 V

8.12 (a) $v(t) = 40e^{-32,000t} \sin 24,000t$ V, $t \geq 0$;
(b) $i_c(t) = 8e^{-32,000t}[3 \cos 24,000t - 4 \sin 24,000t]$ mA, $t > 0$

8.14 (a) $i_L(t) = [4 - e^{-30t}(4 \cos 40t + \sin 40t)]$ mA, $t \geq 0$;
(b) 4.42 mA; (c) 68.58 ms

8.16 (a) 8000 Ω; (b) $i(0^+) = -2$ mA; $\dfrac{di}{dt}(0^+) = 200$ A/s;

(c) $v_c(t) = -40,000t\, e^{-50,000t}$ V, $t \geq 0$

8.19 (a) 312.50 mH; (b) $48t\, e^{-16,000t}$ A, $t \geq 0$; (c) 1.104 mA; (d) 62.5 μs

8.22 $v_c(t) = 120e^{-6000t}[8 \cos 8000t + 6 \sin 8000t]$ V, $t \geq 0$

8.24 $v(t) = [\frac{16}{3} e^{-2000t} + \frac{128}{3} e^{-8000t}]$ V, $t > 0$

Chapter 9

9.1 (a) -15 V; (b) -10 V; (c) -4 V; (d) 7 V; (e) $-1.08 \leq v_a \leq 4.92$ V

9.5 (a) 9 V; (b) noninverting

9.7 (a) 6 μW; (b) 2400 pW; (c) 2500; (d) yes; enables the 60-mV source
to control 2500 times as much power in the load.

9.10 (a) -99.965; (b) 349.88 μV; (c) 1000.35 Ω; (d) -100, 0 V, 1000 Ω

9.12 $v_t = -47.46v_s$, $R_t = 2.033$ kΩ;
(a) 2.033 kΩ; (b) 1978.84 Ω

9.16 $R_B = 12$ kΩ; $R_C = 500$ Ω; $R_D = 4$ kΩ

Chapter 10

10.1 (a) 40 V; (b) 314.16 rad/s; (c) 50 Hz; (d) 60°; (e) 1.05 rad;
(f) 20 ms; (g) 6.67 ms; (h) $40 \cos 100\pi t$ V; (i) 8.33 ms; (j) 11.67 ms

10.3 (a) 100 Hz; (b) $85.06 \cos(200\pi t - 18°)$ V

10.6 (a) $y = 116.62 \cos(800t + 150.96°)$;
(b) $y = 53.85 \cos(300t + 8.20°)$

10.8 (a) 62,831.85 rad/s; (b) 90°; (c) $-159.15\ \Omega$; (d) 0.10 μF;
(e) $-j159.15\ \Omega$

10.11 $Z_{ab} = (60 + j45)\ \Omega = 75\underline{/36.87°}\ \Omega$

10.17 (a) $\mathbf{I}_2 = 4.24\underline{/-45°}$ A, $\mathbf{I}_3 = 3\underline{/90°}$ A, $\mathbf{V}_s = 18.25\underline{/80.54°}$ V;
(b) $i_2 = 4.24\cos(10^5t - 45°)$ A,
$i_3 = 3\cos(10^5t + 90°)$ A,
$v_s = 18.25\cos(10^5t + 80.54°)$ V

10.19 $v = 180\sin 400t$ V

10.21 $\mathbf{V}_0 = (20 - j40)$ V $= 44.72\underline{/-63.43°}$ V

10.23 $v_1 = 23.06\cos(10^4t - 76.97°)$ V,
$v_2 = 24.38\cos(10^4t - 63.42°)$ V

10.25 $\mathbf{I}_1 = 6.013\underline{/93.74°}$ A, $\mathbf{I}_2 = 4.019\underline{/-95.60°}$ A, $\mathbf{I}_3 = 7.756\underline{/-145.57°}$ A,
$\mathbf{I}_4 = 6.013\underline{/3.74°}$ A

10.29 $\mathbf{V}_t = \mathbf{V}_{ab} = 3.09\underline{/163.19°}$ V; $Z_t = 9.052\underline{/-25.19°}$

10.33 (a) $v_0 = 10.18\cos(20,000t - 45°)$ V; (b) 6.65 V

Chapter 11

11.1 (a) $P = 433.01$ W (abs), $Q = 240$ VAR (del);
(b) $P = 433.01$ W (abs), $Q = 250$ VAR (abs);
(c) $P = 383.02$ W (abs), $Q = 321.39$ VAR (del);
(d) $P = 211.31$ W (del); $Q = 453.15$ VAR (del);
(e) $P = 353.55$ W (abs); $Q = 353.55$ VAR (del)

11.3 (a) 1800 W; (b) -200 W; (c) 800 W; (d) 600 VAR; (e) absorb;
(f) 0.8; (g) 0.6

11.5 1250 W

11.7 $(1.875 + j0.625)\ \Omega$

11.11 (a) $(192 - j192\ \Omega$; (b) 4.69 W

11.12 (a) 360 mW; (b) 4000 Ω, 0.1 μF; (c) 443.08 mW, yes; (d) 450 mW;
(e) 4000 Ω, 0.0667 μF; (f) yes

11.14 (a) 200; (b) 437.48 mW

11.15 (a) 657.44 VARS; (b) delivers; (c) 0.1208 lead; (d) 569.44 A;
(e) 416.67 A

Chapter 12

12.1 (a) Balanced, abc; (b) balanced, acb;
(c) unbalanced, unequal phase displacements;
(d) unbalanced, unequal amplitudes;
(e) unbalanced, unequal phase displacements;
(f) unbalanced, unequal frequencies

12.3 $v_{AB} = 293.95\cos(\omega t + 56°)$,
$v_{BC} = 293.95\cos(\omega t - 64°)$,
$v_{CA} = 293.95\cos(\omega t + 176°)$

12.5 (a) Unbalanced, because load impedances are not equal;
(b) $\mathbf{I}_0 = 9.96\underline{/-9.79°}$ A (rms)

12.7 (a) 75 A (rms); (b) 449.03 V (rms)

12.9 (a) 60.93 A (rms); (b) 23.09 A (rms); (c) 24 A (rms);
(d) 2192.86 V (rms)

12.14 346.41 V (rms)

12.15 $(4353.75 + j3510)$ kva

12.18 (a) 2408.48 V rms; (b) 38.38 μF
12.21 (a) W_1 = 139,196.31 W, W_2 = 37,297.54 W;
(b) W_1 = 37,297.54, W_2 = 139,196.31 W
12.22 Z = 25.6 + j19.2 Ω
12.24 (a) 12,470.77 W; (b) 21,600 VAR

Chapter 13

13.1 (a) 0.8; (b) 9 mH; (c) 3
13.3 The terminal connected to the + terminal of the voltmeter should be marked
with a dot.
13.7 (a) 0.25; (b) (1.44 − j1.92) Ω; (c) (14.56 + j46.08) Ω; (d) 13.85 V;
(e) 58.24%; (f) 4.80%
13.9 (a) $\mathbf{V}_t$ = 64.40$\underline{/26.57°}$ V, Z_t = 190.72 + j418.56 Ω;
(b) 190.72 − j418.56 Ω; (c) 5.44 W; (d) 21.18%
13.11 $\mathbf{I}_a$ = 6A, $\mathbf{I}_b$ = 1.5 A, $\mathbf{I}_c$ = 7.5 A
13.13 800$\underline{/45°}$ Ω
13.14 775.07 W
13.15 (a) 0.2; (b) 28.8 mW; (c) 2.4 V
13.20 (a) 10 H, 0.1 H, 0.9 H; (b) k = 0.9
13.22 (a) 42.51 V; (b) verification

Chapter 14

14.1 (a) 40 kΩ, 2.5 mH; (b) 969.24 krad/s, 1.032 Mrad/s
14.3 (a) 160 pF; (b) 32; (c) 6.4; (d) 78.125 krad/s, 15.625 krad/s;
(e) 160 V; (f) 12.8 mA
14.5 (a) 18.24 V, 17.52 V; (b) 17.89 V, 17.89 V; (c) − 1.9%, 2.11%
14.7 250 Ω, 20 μH, 0.2 μF, 0.80 A
14.9 (a) 10 kΩ, 684.48 μH, 1.46 nF; (b) 14.61; (c) 68.45 krad/s; (d) 11.69;
(e) 85.56 krad/s
14.11 (a) 6.25, 16 krad/s, 100 krad/s, 92.32 krad/s, 108.32 krad/s; (b) 625 V;
(c) 5, 20 krad/s, 100 krad/s, 90.50 krad/s, 110.50 krad/s, 500 V
14.13 (a) 92.08 Ω, 10.13 pF; (b) 104.72
14.15 (a) 9.9875 Mrad/s; (b) 20 V; (c) 10.0050 Mrad/s; (d) 20.025 V;
(e) 700 krad/s; (f) 19.97; (g) 14.27
14.18 (a) 4 V; (b) 3.03 V; (c) 3.88 V
14.22 (a) 30.66 nF, 10.09 nF;
(b) 5.31 cos 10,000πt + 0.0801 cos (20,000πt + 1.19°) V
14.23 (a) $C = L/(\omega^2 L^2 + R^2)$; (b) 7.72 nF; (c) 0.2883; (d) 1.0
14.26 (a) 25 mH, 40 F; (b) 50 Ω, 0.25 μH, 0.16 μF
14.27 (a) $\sqrt{99}$ rad/s, $\sqrt{99}$, 1 rad/s; (b) 10^4, 50, 251.89, 1 kΩ, 19.90 mH, 199 pF;
(c) 9.95, 50, 251.89 rad/s
14.29 (a) 1 rad/s; (b) 10 V; (c) 0.905 rad/s, 1.105 rad/s; (d) 0.2 rad/s;
(e) 5; (f) 1 kΩ, 1 nF

Chapter 15

15.1 (a) $f(t)$ = $(t + 10)u(t + 10) − 2tu(t) + (t − 10)u(t − 10)$;
(b) $f(t)$ = $8[−(t + 3)u(t + 3) + (t + 2)u(t + 2) + (t + 1)u(t + 1)$
$− (t − 1)u(t − 1) − (t − 2)u(t − 2) + (t − 3)u(t − 3)]$

15.6 (a) $1/(s + a)^2$;   (b) $s/(s^2 + \omega^2)$;   (c) $[s \cos \theta - \omega \sin \theta]/(s^2 + \omega^2)$;
(d) $1/(s^2 - 1)$
15.9 (a) $\omega s/(s^2 + \omega^2)$;   (b) $s^2/(s^2 + \omega^2)$;   (c) 2
15.10 (a) $1/s(s + a)$;   (b) $1/s^3$
15.14 (a) $\omega s/[(s + a)^2 + \omega^2]$;   (b) $(s + a)/s[(s + a)^2 + \omega^2]$
15.21 (a) $[9 - 6e^{-2t} - 3e^{-8t}]u(t)$;
(b) $[1 - 10e^{-3t} + 45e^{-4t} - 36e^{-5t}]u(t)$;
(c) $[30 + 50e^{-6t} \cos (8t - 126.87°)]u(t)$;
(d) $[80 - 200te^{-5t} - 80e^{-5t}]u(t)$;
(e) $[20\sqrt{5}te^{-5t} \cos (10t - 63.43°) + 2e^{-5t} \cos (10t + 90°)]u(t)$;
(f) $10\delta(t) + [36 + 4e^{-5t}]u(t)$

Chapter 16

16.1 $I_n = -I_0/s, Z_n = sL$
16.3 $V_t = V_0/s, Z_t = 1/sC$
16.5 (a) $2 \times 10^6(s + 6000)/(s^2 + 6000s + 25 \times 10^6)$;
(b) $-3000 - j4000, -3000 + j4000, -6000$
16.8 (a) $-8/(s + 2500), -400/(s + 2500)$;   (b) $-8e^{-2500t}u(t)$ A, $-400e^{-2500t}u(t)$ V
16.10 (a) $160s(s + 30,000)/(s + 80,000)(s^2 + 36 \times 10^8)$;
(b) $64e^{-80,000t} + 107.33 \cos (60,000t + 26.57°)$
16.14 (a) $[-5 + 5.02e^{-160t}]u(t)$ V;   (b) 5.02 μC
16.16 (a) $V_t = 10^4/[s(s + 1000)], Z_t = 500(s + 2000)/[s + 1000]$;
(b) $[10e^{-1000t} \sin 1000t]u(t)$, V
16.19 (a) $[0.429 + 6.836e^{-6.828t} - 3.265e^{-1.172t}]u(t)$, A;
(b) $[0.643 + 1.416e^{-6.828t} + 3.941e^{-1.172t}]u(t)$, A;
(c) $[1.071 + 8.252e^{-6.828t} + 0.676e^{-1.172t}]u(t)$, A;
(d) $[1.286 - 82.524e^{-6.828t} - 6.762e^{-1.172t}]u(t)$, V
16.23 (a) $[9.6 - 4.8e^{-25t}]u(t)$, V; $28.8\delta(t) + [480 + 480e^{-25t}]u(t)$, μA
16.26 $[2 - e^{-10t}]u(t)$, V
16.27 $11s$
16.28 $[10e^{-0.8t} \sin 0.6t]u(t)$, V

Chapter 17

17.1 $[24 + 19,149.70\ e^{-1500t} \cos (500t - 90.19°)]u(t)$ V
17.3 (a) $100/(s + 100), -100$;   (b) $s/(s + 100), -100, 0$;
(c) $s/(s + 500), -500, 0$;   (d) $500/(s + 500), -500$;
(e) $200/(s + 250), -250$
17.5 (a) $(1 - s)/2(1 + s)$;   (b) $-1, 1$
17.7 (a) $y(t) = 100t$, $0 \le t \le 5$;
$y(t) = 100(10 - t)$, $5 \le t \le 10$;
$y(t) = 0$, $t \le 0$ and $t \ge 10$;
(b) $y(t) = 50t$, $0 \le t \le 5$;
$y(t) = 250$, $5 \le t \le 10$;
$y(t) = 50(15 - t)$, $10 \le t \le 15$;
$y(t) = 0$, $t \le 0$ and $t \ge 15$;
(c) $y(t) = 100t$, $0 \le t \le 1$;
$y(t) = 100$, $1 \le t \le 5$;
$y(t) = 100(6 - t)$, $5 \le t \le 6$;
$y(t) = 0$, $t \le 0$ and $t \ge 6$

17.8 13.5, 108, 175.50, 584, 534.50, 208

17.11 (a) $v_0 = [20 - 20e^{-100t}]u(t)$ V, $0 \le t \le 10$ ms;

$v_0 = 20[e^{-100(t-0.01)} - e^{-100t}]u(t - 0.01)$ V, $t \ge 10$ ms

17.14 $25 \cos (50t + 60°)$ V

17.15 (a) $-12.5(s + 40)/(s + 100)$;   (b) $114.02 \cos (200t + 195.26°)$ V

17.16 (a) $2 \times 10^6 s/(s^2 + 60,000s + 25 \times 10^8)$;

(b) $2.5 \times 10^6 e^{-30,000t} \cos (40,000t + 36.87°)$ V;

(c) $156.05 \cos (40,000t + 29.56°)$ mV

17.21 (a) 0, 1400 rad/s;   (b) 989.95 rad/s;   (c) 5.02

17.23 (a) 80 Ω;   (b) -2.18 dB

17.24 $1000(s + 1)/(s + 10)(s + 100)$

Chapter 18

18.3 (a) $v(t) = \dfrac{4V_m}{\pi} \displaystyle\sum_{n=1,3,5,}^{\infty} \dfrac{1}{n} \sin n\omega_0 t$;

(b) $v(t) = \dfrac{2V_m}{\pi} + \dfrac{4V_m}{\pi} \displaystyle\sum_{n=1}^{\infty} \dfrac{\cos n\omega_0 t}{(1 - 4n^2)}$;

(c) $v(t) = \dfrac{V_m}{\pi} + \dfrac{V_m}{2} \sin \omega_0 t + \dfrac{2V_m}{\pi} \displaystyle\sum_{n=2,4,6,}^{\infty} \dfrac{\cos n\omega_0 t}{(1 - n^2)}$

18.8 (a) 15, 707.96 rad/s, 78,539.82 rad/s;   (b) 2500 Hz, 12,500 Hz;

(c) 0, 7.5 V;

(d) $a_k = b_k = 0$, k even;

$a_k = -\dfrac{30}{\pi k} \sin k \dfrac{\pi}{2}$, k odd;

$b_k = \dfrac{90}{\pi k}$, k odd;

$a_k = \dfrac{60}{\pi k} \sin \dfrac{\pi k}{4}$;

$b_k = 0$ for all k;

(e) $\dfrac{30}{\pi} \displaystyle\sum_{n=1,3,5,}^{\infty} \left[-\dfrac{1}{n} \sin \dfrac{n\pi}{2} \cos n\omega_0 t + \dfrac{3}{n} \sin n\omega_0 t \right]$ V;

$7.5 + \dfrac{60}{\pi} \displaystyle\sum_{n=1}^{\infty} \dfrac{1}{n} \sin \dfrac{n\pi}{4} \cos n\omega_0 t$ V

18.13 (a) 62.5 Hz;   (b) no;   (c) yes;   (d) yes;   (e) yes;

(f) 0, 0, 0, 11.46×10^{-3} A, 0

18.15 $\dfrac{2V_m}{\pi} \displaystyle\sum_{n=1,3,5,}^{\infty} \left[\dfrac{1}{n} \sqrt{\dfrac{4}{\pi^2 n^2} + 1} \right] \cos (n\omega_0 t + \theta_n + 180°)$,

where $\tan \theta_n = n\pi/2$

18.20 $95.70 \sin (\omega_0 t + 43.19°) + 38.80 \sin (3\omega_0 t - 27.51°) + 15.91 \sin (5\omega_0 t - 52.70°)$

$+ 8.35 \sin (7\omega_0 t - 63.54°) + \cdots$ mA

18.23 $45.56 - 27.82 \sin (\omega_0 t - 15.94°) - 12.56 \sin (2\omega_0 t - 29.74°) -$

$7.33 \sin (3\omega_0 t - 40.59°) - \cdots$ V

18.25 (a) 17.74 A;   (b) 17.89 A;   (c) 17.95 A;   (d) 18.14 A

18.26 (a) 915 W;   (b) 223.66 V;   (c) 8.77 A

18.27 3.46 W

18.28 9.01%

18.31 $-j\dfrac{32}{\pi^2}\displaystyle\sum_{\substack{n=-\infty \\ (\text{odd})}}^{n=\infty}\dfrac{1}{n^2}\sin\dfrac{n\pi}{2}\,e^{jn\omega_0 t}$

18.32 3.49 W

18.36 (c) 18.94;   (d) 22.68;   (e) $18.94 < 22.68$

Chapter 19

19.1 (a) $-j4\pi A\sin 2\omega/[\pi^2 - 4\omega^2]$;   (b) $\dfrac{4A}{\omega^2\tau}\left(1 - \cos\dfrac{\omega\tau}{2}\right)$

19.13 (a) $\dfrac{\tau}{2}\left[\dfrac{\sin\left[(\omega + \omega_0)(\tau/2)\right]}{(\omega + \omega_0)(\tau/2)} + \dfrac{\sin[(\omega - \omega_0)(\tau/2)]}{(\omega - \omega_0)(\tau/2)}\right]$;

    (b) $F(\omega) \rightarrow \pi[\delta(\omega + \omega_0) + \delta(\omega - \omega_0)]$

19.15 (a) $[3.75 + 3.75\,\text{sgn}(t) + 2.5e^{-4t}u(t)]$ A

19.16 (a) $[40\,\text{sgn}(t) - 80e^{-2.5t}u(t)]$ V

19.17 (a) $[\frac{4}{3}e^{-t} - e^{-2t}]u(t) + \frac{1}{3}e^{2t}u(-t)$ V;

    (b) $\frac{1}{3}$ V;   (c) $\frac{1}{3}$ V;   (d) $[\frac{4}{3}e^{-t} - e^{-2t}]u(t)$ V;   (e) yes

19.18 (a) $\dfrac{1}{\pi}\cdot\dfrac{1}{1 + t^2}$;   (b) $\dfrac{1}{2\pi}$ J;   (c) $\dfrac{1}{2\pi}$ J;   (d) 1.15 rad/s

19.20 (a) 18.17%;   (b) 27.23%;   (c) 10.57%

Chapter 20

20.1 4 Ω, 0.8, −0.8, (8/75)℧; 0.1℧, −0.75, 0.75, 3.75 Ω

20.3 5/3, (40/3) Ω, (2/15)℧, 5/3

20.5 -4×10^{-4}, 20 Ω, -5×10^{-7}℧, −0.02

20.6 0.1℧, −0.04℧, −0.04℧, 0.056℧

20.7 1.951/66.25°, 3.57/0° Ω, 0.7576/45° , 1.071/0°

20.12 $(s + 1)/[2s(s + 2)]$, $-1/[2s(s + 2)]$, $-1/[2s(s + 2)]$, $(s + 1)/[2s(s + 2)]$

20.20 (a) −10 V;   (b) 5 mW;   (c) 40 kΩ;   (d) 7.8125 mW;

    (e) −6.25 mV, 31.25 μA;   (f) −97.66 nW

20.21 (b) $[28 + 28.28e^{-1000t}\cos(7000t + 171.87°)]u(t)$ V

20.22 149.42 Ω

20.24 (b) $10^5/3$

INDEX

INDEX